THE COLORING REVIEW GUIDE TO

HUMAN ANATOMY

THE COLORING REVIEW GUIDE TO

HUMAN ANATOMY

HOGIN McMURTRIE   JAMES KRALL RIKEL
McMurtrie Design   *Pierce College*

WCB   **Wm. C. Brown Publishers**

Book Team

Editor *Edward G. Jaffe*
Developmental Editor *Colin H. Wheatley*
Production Editor *Sherry Padden*
Visuals/Design Consultant *Marilyn M. Phelps*
Designer *Mark Elliot Christianson*
Art Editor *Janice M. Roerig*

 Wm. C. Brown Publishers

President *G. Franklin Lewis*
Vice President, Publisher *George Wm. Bergquist*
Vice President, Publisher *Thomas E. Doran*
Vice President, Operations and Production *Beverly Kolz*
National Sales Manager *Virginia S. Moffat*
Advertising Manager *Ann M. Knepper*
Production Editorial Manager *Colleen A. Yonda*
Production Editorial Manager *Julie A. Kennedy*
Publishing Services Manager *Karen J. Slaght*
Manager of Visuals and Design *Faye M. Schilling*

Electronic page makeup by Hans & Cassady, Inc.

Printed in the United States of America by Wm. C. Brown Publishers,
2460 Kerper Boulevard, Dubuque, IA 52001

To my wife, Denise, for her unwavering love,
support, and courage from flagfall to finish . . . and
to J.B., for that plane flight back from San Francisco

Hogin McMurtrie

Table of Contents

Unit 4 Regulation and Maintenance of the Human Body

9 Cardiovascular System → The Blood, Heart, and Blood Vessels
- Distribution of oxygen and nutrients to the cells
- Carrying of carbon dioxide and wastes from the cells
- Maintenance of body acid-base balance
- Protection against disease
- Prevention of hemorrhage (blood clot formation)
- Regulation of body temperature

10 Lymph Vascular System → Lymph, Lymph Nodes, Vessels, and Glands (ex: SPLEEN, TONSILS, and THYMUS GLAND)
- Return of protein and fluid to the blood vascular system
- Transportation of fats from the digestive system to the blood vascular system
- Filtration of blood
- Production of white blood cells
- Protection against disease

11 Respiratory System → The Lungs, and a series of Passageways in and out of the Lungs
- Supplies oxygen
- Elimination of carbon dioxide
- Regulation of body acid-base balance

12 Digestive System → A Tubular Passageway and Associated Organs (ex: SALIVARY GLANDS, LIVER, GALLBLADDER, and PANCREAS)
- Physical and chemical breakdown of food for cell usage
- Elimination of solid wastes

13 Urinary System → Organs of Urine Production, Collection, and Elimination
- Regulation of blood chemical composition
- Elimination of wastes
- Regulation of blood chemical composition
- Regulation of fluid/electrolyte balance and volume
- Maintenance of body acid-base balance

Unit 5 Continuance of the Human Species

14 Reproductive System → Organs (TESTES and OVARIES) for Production of Reproductive Cells (SPERM and OVA), and Organs for Transportation and Storage of Those Cells
- Reproduction of the organism
- Perpetuation of the species
- Passage of genetic material from generation to generation

PREFACE

Every beginning human anatomy student needs to develop a system of study in order to be successful. This book is designed to facilitate learning by providing useful tools in an effective format. It serves as a framework summarizing material for the instructor's presentation, the primary text, dissection experience, class notes, or outside study. The book's design emphasizes the connections between facts and structures through specific student activities—most notably, *learning through the process of color association*. As students color code each drawing, highlight pertinent information on each plate, add their own notes from the primary text and lectures, become familiar with the summary charts and tables, and learn the proper pronunciation of each term, the book becomes as much the student's creation as the authors'.

This personalized document of the students' knowledge of human anatomy can continue to serve as a useful integrated tool for future study and course work. Thus, not only can *The Coloring Review Guide to Human Anatomy* be used with any textbook, but serves to guide the undergraduate, graduate, and allied health professional student in preparation for written, oral, board, and licensing examinations.

The text is organized into 5 major units (divided into 14 chapters) with the major body systems summarized in the Table of Contents:

Unit 1 Orientation and Organization
of the Human Body

Unit 2 Support and Movement
of the Human Body

Unit 3 Integration and Control
of the Human Body

Unit 4 Regulation and Maintenance
of the Human Body

Unit 5 Continuance of the Human Species

The two chapters in Unit 1 establish a foundation for the terminology and relationships to come in following chapters and should be studied first. Units 2—5 include chapters covering the 11 major systems of the human body, plus a chapter on the articulations associated with the Skeletal and Muscular Systems. Each of the body systems may be studied in any order, adapting to any course sequence. However, each chapter, arranged in a consistent format of features followed by activities, should be studied from beginning to end.

The text's learning system features the following study aids:

1. A consistent vertical format of upper and lower pages divided by a spiral binding visually separates the names and terms from their designated structures. In general, the page above the spiral contains the terms and accompanying text, and the page below, the illustration(s). This separation is valuable in organizing learning and also enables the upper and lower pages to be used as "flash cards."

2. The vertical format also allows both right- and left-handed students to comfortably color and make notes. The spiral binding enables each page to lie surface flat for optimal ease of coloring.

3. Each chapter begins with a brief table of contents, introductory paragraphs providing an overview of the system, a simple illustration of the system as a whole, components and functions of the system, and coloring guidelines for the chapter.

4. Each chapter proceeds with a general overview plate(s), followed by plates of specific areas beginning with the most superior region (head, neck) to the most inferior region (leg, foot).

5. Each chapter ends with summary charts and tables providing pertinent information cross-referenced to specific plates.

6. The book concludes with two appendices: A bibliography for further student references and a pronunciation guide/index that cross-references each term to every plate on which that term appears.

7. The longer chapters (Skeletal and Muscular) end with System Review Plates to aid the student in seeing the system as a working whole rather than as separate parts.

The authors wish to acknowledge their gratitude and indebtedness to Edward G. Jaffe for his enthusiasm, uncompromising vision, and determination in bringing this project to fruition; to Carol Mills, for her continued support, encouragement, and commitment to excellence; to the helpful professional criticisms, reviews and suggestions of

Betty Elliott *Truckee Meadows Community College*
Victor P. Eroschenko *University of Idaho*
Harry Reasor *Miami Dade Community College*
Harold R. Milliken *Loma Linda University*
Anne Funkhouser *University of the Pacific*

and to our parents, who gave us the spirit and value of education.

Hogin McMurtrie
James Krall Rikel, Ph.D.

Before You Begin—*Study the following page carefully. It clearly illustrates and explains how to best use the carefully designed learning system found in this coloring review guide.*

UPPER (TOP) PAGE

CHAPTER NUMBER (number of CHAPTER SUBDIVISION follows the decimal point)

CHAPTER TITLE and SUBDIVISION TITLE (followed by a SUB-HEADING)

•Each CHAPTER is divided into SUB–DIVISIONS, which follow a logical sequence. Each CHAPTER should be studied from the beginning to the end in the order presented.

The left column of each UPPER PAGE (and sometimes the LOWER) contains a general outline or description of the structures emphasized in the SUBDIVISION and how they relate to one another.

Pertinent information may be highlighted from the upper and/or lower text.

★ The text will indicate (where applicable) reference to END-OF-CHAPTER CHARTS, which contain a more detailed synopsis of subject matter. CHARTS also appear on UPPER and LOWER PAGES.

★ As an aid to learning human anatomy through the easy process of COLOR ASSOCIATION (COLOR LINKAGE), emphasis is placed on the concept of VISUALIZATION over ROTE-MEMORIZATION:

■ NAME OF A STRUCTURE is found in BOLD TYPEFACE within a RECTANGULAR BOX. Choose a color and color within the borders of the box. A REFERENCE NUMBER precedes the BOX.

■ Locate the same REFERENCE NUMBER in the ILLUSTRATION on the LOWER PAGE, which pinpoints the STRUCTURE. Use the SAME COLOR from the above box and color the STRUCTURE.
■ NAME and STRUCTURE are now VISUALLY RELATED through COLOR LINKAGE.

★ When colored, WHOLE ILLUSTRATIONS and WHOLE BODY SYSTEMS (CHAPTERS) take on an EXPRESSION of your OWN INDI-VIDUALITY, enhancing RETENTION of SUBJECT MATTER.

Repeat of CHAPTER NUMBER and SUBDIVISION NUMBER (each UPPER PAGE and its corresponding LOWER PAGE work together as a WHOLE UNIT).

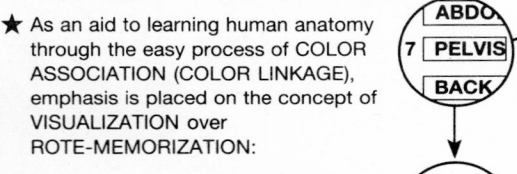

UNIT TITLE and PAGE NUMBER of the book

★ Suggested COLORING NOTES, but you may use any colors you wish. (Consistent colors throughout the book: arteries (red), veins (blue), lymphatic vessels (green), muscles (reddish-brown), fat (yellow).

★ In general, color larger areas with lighter colors, and smaller areas with darker colors (when possible).

★ The authors suggest the use of COLORED PENCILS over FELT-TIP PENS (wax crayons are too coarse):
■ Horizontal, rectangular boxes surrounding the NAME of the STRUCTURE are to be colored in, and darker felt-tip pen colors could obliterate the name or bleed.
■ Colored pencils increase the color palette, since you can create many different color densities from the same pencil by applying lighter or heavier pressure.

A view based on the STANDARD ANATOMICAL POSITION is indicated for each illustration in order to facilitate orientation of the object in space.

★ Complete Chapter 1 (Orientation) and Chapter 2 (Organization) before studying any body system (Chapters 3–14)

Reference numbers are located either inside a structure to be colored, or precede a ruled line that points to the structure.

★ Border lines of each structure to be colored are in heavy outline, facilitating visual separation of areas.

UNIT TITLE and PAGE NUMBER of the book

(UNITS combine CHAPTERS (SYSTEMS) with similar FUNCTIONS – – See Table of Contents and Back outside cover)

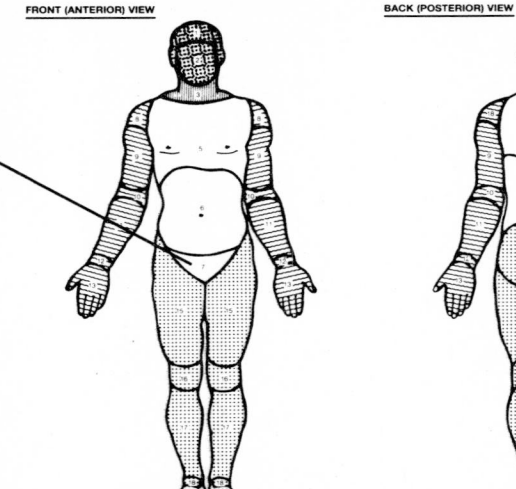

LOWER (BOTTOM) PAGE

THE COLORING REVIEW GUIDE TO

HUMAN ANATOMY

UNIT 1: ORIENTATION AND ORGANIZATION

Chapter 1 ⬛**Orientation**

★ *SAGITTAL, CORONAL, and TRANSVERSE PLANES can be conceived
as a SET OF MUTUALLY PERPENDICULAR PLANES used to depict
structural arrangement*

Color in the body planes after coloring the next pages

SUPERIOR (TOP) Surface
of Object Being Viewed

HORIZONTAL
(TRANSVERSE)
(CROSS-SECTIONAL)
PLANE

FRONTAL
(CORONAL) PLANE

FRONT (ANTERIOR)
Surface of Object
Being Viewed

PARASAGITTAL
PLANE

MIDSAGITTAL
(MEDIAN) PLANE

Concept of Directional Terms as "Viewing Planes"

★ An interesting concept to consider in order to facilitate mental
orientation when viewing an object (or the whole body), is to visually
place the object mentally inside of a transparent, 3-dimensional box
– the faces or sides of the box become the VIEWING PLANES
through which one views the object from different positions around

it. Thus, for example, in the illustrations below, when viewing the
body through the FRONT or ANTERIOR VIEWING PLANE of the box,
that is an ANTERIOR VIEW. When viewing the body through the
dotted BACK or POSTERIOR VIEWING PLANE, one has a
POSTERIOR VIEW of the object, etc... .

THE STANDARD ANATOMICAL POSITION

- **Body Erect, Facing Observer**
- **Eyes Directed Forward**
- **Arms at Sides of Body, Palms of Hands Turned Forward**
- **Feet Placed Flat on Floor and Parallel**

DIRECTIONAL TERMS are used to precisely locate various structures, organs, surfaces, and regions in relation to one another. BODY PLANES are used as points of reference in viewing a structure, by which positions of parts of the body are indicated. A body plane is a flat surface formed by a cut (imaginary or real) through the body or part of it. All directional terms and body planes are always relative to the Standard Anatomical Position.

Color in the arrows (for directional terms) and the body planes after coloring the next pages.

ANTERIOR VIEW

RIGHT LATERAL VIEW

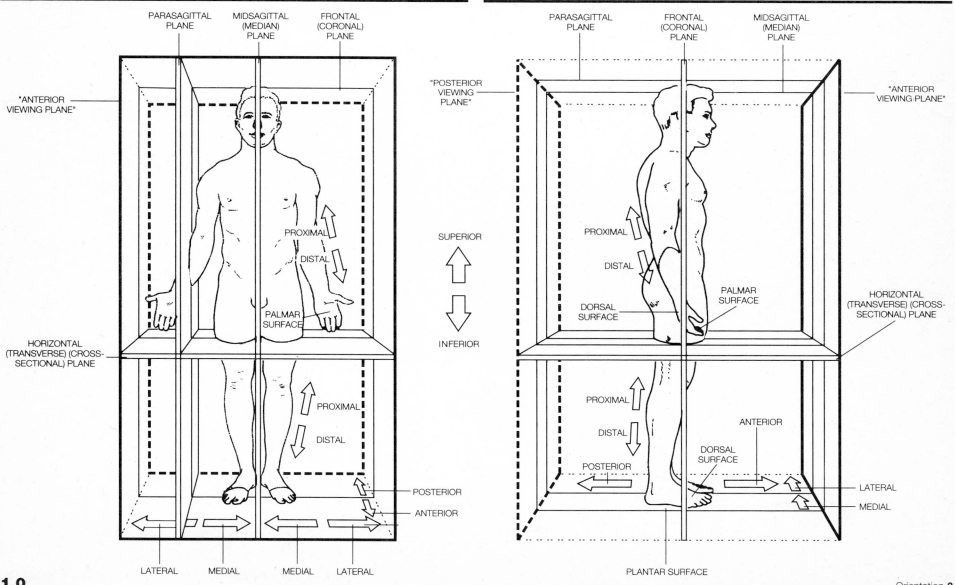

ORIENTATION: PLANES OF REFERENCE AND DESCRIPTIVE TERMINOLOGY

You will be viewing many body structures in section: i.e., looking at the FLAT SURFACE or BODY PLANE that results from a cut made through the 3-dimensional body structure.

**Coloring guideline: 1= light orange
2 = light red 3 = yellow   4 = light blue**

Body Planes

(Imaginary flat surfaces that pass through the body)

1 **MIDSAGITTAL (MEDIAN)** } SAGITTAL PLANES
2 **PARASAGITTAL**
3 **FRONTAL (CORONAL)**
4 **HORIZONTAL (TRANSVERSE) (CROSS-SECTIONAL)**

Body Planes

1 **MIDSAGITTAL (MEDIAN)**   Midline plane of the body, passes lengthwise, extends vertically
Divides body into right and left halves: the plane of symmetry.

2 **PARASAGITTAL**   Plane parallel to midsagittal plane, on either side of it.
Passes lengthwise, extends vertically.
Divides body into unequal right and left portions.

3 **FRONTAL (CORONAL)**   Plane perpendicular to sagittal planes.
Passes lengthwise, extends vertically.
Divides body into equal or unequal front and back portions.

4 **HORIZONTAL (TRANSVERSE) (CROSS-SECTIONAL)**   Plane perpendicular to midsagittal-sagittal planes.
Passes breadthwise, extends horizontally.
Divides body into upper (superior) and lower (inferior) portions.
**Cross/transverse sections are perpendicular to the long axis of structures and may not be horizontal (i.e., oblique).

Directional Terms of the Body

(Used to explain the exact location of various body structures relative to each other)

5 **SUPERIOR (CRANIAD)**   CEPHALIC UPPER
6 **INFERIOR (CAUDAD)**   LOWER
7 **ANTERIOR (VENTRAL)**   FRONT
8 **POSTERIOR (DORSAL)**   BACK
9 **PROXIMAL**
10 **DISTAL**
11 **MEDIAL**
12 **LATERAL**

{ = relative to one another

Directional Terms

5 **SUPERIOR (CRANIAD)**   Any structure that is above another structure (nearer the head) is said to be SUPERIOR to it.

6 **INTERIOR (CAUDAD)**   Any structure that is lower than another structure (toward the feet) is said to be INFERIOR to it.

7 **ANTERIOR (VENTRAL)**   Any structure that is in front of another structure is said to be ANTERIOR to it.

8 **POSTERIOR (DORSAL)**   Refers to a structure that is in back of another.
Toward the back.

9 **PROXIMAL**   Refers to any structure that is closer to the body mass (or midsagittal plane) than another structure.

10 **DISTAL**   Refers to a structure that is farther away from the body mass (or midsagittal plane) than another structure.

11 **MEDIAL**   Refers to a structure closer to the midsagittal plane than another structure.
Toward the midline plane.

12 **LATERAL**   Refers to a structure farther away from the midsagittal plane than another structure.
Towards the sides, away from the midline.

IPSILATERAL   Refers to structures that are on the same side of the body.

CONTRALATERAL   Refers to structures that are on opposite sides of the body.

INTERMEDIATE   Refers to a structure that is between a medial structure and a lateral structure.

INTERNAL (DEEP)   Refers to a structure that is away from the body surface in reference to another structure.

EXTERNAL (SUPERFICIAL)   Refers to a structure that is more toward the body surface in reference to another structure.

VISCERAL   Refers to the covering of an internal organ (viscus).

PARIETAL   Refers to the outer wall of a body cavity.

BODY PLANES (SECTIONS)

Plane   View (section)

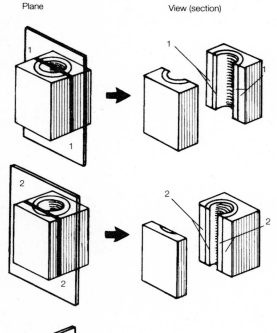

Oblique

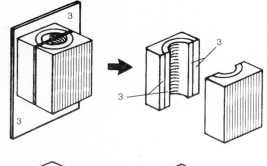

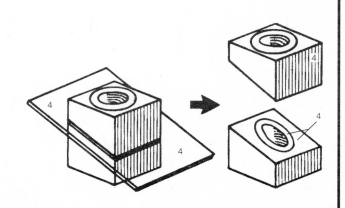

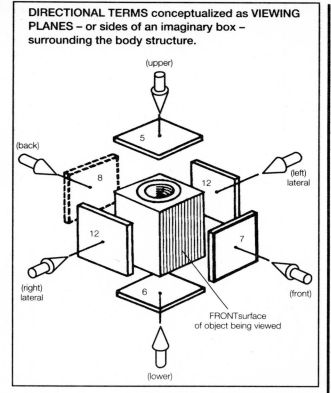

(upper)

(back)

8

12

(left) lateral

12

7

(right) lateral

6

(front)

FRONT surface of object being viewed

(lower)

Directional Terms of the limbs only
ANTERIOR VIEW (left limbs)

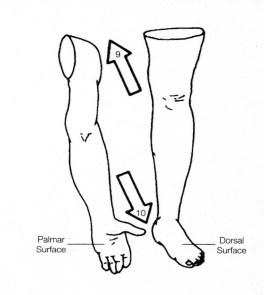

9

10

Palmar Surface

Dorsal Surface

POSTERIOR VIEW (left limbs)

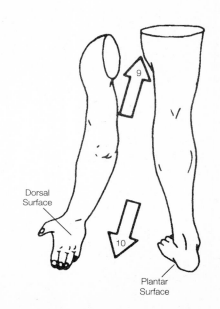

9

10

Dorsal Surface

Plantar Surface

ORIENTATION: BODY REGIONS
The 5 Major Body Regions and Generalized Body Regions

The human body is divided into 5 major regions, all identifiable on the surface:
- HEAD
- NECK
- TRUNK (TORSO)
- UPPER EXTREMITY (2)
- LOWER EXTREMITY (2)

These 5 body regions can be subdivided into more generalized and specific localized areas.
- Each region contains internal structures with specific anatomical names. Learning the specific regional terminology provides a solid basis for later learning the names of the underlying structures and identifying their location

■ **HEAD** (CAPUT) contains the BRAIN and SPECIAL SENSE ORGANS
- Provides openings into the RESPIRATORY SYSTEM and DIGESTIVE SYSTEM

■ **NECK** (COLLUM)
- Tremendous complexity of neck musculature provides a wide variety of movements for the HEAD.
  - Connects HEAD to the THORAX

4 NECK REGIONS contain major organs:

ANTERIOR REGION (CERVIX): contains portions of the RESPIRATORY tract (TRACHEA) and DIGESTIVE tract (ESOPHAGUS), the LARYNX or VOICE BOX, MAJOR BLOOD VESSELS to and from the HEAD, HYOID BONE, NERVES, and THYROID and PARATHYROID GLANDS.

RIGHT and LEFT LATERAL REGIONS contain MAJOR NECK MUSCLES and LYMPH NODES.

POSTERIOR REGION (NUCHA) contains the SPINAL CORD, CERVICAL VERTEBRAE.

■ **TRUNK** (TRUNCUS) or TORSO is frequently divided into an upper THORAX (PECTUS) or CHEST, and a lower ABDOMEN (VENTER), separated by a muscular internal DIAPHRAGM. The TORSO is a major site of vital visceral organs:

THORAX contains the bony RIB CAGE within which lies continuations of the TRACHEA and ESOPHAGUS, the LUNGS, HEART, MAJOR BLOOD and LYMPHATIC VESSELS, and THYMUS GLAND (outside the RIB CAGE are the BREASTS and LYMPH NODES.)

ABDOMEN contains lower ESOPHAGUS, STOMACH, SMALL INTESTINE, LARGE INTESTINE (major portion), LIVER, GALLBLADDER, SPLEEN, PANCREAS, KIDNEYS, and ADRENAL GLANDS

2 other divisions of the TORSO include the BACK (DORSUM) and THE PELVIS.

PELVIS contains LARGE INTESTINE (terminal portion), URINARY BLADDER, and INTERNAL REPRODUCTIVE ORGANS. The floor of the PELVIS is called the PERINEUM and incudes the EXTERNAL GENITALIA.

■ **UPPER EXTREMITIES** are adaptive for freedom of movement.

■ **LOWER EXTREMITIES** are important for locomotion and bearing weight.

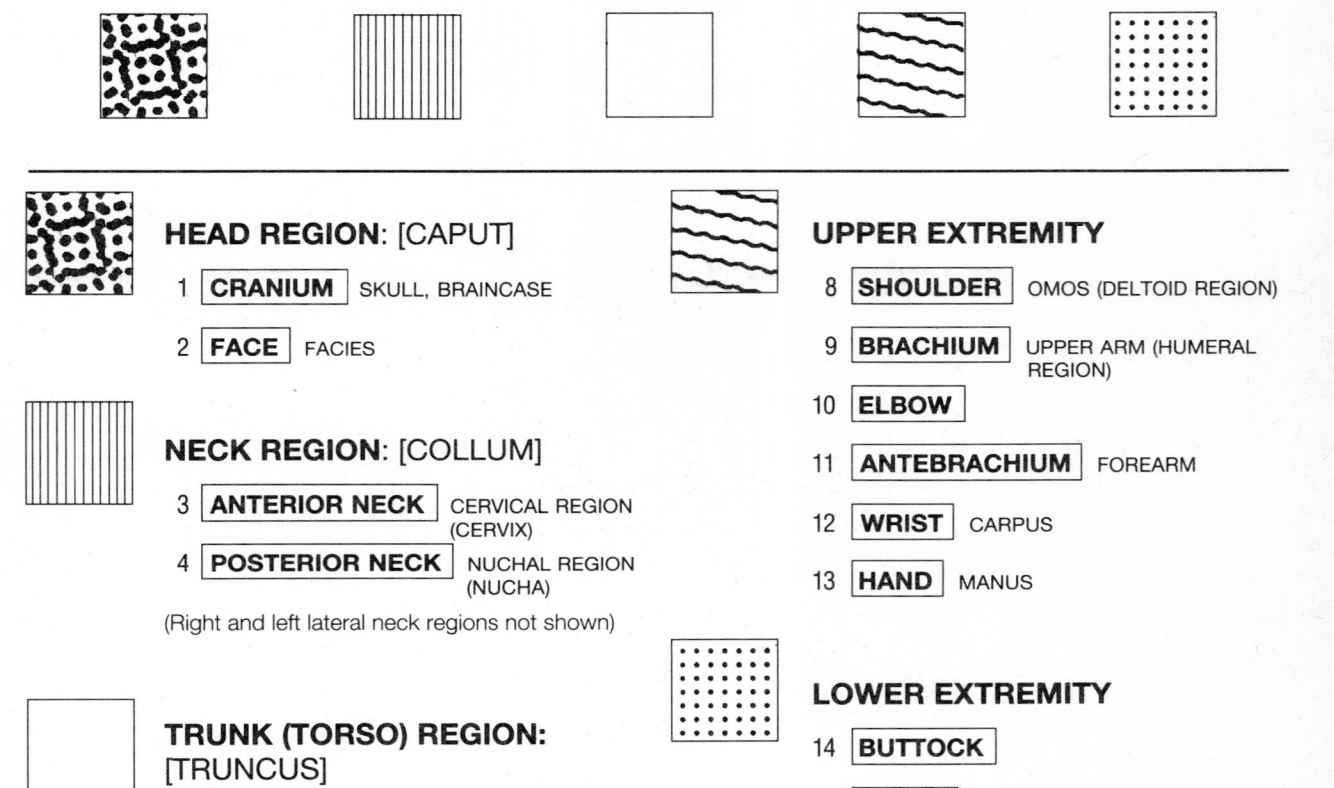

HEAD [CAPUT]   **NECK [COLLUM]**   **TRUNK [TRUNCUS]**   **UPPER EXTREMITY (2)**   **LOWER EXTREMITY (2)**

HEAD REGION: [CAPUT]

1 **CRANIUM**   SKULL, BRAINCASE

2 **FACE**   FACIES

NECK REGION: [COLLUM]

3 **ANTERIOR NECK**   CERVICAL REGION (CERVIX)

4 **POSTERIOR NECK**   NUCHAL REGION (NUCHA)

(Right and left lateral neck regions not shown)

TRUNK (TORSO) REGION: [TRUNCUS]

5 **THORAX**   PECTUS (CHEST/PECTORAL REGION)

6 **ABDOMEN**   VENTER

7 **PELVIS**

BACK (DORSUM) : POSTERIOR THORAX AND POSTERIOR ABDOMEN

UPPER EXTREMITY

8 **SHOULDER**   OMOS (DELTOID REGION)

9 **BRACHIUM**   UPPER ARM (HUMERAL REGION)

10 **ELBOW**

11 **ANTEBRACHIUM**   FOREARM

12 **WRIST**   CARPUS

13 **HAND**   MANUS

LOWER EXTREMITY

14 **BUTTOCK**

15 **THIGH**   UPPER LEG (FEMUR/FEMORAL REGION)

16 **KNEE**

17 **LEG**   CRUS

18 **ANKLE**   TARSUS

19 **FOOT**   PES

GENERALIZED BODY REGIONS

FRONT (ANTERIOR) VIEW

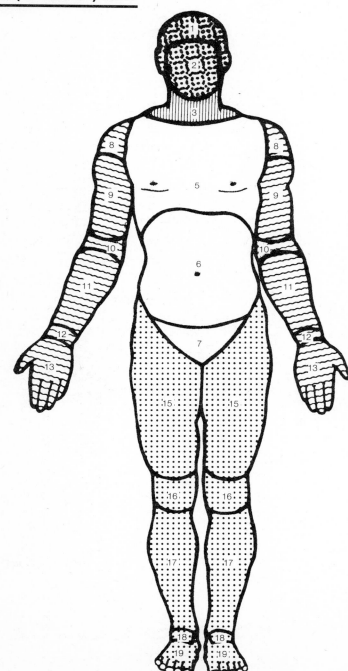

BACK (POSTERIOR) VIEW

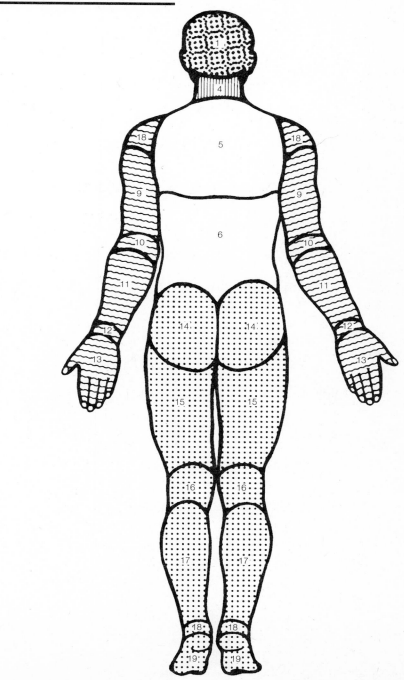

ORIENTATION: BODY REGIONS
Specific Body Regions/Abdominopelvic Regions and Quadrants

■ **HEAD REGION**
   1 CRANIUM
   2 FACE

■ **NECK REGION**
   3 ANTERIOR NECK
   4 POSTERIOR NECK

■ **TRUNK REGION**
   5 THORAX
   6 ABDOMEN } see bottom page
   7 PELVIS

   BACK

■ **UPPER EXTREMITY** (2)
   8 SHOULDER
   9 BRACHIUM
   10 ELBOW
   11 ANTEBRACHIUM
   12 WRIST
   13 HAND

■ **LOWER EXTREMITY** (2)
   14 BUTTOCK
   15 THIGH
   16 KNEE
   17 LEG
   18 ANKLE
   19 FOOT

SPECIFIC BODY REGIONS

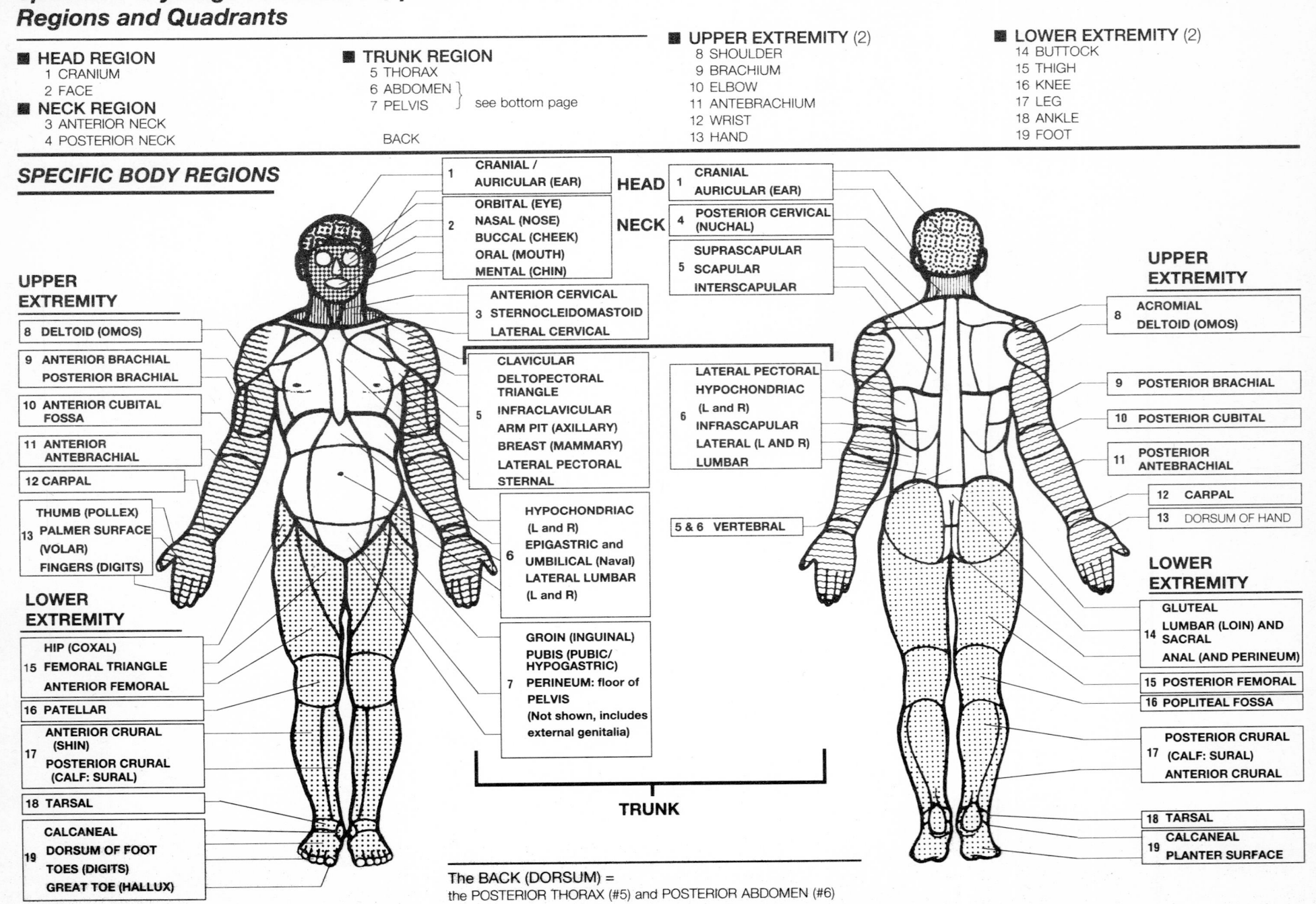

UPPER EXTREMITY

8 DELTOID (OMOS)

9 ANTERIOR BRACHIAL
  POSTERIOR BRACHIAL

10 ANTERIOR CUBITAL FOSSA

11 ANTERIOR ANTEBRACHIAL

12 CARPAL

13 THUMB (POLLEX)
   PALMER SURFACE (VOLAR)
   FINGERS (DIGITS)

LOWER EXTREMITY

15 HIP (COXAL)
   FEMORAL TRIANGLE
   ANTERIOR FEMORAL

16 PATELLAR

17 ANTERIOR CRURAL (SHIN)
   POSTERIOR CRURAL (CALF: SURAL)

18 TARSAL

19 CALCANEAL
   DORSUM OF FOOT
   TOES (DIGITS)
   GREAT TOE (HALLUX)

1 CRANIAL / AURICULAR (EAR)

2 ORBITAL (EYE)
  NASAL (NOSE)
  BUCCAL (CHEEK)
  ORAL (MOUTH)
  MENTAL (CHIN)

3 ANTERIOR CERVICAL
  STERNOCLEIDOMASTOID
  LATERAL CERVICAL

5 CLAVICULAR
  DELTOPECTORAL TRIANGLE
  INFRACLAVICULAR
  ARM PIT (AXILLARY)
  BREAST (MAMMARY)
  LATERAL PECTORAL
  STERNAL

6 HYPOCHONDRIAC (L and R)
  EPIGASTRIC and UMBILICAL (Naval)
  LATERAL LUMBAR (L and R)

7 GROIN (INGUINAL)
  PUBIS (PUBIC/HYPOGASTRIC)
  PERINEUM: floor of PELVIS
  (Not shown, includes external genitalia)

HEAD

NECK

1 CRANIAL
  AURICULAR (EAR)

4 POSTERIOR CERVICAL (NUCHAL)

5 SUPRASCAPULAR
  SCAPULAR
  INTERSCAPULAR

6 LATERAL PECTORAL
  HYPOCHONDRIAC (L and R)
  INFRASCAPULAR
  LATERAL (L AND R)
  LUMBAR

5 & 6   VERTEBRAL

UPPER EXTREMITY

8 ACROMIAL
  DELTOID (OMOS)

9 POSTERIOR BRACHIAL

10 POSTERIOR CUBITAL

11 POSTERIOR ANTEBRACHIAL

12 CARPAL

13 DORSUM OF HAND

LOWER EXTREMITY

14 GLUTEAL
   LUMBAR (LOIN) AND SACRAL
   ANAL (AND PERINEUM)

15 POSTERIOR FEMORAL

16 POPLITEAL FOSSA

17 POSTERIOR CRURAL (CALF: SURAL)
   ANTERIOR CRURAL

18 TARSAL

19 CALCANEAL
   PLANTER SURFACE

TRUNK

The BACK (DORSUM) =
the POSTERIOR THORAX (#5) and POSTERIOR ABDOMEN (#6)
(includes vertebral region)

9 ABDOMINOPELVIC REGIONS:

- used for anatomical studies
- easy location of organs

1	RIGHT HYPOCHONDRIAC		6	LEFT LATERAL	L. LUMBAR
2	EPIGASTRIC		7	RIGHT INGUINAL	R. ILIAC
3	LEFT HYPOCHONDRIAC		8	HYPOGASTRIC	PUBIC
4	RIGHT LATERAL	R. LUMBAR	9	LEFT INGUINAL	L. ILIAC
5	UMBILICAL				

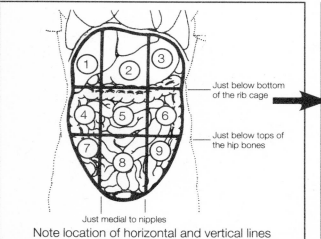

Just below bottom of the rib cage

Just below tops of the hip bones

Just medial to nipples

Note location of horizontal and vertical lines

ABDOMINOPELVIC QUADRANTS:

- used for site location of pain, tumors, other abnormalities

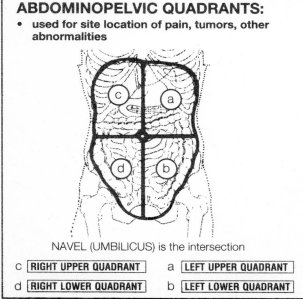

NAVEL (UMBILICUS) is the intersection

| c | RIGHT UPPER QUADRANT | a | LEFT UPPER QUADRANT |
| d | RIGHT LOWER QUADRANT | b | LEFT LOWER QUADRANT |

①
- **Liver** (right lobe)
- **Gallbladder**
- **Right Kidney** (upper 1/3)
- **Right Adrenal Gland**

②
- **Liver** (left lobe, medial part of right lobe)
- **Stomach** (pyloric part and lesser curvature)
- **Small Intestine** (superior and descending duodenum)
- **Pancreas** (body and upper head)
- **2 Adrenal Glands**

③
- **Stomach** (body and fundus)
- **Large Intestine** (Left Colic or Splenic Flexure)
- **Spleen**
- **Pancreas** (tail)
- **Left Kidney** (upper 2/3)
- **Left Adrenal Gland**

④
- **Small Intestine** (Jejunum, Ileum)
- **Large Intestine** (Right Colic or Hepatic Flexure, Ascending Colon, Superior Cecum)
- **Right Kidney** (lower 2/3, lateral portion)

⑤
- **Small Intestine** (Inferior Duodenum, Jejunum, Ileum)
- **Large Intestine** (middle of Transverse Colon)
- **Bifurcation of Abdominal Aorta and Inferior Vena Cava**
- **Kidneys** (Hilar regions)

⑥
- **Small Intestine** (Jejunum, Ileum)
- **Large Intestine** (Descending Colon)
- **Left Kidney** (lower 1/3)

⑦
- **Small Intestine** (Ileum)
- **Large Intestine** (lower end of Cecum)
- **Appendix**

⑧
- **Small Intestine** (Ileum)
- **Large Intestine** (part of Sigmoid Colon and Rectum)
- **Urinary Bladder** (when full)

⑨
- **Small Intestine** (Ileum)
- **Large Intestine** (junction of Descending and Sigmoid Colon)

ORIENTATION: BODY CAVITIES
The 2 Principal Body Cavities:
Cavity Divisions and Body Membranes

Use cool colors (blues, greens, etc.) for the DORSAL BODY CAVITY divisions, and warm colors (orange, red, yellow) for the VENTRAL BODY CAVITY divisions

★ See 7.4, 7.6, 9.4, 11.6, 12.4

★ BODY CAVITIES are CONFINED SPACES WITHIN THE BODY that house and support organs. BODY CAVITIES serve to segregate or confine organs and systems that have related functions. The organs within the BODY CAVITIES are PROTECTED, SUPPORTED, AND COMPARTMENTALIZED (separated) by associated MEMBRANES.

★ The two principal body cavities are the DORSAL BODY CAVITY and VENTRAL BODY CAVITY.

■ **DORSAL BODY CAVITY**
- Includes 2 cavities, the CRANIAL CAVITY and VERTEBRAL CAVITY (CANAL), housing the BRAIN and SPINAL CORD, respectivel
- The major portion of the nervous system (central nervous system, CNS) occupies the dorsal cavity.
- Covering and protective membranes are the 3-layered MENINGES.

■ **VENTRAL BODY CAVITY**
- During development, a body cavity is formed within the TRUNK called the COELOM, which is lined with a membrane secreting lubricating fluid. In time, the COELOM is partitioned by the MUSCULAR DIAPHRAGM into an upper THORACIC (CHEST) CAVITY and a lower ABDOMINOPELVIC CAVITY. The organs contained within the COELOM are called VISCERA or VISCERAL ORGANS.
- Serous membranes line the THORACIC and ABDOMINOPELVIC CAVITIES as PARIETAL MEMBRANES and cover the VISCERA as VISCERAL MEMBRANES.
- Serous membranes secrete a watery SEROUS FLUID for lubrication and they compartmentalize visceral organs so that infections and diseases cannot spread from one compartment to another.

■ **THORACIC CAVITY**
- Contains the principal organs of the RESPIRATORY and CIRCULATORY SYSTEMS: 2 PLEURAL CAVITIES contain the RIGHT and LEFT LUNGS.
- 1 PERICARDIAL CAVITY contains the HEART.
- The area between the 2 LUNGS within the THORACIC CAVITY = MEDIASTINUM.

■ **ABDOMINOPELVIC CAVITY**
- Consists of 2 cavities: The ABDOMINAL CAVITY hosts the primary organs of digestion and the PELVIC CAVITY hosts the internal reproductive organs.

★ Several smaller cavities exist within the HEAD (See 4.6)
- ORAL (BUCCAL) CAVITY contains the TEETH and TONGUE primarily for DIGESTION (secondarily for RESPIRATION). (See 12.5)
- NASAL CAVITY for RESPIRATION. (See 11.2)
- 2 ORBITAL CAVITIES house the eyeballs and associated muscles, nerves, and blood vessels for the SENSATION of VISION. (See 7.19)
- 2 MIDDLE EAR CAVITIES house the ear ossicles (bones) for the SENSATION of HEARING. (See 7.23)

Principal Body Cavities	**I** DORSAL BODY CAVITY	**II** VENTRAL BODY CAVITY	

Dorsal Body Cavity	**Organs and Structures Housed within Cavities**	**3-Layered Membrane**	**Location of Membrane**
A CRANIAL CAVITY	BRAIN	MENINGES (DURA MATER, ARACHNOID, and PIA MATER)	Continuously covers and protects the BRAIN and SPINAL CORD
B VERTEBRAL (SPINAL) CAVITY *VERTEBRAL CANAL*	SPINAL CORD		

Ventral Body Cavity		**Serous Membranes**	**Specific Location of Serous Membranes**
C THORACIC CAVITY *CHEST CAVITY*	RIGHT and LEFT LUNGS	PARIETAL PLEURA	Adheres to the thoracic walls and the thoracic surface of the DIAPHRAGM
1 2 PLEURAL CAVITIES		VISCERAL PLEURA	Adheres to the outer surface of the LUNGS
	MIDDLE MEDIASTINUM: *	PARIETAL PERICARDIUM (PERICARDIAL SAC)	Durable outer covering of the HEART
2 PERICARDIAL CAVITY	HEART	VISCERAL PERICARDIUM (EPICARDIUM)	Adheres to outer surface of the HEART; outer wall layer of the HEART
D ABDOMINOPELVIC CAVITY	STOMACH, SMALL INTESTINE, LARGE INTESTINE (major portion), LIVER, GALL BLADDER, SPLEEN, PANCREAS‡, KIDNEYS‡, ADRENAL GLANDS‡	PARIETAL PERITONEUM	Adheres to the ABDOMINAL WALL
4 ABDOMINAL CAVITY		VISCERAL PERITONEUM	Adheres to the ABDOMINAL VISCERA
5 PELVIC CAVITY	LARGE INTESTINE (terminal portion), URINARY BLADDER Internal Reproductive organs: (♂ :PROSTATE GLAND, SEMINAL VESICLES) (♀ :UTERUS, UTERINE TUBES, OVARIES)	MESENTERY	Double fold of PERITONEUM, which connects the PARIETAL PERITONEUM to the VISCERAL PERITONEUM

BODY MEMBRANES
- 2 basic types: SEROUS MEMBRANES and MUCOUS MEMBRANES
- Both types are composed of thin layers of connective tissue and epithelial tissue
- MUCOUS MEMBRANES secrete a thick, viscid MUCUS for lubrication and protection
- MUCOUS MEMBRANES line the LUMINA (hollow tubular canals) of various organs: ESOPHAGUS, STOMACH, INTESTINES, TRACHEA, UTERUS, ORAL, and NASAL CAVITIES

* *The PERICARDIAL CAVITY, HEART, and PERICARDIUM comprise the MIDDLE MEDIASTINUM. See below.*

‡ *A portion of the PANCREAS, the KIDNEYS, and ADRENAL GLANDS are positioned **behind** the parietal peritoneum, although they are still within the ABDOMINOPELVIC CAVITY. Thus, they are RETROPERITONEAL ORGANS.*

★ Compartmentalization:

4 Distinct Compartments in the THORACIC CAVITY House Organs
2 PLEURAL CAVITIES house the LUNGS
1 PERICARDIAL CAVITY houses the HEART (middle mediastinum)
1 MEDIASTINUM (anterior, posterior, superior, and middle)

3 | The MEDIASTINUM

- The space between the lungs extending from the STERNUM (breastbone) to the VERTEBRAL COLUMN (in the back)
- The MEDIASTINUM houses **ALL** the contents of the THORACIC CAVITY *excluding* the LUNGS and PLEURAE (serous membranes of the lungs)

| S | SUPERIOR MEDIASTINUM | M | MIDDLE MEDIASTINUM |
| P | POSTERIOR MEDIASTINUM | A | ANTERIOR MEDIASTINUM |

Principal Body Cavities

(Right Lateral View)

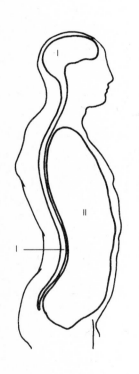

Divisions of Principal Body Cavities

Main Divisions
(Right Lateral View)

Subdivisions
(Anterior View)

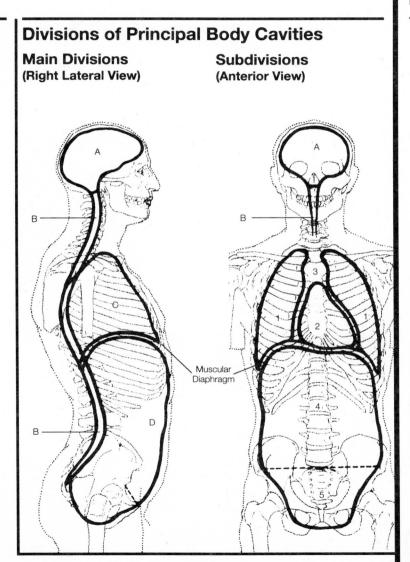

Muscular Diaphragm

Mass of Organs within MEDIASTINUM

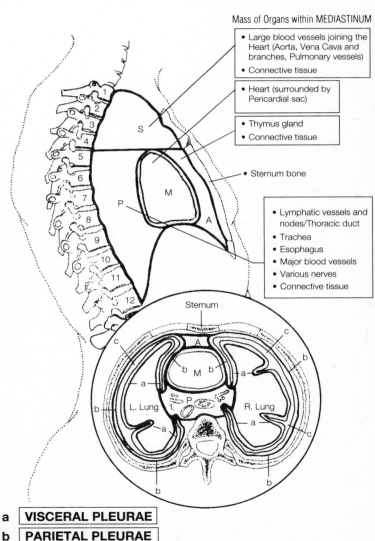

- Large blood vessels joining the Heart (Aorta, Vena Cava and branches, Pulmonary vessels)
- Connective tissue

- Heart (surrounded by Pericardial sac)

- Thymus gland
- Connective tissue

- Sternum bone

- Lymphatic vessels and nodes/Thoracic duct
- Trachea
- Esophagus
- Major blood vessels
- Various nerves
- Connective tissue

Sternum

L. Lung   R. Lung

a	VISCERAL PLEURAE
b	PARIETAL PLEURAE
c	PLEURAL CAVITY

UNIT 1: ORIENTATION AND ORGANIZATION

Chapter **2** 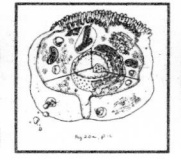 **Organization**

Two Basic Types of Interdependent Bodily Phenomena Requisite for Life

1 ORGANIZATIONAL PHENOMENA: Related to the structural complexity of the organism

• DEVELOPMENT ➤ Cellular differentiation during morphogenesis changes unspecialized tissues into structurally and functionally distinct organs.

• ORGANIZATION ➤ Each body structure is genetically determined and adapted to specific functions.

• ADAPTATION ➤ Narrow tolerance levels result in highly specialized structures with low adaptability.

• SYNTHESIS ➤ Process of forming a complex substance from simple elements or compounds (such as PROTEIN, DNA, RNA, and Fats).

• SECRETION ➤ Specific chemical compounds synthesized within specialized cells for intercellular communication and other purposes.

2 METABOLIC PHENOMENA: The sum total of all dynamic energy and material transformations that occur continuously within the living organism and its cells. ENERGY CHANGES involve the *release* of potential energy of food and its conversion into mechanical work (movement) or heat. MATERIAL CHANGES of substances occur during the life periods of GROWTH, MATURITY, and SENESCENCE (old age).

• ANABOLISM ➤ Building-up process that converts ingested substances into the constituents of protoplasm.

• CATABOLISM ➤ Tearing-down process that breaks down substances into simpler substances, whose end products are usually excreted.

• GROWTH ➤ Growth processes occur both at the cellular and systems level. Newly formed daughter cells need to grow and mature to function. A genetically determined growth process of the organism arises when body structures reflect metabolic needs.

Many of the Essential Life Processes ⟶ are delegated to the ⟶ Separate Systems Level
Active at the Cellular Level

• CONTRACTILITY and MOVEMENT ⟶	MUSCULAR SYSTEM
• INTEGRATION and CONTROL ⟶ • IRRITABILITY and RESPONSIVENESS	NERVOUS SYSTEM ENDOCRINE SYSTEM (INTEGUMENTARY SYSTEM)
• CIRCULATION and TRANSPORT ⟶	CARDIOVASCULAR SYSTEM LYMPH VASCULAR SYSTEM
• RESPIRATION ⟶	RESPIRATORY SYSTEM
• NUTRITION: INGESTION, DIGESTION, and ABSORPTION ⟶	DIGESTIVE SYSTEM
• EXCRETION ⟶	URINARY SYSTEM (RESPIRATORY SYSTEM) (INTEGUMENTARY SYSTEM)
• REPRODUCTION ⟶	REPRODUCTIVE SYSTEM
CELL (PLASMA) MEMBRANE ⟶	INTEGUMENTARY SYSTEM
CYTOPLASMIC LATTICE ⟶	SKELETAL SYSTEM
NUCLEUS ⟶	CENTRAL NERVOUS SYSTEM (BRAIN)

LEVELS OF STRUCTURAL ORGANIZATION AND COMPLEXITY WITHIN THE HUMAN BODY

THE ENTIRE CONSTITUENTS OF THE HUMAN BODY:
- **Cells Organized into Tissues and Organs (and Systems)**
- **Connective Tissue Fibers (Products of Cells)**
- **Fluid**

★ **ENVIRONMENTAL FACTORS UPON WHICH LIFE IS TOTALLY DEPENDENT**
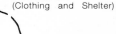
WATER
OXYGEN
FOOD
HEAT
PRESSURE
PROTECTION
(Clothing and Shelter)

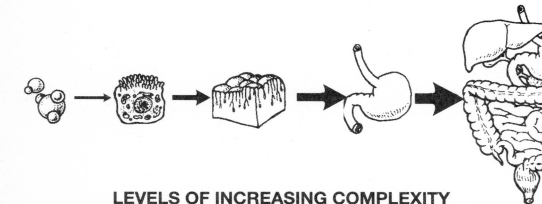

LEVELS OF INCREASING COMPLEXITY
Each level represents an association of units from the preceding level

| CHEMICAL | → | CELLULAR | → | TISSUE | → | ORGAN | → | SYSTEM |

CHEMICAL
- All chemical substances essential for life maintenance
- Composed of atoms joined together to form various molecular structures

CELLULAR
- The basic structural and functional units of the organism
- Composed of chemical substances called protoplasm
- Small protoplasmic organelles within the cell carry out specific functions

TISSUE
- Aggregated groups or layers of similar specialized cells (and their intercellular substance) whose individual, unique structure is directly related to the performance of a specific function
- Similar cells secrete a nonliving MATRIX (liquid, semisolid, or solid), which spaces them uniformly and binds them together as tissue. (Secretory Cells, however, do not have a binding matrix, and are solitary amidst another tissue.)

ORGAN
- Structures of definite form and function
- Composed of two or more tissues joined together to form a structure that has a particular function

SYSTEM
- An association of organs working together to perform a common function

ORGANISM
- All body systems functioning together constitute one living individual. The systems do not work independently, but depend on a highly coordinated effort working as a whole

ORGANIZATION: THE CHEMICAL AND CELLULAR LEVEL
The Generalized Cell and Its Components

N = bright yellow a = flesh
C = light blue 8 = green
4 = orange 1 = light purple
Other organelles = your choice
★ See Chart #1

★ THE CELL: THE BASIC STRUCTURAL AND FUNCTIONAL UNIT OF LIFE.

The PHENOMENA that distinguish living from non-living things are exemplified in the basic structural and functional activities of microscopic units called CELLS. The human being is composed of between 60 –100 trillion living cells, with wide variation of structure and function, only a few hundred specific kinds exist. However, all cells have certain *features in common:*

• All cells are composed entirely of chemical substances called PROTOPLASM.

• A generalized cell consists of a relatively fixed percentage of protoplasmic compounds that are arranged into small func tional structures called ORGANELLES, each one a specific working component of the cell. (See Chart #1)

• PROTOPLASM is divided into a single NUCLEUS and the CYTOPLASM. The NUCLEUS is the hereditary and control center of the cell, without which the cell dies. The CYTO PLASM is all the protoplasm of the cell outside the nucleus.

★ • BASIC CONSTITUENTS OF PROTOPLASM
Protoplasm is made up of certain ELEMENTS in CHEMICAL COMBINATION.
There are 4 ELEMENTS that make up most of the BODY WEIGHT. (92.2%):

(O)	OXYGEN	62.2%
(C)	CARBON	18.0%
(H)	HYDROGEN	9.4%
(N)	NITROGEN	2.6%

■ Most of the (O) + (H) = WATER
■ (O) + (H) + (C) = CARBOHYDRATES and FATS (Both are the MAIN SOURCES of ENERGY in living protoplasm)
■ (O) + (H) + (C) + (N) = PROTEINS (Main building constituent of protoplasm. The Basic Function of All Cells is the Production of PROTEIN)

• General Fixed Percentage of Protoplasmic Compounds of the Cell:

    WATER 80%
    PROTEINS 15%
    LIPIDS (FATS) 3%
    CARBOHYDRATES 1%
    MINERALS and NUCLEIC ACIDS 1%

★ CELLS arise only from pre-existing cells.
• New cells arise from CELL DIVISION, either by MITOSIS or MEIOSIS.
• (See Chart #1 and 14.11)
★ Growth and development arise from :
1. Increase and numbers of cells.
2. Differentiation of cells into different types of tissues.

PRINCIPAL PARTS of a GENERALIZED CELL ⟶ PRINCIPAL PARTS of a CELL

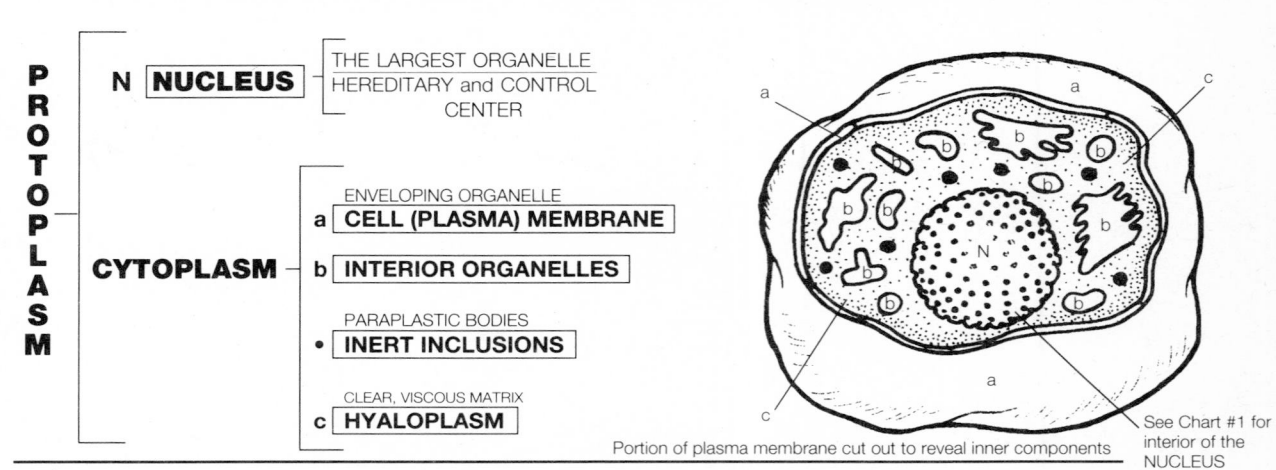

P R O T O P L A S M

N **NUCLEUS** — THE LARGEST ORGANELLE HEREDITARY and CONTROL CENTER

CYTOPLASM

ENVELOPING ORGANELLE
a **CELL (PLASMA) MEMBRANE**
b **INTERIOR ORGANELLES**

PARAPLASTIC BODIES
• **INERT INCLUSIONS**

CLEAR, VISCOUS MATRIX
c **HYALOPLASM**

Portion of plasma membrane cut out to reveal inner components

See Chart #1 for interior of the NUCLEUS

PRINCIPAL INTERIOR ORGANELLES:
Functional Compartmentalization within the Watery Hyaloplasm

1 **ENDOPLASMIC RETICULUM (ER)**
2 **RIBOSOMES** MICROSOMES
3 **GOLGI APPARATUS** GOLGI COMPLEX
4 **MITOCHONDRIA**
5 **LYSOSOMES** CONTAIN ENZYMES
6 **VACUOLES**
7 **PEROXISOMES**
8 **CENTROSOME** CENTROSPHERE
N **NUCLEUS**
   N_1 **NUCLEOLUS**
   CYTOPLASMIC LATTICE
9 **MICROTUBULES**
10 **MICROFILAMENTS**
   MICROFIBRILS

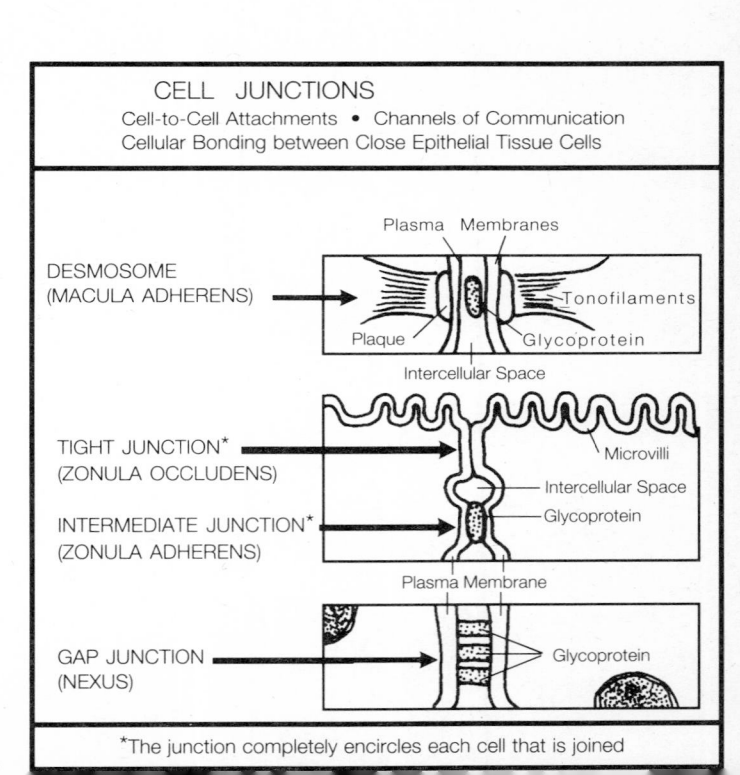

CELL JUNCTIONS
Cell-to-Cell Attachments • Channels of Communication
Cellular Bonding between Close Epithelial Tissue Cells

Plasma Membranes

DESMOSOME (MACULA ADHERENS)
Plaque   Glycoprotein   Tonofilaments
Intercellular Space

TIGHT JUNCTION* (ZONULA OCCLUDENS)
Microvilli
Intercellular Space
Glycoprotein

INTERMEDIATE JUNCTION* (ZONULA ADHERENS)
Plasma Membrane

GAP JUNCTION (NEXUS)
Glycoprotein

*The junction completely encircles each cell that is joined

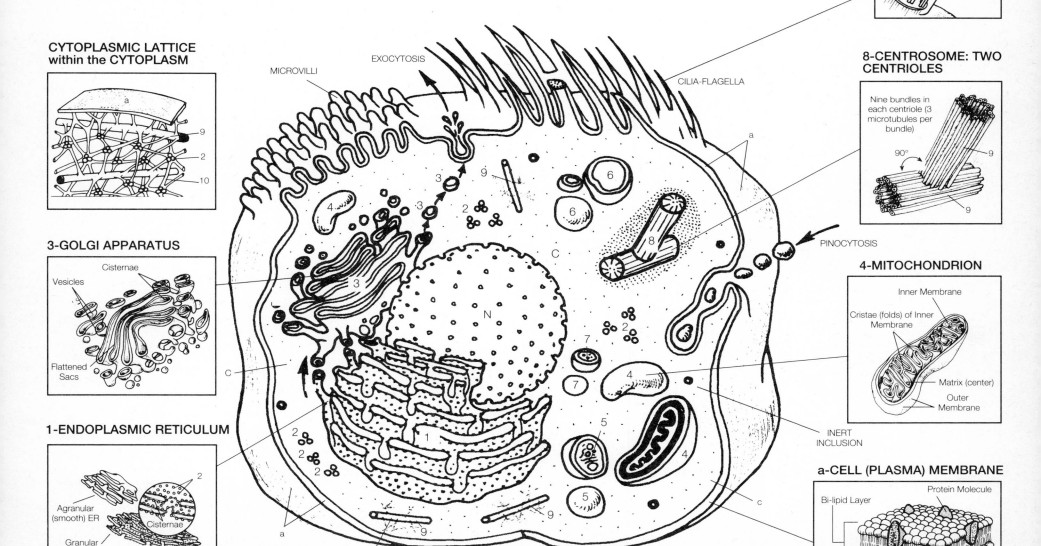

★ The structure and form of a cell are closely correlated with its functions. Cells of one tissue differ from those of other tissues, depending on the specialized function they perform (in accordance with their modified structure).

★ A cell has the POWER OF EXERCISING THE VITAL PROCESSES OF LIFE required for the maintenance of the life of the human organism as a whole. In that sense, the cell can be viewed as a MICROCOSM of the MACROCOSMIC world of

HUMANS. The essential life processes that are carried on at the cellular level are, in the human organism, delegated to the SEPARATE WHOLE SYSTEMS, which function together to constitute the living individual. (Refer to the LIFE REQUISITES chart on 2.0). Before entering into a detailed chapter-by-chapter journey through the separate body systems (and the organs grouped into those systems), we will familiarize ourselves with the ORGANIZATION OF THE 4 PRIMARY types of BODY TISSUES and their principal subdivisions (See 2.2).

CILIA-FLAGELLA

Nine bundles in each cilia (2 microtubules per bundle) + 2 in middle

9

CYTOPLASMIC LATTICE within the CYTOPLASM

a
9
2
10

3-GOLGI APPARATUS

Vesicles
Cisternae
Flattened Sacs

1-ENDOPLASMIC RETICULUM

2
Agranular (smooth) ER
Cisternae
Granular (rough) ER
a

8-CENTROSOME: TWO CENTRIOLES

Nine bundles in each centriole (3 microtubules per bundle)
90°
9
9

4-MITOCHONDRION

Inner Membrane
Cristae (folds) of Inner Membrane
Matrix (center)
Outer Membrane

a-CELL (PLASMA) MEMBRANE

Bi-lipid Layer
Protein Molecule

MICROVILLI
EXOCYTOSIS
CILIA-FLAGELLA
a
PINOCYTOSIS
INERT INCLUSION
c

(Size of microtubules in relation to other parts is exaggerated)
(Size of All Parts Exaggerated)

ORGANIZATION: THE TISSUE LEVEL
Histology: Classification of Tissues

Color the names of the four principal groups of tissues in the first box, and use the same colors for the names as they repeat in the lower left vertical rectangular boxes.

★ See Charts #2, #3, #4

★ HISTOLOGY is the study of tissues within body organs. TISSUES are aggregated groups or layers of similar specialized cells whose unique structure is directly related to the performance of a specific function (or set of functions).

TISSUES of the body are classified into 4 principal groups, with 11 basic kinds of tissues derived from them. (Note the difference between the 4 kinds of CONNECTIVE TISSUE and their different TYPES.)
Classification of tissues is based on:
• EMBRYONIC DEVELOPMENT (derived from one or all of the 3 layers of the embryo)—ECTODERM, MESODERM, ENDODERM
• STRUCTURAL ORGANIZATION
• FUNCTIONAL PROPERTIES

Two Highly Specialized Groups
MUSCLE TISSUE (3 kinds)
NERVOUS TISSUE (2 kinds)
Two All-Purpose, Wide-Variety Groups
EPITHELIAL (LINING) TISSUES (2 kinds)
CONNECTIVE (SUPPORTING) TISSUES (4 kinds)

★ | MUSCLE TISSUE |
• Derived from MESODERM.
• Outstanding characteristic is the ability to shorten or *contract,* effecting movement of the body or an organ within the body for movement of materials.
• So specialized for contraction that the tissue cannot replicate after birth.
• Muscle cells (fibers) are elongated in the direction of contraction.
• Little intercellular matrix with the fibers lying close together.
• Three kinds: SMOOTH, CARDIAC, SKELETAL
Properties of Action
  • CONTRACTILITY • CONDUCTIVITY
  • IRRITABILITY • ELASTICITY

★ | NERVOUS TISSUE |
• Derived from ECTODERM.
• Initiates and conducts IMPULSES for coordination of body activities.
• Two kinds: NEURONS and NEUROGLIA
NEURONS are specialized in:
• IRRITABILITY • CONDUCTIVITY • INTEGRATION
(NEURONS are incapable of mitosis)
NEUROGLIA are 5 times as abundant as neurons, and they support and assist neurons. They are NONRECEPTIVE and NONCONDUCTIVE. There are 6 types of NEUROGLIA (See 7.1)

★ **Master Tissues of the Body:**
NERVOUS TISSUE and SKELETAL MUSCLE
• These tissues work together in receiving stimuli from the external and internal environment and then reacting to them. They are the most highly specialized tissues in the body.

★ **Four Principal Groups of Tissues**

Two Highly Specialized and Modified
| MUSCLE TISSUES |
| NERVOUS TISSUES |

Two All-Purpose, Wide-Variety
| EPITHELIAL TISSUES |
| CONNECTIVE TISSUES |

Components of Tissue

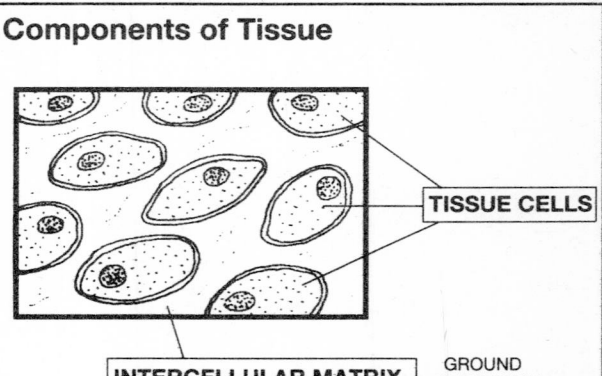

TISSUE CELLS

INTERCELLULAR MATRIX — GROUND SUBSTANCE

MUSCLE TISSUE (3 KINDS)

| SMOOTH MUSCLE | CARDIAC MUSCLE | SKELETAL MUSCLE |

(VISCERAL, PLAIN)

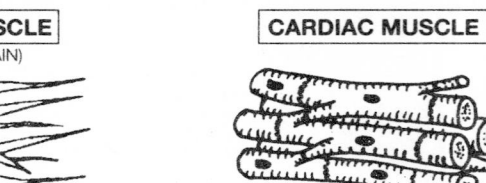
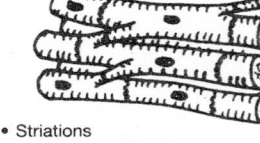

SMOOTH MUSCLE:
• Spindle-shaped, single nucleus
• Nonstriated
• Least specialized
• Slow, rhythmical contraction
• No voluntary control (Involuntary)
• Found in walls of hollow internal organs (VISCERA and BLOOD VESSELS)

CARDIAC MUSCLE:
• Striations
• More highly specialized
• Rapid, rhythmical contraction (spreads through whole muscle mass)
• No voluntary control (Involuntary)
• Found only in HEART WALL (MYOCARDIUM)
• INTERCALATED DISCS adhere cells to cells
• Branching, but no crossing of protoplasm

SKELETAL MUSCLE:
• Cylinder-shaped, Multinucleated
• Marked Striations
• Most highly specialized
• Very rapid, powerful contractions (of individual fibers)
• Conscious control (voluntary)
• Found in all SKELETAL MUSCLES (HEAD, TRUNK, LIMBS) TONGUE, PHARYNX, and UPPER ESOPHAGUS
• Thick bundles, no branching

NERVOUS TISSUE (2 KINDS)

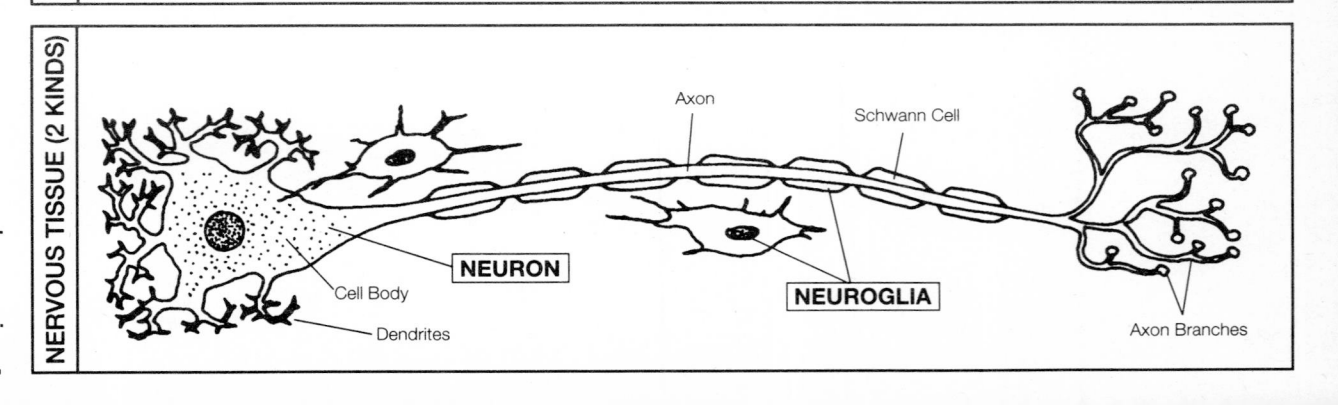

Axon

Schwann Cell

| NEURON |

| NEUROGLIA |

Cell Body

Dendrites

Axon Branches

★ EPITHELIA (See 2.3)

- Derived from ECTODERM, MESODERM, and ENDODERM (the 3 layers of the embryo)
- Cover body and organ surfaces
- Line body cavities and lumina
- Form various glands
- Form the EPIDERMIS of the SKIN
- Form the surface layer of mucous and serous membranes
- Consist of one or more cellular layers. One layer = SIMPLE EPITHELIA
- Multilayered = STRATIFIED EPITHELIA
- Tightly packed cells with *little* INTERCELLULAR MATRIX between them
- Typically avascular (without blood vessels)
- Rest on a basement membrane
- Two kinds: COVER and LINING EPITHELIUM and GLANDULAR

General Functions
- PROTECTION
- ABSORPTION
- EXCRETION
- SECRETION

Specialized Functions
- MOVEMENT OF SUBSTANCES THROUGH DUCTS
- PRODUCTION OF GERM CELLS
- RECEPTION OF STIMULI
- EXCELLENT ABILITY TO REGENERATE (as frequently as every 24 hours)

★ CONNECTIVE TISSUE (See 2.4)

- Derived from MESODERM
- All other connective tissues differentiate from an EMBRYONIC connective tissue called MESENCHYME: MESENCHYMAL CELLS in a SEMISOLID, JELLYLIKE MATRIX with fine fibrils
- Connective tissues are located throughout the body
- 4 kinds of connective tissues: CONNECTIVE TISSUE PROPER, CARTILAGE, BONE, and BLOOD

General Functions
- SUPPORT and BIND OTHER TISSUES, and TISSUES and PARTS
- STORE NUTRIENTS
- MANUFACTURE PROTECTIVE and REGULATORY MATERIALS

★
- The bulk of connective tissue consists of intercellular matrix, whose nature gives each type of connective tissue its unique properties.
- The cells of connective tissue are few in number compared with the matrix substance.
- Connective tissues are highly vascular (except for cartilage).

★ 5 types of CONNECTIVE TISSUE PROPER:
- LOOSE (AREOLAR)
- ADIPOSE (FATTY)
- DENSE FIBROUS (REGULAR and IRREGULAR)
- ELASTIC
- RETICULAR

★ 2 types of DENSE CONNECTIVE TISSUE:
- CARTILAGE
- BONE

★ 1 type of CONNECTIVE TISSUE free–floating in a FLUID MATRIX
- BLOOD

2.2

COVER and LINING EPITHELIUM (2 TYPES)

EPITHELIA (2 KINDS)

SIMPLE (SINGLE LAYER)

 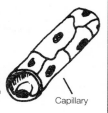

BASEMENT MEMBRANE:
- GLYCOPROTEIN from the epithelial cells
- COLLAGEN and RETICULAR FIBER mesh from underlying connective tissue

Capillary

STRATIFIED (MULTILAYERED)

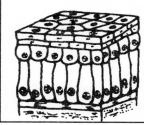

GLANDULAR

CONNECTIVE TISSUE (4 KINDS)

CONNECTIVE TISSUE PROPER (5 TYPES)

4 TYPES of CELLS in a SEMISOLID MATRIX with thick collagenous and elastic fibers

LOOSE (AREOLAR)　　ADIPOSE (FATTY)　　DENSE FIBROUS　　ELASTIC

1 TYPE of CELLS in a SEMISOLID MATRIX with fine network of extracellular reticular fibers

RETICULAR

DENSE CONNECTIVE TISSUE (2 TYPES): CARTILAGE and BONE (OSSEOUS TISSUE)

CARTILAGE (3 TYPES)
CELLS in a SOLID ELASTIC MATRIX (with fibers)

HYALINE

FIBROCARTILAGE　　ELASTIC

BONE (2 TYPES)
CELLS in a SOLID RIGID MATRIX

impregnated with Calcium and Magnesium salts

- Forms rigid body framework
- 2 types: COMPACT and SPONGY

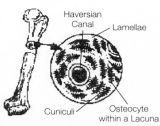

Haversian Canal
Lamellae
Osteocyte within a Lacuna
Cuniculi

BLOOD
CELLS in a FLUID MATRIX

(fibrils not present until clotting occurs)

- Cells float free in the matrix

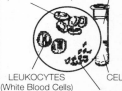

ERYTHROCYTES (Red Blood Cells)
PLASMA
LEUKOCYTES (White Blood Cells)
CELLS
THROMBOCYTES (Platelets)

ORGANIZATION: THE TISSUE LEVEL
All-Purpose, Wide-Variety: Epithelial Tissue

★ See Chart #3 for Type, Structure, Function, and Location of Epithelial Tissue

Choose you color palette

The EPITHELIUM is the layer of cells forming the Epidermis of the Skin (See 3.1) and the surface layer of mucous and serous membranes (See 1.4). The cells rest on a *Basement Membrane* and lie close to each other with little intercellular material between them.
The EPITHELIUM may be either SIMPLE (Consisting of a single layer) or STRATIFIED (Consisting of several layers).
• EPITHELIUM CELLS may be FLAT (SQUAMOUS), CUBE-SHAPED (CUBOIDAL) or CYLINDRICAL (COLUMNAR).

• Modified forms of epithelium include CILIATED, PSEUDOSTRATIFIED, GLANDULAR, and NEUROEPITHELIUM
• NEUROEPITHELIUM is a specialized epithelial structure forming the fermintation of a nerve of special sense. (Gustatory cells, olfactory cells, inner ear hair cells, and rods and cones of the retina. See 7.18-7.23)
• GLANDULAR EPITHELIUM lies in clusters deep to the covering and lining EPITHELIUM. A GLAND may consist of one cell (unicellular GOBLET cells, which secrete mucous) or 2 groups of highly specialized multicellular epithelium that secrete substances into ducts or into the blood:

I. EXOCRINE GLANDS
Secrete their products into ducts or tubes that empty at the surface of the Covering and Lining Epithelium (at the skin surface or into the cavity or lumen of a hollow organ): ENZYMES, OIL, or SWEAT.
II. ENDOCRINE GLANDS
Secrete their products directly into the blood: HORMONES.

---

2 General Kinds of EPITHELIAL TISSUE
- ■ COVERING and LINING EPITHELIUM
- ■ GLANDULAR EPITHELIUM

■ COVERING and LINING EPITHELIUM

SIMPLE
PSEUDOSTRATIFIED* } ONE LAYER
STRATIFIED } MULTILAYERED

* PSEUDOSTRATIFIED is actually SIMPLE epithelium because each cell is in contact with the BASEMENT MEMBRANE.

Lower Surface Support of COVERING and LINING EPITHELIUM
a | BASEMENT MEMBRANE |

b | UNDERLYING SUPPORTIVE CONNECTIVE TISSUE |

■ GLANDULAR EPITHELIUM
EXOCRINE GLANDS
ENDOCRINE GLANDS

Component Structure of GLANDULAR EPITHELIUM
▣ c | DUCT OF GLAND |

▣ d | SECRETORY PORTION |

COVERING and LINING EPITHELIUM

I. SIMPLE (1 LAYER)
(1) | SIMPLE SQUAMOUS |
(2) | SIMPLE CUBOIDAL |
(3) | SIMPLE COLUMNAR |

II. PSEUDOSTRATIFIED (1 LAYER) *
(4) | PSEUDOSTRATIFIED CILIATED COLUMNAR WITH GOBLET CELLS |

III. STRATIFIED (MULTILAYERED)
(5) | STRATIFIED SQUAMOUS | ‡
(6) | STRATIFIED CUBOIDAL |
(7) | STRATIFIED COLUMNAR |
(8) | STRATIFIED TRANSITIONAL |

‡ In the skin, STRATIFIED SQUAMOUS TISSUE CELLS contain the PIGMENT KERATIN.

• SQUAMOUS EPITHELIUM is classified either as ENDOTHELIUM (which lines the blood vessels and the heart) or MESOTHELIUM (which lines the serous cavities)

GLANDULAR EPITHELIUM

I. EXOCRINE GLANDS
Secretion reaches an epithelial surface either directly or through a DUCT

Structural Classification:
9 | UNICELLULAR GOBLET CELL |

MULTICELLULAR
10 | SIMPLE |

TUBULAR
BRANCHED TUBULAR
COILED TUBULAR
ACINAR
TUBULOACINAR

11 | COMPOUND |

TUBULAR
ACINAR
TUBULOACINAR

Functional Classification:
(12) | HOLOCRINE | gland and/or its secretion consists of altered cells of the same gland
(13) | MEROCRINE | glandular cell remains intact during elaboration and discharge of secretion
(14) | APOCRINE | secretory cells contribute part of their protoplasm to the material secreted

II. ENDOCRINE GLANDS:
Secretion is directly into the BLOODSTREAM (See Chapter 8)

COVERING and LINING EPITHELIUM

I. SIMPLE (1 Layer)

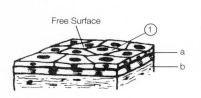

Free Surface

① a b

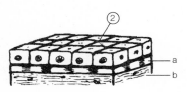

② a b

Unicellular Exocrine Gland: Goblet Cell

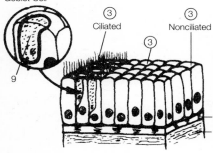

③ Ciliated ③ Nonciliated ③

9 a b

II. PSEUDOSTRATIFIED (1 layer)

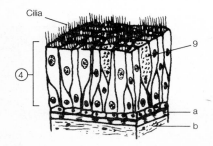

Cilia

④ 9 a b

III. STRATIFIED (2 or more layers)

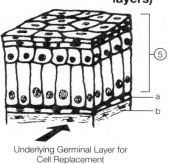

⑤ a b

Underlying Germinal Layer for Cell Replacement

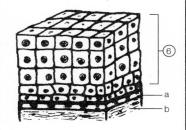

⑥ a b

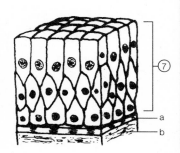

⑦ a b

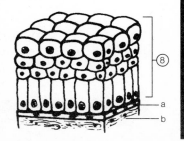

⑧ a b

GLANDULAR EPITHELIUM

I. EXOCRINE GLANDS

10 SIMPLE

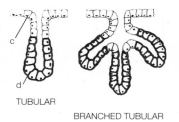

c d

TUBULAR

BRANCHED TUBULAR

COILED TUBULAR

ACINAR

TUBULOACINAR

c d

11 COMPOUND

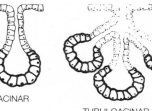

c

TUBULAR

ACINAR

TUBULOACINAR

c d

FUNCTIONAL CLASSIFICATION

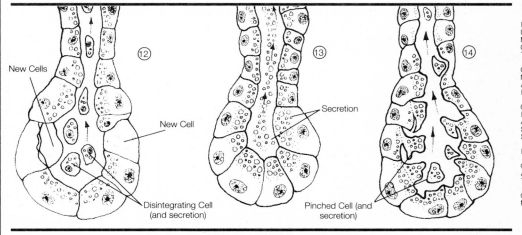

New Cells

⑫

New Cell

⑬ Secretion

⑭

Disintegrating Cell (and secretion)

Pinched Cell (and secretion)

II. ENDOCRINE GLANDS

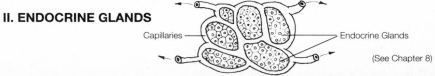

Capillaries

Endocrine Glands

(See Chapter 8)

ORGANIZATION: THE TISSUE LEVEL
All-Purpose, Wide-Variety: Connective Tissue

Color in the left rectangular boxes.
Use a consistent color scheme for
the common elements.
★ See Chart #4

The four general kinds of connective tissue
(CONNECTIVE TISSUE PROPER, CARTILAGE, BONE, and BLOOD)
are all derived from embryonic mesenchyme:

■ *BONE (OSSEOUS) TISSUE: The hardest and most rigid*
connective tissue. For details, see 4.1.

■ *BLOOD (VASCULAR) TISSUE: Specialized, viscous*
connective tissue, only connective tissue with a fluid
matrix. For details, see 9.1.

In this section, we will only discuss CONNECTIVE TISSUE PROPER and CARTILAGE

Transverse
section of a 4-
week embryo

MESENCHYME: EMBRYONIC
CONNECTIVE TISSUE:

a MESENCHYMAL CELLS

b JELLYLIKE MATRIX

- Derived from mesoderm
- Migrates and gives rise to all other kinds of
  connective tissue
- Cytoplasmic processes of the mesenchymal cells
  are drawn out to touch neighboring cells
- MUCOUS CONNECTIVE TISSUE is a type found
  only in the umbilical cord of the fetus

■ CONNECTIVE TISSUE PROPER: SEMISOLID MATRIX

1 LOOSE (AREOLAR)

Also called LOOSE FIBROUS
PACKING TISSUE between organs
Forms sheaths around muscles, nerves, and blood
   vessels

2 ADIPOSE (FATTY)

Provides a protective cushion for organs
The insulating layer in deep skin
Cytoplasm of fat cells is displaced by STORED FAT
RESERVES

3 DENSE FIBROUS

Very strong
Inelastic, yet pliable
(e.g., ligaments, capsules of joints, heart valves)

4 ELASTIC

Very strong
Flexible and extensible
(e.g., in walls of blood vessels and air passages)

5 RETICULAR

Reticular cells are phagocytic (capable of ingesting
   particles)
Reticular fibers form a 3-dimensional net for the
   framework of the: spleen, lymph nodes, bone marrow

■ CARTILAGE: SOLID ELASTIC MATRIX

6 HYALINE (GRISTLE)

Firm, yet resilient
Transition stage in bone formation
(e.g., in air passages, ends of bones at joints)

7 FIBROCARTILAGE

Also called WHITE FIBROCARTILAGE
Tough
Resistant to stretching
Shock-absorbing function between vertebrae

8 ELASTIC

Also called YELLOW FIBROCARTILAGE
Flexible, resilient
(e.g., in the larynx and ear)

Common Elements of CONNECTIVE TISSUE PROPER

COLLAGEN (WHITE) FIBERS	ELASTIC (YELLOW) FIBERS
RETICULAR FIBERS	FIBROBLASTS (FIBROCYTES)
MACROPHAGES (HISTIOCYTES)	PLASMA CELLS
FAT CELLS (ADIPOCYTES)	MATRIX

Common Elements of CARTILAGE

COLLAGEN (WHITE) FIBERS	ELASTIC (YELLOW) FIBERS
LACUNA	CHONDROCYTES
MATRIX	

C O N N E C T I V E   T I S S U E   P R O P E R

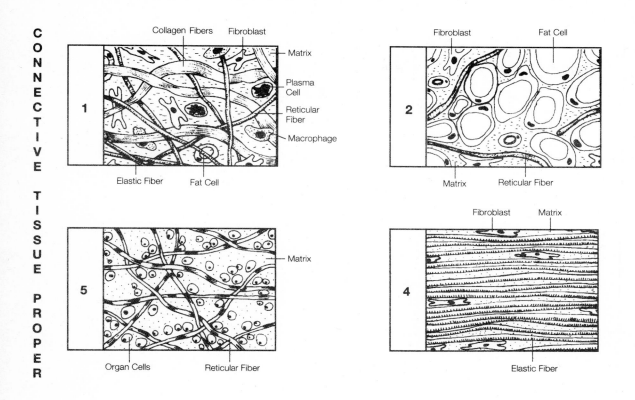

1
- Collagen Fibers
- Fibroblast
- Matrix
- Plasma Cell
- Reticular Fiber
- Macrophage
- Elastic Fiber
- Fat Cell

2
- Fibroblast
- Fat Cell
- Matrix
- Reticular Fiber

5
- Matrix
- Organ Cells
- Reticular Fiber

4
- Fibroblast
- Matrix
- Elastic Fiber

R E G U L A R

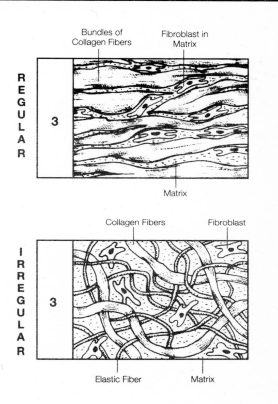

3
- Bundles of Collagen Fibers
- Fibroblast in Matrix
- Matrix

I R R E G U L A R

3
- Collagen Fibers
- Fibroblast
- Elastic Fiber
- Matrix

C A R T I L A G E

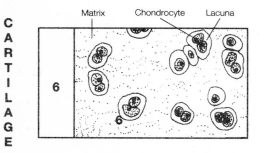

6
- Matrix
- Chondrocyte
- Lacuna

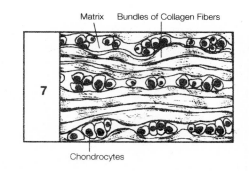

7
- Matrix
- Bundles of Collagen Fibers
- Chondrocytes

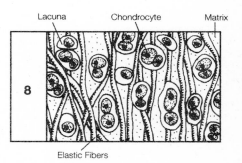

8
- Lacuna
- Chondrocyte
- Matrix
- Elastic Fibers

Chapter 2: *ORGANIZATION* : CHARTS

Chart #1 : **The Generalized Cell and Its Components (See 2.1)**

Protoplasm = I The Nucleus + II The Cytoplasm

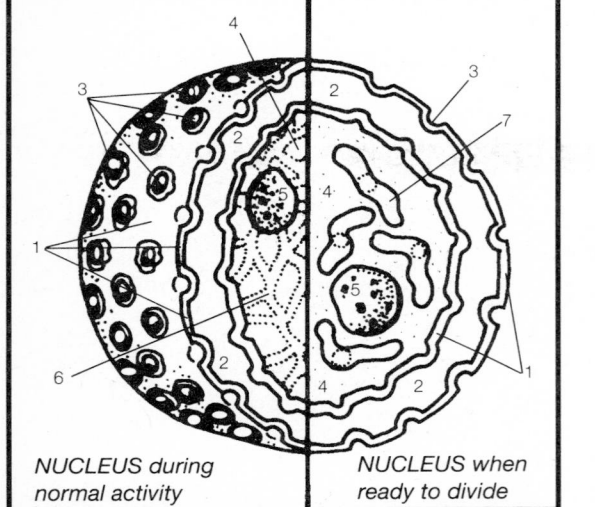

INTERPHASE   MITOSIS

NUCLEUS during normal activity   *NUCLEUS when ready to divide*

★ Pores not seen joined in the illustration to form passageway

I. The Nucleus:

- The supervisor of cell activity—exercises control of cellular development and function.
- The vital body in the protoplasm of the cell, containing the chromosomes.
- The essential agent in growth, metabolism, reproduction, and transmission of cell characteristics.

1 DOUBLE NUCLEAR MEMBRANE - Double layer of protein that acts as a boundary for the NUCLEUS and as a semipermeable membrane.
2 PERINUCLEAR CISTERNA - Space between layers of the nuclear membrane (envelope).
3 NUCLEAR PORES - The two membranes of the nuclear envelope join at regular intervals to form NUCLEAR PORES for migration of ribosomal RNA into the HYALOPLASM.
4 NUCLEOPLASM - Also called KARYOLYMPH, the specialized protoplasmic substance of the nucleus.
5 NUCLEOLUS - Spheroid body within the nucleus where RNA (ribonucleic acid), especially ribosomal RNA, is produced and stored.
6 CHROMATIN - Substance containing the genetic material DNA (deoxyribonucleic acid), histones, and other proteins. It is present in the nondividing or normal metabolic phase of the cell (INTERPHASE). It is a dispersed form of the chromosomes, a threadlike, extended mass of DNA.
7 CHROMOSOMES - Rodlike, "packaged" DNA that becomes microscopically visible in the dividing phase of the cell (MITOSIS). The replication of the DNA molecule in preparation for cell division takes place during mid-INTERPHASE. On the chromosomes are the GENES, the carriers of inheritable characteristics.

II. The Cytoplasm • The contents (protoplasm) of the cell outside the nucleus

Cell Structures and Organelles

CELL (PLASMA) MEMBRANE
(Exterior Enveloping Organelle
of the entire Cell)

Function and Description

- Thin layer of proteins and phospholipids that acts as a boundary for each cell and as a selectively permeable filter.
- Modifications of the cell membrane include MICROVILLI and CILIA-FLAGELLA.
- PROCESSES OF THE CELL MEMBRANE - Methods for ingestion of foreign particles into the cell include ENDOCYTOSIS, PINOCYTOSIS, and PHAGOCYTOSIS. Methods for the discharge of particles out of the cell include EXOCYTOSIS and EMIOCYTOSIS.

Interior Organelles

ENDOPLASMIC RETICULUM (ER)

Channels formed by a pair of parallel membranes that are continuous with the plasma and nuclear membranes. It also serves as a platform for protein synthesis by ribosomes.

RIBOSOMES

Spheroid bodies attached to the surface of the endoplasmic reticulum. Specific site of protein synthesis.

GRANULAR CISTERNAE

Rough areas along the endoplasmic reticulum where ribosomes attach.

Chart #1 : (cont'd)

AGRANULAR CISTERNAE	Smooth areas along the endoplasmic reticulum where there are no ribosomes attached.
GOLGI APPARATUS	Four to eight stacked saclike structures. Involved in the secretion of proteins formed at the granular cisternae of the endoplasmic reticulum and of lipids formed at the agranular cisternae of the endoplasmic reticulum. Vesicles—Terminal ends of the Golgi apparatus that pinch off and contain proteins and/or lipids. Cisternae—Stacked elements of the Golgi apparatus containing cavities.
MITOCHONDRION	Rod-shaped structure involved in the conversion of stored energy into ATP. Outer Membrane—Serves as a boundary for the mitochondrion. Inner Membrane—Folded membrane that creates a large surface area for energy conversion reactions to occur. Folds are known as cristae. Matrix—Central region of mitochondrion.
LYSOSOMES	Spheroid organelles containing digestive enzymes that digest foreign material and nutrients. If part of the cell is damaged, the limiting membrane around the lysosome ruptures, releasing the enzymes that help to digest the injured tissue.
VACUOLE	Any cavity or space within the cytoplasm surrounded by membrane.
PEROXISOMES	Microbodies present in vertebrate animal cells, containing a great number and wide variety of enzymes important in cell metabolism.
CENTROSOME *(CENTROSPHERE)*	Dense area of cytoplasm that is generally spheroid and contains 2 centrioles.
CENTRIOLE *(Structures within the CENTROSOME)*	Paired cyclindrical structures comprised of a ring of nine evenly spaced bundles. Each bundle consists of three microtubules. During cell division the centrioles organize spindle fibers.

Paraplastic Bodies

INERT INCLUSIONS	Lifeless, temporary, principally organic molecules involved in overall body functions. Includes: 1. FOOD SUBSTANCES (Stored fuel): FAT GLOBULES, GLYCOGEN GRANULES, PROTEIN MASSES. 2. PIGMENT GRANULES (ex: Melanin). 3. SECRETORY GRANULES (MUCUS) and WASTE MATERIALS (by-products of cell activities), both awaiting expulsion from the cell. 4. CRYSTALS of various substances. 5. CHROMOPHILIC SUBSTANCE (Nissl bodies, found in neuron cell bodies).

Chart #1 : (cont'd)

Clear, Viscous Matrix

HYALOPLASM — The fluid portion of cytoplasm, 75–90% water, called BASIC PROTOPLASM, or the BASIC GROUND SUBSTANCE.

Cytoplasmic Lattice

Irregular, three-dimensional network that pervades the cytoplasm.

MICROFILAMENTS — Submicroscopic flexible strands of the protein actin. Involved in the constant directional movement of organelles.

MICROTUBULES — Hollow cylinders of protein subunits called tubulins. The tubules are involved in: organelle movement, DNA division, and cell shape, and are part of cilia and flagella.

Cell Junctions

Cell-to-cell attachments and channels of communication.

DESMOSOME (MACULA ADHERENS) — Small thickening at the middle of an intracellular bridge, frequently found in the stratified epithelium of skin and the small intestine.

TIGHT JUNCTION (ZONULA OCCLUDENS) — Points along the junction between adjacent cells in which the plasma membranes have been fused, thereby occluding the intracellular channel. Common in the small intestine where material must be directed through the cell for filtration and absorption.

INTERMEDIATE JUNCTION (ZONULA ADHERENS) — Located near a tight junction in the small intestine epithelium. These junctions are similar to the desmosomes but are less dense. Provides strength in the intracellular junctions.

GAP JUNCTION (NEXUS) — Point along intracellular junction where two plasma membranes come in close proximity. The gap is bridged by membrane proteins that provide for a low resistance electrical bridge between smooth and cardiac muscle cells. The structure is also found in epithelial cells.

Cell Division (★ See 14.11)

MEIOSIS — Specialized germ cells (SPERMATOZOA and OVA) contain in their nuclei the genes for hereditary characteristics. In MEIOSIS, 2 successive divisions of the nucleus of germ cells produce cells that contain 1/2 the number of chromosomes present in all other (SOMATIC) cells.

MITOSIS — The other type of cell division, in which each newly formed daughter cell contains the same number of chromosomes as the parent cell [46 chromosomes (23 pairs) per cell in humans]. It is the process by which the body grows, and by which somatic cells are replaced.

Chart #2 : Four Principal Groups of Tissue (See 2.2, 2.3, 2.4)

	TYPE	COMPOSITION	SOCIAL STRUCTURE	INTERCELLULAR MATRIX	VASCULAR SUPPLY	EMBRYOLOGICAL DEVELOPMENT	LOCATION	FUNCTION
A L L - P U R P O S E / W I D E - V A R I E T Y	**EPITHELIAL**	SIMPLE EPITHELIA (1-layer) / STRATIFIED EPITHELIA (Multilayered) / GLANDULAR EPITHELIA	closely packed	very little	avascular	ECTODERM MESODERM ENDODERM	• covers free body surface (outer layer) of skin and visceral organs • lines body cavities and lumina (inner layers) • forms glands (secretory portions)	• protection • absorption • secretion
	CONNECTIVE	in the adult, composed of CONNECTIVE TISSUE PROPER, CARTILAGE, OSSEOUS (BONE) TISSUE, and VASCULAR (BLOOD) TISSUE	loose, widely scattered	• more matrix than cells • matrix determines quality of tissue	highly vascular (except to dense fibrous and cartilage)	MESODERM	• found throughout the body • does not occur on free surfaces of the body or body cavities	• binds organs together • protection • structural support • metabolic support • repair of body organs
H I G H L Y M O D I F I E D a n d S P E C I A L I Z E D	**MUSCLE TISSUES**	SMOOTH MUSCLE / CARDIAC MUSCLE / SKELETAL MUSCLE	least specialized / more specialized	very little	extensive vascular supply	MESODERM	• walls of hollow internal organs (SMOOTH) • heart muscle (CARDIAC) • covers and spans skeletal joints via the tendons (SKELETAL)	• CONTRACTION • body locomotion • movement of materials through the body • movement of relative body parts • maintenance of body posture • heat production
	NERVOUS TISSUES	NEURONS NEUROGLIA (5–1 ratio to neurons)	MASTER TISSUES OF THE BODY • NERVOUS TISSUE and SKELETAL MUSCLE • The 2 most highly specialized tissues • Highly integrated	very little	extensive vascular supply	ECTODERM	• centered in the brain and spinal cord, with peripheral branches throughout the body	• CONDUCTION • responds to stimuli; initiates and conducts nerve (electrical) impulses • coordinates body activities—regulates organs and glands • stores memories thinking • Neuroglia bind and support neurons

Chart #3 : Epithelial Tissue: Covering and Lining Epithelium, and Glandular Epithelium (See 2.3)

Covering and Lining Epithelium

Type	Structure	Function	Location
I. SIMPLE SQUAMOUS	Single layer of tilelike cells	Absorption, secretion, and filtration. Reduces friction between surfaces.	Covers and lines the organs, skin, cavities, and hollow structures.
			Called ENDOTHELIUM inside the heart, lymphatic, and blood vessels. Called MESOTHELIUM when covering viscera and ventral body cavity.
CUBOIDAL	Single layer of square cube-shapedcells	Secretion and absorption	Lines: renal nephron, ducts of glands, capsule of eye, and pigmented layer of retina; covers ovary.
	SIMPLE CUBOIDAL ⟶	Nonsecretory; protects underlying structures	Small ducts, e.g., salivary glands
	SECRETORY CUBOIDAL ⟶	1. In serous gland, forms watery secretion containing e.g., digestive enzymes	Many glands, e.g., serous-secreting portion of a mixed salivary gland
		2. In mucous gland, forms viscous lubricant, mucus	Mucous glands, e.g., mixed salivary gland
COLUMNAR	Single layer of rectangular cells	Secretion and absorption	Lines the digestive tract from the stomach through the colon. Ducts of glands and gall bladder.
	SIMPLE COLUMNAR ⟶	Protects and lines	Large ducts of the kidney
	SECRETORY COLUMNAR		
	Mucous Columnar ⟶	1. Secretes mucus to protect stomach from its own acid and digestive enzymes	Surface lining of the stomach
	Goblet Columnar ⟶	2. Goblet cells secrete a viscous, lubricant protective mucus	e.g., Lining of the intestine
	STRIATED or BRUSH-BORDERED COLUMNAR ⟶	Absorbs foodstuffs	Small intestine
	Ciliated Columnar ⟶	Cilia form currents to waft ovum toward the womb	Uterine tube

Chart #3 : (cont'd)

Covering and Lining Epithelium

Type	Structure	Function	Location
II. PSEUDOSTRATIFIED CILIATED COLUMNAR (with Goblet Cells)	All cells attached to basement membrane. Not all cells reach surface. Nuclei are at different levels.	Secretion and surface movement by cilia. In respiratory passages, cilia form currents to move particles trapped by secreted mucus toward mouth.	Lines ducts of many glands, male urethra, and eustachian tube; ciliated variety with goblet cells lines most of bronchial tree; male reproductive tract (epididymis and vas deferens) has pseudostratified columnar epithelium covered with stereocilia. No cilia in the urethra.
III. STRATIFIED (COMPOUND)	Many layers of true stratified cells:		
SQUAMOUS	Multiple layers of squamous cells and other cells	Protection	Lines mouth, esophagus, part of epiglottis, male urethra, and vagina. Covers skin.
	UNCORNIFIED STRATIFIED SQUAMOUS	Most resistant type of epithelium; protects against friction.	Surface locations subject to great wear and tear (internal surfaces: e.g., mouth and esophagus)
	CORNIFIED STRATIFIED SQUAMOUS	Protects against wear and tear, evaporation, and temperature extremes	External skin surfaces; cornified (dead cell) layers are thickest on soles of feet and palms of hands
CUBOIDAL	Multiple layers of cuboidal cells and others	Protection	Ducts of sweat glands, fornix of conjunctiva, cavernous urethra, and epiglottis
COLUMNAR	Multiple layers of columnar cells and others	Protection and secretion	Lines part of male urethra and large excretory ducts
TRANSITIONAL	Similar to stratified squamous tissue but superficial cells are larger and more spheroid. Has deep columnar middle cuboidal and superficial squamous cells.	Permits distension. In urinary passages, prevents penetration of urine into underlying passages.	Lines urinary bladder. Lines male urethra (gradually changes to STRATIFIED SQUAMOUS). Lines calyces, pelvis, and ureter.

Chart #3 : (cont'd)

Glandular Epithelium

I. Exocrine Glands

Glands that secrete their products onto a free surface or into ducts. The secretions are then carried by the ducts into body cavities, lumens of organs, or to the body's surface.

Structural Classification

Type	Function	Example
UNICELLULAR	Lubrication and protection	Goblet cells of respiratory, digestive, reproductive, and urinary systems
MULTICELLULAR	Temperature regulation, protection, lactation, digestion	Sweat glands, sebaceous glands, mammary glands, digestive glands
TUBULAR	Digestion	Small intestinal glands
BRANCHED TUBULAR	Digestion, reproduction	Gastric glands and uterine glands
COILED TUBULAR	Temperature regulation	Some sweat glands
ACINAR	Semen production	Seminal vesicles
TUBULOACINAR	Lubrication of skin	Sebaceous glands
COMPOUND		
TUBULAR	Semen production and digestion	Bulbourethral (Cowper's) gland and liver
ACINAR	Lactation and digestion (starch)	Mammary gland and salivary glands (sublingual and submandibular)
TUBULOACINAR	Digestion	Salivary gland (parotid) and pancreas

Functional Classification

Type	Secretion Type	Example
HOLOCRINE	With secretion, entire cell is lost along with secretion	Sebaceous glands
MEROCRINE	Secretion occurs on demand (as needed) with no cell destruction	Some sweat glands, salivary glands, and pancreas
APOCRINE	Secretions accumulate near end of gland cell. Part of cell is discharged with secretion.	Mammary glands

II. Endocrine Glands
(See Chapter 8)

Ductless glands that secrete their products into the extracellular space around the secretory cells. The secretions are then passed into the capillaries to be transported in the blood.

Example

Pituitary, thyroid, parathyroids, adrenals, pancreas, ovaries, testes, pineal, thymus

Chart #4 : Connective Tissue (See 2.4)

2 of the 4 kinds of connective tissue:
Connective Tissue Proper, and Cartilage
(Bone See 8.1, Blood See 9.1)

I. CONNECTIVE TISSUE PROPER

	Descriptive Components	Function	Location
LOOSE (AREOLAR)	• Collagen (white) fibers • Elastic (yellow) fibers • Reticular fibers • Fibroblasts • Macrophages • Plasma cells • Adipocytes (fat cells) • Mast cells (when it surrounds blood vessels has anticoagulant heparin) • Large semifluid matrix	• Binding and packing material providing strength, elasticity, and support • Binds skin to underlying muscles • Provides nutrients to the skin	• Mucous membranes • Surrounds muscle fasciae • Surrounds blood vessels • Surrounds nerves
ADIPOSE (FATTY)	• A specialized type of loose fibrous connective • Adipocytes (fat cells) derived from fibroblasts; store large fat droplets in the cytoplasm • Fibroblasts • Reticular fibers • Medium semifluid matrix	• Food energy reserve • Supports, protects, and insulates various organs • Reduces heat loss through the skin	• Hypodermis of the skin • Concentrated around the kidneys • Surface of the heart • Marrow of long bones • Padding around joints • Breasts of mature female
DENSE FIBROUS	• Collagen (white) fibers densely packed in bundles • Fibroblasts in small matrix between bundles	• Provides strong attachment and flexible support between various structures	• Tendons—attach muscles to bones, transfer forces of contraction • Ligaments—connect bone to bone across articulation • Aponeuroses—flat bands connect one muscle with another muscle or bone • Membranes around various organs • Perichondrium—surrounds cartilage • Fasciae
ELASTIC	• Elastic (yellow) fibers • Fibroblasts in small matrix between irregular fiber arrangement	• Allows stretching of various organs (up to 1 1/2 X their original lengths) • Will snap back to former size	• Walls of large arteries • Cartilage of larynx • Lung tissue • Walls of trachea • Walls of bronchial tubes • True vocal cords • Between vertebrae as the ligamenta flava
RETICULAR	• 3-dimensional network of reticular fibers • Large jellylike matrix with phagocytic cells	• Forms framework or stroma of various organs • Binds together smooth muscle tissue	• Liver • Spleen • Lymph nodes • Thymus gland • Bone marrow

II. CARTILAGE

	Descriptive Components	Function	Location
HYALINE (GRISTLE)	• Chondrocytes (within Lacunas) • Extremely fine collagenous fibers • Glossy, glassy appearance • Homogeneous matrix	• Most abundant cartilage in the body • Most of the bones in the body start out as hyaline cartilage • Provides movement at joints • Flexibility • Support	• Articular surfaces of bones • Reinforces nose • Costal cartilage—forms a bridge between ends of 1st 10 ribs and the sternum • Parts of larynx, trachea, bronchi, and bronchial tubes
FIBROCARTILAGE	• Chondrocytes (within Lacunas) • Bundles of collagen fibers • Small matrix	• Support and fusion • Durability to withstand tension and compression	• Symphysis pubis—where the 2 pelvic bones articulate • Intervertebral discs—cartilage between the vertebrae • Wedges within the knee joint
ELASTIC	• Elastic (yellow) fibers • Chondrocytes (within Lacunas) • Homogeneous matrix	• Provides flexibility and support while maintaining strength and shape	• External ear • Eustachian (auditory) canals • Epiglottis • Parts of larynx

Chapter 3 Integumentary System

Formula for estimating the percentage of body surface areas, particularly helpful in judging the portion of skin that has been burned.

"RULE OF 9'S": EXTENT AND SEVERITY OF BURNS

THE INTEGUMENT (SKIN)

The Major Body Regions as a Percentage (%) of Skin Surface   Total %

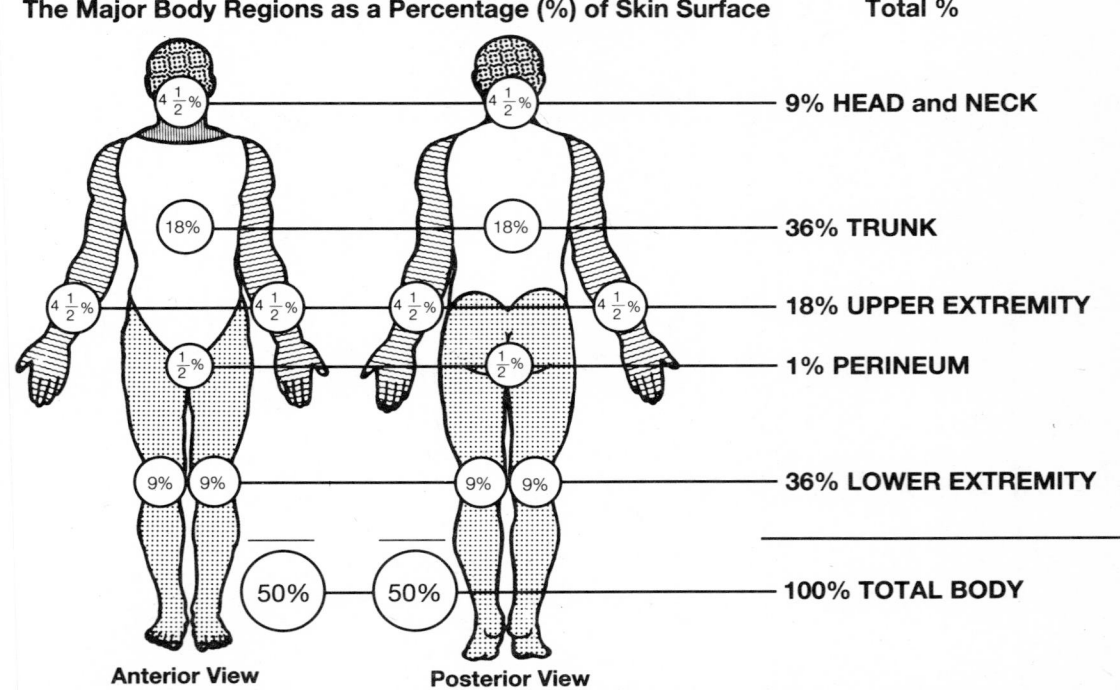

$4\frac{1}{2}$%		9% HEAD and NECK
18%		36% TRUNK
$4\frac{1}{2}$%		18% UPPER EXTREMITY
$\frac{1}{2}$%		1% PERINEUM
9% 9%		36% LOWER EXTREMITY
50% 50%		100% TOTAL BODY

Anterior View   **Posterior View**

DEGREE OF BURN	SKIN DEPTH	LAYER DAMAGE	SKIN SEVERITY
1 FIRST DEGREE	Minimal	Outer epidermis; Shedding	Redness, Pain, Edema (Swelling)
2 SECOND DEGREE	Superficial-Deep Partial	Epidermis and Dermis	Blistering
3 THIRD DEGREE	Full (underlying muscle)	Epidermis and Dermis destroyed	Charring, Coagulation

THE INTEGUMENTARY SYSTEM, which accounts for close to 7% of the total body weight, consists of the SKIN and its associated structures (HAIR, NAILS, SWEAT and OIL GLANDS), integrated with millions of sensory receptors from the Nervous System and an incredibly complex vascular network from the Blood Vascular System. THE SKIN IS THE LARGEST ORGAN IN THE BODY AND IS ESSENTIAL FOR THE SURVIVAL OF THE HUMAN ORGANISM. The skin is often considered a simple covering that basically protects the body and keeps all of its parts together; but it is much more than that. It is an amazingly complex, dynamic organ that interfaces between the constantly changing external environment and the internal environment; its metabolic activity must maintain homeostasis.

★ **See Chart #1: Functions of the Skin**

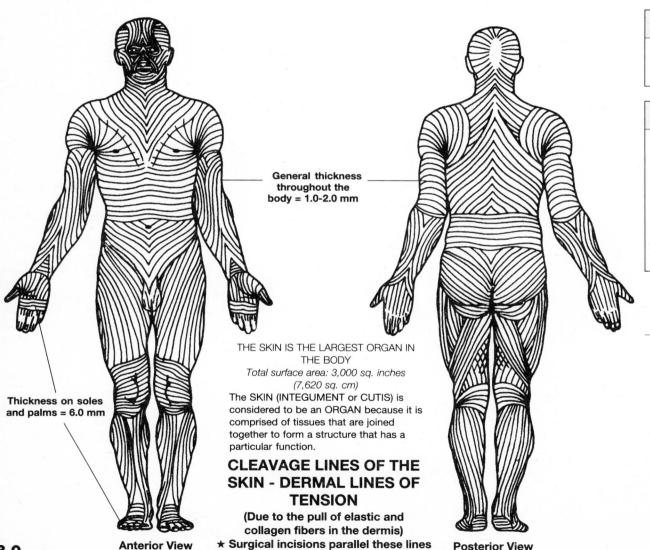

General thickness throughout the body = 1.0-2.0 mm

Thickness on soles and palms = 6.0 mm

THE SKIN IS THE LARGEST ORGAN IN THE BODY
Total surface area: 3,000 sq. inches (7,620 sq. cm)
The SKIN (INTEGUMENT or CUTIS) is considered to be an ORGAN because it is comprised of tissues that are joined together to form a structure that has a particular function.

CLEAVAGE LINES OF THE SKIN - DERMAL LINES OF TENSION

(Due to the pull of elastic and collagen fibers in the dermis)
★ **Surgical incisions parallel these lines**

Anterior View

Posterior View

3.0

System Components

THE SKIN AND ITS STRUCTURAL DERIVATIVES (SUCH AS HAIR, NAILS, SWEAT AND OIL GLANDS)

System Function

- **EXTERNAL SUPPORT OF THE BODY**
- **REGULATION OF BODY TEMPERATURE**
- **PROTECTION OF THE BODY**
- **ELIMINATION OF WASTES**
- **RECEPTION OF CERTAIN STIMULI (SUCH AS TEMPERATURE, PRESSURE, PAIN, TOUCH, AND VIBRATION)**

Chapter Coloring Guidelines

- *You may want to choose your color palette by comparing any colored illustrations from your main textbook.*
- *Arteries (red)*
- *Veins (blue)*
- *Epidermis (fleshy, warm pinks, or browns)*
- *Dermis (light grays)*
- *Fatty Tissue in Hypodermis (yellow)*
- *Sweat and Oil Glands (light or pale greens)*
- *Hair (light browns)*

INTEGUMENTARY SYSTEM: THE INTEGUMENT (SKIN)
Three Principal Layers of Skin (Cutis)

In general, color the EPIDERMIS the color of your skin with close variations for the different layers of epidermis.
Dermis (light grays) Hypodermis (yellow)

★ See Chart #1

The 3 Cutaneous Layers: EPIDERMIS, DERMIS, HYPODERMIS

The 3 principal layers of the skin are established by the eleventh week of embryonic development:

■ The EPIDERMIS is the thinner, outermost layer (cuticle), and is composed of 4–5 sublayers, which are continually formed from within outward (from STRATUM GERMINATIVUM to STRATUM CORNEUM). (The EPIDERMIS and associated structures are derived from the ECTODERM GERM LAYER.)

■ The considerably thicker DERMIS lies directly under the EPIDERMIS and consists of 2 sublayers.

■ The deepest layer is this HYPODERMIS and consists of a single layer. The DERMIS and HYPODERMIS are derived from the MESODERM GERM LAYER.

I. **EPIDERMIS:** STRATIFIED SQUAMOUS EPITHELIUM

a **STRATUM CORNEUM** HORNY LAYER

b **STRATUM LUCIDUM** *

c **STRATUM GRANULOSUM** †

d **STRATUM SPINOSUM** STRATUM MALPIGHII †

e **STRATUM GERMINATIVUM** STRATUM BASALE †

II. **DERMIS** CORIUM: LOOSE CONNECTIVE TISSUE CONTAINING DENSE COLLAGENOUS and ELASTIC FIBERS

f **PAPILLARY LAYER 1/5:** CONTAINS LOOPS OF CAPILLARIES

g **RETICULAR LAYER 4/5:** FIBERS INTERLACE LIKE A NET

The Subcutaneous Layer

III. **HYPODERMIS** SUPERFICIAL FASCIA: ATTACHES RETICULAR REGION TO UNDERLYING BONE and MUSCLE

***STRATUM LUCIDUM** not found in hairy skin, only in thick skin of palms of hands and soles of feet.*
*† **STRATUM GRANULOSUM, STRATUM SPINOSUM, and STRATUM GERMINATIVUM** are the living portions of the **EPIDERMIS**.*

EPIDERMAL SUBLAYERS

	Cell Rows	Cell Type	Nuclei	Substance Formation	
a	25–30	Flat, scalelike, dead; cornified	None	Keratin	Keratinization waterproofing
*b	3–5	Flat, dead, clear, translucent	Nonvisible, degenerate	Eleidin	
†c	1–5	Flattened, granular	Shriveled, degenerate	Keratohyalin	
†d	8–10	Polyhedral, with spinelike projections	Central oval, large	MELANO–CYTES with branching processes produce dark-brown pigment MELANIN	PIGMENTATION *Amount* of melanin produced is directly related to the gradual exposure to sunlight.
†e	1	Cuboidal and columnar	Mitotic		Melanin is a proteinaceous pigment protective against the sun's ultraviolet rays, promoting tanning of the skin.

GENERAL CHARACTERISTICS OF SKIN
VARIATION OF THICKNESS:
• Average—1.0–2.0 mm (Epidermis = 0.07–0.12 mm)
• Thickest—6.0 mm (Soles of Feet, Epidermis = 1.4 mm) (Palms of Hands, Epidermis = 0.8 mm)
• Thinnest—0.5 mm (Eyelids, External Genitalia, Tympanum)

In general, the skin is thicker on the DORSAL SURFACE OF A PART OF THE BODY THAN ON THE VENTRAL SURFACE (except in the Feet and Hands—the Palmar Surface of the Hand and the Plantar Surface of the Foot are both thicker than their dorsal aspects due to exposed wear and abrasion).

VARIATION OF TEXTURE:
• ROUGH and CALLOUS—Elbow and Knuckle Joints
• SOFT and SENSITIVE—Eyelids, External Genitalia, Nipples

VARIATION IN NORMAL SKIN COLOR:
• Genetically determined.
• Reflects a combination of 3 pigments in the skin: MELANIN, CAROTENE, HEMOGLOBIN
• (MELANIN = dark brown, CAROTENE = yellowish, HEMOGLOBIN in the blood = pinkish tones)

I. EPIDERMIS
- 30–50 cells thick (mostly **STRATUM CORNEUM**)
- **Stratification of cells forms a dense protective barrier)**
- **Keratinized and cornified**
- **Innermost STRATUM GERMINATIVUM constantly divides mitotically to replace the rest of the EPIDERMIS as it wears away**

II. DERMIS (CORIUM)
- **The true skin**
- **Upper papillary layer is in contact with the EPIDERMIS**
- **Lower reticular layer is in contact with the HYPODERMIS**
- **Loose connective tissue contains numerous capillaries, lymphatics, nerve endings, hair follicles, sebaceous glands, sweat glands and their ducts, and smooth muscle fibers**

III. HYPODERMIS
- **Deepest subcutaneous layer**
- **Loose fibrous connective tissue contains adipose (fatty) cells, interlaced with blood vessels**
- **Binds DERMIS to underlying organs, stores lipids, insulates and cushions the body, regulates temperature, and aids sexual attraction in mature females (soft contouring)**

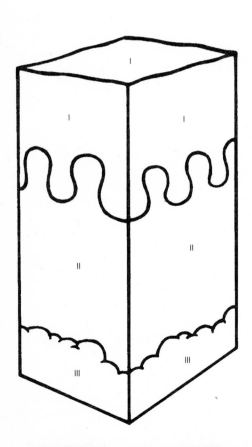

PRINCIPAL SKIN LAYERS

EPIDERMAL
BOTTOM LAYER (1 row)

e STRATUM GERMINATIVUM (BASALE)

- Continued cell division
- As cells multiply, they push upward to become cells of upper layers, away from the blood supply
- Nuclei degenerate, cells die
- Eventually, dead cells are shed off from top layer (a): (STRATUM CORNEUM)

(≈6–8 weeks to move from e to a)

- Elastic fibers here provide skin tone
- Reticular fibers form a strong meshwork
- Collagenous fibers (along with the upper elastic fibers) pull in definite directions to form the dermal lines of tension on the skin's surface
- Collagenous and elastic fibers here reinforce HYPODERMIS

EPIDERMAL
TOP LAYER (25–30 rows)

a STRATUM CORNEUM

- Continually shed and replaced
- Completely filled with keratin
- Barrier against light waves, heat waves, bacteria, chemicals

3% thicker in females than in males (due to greater lipid deposits in adipose cells)

SUBLAYERS IN SECTION OF THICK SKIN
(SOLE of FOOT or PALM of HAND)

INTEGUMENTARY SYSTEM: THE INTEGUMENT (SKIN)
Structural Organization of Skin: Epidermal Derivatives of Hair and the Integumentary Glands

HAIR and INTEGUMENTARY GLANDS form from the EPIDERMIS, and along with NAILS (See 3.3) they constitute the 3 EPIDERMAL SKIN DERIVATIVES
HAIR HAS LIMITED FUNCTIONAL VALUE:
- Protection
- Distinguish individuals
- Sexual ornament of attraction

■ During midfetal life, the EPIDERMIS develops downgrowths into the DERMIS called FOLLICLES, from which the actual hair grows (root and shaft).

■ The FOLLICLES surround the HAIR

■ Each single hair develops from STRATUM GERMINATIVUM cells within the BULB OF THE FOLLICLE, which is the enlarged base of the root.

■ Nutrients from dermal blood vessels in the DERMAL PAPILLA of the BULB foster hair growth.

■ As the hair grows and cells divide, they are pushed away from the blood supply. Results are death of the hair cells and keratinization (See 3.1).

■ Each hair lost is replaced by a new hair from the BULB OF THE FOLLICLE that pushes the old hair out.

■ HAIR LIFE SPAN:
Eyelash 3–4 months
Scalp hair 3–4 years

■ ALL GLANDS OF THE SKIN are located in the DERMIS, even though derived from the EPIDERMIS.

■ INTEGUMENTARY GLANDS HAVE EXTREMELY IMPORTANT FUNCTIONAL VALUE:
- Body defense
- Perspiration (sweat) for excretion of wastes and evaporative cooling
- Maintenance of homeostasis

The OILY SEBACEOUS GLANDS develop from the FOLLICULAR EPITHELIUM OF THE HAIR (EXTERNAL ROOT SHEATH, see HAIR ROOT illustration). These Holocrine (Secretory) glands and specialized smooth muscles—the ARRECTOR PILI MUSCLES (which raise the hair under the influence of cold or fright)—are both attached to the HAIR FOLLICLE.

★ Note the complex vascular system in the HYPODERMIS and the upper RETICULAR LAYER of the DERMIS. Also note the 2 major locations for CAPILLARY LOOPS—in the PAPILLARY LAYER of the DERMIS and in the DERMAL PAPILLA of each HAIR ROOT.

I. EPIDERMIS (a–e from 3.1)

I. Epidermal Derivatives:
HAIR, INTEGUMENTARY GLANDS, NAILS

HAIR
1 SHAFT
2 ROOT
3 FOLLICLE
4 BULB OF FOLLICLE
MATRIX + DERMAL PAPILLA

Medulla / Cortex / Cuticle
Cross-section of a HAIR SHAFT

Longitudinal Section of the Bottom of a HAIR ROOT (in Deep Dermis)
(Arrows represent direction of growth: Outside Down—Inside Up)

INTEGUMENTARY GLANDS
5 SEBACEOUS GLAND | OIL GLAND

SUDORIFEROUS GLANDS
6 APOCRINE SWEAT GLAND
7 ECCRINE SWEAT GLAND

SPECIALIZED SUDORIFEROUS GLANDS
- CERUMINOUS SWEAT GLANDS in external auditory meatus produce CERUMEN (EARWAX) when secretions combine with oil from SEBACEOUS GLANDS
- MAMMARY GLANDS in the breasts secrete milk during lactation periods, under the influence of Pituitary and Ovarian hormones.

II. DERMIS
PAPILLARY LAYER
8 PAPILLAE
RETICULAR LAYER
9 COLLAGENOUS, ELASTIC, and RETICULAR FIBERS
10 ARRECTOR PILI MUSCLE

III. HYPODERMIS
11 LOOSE FIBROUS CONNECTIVE TISSUE
12 ADIPOSE TISSUE
13 ARTERY
14 VEIN
15 LYMPHATIC VESSEL
16 MOTOR NERVE

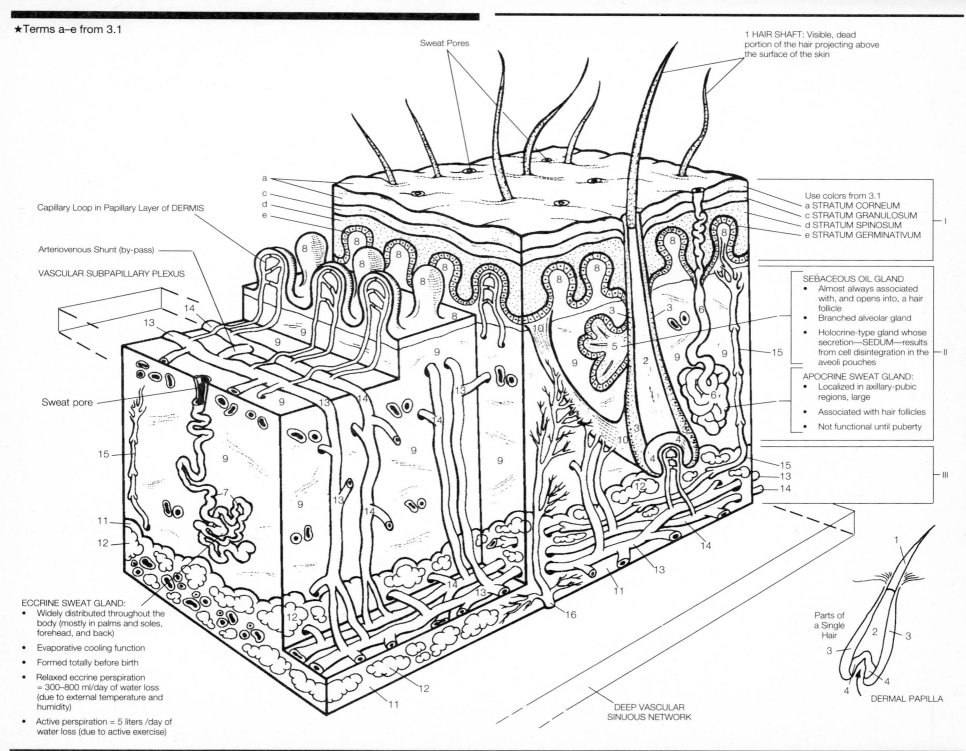

Sweat Pores

1 HAIR SHAFT: Visible, dead portion of the hair projecting above the surface of the skin

a
c
d
e

Use colors from 3.1
a STRATUM CORNEUM
c STRATUM GRANULOSUM
d STRATUM SPINOSUM
e STRATUM GERMINATIVUM

I

Capillary Loop in Papillary Layer of DERMIS

Arteriovenous Shunt (by-pass)

VASCULAR SUBPAPILLARY PLEXUS

SEBACEOUS OIL GLAND
• Almost always associated with, and opens into, a hair follicle
• Branched alveolar gland
• Holocrine-type gland whose secretion—SEDUM—results from cell disintegration in the aveoli pouches

II

APOCRINE SWEAT GLAND:
• Localized in axillary-pubic regions, large
• Associated with hair follicles
• Not functional until puberty

Sweat pore

III

ECCRINE SWEAT GLAND:
• Widely distributed throughout the body (mostly in palms and soles, forehead, and back)
• Evaporative cooling function
• Formed totally before birth
• Relaxed eccrine perspiration = 300–800 ml/day of water loss (due to external temperature and humidity)
• Active perspiration = 5 liters /day of water loss (due to active exercise)

DEEP VASCULAR SINUOUS NETWORK

Parts of a Single Hair

DERMAL PAPILLA

INTEGUMENTARY SYSTEM: THE INTEGUMENT (SKIN)
Epidermal Derivatives (Nails)/ Innervation of the Skin

Refer to 3.1 for colors used in epidermal layers
Regardless of skin color, all NAIL BODIES should
appear in light pink (indicative of the rich vascular
capillaries beneath the translucent nail).

★ See Chart # 2

★ ■ NAILS (as well as HAIR) are structural features of the integument and have limited functional value.

- Protection of digits
- Grasping of small objects

★ ■ Exclusively found on the DISTAL DORSAL SURFACE OF EACH FINGER and TOE. (Thus, there are 20 separate nails.)

■ NAILS are formed from the hard, keratinized cells of the EPIDERMIS, namely, the outer STRATUM CORNEUM.

■ Hardness is due to a dense, parallel arrangement of KERATIN FIBERS between the cells.

★ ■ Main visible part of the NAIL—the NAIL BODY—rests on a NAIL BED (actually STRATUM SPINOSUM) and appears pink due to underlying vascular tissue.

■ The white, semilunar area at the proximal end of the nail body = LUNULA. (It is white because the vascular tissue underneath does not show through.)

■ The FREE EDGE is that part of the nail that projects beyond the distal portion of the digit. It is attached to the undersurface by the HYPONYCHIUM, a thickened area of stratum corneum.

■ The NAIL ROOT is the portion of the nail hidden in the NAIL GROOVE.

■ The EPONYCHIUM is a narrow band of stratum corneum that covers above the NAIL ROOT. It extends from the lateral border (margin) of the NAIL WALL and occupies the proximal border of the NAIL.

★ ■ The NAIL FOLD is a fold of skin extending around the lateral and proximal borders. The furrow between the NAIL FOLD and the NAIL BED is the NAIL GROOVE.

★ GROWTH AREAS OF THE NAIL: THE MATRIX and THE LUNULA

Growth is due to a transformation of the SUPERFICIAL CELLS OF THE MATRIX into hardened, transparent NAIL CELLS (hardened stratum corneum).

These cells start in the NAIL ROOT and are pushed forward over the stratum spinosum (and stratum germinativum) of the NAIL BED as the NAIL BODY. Rate of growth is 1 mm (0.04 inches)/week.

Parts of a Single Nail

1. FREE EDGE
2. NAIL BODY
3. NAIL ROOT

Supportive and Surrounding Areas

Areas of STRATUM CORNEUM

4. NAIL FOLD
5. HYPONYCHIUM QUICK
6. EPONYCHIUM CUTICLE (SPLIT CAUSES "HANGNAIL")

Areas of STRATUM SPINOSUM and STRATUM GERMINATIVUM

7. NAIL BED
8. MATRIX
9. LUNULA
} GROWTH AREAS OF THE NAIL

INNERVATION of the SKIN

(See also Chapter 7: 7.2, 7.18, and Chart #7)

★ There are an abundance of sensory nerve receptors in the skin that convey sensory impulses through afferent (sensory) nerves of the PERIPHERAL NERVOUS SYSTEM to the SENSORIUM—that portion of the BRAIN that functions as a center for SENSATIONS. See Chart #2 in this chapter for the types of CUTANEOUS SENSORY RECEPTORS, their DETECTION FUNCTIONS, and SOMATIC SENSATIONS.

I. EPIDERMIS

1. FREE NERVE ENDINGS
2. MERKEL'S DISCS FOUND ONLY IN HAIRLESS SKIN

II. DERMIS

PAPILLARY LAYER

3. MEISSNER'S CORPUSCLES

RETICULAR LAYER

4. BULBS OF KRAUSE
5. ORGANS OF RUFFINI

III. DERMIS AND HYPODERMIS

6. PACINIAN CORPUSCLES
7. GOLGI-MAZZONI ENDINGS
8. HAIR ROOT PLEXUSES
9. SENSORY NERVE SENSORY "AXONS"

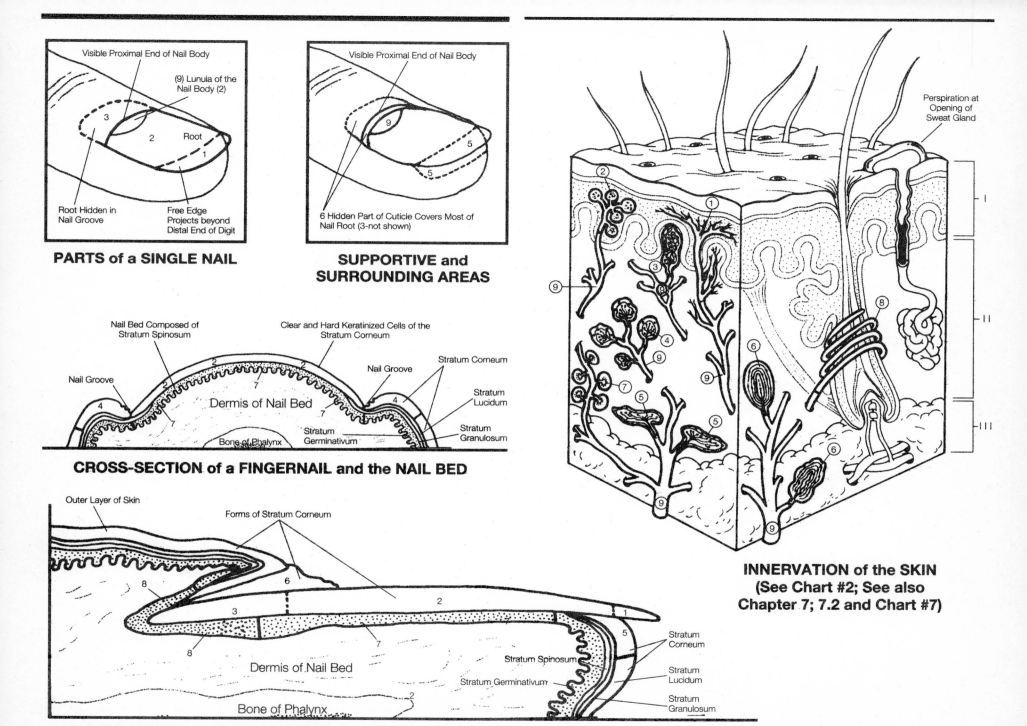

PARTS of a SINGLE NAIL

Visible Proximal End of Nail Body

(9) Lunula of the Nail Body (2)

Root

Root Hidden in Nail Groove

Free Edge Projects beyond Distal End of Digit

SUPPORTIVE and SURROUNDING AREAS

Visible Proximal End of Nail Body

6 Hidden Part of Cuticle Covers Most of Nail Root (3-not shown)

CROSS-SECTION of a FINGERNAIL and the NAIL BED

Nail Bed Composed of Stratum Spinosum

Clear and Hard Keratinized Cells of the Stratum Corneum

Nail Groove

Nail Groove

Stratum Corneum

Stratum Lucidum

Dermis of Nail Bed

Stratum Germinativum

Stratum Granulosum

Bone of Phalynx

SAGITTAL SECTION of a FINGERNAIL and the NAIL BED

Outer Layer of Skin

Forms of Stratum Corneum

Dermis of Nail Bed

Stratum Spinosum

Stratum Germinativum

Bone of Phalynx

Stratum Corneum

Stratum Lucidum

Stratum Granulosum

Perspiration at Opening of Sweat Gland

INNERVATION of the SKIN
(See Chart #2; See also
Chapter 7; 7.2 and Chart #7)

Chapter 3: *INTEGUMENTARY SYSTEM:* CHARTS

Chart #1 : Functions of the Skin (See 3.0, 3.1)

INTERACTION OF OTHER BODY SYSTEMS WITH THE INTEGUMENTARY SYSTEM	
MUSCULAR SYSTEM	• Emotion conveyed through facial expressions
CIRCULATORY SYSTEM	• Blushing involves vasodilation of cutaneous arteries • Extensive interaction helps maintain homeostasis
NERVOUS SYSTEM	• Muscles or glands within the DERMIS act as specialized integumentary EFFECTORS which respond to EFFERENT (MOTOR) impulses transmitted through the ANS (AUTONOMIC NERVOUS SYSTEM) • Countless AFFERENT (SENSORY) RECEPTORS respond to temperature pressure, pain, tactile (touch), and vibratory sensations. (Greatest concentration in the palms, soles, lips, external genitalia)
ENDOCRINE SYSTEM	• Certain hormones alter the function and appearance of the integument
IMMUNE SYSTEMS	• Systems that provide body immunity interact with the integument

• SOCIAL RECOGNITION AND COMMUNICATION

• DYNAMIC INTERFACE BETWEEN EXTERNAL AND INTERNAL ENVIRONMENT

PROTECTION OF EPIDERMIS — BY STRATIFICATION *

• COVERS THE BODY (protection from external injury)
  • Exposed outer layer is tough and cornified, and calluses are formed here in response to friction and outward cell growth. Keratin toughens epidermis.
• KEEPS THE BODY TOGETHER (external body support)
• PROTECTION OF UNDERLYING TISSUE FROM:
  ■ Bacterial invasion
    • Sebum is acidic and antiseptic. Lipids keep epidermis from cracking. Rapid mitotic rate and cell shedding minimizes entry of pathogens.
  ■ Drying out
    • Sebum provides oily fibers; keratin toughens epidermis. Basement membrane seals epidermis.
  ■ Harmful ultraviolet rays
    • Melanin absorbs solar radiation; scalp hair disperses light.

MAINTENANCE OF — BODY HOMEOSTASIS

• REGULATION AND CONTROL OF BODY TEMPERATURE
  ■ DERMIS:
    • Cooling through vasodilation and sweating. Warming through vasoconstriction and shivering.
  ■ HYPODERMIS:
    • Insulation from lipids
• PREVENTS EXCESSIVE LOSS OF INORGANIC MATERIALS
  ■ Wound–healing after blood loss (dermal vasoconstriction, blood coagulation, temporary scab, collagenous scar tissue.)
• RECEPTION OF ENVIRONMENTAL STIMULI
  ■ Sensory afferent receptors for:
    • Temperature • Pressure
    • Pain • Touch • Vibration
• ELIMINATION OF WASTES
  ■ Excretion of water and salts
• STORAGE OF CHEMICAL COMPOUNDS
• SYNTHESIZES SEVERAL COMPOUNDS (including Vitamin D in the DERMIS, in the presence of ultraviolet light) KERATIN, MELANIN, and CAROTENE synthesized in the EPIDERMIS

* Stratification of the Epidermis forms a dense barrier of protection, and a nearly impenetrable barrier against pathogens.

Chart #2 : Cutaneous Sensory Receptors in the Skin (See 3.3)

- Cutaneous sensory receptors for the 6 basic skin sensations (TOUCH, PRESSURE, PAIN, WARMTH, COLD, VIBRATION)
- TICKLING, ITCHING, SOFTNESS, HARDNESS, and WETNESS are due to 2 or more stimulations of these receptors and to a *blending* of the sensations in the brain
  - 1–7 are found in HAIRLESS PARTS of the SKIN, are not uniformly distributed over the whole body surface (e.g., "TOUCH" endings are very numerous in the hands (palms) and feet (soles), but much less frequent in the skin of the back.)
  - 8 form networks around HAIR SHAFTS in HAIRY BODY SURFACES

CUTANEOUS SENSORY RECEPTORS

	CLASSIFICATION	TYPE	DETECTION FUNCTION	SOMATIC SENSATION
1	FREE NERVE ENDINGS	Tactile, Pressure and Pain receptors	Pressure changes, Tissue damage	Touch, Pressure, Pain
2	MERKEL'S DISCS	Tactile receptors	Light motion on skin surface	Light touch
3	MEISSNER'S CORPUSCLES	Mechanoreceptors	Light motion on skin surface	Light touch, Texture, Low frequency vibrations
4	BULBS OF KRAUSE	Thermoreceptors	Temperature changes	Cold
5	ORGANS OF RUFFINI	Thermoreceptors	Temperature changes	Heat, Pain
6	PACINIAN CORPUSCLES	Mechanoreceptors	Pressure changes	Deep pressure, High frequency vibrations
7	GOLGI-MAZZONI ENDINGS	Mechanoreceptors	Pressure changes	Deep pressure
8	HAIR ROOT PLEXUSES	Tactile receptors	Hair movement	Light touch

(See also Chapter 7 Nervous System: 7.2, 7.18, and Chart #7)

UNIT 2: SUPPORT AND MOVEMENT

Chapter 4 / Skeletal System

The SKELETAL SYSTEM in the adult human organism consists of 206 individual bones (and related cartilage) uniquely arranged into the flexible, but strong framework of the body. Each bone is an organ unto itself, and all bones collectively work together to provide the total functioning of the skeletal system. The SKELETAL SYSTEM must not be considered as an isolated system, but one that supports and protects all of the systems of the body, and in particular has close association with the CARDIOVASCULAR, MUSCULAR, and RESPIRATORY SYSTEMS:

•It helps serve the CIRCULATORY SYSTEM in production of blood cells in well-protected areas of bone.

•It helps serve the MUSCULAR SYSTEM in a 2-fold manner:

1. Bony projections provide attachments for muscles that span movable joints, optimizing the muscles' power to move the bones; the skeletal system serves as the basis for the complexity of body movement through space.

2. Bone acts as a storehouse for calcium needed for muscle contraction.

•Bones serve the RESPIRATORY SYSTEM by forming passageways in the nasal cavity that help clean, moisten, and warm inhaled air; bones of the thorax are specially shaped and positioned to serve the chest in expansion during inhalation.

Bones also serve as landmarks for students of anatomy (as well as surgeons). Frequent, misleading conceptions of the hard, dry, lifeless material of laboratory study must be properly replaced by a conception of bone as vital living substance and as a site of enormous metabolic activity, performing many crucial body functions, including support, protection, form and shape, body movement, blood cell production, and storage of mineral nutrients.

The skeleton is divided for purposes of study into 2 portions: the AXIAL SKELETON and the APPENDICULAR SKELETON:

•The AXIAL SKELETON consists of the bones related to the axis of the body that support and protect the head, neck, and trunk.

•The APPENDICULAR SKELETON consists of the bones of the upper and lower extremities and the girdles that anchor them to the axial skeleton.

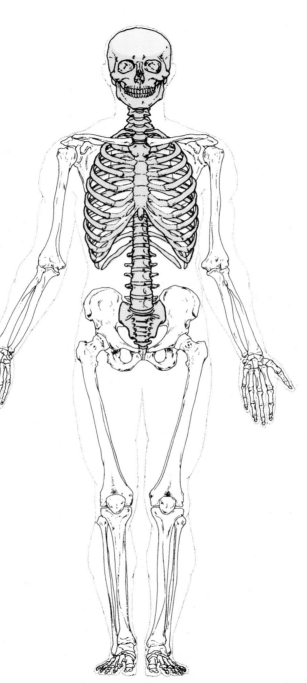

Anterior View

System Components

•THE BONES OF THE BODY AND

•THEIR ASSOCIATED CARTILAGES

System Function

•INTERNAL SUPPORT OF THE BODY

•PROTECTION OF THE BODY

•BODY LEVERAGE

•PRODUCTION OF BLOOD CELLS

•STORAGE OF MINERALS

☐ = AXIAL SKELETON

☐ = APPENDICULAR SKELETON

Chapter Coloring Guidelines

•*You may want to choose your color palette by comparing any colored illustrations from your main textbook.*

•*Axial Skeleton = warm colors*

•*Appendicular Skeleton = cool colors*

•*Note: 4.4–4.7 (CRANIAL and FACIAL Bones) Use the same colors for these bones on all pages.*

•*Observe the direct correlation between the bones of the Upper Limb and the Lower Limb in the Appendicular Skeleton (Humerus-Femur; Ulna-Tibia, etc.) You may want to color correlated bones the same color to enhance understanding of these relationships.*

SKELETAL SYSTEM: GENERAL ORGANIZATION
Gross Bone Anatomy

★ The SKELETON SYSTEM consists of 2 types of connective tissue:
- CARTILAGE
- BONE (OSSEOUS TISSUE)

Bone is the most rigid of all connective tissue, and unlike cartilage, is the site of great metabolic activity enhanced by a large vascular supply. Bone is not to be conceived as the dry, lifeless structure studied in the anatomy laboratory.

★ The process of bone formation is called OSSIFICATION or OSTEOGENESIS. Bone is formed from the human embryo "skeleton" by either fibrous membranes or hyaline cartilage. In the young child, bone tissue has completely replaced cartilage, except in 2 regions:

ARTICULAR CARTILAGE—covers the articular surfaces of the 2 EPIPHYSES, the proximal and distal ends of a typical long bone.

EPIPHYSEAL LINE—remnant of a cartilaginous epiphyseal plate between each EPIPHYSIS and the DIAPHYSIS (the shaft, or long portion of a typical bone). During growth, the EPIPHYSEAL PLATE is a region of mitotic activity responsible for elongation of bone.

★ The PERIOSTEUM—a dense, white fibrous tissue covering the surface of bone *not* covered by articular cartilage.
- Essential for diametric (width) bone growth, repair, and nutrition.
- Point of attachment for ligaments and muscle tendons.
  2 LAYERS OF PERIOSTEUM (See 4.2)
  - INNER OSTEOGENIC LAYER
    - Covers compact bone
    - Elastic fibers, blood vessels
    - OSTEOBLASTS—specialized cells form new bone and help in repair
  - OUTER FIBROUS LAYER
    - fibrous, connective tissue
    - blood vessels, lymphatic vessels, and nerves that pass into the bone

★ Along the DIAPHYSIS are nutrient foramina that allow passage of nutrient vessels into the bone.

Typical Structure of a Long Bone

1 **DIAPHYSIS:** SHAFT ⟶ A **MEDULLARY CAVITY WITH YELLOW MARROW**

2 **2 EPIPHYSES:** PROXIMAL EPIPHYSIS AND DISTAL EPIPHYSIS ⟶ B **CANCELLOUS SPONGY BONE WITH RED MARROW**

3 **EPIPHYSEAL LINE/PLATE**

4 **NUTRIENT ARTERY**

5 **ARTICULAR CARTILAGE**

6 **PERIOSTEUM**

★ In the DIAPHYSIS (shaft) of a typical long bone, the COMPACT BONE forms a cylinder that surrounds a central cavity, called the MEDULLARY CAVITY :
- Also called the MARROW CAVITY, it contains fatty YELLOW MARROW (consists of fat cells and a few scattered blood cells).
- Lined with a layer of connective tissue called the ENDOSTEUM. The endosteum also contains osteoblasts. In addition, scattered OSTEOCLASTS aid in bone removal. (See 4.2)

2 Types of BONE TISSUE, Based on Porosity

C **COMPACT,** DENSE BONE   OUTER LAYER

B **CANCELLOUS,** SPONGY BONE   INNER LAYER

There are 2 types of bone tissue, and most bones have both types:
- CANCELLOUS SPONGY BONE
- COMPACT, DENSE BONE

SPONGY BONE
- Contains many large spaces filled with RED MARROW (responsible for producing red blood cells, some white blood cells and platelets)
- Highly porous, highly vascular inner portion of bone. Makes bone lighter
- Makes up most of the epiphyses of long bones and almost all of short, flat, irregularly shaped bones

COMPACT BONE
- Contains few spaces
- Deposited in a layer over spongy bone
- This dense layer is thicker along the DIAPHYSIS than around the EPIPHYSES
- Hard, outer layer provides protection, durable strength, and support; resists weight stress.
- Adult, compact bone has a concentric ring structure. (Spongy bone does not.)

EXAMPLE
OF A
TYPICAL LONG BONE:
LEFT HUMERUS
(ANTERIOR VIEW)

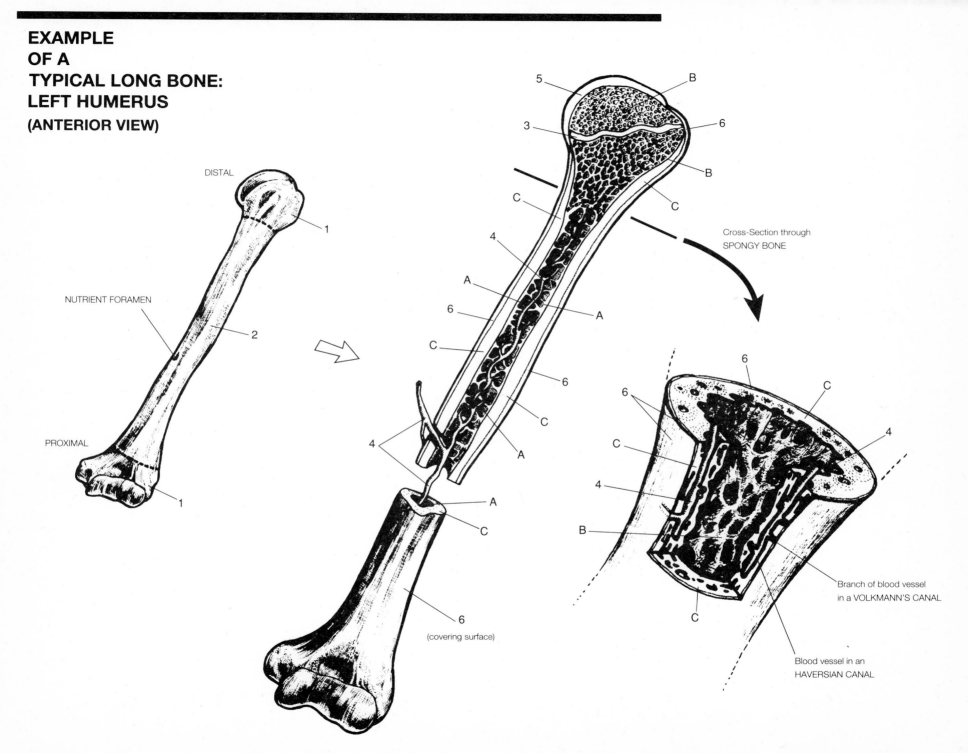

DISTAL

NUTRIENT FORAMEN

PROXIMAL

1

2

1

5

B

3

6

B

C

C

4

A

A

6

C

6

C

A

4

A

C

6

(covering surface)

Cross-Section through
SPONGY BONE

6

6

C

C

C

4

B

4

C

Branch of blood vessel
in a VOLKMANN'S CANAL

Blood vessel in an
HAVERSIAN CANAL

SKELETAL SYSTEM : GENERAL ORGANIZATION
Microscopic Bone Anatomy

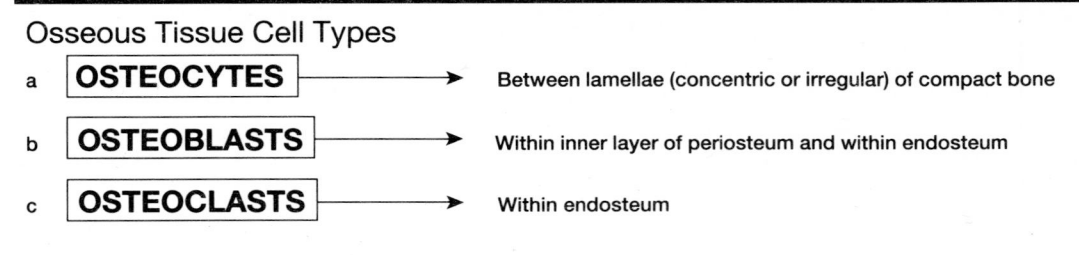

a = light blue   b = blue   c = purple
9 = flesh   10 = cream   11 = yelllow–orange

★ Blood Vessels and nerves from the periosteum penetrate compact bone through VOLKMANN'S CANALS, which connect with the MEDULLARY CAVITY and the longitudinal HAVERSIAN CANALS.

•Around the HAVERSIAN CANALS are concentric rings of hard, calcified intercellular substance called HAVERSIAN LAMELLAE.

•Small spaces between the lamellae are called LACUNAE and each of these spaces contain an OSTEOCYTE, or bone cell.

•Radiating in all directions from each LACUNA are minute canals called CANALICULI forming an intricate branching network throughout the bone for the supply of nutrients to and removal of wastes from the OSTEOCYTES.

•OSTEOCYTES are actually mature OSTEOBLASTS that have lost their ability to produce new bone tissue, but they are in the center of metabolic activity.

★ •Each HAVERSIAN CANAL with its concentric system of surrounding lamellae, lacunae, osteocytes, and canaliculi is called an HAVERSIAN SYSTEM (OSTEON).
Haversian systems are characteristic of adult bone.
•Between the cylindrical, longitudinal Haversian systems is found INTERSTITIAL LAMELLAE containing irregular arrangements of osteocytes, lacunae, and canaliculi.

★ 2 TYPES OF BONE DEVELOPMENT
 ■ INTRAMEMBRANOUS OSSIFICATION
  •"Direct" formation of bony tissue from tissue lying between 2 fibrous layers
  •Flat bones of the roof of the skull, parts of the lower jaw (mandible), part of the clavicles
 ■ INTRACARTILAGINOUS or ENDOCHONDRAL OSSIFICATION
  •"Indirect", two-step process:
   1. Hyaline cartilage model is formed.
   2. Model is replaced by bony tissue.
  •Most of the bones of the body, including parts of the skull

Osseous Tissue Cell Types

a | OSTEOCYTES | ⟶ Between lamellae (concentric or irregular) of compact bone

b | OSTEOBLASTS | ⟶ Within inner layer of periosteum and within endosteum

c | OSTEOCLASTS | ⟶ Within endosteum

Spaces Containing Osteocytes

7 | LACUNAE | (singular = LACUNA)

Canal Network Branching from Lacunae

8 | CANALICULI |

Calcified, Intercellular Substance
Concentric:

9 | HAVERSIAN LAMELLAE |

Irregular:

10 | INTERSTITIAL LAMELLAE |

PERIOSTEUM

6a | OUTER, FIBROUS LAYER |

6b | INNER, OSTEOGENIC LAYER |

11 | ENDOSTEUM |

★ **HAVERSIAN SYSTEMS (OSTEONS)** in Compact Bone

Each OSTEON contains:
•HAVERSIAN CANAL (and VOLKMANN'S CANALS)
•Concentric LAMELLAE
•Surrounding Concentric Layers of OSTEOCYTES
•LACUNAE
•CANALICULI

AREAS BETWEEN OSTEONS

•Interstitial LAMELLAE not connected to the HAVERSIAN SYSTEMS
•Irregularly arranged OSTEOCYTES, LACUNAE, and CANALICULI

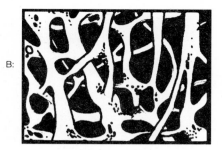

B:

BONY TRABECULAE (SPONGY BONE)
found mostly in the epiphyseal
ends of a long bone and making up
most of the osseous tissue of
short bones

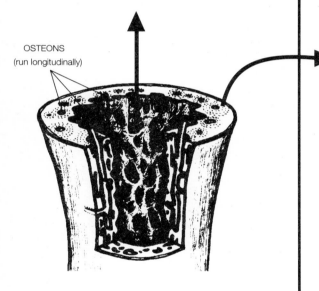

OSTEONS
(run longitudinally)

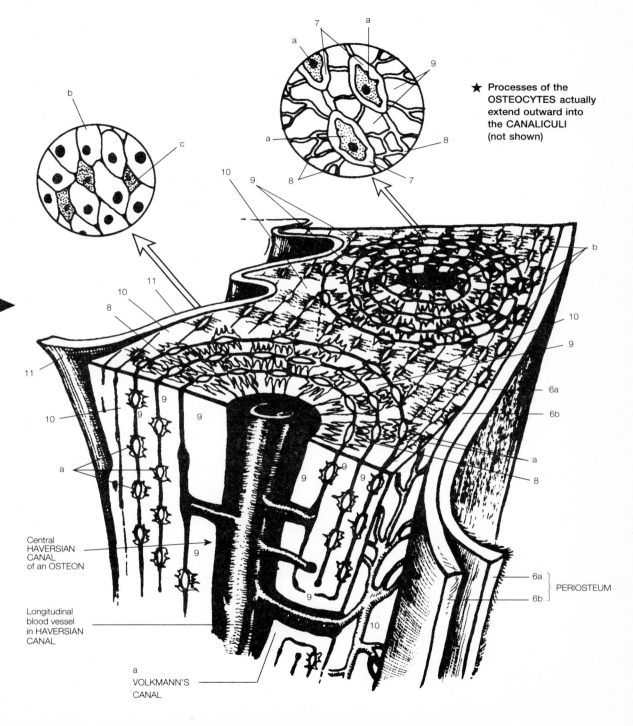

b

c

a

7

a

9

a

a

8

7

8

★ Processes of the
OSTEOCYTES actually
extend outward into
the CANALICULI
(not shown)

10

9

11

10

8

b

11

9

10

10

9

9

6a

6b

9

9

9

9

a

a

8

Central
HAVERSIAN
CANAL
of an OSTEON

9

6a
6b

Longitudinal
blood vessel
in HAVERSIAN
CANAL

9

PERIOSTEUM

10

a
VOLKMANN'S
CANAL

4.2

SKELETAL SYSTEM: GENERAL ORGANIZATION
Superficial Survey of the Skeleton

There is no coloring necessary here, unless you want to backtrack later and use the same colors you chose for the following pages.

See Chart #1 Surface Features and Markings of Bones

★ **2 MAJOR DIVISIONS OF THE SKELETAL SYSTEM**
■ THE AXIAL SKELETON (80 BONES)
■ THE APPENDICULAR SKELETON (126 BONES)
The normal adult human skeleton consists of 206 BONES (80 + 126 = 206).

■ THE AXIAL SKELETON consists of bones that lie around the AXIS of the body. (The AXIS is an imaginary longitudinal line that runs through the head and down to the space between the feet. This straight line runs vertically along the body's center of gravity and is considered the CENTER OF THE HUMAN BODY.)

HEAD and NECK and TRUNK
- BONES OF THE SKULL
- BREASTBONE (STERNUM)
- BACKBONE (VERTEBRAL COLUMN)
- RIBS

■ THE APPENDICULAR SKELETON consists of the bones of the upper and lower extremities (the FREE APPENDAGES) and the bones (GIRDLES) that connect them to the axial skeleton.
- PECTORAL GIRDLE and UPPER LIMBS
- PELVIC GIRDLE and LOWER LIMBS

★ **FUNCTIONS OF THE SKELETAL SYSTEM**
MECHANICAL FUNCTIONS:
■ SUPPORT OF THE BODY
Supports softer tissues and organs of the body that attach to it via connective tissues.
Maintains body form and erect posture.
■ PROTECTION OF THE BODY
Protection of delicate structures.
CRANIUM encloses the brain. VERTEBRAL COLUMN encloses the spinal cord. RIB CAGE protects heart, lungs, great blood vessels, liver, and spleen.
PELVIC CAVITY supports and protects pelvic viscera.
■ BODY LEVERAGE and MOVEMENT
Bones act as anchors for the attachment of most skeletal muscles. When muscles contract, the bones act as LEVERS with the joints (pivots) causing body movement.
METABOLIC FUNCTIONS:
■ BLOOD CELL PRODUCTION
The RED BONE MARROW in the spongy region of bone produces red blood cells, white blood cells, and platelets in the process of HEMOPOIESES.
(Approximately 1 million red blood cells are produced every second to replace those destroyed by the liver!)
■ STORAGE OF MINERALS
The entire skeleton provides a storehouse for CALCIUM and PHOSPHORUS. (Close to 90% of both of these inorganic salts are stored in the bones and teeth, and if insufficient in diet, may be withdrawn from the bones until replaced by proper nutrition.) Lesser amounts of MAGNESIUM and SODIUM SALTS are also stored in bone tissue.

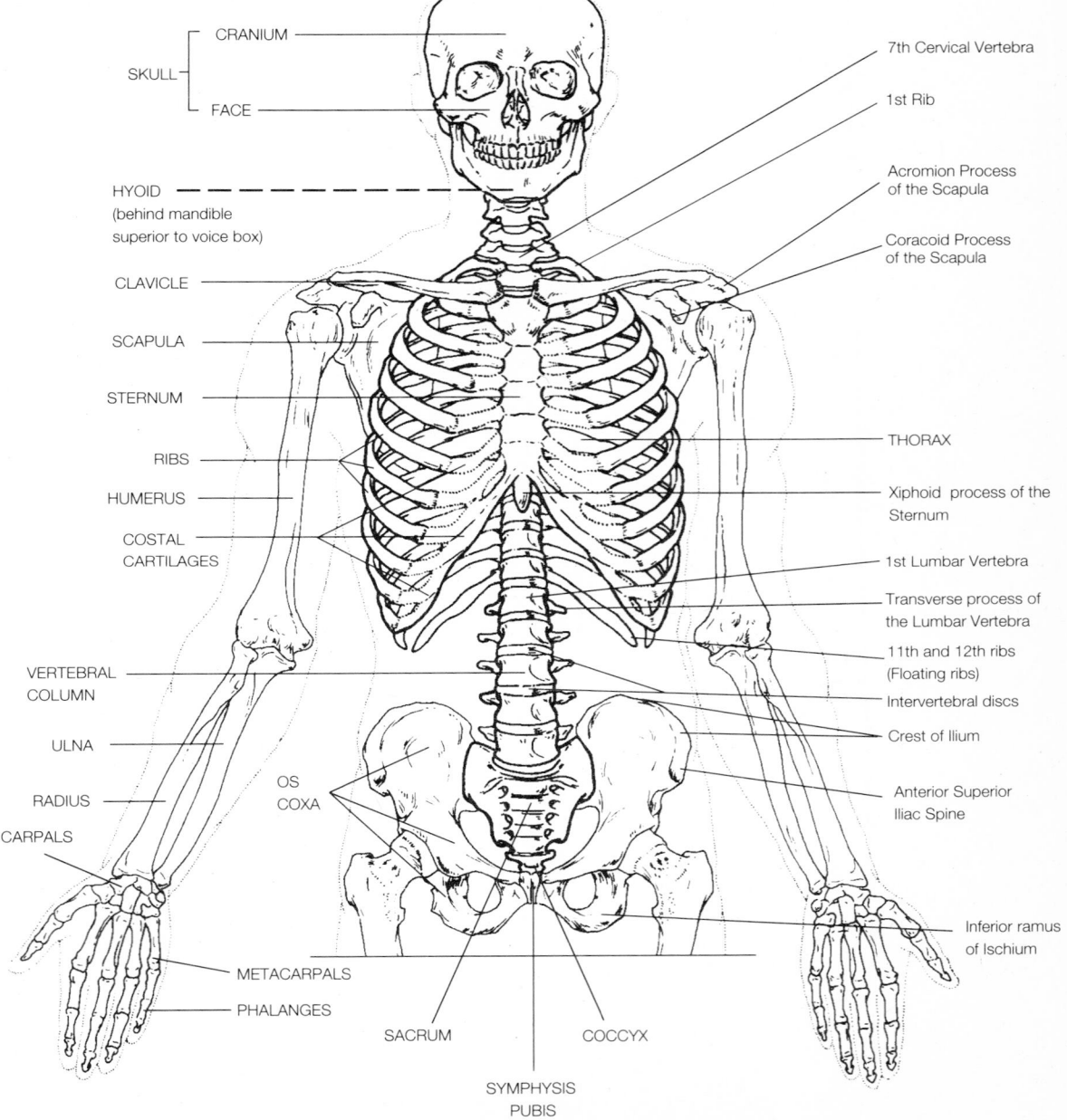

SKULL — CRANIUM — FACE

7th Cervical Vertebra

1st Rib

Acromion Process of the Scapula

Coracoid Process of the Scapula

HYOID (behind mandible superior to voice box)

CLAVICLE

SCAPULA

STERNUM

RIBS

HUMERUS

COSTAL CARTILAGES

THORAX

Xiphoid process of the Sternum

1st Lumbar Vertebra

Transverse process of the Lumbar Vertebra

11th and 12th ribs (Floating ribs)

Intervertebral discs

Crest of Ilium

VERTEBRAL COLUMN

ULNA

RADIUS

CARPALS

OS COXA

METACARPALS

PHALANGES

SACRUM

COCCYX

SYMPHYSIS PUBIS

Anterior Superior Iliac Spine

Inferior ramus of Ischium

★ It is important to remember that EACH bone in the skeletal system is an organ itself. In addition to the principal component of OSSEOUS TISSUE, every bone consists of other types of tissue–NERVOUS, VASCULAR, CARTILAGINOUS–that all coordinate in elaborating each bone's structure and function.

★ Recall that bone tissue consists of widely scattered separated cells surrounded by a very large amount of intercellular substance. Unique among all other connective tissue, osseous tissue contains abundant mineral salts (CALCIUM PHOSPHATE and CALCIUM CARBONATE) in its intercellular substance. Together, these salts are referred to as HYDROXYAPATITES. As bone grows, the framework is formed by COLLAGENOUS FIBERS. As these salts are deposited in the framework of the intercellular substance, the tissue hardens, or becomes OSSIFIED, giving bone its rigid character.

    WEIGHT OF BONE
        HYDROXYAPATITES = 67%
        COLLAGENOUS FIBERS = 33%

★ 4 PRINCIPAL TYPES OF BONE

Rather than size, the bones of the skeleton are classified according to 4 *shapes:*

■ LONG BONES
   • Longer than they are wide
   • 1 diaphysis and 2 epiphyses
   • Slightly curved for strength
   • Function as levers
(most of the bones of the upper and lower extremities)

■ SHORT BONES
   • Nearly equal in length and width
   • Mostly spongy bone, with a thin outer layer of compact bone
   • Somewhat cube-shaped
   • Transfer forces in confined spaces
[the carpal (wrist) bones and ankle (tarsal) bones]

■ FLAT BONES
   • Thin and broad
   • 2 more or less parallel plates of compact bone enclosing a layer of spongy bone
   • Dense surface for protection of underlying organs and muscle attachment
[the cranium, ribs, pectoral girdle (clavicles and scapulae)]

■ IRREGULAR BONES
   • Complex, varied shapes
   • Many surface markings for muscle attachment or articulation
   • Varied amounts of spongy and compact bone
(the vertebrae, sphenoid and ethmoid bones of the skull, certain facial bones)

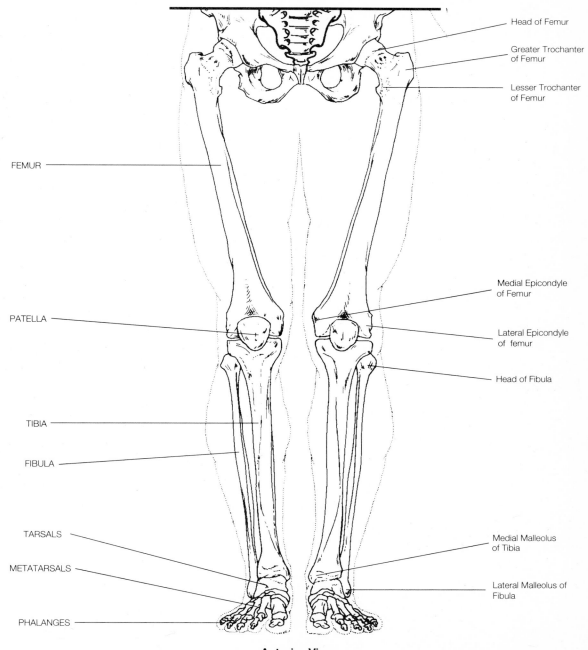

Head of Femur

Greater Trochanter of Femur

Lesser Trochanter of Femur

FEMUR

Medial Epicondyle of Femur

PATELLA

Lateral Epicondyle of femur

Head of Fibula

TIBIA

FIBULA

TARSALS

Medial Malleolus of Tibia

METATARSALS

Lateral Malleolus of Fibula

PHALANGES

Anterior View

SKELETAL SYSTEM: AXIAL SKELETON
Bones of the Adult Skull: Cranial and Facial
Anterior and Posterior Views

See Chart #2 Openings in Bones and
the Structures Passing Through Them
★ See also 6.4

★ **THE ADULT HUMAN SKULL**
- Contains 22 BONES
- Rests on the superior end of the vertebral column
- Each bone of the skull articulates the adjacent bones
- Contains several cavities that house the BRAIN and SENSORY ORGANS
- Consists of 8 CRANIAL BONES and 14 FACIAL BONES

★ **THE 8 CRANIAL BONES: CRANIUM**
- Enclose and protect the BRAIN
- Enclose and protect the SENSORY ORGANS of SIGHT, HEARING, and BALANCE (See Chapter 7, SENSE ORGANS)
- All interlocked by SUTURES

■ THE FRONTAL BONE forms:

- Forehead
- Roof of the nasal cavity
- Superior arch of the BONY ORBITS (which house the eyeballs)
- Anterior part of CRANIAL FLOOR (After birth, left and right halves of the frontal bone unite by a suture, which disappears usually by age 6.)

■ TWO PARIETAL BONES form:
- Superior (upper) sides of the CRANIUM
- Roof of the CRANIUM (CRANIAL CAVITY)

■ TWO TEMPORAL BONES form:
- Inferior (lower) sides of the CRANIUM
- Part of the CRANIAL FLOOR (Structurally, each temporal bone has 4 parts. See 4.10)

■ THE OCCIPITAL BONE forms:
- Posterior (back) portion of the skull
- Prominent portion of the base of the CRANIUM

■ THE SPHENOID BONE forms:
- Anterior base of the CRANIUM, situated at the middle part of the base of the skull.
- The "keystone" of the CRANIAL FLOOR, it *ARTICULATES with all other cranial bones.*
- Part of the floor and sidewalls of the eye sockets
- Inferior PTERYGOID PROCESSES help form the lateral walls of the NASAL CAVITY

■ THE ETHMOID BONE forms:
- The roof of the NASAL CAVITY, situated in the anterior part of the floor of the cranium between the orbits
- A middle inferior projection—THE PERPENDICULAR PLATE forms a major part of the NASAL SEPTUM that separates the NASAL CAVITY into 2 chambers

The Adult Skull Contains **22 BONES: 8 CRANIAL and 14 FACIAL**

(Does not include the single HYOID BONE or the 6 OSSICLES of the Inner Ear)

8 CRANIAL BONES

	Bone	Number of Each Bone
1	**FRONTAL**	1
2	**PARIETAL**	2
3	**TEMPORAL**	2
4	**OCCIPITAL**	1
5	**SPHENOID**	1
6	**ETHMOID**	1

Total Number of Cranial Bones = **8**

H **HYOID BONE**

★ **THE HYOID BONE**
- Unique among *all* bones of the skeletal system in that it is the only bone that does not articulate directly to any other bone
- Suspended from the STYLOID PROCESS OF THE TEMPORAL BONE by the STYLOHYOID LIGAMENTS
- Located in the neck between the mandible and larynx (voice box), just superior to the larynx
- Supports the TONGUE
- Provides attachment for several TONGUE and NECK MUSCLES and LIGAMENTS

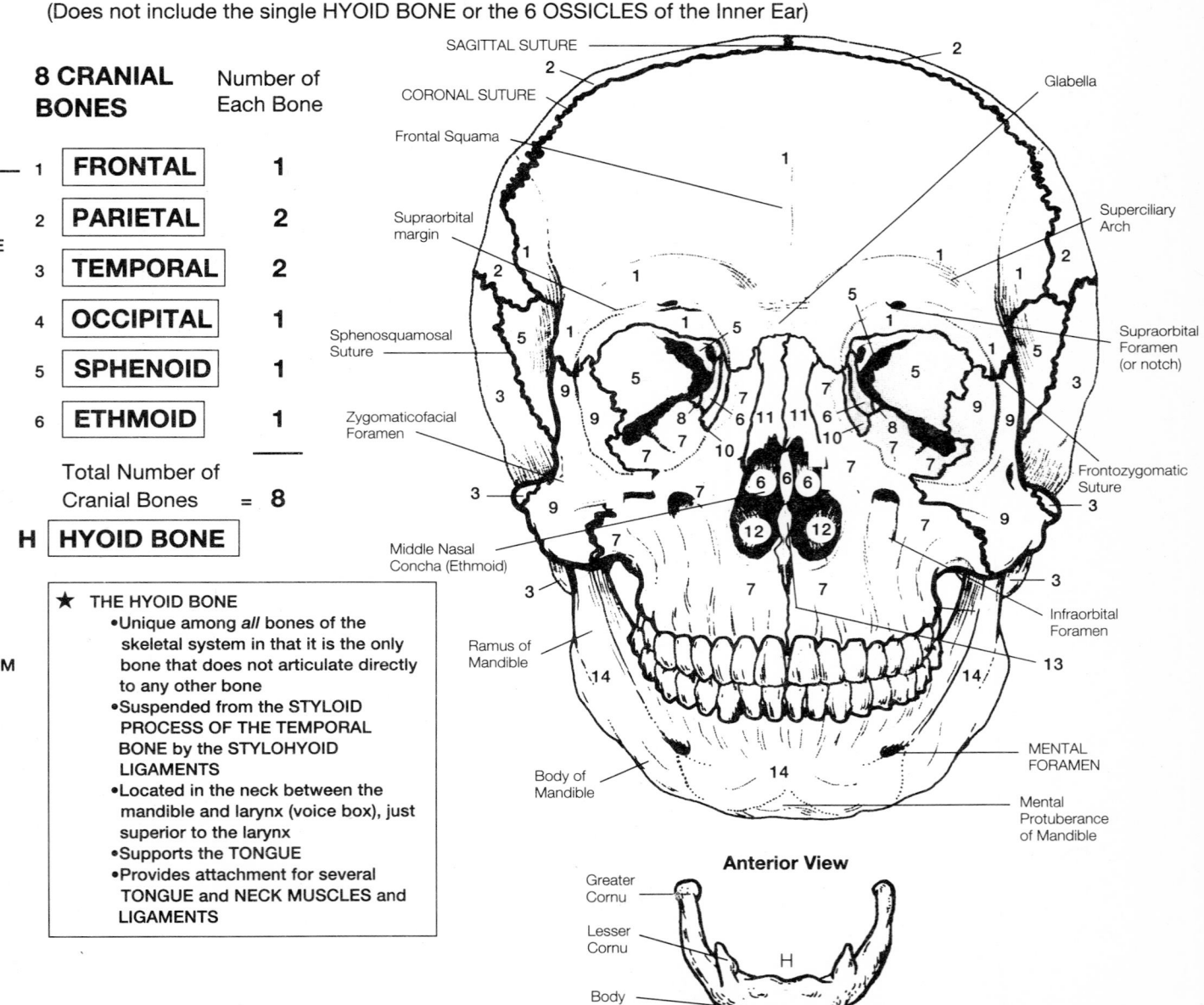

Anterior View

★ The 14 FACIAL BONES
- •Form the framework and basic shape for face
- •Support the teeth
- •The bones of the skull not in contact with the BRAIN
- •All are firmly interlocked by sutures, (except the MANDIBLE (LOWER JAW), with one another and the CRANIAL BONES
- •Attachments for various muscles that move the jaw and cause facial expressions

■ THE MAXILLA BONE
- •Articulates with every bone of the face (except the MANDIBLE) (See 4.8)

■ The TWO PALATINE BONES (See 4.8)

■ THE TWO ZYGOMATIC BONES form:
- •The cheekbones of the face
- •Part of the outer wall and floor of the orbits (Posterior projection called the TEMPORAL PROCESS articulates with anterior projection of the TEMPORAL BONE, forming the ZYGOMATIC ARCH.)

■ THE TWO LACRIMAL BONES form:
- •Anterior part of the medial wall of each orbit (Smallest bones of the face; thin, and shaped like a fingernail) (Tear ducts pass into the nasal cavity through the LACRIMAL FOSSAE)

■ THE TWO NASAL BONES FORM:
- •Bridge of the nose (superior part) (small, oblong bones fuse together) (See 4.9)

■ THE TWO INFERIOR NASAL CONCHAE (See 4.8)

■ THE VOMER BONE (See 4.9)

■ THE MANDIBLE forms:
- •The LOWER JAW
- ★ The ONLY MOVABLE BONE OF THE SKULL
- ★ Does not form from the fusion of 2 bones

BODY OF MANDIBLE
- •Horseshoe-shaped front
- •2 horizontal lateral sides

RAMI OF MANDIBLE
- •2 vertical extensions from posterior portion of the body (single = RAMUS)

ANGLES OF MANDIBLE
- •Where each RAMUS meets the BODY

14 FACIAL BONES

			Number of Each Bone
7	MAXILLA	UPPER JAW	2
8	PALATINE 2		2
9	ZYGOMATIC	MALAR (CHEEKBONE)	2
10	LACRIMAL		2
11	NASAL		2
12	INFERIOR NASAL CONCHA		2
13	VOMER		1
14	MANDIBLE	LOWER JAW	1
	Total Number of Facial Bones =		14

★ Not considered in the typical adult count of 206 total bones are WORMIAN, or SUTURAL, BONES within the joints (sutures) of the skull, whose number varies considerably from person to person

Superior Temporal Line

SAGITTAL SUTURE

LAMBDOIDAL SUTURE

External Occipital Protuberance

SQUAMOSAL SUTURE

Superior Nuchal Line

Inferior Nuchal Line

Mastoid Process (Temporal Bone)

Styloid Process (Temporal Bone)

Pterygoid Tuberosity of Mandible

Mylohyoid Line

Teeth

MASTOID FORAMEN

Posterior View

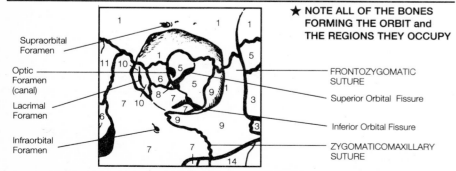

DETAIL of LEFT ORBIT (Anterior View)

Supraorbital Foramen

Optic Foramen (canal)

Lacrimal Foramen

Infraorbital Foramen

★ NOTE ALL OF THE BONES FORMING THE ORBIT and THE REGIONS THEY OCCUPY

FRONTOZYGOMATIC SUTURE

Superior Orbital Fissure

Inferior Orbital Fissure

ZYGOMATICOMAXILLARY SUTURE

SUTURES
- •Immovable articulations that are found ONLY between skull bones
- •Most of the names of the sutures are descriptive of the bones they connect

4 PROMINENT SKULL SUTURES
- ■ CORONAL SUTURE ——➤ between the FRONTAL BONE and the 2 PARIETALS
- ■ SAGITTAL SUTURE ——➤ between the 2 PARIETAL BONES
- ■ LAMBDOIDAL SUTURE ➤ between the occipital BONE and the 2 PARIETALS
- ■ SQUAMOSAL SUTURE ➤ between the 2 TEMPORAL BONES and the 2 PARIETALS (Squamosal refers to the thin, large expanded portion of the TEMPORAL BONE)

Repeat color selections from 4.4
★ See Chart #2 Openings in Bones and the
Structures Passing Through Them
★ See also 6.4

SKELETAL SYSTEM: AXIAL SKELETON
Bones of the Adult Skull: Cranial and Facial
Left Lateral and Inferior Views

22 BONES OF THE SKULL:

8 CRANIAL BONES

		Number of Each Bone
1	**FRONTAL**	1
2	**PARIETAL**	2
3	**TEMPORAL**	2
4	**OCCIPITAL**	1
5	**SPHENOID**	1
6	**ETHMOID**	1
		8

14 FACIAL BONES

		Number of Each Bone
7	**MAXILLA**	2
8	**PALATINE**	2
9	**ZYGOMATIC**	2
10	**LACRIMAL**	2
11	**NASAL**	2
12	**INFERIOR NASAL CONCHA**	2
13	**VOMER**	1
14	**MANDIBLE**	1
		14

| H | **HYOID BONE** | |

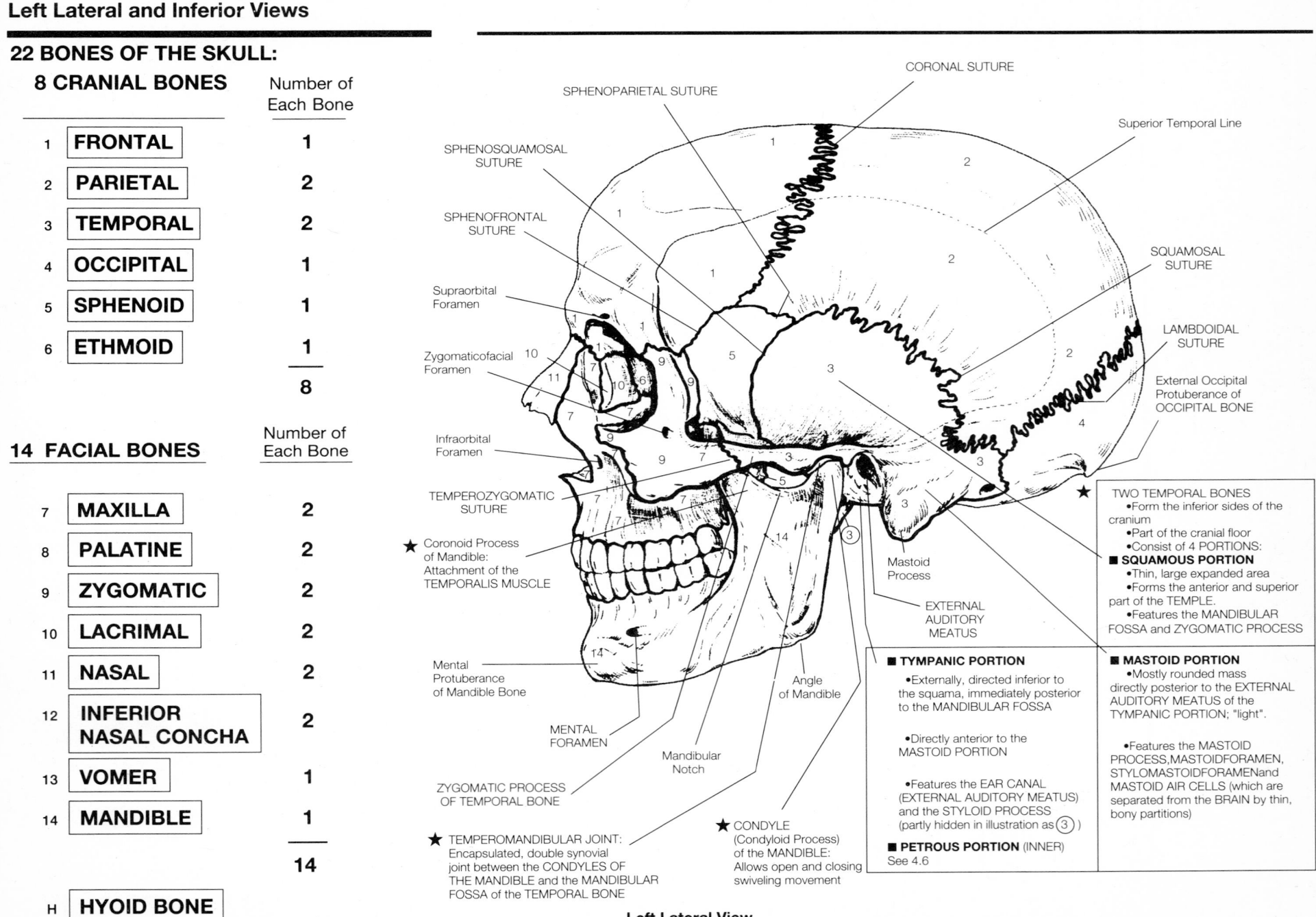

CORONAL SUTURE

SPHENOPARIETAL SUTURE

Superior Temporal Line

SPHENOSQUAMOSAL SUTURE

SPHENOFRONTAL SUTURE

SQUAMOSAL SUTURE

Supraorbital Foramen

LAMBDOIDAL SUTURE

Zygomaticofacial Foramen

External Occipital Protuberance of OCCIPITAL BONE

Infraorbital Foramen

TEMPEROZYGOMATIC SUTURE

★ Coronoid Process of Mandible: Attachment of the TEMPORALIS MUSCLE

Mastoid Process

★ TWO TEMPORAL BONES
•Form the inferior sides of the cranium
•Part of the cranial floor
•Consist of 4 PORTIONS:
■ **SQUAMOUS PORTION**
•Thin, large expanded area
•Forms the anterior and superior part of the TEMPLE.
•Features the MANDIBULAR FOSSA and ZYGOMATIC PROCESS

EXTERNAL AUDITORY MEATUS

Mental Protuberance of Mandible Bone

Angle of Mandible

■ **TYMPANIC PORTION**
•Externally, directed inferior to the squama, immediately posterior to the MANDIBULAR FOSSA

•Directly anterior to the MASTOID PORTION

•Features the EAR CANAL (EXTERNAL AUDITORY MEATUS) and the STYLOID PROCESS (partly hidden in illustration as ③)

■ **PETROUS PORTION** (INNER)
See 4.6

■ **MASTOID PORTION**
•Mostly rounded mass directly posterior to the EXTERNAL AUDITORY MEATUS of the TYMPANIC PORTION; "light".

•Features the MASTOID PROCESS, MASTOID FORAMEN, STYLOMASTOID FORAMEN and MASTOID AIR CELLS (which are separated from the BRAIN by thin, bony partitions)

MENTAL FORAMEN

Mandibular Notch

ZYGOMATIC PROCESS OF TEMPORAL BONE

★ TEMPEROMANDIBULAR JOINT:
Encapsulated, double synovial joint between the CONDYLES OF THE MANDIBLE and the MANDIBULAR FOSSA of the TEMPORAL BONE

★ CONDYLE
(Condyloid Process) of the MANDIBLE:
Allows open and closing swiveling movement

Left Lateral View

5 SPHENOID BONE :

"KEYSTONE" OF THE CRANIAL FLOOR
ARTICULATES WITH all OTHER CRANIAL BONES
"A BAT WITH OUT-STRETCHED WINGS"

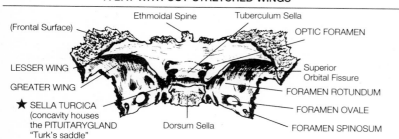

(Frontal Surface)
Ethmoidal Spine
Tuberculum Sella
OPTIC FORAMEN
LESSER WING
GREATER WING
Superior Orbital Fissure
FORAMEN ROTUNDUM
★ SELLA TURCICA (concavity houses the PITUITARY GLAND "Turk's saddle")
Dorsum Sella
FORAMEN OVALE
FORAMEN SPINOSUM

Superior View (see 4.6, bottom)

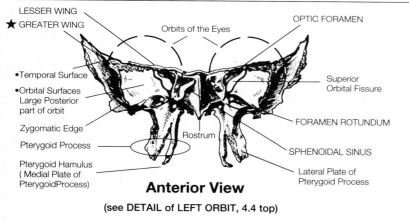

LESSER WING
★ GREATER WING
Orbits of the Eyes
OPTIC FORAMEN
•Temporal Surface
•Orbital Surfaces Large Posterior part of orbit
Zygomatic Edge
Rostrum
Superior Orbital Fissure
FORAMEN ROTUNDUM
Pterygoid Process
SPHENOIDAL SINUS
Pterygoid Hamulus (Medial Plate of PterygoidProcess)
Lateral Plate of Pterygoid Process

Anterior View

(see DETAIL of LEFT ORBIT, 4.4 top)

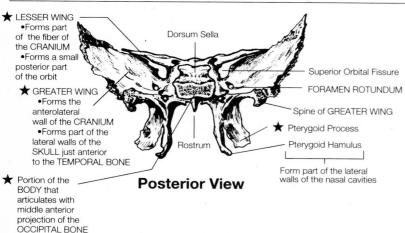

★ LESSER WING
•Forms part of the fiber of the CRANIUM
•Forms a small posterior part of the orbit
Dorsum Sella
★ GREATER WING
•Forms the anterolateral wall of the CRANIUM
•Forms part of the lateral walls of the SKULL just anterior to the TEMPORAL BONE
Superior Orbital Fissure
FORAMEN ROTUNDUM
Spine of GREATER WING
★ Pterygoid Process
Pterygoid Hamulus
Rostrum
★ Portion of the BODY that articulates with middle anterior projection of the OCCIPITAL BONE
Form part of the lateral walls of the nasal cavities

Posterior View

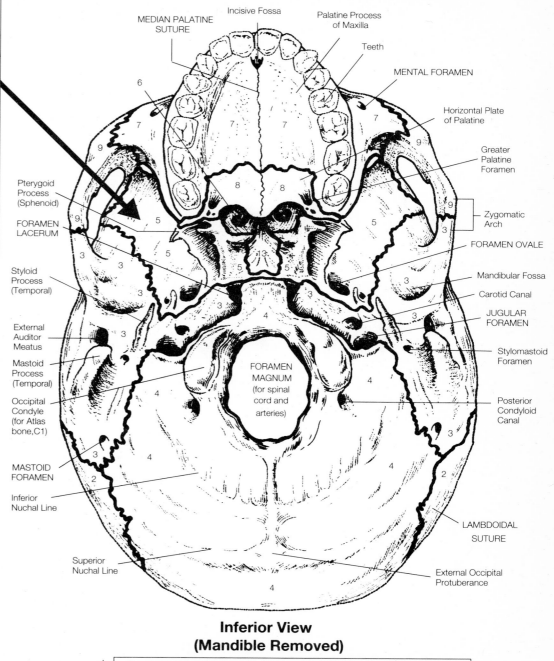

MEDIAN PALATINE SUTURE
Incisive Fossa
Palatine Process of Maxilla
Teeth
MENTAL FORAMEN
Horizontal Plate of Palatine
Greater Palatine Foramen
Zygomatic Arch
FORAMEN OVALE
Mandibular Fossa
Carotid Canal
JUGULAR FORAMEN
Stylomastoid Foramen
Posterior Condyloid Canal
LAMBDOIDAL SUTURE
External Occipital Protuberance

Pterygoid Process (Sphenoid)
FORAMEN LACERUM
Styloid Process (Temporal)
External Auditor Meatus
Mastoid Process (Temporal)
Occipital Condyle (for Atlas bone,C1)
MASTOID FORAMEN
Inferior Nuchal Line
Superior Nuchal Line

FORAMEN MAGNUM (for spinal cord and arteries)

Inferior View
(Mandible Removed)

★ The FORAMEN MAGNUM is the large hole in the inferior part of the cipital bone through which the spinal cord attaches to the brain. On either side of this large foramen are oval processes with convex surfaces—the OCCIPITAL CONDYLES. They articulate with the ATLAS (1st cervical vertebra).

SKELETAL SYSTEM: AXIAL SKELETON
Bones of the Adult Skull: Cranial and Facial
Right Midsagittal and Superior Views

Repeat color selections from 4.4 and 4.5.
★ See Chart #2 Openings in Bones and the
Structures Passing Through Them

22 BONES OF THE SKULL :

8 CRANIAL BONES	Number of Each Bone
1 FRONTAL	1
2 PARIETAL	2
3 TEMPORAL	2
4 OCCIPITAL	1
5 SPHENOID	1
6 ETHMOID	1
	8

14 FACIAL BONES	Number of Each Bone
7 MAXILLA	2
8 PALATINE	2
9 ZYGOMATIC	2
10 LACRIMAL	2
11 NASAL	2
12 INFERIOR NASAL CONCHA	2
13 VOMER	1
14 MANDIBLE	1
	14

CORONAL SUTURE

SPHENOPARIETAL SUTURE

SELLA TURCICA OF SPHENOID BONE (Concavity houses the PITUITARY GLAND)

Grooves for Middle Meningeal Vessels

SPHENOSQUAMOSAL SUTURE

SQUAMOSAL SUTURE

LAMBDOIDAL SUTURE

SPHENOFRONTAL SUTURE

FRONTAL SINUS

Groove for the TRANSVERSE SINUS

Crista Galli of the Ethmoid

Cribriform Plate of the Ethmoid

Perpendicular Plate of Ethmoid (Superior portion of NASALSEPTUM)

Anterior Nasal Spine

Incisive Canal

★ ALVEOLAR PROCESSES of both MAXILLA (7) and MANDIBLE (14): Sets of bony sockets into which the teeth are set

OCCIPITO-MASTOIDAL SUTURE

HYPOGLOSSAL FORAMEN

INTERNAL AUDITORY MEATUS

Styloid Process of Temporal Bone

SPHENOIDAL SINUS (of 5)

MANDIBULAR FORAMEN

Pterygoid Process of Sphenoid Bone

Palate

TEMPORAL BONE
4 PORTIONS
■ SQUAMOUS PORTION
■ PETROUS PORTION
•Internally, forms part of the floorof the cranial cavity;"dense"
FEATURES
•Structures of the middle and inner ear (including the 3 OSSICLE BONES per ear)
•CAROTID CANAL (allows blood into the BRAIN via the INTERNALCAROTID ARTERY)
•JUGULAR FORAMEN (lets blood drain from the BRAIN via the INTERNAL JUGULAR VEIN)
•INTERNAL AUDITORY MEATUS
■ TYMPANIC PORTION
■ MASTOID PORTION (See 4.5)

Right Midsagittal View
(View of Inner Right Cranial Cavity)

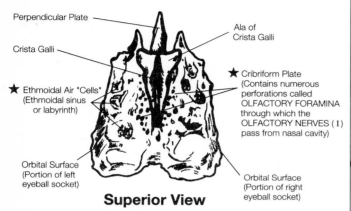

Perpendicular Plate

Crista Galli

★ Ethmoidal Air "Cells"
(Ethmoidal sinus
or labyrinth)

Orbital Surface
(Portion of left
eyeball socket)

Ala of
Crista Galli

★ Cribriform Plate
(Contains numerous
perforations called
OLFACTORY FORAMINA
through which the
OLFACTORY NERVES (I)
pass from nasal cavity)

Orbital Surface
(Portion of right
eyeball socket)

Superior View

★ CRISTA GALLI: An upper spine of the PERPENDICULAR PLATE,
projects superiorly into the cranial cavity.
It is an attachment for the MENINGES covering
the BRAIN.

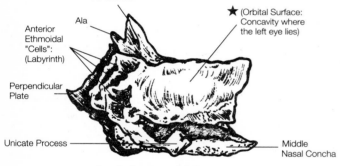

Anterior
Ethmoidal
"Cells":
(Labyrinth)

Ala

Perpendicular
Plate

Unicate Process

★ (Orbital Surface:
Concavity where
the left eye lies)

Middle
Nasal Concha

Left Lateral View
(See 4.7)

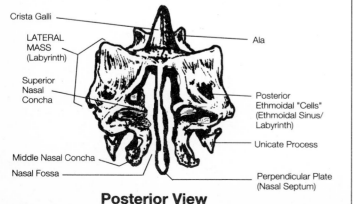

Crista Galli

LATERAL
MASS
(Labyrinth)

Superior
Nasal
Concha

Middle Nasal Concha

Nasal Fossa

Ala

Posterior
Ethmoidal "Cells"
(Ethmoidal Sinus/
Labyrinth)

Unicate Process

Perpendicular Plate
(Nasal Septum)

Posterior View

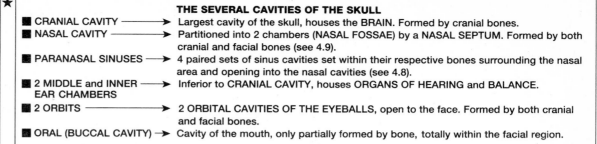

THE SEVERAL CAVITIES OF THE SKULL

■ CRANIAL CAVITY ⟶ Largest cavity of the skull, houses the BRAIN. Formed by cranial bones.

■ NASAL CAVITY ⟶ Partitioned into 2 chambers (NASAL FOSSAE) by a NASAL SEPTUM. Formed by both cranial and facial bones (see 4.9).

■ PARANASAL SINUSES ⟶ 4 paired sets of sinus cavities set within their respective bones surrounding the nasal area and opening into the nasal cavities (see 4.8).

■ 2 MIDDLE and INNER EAR CHAMBERS ⟶ Inferior to CRANIAL CAVITY, houses ORGANS OF HEARING and BALANCE.

■ 2 ORBITS ⟶ 2 ORBITAL CAVITIES OF THE EYEBALLS, open to the face. Formed by both cranial and facial bones.

■ ORAL (BUCCAL CAVITY) ⟶ Cavity of the mouth, only partially formed by bone, totally within the facial region.

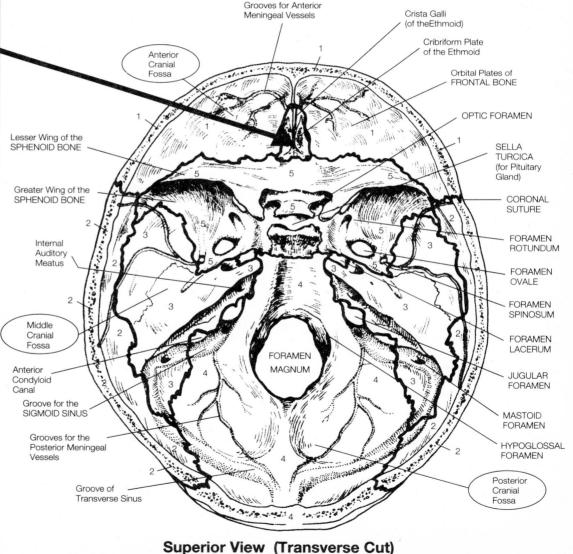

Grooves for Anterior
Meningeal Vessels

Crista Galli
(of theEthmoid)

Cribriform Plate
of the Ethmoid

Orbital Plates of
FRONTAL BONE

OPTIC FORAMEN

SELLA
TURCICA
(for Pituitary
Gland)

CORONAL
SUTURE

FORAMEN
ROTUNDUM

FORAMEN
OVALE

FORAMEN
SPINOSUM

FORAMEN
LACERUM

JUGULAR
FORAMEN

MASTOID
FORAMEN

HYPOGLOSSAL
FORAMEN

Posterior
Cranial
Fossa

Anterior
Cranial
Fossa

Lesser Wing of the
SPHENOID BONE

Greater Wing of the
SPHENOID BONE

Internal
Auditory
Meatus

Middle
Cranial
Fossa

Anterior
Condyloid
Canal

Groove for the
SIGMOID SINUS

Grooves for the
Posterior Meningeal
Vessels

Groove of
Transverse Sinus

FORAMEN
MAGNUM

Superior View (Transverse Cut)
(Floor of Cranial Cavity/Base of Skull)

SKELETAL SYSTEM: AXIAL SKELETON
Bones of the Adult Skull:
Lateral Walls and Medial Nasal Septum of the Nasal Cavity/ External Nose

1 = pink   5 = yellow   6 = light blue
7 = flesh   8 = light range   10 = yellow-orange
11 = light brown   12 = blue-green   13 = light red
CARTILAGES = any colors except cream, green, or grey.
★ See also 11.2

★ The NASAL CAVITY is the cavity in the SKULL between the floor of the cranium and the roof of the mouth. The ETHMOID BONE is located in the anterior portion of the floor of the cranium between the orbits, where it forms the ROOF OF THE NASAL CAVITY.

•A major portion of the RIGHT and LEFT LATERAL WALLS OF THE NASAL CAVITY is composed of the inner portion of the LATERAL MASSES (LABYRINTHS) OF THE ETHMOID BONE (See 4.6). Portions of the FRONTAL and SPHENOID cranial bones, along with portions of many facial bones, help to complete the lateral walls of the NASAL CAVITY.

•Two scroll-shaped plates from both inner lateral walls of the ETHMOID BONE project into the NASAL CAVITY, the SUPERIOR and MIDDLE NASAL CONCHAE.

•An inferior middle projection of the ETHMOID BONE, THE PERPENDICULAR PLATE helps to form a MEDIAL SEPTUM that splits the NASAL CAVITY into 2 separate chambers:

★ NASAL SEPTUM: MEDIAL WALL of the NASAL CAVITY. Partitions the NASAL CAVITY into 2 chambers or NASAL FOSSAE, and is formed of both bone and cartilage.
SUPPORTING FRAMEWORK OF NASAL SEPTUM
•The median COLUMNA (SEPTAL CARTILAGE) of the nose extends inward to form the anterior part of the septum.
•The thin, plowshare-shaped VOMER bone forms the inferior and posterior part of the nasal septum.
•The PERPENDICULAR PLATE of THE ETHMOID BONE articulates with the VOMER BONE and COLUMNA, and forms the superior portion of the nasal septum.
An inner portion of GREATER ALAR CARTILAGE, NASAL and VOMERONASAL CARTILAGE, and NASAL CRESTS of the NASAL, SPHENOID, PALATINE, and MAXILLA BONES complete the NASAL SEPTUM.

RIGHT AND LEFT LATERAL WALLS OF THE NASAL CAVITY

Associated CRANIAL BONES

1 | **FRONTAL BONE** |

5 | **SPHENOID BONE** |

6 | **ETHMOID BONE** |

Associated FACIAL BONES

7 | **MAXILLA BONE** | UPPER JAW

8 | **PALATINE BONE** |

10 | **LACRIMAL BONE** |

11 | **NASAL BONE** |

12 | **INFERIOR NASAL CONCHA** | *

* The fragile INFERIOR NASAL CONCHA are not part of the ETHMOID BONE, as are the SUPERIOR and MIDDLE NASAL CONCHA. They are all scroll-shaped bones that form portions of the lateral walls of the nasal cavities and project into the cavities. All of these 3 nasal conchae allow for circulation and filtration of air before it passes into the Lungs.

NASAL SEPTUM (MEDIAL WALL OF THE NASAL CAVITY)

Associated SKULL BONES

5 | **NASAL CREST OF THE SPHENOID BONE** |

6 | **PERPENDICULAR PLATE OF THE ETHMOID BONE** | *

11 | **NASAL CREST OF THE NASAL BONES** |

13 | **VOMER BONE** | *

Associated CARTILAGES

15 | **SEPTAL CARTILAGE** | *COLUMNA

16 | **GREATER ALAR CARTILAGE** |

19 | **NASAL CARTILAGE** |

20 | **VOMERONASAL CARTILAGE** |

*Major components of the NASAL SEPTUM

★ THE EXTERNAL NOSE
•The framework of the nose is formed of both bone and cartilage, but mostly cartilage.
2 NASAL BONES
•Fuse together to form the superior part of the "bridge" of the nose that lies between the orbits.
•The orifices of the nose, the NOSTRILS or NARES, are separated from one another by a median septum called the COLUMNA (SEPTAL CARTILAGE).

EXTERNAL NOSE

11 | **NASAL BONES** | ■

F | **FIBRO-FATTY TISSUE** |

■ Although not part of the NOSE, the FRONTAL PROCESSES OF THE 2 FUSED MAXILLAE BONES form the posterolateral support of the NOSE near the NASAL BONES.

CARTILAGES of External Nose

15 | **SEPTAL CARTILAGE** | COLUMNA

16 | **GREATER ALAR CARTILAGE** |

17 | **LESSER ALAR CARTILAGE** |

18 | **LATERAL NASAL CARTILAGE** |

LATERAL WALLS OF THE NASAL CAVITY

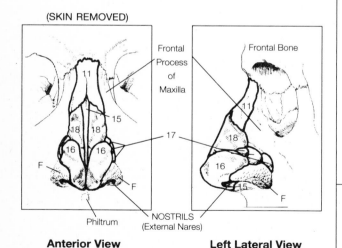

Cribriform Plate (of Ethmoid)
Crista Galli of Ethmoid
Middle Nasal Concha (of Ethmoid)
Superior Nasal Concha (of Ethmoid)
Sella Turcica
FRONTAL SINUS
SPHENOIDAL SINUS
SPHENOPALATINE FORAMEN
MAXILLARY SINUS (Antrum of Highmore)

Right Lateral Wall (Medial View)
(Left wall and perpendicular plate of the Ethmoid bone cut away to expose the Ethmoid's right wall)

EXTERNAL NOSE

(SKIN REMOVED)

Frontal Process of Maxilla
Frontal Bone
Philtrum
NOSTRILS (External Nares)

Anterior View   **Left Lateral View**

MEDIAL NASAL SEPTUM OF NASAL CAVITY

★ **See Respiratory System, Chapter 11, 11.2 for more details of the Openings of the Paranasal Sinuses into the Nasal Cavity.**

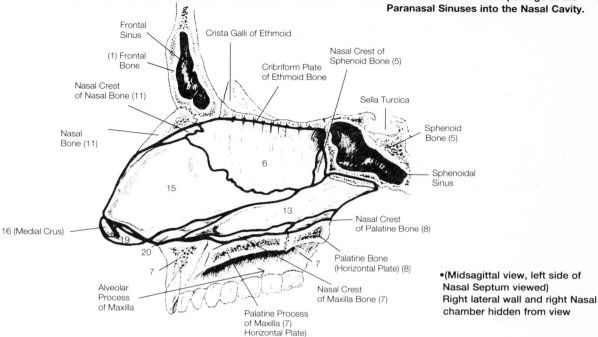

Frontal Sinus
Crista Galli of Ethmoid
Nasal Crest of Sphenoid Bone (5)
(1) Frontal Bone
Cribriform Plate of Ethmoid Bone
Sella Turcica
Nasal Crest of Nasal Bone (11)
Sphenoid Bone (5)
Nasal Bone (11)
Sphenoidal Sinus
16 (Medial Crus)
Nasal Crest of Palatine Bone (8)
Palatine Bone (Horizontal Plate) (8)
Alveolar Process of Maxilla
Nasal Crest of Maxilla Bone (7)
Palatine Process of Maxilla (7) Horizontal Plate)

•(Midsagittal view, left side of Nasal Septum viewed)
Right lateral wall and right Nasal chamber hidden from view

7 MAXILLA BONE

The paired **MAXILLARY BONES** unite to form the single upper jawbone, the **MAXILLA BONE**.
■ Articulates with *every bone of the face* except the lower jawbone, or **MANDIBLE**
■ Fusion complete before birth
FORMATIONS OF THE MAXILLA:
•Part of the lateral walls and floor of the nasal cavities
•Part of the roof of the mouth, comprising most of the hard palate
•Part of the floors of the orbits (See 4.4)

Orbital Surface (Portion of the right eyeball socket)
Frontal Process
Lacrimal Groove
MAXILLARY SINUS
Middle Meatus
Inferior Meatus
Zygomatic Process
INFRAORBITAL FORAMEN
Nasal Notch
INCISIVE CANAL
Greater Palatine Groove
Palatine Process
Horizontal Plate
Alveolar Process

R. Lateral View   **R. Medial View**

8 PALATINE BONES

•L-shaped bones whose horizontal processes unite to form a **NASAL CREST**. These **HORIZONTAL PLATES** form the posterior portion of the hard palate. (This hard palate separates the upper nasal cavity from the lower oral cavity.)
•Small part of the lateral walls and floor of the nasal cavities
•Small part of the floors of the orbit (See 4.4, bottom)

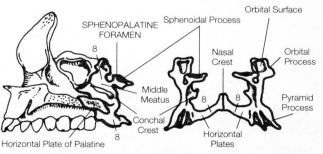

SPHENOPALATINE FORAMEN
Sphenoidal Process
Orbital Surface
Nasal Crest
Orbital Process
Middle Meatus
Conchal Crest
Horizontal Plates
Pyramid Process
Horizontal Plate of Palatine

In Situ   **Posterior View (both bones)**

SKELETAL SYSTEM: AXIAL SKELETON
The Vertebral Column and Vertebrae

★ THE VERTEBRAL COLUMN
- Commonly called the "backbone"
- The backbone (or spine) together
with the SPINAL CORD = SPINAL COLUMN
26 MOVABLE VERTEBRAE
■ 24 SEPARATE VERTEBRAE separated
by fibrocartilaginous INTERVERTEBRAL
DISCS. (The drum-shaped BODY or
CENTRUM of each VERTEBRA is in contact
with the DISCS at each end.) Of the 24,
there are 7 CERVICAL VERTEBRAE, 12 THORACIC
VERTEBRAE, and 5 LUMBAR VERTEBRAE.

■ 1 SACRUM develops as 5 distinct
bones that fuse together.

■ 1 COCCYX consists of 3–5 (usually 4) more or
less fused rudimentary "tailbones".

★ FUNCTIONS OF THE SPINE
- Supports the head
- Supports the thorax by serving as a point of
attachment for the ribs (and the muscles of the back)
- Supports the girdles (pectoral and pelvic girdles
that attach the appendages to the trunk)
- Houses and protects the spinal cord
- Permits passage of the spinal nerves
- A strong, flexible rod permits freedom of movement
(anteriorly, posteriorly, and laterally)
- INTERVERTEBRAL DISCS lend flexibility and absorb
stress of movement

★ LINKAGE OF THE VERTEBRAE
- Intervertebral discs ⎫
- Interlocking processes ⎬ permits:
- Binding ligaments ⎭
■ LIMITED MOVEMENT BETWEEN VERTEBRAE
■ EXTENSIVE MOVEMENTS OF THE ENTIRE
VERTEBRAL COLUMN

★ OPENINGS OF THE VERTEBRAL COLUMN
- VERTEBRAL FORAMEN of each VERTEBRA all
add up to form the SPINAL CANAL through which
the SPINAL CORD traverses.
- Between the vertebrae are openings called
INTERVERTEBRAL FORAMINA, through which spinal
nerves traverse as they branch off from the spinal cord.
- TRANSVERSE FORAMINA in the transverse
processes of the CERVICAL VERTEBRA allow passage
of the VERTEBRAL VESSELS, which transfer blood to
and from the BRAIN.

The VERTEBRAL COLUMN is composed of **26 MOVABLE VERTEBRAE**
(It Is Actually Composed of **33 INDIVIDUAL VERTEBRAE**, But
5 Are Fused into **1 SACRUM** And **4** Are More or Less Fused into **1 COCCYX**)

The 4 Adult SPINAL CURVATURES:

CERVICAL CURVATURE: CONVEX ANTERIORLY

C | **7 CERVICAL VERTEBRAE** |

THORACIC CURVATURE: CONCAVE ANTERIORLY

T | **12 THORACIC VERTEBRAE** |

LUMBAR CURVATURE: CONVEX ANTERIORLY

L | **5 LUMBAR VERTEBRAE** |

PELVIC CURVATURE: CONCAVE ANTERIORLY

S | **1 SACRUM** |

Co | **1 COCCYX** |

= **26 VERTEBRAE**

★ ■ The FETAL SPINE has only 1 curve (concave anteriorly)
■ Thus, the THORACIC and PELVIC CURVATURES in
the adult spine are designated as PRIMARY CURVES
■ Therefore, the opposite CERVICAL and LUMBAR
CURVATURES are designated as SECONDARY CURVES

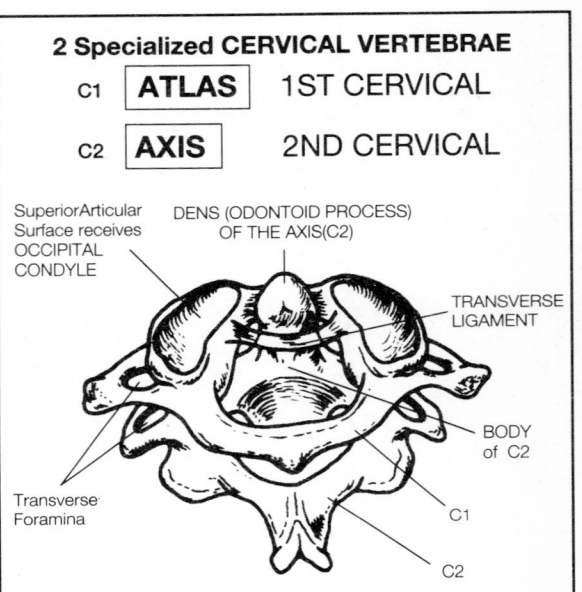

2 Specialized CERVICAL VERTEBRAE

C1 | ATLAS | 1ST CERVICAL

C2 | AXIS | 2ND CERVICAL

SuperiorArticular
Surface receives
OCCIPITAL
CONDYLE

DENS (ODONTOID PROCESS)
OF THE AXIS(C2)

TRANSVERSE
LIGAMENT

BODY
of C2

Transverse
Foramina

C1

C2

(Posterior View)

ATLAS
- A ring of bone with ANTERIOR and POSTERIOR
ARCHES
- 2 large lateral masses
- SUPERIOR ARTICULAR SURFACES of the LATERAL
MASSES receive the OCCIPITAL CONDYLES of the skull,
thus supporting the head and permit nodding of the head
- Has *no* BODY and *no* SPINAL PROCESS
- TUBERCLES for attachment of TRANSVERSE
LIGAMENT

AXIS
- DENS (ODONTOID PROCESS) attached to the
superior aspect of the BODY. It is regarded as *what
remains of* the ATLAS BODY that has become attached
to the AXIS
- TRANSVERSE LIGAMENT OF THE ATLAS passes
behind the DENS to create a PIVOT JOINT permitting
ROTATION OF THE SKULL.

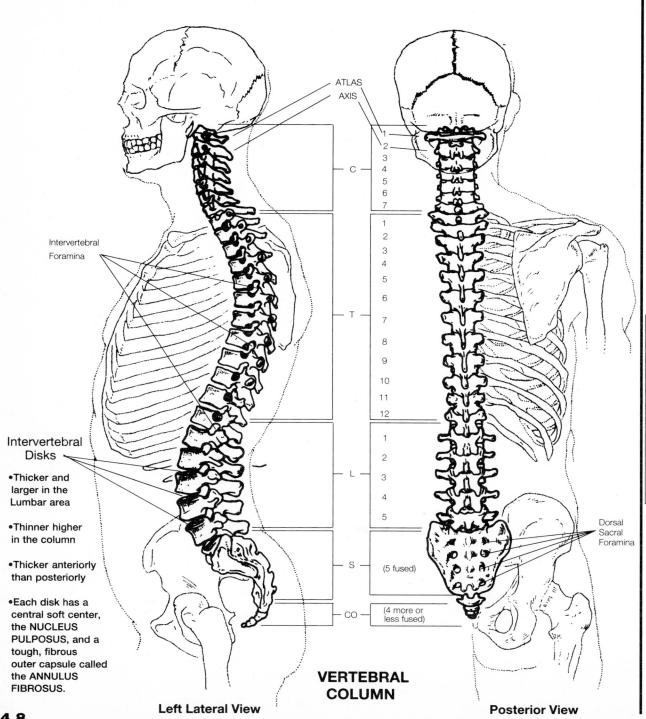

GENERAL STRUCTURE OF A TYPICAL VERTEBRA
(no coloring necessary)

POSTERIOR

NEURAL ARCH

ANTERIOR

Spinous Process

LAMINA

Superior Articular Process (and facet) of the PEDICLE

Transverse Process

PEDICLE

★ VERTEBRAL FORAMEN (encloses spinal cord)

Body (Centrum)

Superior View

■ **BODY (CENTRUM)**
  • Weight-bearing part of a VERTEBRA adapted to withstand compression; INTERVERTEBRAL DISCS contact on each end of the BODY.
■ **NEURAL ARCH**
  • Affixed to the posterior surface of the BODY
  • Composed of 2 short, thick PEDICLES and 2 arched LAMELLAE
■ **VERTEBRAL FORAMEN**
  • Hollow space formed by the BODY and the NEURAL ARCH
  • Spinal cord passes through all the VERTEBRAL FORAMINA
■ **7 PROCESSES OF THE NEURAL ARCH**
  • 1 SPINAL PROCESS
  • 2 TRANSVERSE PROCESSES
  • 2 SUPERIOR ARTICULAR PROCESSES
  • 2 INFERIOR ARTICULAR PROCESSES

★ Superior articular process (and facet) interlocks with the inferior articular process of the adjacent anterior vertebra

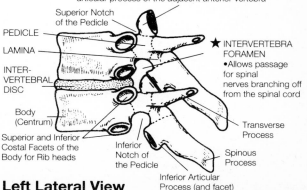

Superior Notch of the Pedicle

PEDICLE

LAMINA

INTER-VERTEBRAL DISC

★ INTERVERTEBRA FORAMEN
• Allows passage for spinal nerves branching off from the spinal cord

Body (Centrum)

Superior and Inferior Costal Facets of the Body for Rib heads

Inferior Notch of the Pedicle

Transverse Process

Spinous Process

Inferior Articular Process (and facet)

Left Lateral View

ATLAS
AXIS

C
1
2
3
4
5
6
7

T
1
2
3
4
5
6
7
8
9
10
11
12

L
1
2
3
4
5

S (5 fused)

CO (4 more or less fused)

Intervertebral Foramina

Intervertebral Disks

• Thicker and larger in the Lumbar area

• Thinner higher in the column

• Thicker anteriorly than posteriorly

• Each disk has a central soft center, the NUCLEUS PULPOSUS, and a tough, fibrous outer capsule called the ANNULUS FIBROSUS.

Dorsal Sacral Foramina

VERTEBRAL COLUMN

Left Lateral View

Posterior View

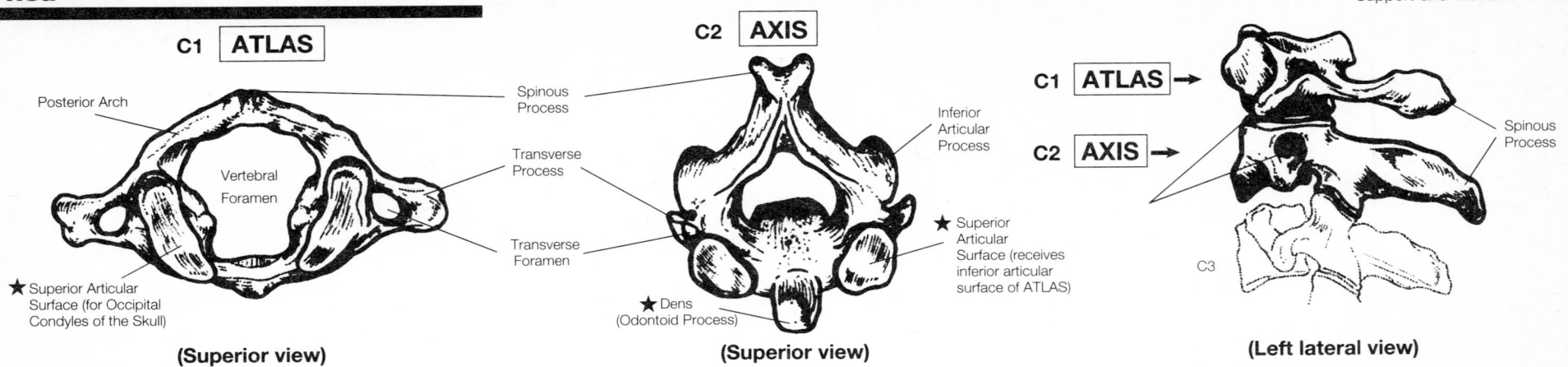

C1 ATLAS

Posterior Arch

Vertebral Foramen

★ Superior Articular Surface (for Occipital Condyles of the Skull)

(Superior view)

C2 AXIS

Spinous Process

Transverse Process

Transverse Foramen

Inferior Articular Process

★ Superior Articular Surface (receives inferior articular surface of ATLAS)

★ Dens (Odontoid Process)

(Superior view)

C1 ATLAS →

C2 AXIS →

Spinous Process

C3

(Left lateral view)

REGIONAL CHARACTERISTICS

Superior View (Posterior aspect on top)

Left Lateral View

c 7 CERVICAL

- ■ Second smallest vertebrae (next to coccyx)
- ■ Form flexible framework of neck region
- ■ Support the head
- ■ Compressed bodies; osseous tissue is more dense than any other vertebral region.
- ■ TRANSVERSE FORAMINA in the base of the transverse processes (passage of vertebral vessels which transfer blood to and from the brain)
- ■ BIFID SPINOUS PROCESSES (C2-C6), for attachment of the NUCHAL LIGAMENT. This strong ligament adds added support at the back of the skull (■ The ATLAS and AXIS have been discussed.)

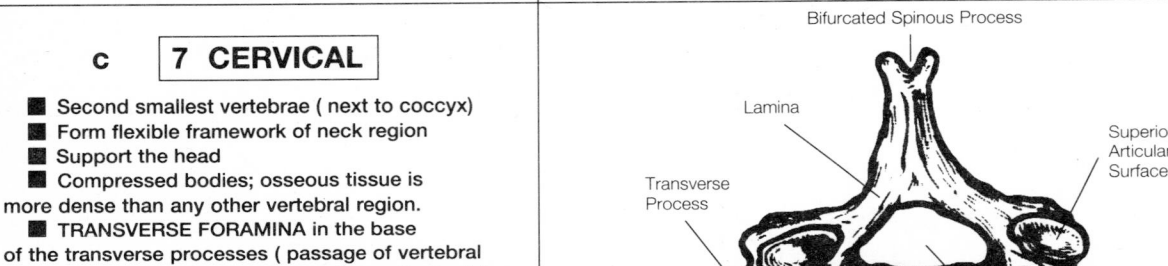

Bifurcated Spinous Process

Lamina

Transverse Process

Superior Articular Surface

Body of Transverse Process

Transverse Foramina

Body (Centrum)

VERTEBRAL FORAMEN

(Anterior ◄———► Posterior)

Superior Articular Process

Bifurcated Spinous Process

Body

Inferior Articular Process

★ C7, the VERTEBRA PROMINENS
  • Elongated spinous process which is *not* bifurcated

T 12 THORACIC

- ■ Serve as Attachments for the ribs, which form the posterior anchor of the rib cage.
- ■ Facets or demifacets on the body for rib attachment:
  - •T1 ——→ a superior facet and an inferior demifacet on both sides
  - •T2-T8 ——→ 2 demifacets on each side (so each head of a rib articulates with 2 vertebrae)
  - •T9 ——→ a single demifacet on each side
  - •T10-T12 ➤ facets on each side
- ■ Facets on the transverse processes for rib attachment:
  - •T2-T10 ➤ have facets on their transverse processes for articulation with the tubercles of the ribs
- ■ Increase in size from T1–T12
- ■ Each thoracic vertebra has a characteristic long spinous process (slopes obliquely downward)

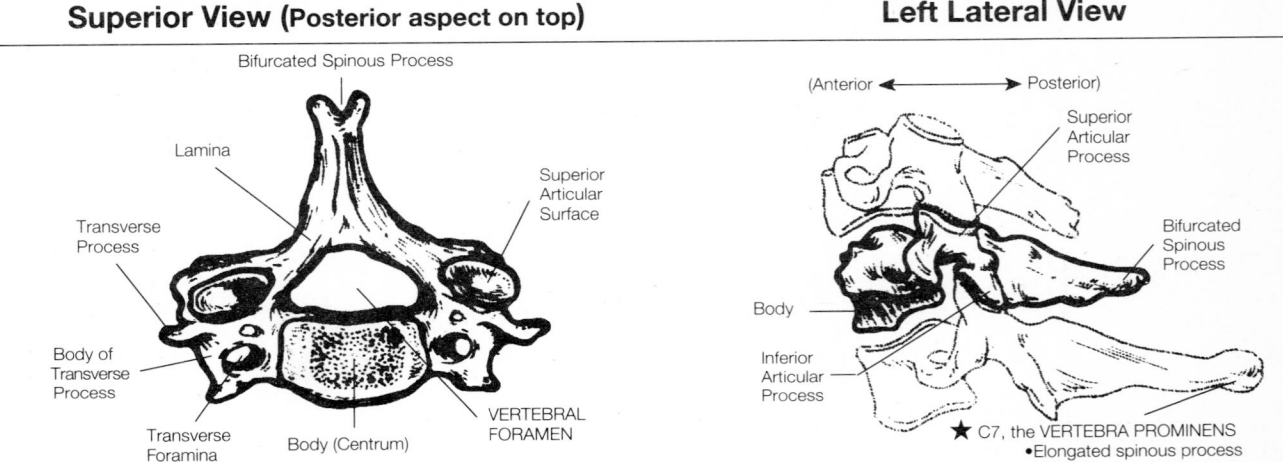

Spinous Process

★ Superior Articular Surface (articulates with the Inferior Articular Surface of the upper adjacent vertebra)

Transverse Process

Lamina

Pedicle

Demifacet for portion of a head of a Rib

Facet for Tubercle of Rib

VERTEBRAL FORAMEN

Body (Centrum)

Intervertebral Foramen

Body

Demifacets for head of "A"

Rib

Inferior Articular Process

Superior Articular Surface Transverse Process

Facet for Tubercle of Rib "B" above Rib "A"

Facet for Tubercle of Rib "A"

Spinous Process

(See 4.9)

5 LUMBAR

- ■ Largest vertebrae of the vertebral column
- ■ Heavy bodies
- ■ Thick, blunt spinous processes (for attachment of powerful back muscles)
- ■ ARTICULAR PROCESSES clasp one another:
  - •Superior articular processes are directed MEDIALLY (and usually SUPERIORLY)
  - •Inferior articular processes are directed LATERALLY (and usually INFERIORLY)
- ■ Lack of transverse foramina (as in CERVICALS)
- ■ Lack of facets for ribs (as in THORACICS)
- ■ Processes are termed "mamillary"

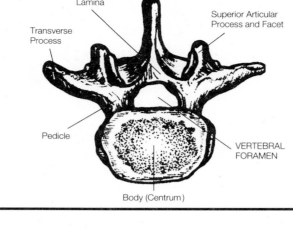

Spinous Process
Lamina
Transverse Process
Superior Articular Process and Facet
Pedicle
VERTEBRAL FORAMEN
Body (Centrum)

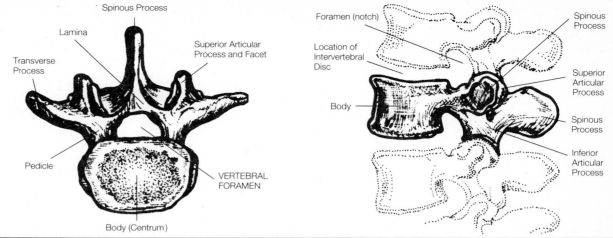

Foramen (notch)
Spinous Process
Location of Intervertebral Disc
Superior Articular Process
Body
Spinous Process
Inferior Articular Process

S

1 SACRUM

- ■ Composed of 5 fused and flattened vertebrae (total fusion after age 26)
- ■ Provides strong foundation for pelvic girdle
- ■ Triangular, wedge-shaped
- ■ Spinous processes of the 5 vertebrae fuse to form the MEDIAN SACRAL CREST
- ■ DORSAL SACRAL FORAMINA on either side of the crest pass the POSTERIOR DIVISIONS OF THE SACRAL NERVES from the spinal cord
- ■ VERTEBRAL CANAL (flattened): (for spinal cord)
  - • Opens superiorly at the SACRAL CANAL
  - • Opens inferiorly at HIATUS OF THE SACRAL CANAL (bounded by the SACRAL CORNU)
- Lateral processes, or ALAE (which bear the 4 pairs of DORSAL SACRAL FORAMINA) have on their upper sides the AURICULAR ARTICULATING SURFACES for the OS COXAE (HIPBONES)
- ■ Large SUPERIOR ARTICULAR PROCESSES for articulation with the 5th LUMBAR VERTEBRA (L5)
- ■ SACRAL PROMONTORY on the superior border of the anterior surface (important obstetrical landmark)
- ■ Anterior surface is smooth and forms the posterior surface of the PELVIC CAVITY
- ■ Anteriorly, the TRANSVERSE LINES OF FUSION can be easily seen
- ■ VENTRAL PELVIC FORAMINA on either side of each transverse line pass the ANTERIOR DIVISIONS OF THE SACRAL NERVES from the spinal cord

Co

1 COCCYX

- ■ 3 – 5 rudimentary, reduced bodies, more or less fused
- ■ CO 1 carries COCCYGEAL CORNU, which butt with the SACRAL CORNU

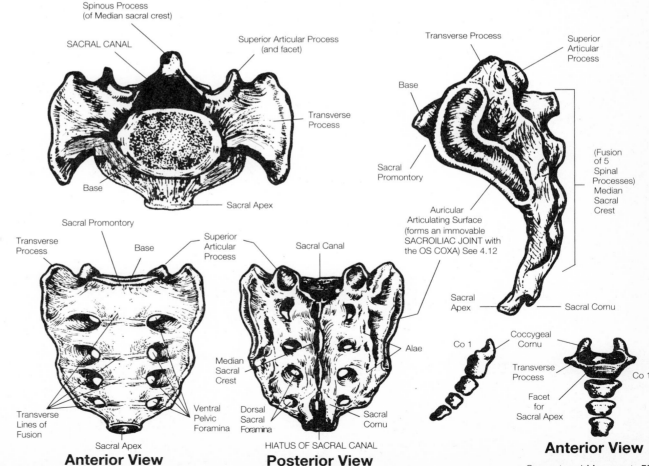

Spinous Process (of Median sacral crest)
SACRAL CANAL
Superior Articular Process (and facet)
Transverse Process
Base
Sacral Apex

Transverse Process
Superior Articular Process
Base
Sacral Promontory
Auricular Articulating Surface (forms an immovable SACROILIAC JOINT with the OS COXA) See 4.12
Sacral Apex
Sacral Cornu
(Fusion of 5 Spinal Processes) Median Sacral Crest

Sacral Promontory
Transverse Process
Base
Superior Articular Process
Sacral Canal
Median Sacral Crest
Alae
Sacral Cornu
Transverse Lines of Fusion
Ventral Pelvic Foramina
Dorsal Sacral Foramina
Sacral Apex
HIATUS OF SACRAL CANAL

Co 1
Coccygeal Cornu
Transverse Process
Facet for Sacral Apex
Co 1

Anterior View
Posterior View
Anterior View

S = yellow-orange R = cream
CC = light grey a = yellow
b = orange c = red
TR = flesh FR = light brown

SKELETAL SYSTEM: AXIAL SKELETON
The Rib (Thoracic) Cage: The Skeletal Portion of the Thorax

★ See also 4.8a, 6.9

★ THE RIB (THORACIC) CAGE
- •Conical shape with a narrow, superior *inlet* and a broad, inferior *outlet*
- •Compressed ANTEROPOSTERIORLY
- •Encloses and protects the thoracic viscera (Recall that the THORAX, or CHEST, is that region of the body between the base of the neck superiorly and the diaphragm inferiorly.)
- •Very flexible; involved directly in the mechanics of breathing (see 11.9).
- •Supports the PECTORAL GIRDLE and UPPER EXTREMITIES

★ STERNUM (BREASTBONE)
- •Elongated, flattened bony plate
- •Forms the ANTERIOR MIDLINE OF THE UPPER THORAX
- •15 cm (6 in.) in length
- •Attaches to the COSTAL CARTILAGES of the 1st–7th RIBS

★ THE 12 PAIRS OF RIBS
- •Each pair attaches POSTERIORLY to A THORACIC VERTEBRA
- •ANTERIORLY, the FIRST 7 PAIRS OF RIBS are called TRUE RIBS because they attach (with their costal cartilages) directly to the STERNUM
- •The LAST 5 PAIRS OF RIBS are called FALSE RIBS because they have no direct attachment or no attachment at all:
  - ■ Anteriorly, the first 3 pairs of FALSE RIBS—8th, 9th, 10th attach to the COSTAL CARTILAGE OF the 7th PAIR
  - ■ The last 2 pairs of FALSE RIBS have no association with the STERNUM at all, and are called FLOATING RIBS,11th and 12th PAIR

★ GENERAL MORPHOLOGY OF A TYPICAL RIB
- ■ HEAD
  - •Articulates with the bodies of 2 adjacent vertebrae by 2 DEMIFACETS separated by an INTERARTICULAR CREST
- ■ NECK
- ■ TUBERCLE : knoblike process lateral to HEAD
  - •Large ARTICULATING PORTION attaches to the facet of the TRANSVERSE PROCESS of the THORACIC VERTEBRA
  - •Small NONARTICULATING PORTION serves as the attachment for the COSTOTRANSVERSE LIGAMENT connecting the RIB to the TRANSVERSE PROCESS of the THORACIC VERTEBRA
- ■ BODY (SHAFT)
  - •Composed of parts of different curvature meeting at the ANGLE
  - •Curves downward and forward
  - •Cannot be flattened on a surface
  - •SUPERIOR SURFACE is blunt or rounded and the INFERIOR SURFACE is sharper. Both surfaces afford attachment for INTERCOSTAL MUSCLES (except the 1st RIB).

The RIB (THORACIC) CAGE:

S	STERNUM	BREASTBONE
R	12 PAIRED RIBS = 24	
CC	COSTAL CARTILAGES	
T	12 THORACIC VERTEBRAE	

The STERNUM

a	MANUBRIUM
b	BODY
c	XIPHOID PROCESS	(cartilage in youth)

The RIBS   (A RIB = A COSTA)
There Are A Total of
24 RIBS, 12 Paired:

TR	7 PAIRED TRUE RIBS
FR	5 PAIRED FALSE RIBS

Thus, There Are
14 TRUE RIBS and
10 FALSE RIBS: of These
10, The Last 4 Are **FLOATING
FALSE RIBS (2 Pair)**

★ (Upper Demifacet concave surface) for articulation with the body of the adjacent anterior Vertebra (to the Inferior Costal Facet)

Interarticular Crest

★ (Lower Demifacet concave surface) for articulation with the body of the THORACIC Vertebra (to the Superior Costal Facet)

Distal Sternal End (attaches to Costal Cartilage)

Nonarticular portion of TUBERCLE (attachment of Costotransverse Ligament)

HEAD

NECK

BODY

ANGLE

Intercostal Groove for passage of the Intercostal Vessels and Nerves

Articular facet of TUBERCLE (articulates with the facet of the Transverse Process of the THORACIC Vertebrae)

R: A TYPICAL RIGHT RIB (Viewed from Below and Behind)
(See 4.8a)

THE STERNUM

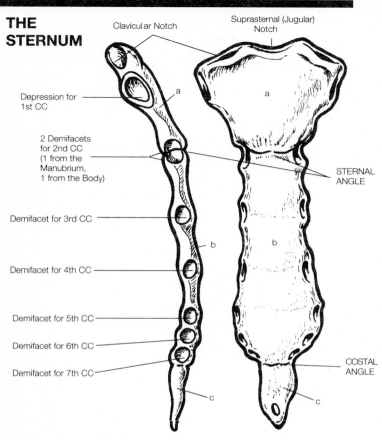

Clavicular Notch

Suprasternal (Jugular) Notch

Depression for 1st CC

2 Demifacets for 2nd CC (1 from the Manubrium, 1 from the Body)

STERNAL ANGLE

Demifacet for 3rd CC

Demifacet for 4th CC

Demifacet for 5th CC

Demifacet for 6th CC

Demifacet for 7th CC

COSTAL ANGLE

a

b

c

(Right Lateral View)   (Anterior View)

3 BASIC PORTIONS OF THE STERNUM

a ■ **UPPER MANUBRIUM**
- •Superior, triangular portion
  - •JUGULAR (SUPRASTERNAL) NOTCH—a depression on the superior surface
  - •CLAVICULAR NOTCHES—on each side of the jugular notch, they articulate with the medial ends of the CLAVICLES
  - •Articulates with the 1st and 2nd RIBS

b ■ **CENTRAL BODY**
- •Middle, largest portion (the GLADIOLUS)
- •Attaches to the COSTAL CARTILAGES of the 2nd–10th RIBS
- •STERNAL ANGLE (ANGLE OF LOUIS)—a depression between the MANUBRIUM and the BODY at the level of the 2nd RIB,— PALATABLE SURFACE LANDMARK

c ■ **LOWER XIPHOID PROCESS (ENSIFORM PROCESS)**
- •Inferior, smallest portion
- •No RIBS attached
- •Attaches to some abdominal muscles
- •COSTAL ANGLE—formed where the two costal margins meet at the XIPHOID PROCESS

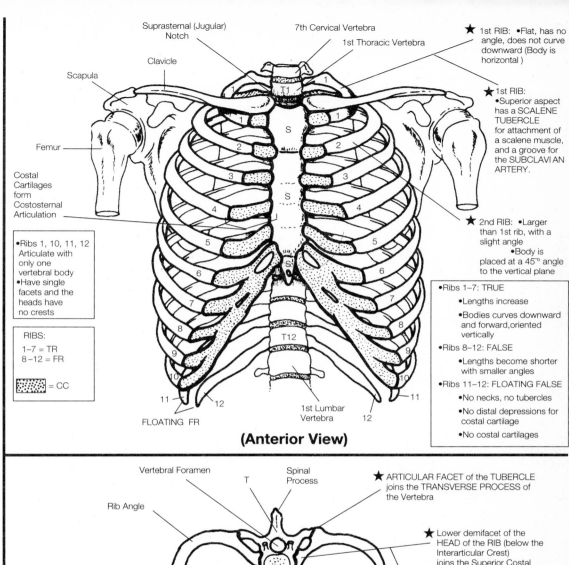

Suprasternal (Jugular) Notch

7th Cervical Vertebra

1st Thoracic Vertebra

Clavicle

Scapula

Femur

Costal Cartilages form Costosternal Articulation

•Ribs 1, 10, 11, 12 Articulate with only one vertebral body
•Have single facets and the heads have no crests

RIBS:
1–7 = TR
8–12 = FR

▨ = CC

★ 1st RIB: •Flat, has no angle, does not curve downward (Body is horizontal)

★ 1st RIB: •Superior aspect has a SCALENE TUBERCLE for attachment of a scalene muscle, and a groove for the SUBCLAVIAN ARTERY.

★ 2nd RIB: •Larger than 1st rib, with a slight angle •Body is placed at a 45° angle to the vertical plane

•Ribs 1–7: TRUE
  •Lengths increase
  •Bodies curves downward and forward, oriented vertically
•Ribs 8–12: FALSE
  •Lengths become shorter with smaller angles
•Ribs 11–12: FLOATING FALSE
  •No necks, no tubercles
  •No distal depressions for costal cartilage
  •No costal cartilages

FLOATING FR

1st Lumbar Vertebra

(Anterior View)

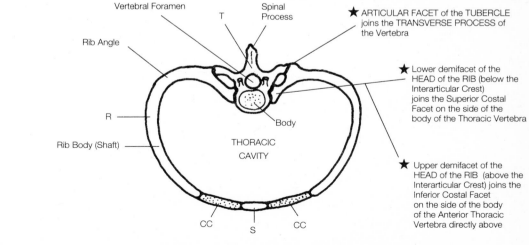

Vertebral Foramen

Spinal Process

Rib Angle

Rib Body (Shaft)

THORACIC CAVITY

Body

CC   S   CC

★ ARTICULAR FACET of the TUBERCLE joins the TRANSVERSE PROCESS of the Vertebra

★ Lower demifacet of the HEAD of the RIB (below the Interarticular Crest) joins the Superior Costal Facet on the side of the body of the Thoracic Vertebra

★ Upper demifacet of the HEAD of the RIB (above the Interarticular Crest) joins the Inferior Costal Facet on the side of the body of the Anterior Thoracic Vertebra directly above

In the CROSS - SECTIONAL VIEW, THE RIB CAGE APPEARS "KIDNEY - SHAPED"

THE RIB (THORACIC CAGE)
(Superior Schematic)

SKELETAL SYSTEM: APPENDICULAR SKELETON
Bones of the Upper Limb /Extremity
Pectoral Girdle, Brachium (Arm), and Forearm

1 = light blue 2 = light green
3 = yellow 4 = orange 5 = red
★ See 6.12-6.15

The PECTORAL GIRDLE:
2 CLAVICLES
AND
2 SCAPULAE

① **CLAVICLE** COLLARBONE

② **SCAPULA** SHOULDER BLADE

★ PECTORAL GIRDLE
•The 4 bones of the PECTORAL GIRDLE attach the bones of the upper extremities to the axial skeleton.
•Each of the 2 SHOULDER GIRDLES consist of 1 CLAVICLE and 1 SCAPULA.
•The main function of the PECTORAL GIRDLE is to provide for the attachment of many muscles that move the arm (brachium) and the forearm.
•The PECTORAL GIRDLE is not a COMPLETE GIRDLE, because it attaches to the axial skeleton at only one point—at the STERNUM (STERNOCLAVICULAR JOINT). The PECTORAL GIRDLE has *no* attachment to the vertebral column.
•The PECTORAL GIRDLE is *not* weight-bearing and the joints are delicate and not very stable. This allows the great mobility of the appendages and freedom of movement in many directions.

★ CLAVICLE (anterior component)
•Long, slender, S-shaped
•Binds the shoulder to the axial skeleton
•Responsible for maintaining the normal height of the shoulder
•Allows freedom of movement by positioning the shoulder joint away from the trunk.

★ SCAPULA (posterior component)
•Large, flat, triangular
•Some 15 muscles attach to the processes and fossae of the scapula
•Lies over the posterosuperior aspect of the thorax; has 3 borders.

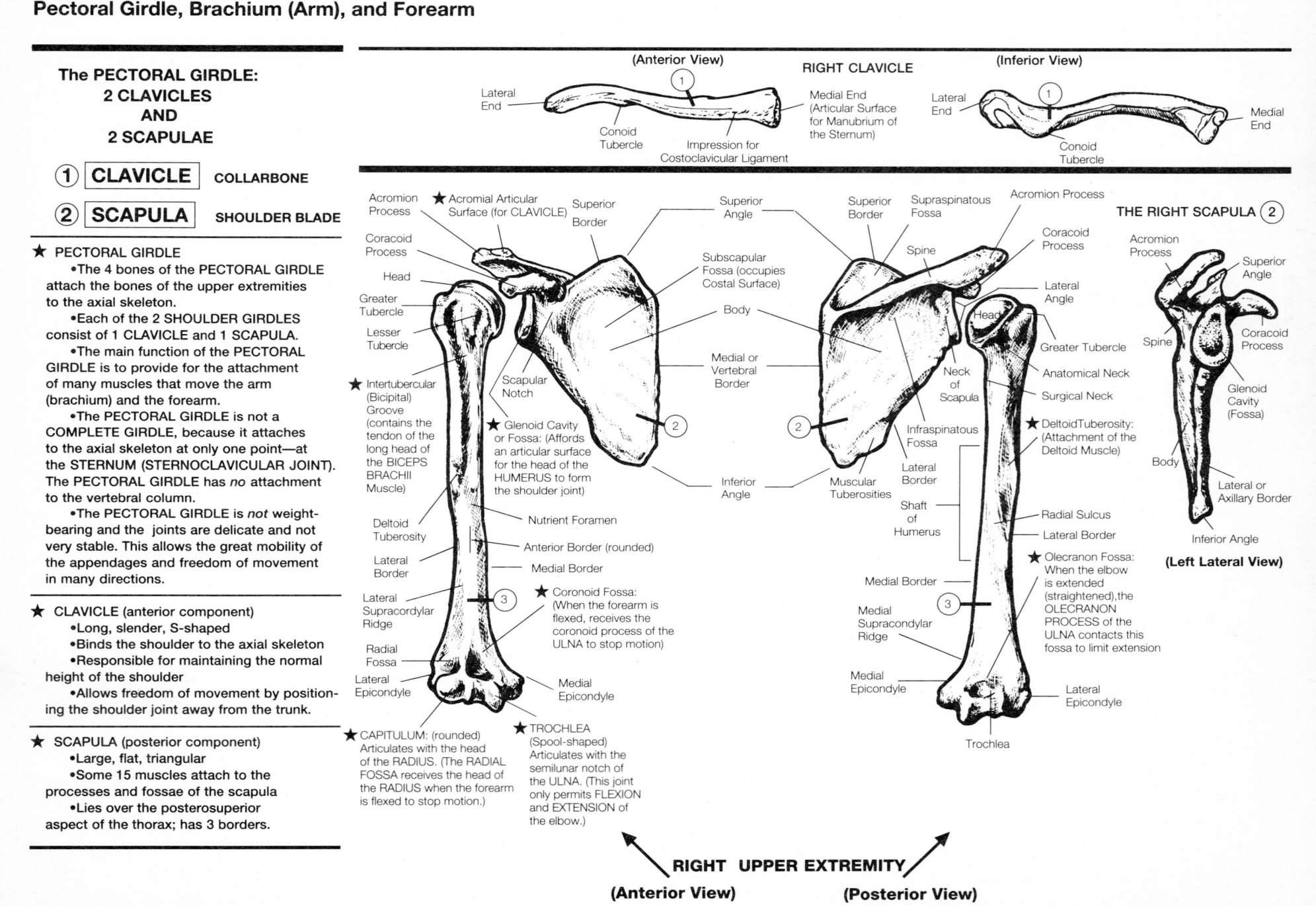

(Anterior View) RIGHT CLAVICLE **(Inferior View)**

Lateral End — Medial End (Articular Surface for Manubrium of the Sternum) — Conoid Tubercle — Impression for Costoclavicular Ligament — Lateral End — Medial End — Conoid Tubercle

THE RIGHT SCAPULA ②

Acromion Process — Superior Angle — Coracoid Process — Spine — Glenoid Cavity (Fossa) — Body — Lateral or Axillary Border — Inferior Angle

(Left Lateral View)

Acromion Process — ★ Acromial Articular Surface (for CLAVICLE) — Coracoid Process — Head — Greater Tubercle — Lesser Tubercle — ★ Intertubercular (Bicipital) Groove (contains the tendon of the long head of the BICEPS BRACHII Muscle) — Scapular Notch — Superior Border — Superior Angle — Subscapular Fossa (occupies Costal Surface) — Body — Medial or Vertebral Border — Inferior Angle

★ Glenoid Cavity or Fossa: (Affords an articular surface for the head of the HUMERUS to form the shoulder joint) — Nutrient Foramen — Anterior Border (rounded) — Medial Border — Deltoid Tuberosity — Lateral Border — Lateral Supracordylar Ridge — Radial Fossa — Lateral Epicondyle

★ Coronoid Fossa: (When the forearm is flexed, receives the coronoid process of the ULNA to stop motion) — Medial Epicondyle

★ CAPITULUM: (rounded) Articulates with the head of the RADIUS. (The RADIAL FOSSA receives the head of the RADIUS when the forearm is flexed to stop motion.)

★ TROCHLEA (Spool-shaped) Articulates with the semilunar notch of the ULNA. (This joint only permits FLEXION and EXTENSION of the elbow.)

Superior Border — Superior Angle — Supraspinatous Fossa — Acromion Process — Coracoid Process — Spine — Lateral Angle — Head — Greater Tubercle — Anatomical Neck — Neck of Scapula — Surgical Neck — Infraspinatous Fossa — Lateral Border — Muscular Tuberosities — Shaft of Humerus — Radial Sulcus — Lateral Border — ★ DeltoidTuberosity: (Attachment of the Deltoid Muscle) — Medial Border — Medial Supracondylar Ridge — Medial Epicondyle — Trochlea — Lateral Epicondyle

★ Olecranon Fossa: When the elbow is extended (straightened),the OLECRANON PROCESS of the ULNA contacts this fossa to limit extension

RIGHT UPPER EXTREMITY
(Anterior View)   **(Posterior View)**

Longer Bones of the Upper Extremities:
HUMERUS, ULNA, AND RADIUS

★ See 6.13-6.15

Singular Bone of the BRACHIUM (ARM)
③ **HUMERUS**

2 Skeletal Structures of the FOREARM
④ **ULNA** (Medial)

⑤ **RADIUS** (Lateral)

★ THE UPPER EXTREMITIES
•The 2 appendages (RIGHT and LEFT) attach at the shoulder, consisting of an ARM, A FOREARM, A WRIST, and A HAND.
•Together, the RIGHT and LEFT UPPER EXTREMITIES contain 60 BONES. Therefore, each upper extremity contains 30 bones (27 of which comprise the WRIST and HAND. See 4.13). The 3 remaining are the longer bones of the upper extremity:
•1 HUMERUS bone in the UPPER ARM (BRACHIUM)
•1 ULNA and 1 RADIUS in the FOREARM

★ HUMERUS (see upper page)
•Longest, largest bone of the upper extremity
•Articulates proximally with the GLENOID FOSSA (CAVITY) of the SCAPULA by a large, rounded HEAD
•Articulates distally with the ULNA and RADIUS by 2 rounded CONDYLES:
■ Medial condyle is the TROCHLEA which is spool-shaped and articulates with the ULNA with a pullylike surface.
■ Lateral condyle is the CAPITULUM, a rounded knob that receives the RADIUS

★ ULNA
•Medial bone of the forearm. located on the small finger side
•Proximally, contains 2 processes (a prominent posterior OLECRANON PROCESS and a pointed anterior CORONOID PROCESS)
•Between these 2 processes is the SEMILUNAR NOTCH that articulates with the TROCHLEA of the HUMERUS
•Distally, an expanded HEAD is separated from the WRIST BONES by a disc of fibrocartilage

★ RADIUS
•Lateral bone of the forearm, located on the thumb side
•Proximally, a disc-shaped HEAD articulates with the CAPITULUM of the HUMERUS and the RADIAL NOTCH of the ULNA
•Distally, an expanded concave CARPAL SURFACE articulates with the WRIST BONES

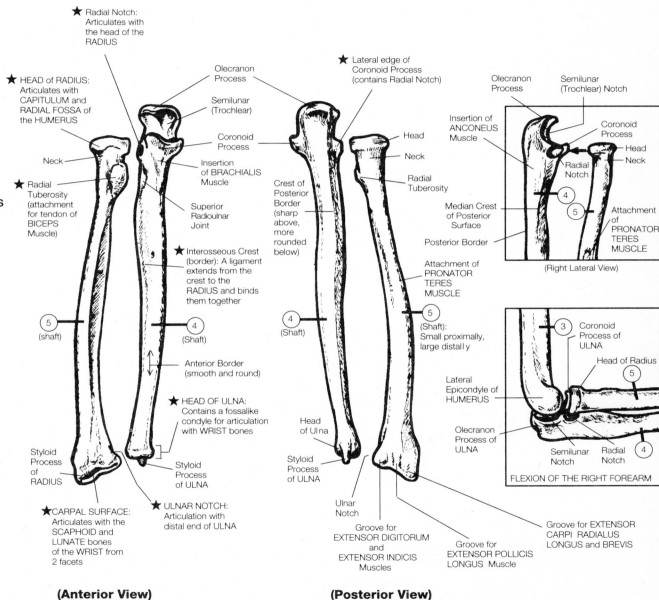

(Anterior View)   **(Posterior View)**

RIGHT UPPER EXTREMITY

SKELETAL SYSTEM: APPENDICULAR SKELETON
Bones of the Upper Limb/Extremity
The Wrist and the Hand

a = blue   b = green   c = purple
d = cream   e = flesh   f = ochre g = brown
h = red   i = yellow   j = orange
★ See also 6.16

★ THE WRIST
•The 8 CARPAL BONES OF THE WRIST are arranged in 2 transverse rows of 4 bones each:
■ PROXIMAL ROW
•Presents a curved surface that articulates with the distal and concave ends of the RADIUS and ULNA
•From Lateral (thumb side) to Medial (small finger side):
  ❏ SCAPHOID—shaped like a boat
  ❏ LUNATE—half-moon shaped surface
  ❏ TRIQUETRUM—shaped like a right triangle
  ❏ PISIFORM—shaped like a pea. Forms in a tendon as a sesamoid bone.

■ DISTAL ROW
•Bears surfaces for articulation with the METACARPAL BONES of the HAND
•From Lateral to Medial side:
  ❏ TRAPEZIUM—"a little table"
  ❏ TRAPEZOID—"table-shaped"
  ❏ CAPITATE—has rounded head
  ❏ HAMATE—has a curved, hooklike process called the HAMULUS

★ THE HAND
•Sometimes the WRIST is included in the definition of the HAND—that part of the body attached to the forearm at the wrist.
■ METACARPUS (OSSA METACARPALIA)
  •Body of Palm of the HAND
  •5 METACARPAL BONES, numbered I–V from the thumb (lateral) side
  •Each metacarpal bone has an enlarged proximal BASE with a concave surface for articulation with the CARPALS
  •Distal round HEADS of the bones form the knuckles when the digits are flexed and articulate with the proximal phalanges.
■ PHALANGES
  •14 phalanges form a total of 5 digits or fingers. 4 fingers each carry 3 phalanges (proximal or 1st, middle or 2nd, and distal or 3rd or terminal). The thumb has only 2 phalanges.

Distal Portion of the UPPER EXTREMITY: The WRIST and the HAND

One WRIST and one HAND contain **27 BONES: 8 CARPAL (WRIST) BONES** and **19 BONES** of the **HAND**
Of the 19 HAND BONES, There are **5 METACARPAL (PALM) BONES** and **14 PHALANGES**
Thus, both WRISTS and HANDS contain **54 BONES: 16 CARPALS, 10 METACARPALS** and **28 PHALANGES**

The WRIST (CARPUS): 8 BONES

RADIAL SIDE (Lateral)

PROXIMAL ROW

a   **SCAPHOID**   NAVICULAR (OS SCAPHOIDEUM)

b   **LUNATE**   SEMILUNAR

c   **TRIQUETRUM**   TRIANGULAR/CUNEIFORM (TRIQUETRAL)

ULNAR SIDE (Medial)

d   **PISIFORM**

The HAND (MANUS): 19 BONES

PALM (METACARPUS)

i   **5 METACARPAL BONES {I–V}**

FINGERS (DIGITS)

j   **14 PHALANGES** (single = PHALANX)

RADIAL SIDE (Lateral)

DISTAL ROW

e   **TRAPEZIUM**   GREATER MULTANGULAR (OS TRAPEZIUM)

f   **TRAPEZOID**   LESSER MULTANGULAR

g   **CAPITATE**

ULNAR SIDE (Medial)

h   **HAMATE**

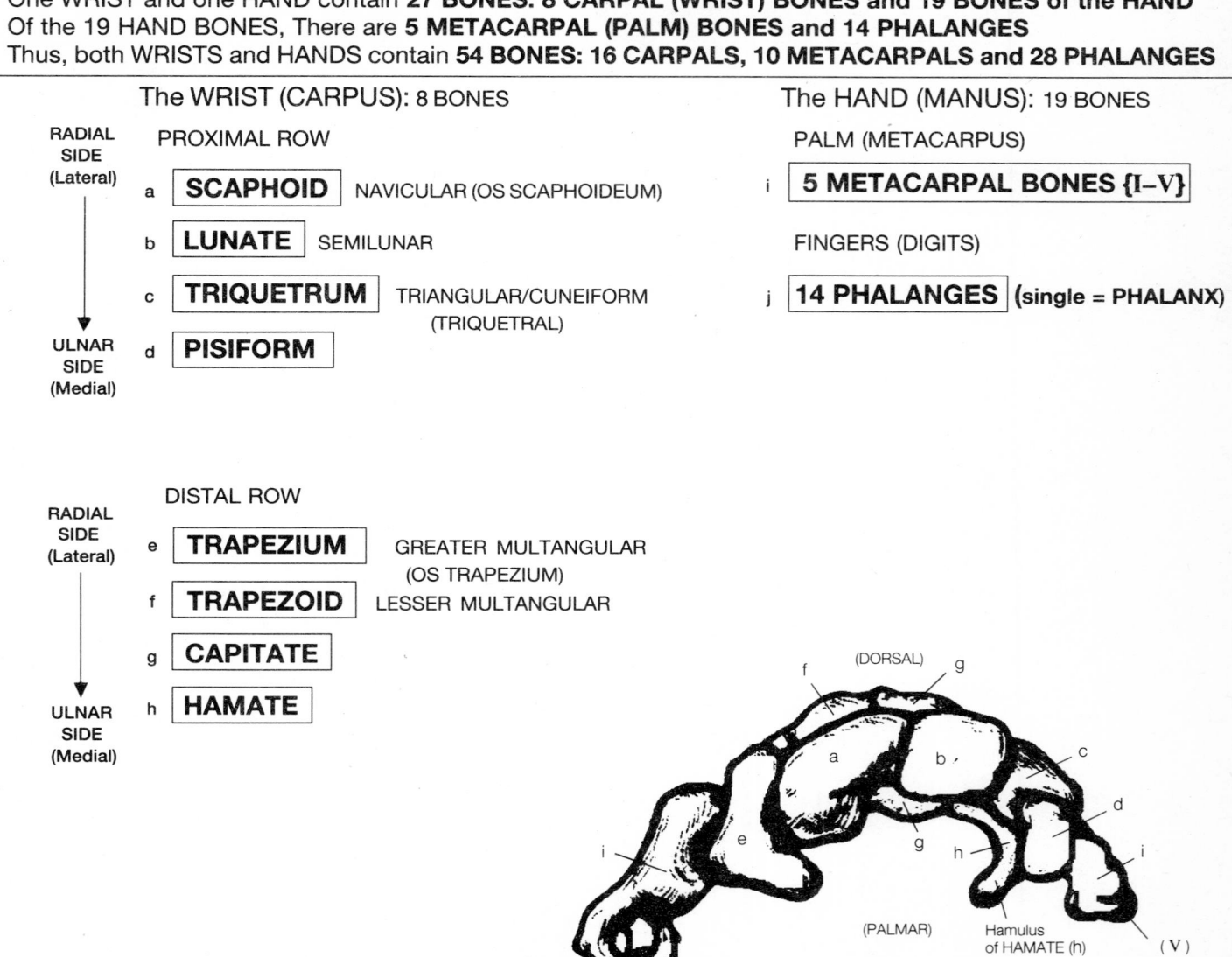

Superior View of Right Wrist

(DORSAL)

(PALMAR)   Hamulus of HAMATE (h)

(I)
1st Metacarpal (Thumb Side)

(V)
5th Metacarpal

★ The HAND is the expression of an incredible structural organization and is able to sustain a great amount of abuse, notwithstanding its complexity.

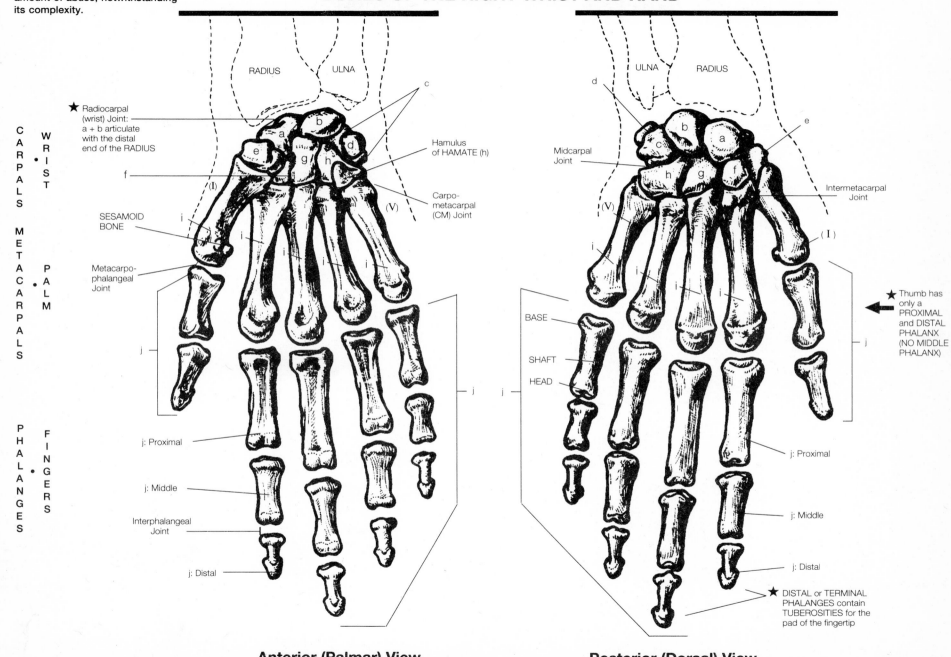

★ Radiocarpal (wrist) Joint: a + b articulate with the distal end of the RADIUS

RADIUS

ULNA

c

Hamulus of HAMATE (h)

f

(I)

(V)

Carpo-metacarpal (CM) Joint

SESAMOID BONE

Metacarpo-phalangeal Joint

j

i

j: Proximal

j: Middle

Interphalangeal Joint

j: Distal

Anterior (Palmar) View

Palm Side Up

ULNA

RADIUS

d

b

c

a

Midcarpal Joint

h

g

f

e

Intermetacarpal Joint

(V)

(I)

BASE

SHAFT

HEAD

★ Thumb has only a PROXIMAL and DISTAL PHALANX (NO MIDDLE PHALANX)

j

j: Proximal

j: Middle

j: Distal

★ DISTAL or TERMINAL PHALANGES contain TUBEROSITIES for the pad of the fingertip

Posterior (Dorsal) View

Back of the Hand

C A R P A L S • W R I S T

M E T A C A R P A L S • P A L M

P H A L A N G E S • F I N G E R S

SKELETAL SYSTEM: APPENDICULAR SKELETON
Bones of the Lower Limb/Extremity
Pelvic Girdle and Pelvis

a = light green b = light purple c = light blue
1 = yellow-green S.P = cream S = red
Co = orange cartilage = grey

★ See also 6.9 , 6.11 , & 14.9

★ The PELVIC GIRDLE is formed by the union of 2 OSSA COXAE (also called COXAL BONES, INNOMINATE BONES or HIP BONES):
■ Anteriorly, the PUBIC PORTION OF EACH COXAL BONE presents a roughened SYMPHYSEAL SURFACE that forms a joint with the opposite bone held by fibrocartilage called the SYMPHYSIS PUBIS.
■ Posteriorly, the ILIAC PORTION OF EACH COXAL BONE presents a medial roughened AURICULAR SURFACE that forms the SACROILIAC JOINTS with the SACRUM of the VERTEBRAL COLUMN.

★ Fetally, each OS COXA develops as three separate bones:
■ ILIUM (Large, superior)
■ ISCHIUM (inferior and posterior)
■ PUBIS (inferior and anterior)
These 3 bones are fused together in the adult, but they are considered separately for descriptive purposes

★ The 3 bones ossify (on the LATERAL SURFACE of each OS COXA) at a large, circular depression (socket) called the ACETABULUM, which receives the head of the FEMUR to form the HIP JOINT.
3 ACETABULAR FEATURES:
•ACETABULAR FOSSA
•LUNATE SURFACE
•ACETABULAR NOTCH (inferior)

★ FUNCTIONS OF THE PELVIC GIRDLE
•Provides strong, stable support for the lower extremities, which carry the weight of the body.
•Supports the weight of the body through the vertebral column.
•In its attachment to the vertebral column, permits an upright posture and locomotion on two legs (bipedal) in contrast to 4-legged locomotion of other mammals.
•Supports and protects the lower viscera, including:
•The urinary bladder
•The reproductive organs
•Developing fetus during pregnancy

The PELVIC GIRDLE = Unification of 2 HIPBONES (INNOMINATE BONES) (or 2 **OSSA COXAE**)

A Single **OS COXA** (HIPBONE): Consists of 3 Fused Bones in the Adult:

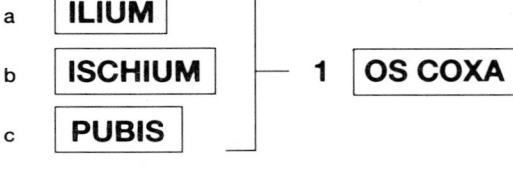

a | ILIUM |
b | ISCHIUM | — 1 | OS COXA |
c | PUBIS |

Attachments *to* the PELVIC GIRDLE :
One Anterior Attachment (Joint)

s.p. | SYMPHYSIS PUBIS |

Two Posterior Attachments (Sacroiliac Joints)

S | SACRUM | attachments are the 2 sacral auricular surfaces

Co | COCCYX | attached to the bottom of the sacrum

RIGHT LATERAL VIEW OF AN OS COXA OF A CHILD SHOWING LIMITS OF THE 3 BONES (Note formation of an ACETABULUM socket at the point of fusion of all 3 bones.)

Iliac Crest (cartilaginous)

Wing of ILIUM

Body of ILIUM

ACETABULUM SOCKET

Ischial Tuberosity (cartilaginous)

Symphysis Pubis (cartilage)

OBTURATOR FORAMEN

THE PELVIS = BASINLIKE BONY STRUCTURE composed of:
■ PELVIC GIRDLE (2 OSSA COXAE)
AND ■ SACRUM and COCCYX (of vertebral column)
AND ■ SYMPHYSIS PUBIS (fibrocartilage and other uniting ligaments)

ISCHIOPUBIC RAMUS =
■ RAMUS OF THE ISCHIUM (ISCHIAL RAMUS)
■ INFERIOR PUBIC RAMUS

★ The GLUTEAL muscles attach to the ILIUM between the 3 gluteal lines (posterior, anterior, inferior).

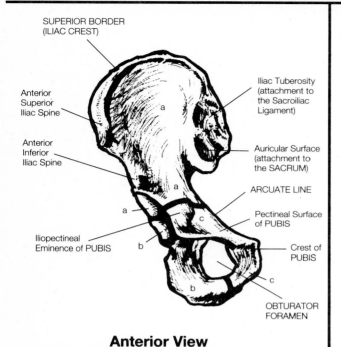

SUPERIOR BORDER (ILIAC CREST)

Anterior Superior Iliac Spine

Anterior Inferior Iliac Spine

Iliopectineal Eminence of PUBIS

Iliac Tuberosity (attachment to the Sacroiliac Ligament)

Auricular Surface (attachment to the SACRUM)

ARCUATE LINE

Pectineal Surface of PUBIS

Crest of PUBIS

OBTURATOR FORAMEN

Anterior View

Wing of ILIUM

External Iliac Fossa

Iliac Crest

Anterior or Middle Gluteal Line

Posterior Gluteal Line

Greater Sciatic Notch

ACETABULUM

Lip
Fossa
Lunate Surface
Notch

Ischial Tuberosity

Inferior Gluteal Line

Anterior Superior Iliac Spine

Body of ILIUM

Anterior Inferior Iliac Spine

Inferior Pubic Ramus

OBTURATOR FORAMEN

Ramus of ISCHIUM

Right Lateral View

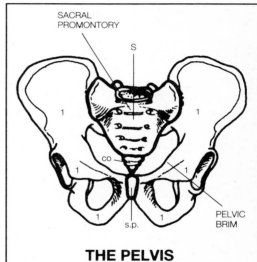

SACRAL PROMONTORY

S

1

1

co

1

1

s.p.

PELVIC BRIM

THE PELVIS

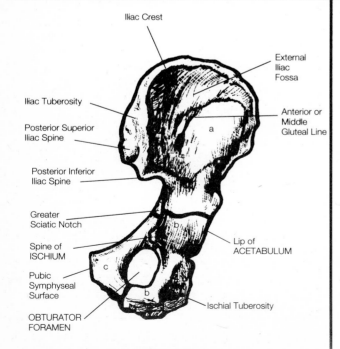

Iliac Crest

Iliac Tuberosity

Posterior Superior Iliac Spine

Posterior Inferior Iliac Spine

Greater Sciatic Notch

Spine of ISCHIUM

Pubic Symphyseal Surface

OBTURATOR FORAMEN

External Iliac Fossa

Anterior or Middle Gluteal Line

Lip of ACETABULUM

Ischial Tuberosity

Posterior View

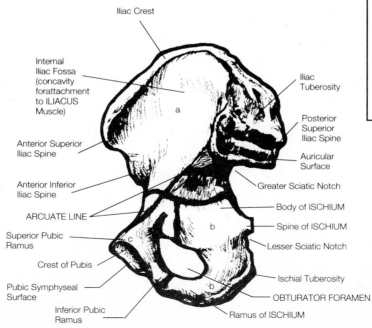

Iliac Crest

Internal Iliac Fossa (concavity forattachment to ILIACUS Muscle)

Anterior Superior Iliac Spine

Anterior Inferior Iliac Spine

ARCUATE LINE

Superior Pubic Ramus

Crest of Pubis

Pubic Symphyseal Surface

Inferior Pubic Ramus

Iliac Tuberosity

Posterior Superior Iliac Spine

Auricular Surface

Greater Sciatic Notch

Body of ISCHIUM

Spine of ISCHIUM

Lesser Sciatic Notch

Ischial Tuberosity

OBTURATOR FORAMEN

Ramus of ISCHIUM

Medial View

SKELETAL SYSTEM: APPENDICULAR SKELETON
Bones of the Lower Limb/Extremity
Characteristics of the Male and Female
Pelves/Clinical Divisions of the Pelvis

There is no coloring required for the male & female pelves unless you want to color in the components of the pelvis from 4.12. Study the chart along with the illustrations.

★ See also 6.9 , 6.11 , and 14.9

★ MALE and FEMALE PELVES
•Significant structural differences exist between male and female pelves, particularly in relation to pregnancy and childbirth.

★ Below are some general characteristics of the FEMALE PELVIS that differ from the MALE PELVIS:

P E L V I S	■ Joints of the pelvic girdle stretch during pregnancy and parturition ■ Entire pelvic girdle titled forward ■ Greater (false) pelvis is shallower ■ Distance between anterior superior iliac spine is wider ■ Lesser (true) pelvis is spherical and wider ■ Pelvic inlet is larger and more oval ■ Pelvic outlet is comparatively large
P U B I S	■ Obturator foramen is more oval and triangular (compared to round) ■ Symphysis pubis is shallower (less deep) ■ Pubic arch is >90°, wider, and more rounded ■ Inferior ramus of pubis does not have an everted surface
I L I U M	■ Ilium is less vertical ■ Iliac fossa is shallow (male=deep) ■ Iliac crest is less curved ■ Greater sciatic notch of the ilium is wide (male=narrow)
I S C H I U M	■ Ischial spine is turned inward less than the male ■ Ischial tuberosity is turned outward (male= turned inward)
S A C R U M	■ Superior surface of sacral body only spans 1/3 the width of the sacrum (male spans 1/2) ■ Auricular surface of the sacrum only extends down to S3 (male extends well beyond S3) ■ Sacrum is short, wide, flat and curving forward in the lower part (male=long, narrow, smooth concavity)
G E N E R A L	■ Acetabulum is small (male=large) ■ Joint surfaces are small (male=large) ■ Muscle attachments are indistinct (male=well-marked) ■ General bone structure is lighter and thinner (male=heavy and thick)/

★ PELVIC INLET → The upper pelvic *entrance of the LESSER (TRUE) PELVIS*
★ PELVIC OUTLET → The lower pelvic opening (border) of the LESSER (TRUE) PELVIS

"HEART-SHAPED" MALE ♂ PELVIS

"KIDNEY-SHAPED" FEMALE ♀ PELVIS

Clinical Divisions of the PELVIS

FP | GREATER, OR FALSE, PELVIS | SUPERIOR

TP | LESSER, OR TRUE, PELVIS | INFERIOR

Dividing Curved, Bony Rim between the False and True PELVES:

PB | PELVIC BRIM

Diameters of the PELVIS

X | TRANSVERSE DIAMETER OF THE PELVIC INLET

Y | ANTEROPOSTERIOR/ TRUE CONJUGATE DIAMETER

Components of the PELVIS

a | ILIUM 2 ⎤
b | ISCHIUM 2 ⎬— 2 OSSA COXAE
c | PUBIS 2 ⎦

s.p. | SYMPHYSIS PUBIS

S | SACRUM

CO | COCCYX

PELVIC BRIM:
■ Curved, bony rim that passes INFERIORLY from the SACRAL PROMONTORY to the UPPER MARGIN OF THE SYMPHYSIS PUBIS
■ Divides the PELVIS into 2 portions, a GREATER and LESSER PELVIS
■ Surrounds the PELVIC INLET OF THE LESSER PELVIS as a boundary

PELVIC INLET OF THE LESSER (TRUE) PELVIS
■ The circumference of the PELVIC BRIM bounds the PELVIC INLET

PELVIC OUTLET OF THE LESSER (TRUE) PELVIS
■ The lower circumference of the lesser pelvis bounds the PELVIC OUTLET
■ Formed by the COCCYX, ISCHIAL PROTUBERANCES, ASCENDING RAMI OF THE ISCHIA, DESCENDING RAMI OF THE OSSA PUBIS and THE SACROSCIATIC LIGAMENTS

ARCUATE (ILIOPECTINEAL) LINE

SACRAL PROMONTORY

PB

UPPER MARGIN OF SYMPHYSIS PUBIS

FP: (between the Iliac Wings)

TP: (enclosed mostly by the Ischial and Pubic Bones)

X: between the widest points of the ARCUATE LINE (12.5 cm)

Y: Line drawn between the SACRAL PROMONTORY and THE UPPER MARGIN OF THE SYMPHYSIS PUBIS (11.5 cm)

Subpubic Angle

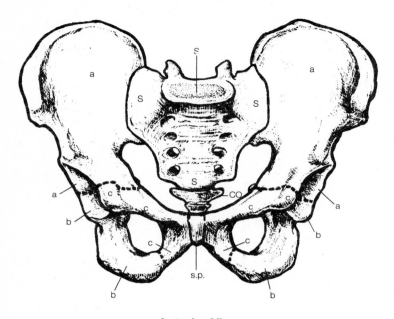

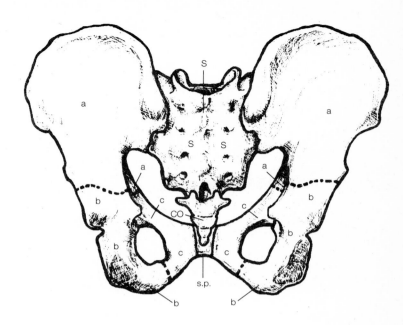

**MALE
PELVIS**

Anterior View

Posterior View

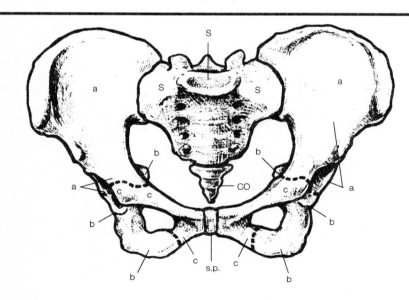

**FEMALE
PELVIS**

Anterior View

Posterior View

SKELETAL SYSTEM: APPENDICULAR SKELETON
Bones of the Lower Limb/Extremity
The Thigh, Knee, and Leg

Longer Bones of the LOWER EXTREMITY:
FEMUR, TIBIA AND FIBULA

★ Singular Bone of the Thigh: Proximal Portion of the LOWER EXTREMITY

② **FEMUR** THIGHBONE

■ The LONGEST, HEAVIEST and STRONGEST BONE of the Body
■ Extends from the HIP to the KNEE
■ The proximal, rounded HEAD articulates with the ACETABULUM OF THE COXAL BONE
■ FOVEA CAPITIS
  • A roughened, shallow pit on the lower center of the rounded HEAD
  • Point of attachment for the LIGAMENTUM TERES, which helps support the HEAD against the ACETABULUM
  •It also carries blood vessels into the bone.
■ SHAFT
  • Bows medially to approach the FEMUR of the opposite thigh
  • This convergence results in the knee joints being brought nearer to the body's line of gravity. (> in women)
  • Contains several important structures for muscle attachment
■ The distal end of the FEMUR is expanded for articulation with the TIBIA and includes the MEDIAL CONDYLE and LATERAL CONDYLE.

★ SESAMOID BONE OF THE KNEE:

③ **PATELLA** KNEE CAP

■ Positioned on the anterior side of the knee joint; small and triangular.
■ Being a sesamoid bone, it forms in the tendon of the QUADRICEPS FEMORIS MUSCLE as a response to stress.
■ Protects the knee joint
■ Strengthens the QUADRICEPS TENDON
■ Increases leverage of the QUADRICEPS FEMORIS MUSCLE
■ Changes direction of muscle pull on the TIBIA to a more efficient angle

SIMILAR TO THE UPPER EXTREMITIES, THE LOWER EXTREMITIES CONTAIN 60 BONES:
Each Extremity Contains 30 BONES (26 of Which Are in the ANKLE and FOOT)

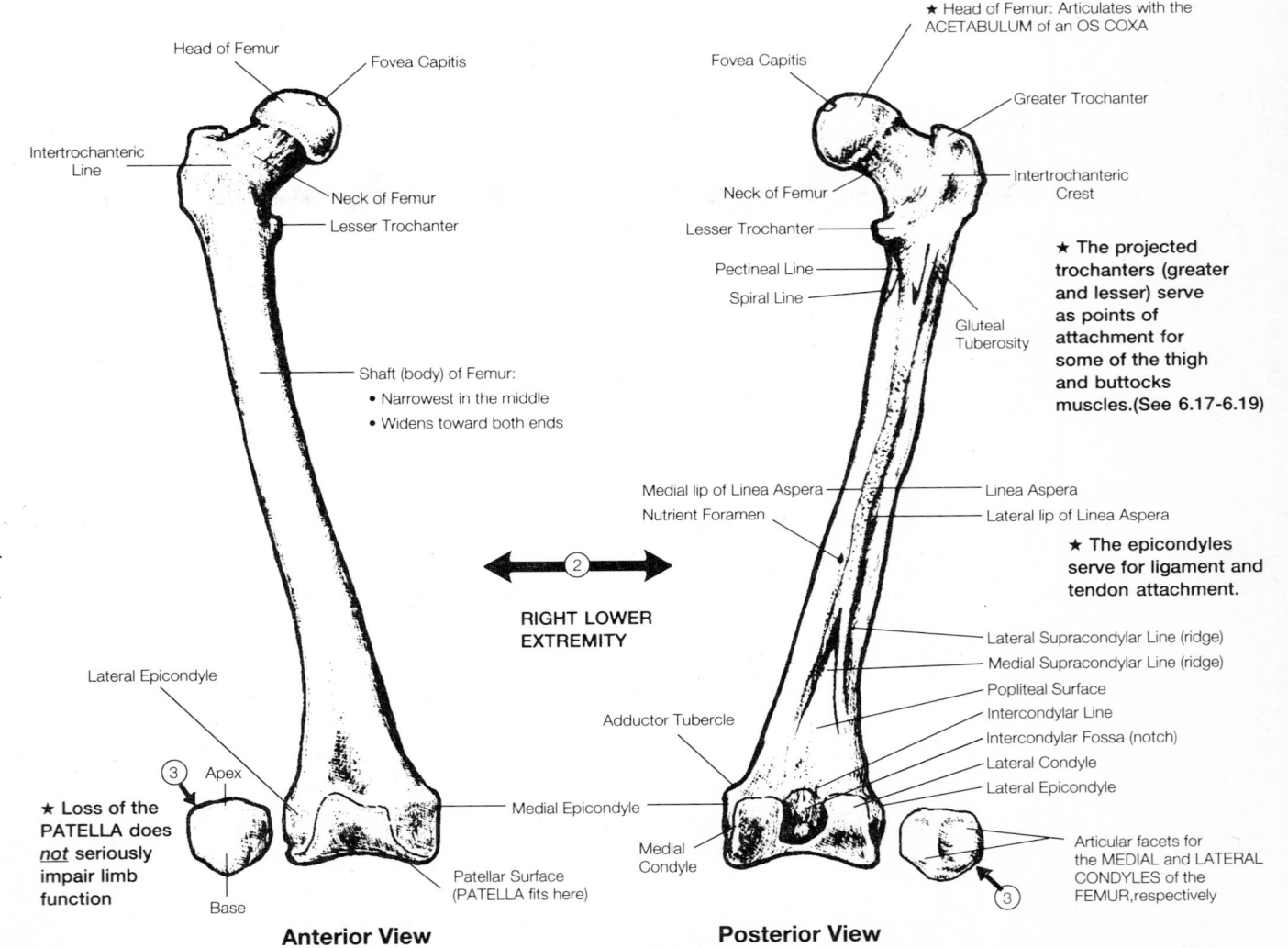

Head of Femur
Fovea Capitis
Intertrochanteric Line
Neck of Femur
Lesser Trochanter
Shaft (body) of Femur:
  • Narrowest in the middle
  • Widens toward both ends

② RIGHT LOWER EXTREMITY

Lateral Epicondyle
③ Apex
★ Loss of the PATELLA does *not* seriously impair limb function
Base
Medial Epicondyle
Patellar Surface (PATELLA fits here)
Anterior View

★ Head of Femur: Articulates with the ACETABULUM of an OS COXA
Fovea Capitis
Greater Trochanter
Neck of Femur
Intertrochanteric Crest
Lesser Trochanter
Pectineal Line
Spiral Line
Gluteal Tuberosity
★ The projected trochanters (greater and lesser) serve as points of attachment for some of the thigh and buttocks muscles.(See 6.17-6.19)
Medial lip of Linea Aspera
Nutrient Foramen
Linea Aspera
Lateral lip of Linea Aspera
★ The epicondyles serve for ligament and tendon attachment.
Lateral Supracondylar Line (ridge)
Medial Supracondylar Line (ridge)
Popliteal Surface
Intercondylar Line
Intercondylar Fossa (notch)
Lateral Condyle
Lateral Epicondyle
Adductor Tubercle
Medial Condyle
Articular facets for the MEDIAL and LATERAL CONDYLES of the FEMUR,respectively
③
Posterior View

★ 2 SKELETAL STRUCTURES of the Leg, Between the Knee and the Ankle:

④ | TIBIA | SHINBONE

⑤ | FIBULA | "SPLINT BONE"

★TIBIA

■ The larger of the 2 bones of the leg
■ PROXIMAL END
•This medial bone bears the major portion of the weight of the leg, where it articulates proximally at the knee joint.
•The MEDIAL and LATERAL condyles of the TIBIA articulate with the FEMORAL CONDYLES.
• The LATERAL CONDYLE of the TIBIA carries a FACET for articulation with the HEAD OF THE FIBULA.
■ SHAFT
• Triangular in cross-section (with the triangular base posterior)
• Contains 3 surfaces (posterior, medial, and lateral) and 3 borders (anterior, medial, and interosseous)
• On the proximal, anterior portion of the SHAFT lies the TIBIAL TUBEROSITY for attachment of the PATELLAR LIGAMENT
■ DISTAL END
• Bears an INFERIOR ARTICULAR SURFACE with a medially oriented MEDIAL MALLEOLUS for articulation with the TALUS (an ankle bone)
• On the distal-lateral end, there is a FIBULAR NOTCH for articulation with the FIBULA

★ FIBULA

• Long, narrow bone parallel and lateral to the TIBIA, considerably smaller than the TIBIA.
• Fragility lends little to weight-bearing or support.
• Extends the area for muscle attachment (along with the INTEROSSEOUS MEMBRANE that connects it to the TIBIA); the medial surface of the shaft is obviously grooved to serve this function.
• The distal end of the bone bears a pointed, prominent knob called the LATERAL MALLEOLUS, which bears a facet for the TALUS on its medial aspect.

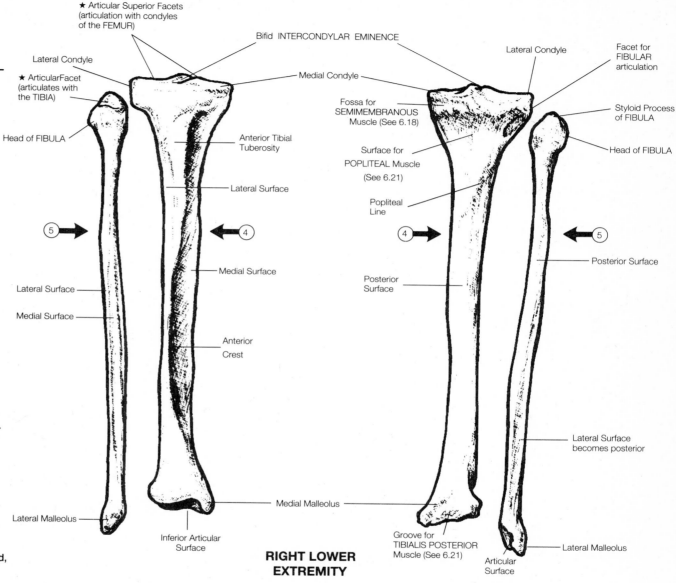

RIGHT LOWER EXTREMITY

Anterior View

Posterior View

★ The TIBIA and FIBULA together form a cuplike articulating surface for the ANKLE. The MEDIAL MALLEOLUS (TIBIA) and LATERAL MALLEOLUS (FIBULA) are positioned on either side of the TALUS BONE to offer STABILIZATION TO THE ANKLE JOINT. (See 4.15)

4.14

d = blue   e = green   f = purple   g = flesh
h = red   i = yellow   j = orange
★ See also 6.22, 6.23

SKELETAL SYSTEM: APPENDICULAR SKELETON
Bones of the Lower Limb/Extremity
The Ankle and the Foot

★ **THE ANKLE**
  • Of the 7 TARSAL BONES that make up the ANKLE, *only* the TALUS BONE is involved in forming the ANKLE JOINT;
  ■ TALUS–contains large superior and lateral ARTICULAR SURFACES that connect to the TIBIA and FIBULA

  ■ CALCANEUS–largest of the Tarsal Bones; forms the heel of the foot

  ■ NAVICULAR–boat-shaped; is anterior to the TALUS

4 DISTAL BONES OF THE ANKLE
  • Series of bones that articulate with the METATARSALS
  • From the medial to the lateral side, they are the 1st, 2nd, 3rd CUNEIFORMS (wedge-shaped) and the CUBOID (which is anterior to the CALCANEUS)

★ **THE FOOT**
  ■ METATARSUS
  • 5 METATARSAL BONES of the SOLE of the foot, numbered (I–V) from the medial side (big toe side) to the lateral side
  ■ METATARSAL (I) is the largest and carries most of the weight-bearing function of the metatarsals
  • BALL OF THE FOOT is formed by the HEADS of METATARSALS (I) and (II)
  • METATARSALS (II–V) are slender and function like an "outrigger", providing stability
  • The proximal *bases* of METATARSALS (I–III) articulate proximally with the 3 CUNEIFORMS
  • The distal *heads* of all the METATARSALS articulate distally with the PROXIMAL PHALANGES

  ■ 14 PHALANGES OF THE TOES
  • Arranged like those of hands, with PROXIMAL PHALANGES being the longest. MIDDLE and DISTAL PHALANGES are very short

DISTAL PORTION OF THE LOWER EXTREMITY: THE ANKLE AND THE FOOT

• **ONE ANKLE AND ONE FOOT CONTAIN 26 BONES:**
    7 TARSAL (ANKLE) BONES AND 19 BONES OF THE FOOT

• **OF THE 19 FOOT BONES,**
    THERE ARE 5 METATARSAL (SOLE) BONES AND 14 PHALANGES

• **THUS, BOTH ANKLES AND FEET CONTAIN 52 BONES:**
    14 TARSALS, 10 METATARSALS, AND 28 PHALANGES

The ANKLE (TARSUS): 7 BONES

PROXIMAL ROW
D NAVICULAR   SCAPHOID

E TALUS   ASTRAGALUS

F CALCANEUS   "HEELBONE" (OS CALCIS)

DISTAL ROW
G 3 CUNEIFORMS   1st, 2nd, 3rd

H CUBOID   OS CUBOIDEUM

The FOOT (PES)

Bones of the SOLE: The PLANTAR SURFACE of the FOOT:

METATARSUS:

i 5 METATARSAL BONES { I–V }

TOES (DIGITS)

j 14 PHALANGES

Arches of the FOOT

L.A. LONGITUDINAL ARCH
      MEDIAL PART
      PROXIMAL PART

T.A. TRANSVERSE ARCH

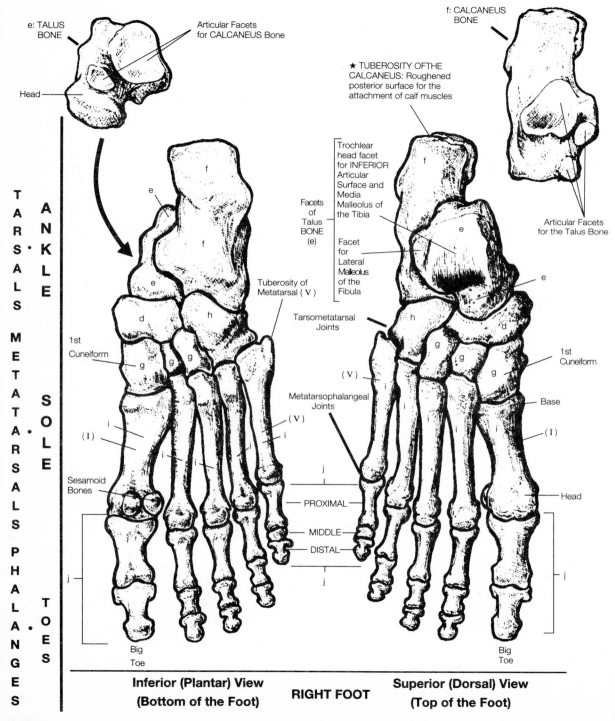

e: TALUS BONE

Head

Articular Facets for CALCANEUS Bone

f: CALCANEUS BONE

★ TUBEROSITY OF THE CALCANEUS: Roughened posterior surface for the attachment of calf muscles

Articular Facets for the Talus Bone

Trochlear head facet for INFERIOR Articular Surface and Media Malleolus of the Tibia

Facets of Talus BONE (e)

Facet for Lateral Malleolus of the Fibula

f

e

f

e

h

d

Tuberosity of Metatarsal (V)

Tarsometatarsal Joints

1st Cuneiform

1st Cuneiform

g g g

g

g g

g

Metatarsophalangeal Joints

(V)

Base

(V)

(I)

i

i

i

i

i

(I)

Sesamoid Bones

Head

i i i i i

j

PROXIMAL

MIDDLE

DISTAL

j

j

j

Big Toe

Big Toe

Inferior (Plantar) View

(Bottom of the Foot)

RIGHT FOOT

Superior (Dorsal) View

(Top of the Foot)

T A R S A L S   M E T A T A R S A L S   P H A L A N G E S

A N K L E   ·   S O L E   ·   T O E S

4.15

2 ARCHES OF THE FOOT

■ Formed by the STRUCTURE and ARRANGEMENT OF THE BONES

■ Held in place by ligaments and tendons, but are not rigid

■ Arches yield when force (weight) is applied and spring back "to normal" when the force is lifted

■ The arches SUPPORT THE WEIGHT OF THE BODY and PROVIDE LEVERAGE WHEN WALKING

↓

LONGITUDINAL ARCH

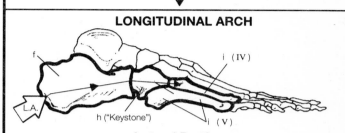

f

i (IV)

L.A.

h ("Keystone")

i (V)

Lateral Portion

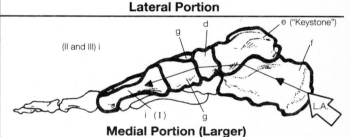

g   d

e ("Keystone")

(II and III) i

f

i (I)

g

L.A.

Medial Portion (Larger)

TRANSVERSE ARCH

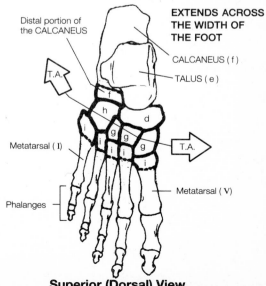

Distal portion of the CALCANEUS

EXTENDS ACROSS THE WIDTH OF THE FOOT

CALCANEUS (f)

TALUS (e)

T.A.

f

h

d

i

g g

Metatarsal (I)

i   g   g

i

T.A.

Metatarsal (V)

Phalanges

i

Superior (Dorsal) View

4.16

Support and Movement **74**

Review the chapter and
utilize the illustrations
any way you wish for study.
(See comment at the bottom
of the bottom page.)

SKELETAL SYSTEM: AXIAL AND APPENDICULAR SKELETON
REVIEW OF THE WHOLE SKELETAL SYSTEM

CLASSIFICATION OF BONES OF THE ADULT SKELETAL SYSTEM

AXIAL SKELETON				See Chapter 7, Sensory Organs.)			
				UPPER LIMB	APPENDICULAR SKELETON	LOWER LIMB	
Name	Total Number of Bones	Name	Total Number of Bones	Name	Total Number of Bones	Name	Total Number of Bones
SKULL	→ [22]	MIDDLE EAR OSSICLES* → [6]		PECTORAL (SHOULDER) GIRDLE → [4]		PELVIC GIRDLE → [2]	
		MALLEUS	2	**CLAVICLE**	2	2 **OSSA COXAE**‡ (each OS coxa	2
CRANIAL	→ 8	**INCUS**	2	**SCAPULA**	2	contains 3 fused bones)	
FRONTAL	1	**STAPES**	2				
PARIETAL	2						
TEMPORAL	2	HYOID → [1]	1				
OCCIPITAL	1						
SPHENOID	1						
ETHMOID	1	VERTEBRAL COLUMN → [26]					
		CERVICAL	7	UPPER EXTREMITY → [60]		LOWER EXTREMITY → [60]	
FACIAL	→ 14	**THORACIC**	12	**HUMERUS**	2		
MAXILLA	2	**LUMBAR**	5			**FEMUR**	2
PALATINE	2	**SACRUM** (5 fused bones)	1	**ULNA**	2	**PATELLA**	2
ZYGOMATIC	2	**COCCYX** (3–5 fused bones)	1	**RADIUS**	2	**TIBIA**	2
LACRIMAL	2			**CARPAL**	16	**FIBULA**	2
NASAL	2	THORAX → [25]		**METACARPAL**	10	**TARSAL**	14
INFERIOR NASAL CONCHA	2	**STERNUM**	1	**PHALANX**	28	**METATARSAL**	10
VOMER	1	**RIBS** (12 paired)	24			**PHALANX**	28
MANDIBLE (only movable bone of skull)	1						
	22		58		64		62

TOTAL NUMBER OF AXIAL BONES = 80

BONES OF THE AXIAL SKELETON:
- Form the axis of the body
- Support and protect the organs of the HEAD, NECK, and TORSO
- SKULL BONES, BREASTBONE (STERNUM), BACKBONE (VERTEBRAL COLUMN), RIBS (RIB CAGE), HYOID BONE

(*The MIDDLE EAR OSSICLES are technically considered a
separate group of bones—apart from the AXIAL and APPENDICULAR
skeletons, but placed in AXIAL classification for convenience.

TOTAL NUMBER OF APPENDICULAR BONES = 126

BONES OF THE APPENDICULAR SKELETON:
- Free appendages (UPPER and LOWER EXTREMITIES)
- Body girdles (PECTORAL and PELVIC GIRDLES) anchor the free appendages to the AXIAL SKELETON

(‡Each OS COXA actually consists of 3 fused bones in the adult:
1 ILIUM, 1 ISCHIUM and 1 PUBIS)

TOTAL NUMBER OF BONES IN THE ADULT SKELETAL SYSTEM = 206

GENERAL COMPARISON OF MALE AND FEMALE SKELETONS

- Bones of the male are in general larger and heavier than those of the female
- In the male, articular ends are thicker in relation to the shafts than in the female
- In the male, certain muscles are larger than those of the female
- In the male, there are larger tuberosities, lines, and ridges for the attachment of the larger muscles
- The pelvis displays many significant structural differences between the male and female skeletons (See 4.13)

★ • Recall that the PECTORAL GIRDLE
and BOTH UPPER EXTREMITIES
are structurally adapted for:
  • FREEDOM OF MOVEMENT
  • EXTENSIVE MUSCLE ATTACHMENT
They are *not* adapted to bearing
weight and, therefore, in relation
to the PELVIC GIRDLE and LOWER
EXTREMITIES are more delicate.
Together both UPPER EXTREMITIES
including the PECTORAL GIRDLE
contain 64 BONES.

■ The PECTORAL GIRDLE is composed
of 2 CLAVICLES and 2 SCAPULAE
• The 2 CLAVICLES attach the
PECTORAL GIRDLE to the AXIAL
SKELETON ONLY at the STERNUM.
(The GIRDLE is not connected
to the VERTEBRAL COLUMN,
thus permitting a wide variety
of movement. Thus, it is not a
complete girdle.)
• The 2 SCAPULAE of the
PECTORAL GIRDLE attach the
upper extremity on each side
of the body to the trunk.
• Thus, each UPPER EXTREMITY contains
30 bones (both combined = 60 bones).

★ Recall that the PELVIC GIRDLE
and BOTH LOWER EXTREMITIES
are structurally adapted for:
  • SUPPORT
  • LOCOMOTION
They are adapted to bearing
weight and, therefore, in relation
to the PECTORAL GIRDLE and UPPER
EXTREMITIES are much stronger.
Together both LOWER EXTREMITIES,
including the PELVIC GIRDLE,
contain 62 bones.
• The PELVIC GIRDLE is formed
by 2 OSSA COXAE united anteriorly
by the SYMPHYSIS PUBIS and
posteriorly by the SACRUM.
Therefore, it is a complete girdle
(as opposed to the PECTORAL GIRDLE,
which is only attached anteriorly).
• Thus, each LOWER EXTREMITY contains
30 bones (both combined = 60 bones).

★ An interesting exercise is to
compare the PECTORAL GIRDLE
and UPPER EXTREMITIES with the
PELVIC GIRDLE and LOWER EXTREMITIES.
Use a similar color scheme
for both and then study their
relationships in terms of structure and function.

4.16

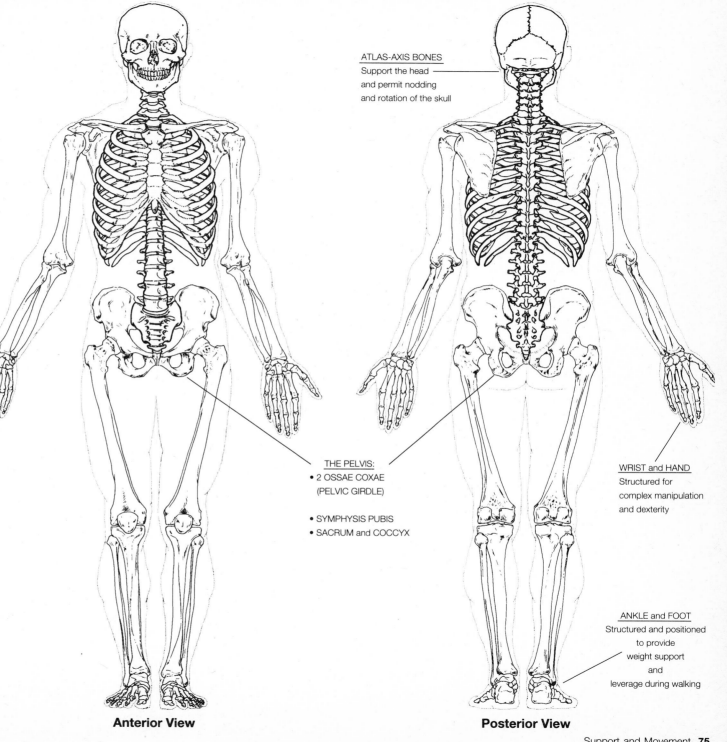

Anterior View

ATLAS-AXIS BONES
Support the head
and permit nodding
and rotation of the skull

THE PELVIS:
• 2 OSSAE COXAE
  (PELVIC GIRDLE)

• SYMPHYSIS PUBIS
• SACRUM and COCCYX

WRIST and HAND
Structured for
complex manipulation
and dexterity

ANKLE and FOOT
Structured and positioned
to provide
weight support
and
leverage during walking

Posterior View

Chapter 4 : *SKELETAL SYSTEM* : CHARTS

Chart #1 : Surface Features and Markings of Bone (See 4.3)

Protrusions (Processes): Any prominent projection	Definition	Example
CONDYLE*	A relatively large, rounded protuberance that forms articulations	Occipital condyle of the occipital bone
EPICONDYLE	A prominence on or above a condyle	Medial epicondyle of the femur
CREST	A prominent ridge or border	Iliac crest of the os coxa
HEAD*	A rounded projection separated from the main part of the bone by a constricted region (the neck)	Head of the femur
TUBERCLE	A small, rounded projection	Greater tubercle of the humerus
TUBEROSITY	A large, rounded process often with a roughened surface	Radial tuberosity of the radius
TROCHANTER	A large, blunt process found only on the femur	Greater trochanter of the femur
SPINE (Spinous Process)	A sharp, pointed process	Spinous process of the vertebrae
LINE	A low, less prominent ridge	Superior temporal line of the parietal bones
Flat Surfaces, Depressions, and Openings	**Definition**	**Example**
ALVEOLUS	A deep pit or socket	Alveoli for teeth in the maxilla
FACET*	A flat or shallow smooth surface for articulations	Costal facet of a thoracic vertebrae
FOVEA	A small, shallow depression or pit	Fovea capitis of the femur
GROOVE or SULCUS	A furrow that accommodates a blood vessel, nerve, tendon, or other soft structure	Costal groove of a thoracic rib Intertubercular sulcus of the humerus
FOSSA	A depression or groove in or on a bone	Mandibular fossa of the temporal bone
NOTCH	Deep indentation in the edge of a bone	Semilunar notch of the ulna
NECK	The narrow section of bone between the head and the shaft	Neck of the humerus
SINUS (Paranasal Sinus)	An air-filled cavity within a bone connected to the nasal cavity	Frontal sinus of the frontal bone
FISSURE	A groove or narrow cleft passage between adjacent parts of bone through which blood vessels or nerves pass	Superior orbital fissure of the sphenoid bone
FORAMEN	A hole or perforation through a bone through which blood vessels, nerves, or ligaments pass	Foramen magnum of the occipital bone
MEATUS	A tubelike passageway or canal running within a bone	External auditory meatus of the temporal bone

*Processes that form joints (articulations)

Chart #2 : Openings in Bones and the Structures Passing through Them (See 4.4–4.7, and 4.11)

Foramen, Fossa, Fissure, etc.	Location	Structures Passing Through*
Bones of the Skull		
CAROTID FORAMEN (Canal)	Petrous portion of the temporal bone	Internal carotid artery and sympathetic nerves
CONDYLAR FOSSA (Canal)	Behind the occipital epicondyle	Facial Nerve (VII)* and stylomastoid artery
EXTERNAL ACOUSTIC MEATUS	Tympanic portion of the temporal bone	Air passageway to tympanic membrane (eardrum)
FORAMEN LACERUM	Between petrous portion of the temporal and sphenoid, bounded medially by the sphenoid and occipital	Internal carotid artery and ascending branch of pharyngeal artery
FORAMEN MAGNUM	Occipital bone	Medulla oblongata and its membranes, spinal accessory nerve (XI), and the vertebral and spinal arteries and meninges
FORAMEN OVALE	Greater wing of sphenoid	Mandibular branch of trigeminal nerve (V)
FORAMEN ROTUNDUM	Junction of anterior and medial parts of the body of sphenoid	Maxillary branch of trigeminal nerve (V)
FORAMEN SPINOSUM	Posterior angle of sphenoid	Middle meningeal vessels
GREATER PALATINE FORAMEN	Palatine bone of hard palate (posterolateral angle)	Greater palatine nerve and descending greater palatine vessels
HYPOGLOSSAL CANAL	Anterolateral edge of the occipital condyle	Hypoglossal nerve (XII) and branch of ascending pharyngeal artery
INCISIVE (Stensen, Scarpa) CANAL	Anterior region of hard palate, posterior to the incisor teeth	Stensen, anterior branches of descending palatine vessels and scarpa, nasopalatine nerves
INFERIOR ORBITAL FISSURE	Between maxilla and greater wing of sphenoid	Maxillary branch of trigeminal nerve (V), zygomatic nerve and infraorbital vessels
INFRAORBITAL FORAMEN	Inferior to orbit in maxilla	Infraorbital nerve and artery
INTERNAL ACOUSTIC MEATUS	Tympanic portion of the temporal bone	Facial nerve (VII), Auditory nerve (VIII), internal auditory artery and nervus intermedius
JUGULAR FORAMEN	Between petrous portion of temporal and occipital, posterior to carotid canal	Internal jugular vein, glossopharyngeal nerve (IX), vagus nerve (X), spinal accessory nerve (XI), and sigmoid sinus
LACRIMAL CANAL	Lacrimal bone	Lacrimal (tear) duct
LESSER PALATINE FORAMEN	Posterior to greater palatine foramen in hard palate	Lesser palatine nerves and artery
MANDIBULAR FORAMEN	Medial surface of ramus of mandible	Inferior alveolar nerve and vessels

Foramen, Fossa, Fissure, etc.	Location	Structures Passing Through
MASTOID FORAMEN	Posterior border of the mastoid portion of temporal bone	Branch of occipital artery to dura mater and emissary vein transversing sinus
MENTAL FORAMEN	Below the second premolar tooth on the lateral side of mandible	Mental nerve and vessels
OLFACTORY FORAMINA	Medial portion of the horizontal cribriform plate of ethmoid bone	Olfactory nerves (I)
OPTIC FORAMEN (Canal)	Back of orbit between upper and lower portions of lesser wing of sphenoid	Optic nerves (II) and ophthalmic artery
PETROTYMPANIC FISSURE (Glaserian Fissure)	Tympanic portion of temporal bone	Glaserian artery
STYLOMASTOID FORAMEN	Between styloid and mastoid processes of temporal bone	Facial nerve (VII) and stylomastoid artery
SUPERIOR ORBITAL FISSURE	Between greater and lesser wings of the sphenoid	Oculomotor nerve (III), trochlear nerve (IV), ophthalmic branch of the trigeminal nerve (V) and abducens nerve (VI)
SUPRAORBITAL FORAMEN (Notch)	Frontal bone, supraorbital ridge of orbit	Supraorbital nerve and artery
ZYGOMATICOFACIAL FORAMEN	Anterior (cheek) surface of zygomatic bone	Zygomaticofacial nerve and vessels
ZYGOMATICOORBITAL FORAMEN	Zygomatic bone, orbital surface	Zygomaticotemporal and zygomaticoorbital nerves
ZYGOMATICOTEMPORAL FORAMEN	Zygomatic bone, temporal surface	Zygomaticotemporal nerve

Vertebral Column

DORSAL SACRAL FORAMINA (8 in number)	Dorsal surface of sacrum	Dorsal rami of sacral nerves
FORAMEN TRANSVERSARIUM (Transverse Foramen)	Transverse processes of the 7 cervical vertebrae	Vertebral artery and vein
INTERVERTEBRAL FORAMEN	Between two vertebrae, between the spinous process and the vertebral body	Spinal nerves
VENTRAL SACRAL FORAMINA (8 in number)	Ventral surface of sacrum	Ventral rami of the sacral nerves
SACRAL CANAL	Central cavity of sacrum	Spinal cord and accessory structures
VERTEBRAL FORAMEN	Between body (pedicles) and lamina of each vertebra	Spinal cord and accessory structures

Appendicular Skeleton

Various Bones, NUTRIENT FORAMEN	Generally on shaft of long bones	Nutrient vessels to and from bone marrow

* The roman numerals of the 12 paired cranial nerves indicate the relative position from which the nerves arise from the brain, from front (I) to back (XII).

UNIT 2: SUPPORT AND MOVEMENT

Chapter 5 Articulations

JOINTS BETWEEN BONES

Charts #1-2

Principle of Leverage in Relation to Bones and Joints

• The rigid LEVER **[BONE]** turns about a FULCRUM **[JOINT]** when force is applied **[MUSCLE]** to move the **RESISTING OBJECT.**

Three Classes of Levers

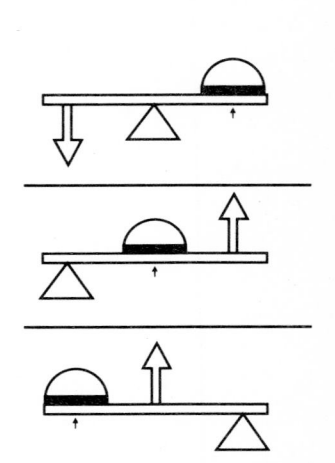

1️⃣ FIRST-CLASS LEVER

  • The FULCRUM (JOINT) is positioned between the FORCE (MUSCLE) and the RESISTANCE

2️⃣ SECOND-CLASS LEVER

  • The RESISTANCE is positioned between the FORCE (MUSCLE) and the FULCRUM (JOINT)

3️⃣ THIRD-CLASS LEVER

  • The FORCE (MUSCLE) is positioned between the FULCRUM (JOINT) and the RESISTANCE

Choose a color for each symbol, and color the illustrations above.

LEVER = BONE

△ PIVOT   FULCRUM = JOINT

EFFORT   FORCE = MUSCLE

RESISTANCE = OBJECT MOVED

JOINTS, or ARTICULATIONS, do not comprise a system
as such. However, they are directly related to the SKELETAL SYSTEM
and specifically to one of the functions of the SKELETAL SYSTEM:
TO PERMIT BODY MOVEMENT. The ARTICULATIONS between the bones
allow movement of body parts, and the range of movement permitted is
determined by the structure of a joint. Articulations are
classified into 3 types
according to STRUCTURE, and
are also classified into 3
types according to FUNCTION
(DEGREE OF MOVEMENT PERMITTED
WITHIN THE JOINT).

ARTHROLOGY
is
the science
concerned
with
the study
of
ARTICULATIONS
(JOINTS)
and
all aspects
of
joint
classifications.

KINESIOLOGY
is an applied,
practical, dynamic science
concerned with the
BIOMECHANICS OF MOVEMENTS
involving certain joints.

JOINTS or ARTICULATIONS
are composed of
FIBROUS CONNECTIVE TISSUE,
CARTILAGE, or LIGAMENTS.
•
Not all joints are flexible
in permitting movement, as
some joints remain rigid for
BODY STABILIZATION
and
MAINTENANCE OF BALANCE.

Anterior View

Posterior View

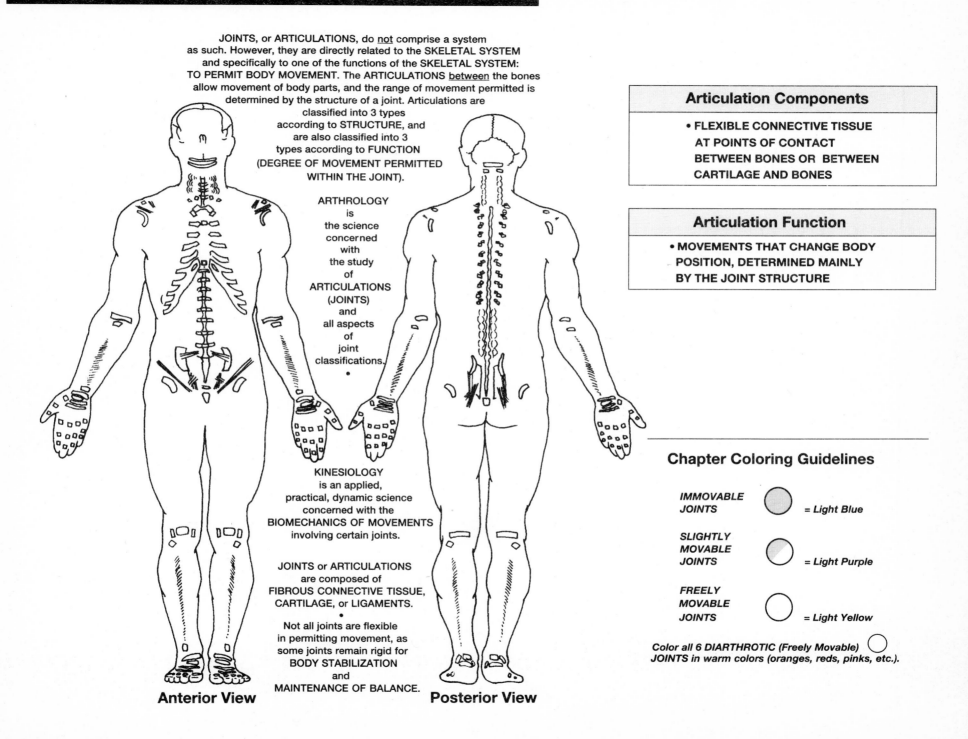

Articulation Components

- **FLEXIBLE CONNECTIVE TISSUE
  AT POINTS OF CONTACT
  BETWEEN BONES OR BETWEEN
  CARTILAGE AND BONES**

Articulation Function

- **MOVEMENTS THAT CHANGE BODY
  POSITION, DETERMINED MAINLY
  BY THE JOINT STRUCTURE**

Chapter Coloring Guidelines

*IMMOVABLE
JOINTS*   = *Light Blue*

*SLIGHTLY
MOVABLE
JOINTS*   = *Light Purple*

*FREELY
MOVABLE
JOINTS*   = *Light Yellow*

*Color all 6 DIARTHROTIC (Freely Movable)
JOINTS in warm colors (oranges, reds, pinks, etc.).*

5.0

ARTICULATIONS : JOINTS BETWEEN BONES
Arthrology: *Classification of Joints*

◯ = light blue (Color 3 kinds in cool colors)

◯ = light purple (Color 2 kinds in neutral colors)

◯ = light yellow (Color 6 kinds in warm colors)

★ See Chart #1 and #2, 5.4

★ FUNCTIONAL CLASSIFICATION
OF JOINTS : 3 TYPES OF JOINTS
Basis of Classification:
- Degree of Movement
  Permitted within a JOINT

DEGREE OF
JOINT MOVEMENT

◯ **SYNARTHROSES**   IMMOVABLE
                              (NONE)
    1   SUTURES
    2   GOMPHOSES
    4   SYNCHONDROSES

◯ **AMPHIARTHROSES**   SLIGHTLY
                              MOVABLE
    3   SYNDESMOSES
    5   SYMPHYSES

◯ **DIARTHROSES**   FREELY
SYNOVIAL JOINTS   MOVABLE

    H   HINGE
    P   PIVOT
    G   GLIDING
    E   ELLIPSOID
    S   SADDLE
    BS BALL-AND-SOCKET

★ STRUCTURAL CLASSIFICATION
OF JOINTS : 3 TYPES OF JOINTS
Basis of Classification:
- Presence or Absence of a Joint Cavity
- Kinds of Supporting Connective Tissue
  Surrounding the Joint

	FUNCTIONAL CLASSIFICATION	JOINTS CLOSED BY JOINT CAPSULE: JOINT CAVITY	CONNECTIVE TISSUE SURROUNDING JOINT and HOLDING BONES TOGETHER
FIBROUS JOINTS			
1 **SUTURES** †	◯		DENSE FIBROUS (thin layer)
2 **GOMPHOSES**	◯	NO	DENSE FIBROUS (ligaments)
3 **SYNDESMOSES**	◯		DENSE FIBROUS (interosseous membrane or Collagenous Fibers)
CARTILAGINOUS JOINTS			
4 **SYNCHONDROSES**	◯	NO	HYALINE CARTILAGE
5 **SYMPHYSES**	◯		FIBROCARTILAGE

LIGAMENTOUS AND
SYNOVIAL JOINTS   : associated with free movement

H **HINGE**	◯	
P **PIVOT**	◯	
G **GLIDING**	◯	YES
E **ELLIPSOID (CONDYLOID)**	◯	ALSO CONTAINS SYNOVIAL FLUID AND SYNOVIAL MEMBRANE
S **SADDLE**	◯	
BS **BALL-AND-SOCKET**	◯	

DENSE FIBROUS
(LIGAMENTS)

† SYNOSTOSES = Sutures where there
is a complete fusion of bones
across the suture line
(usually in older adults)

FUNCTIONAL AND STRUCTURAL CLASSIFICATION OF ARTICULATIONS

STRUCTURAL	FUNCTIONAL

FIBROUS JOINTS (3 KINDS) — -ARTHROSES

1 SUTURES

- SERRATE ➞ Sagittal suture between Parietal Bones of the Skull
- LAP (SQUAMOUS) ➞ Squamous suture beteen Temporal and Parietal Bones
- PLANE (BUTT) ➞ Maxillary suture = the 2 Maxillary Bones form HARD PALATE
- ✝. SYNOSTOSIS ➞ Union of Right and Left Portions of the Frontal Bone

SYN-

2 GOMPHOSES

- Styloid process in Temporal Bone inserted into a socket
- Teeth roots embedded in the Alveoli of the Maxilla and Mandible

SYN-

3 SYNDESMOSES

- Distal and articulation of the Tibia and Fibula
- Articulations between shafts of the Ulna and Radius

AMPHI-

STRUCTURAL	FUNCTIONAL

CARTILAGINOUS JOINTS (2 KINDS) — -ARTHROSES

4 SYNCHONDROSES

- Temporary joints forming the growth lines or EPIPHYSEAL PLATES in the long bones in children
- Occipital-sphenoid joint in children
- Joints ossify when bone growth is complete. Hyaline cartilage of the temporary joint is replaced by bone.
- Joint changes from a SYNCHONDROSES to a SYNOSTOSES

- 1st rib STERNUM
- Sacroiliac joint

SYN-

5 SYMPHYSES

- Symphysis Pubis (between the anterior surfaces of the Coxal Bones).
- Intervertebral discs between the vertebrae

AMTHI-

STRUCTURAL	FUNCTIONAL

SYNOVIAL JOINTS (6 KINDS) - DIARTHROTIC — -ARTHROSES

H

- Bending in only one plane (like a door)
- 1 bone surface is always concave, and the other is always convex

Most common type of DIARTHROSES :
- KNEE (Tibiofemoral) JOINT
- ELBOW (Humeroulnar) JOINT
- Joints of the PHALANGES
- TEMPOROMANDIBULAR JOINT

DI-

P

- Rotation about a central axis
- ATLAS and AXIS JOINT (rotational movement of the head)
- Proximal articulation of the RADIUS and ULNA (rotational movement of the forearm)

DI-

G

- Smooth sliding
- Simplest kind of movement
- COSTOVERTEBRAL JOINTS (between ribs, and vertebrae and vertebral processes)
- Joints between the carpals
- Joints between the tarsals
- STERNOCLAVICULAR JOINT

DI-

E

- Angular movement in 2 directions (Biaxial): up-and-down and side-to-side
- Oval, convex surface-elliptical, concave depression
- RADIOCARPAL JOINT
- ATLANTO-OCCIPITAL JOINT (between the ATLAS BONE and the Occipital Bone of the skull)

DI-

S

- Modified CONDYLOID (ELLIPTICAL) JOINT
- Associated only with the thumb
- Concave surface in one direction and convex surface in another direction
- The articulation of the TRAPEZIUM BONE IN THE CARPUS (WRIST) with the FIRST METACARPAL BONE
- CARPOMETACARPAL JOINT

DI-

BS

- Greatest range of movement of all DIARTHROSES (all planes and rotation)
- Rounded convex surface and cuplike cavity
- HIP JOINTS
- SHOULDER JOINTS

DI-

ARTICULATIONS: DIARTHROTIC-SYNOVIAL JOINTS
Typical Joint Structure

a and b = warm colors c = light brown
d = cream e = light yellow
f = light gray synovial cavity = light blue

★ DIARTHROSES
- Freely movable joints
- Contain SYNOVIAL FLUID, or SYNOVIUM. (Sometimes the term SYNOVIUM refers to the SYNOVIAL MEMBRANE, which secretes the SYNOVIAL FLUID into the SYNOVIAL CAVITY)
- Enclosed by JOINT CAPSULES
- Most complex and varied of the 3 types of joints

★ DUAL FUNCTION OF DIARTHROTIC JOINTS
1. Provide wide range of precise, smooth movements
2. Maintain Stability, Strength (and to a certain extent, Rigidity)

★ JOINT CAPSULE

■ Flexibility of the FIBROUS JOINT CAPSULE permits movement at a DIARTHROTIC JOINT.

■ Great tensile strength of the FIBROUS JOINT CAPSULE resists dislocation. (Fibers arranged in parallel bundles adapted to resist recurrent strain are called LIGAMENTS.)

■ FIBROUS JOINT CAPSULE is attached to the PERIOSTEUM of the ARTICULATING BONES (at a distance away from the ARTICULAR CARTILAGE— which is smooth and caps the ends of the ARTICULATING BONES, and resides within the SYNOVIAL CAVITY filled with SYNOVIAL FLUID.)

★ 3 FACTORS LIMITING MOVEMENT OF A DIARTHROTIC-SYNOVIAL JOINT

1. Structure of the bone (sometimes a process will limit range and "lock" articulation)
2. Strength and tautness of associated structures (Ligaments, Tendons, and Joint Capsules)
3. Size, arrangement, and action of associated muscles spanning the joint

★ CUSHIONED BURSAE near DIARTHROTIC-SYNOVIAL JOINTS
- Flattened, pouchlike synovial membrane sacs filled with synovial fluid
- Located between muscles or within areas where tendon passes over bone
- Cushion muscles, facilitate movement of muscles or tendons over surfaces of bones or ligaments

Typical Structure of a DIARTHROTIC-SYNOVIAL JOINT

| SYNOVIAL CAVITY WITH SYNOVIAL FLUID/SYNOVIUM |
a	SYNOVIAL MEMBRANE	INNER LAYER OF THE JOINT CAPSULE
b	JOINT CAPSULE	FIBROUS ARTICULAR CAPSULE
c	ARTICULAR CARTILAGE
d	ARTICULATING BONES	NEVER COME IN ACTUAL CONTACT WITH ONE ANOTHER
e	ARTICULAR DISK
f	PERIOSTEUM OF BONES

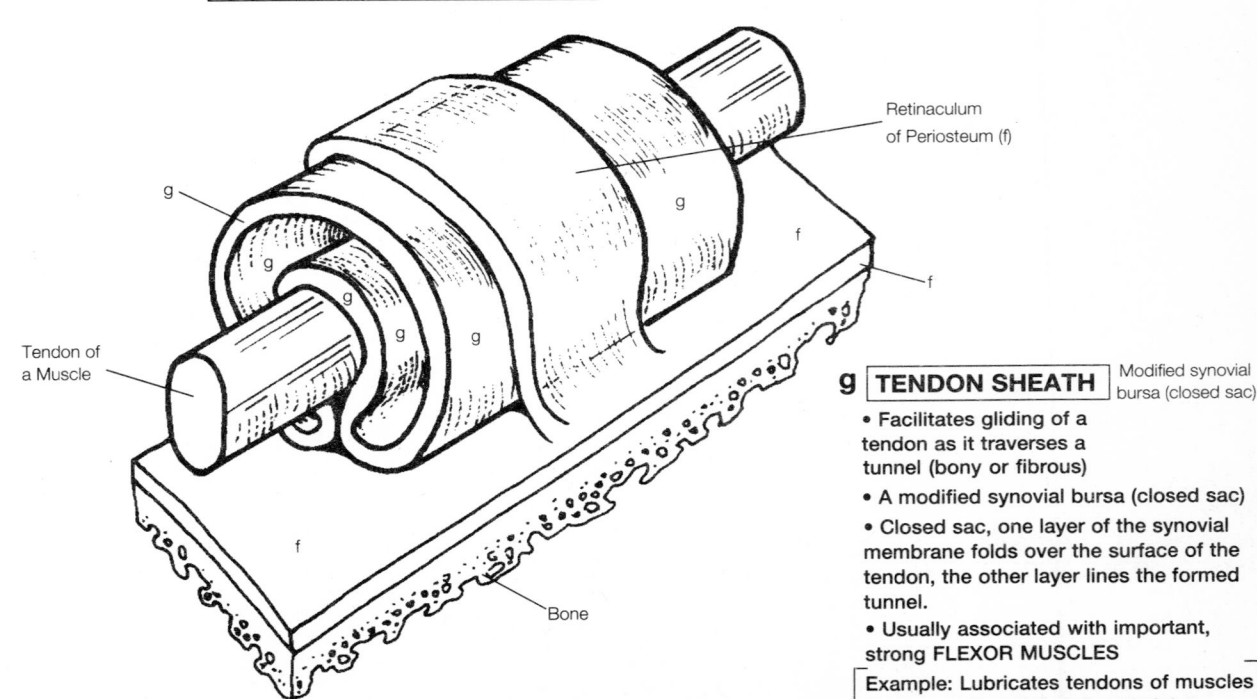

Retinaculum of Periosteum (f)

g

Tendon of a Muscle

Bone

g | TENDON SHEATH | Modified synovial bursa (closed sac)

- Facilitates gliding of a tendon as it traverses a tunnel (bony or fibrous)
- A modified synovial bursa (closed sac)
- Closed sac, one layer of the synovial membrane folds over the surface of the tendon, the other layer lines the formed tunnel.
- Usually associated with important, strong FLEXOR MUSCLES

Example: Lubricates tendons of muscles that cross the wrist and ankle joints

Typical Structure of a DIARTHROTIC-SYNOVIAL JOINT

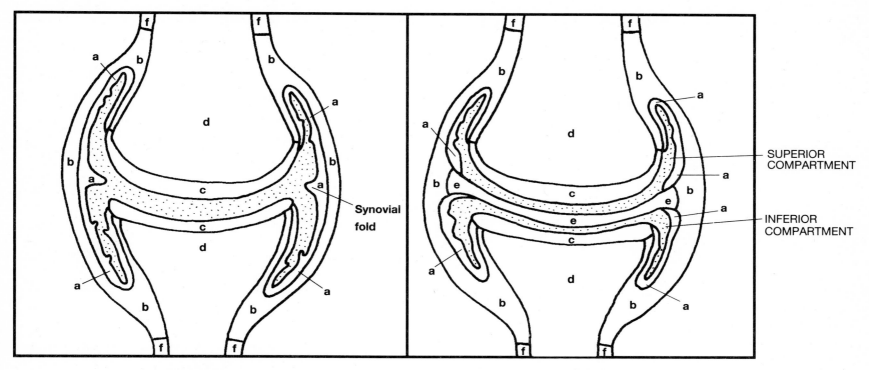

Coronal View

■ All DIARTHROSES are SYNOVIAL JOINTS

■ They are enclosed by a JOINT CAPSULE (b) made of dense fibroelastic tissue, also called an ARTICULAR CAPSULE.

■ A JOINT CAVITY within the CAPSULE creates a SPACE BETWEEN ARTICULATING BONES, also called a SYNOVIAL CAVITY.

■ SYNOVIAL FLUID, SYNOVIUM which fills the SYNOVIAL CAVITY, is secreted by a thin, SYNOVIAL MEMBRANE, which lines the inside of the JOINT CAPSULE. The fluid contains the lubricant HYALURONIC ACID and interstitial fluid from BLOOD PLASMA.

■ SYNOVIAL MEMBRANE:
- Lubricates the joint
- Contains phagocytic cells that remove microbes and debris
- Provides nourishment for the ARTICULAR CARTILAGE (c), which caps both articulating bones. Bones never come in contact with each other.

Coronal View
(with Articulating Disk)

Example: Temporomandibular Joint

■ The encapsulated, DOUBLE SYNOVIAL JOINT between the condyles of the MANDIBLE and the TEMPORAL BONES of the SKULL

■ The only DIARTHROTIC JOINT in the SKULL

■ A combination of an [H] HINGE JOINT and a [G] GLIDING JOINT (See 5.1)

■ The ARTICULAR DISK separates the SYNOVIAL CAVITY into 2 separate compartments

■ Clinically important due to TEMPOROMANDIBULAR JOINT (TMJ) SYNDROME

ARTICULATIONS: DIARTHROTIC-SYNOVIAL JOINTS
A Specific Example: The Knee (Tibiofemoral) Joint

Repeat similar colors from 5.2 a-f for Sagittal section drawing. Subpatellar fat is deep yellow. Muscle is light red or orange. Leave bones white and color in ligaments with cool colors, and muscles with warm colors.

★ See also 4.14

★ The KNEE (PATELLAR) JOINT:
  • Largest joint in the body
  • Most complex joint in the body
  • Most vulnerable joint in the body
  • Hinged DIARTHROSIS
  • Permits FLEXION and EXTENSION
    (limited ROLLING and GLIDING)

Specific Example of a DIARTHROTIC-SYNOVIAL JOINT: The KNEE (PATELLAR) JOINT

1 FIBULAR COLLATERAL LIGAMENT

2 TIBIAL COLLATERAL LIGAMENT

3 ANTERIOR CRUCIATE LIGAMENT

4 POSTERIOR CRUCIATE LIGAMENT

5 TRANSVERSE LIGAMENT

6 PATELLAR LIGAMENT

7 OBLIQUE POPLITEAL LIGAMENT

8 ARCUATE POPLITEAL LIGAMENT

9 LIGAMENT OF WRISBERG

10 LATERAL MENISCUS

11 MEDIAL MENISCUS ⎫ ARTICULAR DISCS (PADS):
  • Attached by their margins to the FIBROUS CAPSULE.
  • Create separate cavities for independent movement of ARTICULATING BONES.

12 SYNOVIAL BURSA REDUCES FRICTION WHERE SKIN RUBS OVER BONE

13 SUBPATELLAR FAT

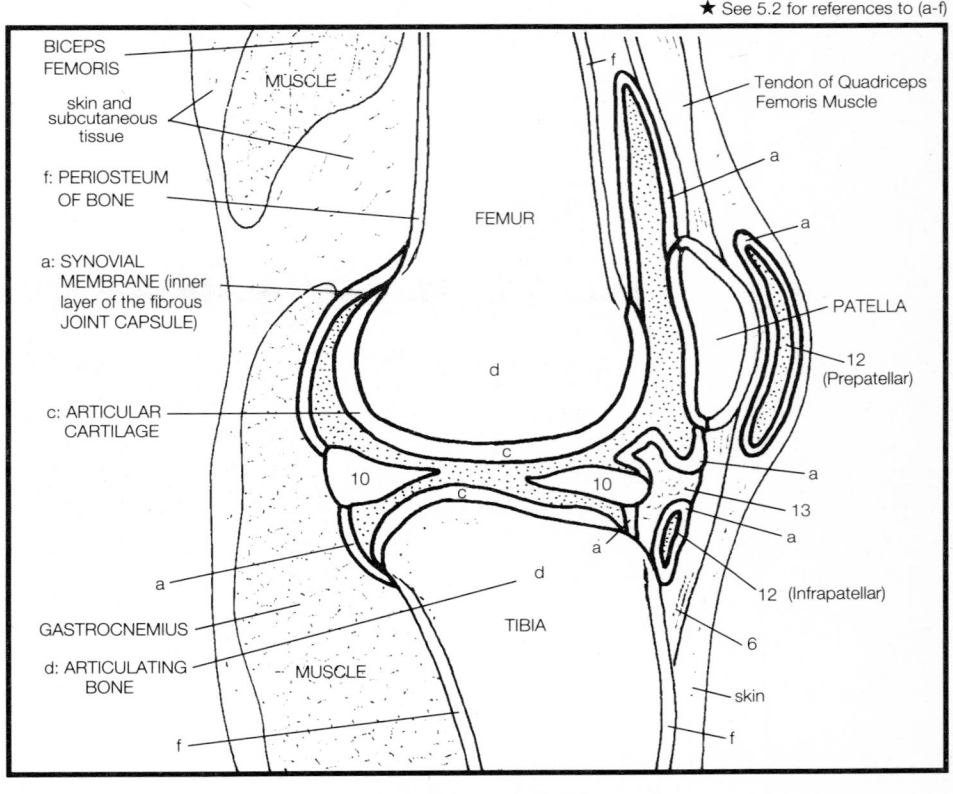

★ See 5.2 for references to (a-f)

BICEPS FEMORIS

MUSCLE

f

Tendon of Quadriceps Femoris Muscle

skin and subcutaneous tissue

f: PERIOSTEUM OF BONE

FEMUR

a

a

a: SYNOVIAL MEMBRANE (inner layer of the fibrous JOINT CAPSULE)

PATELLA

12 (Prepatellar)

d

c: ARTICULAR CARTILAGE

c

10   c   10

a

a

13

a

GASTROCNEMIUS

a

d

12 (Infrapatellar)

d: ARTICULATING BONE

MUSCLE

TIBIA

6

skin

f

f

Right Lateral View
(Sagittal Section)

⋰⋱ = SYNOVIAL CAVITY filled with SYNOVIAL FLUID

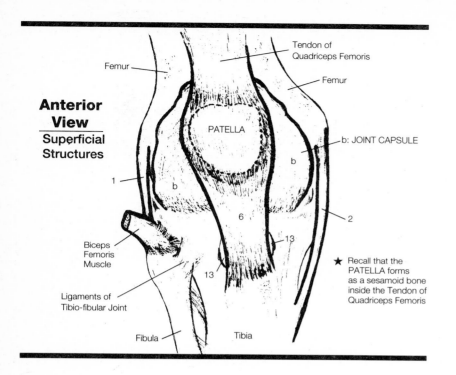

Anterior View
Superficial Structures

Femur

Tendon of Quadriceps Femoris

Femur

PATELLA

b: JOINT CAPSULE

1

b

b

6

2

13

13

Biceps Femoris Muscle

Ligaments of Tibia-fibular Joint

Fibula

Tibia

★ Recall that the PATELLA forms as a sesamoid bone inside the Tendon of Quadriceps Femoris

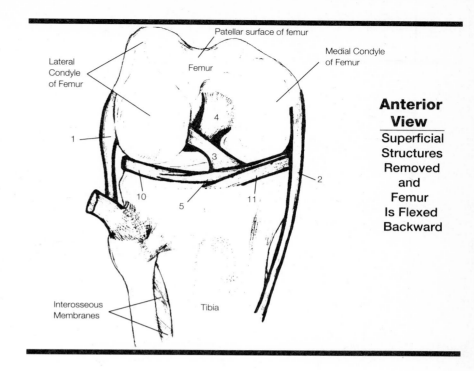

Anterior View
Superficial Structures Removed and Femur Is Flexed Backward

Patellar surface of femur

Lateral Condyle of Femur

Femur

Medial Condyle of Femur

1

4

3

2

10

5

11

Interosseous Membranes

Tibia

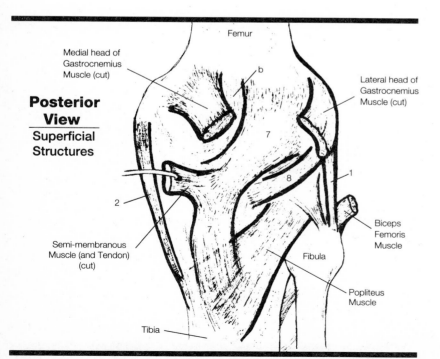

Posterior View
Superficial Structures

Femur

Medial head of Gastrocnemius Muscle (cut)

b

Lateral head of Gastrocnemius Muscle (cut)

7

8

1

2

7

Biceps Femoris Muscle

Semi-membranous Muscle (and Tendon) (cut)

Fibula

Popliteus Muscle

Tibia

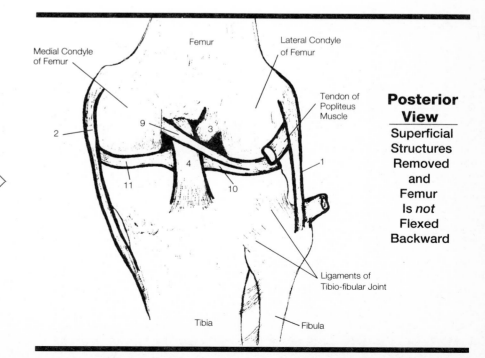

Posterior View
Superficial Structures Removed and Femur Is *not* Flexed Backward

Femur

Medial Condyle of Femur

Lateral Condyle of Femur

Tendon of Popliteus Muscle

2

9

3

1

11

4

10

Ligaments of Tibia-fibular Joint

Tibia

Fibula

ARTICULATIONS: DIARTHROTIC-SYNOVIAL JOINTS
Movements of Joints

Color in opposite colors for opposite movements.
(ex: yellow-purple, orange-blue, red-green, etc.)

★ See Charts #1 and #2, 5.1, 6.3

OPPOSITE
MOVEMENTS =

Limiting Factors of Movement

•Apposition of Soft Parts   •Tension of Associated Ligaments
•Structure of Participating Bones   •Tension, Size, and Arrangement of Associated Muscles

MOVEMENT TERM   **DEFINITION OF MOVEMENT**

ANGULAR
Increase or decrease the joint angle produced by the articulating bone

2 **FLEXION** — Movement that decreases the joint angle on an anterior-posterior plane
(ex: bending the head forward, flexing arm upward)

8 **DORSIFLEXION** — Upward bending of the foot backward at the ankle

3 **EXTENSION** — Movement that increases the joint angle on an anterior-posterior plane
(ex: bending the head backward, extending arm downward)

9 **PLANTAR FLEXION** — Downward bending of the foot (forepart depressed in respect to the ankle)

10 **HYPEREXTENSION** — Extension of a joint beyond the anatomical position

4 **ABDUCTION** — Movement away from the midline of the body

5 **ADDUCTION** — Movement toward the midline of the body

CIRCULAR

6 **ROTATION** — Movement of a bone around its own axis

11 **SUPINATION** — Assuming the supine position; turning wrists so that palms are upward

12 **PRONATION** — Assuming the prone position; turning wrists so that palms are downward
(Both SUPINATION and PRONATION pertain to position of body as a whole)

7 **CIRCUMCUCTION** — Movement at a synovial joint describing a cone-shaped figure, where the
distal end of a bone moves in a circle, while proximal end remains stationary

SPECIALZED MOVEMENTS

13 **INVERSION** — Turning bottom of the feet (soles) inward

14 **EVERSION** — Turning bottom of the feet (soles) outward

15 **PROTRACTION** — To draw forward (ex: jutting of the lower jaw forward)

16 **RETRACTION** — To draw backward (ex: pulling the lower jaw backward behind upper teeth)

17 **ELEVATION** — An upward movement; lifting the shoulders (scapulae) as in a shrug

18 **DEPRESSION** — A downward movement; moving shoulders (scapulae) in a downward direction

★ See 5.1

MONAXIAL MOVEMENT

H HINGE → FLEXION-EXTENSION

P PIVOT → ROTATION

BIAXIAL MOVEMENT

E ELLIPSOIDAL   FLEXION-EXTENSION
                 ABDUCTION-ADDUCTION

S SADDLE

TRIAXIAL MOVEMENT

G GLIDING

BS BALL AND SOCKET →   FLEXION-EXTENSION
                        ABDUCTION-ADDUCTION
                        ROTATION, CIRCUMDUCTION

GLIDING, ANGULAR, AND CIRCULAR MOVEMENTS OF JOINTS

Midline of body

1 GLIDING

Back-and-forth and side-to-side
(one surface over another surface)

Chart #1 : Types of Articulations (See 5.1)

FUNCTIONAL CLASSIFICATION

Type	Structure	Movements and Examples
SYNARTHROSES	ARTICULATIONS IN WHICH BONES ARE HELD TIGHTLY TOGETHER BY FIBROUS CONNECTIVE TISSUE	Immovable Joints
• **SUTURES**	Grooves on edge of bone, fit tightly together; have short connecting fibers. (Synostosis: fused sutures found in older adults.)	(Found in flat bones of skull)
• **GOMPHOSES**	Spike of a bone fits into a socket	(Teeth set within the alveoli of the maxilla and mandible)
• **SYNCHONDROSES**	Joints in which the bones are held together by hyaline cartilage	Epiphyses of long bone; a temporary joint; cartilage replaced by bone in the adult. (Joint between ribs and sternum involving costal cartilage)
AMPHIARTHROSES	BONES UNITED BY CARTILAGE	Slight Movement in Joints
• **SYNDESMOSES**	Joint fibers longer, called ligaments; some give in a slight stretch.	(Joint between distal ends of tibia and fibula)
• **SYMPHYSES**	Articular surface of bones covered with thin layer of hyaline cartilage. Fibrocartilage discs between bones.	[Junction of pubic bones (Symphysis Pubis)]

⬤ DIARTHROSES	FREE MOVEMENT LIMITED BY LIGAMENTS, MUSCLES, TENDONS, OTHER BONES, AND SHAPE OF BONE INVOLVED IN JOINT	Freely Movable Joints
H **HINGE**	Convex surface of one bone articulates with concave surface of the other	Flexion and Extension; (Knee and elbow)
P **PIVOT**	Conical portion of one bone articulates with a depression of another	Rotational around a longitudinal axis of a bone (atlas-axis and proximal radius-ulna joint)
G **GLIDING**	Flattened or slightly curved articulating surfaces	Any direction; intercarpal and intertarsal sliding (side-to-side and back-and-forth)
E **ELLIPSOID** (CONDYLOID)	Oval condyle of one bone articulates with elliptical cavity of another	In two perpendicular planes (biaxial movement); (Radius-carpals joint)
S **SADDLE**	Concave- and convex-shaped surfaces of each articulating bone	Biaxial; movement in most directions; (some limitation preventing circumduction) (Carpal–metacarpal joint of thumb)
BS **BALL-AND-SOCKET**	Rounded, convex head of one bone articulates with cuplike socket of another	Rotation in all planes (flexion, extension, abduction, adduction); shoulder and hip. Permits the greatest range of movement with the least amount of stability. Therefore, joint type is often damaged.

Chart #2 : Relation of Major Joints and Ligaments to the Skeletal System (See 5.1, 5.4)

◯ = Synarthroses
◯ = Amphiarthroses
◯ = Diarthroses

			AXIAL SKELETON	APPENDICULAR SKELETON	
				UPPER EXTREMITY	LOWER EXTREMITY
FIBROUS JOINTS					
1	**SUTURES**	◯	Between (linking) Bones of the Skull		
2	**GOMPHOSES**	◯	Teeth roots ⟷ Alveoli of the Maxilla and Mandible Styloid process of the Temporal Bone		
3	**SYNDESMOSES**	◯		Interosseous Membrane between shafts of the Ulna and Radius (see IM below)	Distal Tibiofibular
CARTILAGINOUS JOINTS					
4	**SYNCHONDROSES** (Hyaline cartilage)	◯	1st Rib ⟷ Sternum Sacroiliac (Epiphysis ⟷ Diaphysis: Bone growth)		
5	**SYMPHYSES** (Fibrocartilage)	◯	• Intervertebral discs • Sacrococcygeal • Symphysis Pubis • Manubrium ⟷ Body of Sternum ($\frac{1}{3}$ synovial)		
DIARTHROTIC-SYNOVIAL JOINTS					
H	**HINGE**	◯	Temperomandibular (double)	Elbow (Humerus ⟷ Ulna and Radius) Interphalangeal	Knee (Tibiofemoral) and WHOLE JOINT Ankle (Tarsal ⟷ Tibia and Fibula) Interphalangeal
P	**PIVOT**	◯	Median Atlanto-Axial	Proximal Radioulnar Distal Radioulnar*	
G	**GLIDING**	◯	Lateral Atlanto-Axial Body of Sternum ⟷ Xiphoid Process Sternocostal (Sternum ⟷ ribs) Costovertebral (heads of ribs) Costotransversal (necks and tubercles of ribs) Vertebral arches Vertebral column ⟷ Pelvis Sacrum ⟷ Ischium	Sternoclavicular (double) Acromioclavicular Intercarpal Carpometacarpal Intermetacarpal	Knee: (Patella ⟷ Femur) Proximal tibiofibular Intertarsal Tarsometatarsal Intermetatarsal
E	**ELLIPSOIDAL** (CONDYLOID)	◯	Atlanto-Occipital	Wrist (Radiocarpal) Metacarpophalangeal	Knee: (Femur-Tibia), two Metatarsophalangeal
S	**SADDLE**	◯		Carpometacarpal Joint of the Thumb	
BS	**BALL-AND SOCKET**	◯		Shourder (Humerus ⟷ Scapula)	Hip (Femur ⟷ Coxal, pubic) *Some
IM	**INTEROSSEOUS MEMBRANES**			Median radioulnar (see 3 above)	Median tibiofibular
FC	**FIBROUS CAPSULES**		example: Vertebral column		

SPECIFIC LIGAMENTS

From Vertebral columi ⟷ Pelvis: G
   Iliolumbar ligament
From Sacrum–Ischium: G
   Sacrotuberous ligament
   Dorsal–sacroilliac ligament
From Pelvic area

From Temperomandibular joint: H
   Stylomandibular ligament
From Vertebra arches: G
   Anterior longitudinal ligament
   Supraspinal ligament
   Interspinal ligament
   Ligamentum nuchae

From Acromioclavicular joint: G
   Coracoclavicular ligament
   Coracoacromial ligament

From Hip joint: BS
   Iliofemoral ligament

* Some authorities contend that the DISTAL RADIOULNAR joint is not a pivot joint, although it is listed as such in Gray's Anatomy.

UNIT 2: SUPPORT AND MOVEMENT

Chapter

6 Muscular System

PROPERTIES	SMOOTH	CARDIAC	SKELETAL
Synonyms	Involuntary Nonstriated	Myocardium	Voluntary Striated
Fibers: Fiber length(μm) Fiber thickness(μm) Shape Markings Connections	50-200 4-8 Spindles No striation Arranged in sheets or layers (may be isolated)	Quadrangular Striation Bifurcated and inter-connected by intercalated discs	25,000 75 Cylinders Marked striation Single and separated
Nuclei(location)	Single(central)	Single(central)	Multiple(peripheral)
Contraction	Very slow	Moderate (contracts without nervous stimulation but regulated by it)	Very quick (contracts only by nervous stimulation)
Effects of cutting a related nerve	Slight	Slight	Complete paralysis

Although there are 3 types of contractile muscle tissue in the body, SKELETAL, CARDIAC, and SMOOTH, the Muscular System in this manual includes mainly a study of the SKELETAL MUSCLES, which give FORM and STABILITY to the body. (See 2.2)

CARDIAC MUSCLE is found only in the heart, which is part of the CARDIOVASCULAR SYSTEM (See chapter 9, 9.4-9.6); it is involuntary and striated.

SMOOTH MUSCLE is generally associated with other (usually hollow) internal organs and glands of involuntary body systems—CARDIOVASCULAR (walls of blood vessels), DIGESTIVE, GENITOURINARY, and RESPIRATORY. It is involuntary and nonstriated.

SKELETAL MUSCLE TISSUE is highly specialized (voluntary and striated), and specifically adapted to contract and relax in order to carry out 3 general functions:

    1) BODY MOTION—whole body movements produced by functional integration with bones and joints, which provide the LEVERAGE and FORMATIVE FRAMEWORK of the Body

    2) HEAT PRODUCTION AND MAINTENANCE OF BODY TEMPERATURE (Muscle tissue produces 40–50% of Heat from cell respiration)

    3) POSTURE and BODY SUPPORT around the flexible joints

Muscle tissue comprises 40–50% of total body weight. Although the 3 types of muscle tissue differ in structure and function, they ALL possess 4 striking characteristics:
- CONTRACTILITY
- EXCITABILITY (IRRITABILITY)
- EXTENSIBILITY
- ELASTICITY
- The outstanding characteristic of muscle tissue is its ability to shorten or contract
- Muscle tissue possesses little intercellular material, hence its cells or fibers lie close together

THE SKELETAL MUSCLE SYSTEM may be divided into 2 main groups associated with the AXIAL and APPENDICULAR SKELETONS (see chapter 4)
▪ MUSCLES OF THE AXIAL SKELETON
- FACIAL EXPRESSION • MASTICATION • EYE, TONGUE, AND NECK MOVEMENT • VERTEBRAL COLUMN AND TRUNK MOVEMENT • RESPIRATION • ABDOMINAL WALL • PELVIC OUTLET

▪ MUSCLES OF THE APPENDICULAR SKELETON
- MUSCLES THAT ACT ON THE GIRDLES •
- MUSCLES THAT MOVE APPENDAGE SEGMENTS:
  UPPER LIMB: PECTORAL GIRDLE-ARM-FOREARM-
          WRIST-HAND-FINGERS
  LOWER LIMB: PELVIC GIRDLE-THIGH-LEG-ANKLE-
          FOOT-TOES

★ Refer to the End-of-Chapter Charts
For *Each* Muscle's:
    ▪ ORIGIN AND INSERTION
    ▪ ACTION AND FUNCTION
    ▪ INNERVATION

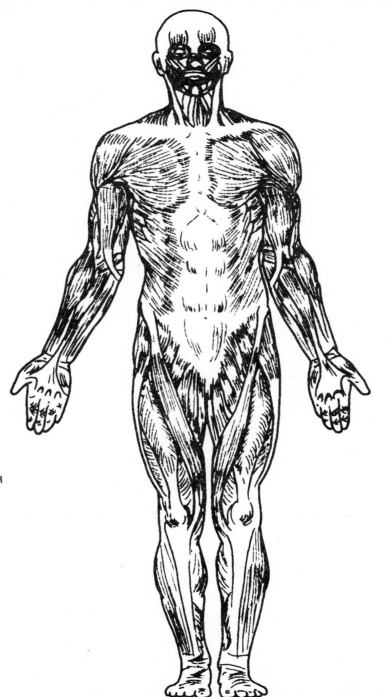

Anterior View

System Components

MUSCLE TISSUE OF THE BODY
(INCLUDING SKELETAL, CARDIAC, AND SMOOTH CONTRACTILE TISSUE)

System Function

- MOVEMENT
- HEAT PRODUCTION
- MAINTENANCE OF BODY POSTURE
- MAINTENANCE OF BODY TEMPERATURE

Chapter Coloring Guidelines

■ *The optimal situation for coloring the muscles would be to use as many warm and fleshy colors as possible (reds, oranges, etc.), but in some situations that is not possible due to the number of muscles on a page.*
■ *Use complementary (opposing) color schemes (red-green, orange-blue, yellow-purple) (warms vs. cools) when dealing with:*
    • *Opposing movements (FLEXION-EXTENSION), contrasting movements (ACTIN-MYOSIN), different layers of muscle, or different groups (extrinsic-intrinsic; suprahyoid-infrahyoid).*
■ *Tendons should be similar in color to their associated muscle, but lighter or clearer in tone to distinguish them.*

MUSCULAR SYSTEM : GENERAL ORGANIZATION
Skeletal Muscle: Structure, Architecture, and Movement Mechanics

When coloring tendons throughout the chapter, try to find a color similar to the associated muscle, but lighter and clearer, or more transparent in tone. Use warm colors for muscles as much as possible.

★ When a muscle contracts, it shortens, bringing its two ends together.
• SKELETAL MUSCLES produce MOVEMENT by exerting PULLING FORCE on TENDONS, which PULL on BONES (mostly TUBULAR BONES).

• TENDONS are white, fibrous cords of dense, regular connective tissue that attach MUSCLE to BONE and are continuous with the outer PERIOSTEUM layer of BONE at the point of attachment (see 4.1)

• The TENDONS pull on the BONES and cross at least one joint, attaching to the ARTICULATING BONES forming the JOINT.

• The muscle that contracts to move the joint is called the PRIME MOVER or AGONIST.

• When a MUSCLE CONTRACTS, one ARTICULATING BONE is drawn toward the other.

• ONE BONE is ordinarily held stationary (other muscles contract in the opposite direction to equalize the exertive forces— these opposing muscles are called ANTAGONISTS. They gradually relax, causing a "braking" control on the movement.)

★ Attachment of MUSCLE TENDON to a STATIONARY BONE = ORIGIN (usually PROXIMAL)

• Attachment of MUSCLE TENDON of the MUSCLE to a MOVABLE BONE = INSERTION (usually DISTAL)

• Muscles that move a body part DO NOT (in general) cover that part.

MECHANICS OF MOVEMENT

★ • BONES act as LEVERS
• JOINTS function as FULCRUMS
• BONE (LEVER) is acted on by 2 FORCES:
① RESISTANCE = Force to be overcome
② EFFORT = Force exerted to overcome RESISTANCE

• LEVERAGE is responsible for a muscle's STRENGTH and RANGE of MOVEMENT.
• STRENGTH increases with distance from a joint
• RANGE OF MOVEMENT decreases with distance from a joint

Structure of SKELETAL MUSCLE
ATTACHMENTS : ORIGIN AND INSERTION

1 | ORIGIN | STATIONARY ATTACHMENT

2 | INSERTION | MOVABLE ATTACHMENT

3 | TENDONS

4 | BELLY (GASTER) | MUSCLE BULK

5 | APONEUROSIS | SHEETLIKE LAYER OF CONNECTIVE TISSUE JOINING A MUSCLE TO THE PART THAT IT MOVES

MUSCLE Architecture

FIBER ARRANGEMENTS: 4 MUSCLE TYPES
PARALLEL, CONVERGENT, PENNATE, CIRCULAR

6 | PARALLEL

7 | CONVERGENT

PENNATE (PINNATE)
8 | UNIPENNATE

9 | BIPENNATE

10 | MULTIPENNATE

11 | CIRCULAR

12 | BICIPITAL

13 | TRIANGULAR

MECHANICS OF MOVEMENT

▭ LEVER (BONE)

△ FULCRUM (JOINT)

⌓ RESISTANCE (WEIGHT) | BODY PART TO BE MOVED

⇧ EFFORT (MUSCLE) | MUSCLE CONTRACTION APPLIED TO BONE AT INSERTION

CLASS OF LEVER		
1st	2nd	3rd
ELEMENT IN BETWEEN	△	⌓	⇧
	NOT MANY ex: (Head resting on the vertebral column)	VERY FEW (ex: Raising body on the toes)	MOST COMMON (ex: Flexing forearm at the elbow)

STRUCTURE OF A SKELETAL MUSCLE

Periosteum of Bone (continuous with Tendons) (See 4.1, 4.2)

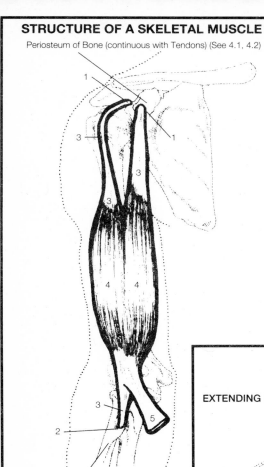

Periosteum of Bone
(continuous with Tendons)

"THE LIMB SYSTEM"

• Skeletal muscles would be powerless to contract (become hard) and relax (become soft) if they could not make full use of the resistance supplied by rigid bone.

• Because these muscles are attached to bones by tendons, they have the firm support necessary to release their energy outward.

• Bones, in turn, become mobile through the joints and ligaments.

• What we may call the "LIMB SYSTEM" comprises all the tubular bones, as these are connected by joints, and moved with the help of muscles.

MUSCLE ARCHITECTURE

6
• Long excursions (contract over a long distance)
• Enduring but weak
• Sometimes with tendinous inscriptions

7
• Fibers converge at the insertion point
• Maximization of contraction

PENNATES (PINNATES)

8   9   10
• Many fibers per unit; strong and short
• Short span of workload
• Provide dexterity

11
• Surround orifices
• Usually act as sphincters

12   13
Types of CONVERGENT MUSCLES according to shape.

EXAMPLES OF THE THREE CLASSES OF LEVERS IN THE BODY

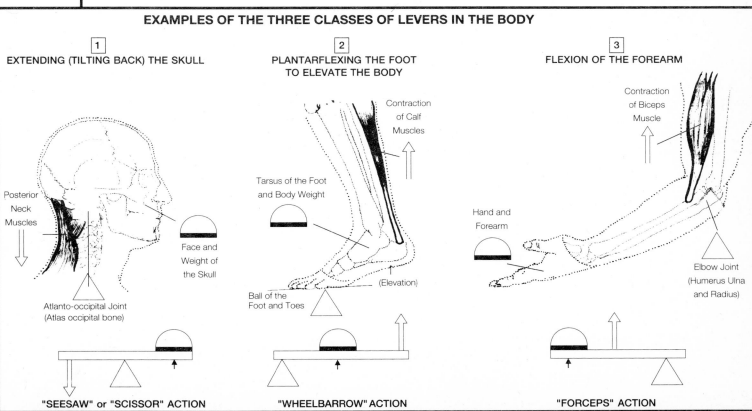

1
EXTENDING (TILTING BACK) THE SKULL

Posterior Neck Muscles

Face and Weight of the Skull

Atlanto-occipital Joint (Atlas occipital bone)

"SEESAW" or "SCISSOR" ACTION

2
PLANTARFLEXING THE FOOT TO ELEVATE THE BODY

Contraction of Calf Muscles

Tarsus of the Foot and Body Weight

Ball of the Foot and Toes

(Elevation)

"WHEELBARROW" ACTION

3
FLEXION OF THE FOREARM

Contraction of Biceps Muscle

Hand and Forearm

Elbow Joint (Humerus Ulna and Radius)

"FORCEPS" ACTION

6.1

MUSCULAR SYSTEM: GENERAL ORGANIZATION
Skeletal Muscle: Tissue Structure and Connective Tissue Sheaths

Use warm colors for 1, 2, and 3.
Use light colors for the connective
tissue sheath coverings.
★ See 7.2, 7.3

• The main, fleshy part of a SKELETAL MUSCLE = BELLY (GASTER). (The entire muscle is wrapped by a large, thick connective tissue sheath = EPIMYSIUM, an extension of the deep fascia.)
• Invaginations of the EPIMYSIUM that divide the muscle into BUNDLES or FASCICLES are called the PERIMYSIUM, which is also an extension of the deep fascia.
• Invaginations of the PERIMYSIUM divide the BUNDLES into each separate MUSCLE CELL or FIBER and are called the ENDOMYSIUM, also an extension of the deep fascia.

• Histologically, a large SKELETAL MUSCLE is composed of many ELONGATED, parallel CYLINDRICAL MULTINUCLEATED CELLS called MUSCLE FIBERS.
• Each muscle fiber (cell) contains small, numerous parallel longitudinally arranged bundles called MYOFIBRILS.
• In turn, each MYOFIBRIL is composed of MYOFILAMENTS (which contain the contractile proteins ACTIN and MYOSIN).
• Each MUSCLE FIBER (CELL) is surrounded by a PLASMA MEMBRANE called a SARCOLEMMA.
• The CYTOPLASMIC interior of the cell is called SARCO-PLASM.
• SARCOPLASMIC RETICULUM is a network of membrane-enclosed tubules (channels) that forms a sleeve around each MYOFIBRIL. The reticulum tubules are essential to the metabolic functions of the cell and are involved with protein synthesis.
• Perpendicular to the SARCOPLASMIC RETICULUM running transversely through the muscle fiber are T-TUBULES, which are internal extensions of the sarcolemma. They penetrate deep into the cell's interior, and also open to the outside of the cell (carrying extracellular fluid in their lumina).
• T-tubules pass through adjacent expanded chambers of sarcoplasmic reticulum, called TERMINAL CISTERNAE. A T-tubule and its two adjacent cisternae comprise a MUSCLE TRIAD. The T-TUBULES (and the RETICULUM) are involved in transmission of the nerve impulse to the muscle fiber.
■ SKELETAL MUSCLE FIBERS only contract when stimulated by SOMATIC MOTOR NEURONS. (See 7.2, 7.3)
• A motor unit consists of a single MOTOR NEURON and the AGGREGATION of MUSCLE FIBERS innervated by the MOTOR NEURON. As the MUSCLE is penetrated, it splays into a number of branching neuron processes called AXONS. The terminal ends of the AXONS (TELODENDRIA) contact the SARCOLEMMA of the individual MUSCLE FIBERS by means of MOTOR END PLATES, forming the NEUROMUSCULAR JUNCTION.
• A nerve impulse causes the release of a NEUROTRANSMIT-TER CHEMICAL across the junction. As it contacts the sarcolemma, physiological activity within the fiber causes contraction.

■ SKELETAL MUSCLE contracts according to an ALL-OR-NONE PRINCIPLE
• When a nerve impulse travels through a MOTOR UNIT, *all* the muscle fibers served by it simultaneously contract to their maxium extent, but only when exposed to a stimulus that reaches a specific threshold strength. (Otherwise, there is no contraction at all.)

Structure of Skeletal Tissue

1 | BELLY (GASTER) |

2 | FASCICLE (BUNDLE) | MUSCLE FASCICULUS

3 | SKELETAL MUSCLE FIBER (CELL) |

COMPONENTS OF MUSCLE FIBER (CELL)

4 | SARCOLEMMA | PLASMA MEMBRANE "MUSCLE HUSK"

5 | SARCOPLASM CYTOPLASM |
a | SARCOPLASMIC RETICULUM |
b | TRANSVERSE (T-) TUBULES |

c | MYOFIBRILS | Composed of MYOFILAMENTS
d | MITOCHONDRIA |

6 | MANY PERIPHERAL NUCLEI |

MYOFIBRIL showing compartmentalization of SARCOMERES

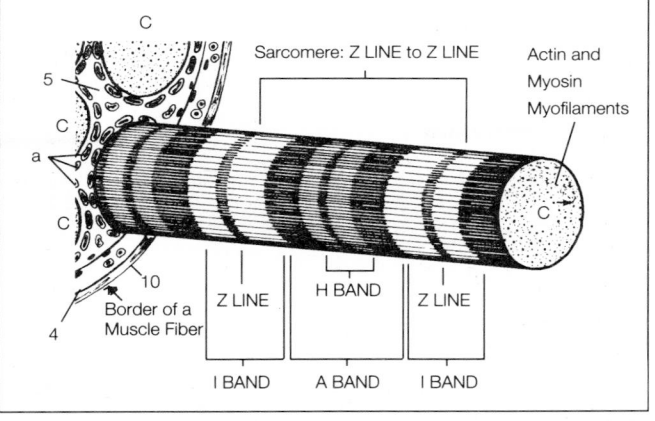

Sarcomere: Z LINE to Z LINE

Actin and Myosin Myofilaments

Z LINE H BAND Z LINE

I BAND A BAND I BAND

Border of a Muscle Fiber

CONTINUOUS CONNECTIVE TISSUE SHEATH COVERINGS OF ASSOCIATED MUSCLE ELEMENTS
• Periosteum covering of attached bones
• Tendons of muscle

7 | EPIMYSIUM | PERIMYSIUM EXTERNUM (INVESTS THE ENTIRE BELLY, OR GASTER)

8 | PERIMYSIUM | PERIMYSIUM INTERNUM (INVESTS EACH BUNDLE, OR FASCICLE)

9 | ENDOMYSIUM | INVESTS EACH MUSCLE CELL (or MUSCLE FIBER)

COMPONENTS OF A MYOFIBRIL
PROTEIN STRANDS
| MYOFILAMENTS |

A | ACTIN | (F-ACTIN) Thin FILAMENTS (50–60 Å)

M | MYOSIN | Thick FILAMENTS (110 Å)

C-B | CROSS-BRIDGES (MYOSIN "HEADS") | LINK WITH THE BINDING SITES ON THE ACTIN DURING CONTRACTION

★ • MYOFILAMENTS (ACTIN and MYOSIN) do not extend the entire length of a MUSCLE FIBER. They are stacked in compartments called SARCOMERES separated by thin, narrow zones of dark material called Z LINES (which are located under the T-TUBULES). (This accounts for the striations of skeletal muscle; nonstriated smooth muscle does not contain sarcomeres.)

• CONTRACTION of muscle tissue results from a sliding movement of ACTIN FILAMENTS in the MYOFIBRILS causing a reduction in the length of the SARCOMERES.
• The alternating dark and light cross-banding striations of skeletal (and cardiac) muscle are due to regular spatial organization of these ACTIN-MYOSIN FILAMENT arrangements:

| I BAND | → Only ACTIN FILAMENTS (Light)

| A BAND | → MYOSIN FILAMENTS (Dark) and overlapping ACTIN FILAMENTS (Light)

CENTRAL REGION OF A BAND
| H BAND | → The portion of MYOSIN FILAMENTS *not* overlapped by ACTIN FILAMENTS

| Z LINES | are in the middle of the I BANDS and constitute the boundary lines of the SARCOMERE UNITS.

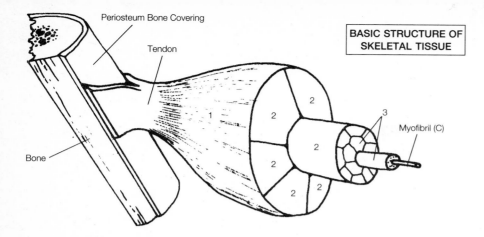

BASIC STRUCTURE OF SKELETAL TISSUE

Periosteum Bone Covering

Tendon

Bone

1

2 2 2 2 2 2 2

3

Myofibril (C)

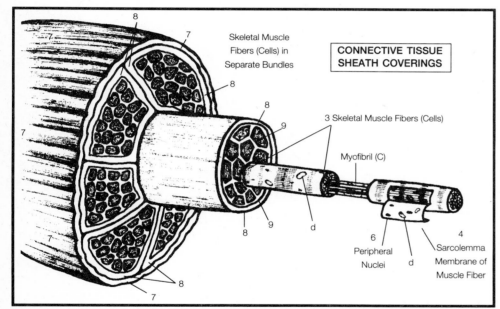

CONNECTIVE TISSUE SHEATH COVERINGS

Skeletal Muscle Fibers (Cells) in Separate Bundles

3 Skeletal Muscle Fibers (Cells)

Myofibril (C)

7 8 8 8 9 9 d 8 7

6
Peripheral Nuclei

d

4
Sarcolemma Membrane of Muscle Fiber

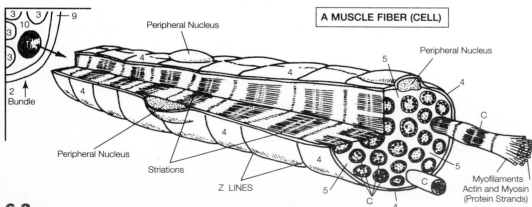

A MUSCLE FIBER (CELL)

Peripheral Nucleus

Peripheral Nucleus

Peripheral Nucleus

Striations

Z LINES

Myofilaments Actin and Myosin (Protein Strands)

3 3 9
3 10
3
2
Bundle

4 4 4 4 4 4 5 5 5 C C 4

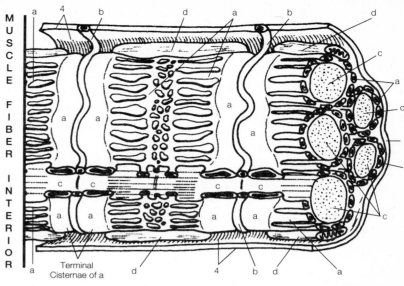

MUSCLE FIBER INTERIOR

a 4 b d a b d
c
a
c
a
a a a
a
4
c
a
a
c c c c
a a
a
4 b d
Terminal Cisternae of a
d 4 b d a

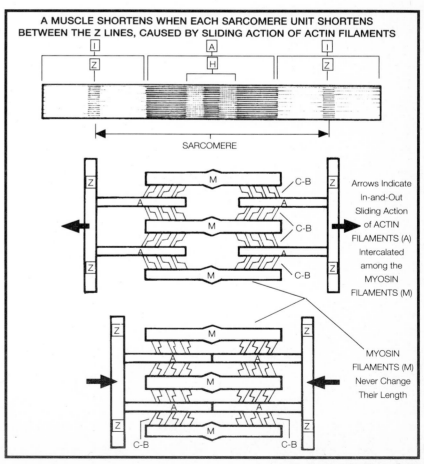

A MUSCLE SHORTENS WHEN EACH SARCOMERE UNIT SHORTENS BETWEEN THE Z LINES, CAUSED BY SLIDING ACTION OF ACTIN FILAMENTS

I A I
Z H Z

SARCOMERE

Z M C-B Z
A A
Z M C-B Z
A C-B
Z M
A
Z M C-B

Arrows Indicate In-and-Out Sliding Action of ACTIN FILAMENTS (A) Intercalated among the MYOSIN FILAMENTS (M)

MYOSIN FILAMENTS (M) Never Change Their Length

Z M Z
A A
Z M Z
A A
Z M Z
C-B C-B

MUSCULAR SYSTEM: GENERAL ORGANIZATION
Superficial Survey: Muscular Groups
According to Function

*To represent antagonistic (opposing) muscle action,
use complementary (opposing) color schemes:
(red-green, orange-blue, yellow-purple, etc.).
Do not color sheaths, bands, bones, or
aponeuroses — leave them white.*

★ See 5.4

■ There are over 650 skeletal muscles in
the body. This chapter includes the
PRIMARY SKELETAL MUSCLES.

■ Most of the muscles occur in PAIRS, in
mirror images, on either side of a midsagittal
symmetrical plane through the upright body.
(Right side is a mirror image of the left side)

■ There are only 5 single muscles!! (not paired)

■ Muscles generally contract as FUNCTIONAL
GROUPS rather than individually.

• SYNERGISTIC MUSCLES = Muscles that
function in cooperation and coordination
with one another to achieve a particular
movement(s).

• ANTAGONISTIC MUSCLES = Muscles that
function in opposition to one another to achieve
a particular movement(s). They are generally
located on opposite sides of a limb.

■ FLEXOR-EXTENSOR ACTION AND
ABDUCTOR-ADDUCTOR ACTION are
examples of ANTAGONISTIC MUSCLE
GROUPS.

■ Relate these muscle groups:
• FRONT OF BODY-BACK OF BODY
• UPPER LIMBS-LOWER LIMBS

BASIS OF NAMING SKELETAL MUSCLES
ACCORDING TO PRIMARY FUNCTIONS ‡

1 | FLEXORS | ⎱ Opposite action
2 | EXTENSORS | ⎰ (ANTAGONISTIC)

3 | ABDUCTORS | ⎱ Opposite action
4 | ADDUCTORS | ⎰ (ANTAGONISTIC)

5 | ROTATORS | LATERAL ROTATORS ⎱ Opposite action
             | MEDIAL ROTATORS ⎰ (ANTAGONISTIC)

6 | SCAPULA STABILIZERS

‡ Reference numbers in parentheses (1–6)
on the illustrations indicate an additional and/or
secondary function. Color them as dots of color
on top of the primary colors.

★ DEFINITIONS OF PRIMARY FUNCTIONS AND
ACTIONS IN NAMING SKELETAL MUSCLES

1 FLEXORS
• Bend a part in proximal direction
• Decrease anterior angle at a joint (generally)
• Bend a limb at a joint

2 EXTENSORS
• Decrease anterior angle at a joint (generally)
• Straighten a limb at a joint

3 ABDUCTORS
• Move a bone *away* from the midline

4 ADDUCTORS
• Move a bone *toward* or *closer* to the midline

5 ROTATORS
• Move a bone around its longitudinal axis

★ 8 LOGICAL DERIVATIONS FOR THE NAMING OF
SKELETAL MUSCLES (WITH EXAMPLES GIVEN)

■ **FUNCTION**
FLEXOR, EXTENSOR, ABDUCTOR, ADDUCTOR,
PRONATOR, LEVATOR

■ **SHAPE**
RHOMBOIDEUS, TRAPEZIUS

■ **SIZE**
MAXIMUS, MEDIUS, MINIMUS, LONGUS, BREVIS

■ **LOCATION**
PECTORALIS, INTERCOSTAL, BRACHII

■ **RELATIVE POSITION**
LATERAL, MEDIAL, ABDOMINAL, FEMORAL,
INTERNAL, EXTERNAL

■ **FIBER ORIENTATION**
RECTUS (STRAIGHT), OBLIQUE, TRANSVERSE

■ **ATTACHMENT-ORIGIN AND INSERTION**
ZYGOMATICUS, TEMPORALIS, NASALIS,
STERNOCLEIDOMASTOID, STYLOHYOID

■ **NUMBER OF ORIGINS**
TRICEPS, BICEPS

(NOT Shown) :

DEFINITIONS OF OTHER FUNCTIONS AND
ACTIONS IN NAMING SKELETAL MUSCLES
SPECIALIZED FLEXORS

■ **DORSIFLEXORS** ⎫
• Flex the Foot (upward)   ⎪ O
 at the Ankle Joint   ⎪ P
SPECIALIZED EXTENSORS   ⎪ P
■ **PLANTAR FLEXORS**   ⎬ O
• Extend the Foot (downward)   ⎪ S
 at the Ankle Joint   ⎪ I
                                  ⎪ T
                                  ⎪ E
                                  ⎭ S

SPECIALIZED ROTATORS

■ **SUPINATORS** ⎫
• Turn the palm upward or   ⎪ O
 anteriorly (forward)   ⎪ P
■ **PRONATORS**   ⎬ P
• Turn the palm downward or   ⎪ O
 posteriorly (backward)   ⎪ S
                                  ⎭ I
                                    T
                                    E
                                    S

■ **LEVATORS** ⎫
• ELEVATION   ⎪ O
• LIFTING   ⎪ P
• Produce an upward movement   ⎬ P
■ **DEPRESSORS**   ⎪ O
• Produce a downward movement   ⎭ S
                                    I
                                    T
                                    E
                                    S

■ **INVERTORS** ⎫
• Turn the sole of the foot inward ⎪ O
■ **EVERTORS**   ⎬ P
• Turn the sole of the foot outward⎪ P
                                   ⎭ O
                                     S
                                     I
                                     T
                                     E
                                     S

■ **SPHINCTERS**
• Decrease the size of an opening

■ **TENSORS**
• Make a body part more rigid to
 increase tension

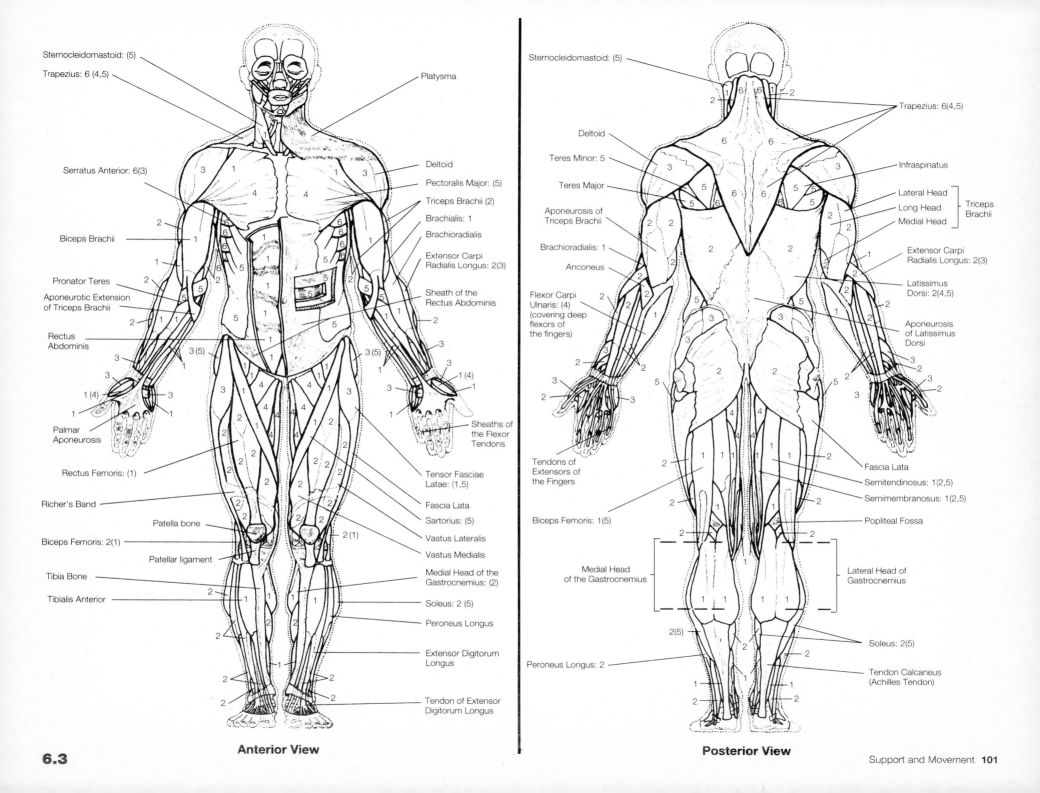

Anterior View

Sternocleidomastoid: (5)

Trapezius: 6 (4,5)

Serratus Anterior: 6(3)

Biceps Brachii

Pronator Teres

Aponeurotic Extension of Triceps Brachii

Rectus Abdominis

Palmar Aponeurosis

Rectus Femoris: (1)

Richer's Band

Biceps Femoris: 2(1)

Patella bone

Patellar ligament

Tibia Bone

Tibialis Anterior

Platysma

Deltoid

Pectoralis Major: (5)

Triceps Brachii (2)

Brachialis: 1

Brachioradialis

Extensor Carpi Radialis Longus: 2(3)

Sheath of the Rectus Abdominis

Sheaths of the Flexor Tendons

Tensor Fasciae Latae: (1,5)

Fascia Lata

Sartorius: (5)

Vastus Lateralis

Vastus Medialis

Medial Head of the Gastrocnemius: (2)

Soleus: 2 (5)

Peroneus Longus

Extensor Digitorum Longus

Tendon of Extensor Digitorum Longus

Posterior View

Sternocleidomastoid: (5)

Deltoid

Teres Minor: 5

Teres Major

Aponeurosis of Triceps Brachii

Brachioradialis: 1

Anconeus

Flexor Carpi Ulnaris: (4) (covering deep flexors of the fingers)

Tendons of Extensors of the Fingers

Biceps Femoris: 1(5)

Medial Head of the Gastrocnemius

Peroneus Longus: 2

Trapezius: 6(4,5)

Infraspinatus

Lateral Head
Long Head
Medial Head
} Triceps Brachii

Extensor Carpi Radialis Longus: 2(3)

Latissimus Dorsi: 2(4,5)

Aponeurosis of Latissimus Dorsi

Fascia Lata

Semitendinosus: 1(2,5)

Semimembranosus: 1(2,5)

Popliteal Fossa

Lateral Head of Gastrocnemius

Soleus: 2(5)

Tendon Calcaneus (Achilles Tendon)

MUSCULAR SYSTEM: MUSCLES OF THE AXIAL SKELETON
Muscles of the Head: Facial Expression and Mastication

Use warm colors for MUSCLES OF MASTICATION, and superficial muscles.
Use cool colors for deeper muscles.
★ See 7.19

★ ■ **MUSCLES OF FACIAL EXPRESSION**
• Originate on the different facial bones

• Insert into the dermis (second major layer of the skin)

• When these muscles contract, surface features are affected, and various emotions are expressed.

• An incredible array of complex facial expressions are very important in social communication, especially speechless communication.

★ Muscles of the EYES (ORBITS) are discussed in 7.19, but their origin, insertion, action, and innervation will be found in the charts at the end of this chapter.

† ORBICULARIS OCULI (5) is a sphincter muscle that encircles the eye. It permits BLINKING, WINKING, and SQUINTING. Also compresses the LACRIMAL GLAND which keeps the eyeball continuously moistened. (See 7.19).

MUSCLES of FACIAL EXPRESSION:

THE SCALP MUSCLES
EPICRANIUS (OCCIPITOFRONTALIS)
1 FRONTALIS FRONTALIS BELLIES
2 OCCIPITALIS

THE FACIAL MUSCLES
THE EYELIDS
4 CORRUGATOR CORRUGATOR SUPERCILII
5 ORBICULARIS OCULI † ORBICULARIS PALPEBRARUM (PARS PALPEBRALIS)

THE NOSE
6 PROCERUS PYRAMIDALIS NASI

NASALIS
7 TRANSVERSE NASALIS
8 ALAR NASALIS
9 DEPRESSOR SEPTI DEPRESSOR ALAE NASI

THE MOUTH
10 LEVATOR LABII SUPERIORIS •
11 LEVATOR LABII SUPERIORIS ALAEQUE NASI
12 LEVATOR ANGULI ORIS CANINUS
13 ZYGOMATICUS MAJOR •
14 ZYGOMATICUS MINOR •
15 RISORIUS • "LAUGHING MUSCLE": PARTLY A CONTINUATION OF THE PLATYSMA (See 6.6)
16 DEPRESSOR LABII INFERIORIS QUADRATUS LABII INFERIORIS (QUADRATUS MENTI)
17 DEPRESSOR ANGULI ORIS TRIANGULARIS
18 MENTALIS LEVATOR LABII INFERIORIS(LEVATOR MENTI)
19 TRANSVERSUS MENTI
20 ORBICULARIS ORIS "KISSING MUSCLE"
21 BUCCINATOR

THE EARS: EXTRINSIC EAR MUSCLES (See charts for Intrinsic Ear Muscles)
26 AURICULARIS ANTERIOR ATTRAHENS AUREM
27 AURICULARIS POSTERIOR RETRAHENS AUREM
28 AURICULARIS SUPERIOR ATTOLENS AUREM

DEEP MUSCLES OF MASTICATION

Temporalis Muscle (reflected)

Zygomatic Process of Temporal Bone (cut)

Mandibular Ramus (cut)

Zygomatic Bone (cut)

DEEP MUSCLES OF MASTICATION
(Coronal section of skull, immediately behind the jaw, Posterior View)

Lateral Pterygoid Plate of Sphenoid Bone

Medial Pterygoid Plate of Sphenoid Bone

Body of Mandible

MUSCLES OF MASTICATION
(4 PAIRS)
22 TEMPORALIS TEMPORAL

23 MASSETER

24 MEDIAL PTERYGOID (PTERYGOIDEUS MEDIALIS)

25 LATERAL PTERYGOID (PTERYGOIDEUS LATERALIS)

• 10, 13, 14, and 15 insert at different locations on ORBICULARIS ORIS (20) and cause the expressions of smiling and laughing.

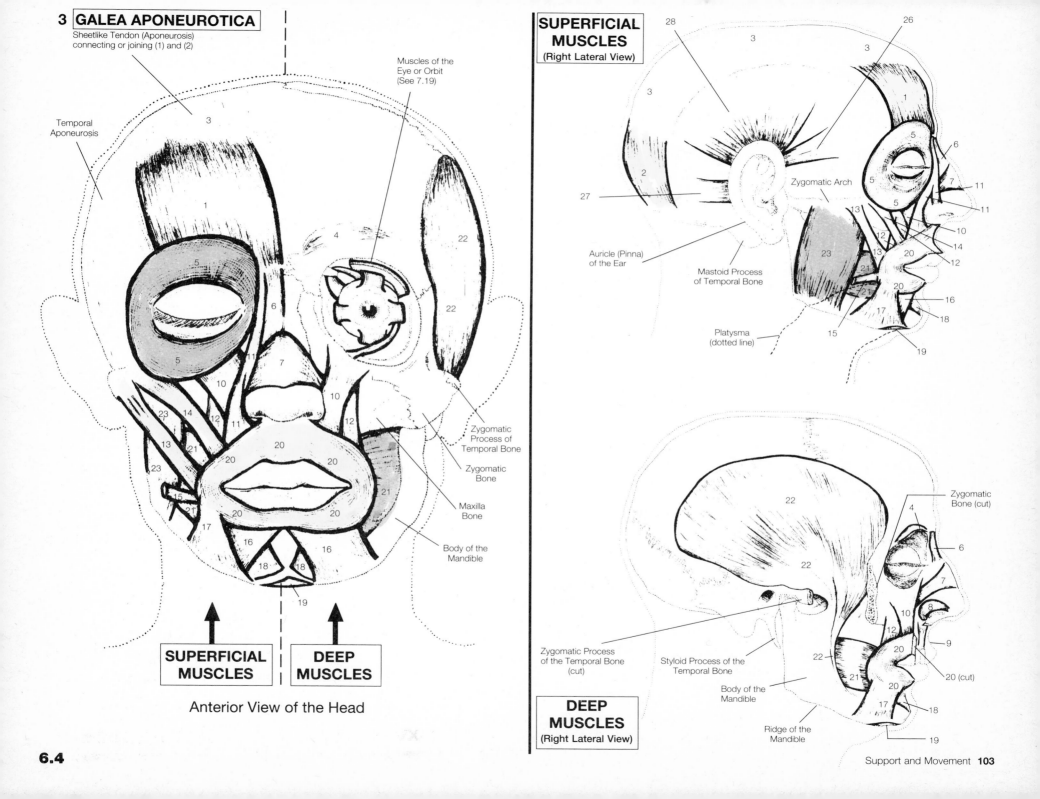

3 GALEA APONEUROTICA

Sheetlike Tendon (Aponeurosis) connecting or joining (1) and (2)

Temporal Aponeurosis

Muscles of the Eye or Orbit (See 7.19)

3

1

4

5

5

6

7

10

11

12

11

14

23

13

21

23

15

21

17

20

20

20

20

20

16

18

18

19

22

22

Zygomatic Process of Temporal Bone

Zygomatic Bone

Maxilla Bone

Body of the Mandible

SUPERFICIAL MUSCLES

DEEP MUSCLES

Anterior View of the Head

SUPERFICIAL MUSCLES
(Right Lateral View)

28

26

3

3

3

1

2

5

6

7

11

5

5

11

27

13

10

14

12

13

20

12

23

21

16

21

20

18

17

19

15

Zygomatic Arch

Auricle (Pinna) of the Ear

Mastoid Process of Temporal Bone

Platysma (dotted line)

DEEP MUSCLES
(Right Lateral View)

22

22

22

22

4

6

7

8

10

12

9

20

20 (cut)

21

20

17

18

19

Zygomatic Bone (cut)

Zygomatic Process of the Temporal Bone (cut)

Styloid Process of the Temporal Bone

Body of the Mandible

Ridge of the Mandible

Pharynx = warm colors
Larynx = cool colors
★ See 11.2, 11.4, 12.5, 12.7, 12.8

MUSCULAR SYSTEM: MUSCLES OF THE AXIAL SKELETON
Muscles of the Pharyngeal and Laryngeal Regions

Muscles of the Pharyngeal and Laryngeal regions include those of the PHARYNX, LARYNX, TONGUE, and SOFT PALATE.

The PHARYNX and LARYNX are located in the anterior region of the neck, or cervix:
★ ■ PHARYNX
    • Supporting walls are composed of skeletal muscles, the CONSTRICTORS.
    • STYLOPHARYNGEUS connects the styloid process of the Temporal bone to the PHARYNX. This muscle elevates and dilates the PHARYNX.
    • SALPINGOPHARYNGEUS raises the upper third division of the PHARYNX, the NASOPHARYNX. (See 11.2, 12.8)

★ ■ LARYNX
    • Intrinsic muscles of the LARYNX are shown here
    • Extrinsic muscles associated with the LARYNX include the INFRAHYOID MUSCLES, located below the HYOID BONE. (See 6.6)

★ ■ TONGUE (See 12.7)
    • A thick, highly specialized muscular organ covered with a mucous membrane.
    INTRINSIC TONGUE MUSCLES
    • Muscles that are within the tongue and comprise the tongue. They change the shape and size of the tongue for swallowing, and cause motility of the tongue.

    EXTRINSIC TONGUE MUSCLES (12-14)
    • Muscles that *insert on* the tongue, but have their origins on structures away from the tongue.
    • Cause gross tongue movement (side-to-side, in-and-out)
    • Maneuver food for chewing
    • Shape food into a rounded bolus
    • Forces food to the back of the mouth

12 | **GENIOGLOSSUS** | Depresses tongue, thrusts it forward

13 | **HYOGLOSSUS** | Depresses sides of tongue

14 | **STYLOGLOSSUS** | Antagonist to #12 Retracts and elevates tongue

15 | **STYLOHYOID**

Does not attach directly to the tongue. Has an indirect effect on the tongue as it elevates the HYOID BONE, where it inserts. (See 6.6)

(P) | **PHARYNX** THROAT, OR GULLET (See 11.2, 12.8)
• Common passageway for both the Respiratory (air) and Digestive (food) systems. Passageway for air from the Nasal (and oral) cavity to the LARYNX, and for food from the oral cavity to the ESOPHAGUS.

ASSOCIATED MUSCLES
(1) | **STYLOPHARYNGEUS**

(2) | **SALPINGOPHARYNGEUS**

SKELETAL SUPPORTING WALL
(3) | **INFERIOR CONSTRICTOR** | LARYNGOPHARYNGEUS (CONSTRICTOR PHARYNGIS INFERIOR)

(4) | **MEDIAL CONSTRICTOR** | HYPOPHARYNGEUS (CONSTRICTOR PHARYNGIS MEDIUS)

(5) | **SUPERIOR CONSTRICTOR** | CEPHALOPHARYNGEUS (CONSTRICTOR PHARYNGIS SUPERIOR)

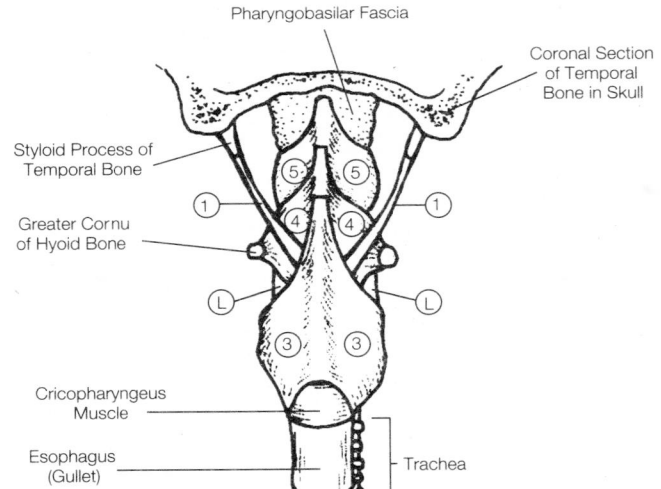

(Posterior View)
of PHARYNX

(L) | **LARYNX** VOICE BOX (See 11.4)
Musculocartilaginous structure comprising the enlarged upper end of the TRACHEA (WINDPIPE), just below the root of the TONGUE
The organ of voice (See Chapter 11, Respiratory System)

INTRINSIC MUSCLES
6 | **ARYEPIGLOTTIC** | ARYEPIGLOTTICUS

7 | **THYROEPIGLOTTIC** | THYREOEPIGLOTTICUS (THYROEPIGLOTTICUS)

8 | **ARYTENOID** | ARYTENOIDEUS

9 | **THYROARYTENOID** | THYREOARYTENOIDEUS

10 | **CRICOARYTENOID** | CRICOARYTENOIDEUS

11 | **CRICOTHYROID** | CRICOTHYROIDEUS

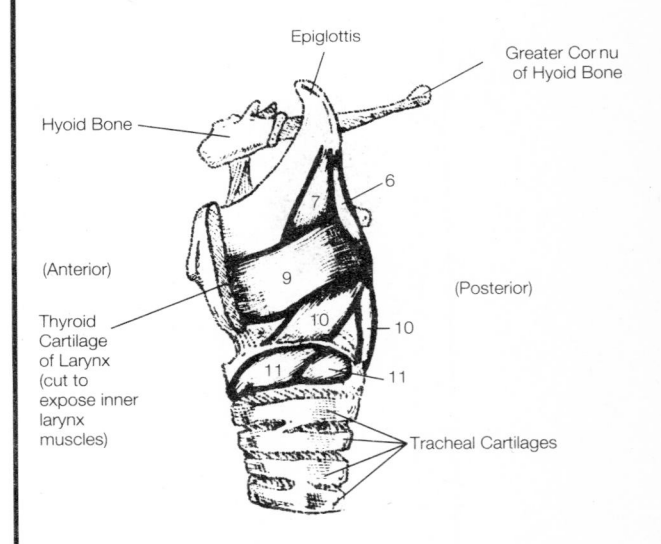

(Left Lateral View)
of LARYNX

MUSCLES ASSOCIATED WITH THE TONGUE

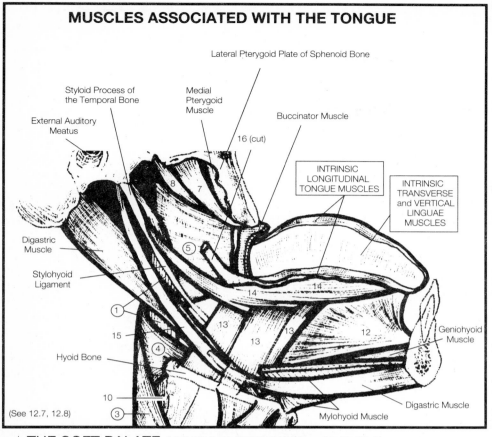

Lateral Pterygoid Plate of Sphenoid Bone

Styloid Process of the Temporal Bone

Medial Pterygoid Muscle

Buccinator Muscle

External Auditory Meatus

16 (cut)

INTRINSIC LONGITUDINAL TONGUE MUSCLES

INTRINSIC TRANSVERSE and VERTICAL LINGUAE MUSCLES

Digastric Muscle

Stylohyoid Ligament

Geniohyoid Muscle

Hyoid Bone

Digastric Muscle

Mylohyoid Muscle

(See 12.7, 12.8)

★ THE SOFT PALATE (VELUM PLATINUM) (See 12.5, 12.6)

• The fleshy, posterior portion of the roof of the mouth (the anterior portion of the roof of the mouth is the bony HARD PALATE (See 4.7 bottom)

• Extends posteriorly from the PALATINE BONES and ends at the V-shaped, soft UVULA.

• Partially separates the MOUTH and the PHARYNX

• A muscular arch lined with a mucous membrane that acts as a partial partition separating the MOUTH and the PHARYNX. (Both PALATOPHARYNGEUS MUSCLES form this muscular PALATOPHARYNGEUS ARCH in the rear of the oral cavity.)

• PALATOGLOSSUS helps to elevate the tongue

16	PALATOGLOSSUS	GLOSSOPALATINUS (TO TONGUE)
17	PALATOPHARYNGEUS	PHARYNGOPALATINUS
18	TENSOR VELI PALATINI	TENSOR PALATI
19	LEVATOR VELI PALATINI	LEVATOR PALATI
20	MUSCULUS UVULAE	UVULA

6.5

MUSCLES OF THE PHARYNX, LARYNX, AND SOFT PALATE
(Constrictor muscles dissected and pulled back to expose interior)

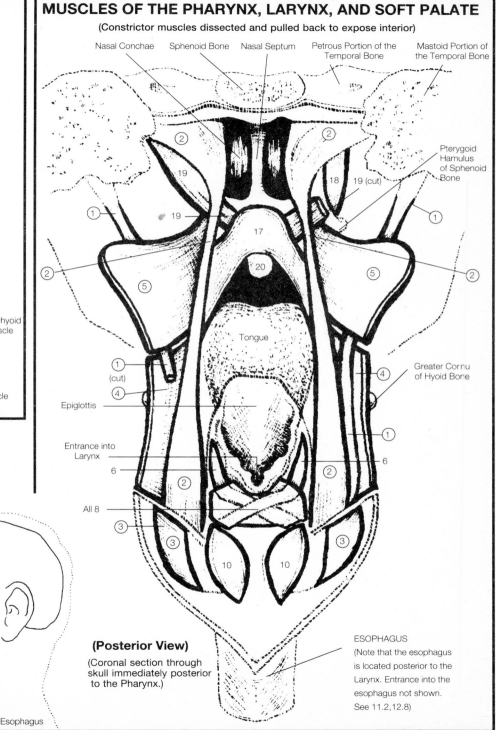

Nasal Conchae   Sphenoid Bone   Nasal Septum   Petrous Portion of the Temporal Bone   Mastoid Portion of the Temporal Bone

Pterygoid Hamulus of Sphenoid Bone

Tongue

Greater Cornu of Hyoid Bone

Epiglottis

Entrance into Larynx

All 8

(Posterior View)

(Coronal section through skull immediately posterior to the Pharynx.)

Hard Palate

Soft Palate

Nasal Cavity

Tongue

Oral Cavity

Esophagus

ESOPHAGUS
(Note that the esophagus is located posterior to the Larynx. Entrance into the esophagus not shown. See 11.2,12.8)

MUSCULAR SYSTEM: MUSCLES OF THE AXIAL SKELETON
Muscles of the Neck:
Cervicals, Hyoids, and Prevertebrals

★ ■ Muscles of the NECK either:
• Support and move the head
• Or are associated with structures within the neck, such as the LARYNX and PHARYNX (already discussed, see 6.5), and the HYOID BONE.

★ ■ NECK MUSCLES Associated with the HYOID BONE: HYOIDS
• The HYOID BONE is the only bone in the body that does not articulate with any other bone. (See 4.4)Seven paired muscles attach to it (#3, 6–11), and one unpaired muscle attaches to it (#4, the MYLOHYOID).

SUPRAHYOIDS: above the HYOID BONE
• DIGASTRIC 2-bellied muscle of double origin. Can open the mouth (jaw) or elevate hyoid bone
• MYLOHYOID Forms floor of the mouth, elevates floor to aid in swallowing; elevates hyoid bone
• GENIOHYOID Helps to depress the jaw; elevates and advances hyoid bone
• STYLOHYOID See 6.5
• HYOGLOSSUS See 6.5

INFRAHYOIDS: below the HYOID BONE
• STERNOTHYROID Pulls down larynx (depresses thyroid cartilage)
• STERNOHYOID Depresses hyoid
• THYROHYOID Elevates the larynx (thyroid cartilage); depresses hyoid
• OMOHYOID Depresses hyoid

★ ■ THE CERVICAL NECK MUSCLES
THE 2 ANTEROLATERAL CERVICALS:
■ STERNOCLEIDOMASTOIDS
■ Contraction of one muscle – –
• Rotates head to the opposite side (turns head sideways)
• Tilts chin upward to the opposite side
■ Contraction of both muscles
• Head is pulled forward and down (flexes the neck on the chest)
■ Obliquely transects the neck, dividing it into 2 triangles, an ANTERIOR TRIANGLE and a POSTERIOR TRIANGLE. ➡

THE 2 SUPERFICIAL CERVICALS
■ The 2 PLATYSMA muscles are broad, thin sheets of muscle, beginning on the ANTERIOR CHEST and sweeping up each side of the neck over the large STERNO-CLEIDOMASTOIDS. They terminate at the body of the mandible and the corners of the mouth, producing expressions of sadness, screaming, and horror when contracted.

CERVICALS
SUPERFICIAL CERVICAL
1 | **PLATYSMA** | PLATYSMA MYOIDES

ANTEROLATERAL CERVICALS
2 | **STERNOCLEIDOMASTOID** | STERNOCLEIDO–MASTOIDEUS

HYOIDS
SUPRAHYOIDS:
3 | **DIGASTRIC** | DIGASTRICUS (ANTERIOR BELLY) (POSTERIOR BELLY)

4 | **MYLOHYOID** | MYLOHYOIDEUS

5 | **GENIOHYOID** | GENIOHYOIDEUS (not shown, see 6.5)

6 | **STYLOHYOID** | STYLOHYOIDEUS

7 | **HYOGLOSSUS**

INFRAHYOIDS
8 | **STERNOTHYROID** | STERNOTHYREOIDEUS (not shown)

9 | **STERNOHYOID** | STERNOHYOIDEUS

10 | **THYROHYOID** | THYROHYOIDEUS

11 | **OMOHYOID** | OMOHYOIDEUS

PREVERTEBRALS: Muscles of the neck region located immediately in front of the vertebral column.

ANTERIOR PREVERTEBRALS: FLEX AND SUPPORT THE HEAD AND HEAD CERVICAL VERTEBRAE

RECTUS CAPITIS ANTERIOR → Turns and inclines the head
RECTUS CAPITIS LATERALIS → Supports head and inclines it laterally
LONGUS CAPITIS → Flexes head
LONGUS COLLI → Twists and bends neck forward

LATERAL PREVERTEBRALS: LATERALLY BEND (FLEX) THE NECK AND ELEVATE FIRST 2 RIBS

SCALENUS ANTERIOR ⎤
SCALENUS MEDIUS ⎦ Flexes neck, elevates 1st rib

SCALENUS POSTERIOR → Flexes neck, elevates 2nd rib

ANTERIOR PREVERTEBRALS
12 | **RECTUS CAPITIS ANTERIOR** | RECTUS CAPITIS ANTICUS MINOR

13 | **RECTUS CAPITIS LATERALIS**

14 | **LONGUS CAPITIS** | RECTUS CAPITIS ANTICUS MAJOR

15 | **LONGUS COLLI** | 3 PARTS: SUPERIOR OBLIQUE, INFERIOR OBLIQUE, VERTICAL

LATERAL PREVERTEBRALS
16 | **SCALENUS ANTERIOR** | SCALENUS ANTICUS

17 | **SCALENUS MEDIUS**

18 | **SCALENUS POSTERIOR** | SCALENUS POSTICUS

---

2 LARGE TRIANGULAR NECK DIVISIONS

(A) **ANTERIOR TRIANGLE**
• 4 lesser triangles • Salivary glands
• Larynx • Trachea • Thyroid glands
• Various nerves and vessels

(P) **POSTERIOR TRIANGLE**
• 2 lesser triangles
• Various nerves and vessels

6 LESSER TRIANGULAR NECK SUBDIVISIONS

ANTERIOR TRIANGLE

SUBMANDIBULAR TRIANGLE
• Salivary glands

SUPRAHYOID TRIANGLE
• Muscles of the floor of mouth
• Salivary glands and ducts

CAROTID TRIANGLE
• Carotid arteries
• Internal jugular vein
• Vagus nerve

MUSCULAR TRIANGLE
• Larynx
• Trachea
• Thyroid glands
• Carotid sheath

POSTERIOR TRIANGLE

OCCIPITAL TRIANGLE
• Cervical nerve plexus
• Accessory nerve

SUBCLAVIAN TRIANGLE
• Brachial nerve plexus
• Subclavian artery

PLATYSMA: OUTER MUSCULAR LAYER COVERING THE NECK

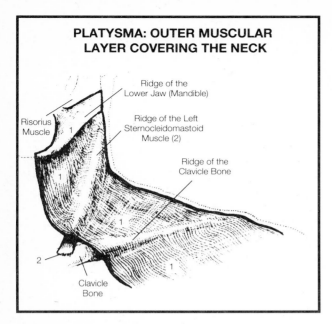

Risorius Muscle

Ridge of the Lower Jaw (Mandible)

Ridge of the Left Sternocleidomastoid Muscle (2)

Ridge of the Clavicle Bone

Clavicle Bone

STERNOCLEIDOMASTOID
- Forms anterior border of POSTERIOR TRIANGLE
- Forms lateral border of ANTERIOR TRIANGLE

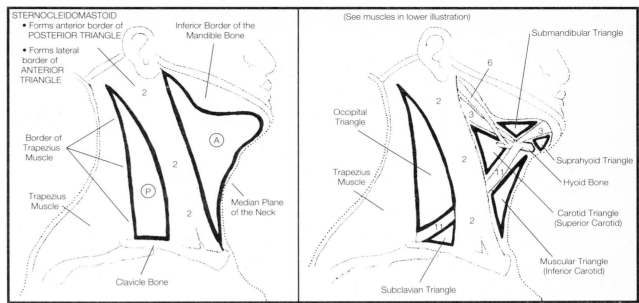

Inferior Border of the Mandible Bone

Border of Trapezius Muscle

Trapezius Muscle

Clavicle Bone

Median Plane of the Neck

(See muscles in lower illustration)

Submandibular Triangle

Occipital Triangle

Trapezius Muscle

Suprahyoid Triangle

Hyoid Bone

Carotid Triangle (Superior Carotid)

Muscular Triangle (Inferior Carotid)

Subclavian Triangle

MUSCLES OF THE ANTERIOR AND LATERAL NECK REGIONS

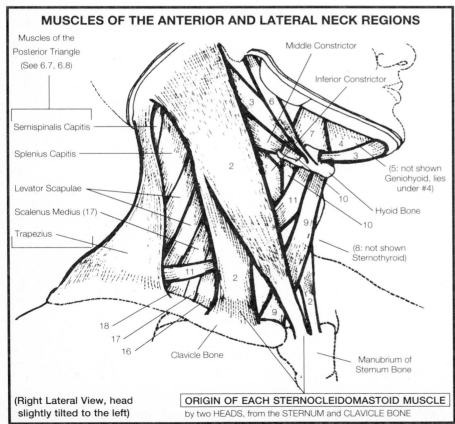

Muscles of the Posterior Triangle (See 6.7, 6.8)

Semispinalis Capitis

Splenius Capitis

Levator Scapulae

Scalenus Medius (17)

Trapezius

Middle Constrictor

Inferior Constrictor

(5: not shown Geniohyoid, lies under #4)

Hyoid Bone

(8: not shown Sternothyroid)

Clavicle Bone

Manubrium of Sternum Bone

(Right Lateral View, head slightly tilted to the left)

ORIGIN OF EACH STERNOCLEIDOMASTOID MUSCLE
by two HEADS, from the STERNUM and CLAVICLE BONE

ANTERIOR AND LATERAL PREVERTEBRAL NECK MUSCLES

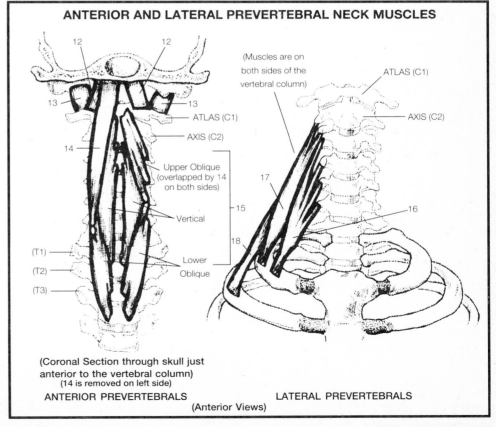

(Muscles are on both sides of the vertebral column)

ATLAS (C1)

AXIS (C2)

ATLAS (C1)

AXIS (C2)

Upper Oblique (overlapped by 14 on both sides)

Vertical

(T1)
(T2)
(T3)

Lower Oblique

(Coronal Section through skull just anterior to the vertebral column)
(14 is removed on left side)

ANTERIOR PREVERTEBRALS

LATERAL PREVERTEBRALS

(Anterior Views)

Erector Spinae = warm colors
Remaining Muscles = cool and neutral colors

MUSCULAR SYSTEM: MUSCLES OF THE AXIAL SKELETON
Deep Muscles of the Vertebral Column
Postvertebral Muscles of the Neck and Back

★ See 4.8, 4.8a

★ The EXTENSOR MUSCLES located on the posterior side of the vertebral column are much stronger and more complex than the muscles that FLEX the vertebral column anterior to it (. . . i.e. RECTUS ABDOMINUS—long, straplike muscles of the anterior abdominal wall).These long extensors have to be stronger than the flexors because EXTENSION (ex.: lifting a weight) moves in RESISTANCE or OPPOSITION to the FORCE OF GRAVITY. (See 6.1)

These deep muscles on the posterior side of the vertebral column that extend, rotate, and abduct the spine can be divided into 2 main groups:SUPERFICIAL GROUP and DEEP GROUP (divided into a Deeper and Deepest Stratum)

■ The SUPERFICIAL GROUP of the SPINE EXTENSORS is a massive group called the ERECTOR SPINAE or SACROSPINALS, extending vertically from the skull to the sacrum.

■ The ERECTOR SPINAE consists of 3 main groups, each of which consists of overlapping strips of muscle:

LONGISSIMUS —— INTERMEDIATE COLUMNS

SPINALIS —— MEDIAL COLUMNS (in contact with the spinal processes of the vertebrae)

ILIOCOSTALIS —— LATERAL COLUMNS

★ POSTERIOR COSTOVERTEBRAL MUSCLES OF THE THORAX
• These muscles overlap the ERECTOR SPINAE, originating on the vertebral column and inserting on the ribs. Their primary action is on the ribcage (Levatores costarum also flexes the vertebral column)

21 LEVATORES COSTARUM

22 SERRATUS POSTERIOR SUPERIOR

23 SERRATUS POSTERIOR INFERIOR

DEEP POSTVERTEBRAL MUSCLES OF THE VERTEBRAL COLUMN (for Action, see charts)

	SUPERFICIAL GROUP THE COLUMNAR MUSCLES: PRINCIPAL MOVERS of the HEAD AND BACK	**DEEPER STRATUM** THE TRANSVERSOSPINAL MUSCLES: ROTATORS of the VERTEBRAL COLUMN	**DEEPEST STRATUM** CONTIGUOUS VERTEBRAE MUSCLES: (SUBOCCIPTAL MUSCLES are *not* PRIME MOVERS, but POSTURAL MUSCLES)
DEEP NECK MUSCLES	"BANDAGE MUSCLES": 1 SPLENIUS CAPITIS 2 SPLENIUS CERVICIS SPLENIUS COLLI		SUBOCCIPTAL MUSCLES: 15 RECTUS CAPITIS POSTERIOR MAJOR R.C. POSTICUS MAJOR
	LONGISSIMUS (INTERMEDIATE COLUMN): 3, 4, 5 3 LONGISSIMUS CAPITIS TRACHELOMASTOID	SEMISPINALIS: 10, 11, 12 10 SEMISPINALIS CAPITIS COMPLEXUS	16 RECTUS CAPITIS POSTERIOR MINOR R.C. POSTICUS MINOR 17 OBLIQUUS CAPITIS INFERIOR 18 OBLIQUUS CAPITIS SUPERIOR
DEEP BACK MUSCLES	4 LONGISSIMUS THORACIS LONGISSIMUS DORSI 5 LONGISSIMUS CERVICIS TRANSVERSALIS COLLI SPINALIS (MEDIAL COLUMN)* 6 SPINALIS THORACIS SPINALIS DORSI ILIOCOSTALIS (LATERAL COLUMN) 7 ILIOCOSTALIS LUMBORUM SACRO-LUMBALIS 8 ILIOCOSTALIS THORACIS ILIOCOSTALIS DORSI (ACCESSORIUS) 9 ILIOCOSTALIS CERVICIS CERVICALIS ASCENDENS	11 SEMISPINALIS THORACIS SEMISPINALIS DORSI 12 SEMISPINALIS CERVICIS SEMISPINALIS COLLI 13 MULTIFIDUS MULTIFIDUS SPINAE 14 ROTATORES ROTATORS SPINAE	19 INTERSPINALES (A SERIES) 20 INTERTRANSVERSARII INTERTRANSVERSALES

(Left vertical label: ERECTOR SPINAE)

*SPINALIS CAPITIS and SPINALIS CERVICIS are not shown. They are inseparable from one another.

★ See also QUADRATUS LUMBORUM (6.9) extends LUMBAR REGION of the spine and causes spinal lateral flexion or abduction.

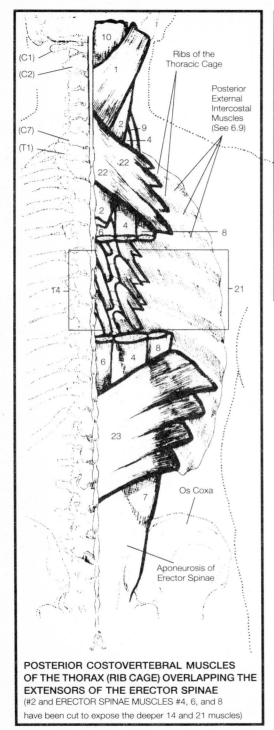

POSTERIOR COSTOVERTEBRAL MUSCLES OF THE THORAX (RIB CAGE) OVERLAPPING THE EXTENSORS OF THE ERECTOR SPINAE

(#2 and ERECTOR SPINAE MUSCLES #4, 6, and 8 have been cut to expose the deeper 14 and 21 muscles)

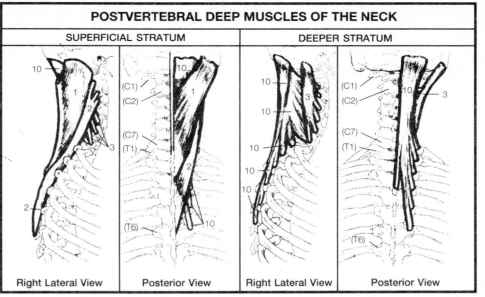

POSTVERTEBRAL DEEP MUSCLES OF THE NECK

SUPERFICIAL STRATUM	DEEPER STRATUM
Right Lateral View / Posterior View	Right Lateral View / Posterior View

DEEPER AND DEEPEST STRATUM OF DEEP BACK MUSCLES

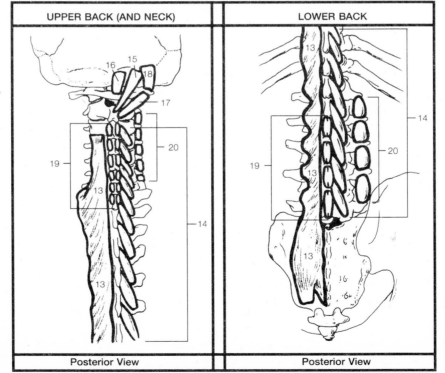

UPPER BACK (AND NECK)	LOWER BACK
Posterior View	Posterior View

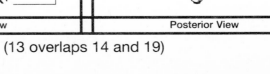

(13 overlaps 14 and 19)

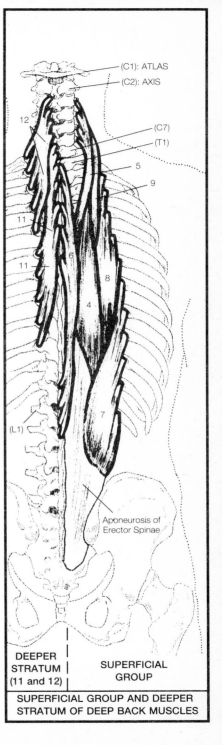

DEEPER STRATUM (11 and 12) | SUPERFICIAL GROUP

SUPERFICIAL GROUP AND DEEPER STRATUM OF DEEP BACK MUSCLES

6.7

MUSCULAR SYSTEM: MUSCLES OF THE AXIAL SKELETON
Trunk Connectors to the Shoulder

Posterior/Dorsal Muscles = warm colors
Ventral Muscles = cool colors
★ See 4.8, 6.12

★ There are 7 paired muscles and 2 superficial triangular back muscles that originate on the axial skeleton and connect to the shoulder. Most of these muscles act on the PECTORAL GIRDLE (the paired SCAPULAE and CLAVICLES), but some of them help to move the BRACHIUM (Upper Arm):

■ AXIAL MUSCLES THAT HELP MOVE THE BRACHIUM
The large superficial LATISSIMUS DORSI back muscle and the paired PECTORALIS MAJOR muscles have their origins on the axial skeleton, are termed axial muscles, but are *not* actually muscles of the axial skeleton.
  • They develop in the FORELIMB
  • They extend to the TRUNK secondarily

The LATISSIMUS DORSI and 8 paired muscles span the shoulder joint to insert on the HUMERUS BONE.

The LATISSIMUS DORSI and the PECTORALIS MAJORS are the only muscles of this group that help move the BRACHIUM that do *not* originate on the SCAPULA (See 6.12 for these 7 muscle pairs)

■ AXIAL MUSCLES THAT ACT ON THE PECTORAL GIRDLE
Most of the muscles that move the Brachium originate on the Scapula, and during brachial movement the scapula has to be held fixed.

★ THE AXIAL MUSCLES OF SCAPULAR STABILIZATION
  • The PECTORAL GIRDLE (both SCAPULAE and CLAVICLES) is an incomplete girdle because it only has an anterior attachment to the Axial Skeleton at the STERNOCLAVICULAR JOINT (Where the Clavicles join the Manubrium of the Sternum at the Clavicular notches)
  • Therefore, the PECTORAL GIRDLE (in particular, the SCAPULAE), requires support from strong straplike muscles
  • 5 paired muscles and one superficial triangular back muscle (TRAPEZIUS) span the shoulder joint from the axial skeleton, insert on the 2 scapulae, and hold them firm (for the muscles that originate on the Scapulae to help move the Brachium, See 6.12)

★ THE AXIAL MUSCLE THAT ACTS ON THE CLAVICLE
  • The paired SUBCLAVIUS muscles depress the Clavicles (thus, drawing the shoulder downward)

■ THE TRAPEZIUS MUSCLE
  • Large, superficial posterior triangular muscle of the back (and neck): A portion of the TRAPEZIUS muscle extends over the posterior neck region, but it is primarily a superficial muscle of the back.
  • Can adduct the scapula, elevate the scapula (ex: shrugging the shoulder), and hyperextend the head.

■ THE LATISSIMUS DORSI MUSCLE
  • Large, flat triangular posterior muscle, often referred to as "the swimmer's muscle," covers the lower half of the thoracic region of the back.
  • Powerful adductor of the arm, drawing it downward and backward while rotating the arm medially.
  • Also extends the arm and retracts the shoulder.

EXTENSORS of the SHOULDER JOINT

SUPERFICIAL MUSCLES of the BACK

POSTERIOR/DORSAL (VERTEBRAL COLUMN)
1 TRAPEZIUS
2 LATISSIMUS DORSI
3 RHOMBOIDEUS MAJOR
4 RHOMBOIDEUS MINOR
5 LEVATOR SCAPULAE   LEVATOR ANGULI SCAPULAE

FLEXORS of the SHOULDER JOINT

VENTRAL (ANTEROLATERAL THORACIC)
6 PECTORALIS MAJOR
7 PECTORALIS MINOR
8 SUBCLAVIUS
9 SERRATUS ANTERIOR   SERRATUS MAGNUS

(CORACOBRACHIALIS is also FLEXOR of the SHOULDER JOINT See 6.20 and 6.21)

AXIAL MUSCLES THAT ACT ON THE PECTORAL GIRDLE

PECTORAL GIRDLE BONES (the SCAPULAE and CLAVICLES) serve as attachment for origin muscles that insert on the BRACHIUM and help move the BRACHIUM and FOREARM. Muscles that move the BRACHIUM originate on the SCAPULA, except for LATISSIMUS DORSI and PECTORALIS MAJOR

Most muscles that act on and stabilize the PECTORAL GIRDLE itself originate on the AXIAL SKELETON and insert on the SCAPULAE. They are divided into ANTERIOR and POSTERIOR GROUPS: (all but one insert on the scapulae)

■ POSTERIOR GROUP
  • TRAPEZIUS
  • MAJOR and MINOR RHOMBOIDS
  • LEVATOR SCAPULAE

■ ANTERIOR GROUP
  • PECTORALIS MINOR
  • SERRATUS ANTERIOR

(The SUBCLAVIUS originates on the AXIAL SKELETON and inserts on the CLAVICLE)

★ See 6.12 for the muscles that both arise from and insert on the scapulae

	ORIGIN	INSERTION
TRAPEZIUS	Superior curved line of occipital bone; nuchal ligament; spinous processes of 7th cervical and all thoracic vertebrae	Clavicle; acromion of scapula and base of spine of scapula
LATISSIMUS DORSI	Broad lumbar aponeurosis that is attached to the iliac crest and to the spines of the sacral, lumbar, and lower thoracic vertebrae	Floor of the bicipital (intertubercular) groove of the humerus
PECTORALIS MAJOR	Clavicle; sternum; 6 upper costal cartilages; aponeurosis of external oblique muscle of abdomen	Crest of the great tubercle of the humerus

SUPERFICIAL MUSCLES OF THE BACK AND NECK

★ (1 overlaps 2, 3, 4, and 5)

Acromion Process
of the Scapula

(C1)

Scapula

Scapula

2

3

Humerus

9

2

2

★ (2 Overlaps parts
of 3 and 9, see the
dotted line, but
note that the lower
part of 1 overlaps
the upper part of 2)

2

Latissimus dorsi
curves around and
inserts on the Anterior
Aspect of the Humerus
(causes Medial, instead
of Lateral, Rotation)

Aponeurosis of
Latissimus Dorsi

Os Coxa

ARM MOVEMENTS OF
PECTORALIS MAJOR
• FLEXES
• ADDUCTS, and
• ROTATES
    MEDIALLY

Posterior View

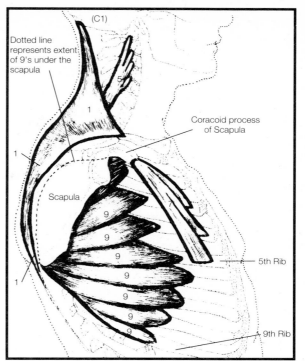

(C1)

Dotted line
represents extent
of 9's under the
scapula

Coracoid process
of Scapula

Scapula

9

9

9

9

9

9

5th Rib

9th Rib

Right Lateral View

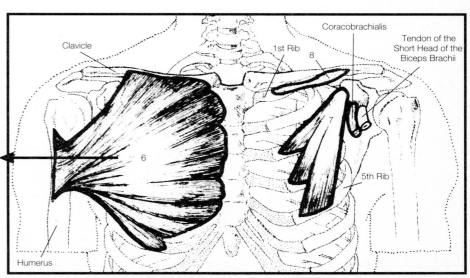

Coracobrachialis

Clavicle

1st Rib

8

Tendon of the
Short Head of the
Biceps Brachii

6

7

5th Rib

Humerus

Anterior View (6 overlaps 7 and 8 and covers 9)

6.8

MUSCULAR SYSTEM: MUSCLES OF THE AXIAL SKELETON
Muscles of the Trunk: The Thorax and the Abdominal Wall (Posterior)

1 = *warm color* 2 = *opposing cool color*
3 = *yellow* 4 = *fleshy* 5 = *light brown*

★ See 4.9, 4.12, 4.13, 11.1, 11.6

★ The THORAX (CHEST) is the closed cavity of the body between the base of the neck and the muscular DIAPHRAGM inferiorly, which contains the HEART, GREAT VESSELS, portions of the ESOPHAGUS and TRACHEA, and the LUNGS.

★ The muscles associated with the THORAX—mainly the DIAPHRAGM and the INTERNAL and EXTERNAL INTERCOSTALS—cause the CONTINUOUS RHYTHMICAL (and usually INVOLUNTARY) CONTRACTIONS associated with the changes in the CAPACITY (AIR VOLUME) of the THORAX during BREATHING.

There are 2 phases to the process of BREATHING:

INSPIRATION (INHALATION) and EXPIRATION (EXHALATION).

•During the ACTIVE PHASE of INSPIRATION, contraction of the DIAPHRAGM and EXTERNAL INTER-COSTAL MUSCLES is prominent.
•During the PASSIVE PHASE of FORCED EXPIRATION, contraction of the INTERNAL INTERCOSTALS and TRANSVERSIS THORACIS (and some ABDOMINAL MUSCLES) is prominent.

★ THE DIAPHRAGM is a dome-shaped musculomembranous wall of skeletal muscle separating the THORACIC CAVITY from the ABDOMINAL CAVITY, with its convexity upward.

•During the ACTIVE phase of INSPIRATION, the DIAPHRAGM contracts and flattens out downward, permitting the descent of the bases of the LUNGS and increase in the CAPACITY of the THORAX. This condition creates an increased "suction pull" on the LUNG TISSUE as the LUNGS EXPAND to fill the THORACIC CAVITY. Because the AIR PRESSURE within the LUNGS is less than the ATMOSPHERIC PRESSURE, AIR is sucked passively into the LUNGS (air pressure always tends to equalize).

•During the passive phase of FORCED EXPIRATION, the DIAPHRAGM relaxes, elevating it and restoring its inverted, domed-basin shape. The CAPACITY of the THORAX decreases, there is less "suction pull" on the LUNG TISSUE, and LUNGS RECOIL. Now the AIR PRESSURE within the LUNGS is greater than the ATMOSPHERIC PRESSURE, and thus AIR is pushed out of the LUNGS into the ATMOSPHERE.

★ Thus, MUSCULAR ACTION is the foundation of the breathing process, with the INTAKE or EXPULSION of AIR following passively.

The MUSCLES of RESPIRATION:
INTERCOSTALS (BETWEEN THE RIBS)

1 | EXTERNAL INTERCOSTALS | INTERCOSTALES EXTERNUS

2 | INTERNAL INTERCOSTALS | INTERCOSTALES INTERNUS

3 | SUBCOSTALS | SUBCOSTALES (INFRACOSTALES)

STERNOCOSTALS (INSIDE ANTERIOR)

4 | TRANSVERSUS THORACIS | TRIANGULARIS STERNI

FLOOR OF THE THORACIC CAVITY

5 | THE DIAPHRAGM

• (ORIGIN = Anteriorly: XIPHOID PROCESS of STERNUM, COSTAL CARTILAGES of LAST 6 RIBS (and INTERCOSTAL SPACES)
  Posteriorly: LUMBAR VERTEBRAE 7th–12th RIBS
• (INSERTION = CENTRAL TENDON)
• The RIGHT half of the DIAPHRAGM rises higher than the LEFT half

STRUCTURES THAT ENCLOSE AND BOUND THE THORAX	
ABOVE	• Upper Ribs • Neck Tissues
AT THE SIDES	• Ribs • Intercostal Muscles
AT THE BACK	• Ribs • Vertebral Column • Intercostal Muscles
IN FRONT	• Ribs • Costal Cartilages • Sternum • Intercostal Muscles
BELOW	• Diaphragm

EXTERNAL INTERCOSTALS
  Outer layer of muscles between the ribs
•ORIGIN = lower margin of each rib (excluding 12th)
•INSERTION = upper margin of next rib
•Fibers directed downward and forward

INTERNAL INTERCOSTALS
•Lie beneath externals
•Fibers directed downward and backward

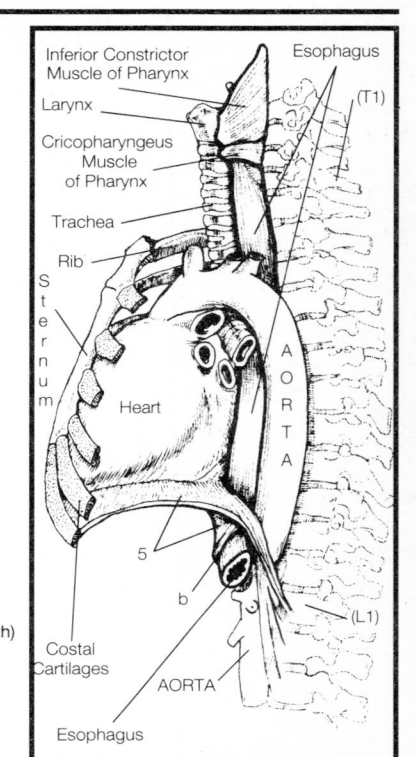

Inferior Constrictor Muscle of Pharynx
Esophagus
(T1)
Larynx
Cricopharyngeus Muscle of Pharynx
Trachea
Rib
Sternum
AORTA
Heart
5
b
(L1)
Costal Cartilages
AORTA
Esophagus

OPENINGS OF THE DIAPHRAGM
a AORTIC HIATUS
b ESOPHAGEAL HIATUS
c VENA CAVAL FORAMEN

TENDONS OF THE DIAPHRAGM
d CENTRAL TENDON
e RIGHT CRUS
f LEFT CRUS

LIGAMENTS OF THE DIAPHRAGM
g MEDIAL ARCUATE LIGAMENT
h LATERAL ARCUATE LIGAMENT
i MEDIAN ARCUATE LIGAMENT

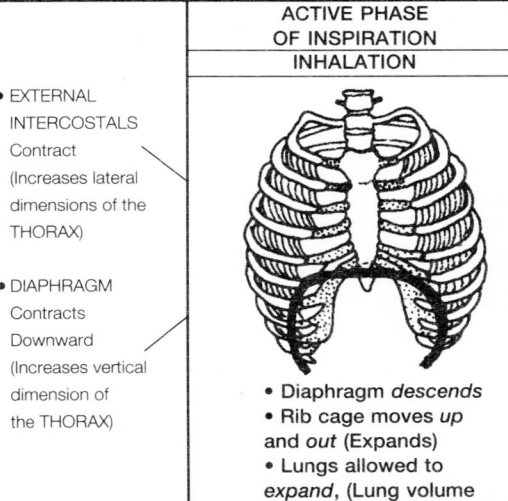

	ACTIVE PHASE OF INSPIRATION INHALATION	PASSIVE PHASE OF FORCED EXPIRATION EXHALATION	
• EXTERNAL INTERCOSTALS Contract (Increases lateral dimensions of the THORAX) • DIAPHRAGM Contracts Downward (Increases vertical dimension of the THORAX)	• Diaphragm *descends* • Rib cage moves *up* and *out* (Expands) • Lungs allowed to *expand*, (Lung volume increased) bringing air in	• Diaphragm *ascends* • Rib Cage moves *down* and *in* (Depressed) • Lungs contract, (Lung volume decreased), forcing air out	• EXTERNAL INTERCOSTALS and DIAPHRAGM Relax • INTERNAL INTERCOSTALS and TRANSVERSIS THORACIS Contract • ABDOMINAL MUSCLES Contract (Aids ascent of DIAPHRAGM)

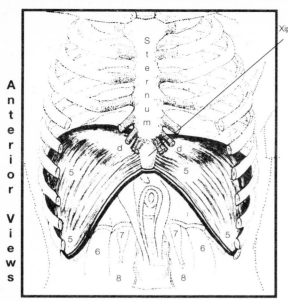

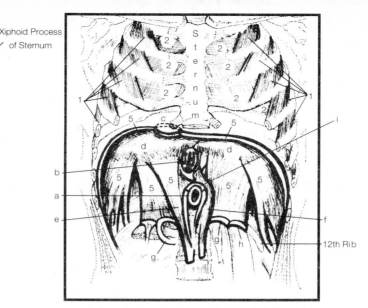

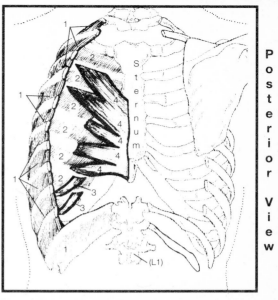

Anterior Views — Xiphoid Process of Sternum

Ribs cut (5–10) to expose DIAPHRAGM (5)

Coronal Section midway through the DIAPHRAGM (5) to expose the back wall and inside of DIAPHRAGM — 12th Rib

Posterior View — (L1)

Scapula removed; posterior portion of thoracic (rib) cage removed to expose muscles on the inner side of the anterior portion of rib cage.

MUSCLES OF THE POSTERIOR ABDOMINAL WALL

★ QUADRATUS LUMBORUM is also considered as a muscle of the VERTEBRAL COLUMN: (See 6.7)
 ■ An EXTENSOR of the LUMBAR REGION of the VERTEBRAL COLUMN when the Right and Left quadratus lumborum muscles CONTRACT TOGETHER.
 ■ LATERAL FLEXOR and ABDUCTOR of the VERTEBRAL COLUMN when these muscles contact SEPARATELY.

★ The PSOAS MAJOR (and MINOR) and the ILIACUS muscles of the POSTERIOR ABDOMINAL WALL are frequently referred to as a single muscle—the ILIOPSOAS. These muscles cooperate in a JOINT SYNERGISTIC ACTION that independent action could not achieve:
 ■ FLEXORS AND ROTATORS OF THE THIGH
 ■ FLEXORS OF THE VERTEBRAL COLUMN ILIOPSOAS are sometimes included as muscles of the HIP (ILIAC-PELVIC) REGION (See 6.17)

6 | **QUADRATUS LUMBORUM**

ILIOPSOAS (PREVERTEBRAL)
7 | **PSOAS MINOR** PSOAS PARVUS
8 | **PSOAS MAJOR** PSOAS MAGNUS
9 | **ILIACUS**

S Y N E R G I S M

	ORIGIN	INSERTION
QUADRATUS LUMBORUM	ILIAC CREST and Transverse processes of lower 3 LUMBAR VERTEBRAE	12th RIB (inferior margin) and Transverse processes of 1st 4 LUMBAR VERTEBRAE
PSOAS MAJOR	Transverse processes of all LUMBAR VERTEBRAE	Both muscles insert on LESSER TROCHANTER of the FEMUR
ILIACUS	ILIAC FOSSA	

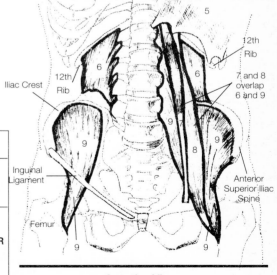

Anterior View

MUSCULAR SYSTEM: MUSCLES OF THE AXIAL SKELETON
Muscles of the Trunk: The Abdominal Wall (Anterolateral)

ANTEROLATERAL ABDOMINAL WALL
4 Pairs of Flat, Sheetlike MUSCLES: (5–8)

3-LAYERED LATERAL ABDOMINAL WALL

5 | EXTERNAL OBLIQUE | OBLIQUUS EXTERNUS ABDOMINIS — SUPERFICIAL

6 | INTERNAL OBLIQUE | OBLIQUUS INTERNUS ABDOMINIS — MIDDLE

7 | TRANSVERSUS ABDOMINIS |
TRANSVERSALIS ABDOMINIS

DEEP

1-LAYERED ANTERIOR ABDOMINAL WALL
8 | RECTUS ABDOMINIS |

9 | PYRAMIDALIS |

INGUINAL REGION (LOWER MEDIAL)

10 | CREMASTER | (OF THE INTERNAL OBLIQUE)

★ THE SHEATH COVERING THE RECTUS ABDOMINIS:

• Composed of APONEUROSES of the 3 Muscle Layers of the LATERAL ABDOMINAL WALL: EXTERNAL OBLIQUE, INTERNAL OBLIQUE, and TRANSVERSUS ABDOMINIS

• APONEUROSES converge on the ANTERIOR MIDLINE of the Abdomen to form the
11 | LINEA ALBA |"WHITE LINE"

• Also in the INGUINAL REGION forms the
12 | INGUINAL LIGAMENT |

■ EXTERNAL OBLIQUE
  • Contracts abdomen and viscera
■ INTERNAL OBLIQUE
  • Compresses viscera, flexes thorax forward
■ TRANSVERSUS ABDOMINIS
  • Compresses abdomen, flexes thorax
■ RECTUS ABDOMINIS
  • Compresses abdomen
  • Flexes lumbar region of vertebral column
■ PYRAMIDALIS
  • Tightens Linea Alba
■ CREMASTER
  • Raises testicle

★ 5 and 6 are right angles to one another

Aponeurosis of External Oblique (5) (overlaps 8 and 6)

★ Rectus Abdominis (8) Flexes the Vertebral Column (mainly lumbar region)

Umbilicus (Navel)

Aponeurosis of Internal Oblique (6) (cut to expose 8)

Inguinal Ligament

Inguinal Canal

Inguinal Falx: Aponeuroses of Internal Oblique (6) and Transversus Abdominis (7)

Superficial Inguinal Ring

Spermatic Cord

ANTERIOR ABDOMINAL WALL OF THE TRUNK
(Anterior View)

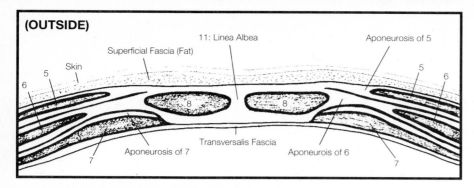

ANTERIOR ABDOMEN (Above Umbilicus)
(Transverse Section, Superior View)

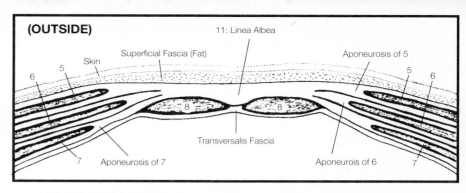

ANTERIOR ABDOMEN (Level of Iliac Crest, just below Umbilicus)
(Transverse Section, Superior View)

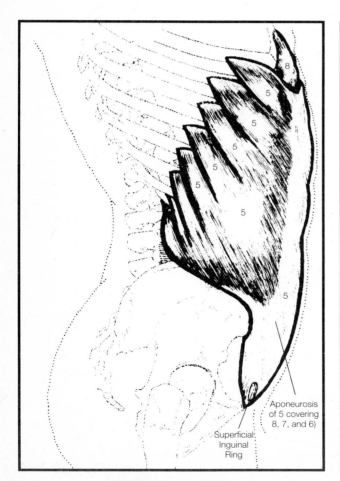

SUPERFICIAL LAYER

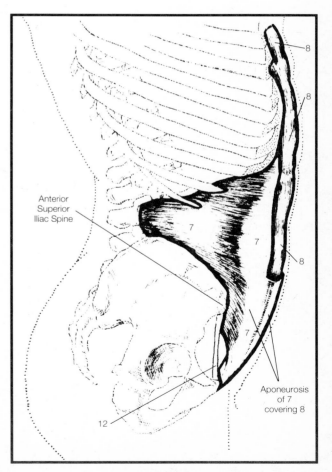

MIDDLE LAYER
(Right Lateral Views)

DEEP LAYER

6.10

MUSCULAR SYSTEM: MUSCLES OF THE AXIAL SKELETON
Muscles of the Trunk: The Pelvic Outlet and the Perineum

Pelvic Diaphragm = warm colors
Urogenital Diaphragm = cool colors

★ See 4.3, 4.9, 4.12, 4.13, 6.17

★ The PERINEUM consists of the structures occupying the PELVIC OUTLET and comprising the PELVIC FLOOR. It is the region containing the external GENITALIA (SEXUAL ORGANS) and the ANAL OPENING.
The PERINEUM is composed of SKIN, PERINEAL FASCIAE, and MUSCLE.

♂ • In the male, the PERINEUM comprises the external region between the SCROTUM and ANUS. (See 14.3)
♀ • In the female, the PERINEUM comprises the external region between the VULVA and ANUS. (See 14.9)

★ The muscles of the PELVIC OUTLET comprise the PELVIC FLOOR (of the PERINEUM). These 10 muscles are composed of two groups of muscular sheets (and their associated perineal fascia): POSTERIOR and ANTERIOR GROUP.
■ POSTERIOR GROUP called the PELVIC DIAPHRAGM consists of 3 muscles.
  • Similar in the male and female.
■ ANTERIOR GROUP called the UROGENITAL DIAPHRAGM, consists of 7 muscles associated with the genitalia
  • Differ greatly in the male and female.

★ ■ PELVIC DIAPHRAGM (POSTERIOR GROUP): LEVATOR ANI and COCCYGEUS
  ■ 2 LEVATOR ANI MUSCLES
  • Help support the pelvic viscera, and help constrict and pull forward the lower part of the rectum (thus aiding in defecation).
  ■ COCCYGEUS MUSCLE
  • Aids LEVATOR ANI in its functions and pulls coccyx forward following defecation. The PELVIC DIAPHRAGM as a whole resists increased intra-abdominal pressure.

★ ■ UROGENITAL DIAPHRAGM (ANTERIOR GROUP)
  (3 superficial muscles, 2 deep muscles, and 2 muscles of the anal region)
  The OBTURATOR INTERNUS and PIRIFORMIS MUSCLES of the PELVIC WALL (Deep Lateral Rotators of the Gluteal Region) are sometimes considered part of the muscles of the PELVIC OUTLET. (See 6.17)

MUSCLES of the PELVIC OUTLET

■ **POSTERIOR GROUP: THE PELVIC DIAPHRAGM**

1 LEVATOR ANI:

1a **PUBOCOCCYGEUS** Anterior portion of LEVATOR ANI

1b **ILIOCOCCYGEUS** Posterior portion of LEVATOR ANI

2 **COCCYGEUS** ISCHIOCOCCYGEUS

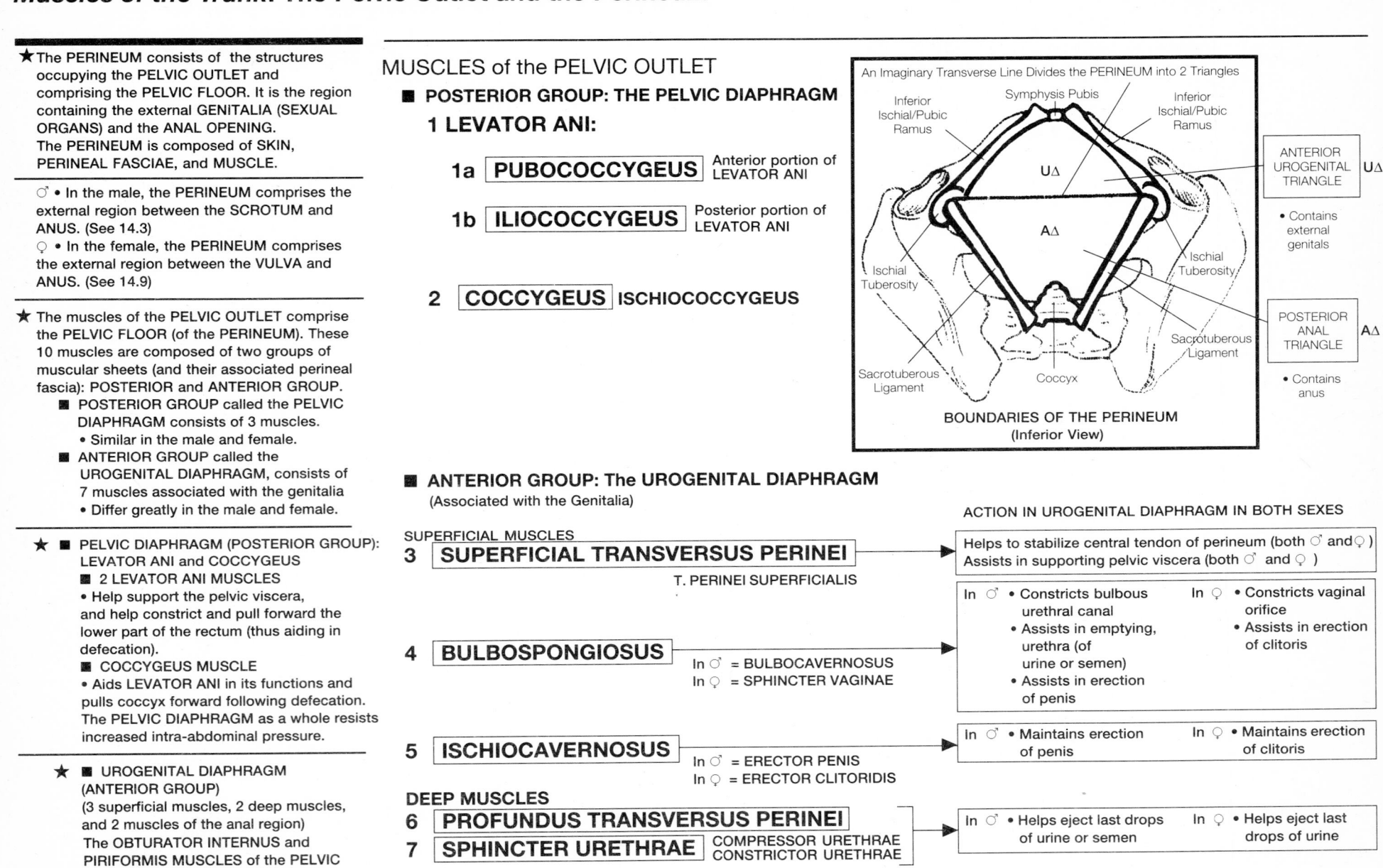

An Imaginary Transverse Line Divides the PERINEUM into 2 Triangles

Symphysis Pubis

Inferior Ischial/Pubic Ramus

Inferior Ischial/Pubic Ramus

UΔ

AΔ

Ischial Tuberosity

Ischial Tuberosity

Sacrotuberous Ligament

Sacrotuberous Ligament

Coccyx

ANTERIOR UROGENITAL TRIANGLE | UΔ
• Contains external genitals

POSTERIOR ANAL TRIANGLE | AΔ
• Contains anus

BOUNDARIES OF THE PERINEUM
(Inferior View)

■ **ANTERIOR GROUP: The UROGENITAL DIAPHRAGM**
(Associated with the Genitalia)

ACTION IN UROGENITAL DIAPHRAGM IN BOTH SEXES

SUPERFICIAL MUSCLES

3 **SUPERFICIAL TRANSVERSUS PERINEI**

T. PERINEI SUPERFICIALIS

→ Helps to stabilize central tendon of perineum (both ♂ and ♀)
Assists in supporting pelvic viscera (both ♂ and ♀)

4 **BULBOSPONGIOSUS**
In ♂ = BULBOCAVERNOSUS
In ♀ = SPHINCTER VAGINAE

→ In ♂ • Constricts bulbous urethral canal
• Assists in emptying, urethra (of urine or semen)
• Assists in erection of penis

In ♀ • Constricts vaginal orifice
• Assists in erection of clitoris

5 **ISCHIOCAVERNOSUS**
In ♂ = ERECTOR PENIS
In ♀ = ERECTOR CLITORIDIS

→ In ♂ • Maintains erection of penis

In ♀ • Maintains erection of clitoris

DEEP MUSCLES

6 **PROFUNDUS TRANSVERSUS PERINEI**

7 **SPHINCTER URETHRAE** COMPRESSOR URETHRAE CONSTRICTOR URETHRAE

→ In ♂ • Helps eject last drops of urine or semen

In ♀ • Helps eject last drops of urine

ANAL REGION

8 **SPHINCTER ANI EXTERNUS** EXTERNAL ANAL SPHINCTER

9 **SPHINCTER ANI INTERNUS** (not shown)

→ • Keeps anal canal (both ♂ and ♀) and orifice closed

MUSCLES OF THE PERINEUM, PELVIC OUTLET, AND PELVIC FLOOR

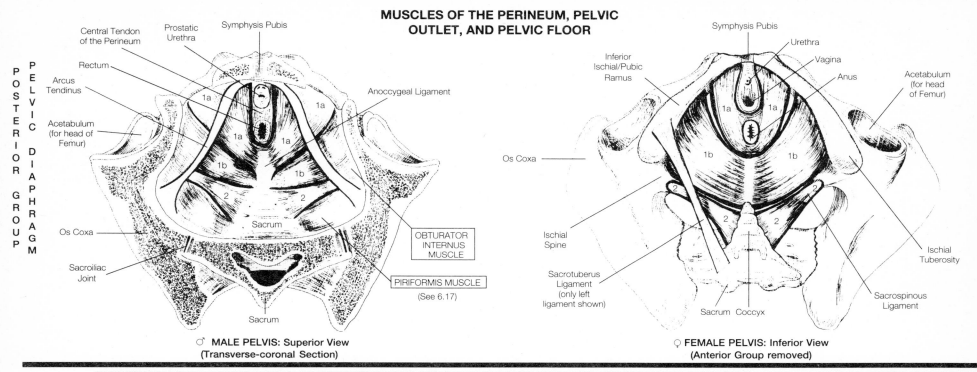

POSTERIOR GROUP — PELVIC DIAPHRAGM

Labels (left diagram):
- Central Tendon of the Perineum
- Prostatic Urethra
- Symphysis Pubis
- Rectum
- Arcus Tendinus
- Acetabulum (for head of Femur)
- Anoccygeal Ligament
- Os Coxa
- Sacroiliac Joint
- Sacrum
- OBTURATOR INTERNUS MUSCLE
- PIRIFORMIS MUSCLE (See 6.17)
- 1a, 1b, 2

♂ **MALE PELVIS: Superior View**
(Transverse-coronal Section)

Labels (right diagram):
- Symphysis Pubis
- Urethra
- Vagina
- Anus
- Inferior Ischial/Pubic Ramus
- Acetabulum (for head of Femur)
- Os Coxa
- Ischial Spine
- Ischial Tuberosity
- Sacrotuberus Ligament (only left ligament shown)
- Sacrospinous Ligament
- Sacrum Coccyx
- 1a, 1b, 2

♀ **FEMALE PELVIS: Inferior View**
(Anterior Group removed)

ANTERIOR GROUP — UROGENITAL DIAPHRAGM

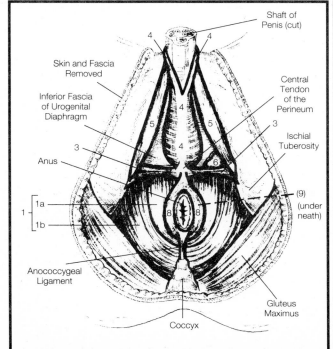

Labels (male perineum):
- Shaft of Penis (cut)
- Skin and Fascia Removed
- Inferior Fascia of Urogenital Diaphragm
- Central Tendon of the Perineum
- Ischial Tuberosity
- Anus
- (9) (underneath)
- Anococcygeal Ligament
- Gluteus Maximus
- Coccyx
- 1, 1a, 1b, 2, 3, 4, 5, 6, 8

♂ **MALE PERINEUM**
(Inferior View)

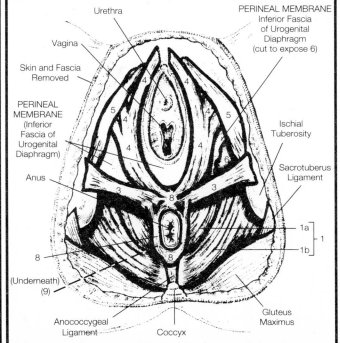

Labels (female perineum):
- Urethra
- PERINEAL MEMBRANE Inferior Fascia of Urogenital Diaphragm (cut to expose 6)
- Vagina
- Skin and Fascia Removed
- PERINEAL MEMBRANE (Inferior Fascia of Urogenital Diaphragm)
- Ischial Tuberosity
- Anus
- Sacrotuberus Ligament
- (Underneath) (9)
- Anococcygeal Ligament
- Coccyx
- Gluteus Maximus
- 1, 1a, 1b, 2, 3, 4, 5, 6, 8

♀ **FEMALE PERINEUM**
(Inferior View)

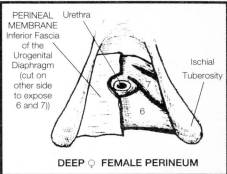

Labels (deep female perineum):
- PERINEAL MEMBRANE Inferior Fascia of the Urogenital Diaphragm (cut on other side to expose 6 and 7))
- Urethra
- Ischial Tuberosity
- 6, 7

DEEP ♀ FEMALE PERINEUM

MUSCULAR SYSTEM: MUSCLES OF THE APPENDICULAR SKELETON
Scapular Connectors to the Shoulder

Coracobrachialis = yellow Abductors = warm colors
Rotators = cool colors (You may use colors from 6.8 on lower
illustration if you so desire)
★ See 4.10, 6.8, 6.13, 6.17

★ Refer back to 6.8. Recall that there are 5 paired muscles and 1 large, triangular superficial back muscle (TRAPEZIUS) that specifically function as SCAPULA STABILIZERS. These muscles originate on the Axial Skeleton, and insert on the SCAPULA, holding the Scapula firm and stable for the muscles that help move the Brachium (Arm).
Review the muscles of Scapular Stabilization:
Posterior Group:
  • TRAPEZIUS
  • RHOMBOIDEUS MAJOR and MINOR
  • LEVATOR SCAPULAE
Anterior Group:
  • PECTORALIS MINOR
  • SERRATUS ANTERIOR

★ There are 8 paired muscles and 1 large, triangular superficial back muscle (LATTISSIMUS DORSI) that span the shoulder joint, insert on the HUMERUS BONE, and help move the Brachium:
  ■ The LATISSMUS DORSI and the paired PECTORALIS MAJOR muscles originate on the Axial Skeleton, and are termed Axial Muscles.
  ■ The remaining 7 paired muscles originate on the Scapula and insert on the Humerus. They are the major muscles involved in the tremendous complexity of movement offered by the arm. (The scapular security offered by the Muscles of Scapular Stabilization offers almost unlimited motility for the arm.)

★ The 7 paired Scapular Muscles of the Shoulder are:
Shoulder Abductors:
  • DELTOID
  • SUPRASPINATUS
Shoulder Rotators:
  • INFRASPINATUS
  • SUBSCAPULARIS
  • TERES MINOR
  • TERES MAJOR
Shoulder Flexor/Adductor:
  • CORACOBRACHIALIS

★ The DELTOID muscle is a thick, powerful muscle involved in a variety of arm movements:
  • Raises arm • Abducts arm
  • Rotates arm • Extends or flexes humerus
It caps the shoulder joint,
ORIGIN = ACROMION PROCESS of the
        SCAPULA, CLAVICLE, SPINE of the
        SCAPULA
INSERTION = DELTOID TUBEROSITY of the
            HUMERUS.
★ Individual muscle groups within the DELTOID (#1) can act separately for different movements.

MUSCLES That Move the Brachium (HUMERUS): ARM
7 SCAPULAR MUSCLES of the SHOULDER

(ORIGINATE on the SCAPULAE
and INSERT on the HUMERUS)

| ABDUCTORS OF THE SHOULDER JOINT |

1 | DELTOID DELTOIDEUS |

2 | SUPRASPINATUS |

| ROTATORS OF THE SHOULDER JOINT |

3 | INFRASPINATUS |

4 | SUBSCAPULARIS |

5 | TERES MINOR |

6 | TERES MAJOR |

| FLEXOR/ADDUCTOR OF SHOULDER JOINT |

7 | CORACOBRACHIALIS |

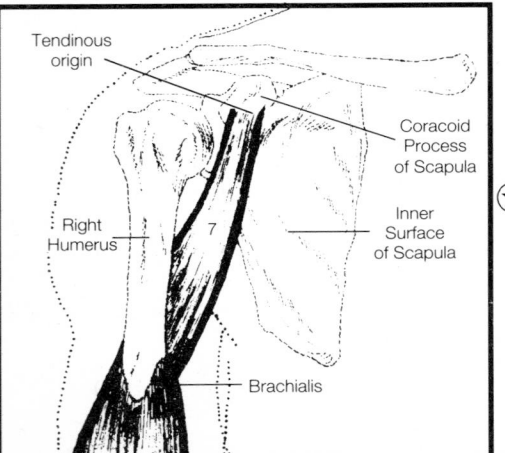

Tendinous origin

Coracoid Process of Scapula

Right Humerus

Inner Surface of Scapula

7

Brachialis

Anterior View

4 MUSCLES OF THE MUSCULOTENDINOUS (ROTATOR) CUFF: (2, 3, 4, 5)

■ There are 4 muscles that originate on the SCAPULAE, span the SHOULDER (GLENOHUMERAL) JOINT, and insert about the HEAD of the HUMERUS.
■ These 4 muscles (and their tendons) reinforce the SHOULDER JOINT, whose socket is too flat to offer security to the HEAD of the HUMERUS on its own.
■ The muscles of SCAPULAR STABILIZATION secure the scapula for the stable base of origin for these 4 muscles.
■ In general, the musculotendinous rotator cuff muscles correspond to the action of the deep lateral rotators of the iliac-pelvic region. (See 6.17)
■ The rotator cuff of the shoulder is commonly a problem with baseball pitchers.

2 | SUPRASPINATUS | ABDUCTOR

3 | INFRASPINATUS |
5 | TERES MINOR | } LATERAL ROTATORS

4 | SUBSCAPULARIS | MEDIAL ROTATOR

ARM MOVEMENTS

ABDUCTION:
  • SUPRASPINATUS
  • DELTOID

LATERAL ROTATION:
  • INFRASPINATUS
  • TERES MINOR

MEDIAL ROTATION:
  • DELTOID (Anterior fibers)
  • SUBSCAPULARIS
  • TERES MAJOR
  • LATISSIMUS DORSI } See 6.8
  • PECTORALIS MAJOR

★ Although both the TERES MAJOR and the LATISSIMUS DORSI arise from the back, they cross from the back to the *anterior aspect* of the HUMERUS. These insertions to the bicipital groove of the humerus anteriorly causes MEDIAL ARM ROTATION (See lower illustration)

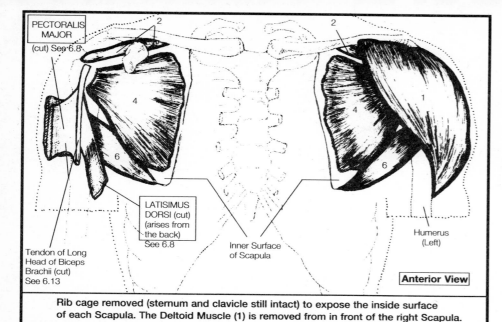

PECTORALIS MAJOR (cut) See 6.8

2

2

LATISIMUS DORSI (cut) (arises from the back) See 6.8

Tendon of Long Head of Biceps Brachii (cut) See 6.13

Inner Surface of Scapula

Humerus (Left)

Anterior View

Rib cage removed (sternum and clavicle still intact) to expose the inside surface of each Scapula. The Deltoid Muscle (1) is removed from in front of the right Scapula.

Humerus (Left)

Posterior View

The Deltoid Muscle (1) is removed from the right arm to expose Scapular Muscles. (Note that the Deltoid caps the shoulder joint in both views.)

Supraglenoid Tubercle of Glenoid Cavity

Acromion Process of Scapula

Short Head of Biceps Brachii

Clavicle Bone

Sternocleidomastoid

Pectoralis Major

Long Head of Biceps Brachii

Coracoid Process of Scapula

Bicipital Groove

Pectoralis Minor

Latissimus Dorsi (★)

Pectoralis Major

Long Head of Triceps

Serratus Anterior

Infraglenoid Tubercle

(★)

Humerus

Anterior View

Sternocleidomastoid

Trapezius

Coracoid Process of Scapula

Levator Scapulae

Rhomboideus Minor

Humerus

Long Head of Triceps Brachii

Rhomboideus Major

Lateral Head of Triceps

AREAS OF ATTACHMENT OF ASSOCIATED MUSCLES

Medial Head of Triceps

(Deltoid Tuberosity)

RIGHT CLAVICLE, SCAPULA, AND HUMERUS:
(★ See 6.13 for Biceps and Triceps Brachii)
(★ See 4.10 for details of Skeletal structure)

Posterior View

MUSCULAR SYSTEM: MUSCLES OF THE APPENDICULAR SKELETON
Muscles of the Upper Limb: The Brachium (Humerus): The Arm

★ The UPPER ARM, or more commonly the ARM, is the PROXIMAL PORTION of the UPPER LIMB or UPPER EXTREMITY.

★ The muscles associated with the BRACHIUM (HUMERUS, or UPPER ARM) are very powerful. They act on the FOREARM at the ELBOW JOINT and THE RADIOULNAR JOINT (assisted by smaller muscles from the FOREARM proper).

■ BICEPS BRACHII = The most prominent muscle (structurally and functionally) of the humerus bone, but has no attachments on the humerus. It lies on the anterior surface of the humerus and FLEXES the FOREARM.
Dual Origin:
• SHORT HEAD
  • ORIGIN = From Coracoid Process of SCAPULA
• LONG HEAD
  • ORIGIN = On superior Tuberosity of the Glenoid Fossa
• INSERTION OF BOTH HEADS: On Radial Tuberosity

■ TRICEPS BRACHII = On the posterior surface of the humerus; Antagonistic to the BICEPS BRACHII, EXTENDS FOREARM
Triple Origin
• LONG HEAD
  • ORIGIN = Infraglenoid Tuberosity of the Scapula – *below glenoid cavity*
• LATERAL HEAD
  • ORIGIN = Posterior surface of Humerus below the great tubercle *lateral surface of humerus*
• MEDIAL HEAD
  • ORIGIN = Humerus below the Radial Groove *medial surface of humerus*
COMMON TENDINOUS INSERTION
• Olecranon Process of Ulna

★ The BRACHIORADIALIS, apart from its FLEXION and SUPINATION action on the forearm, is also shown to play a part of RAPID EXTENSION (countering the centrifugal force produced by EXTENSION).

MUSCLES of the
BRACHIUM (HUMERUS): The ARM
FLEXOR/ADDUCTOR of SHOULDER JOINT

1 **CORACOBRACHIALIS** (See 6.12)
Origin – corocoid process of scapula to insert – shaft of humerus

BRACHIAL MUSCLES That Act on the FOREARM:

2 **BICEPS BRACHII** (2 HEADS/ORIGINS)

3 **BRACHIALIS** *anterior shaft of humerus / Coronoid process of ulna*

4 **BRACHIORADIALIS** SUPINATOR LONGUS
flat on lateral surface of forearm

5 **TRICEPS BRACHII** (3 HEADS/ORIGINS)
origin: lateral epi condyle of humerus to insertion at styloid process of radius

MUSCLES ACTING ON THE FOREARM:
4 FLEXORS OF FOREARM

SYNERGISTIC ACTION OF PRINCIPAL FLEXORS

2 **BICEPS BRACHII**

3 **BRACHIALIS**

4 **BRACHIORADIALIS** ◉

PRONATOR TERES ◉
Origin – medial epicondyle of humerus to lateral surface of radius

2 EXTENSORS OF FOREARM

5 **TRICEPS BRACHII** PRINCIPAL EXTENSOR

6 **ANCONEUS** ◉

3 SUPINATORS OF HAND (AND FOREARM)

SYNERGISTIC ACTION

2 **BICEPS BRACHII**

4 **BRACHIORADIALIS** ◉

7 **SUPINATOR** ◉ *lateral epicondyle of humerus to lateral surface of radius*

2 PRONATORS OF HAND (AND FOREARM)

8 **PRONATOR TERES** ◉

9 **PRONATOR QUADRATUS** ◉
Origin ulna → radius ▱

★ ◉ MUSCLES OF THE FOREARM (See also 6.14 and 6.15)
PRONATOR TERES = Anterior superficial
PRONATOR QUADRATUS = Anterior deep
BRACHIORADIALIS and ACONEUS = Posterior Superficial
SUPINATOR = Posterior deep

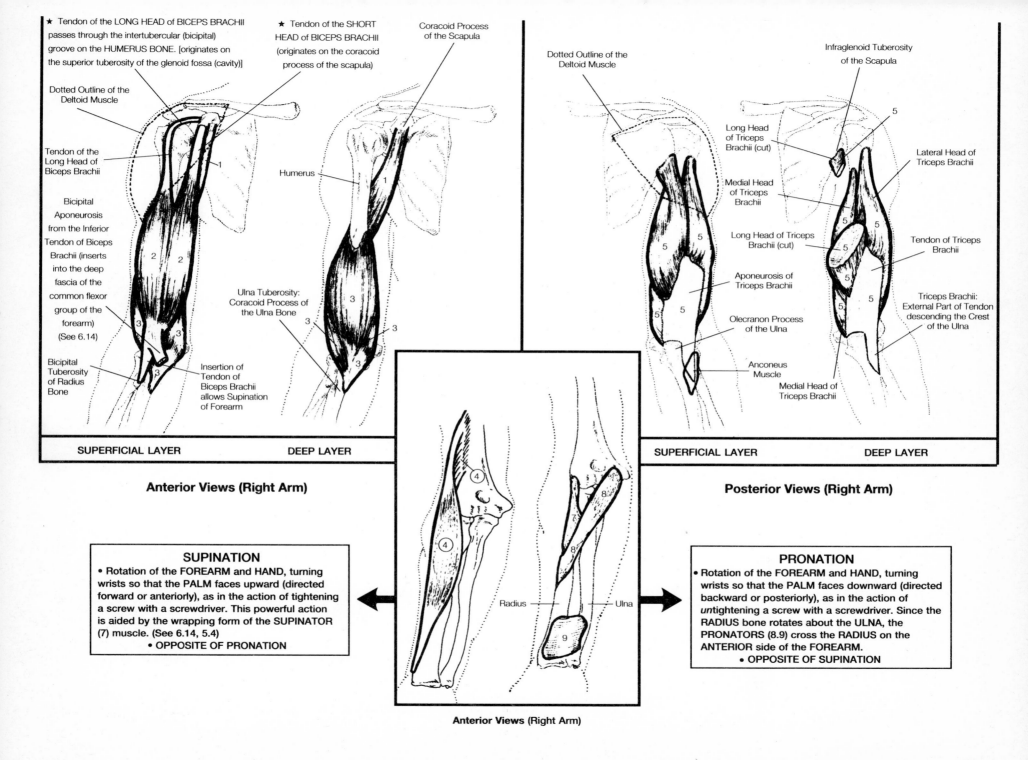

★ Tendon of the LONG HEAD of BICEPS BRACHII passes through the intertubercular (bicipital) groove on the HUMERUS BONE. [originates on the superior tuberosity of the glenoid fossa (cavity)]

★ Tendon of the SHORT HEAD of BICEPS BRACHII (originates on the coracoid process of the scapula)

Coracoid Process of the Scapula

Dotted Outline of the Deltoid Muscle

Infraglenoid Tuberosity of the Scapula

Dotted Outline of the Deltoid Muscle

Tendon of the Long Head of Biceps Brachii

Humerus

Long Head of Triceps Brachii (cut)

Lateral Head of Triceps Brachii

Bicipital Aponeurosis from the Inferior Tendon of Biceps Brachii (inserts into the deep fascia of the common flexor group of the forearm) (See 6.14)

Medial Head of Triceps Brachii

Long Head of Triceps Brachii (cut)

Tendon of Triceps Brachii

Ulna Tuberosity: Coracoid Process of the Ulna Bone

Aponeurosis of Triceps Brachii

Triceps Brachii: External Part of Tendon descending the Crest of the Ulna

Bicipital Tuberosity of Radius Bone

Insertion of Tendon of Biceps Brachii allows Supination of Forearm

Olecranon Process of the Ulna

Anconeus Muscle

Medial Head of Triceps Brachii

SUPERFICIAL LAYER   **DEEP LAYER**

Anterior Views (Right Arm)

SUPERFICIAL LAYER   **DEEP LAYER**

Posterior Views (Right Arm)

Radius — — Ulna

SUPINATION
• Rotation of the FOREARM and HAND, turning wrists so that the PALM faces upward (directed forward or anteriorly), as in the action of tightening a screw with a screwdriver. This powerful action is aided by the wrapping form of the SUPINATOR (7) muscle. (See 6.14, 5.4)
  • OPPOSITE OF PRONATION

PRONATION
• Rotation of the FOREARM and HAND, turning wrists so that the PALM faces downward (directed backward or posteriorly), as in the action of *un*tightening a screw with a screwdriver. Since the RADIUS bone rotates about the ULNA, the PRONATORS (8.9) cross the RADIUS on the ANTERIOR side of the FOREARM.
  • OPPOSITE OF SUPINATION

Anterior Views (Right Arm)

6.13

MUSCULAR SYSTEM: MUSCLES OF THE APPENDICULAR SKELETON
Muscles of the Upper Limb: Muscles of the Forearm:
Anterior (Palmar) Muscles

Color Flexors of the Forearm in warm colors, and the other Anterior Forearm Muscles in light warm colors.
★ See 4.10, 6.13, 6.16

Review the anatomical position of the body (see 1.0)

★ Most of the ANTERIOR (or PALMAR) PORTION of the FOREARM is made up of FLEXOR MUSCLES of the WRIST AND FINGERS (DIGITS).

As a group, these FLEXORS ORIGINATE from the:
- MEDIAL EPICONDYLE of the HUMERUS
- UPPER RADIUS and ULNA
- INTEROSSEUS MEMBRANE OF FOREARM

■ FLEXOR CARPI MUSCLES (2,4), as they cross the wrist, INSERT on:
- DISTAL CARPAL BONES or
- METACARPALS

■ The 2 FINGER (DIGIT) FLEXORS (5,6) INSERT on:
- MIDDLE and DISTAL PHALANGES
- (when the fingers are flexed, they close over the palm as in a fist)

■ PALMARIS LONGUS (3) "INSERTS" or merges with "(ORIGINATES)" the PALMAR APONEUROSIS

The FLEXOR DIGITORUM SUPERFICIALIS (5) (of the superficial group) consists of 3 heads: HUMERAL, ULNAR, and RADIAL

★ The LUMBRICALE MUSCLES OF THE HAND (See 6.16) arise from TENDONS of the FLEXOR DIGITORUM PROFUNDUS (6)

ANTERIOR (PALMAR) MUSCLES of the FOREARM

SUPERFICIAL GROUP

LATERAL

1 | PRONATOR TERES | HUMERAL HEAD and ULNAR HEAD

2 | FLEXOR CARPI RADIALIS | RADIOCARPUS

3 | PALMARIS LONGUS |

MEDIAL

4 | FLEXOR CARPI ULNARIS | HUMERAL HEAD and ULNAR HEAD

5 | FLEXOR DIGITORUM SUPERFICIALIS |
FLEXOR DIGITORUM SUBLIMIS

DEEP GROUP

6 | FLEXOR DIGITORUM PROFUNDUS |

7 | FLEXOR POLLICIS LONGUS |

8 | PRONATOR QUADRATUS |

ANTERIOR FOREARM MUSCLES ACTING ON THE WRIST JOINT, HANDS, AND FINGERS

FLEXORS OF THE WRIST AND HAND

ALL OF THE ANTERIOR FOREARM MUSCLES *EXCEPT PRONATORS*: MAINLY THE SUPERFICIAL GROUP

ABDUCTOR OF THE WRIST AND HAND

2 | FLEXOR CARPI RADIALIS |

ADDUCTOR OF THE WRIST AND HAND

4 | FLEXOR CARPI ULNARIS |

PRONATORS OF THE WRIST AND HAND (AND FOREARM)

1 | PRONATOR TERES |

8 | PRONATOR QUADRATUS |

FLEXORS OF THE FINGERS (DIGITS)

5 | FLEXOR DIGITORUM SUPERFICIALIS |

6 | FLEXOR DIGITORUM PROFUNDUS |

FLEXOR OF THE THUMB

7 | FLEXOR POLLICIS LONGUS |

ROTATORS OF THE FOREARM

1 | PRONATOR TERES |

8 | PRONATOR QUADRATUS |

9 | SUPINATOR | (See 6.13, 6.15)

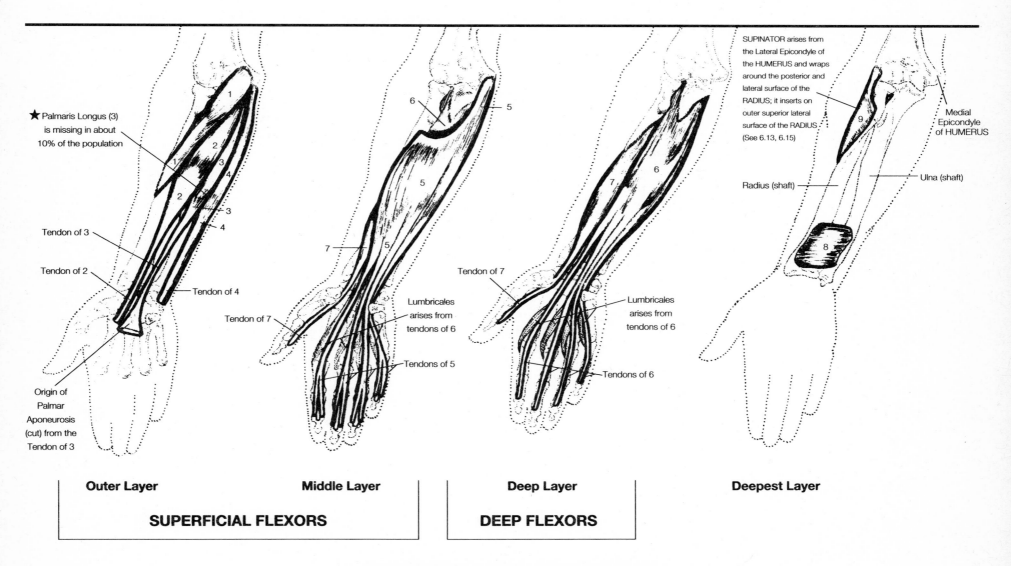

★ Palmaris Longus (3) is missing in about 10% of the population

Tendon of 3

Tendon of 2

Tendon of 4

Origin of Palmar Aponeurosis (cut) from the Tendon of 3

Tendon of 7

Lumbricales arises from tendons of 6

Tendons of 5

Tendon of 7

Lumbricales arises from tendons of 6

Tendons of 6

SUPINATOR arises from the Lateral Epicondyle of the HUMERUS and wraps around the posterior and lateral surface of the RADIUS; it inserts on outer superior lateral surface of the RADIUS (See 6.13, 6.15)

Medial Epicondyle of HUMERUS

Radius (shaft)

Ulna (shaft)

Outer Layer   **Middle Layer**   **Deep Layer**   **Deepest Layer**

SUPERFICIAL FLEXORS   **DEEP FLEXORS**

ANTERIOR (PALMAR) MUSCLES OF THE FOREARM

Anterior Views (Right Arm)

MUSCULAR SYSTEM: MUSCLES OF THE APPENDICULAR SKELETON
Muscles of the Upper Limb: Muscles of the Forearm:
Posterior (Dorsal) Muscles

Color Extensors of the Forearm in cool colors (in opposition to the Forearm Flexors on 6.22). Color the other Posterior Forearm Muscles in light cool colors. Color Tendons of Extensor Muscle the same color as the Extensors. (Lighter if possible.)

★ See 4.10, 6.13, 6.16

★ Most of the POSTERIOR (DORSAL/BACK) PORTION of the FOREARM is made up of EXTENSOR MUSCLES OF THE WRIST and FINGERS (DIGITS), although certainly less massive than the FLEXOR side.

The ANCONEUS is a small superficial forearm muscle that helps EXTEND the forearm itself.

The BRACHIORADIALIS muscle originates dorsally on the supracondylar ridge of the Humerus bone, but curves most of its mass to the anterior (palmer) side of the forearm. It helps to FLEX and SUPINATE the forearm, although it has been shown to play a part in RAPID EXTENSION (countering the centrifugal force produced by FLEXION). (See 6.13)

★ As a group, these EXTENSORS originate from the:
- LATERAL EPICONDYLE of the HUMERUS
- UPPER RADIUS and ULNA
- INTEROSSEUS MEMBRANE of FOREARM

■ EXTENSOR CARPI MUSCLES (2,3,6) INSERT ON:
- DISTAL CARPAL BONES
- or METACARPALS

■ The 3 FINGER (DIGIT) EXTENSORS (4,5,12) INSERT (FORM A TENDON EXPANSION) ON:
- MIDDLE and DISTAL PHALANGES
- The SMALL INTRINSIC MUSCLES of the HAND (see 6.16) INSERT on these EXPANDED TENDONS
- If a fist was made over the palm by FLEXORS, the FINGER EXTENSORS WOULD "pull" the fingers "back out" into the anatomical position.

★ The muscles that extend the hand are located on the posterior side of the forearm, and their tendons extend along the dorsal surface of the hand, mainly over the DORSAL INTEROSSEI onto the phalanges. Note the complex division of the tendon of the EXTENSOR DIGITORUM COMMUNIS into the 4 phalanges (fingers).

★ THE ANATOMICAL "SNUFFBOX"
■ A small depression in the skin at the base of the thumb (laterally)
■ Created by the boundaries of the 2 thumb extensors and the thumb abductor (and their tendons).

POSTERIOR (DORSAL/BACK) MUSCLES of the FOREARM

SUPERFICIAL GROUP

LATERAL ① **BRACHIORADIALIS** SUPINATOR LONGUS

2 **EXTENSOR CARPI RADIALIS LONGUS**

3 **EXTENSOR CARPI RADIALIS BREVIS**

4 **EXTENSOR DIGITORUM COMMUNIS**

5 **EXTENSOR DIGITI MINIMI** EXTENSOR DIGITI QUINTI PROPRIUS

6 **EXTENSOR CARPI ULNARIS**

MEDIAL ⑦ **ANCONEUS**

DEEP GROUP

8 **SUPINATOR** SUPINATOR RADII BREVIS

9 **ABDUCTOR POLLICIS LONGUS** EXTENSOR OSSIS METACARPI POLLICIS

10 **EXTENSOR POLLICIS BREVIS** EXTENSOR PRIMII INTERNODII POLLICIS

11 **EXTENSOR POLLICIS LONGUS** EXTENSOR SECONDII INTERNODII POLLICIS

12 **EXTENSOR INDICIS** EXTENSOR INDICIS PROPRIUS

POSTERIOR FOREARM MUSCLES ACTING ON THE WRIST JOINT, HANDS, AND FINGERS

EXTENSORS OF THE WRIST AND HAND

ALL POSTERIOR FOREARM MUSCLES OF THE SUPERFICIAL GROUP *EXCEPT BRACHIORADIALIS AND ANCONEUS*

ABDUCTORS OF THE WRIST AND HAND
2 EXTENSOR CARPI RADIALIS LONGUS
3 EXTENSOR CARPI RADIALIS BREVIS
6 EXTENSOR CARPI ULNARIS

ADDUCTORS OF THE WRIST AND HAND
10 EXTENSOR POLLICIS BREVIS
11 EXTENSOR POLLICIS LONGUS

EXTENSORS OF THE FINGERS
4 EXTENSOR DIGITORUM COMMUNIS
5 EXTENSOR DIGITI MINIMI
12 EXTENSOR INDICIS

EXTENSORS OF THE THUMB
10 EXTENSOR POLLICIS BREVIS
11 EXTENSOR POLLICIS LONGUS

ABDUCTOR OF THE THUMB
9 ABDUCTOR POLLICIS LONGUS

ROTATOR OF THE FOREARM
8 SUPINATOR (See 6.13, 6.14)

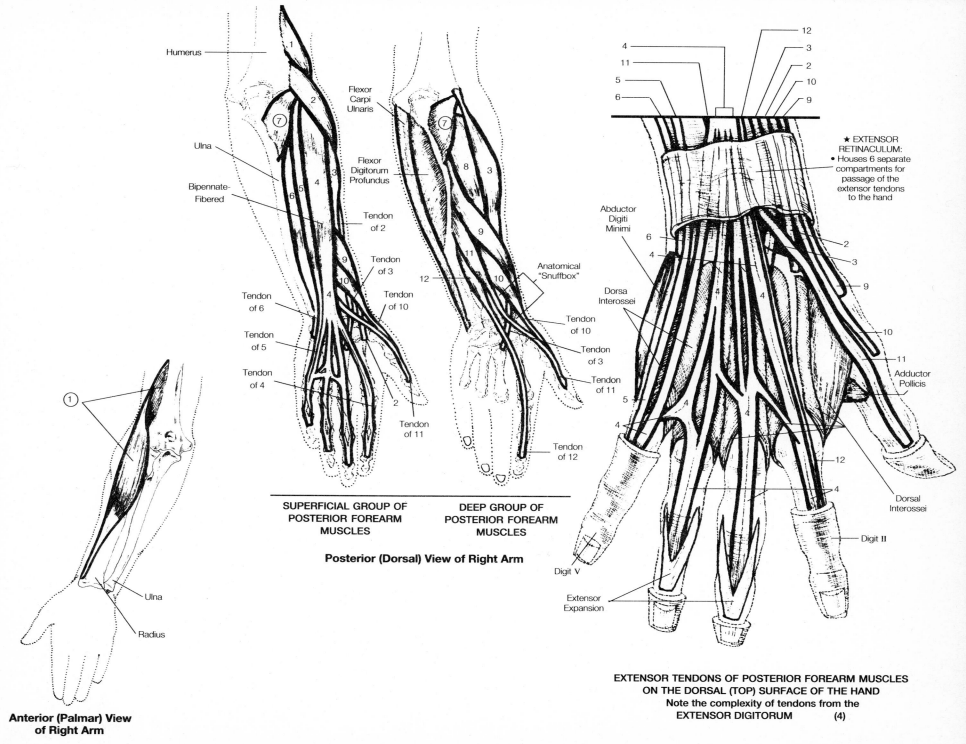

Humerus

Ulna

Bipennate-Fibered

Flexor Carpi Ulnaris

Flexor Digitorum Profundus

Tendon of 2

Tendon of 3

Tendon of 10

Tendon of 6

Tendon of 5

Tendon of 4

Tendon of 11

Tendon of 2

Anatomical "Snuffbox"

Tendon of 10

Tendon of 3

Tendon of 11

Tendon of 12

SUPERFICIAL GROUP OF POSTERIOR FOREARM MUSCLES

DEEP GROUP OF POSTERIOR FOREARM MUSCLES

Posterior (Dorsal) View of Right Arm

★ EXTENSOR RETINACULUM:
• Houses 6 separate compartments for passage of the extensor tendons to the hand

Abductor Digiti Minimi

Dorsa Interossei

Adductor Pollicis

Dorsal Interossei

Digit II

Extensor Expansion

Digit V

EXTENSOR TENDONS OF POSTERIOR FOREARM MUSCLES ON THE DORSAL (TOP) SURFACE OF THE HAND
Note the complexity of tendons from the
EXTENSOR DIGITORUM   (4)

Ulna

Radius

Anterior (Palmar) View of Right Arm

MUSCULAR SYSTEM: MUSCLES OF THE APPENDICULAR SKELETON
Muscles of the Upper Limb: The Hand (Palmar and Dorsal Surfaces)

Thenar Eminence = reds and oranges
Hypothenar Eminence = pinks and purples
Intermediate Group = blues and greens
Palmaris Brevis = yellow

★ See 4.11, 6.14, 6.15

★ (Numbers in parenthesis to the right of rectangular boxes indicate numbers of muscles)
COMPLEX FUNCTIONS OF THE HAND:

FLEXION AND EXTENSION OF HAND AND FINGERS ⎤ **FOREARM MUSCLES**

PRECISE FINGER MOVEMENTS: FLEXION AND EXTENSION coordinated with ABDUCTION AND ADDUCTION ⎤ **SMALL INTRINSIC MUSCLES OF THE HAND**

■ *EXTENSORS OF THUMB*
EXTENSOR POLLICIS BREVIS (6.15)
EXTENSOR POLLICIS LONGUS

■ *OPPOSITION (GRASPING)*
OPPONENS POLLICIS ⟷ OPPONENS DIGITI MINIMI

■ *CIRCUMDUCTION OF THUMB*
OPPONENS POLLICIS → FLEXOR POLLICIS BREVIS
← ABDUCTOR POLLICIS BREVIS ◄

★ INTEROSSEI AND LUMBRICALES of the INTERMEDIATE PALM GROUP insert into the EXTENSOR EXPANSION (along with certain tendons of arm muscles), which aids in complex functions for extending and flexing the fingers.

★ The LUMBRICALES arise from tendons of the FLEXOR DIGITORUM PROFUNDUS (See also 6.14)

★ Even when relaxed, the hand is contracted inward slightly toward the PALMAR SURFACE. This is because the FLEXOR MUSCLES that FLEX the HAND are naturally stronger (and more massive as a whole) than the EXTENSOR MUSCLES that EXTEND the HAND.

INTRINSIC MUSCLES of the HAND (MANUS)

THENAR EMINENCE of the THUMB

1 ABDUCTOR POLLICIS BREVIS

2 OPPONENS POLLICIS

3 FLEXOR POLLICIS BREVIS

HYPOTHENAR EMINENCE of the LITTLE FINGER

4 ABDUCTOR DIGITI MINIMI `ABDUCTOR DIGITI QUINTI`

5 OPPONENS DIGITI MINIMI `OPPONENS DIGITI QUINTI`

6 FLEXOR DIGITI MINIMI BREVIS `FLEXOR DIGITI QUINTI BREVIS`

INTERMEDIATE GROUP of the PALM
(Deep Muscles between Metacarpal Bones)

7 ADDUCTOR POLLICIS `OBLIQUE AND TRANSVERSE HEADS`

8 PALMAR INTEROSSEI `(3) INTEROSSEI VOLARES`

9 DORSAL INTEROSSEI `(4) INTEROSSEI DORSALES MANUS`

10 LUMBRICALES `(4) LUMBRICALES MANUS`

SUBCUTANEOUS MUSCLE
11 PALMARIS BREVIS

ACTION of INTRINSIC HAND MUSCLES

ABDUCTOR OF THE THUMB
1 ABDUCTOR POLLICIS BREVIS

ADDUCTOR OF THE THUMB
7 ADDUCTOR POLLICIS

FLEXORS OF THE THUMB
2 OPPONENS POLLICIS

3 FLEXOR POLLICIS BREVIS

ABDUCTORS OF THE FINGERS
4 ABDUCTOR DIGITI MINIMI
8 PALMAR INTEROSSEI (3)
9 DORSAL INTEROSSEI (4)

ADDUCTORS OF THE FINGERS
8 PALMAR INTEROSSEI (3)
9 DORSAL INTEROSSEI (4)

FLEXORS OF THE FINGERS
6 FLEXOR DIGITI MINIMI BREVIS

10 LUMBRICALES (4)

MUSCLES That Change the Shape of the PALM
5 OPPONENS DIGITI MINIMI

11 PALMARIS BREVIS

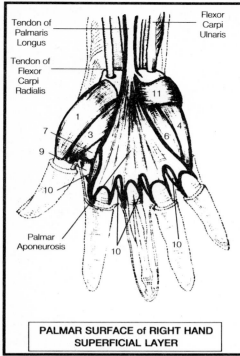

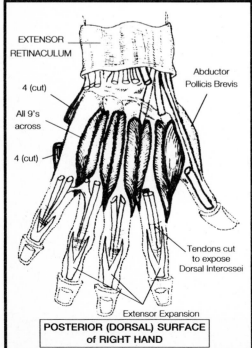

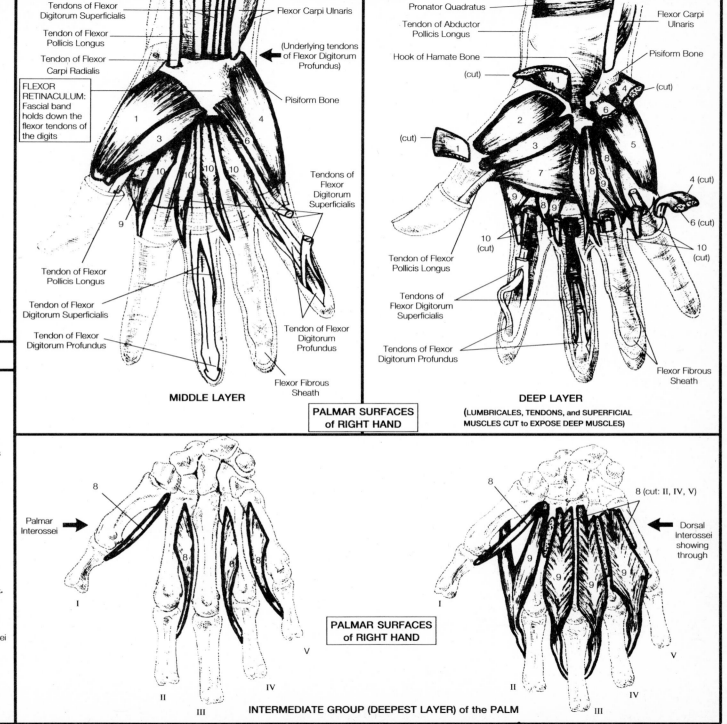

6.17

Support and Movement **128**

1 = yellow   2 = orange
3 = red   4 = light pink
Deep Laterals = cool colors

★ See 4.14, 6.18, 6.19

MUSCULAR SYSTEM: MUSCLES OF THE APPENDICULAR SKELETON
Muscles of the Iliac-Pelvic (Girdle) Region: Buttock (Gluteal) Muscles

★ BUTTOCK MUSCLES act on the HIP JOINT AND EXTEND, ABDUCT, AND ROTATE the THIGH (FEMUR).

■ *GLUTEUS MAXIMUS*
- Powerful thigh extensor (1–2 in. thick)
- Bipedal stance and locomotion
- Running and climbing
- Superficial fascia (fat) adds form
- ORIGIN = Superior curved iliac line and crest, sacrum, coccyx, and the aponeurosis of the back
- INSERTION = Gluteal tuberosity of Femur, (just below greater trochanter), and the ILIOTIBIAL TRACT

■ *GLUTEUS MEDIUS*
- HIP STABILIZER: Keeps hips level during running/walking
- ORIGIN = Lateral surface of the ilium
- INSERTION = Greater trochanter of the Femur

■ *GLUTEUS MINIMUS*
- Smallest and deepest of gluteals
- ORIGIN = Lateral surface of lower part of ilium
- INSERTION = Greater trochanter

■ *TENSOR FASCIAE LATAE* is actually part of the THIGH REGION (being a THIGH FLEXOR), but has similar attachments to the BUTTOCKS
- ORIGIN = Iliac crest
- INSERTION = Tibia by way of the ILIOTIBIAL TRACT and FASCIA LATA

★ *ILIOTIBIAL TRACT*
- A tendinous band that extends down the thigh
FASCIA LATA
- A broad, wide, lateral fascia encasing the thigh muscles
- It is continuous with the ILIOTIBIAL TRACT.The ILIOTIBIAL TRACT and The FASCIA LATA (near the junction of the GLUTEUS MAXIMUS and TENSOR FASCIAE LATAE) are collectively referred to as the FEMORAL APONEUROSIS.

★ • In general, the 6 DEEP LATERAL ROTATORS correspond to the action of the MUSCULOTENDINOUS CUFF OF THE SHOULDER JOINT. (See 6.12)
■ Located directly over the posterior aspect of the HIP JOINT

Superficial Posterior BUTTOCK MUSCLES

1 | GLUTEUS MAXIMUS |

2 | GLUTEUS MEDIUS | ⎤
                              } SYNERGISTIC ACTION
3 | GLUTEUS MINIMUS | ⎦

4 | TENSOR FASCIAE LATAE |
(TENSOR FASCIAE FEMORIS)

SUPERFICIAL POSTERIOR BUTTOCK MUSCLES THAT ACT ON THE HIP JOINT AND THE THIGH (FEMUR)

EXTENSOR OF THIGH
1 | GLUTEUS MAXIMUS |

FLEXOR OF THIGH
4 | TENSOR FASCIAE LATAE |

ABDUCTORS OF THIGH
| ALL SUPERFICIAL POSTERIOR BUTTOCK MUSCLES |

ROTATORS OF THIGH
MEDIAL ROTATORS
| ALL SUPERFICIAL POSTERIOR BUTTOCK MUSCLES *EXCEPT GLUTEUS MAXIMUS* |

6 Deep Lateral BUTTOCK MUSCLES (ROTATORS)

5 | PIRIFORMIS | ⊙

6 | OBTURATOR INTERNUS | ⊙

7 | GEMELLUS SUPERIOR |

8 | GEMELLUS INFERIOR |

9 | QUADRATUS FEMORIS |

10 | OBTURATOR EXTERNUS |

DEEP LATERAL BUTTOCK MUSCLES (AND ILIOPSOAS) THAT ACT ON THE HIP JOINT AND THE THIGH (FEMUR)

LATERAL ROTATORS
ALL 6 DEEP LATERAL BUTTOCK MUSCLES *PLUS GLUTEUS MAXIMUS*

★ **ILIOPSOAS: The compound ILIACUS AND PSOAS MAJOR Muscle**
(not shown here, see 6.9)

• In addition to LATERALLY ROTATING the THIGH, the single synergistic ILIOPSOAS also FLEXES the THIGH and FLEXES THE VERTEBRAL COLUMN.

⊙ PIRIFORMIS and OBTURATOR INTERNUS are sometimes included as muscles of the PELVIC WALL. (See 6.11) They not only LATERALLY ROTATE THE THIGH , but also ABDUCT THE THIGH.

GLUTEAL MUSCLES OF THE ILIAC-PELVIC (GIRDLE) REGION

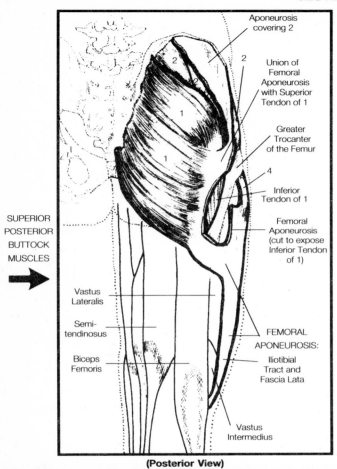

SUPERIOR POSTERIOR BUTTOCK MUSCLES →

Aponeurosis covering 2

Union of Femoral Aponeurosis with Superior Tendon of 1

Greater Trocanter of the Femur

Inferior Tendon of 1

Femoral Aponeurosis (cut to expose Inferior Tendon of 1)

Vastus Lateralis

Semi-tendinosus

Biceps Femoris

FEMORAL APONEUROSIS:
Iliotibial Tract and Fascia Lata

Vastus Intermedius

(Posterior View)

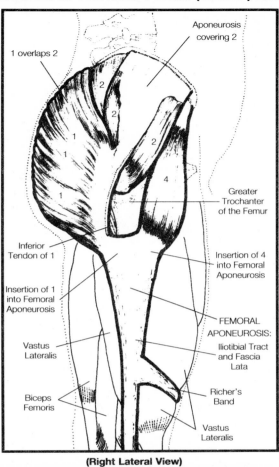

1 overlaps 2

Aponeurosis covering 2

Greater Trochanter of the Femur

Inferior Tendon of 1

Insertion of 1 into Femoral Aponeurosis

Insertion of 4 into Femoral Aponeurosis

Vastus Lateralis

FEMORAL APONEUROSIS:
Iliotibial Tract and Fascia Lata

Biceps Femoris

Richer's Band

Vastus Lateralis

(Right Lateral View)

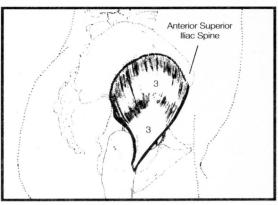

Anterior Superior Iliac Spine

(Right Lateral View)

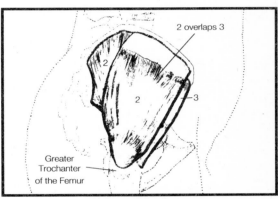

2 overlaps 3

Greater Trochanter of the Femur

(Right Lateral View)

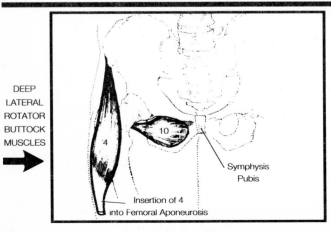

DEEP LATERAL ROTATOR BUTTOCK MUSCLES →

Symphysis Pubis

Insertion of 4 into Femoral Aponeurosis

(Anterior View)

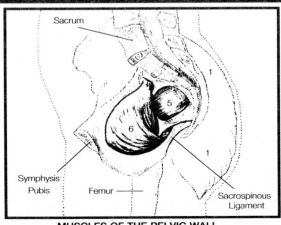

Sacrum

Symphysis Pubis

Femur

Sacrospinous Ligament

MUSCLES OF THE PELVIC WALL
(Midsagittal Section, Medial View)

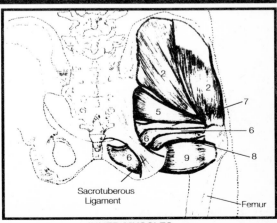

Sacrotuberous Ligament

Femur

INNER MUSCLES
(Posterior View)

6.17

6.18

Support and Movement **130**

1 = yellow   2 = yellow-orange
3, 4, 5 = light cool colors (blues, greens)
6 = orange   7 = red   8 = pink
★ See 4.14

MUSCULAR SYSTEM: MUSCLES OF THE APPENDICULAR SKELETON
Muscles of the Lower Limb: Muscles of the Thigh: *Medial and Posterior Femoral*

★ The THIGH is the proximal portion of the lower extremity, lying *between* the HIP JOINT and the KNEE.

MUSCLES OF THE MEDIAL THIGH
• The MEDIAL FEMORAL THIGH MUSCLES are basically ADDUCTORS, and make up most of the thigh mass.

• Recall that an ADDUCTOR muscle draws toward the medial line of the body.

• The long, thin *GRACILIS* is the most superficial of the MEDIAL THIGH MUSCLES, and is a 2-joint muscle:
   ■ ADDUCTS THE THIGH
   ■ FLEXES THE LEG

■ The *PECTINEUS* is flat and quadrangular, and is the uppermost of the MEDIAL THIGH MUSCLES

■ The ADDUCTOR LONGUS, BREVIS, and MAGNUS muscles are basically SYNERGISTIC in ADDUCTING, FLEXING, and ROTATING the THIGH.

■ Note the HIATUS (opening) in the ADDUCTOR MAGNUS for the passing of the FEMORAL ARTERY and VEIN from anterior thigh to posterior thigh.

THE 3 "HAMSTRINGS": BICEPS FEMORIS, SEMITENDINOSUS, and SEMIMEM-BRANOSUS
★ MUSCLES OF THE POSTERIOR THIGH:
   ■ Act as EXTENSORS of the HIP JOINT (more efficient when the KNEE JOINT is EXTENDED)

   ■ Act as FLEXORS of the KNEE JOINT (more efficient with the HIP JOINT is EXTENDED)

   ■ When raising the leg (as in kicking a ball), the POSTERIOR THIGH MUSCLES *restrict knee extension.*

   ■ Are ANTAGONISTIC to the QUADRICEPS FEMORIS (See 6.19) when flexing the leg.

MEDIAL FEMORAL (ADDUCTOR) THIGH MUSCLES

1 | GRACILIS |
2 | PECTINEUS |
3 | ADDUCTOR LONGUS |
4 | ADDUCTOR BREVIS | SYNERGISTIC ACTION
5 | ADDUCTOR MAGNUS |

MEDIAL THIGH MUSCLES THAT ACT ON THE HIP JOINT (AND THIGH)

ADDUCTORS OF THIGH
| ALL THE MEDIAL FEMORAL THIGH MUSCLES |

FLEXORS OF THIGH
2 | PECTINEUS |
3 | ADDUCTOR LONGUS |
4 | ADDUCTOR BREVIS |

LATERAL ROTATORS OF THIGH
3 | ADDUCTOR LONGUS |
4 | ADDUCTOR BREVIS |
5 | ADDUCTOR MAGNUS |

MEDIAL THIGH MUSCLES THAT ACT ON THE KNEE JOINT (AND LOWER LEG)

FLEXOR OF LEG
1 | GRACILIS |

EXTENSOR OF LEG
5 | ADDUCTOR MAGNUS | (Superficial)

POSTERIOR FEMORAL (FLEXOR) THIGH MUSCLES
THE 3 "HAMSTRINGS":

6 | BICEPS FEMORIS |
7 | SEMITENDINOSUS |
8 | SEMIMEMBRANOSUS |

■ BICEPS FEMORIS
• A superficial LONG HEAD (ORIGIN = ISCHIAL TUBEROSITY)
• A deep, SHORT HEAD (ORIGIN = LINEA ASPERA of FEMUR)
• INSERTION = Head of FIBULA and LATERAL CONDYLE of TIBIA
• Action is complicated, works over both HIP and KNEE JOINTS

■ SEMITENDINOSUS
• Fusiform (ORIGIN = ISCHIAL TUBEROSITY from a COMMON TENDON WITH the BICEPS FEMORIS)
• Long slender tapering tendon (INSERTION = PROXIMAL LATERAL portion of TIBIA)
• Action is complicated, works over both HIP and KNEE JOINTS

■ SEMIMEMBRANOSUS
• Flat (ORIGIN = ISCHIAL TUBEROSITY)
• INSERTION = MEDIAL CONDYLE OF TIBIA
• Action similar to SEMITENDINOSUS

POSTERIOR THIGH MUSCLES THAT ACT ON THE HIP JOINT (AND THIGH)

EXTENSORS OF THIGH
| ALL THE POSTERIOR FEMORAL THIGH MUSCLES |

POSTERIOR THIGH MUSCLES THAT ACT ON THE KNEE JOINT (AND LOWER LEG)

LATERAL ROTATOR OF LEG AND THIGH
6 | BICEPS FEMORIS |

MEDIAL ROTATOR OF LEG AND THIGH
7 | SEMITENDINOSUS |

FLEXORS OF LEG
| ALL THE POSTERIOR FEMORAL THIGH MUSCLES |

MEDIAL FEMORAL (ADDUCTOR) THIGH MUSCLES

POSTERIOR FEMORAL (FLEXOR) THIGH MUSCLES

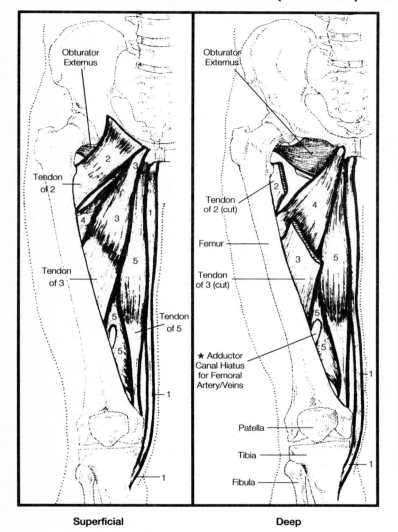

Superficial

Deep

Anterior Views

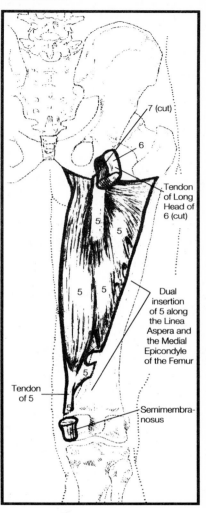

Posterior View

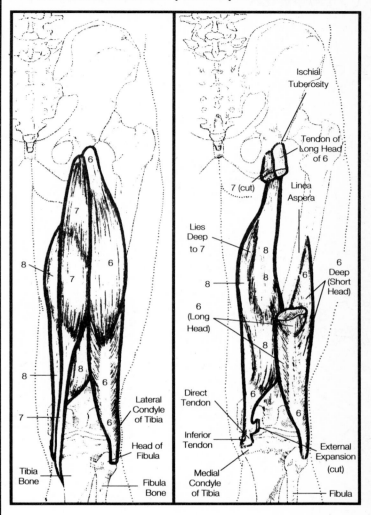

Posterior Views of the 3 "HAMSTRINGS"

MUSCULAR SYSTEM: MUSCLES OF THE APPENDICULAR SKELETON
Muscles of the Lower Limb: Muscles of the Thigh: *Anterior Femoral*

★ QUADRICEPS FEMORIS arises from 4 distinct heads: (4 separate origins)
■ RECTUS FEMORIS
  • ORIGIN = The anterior inferior spine of the ilium and the lip of the acetabulum
■ VASTUS LATERALIS
  • ORIGIN = Greater trochanter and Linea aspera of Femur
■ VASTUS MEDIALIS
  • ORIGIN = Medial surface of the Femur
■ VASTUS INTERMEDIUS
  • ORIGIN = Anterior and lateral surfaces of Femur
  • All 4 muscles of the QUADRICEPS FEMORIS insert on the common TENDON of the QUADRICEPS FEMORIS

★ QUADRICEPS FEMORIS (4 composite muscles) has a common tendon INSERTION ON THE PATELLA and thick PATELLAR TENDON, which attaches to the TIBIAL TUBEROSITY.

★ ■ PATELLAR TENDON is technically a LIGAMENT because the PATELLA is a sesamoid bone that develops inside of the tendon.

■ QUADRICEPS FEMORIS
The *ONLY* EXTENSORS of the KNEE JOINT
■ RECTUS FEMORIS
  • Main FLEXOR of the HIP JOINT (aided by ILIOPSOAS (See 6.9 and 6.17) and SARTORIUS (weak)
  • The only one of the 4 QUADRICEPS that contracts over both HIP and KNEE JOINTS.

★ ■ SARTORIUS can act on both the HIP and KNEE JOINT
  • Flexes and rotates thigh
  • Flexes leg
  • Flexes knee

The SARTORIUS muscle is the LONGEST MUSCLE in the body.
A long, ribbon-shaped thigh muscle that obliquely crosses the anterior aspect of the thigh,
ORIGIN = ANTERIOR SUPERIOR ILIAC SPINE
INSERTION = TIBIAL TUBEROSITY

ANTERIOR FEMORAL (EXTENSOR) THIGH MUSCLES

1 | SARTORIUS | THE "TAILOR'S MUSCLE"

| QUADRICEPS FEMORIS | : QUADRICEPS EXTENSOR FEMORIS
COMPOSITE OF 4 DISTINCT MUSCLES (HEADS)

2 | RECTUS FEMORIS |

3 | VASTUS LATERALIS | VASTUS EXTERNUS

4 | VASTUS MEDIALIS | VASTUS INTERNUS ⎤ SYNERGISTIC ACTION OF EXTENSION

5 | VASTUS INTERMEDIUS | CRUREUS

ANTERIOR THIGH MUSCLES that ACT on the THIGH (and HIP JOINT)

ABDUCTOR OF THIGH (AND HIP)
1 | SARTORIUS |

LATERAL ROTATOR OF THIGH (AND HIP)
1 | SARTORIUS |

FLEXORS OF THIGH (AND HIP)
1 | SARTORIUS | (weak)

2 | RECTUS FEMORIS |

ANTERIOR THIGH MUSCLES THAT ACT ON THE KNEE JOINT (AND LOWER LEG)

FLEXOR OF THE KNEE JOINT/ MEDIAL ROTATOR OF LEG
1 | SARTORIUS |

EXTENSORS OF LEG AT KNEE (KNEE JOINT)
| QUADRICEPS FEMORIS: ALL 4 HEADS |

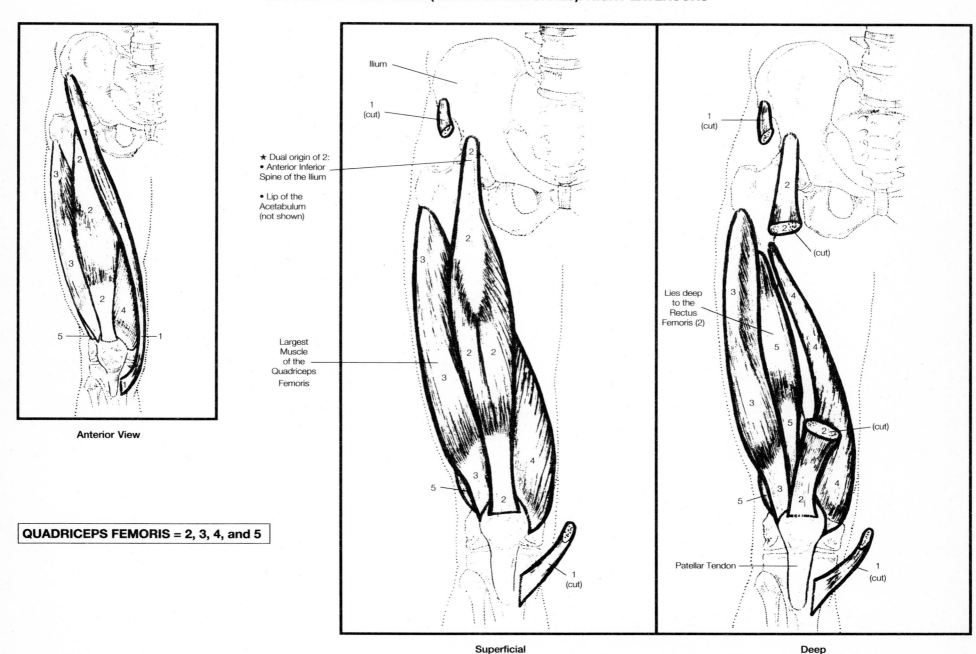

Anterior View

QUADRICEPS FEMORIS = 2, 3, 4, and 5

Ilium

1 (cut)

★ Dual origin of 2:
• Anterior Inferior Spine of the Ilium

• Lip of the Acetabulum (not shown)

Largest Muscle of the Quadriceps Femoris

1 (cut)

Superficial

1 (cut)

Lies deep to the Rectus Femoris (2)

(cut)

(cut)

Patellar Tendon

1 (cut)

Deep

Anterior Views of RIGHT THIGH

6.20

Support and Movement 134

Dorsiflexors = warm colors
Evertors = cool colors
★ See 4.14, 6.21

MUSCULAR SYSTEM: MUSCLES OF THE APPENDICULAR SKELETON
Muscles of the Lower Limb: Muscles of the Leg: *Anterior and Lateral*

CRURAL LEG MUSCLES MOVE THE FOOT:

★ ■ ANTERIOR LEG MUSCLES
Cross Several Joints:
- EXTENSORS of TOES
- DORSIFLEXORS of the ANKLE

■ (TIBIALIS ANTERIOR is also an
INVERTOR of the FOOT as well as
an ankle DORSIFLEXOR)
- Found in the ANTEROLATERAL
portion of the LEG.

■ *TIBIALIS ANTERIOR*
ORIGIN = Upper tibia (under lateral
condyle); interosseous
membrane, intermuscular
septum
INSERTION = Internal cuneiform
and 1st metatarsal

■ *EXTENSOR DIGITORUM LONGUS*
ORIGIN = Under lateral condyle of tibia;
body of fibula
INSERTION = 2-5 phalanges of toes

■ *EXTENSOR HALLUCIS LONGUS*
ORIGIN = Front of fibula shaft;
interosseous membrane
INSERTION = Terminal phalanx of the
Great Toe

■ *PERONEUS TERTIUS* is actually an
extension of EXTENSOR DIGITORUM
LONGUS. Its EVERSION qualities are
suspect.

■ *LATERAL (PERONEAL) LEG MUSCLES*
- Have their tendons passing
laterally and under the FOOT
- EVERTORS OF FEET

■ *PERONEUS LONGUS*
ORIGIN = Head of fibula, and lateral
condyle of tibia
INSERTION = By tendon under to
internal cuneiform
and 1st metatarsal

■ *PERONEUS BREVIS*
ORIGIN = Midportion of front
of fibula shaft
INSERTION = Of 5th metatarsal

★ There is NO MUSCLE ATTACHMENT along
the ANTEROMEDIAL ASPECT or portion
of the LEG along the TIBIAL SHAFT.

MUSCLES of the LEG: CRURAL MUSCLES
ANTERIOR CRURAL MUSCLES: DORSIFLEXORS

1 | TIBIALIS ANTERIOR | TIBIALIS ANTICUS

2 | EXTENSOR DIGITORUM LONGUS |

3 | EXTENSOR HALLUCIS LONGUS | GREAT TOE

4 | PERONEUS TERTIUS |

LATERAL CRURAL MUSCLES: EVERTORS

5 | PERONEUS LONGUS |
6 | PERONEUS BREVIS | — Synergistic Action

ANTERIOR and LATERAL MUSCLES of
the LOWER LEG that ACT on the
ANKLE JOINT, FOOT, and TOES

DORSIFLEXION OF ANKLE AND FOOT
| ALL ANTERIOR CRURAL MUSCLES |

INVERSION OF FOOT
1 | TIBIALIS ANTERIOR |

EVERSION OF FOOT
| ALL PERONEAL MUSCLES:
P. TERTIUS, P. LONGUS, P. BREVIS |

PLANTAR FLEXION OF FOOT
5 | PERONEUS LONGUS |

6 | PERONEUS BREVIS |

ADDUCTION OF FOOT
5 | PERONEUS LONGUS |

ABDUCTION OF FOOT
6 | PERONEUS BREVIS |

EXTENSORS OF THE TOES
2 | EXTENSOR DIGITORUM LONGUS |

3 | EXTENSOR HALLUCIS LONGUS |
(Great Toe)

MUSCLES OF THE LEG: (ANTERIOR AND LATERAL): RIGHT CRURALS

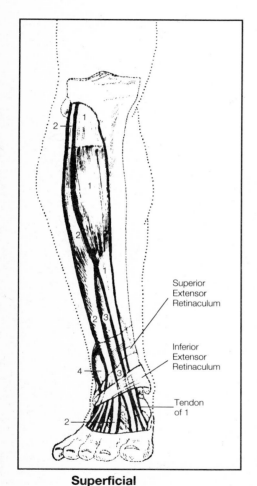

Superficial

ANTERIOR CRURALS
(Anterior Views)

Superior Extensor Retinaculum

Inferior Extensor Retinaculum

Tendon of 1

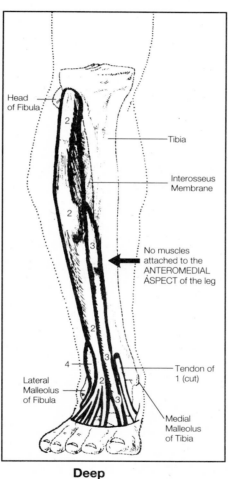

Deep

Head of Fibula

Tibia

Interosseus Membrane

No muscles attached to the ANTEROMEDIAL ASPECT of the leg

Tendon of 1 (cut)

Lateral Malleolus of Fibula

Medial Malleolus of Tibia

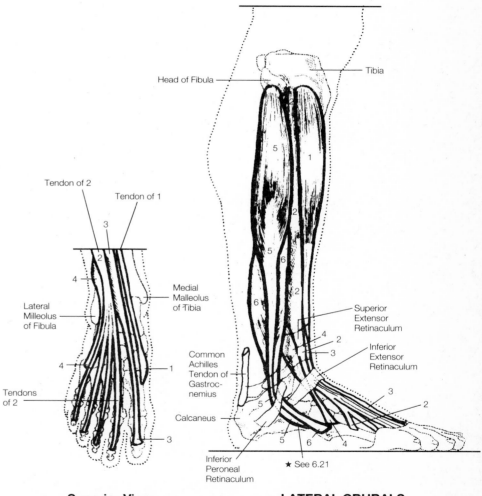

Tendon of 2

Tendon of 1

Lateral Milleolus of Fibula

Medial Malleolus of Tibia

Tendons of 2

Superior View
(Top-Dorsal Surface of the Foot)

Head of Fibula

Tibia

Superior Extensor Retinaculum

Inferior Extensor Retinaculum

Common Achilles Tendon of Gastroc-nemius

Calcaneus

Inferior Peroneal Retinaculum

★ See 6.21

LATERAL CRURALS
(Right Lateral View)

MUSCULAR SYSTEM: MUSCLES OF THE APPENDICULAR SKELETON
Muscles of the Lower Limb: Muscles of the Leg: *Posterior*

Superficial Group = warm colors
Deep Group = cool colors
(Use colors from 6.20 for
PERONEUS LONGUS and
PERONEUS BREVIS)
★ See 4.14, 6.20, 6.22

★ **7 POSTERIOR CRURAL LEG MUSCLES**

■ **3 SUPERFICIAL MUSCLES** called the
TRICEPS SURAE:
  • GASTROCNEMIUS
  • SOLEUS
  • PLANTARIS

■ GASTROCNEMIUS = Major portion of
the Calf muscles has 2 distinct heads
and ORIGINS:
  • ORIGIN OF LATERAL HEAD =
    Posterior surface of Lateral
    Condyle of Femur
  • ORIGIN OF MEDIAL HEAD =
    Posterior surface of Medial
    Condyle of Femur
■ SOLEUS
  • ORIGIN = Head and upper shaft
    of fibula, oblique line of tibia
■ A TRICIPITAL MUSCLE is formed from
the 2 HEADS of the GASTROCNEMIUS
and the SOLEUS
  • INSERTION = Common TENDON
    OF ACHILLES, or TENDO
    CALCANEUS, inserts on the
    CALCANEUS bone of the FOOT,
    (OSCALCIS)
  • Can lift the entire body onto the
    heads of the METATARSAL BONES
    in a special movement = PLANTAR
    FLEXION. This action involves only
    one joint, the ANKLE.
■ PLANTARIS
  • ORIGIN = External supracondyloid
    ridge of femur
  • INSERTION = Inner border of
    Achilles Tendon

■ **4 DEEP MUSCLES**
  • Unique INSERTION of
    Tendons of TIBIALIS POSTERIOR
    allows INVERSION of the FOOT,
    as well as PLANTAR FLEXION.

  • Note how the tendons of FLEXOR
    HALLUCIS LONGUS and FLEXOR
    DIGITORUM LONGUS pass under
    the sole of the foot to insert on the
    terminal phalanges of the toes.

MUSCLES of the LEG: CRURAL MUSCLES
POSTERIOR CRURAL MUSCLES

SUPERFICIAL GROUP: TRICEPS SURAE

1 | GASTROCNEMIUS | 2 **HEADS (LATERAL & MEDIAL)**

2 | SOLEUS |

3 | PLANTARIS |

DEEP GROUP:

4 | POPLITEUS |

5 | FLEXOR HALLUCIS LONGUS |

6 | FLEXOR DIGITORUM LONGUS |

7 | TIBIALIS POSTERIOR | **TIBIALIS POSTICUS**
  (ALSO SUPPORTS ARCHES)

POSTERIOR MUSCLES of the
LEG that ACT on the LEG

  FLEXOR OF KNEE JOINT
  1 | GASTROCNEMIUS |

  FLEXORS OF THE LEG
  1 | GASTROCNEMIUS |
  3 | PLANTARIS |
  4 | POPLITEUS |

  MEDIAL ROTATOR OF LEG
  4 | POPLITEUS |

POSTERIOR MUSCLES of the LEG that ACT
on the ANKLE JOINT, FOOT, and TOES

  DORSIFLEXION OF ANKLE AND FOOT
  5 | FLEXOR HALLUCIS LONGUS |

  INVERSION OF FOOT
  7 | TIBIALIS POSTERIOR |

  EXTENSION OF FOOT
  3 | PLANTARIS |
  6 | FLEXOR DIGITORUM LONGUS |

  PLANTAR FLEXION OF FOOT
  1 | GASTROCNEMIUS |
  2 | SOLEUS |
  3 | PLANTARIS |
  7 | TIBIALIS POSTERIOR |

  ROTATION OF FOOT
  2 | SOLEUS |

  FLEXION OF THE TOES
  6 | FLEXOR DIGITORUM LONGUS |
  5 | FLEXOR HALLUCIS LONGUS |
  (Great Toe)

MUSCLES OF THE LEG (POSTERIOR): RIGHT CRURALS

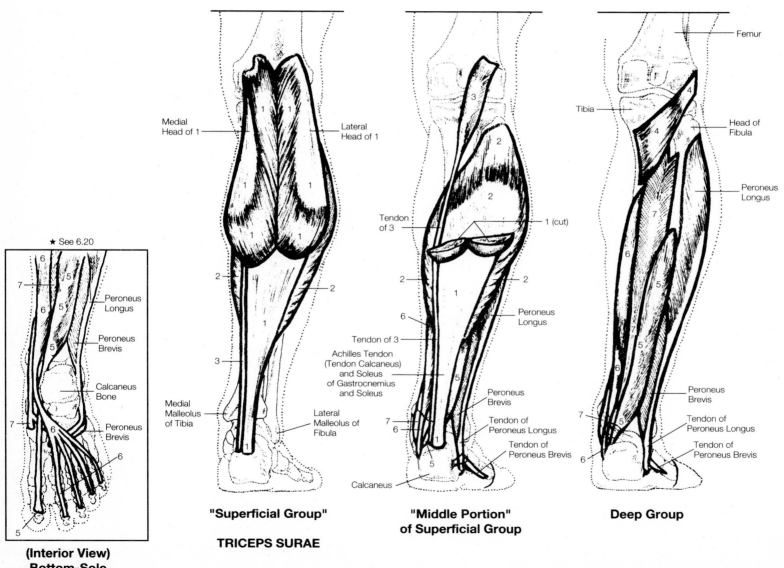

★ See 6.20

**(Interior View)
Bottom-Sole
of the Foot**

Medial
Head of 1

Lateral
Head of 1

Peroneus
Longus

Peroneus
Brevis

Calcaneus
Bone

Peroneus
Brevis

Medial
Malleolus
of Tibia

Lateral
Malleolus of
Fibula

"Superficial Group"

TRICEPS SURAE

Tendon
of 3

1 (cut)

Peroneus
Longus

Tendon of 3

Achilles Tendon
(Tendon Calcaneus)
and Soleus
of Gastrocnemius
and Soleus

Peroneus
Brevis

Tendon of
Peroneus Longus

Tendon of
Peroneus Brevis

Calcaneus

**"Middle Portion"
of Superficial Group**

Femur

Tibia

Head of
Fibula

Peroneus
Longus

Peroneus
Brevis

Tendon of
Peroneus Longus

Tendon of
Peroneus Brevis

Deep Group

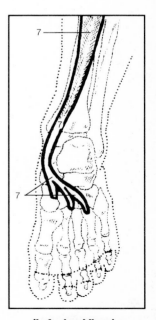

**(Inferior View)
Bottom-Sole
of the Foot**

(Posterior Views)

MUSCULAR SYSTEM: MUSCLES OF THE APPENDICULAR SKELETON
Muscles of the Lower Limb: Muscles of the Foot:
Plantar (Bottom) Layers: Superficial and Middle

① = *yellow*
Superficial Layer = oranges and reds
Middle Layer = pinks and purples
Tendons = light or neutral colors
★ See 4.15, 6.16, 6.21, 6.23

The MUSCLES OF THE FOOT are arbitrarily divided into 3 PLANTAR (BOTTOM) LAYERS, (Superficial, Middle, and Deep) and DORSAL SURFACE MUSCLES. Here we will concentrate on the outermost plantar layers of the sole.

★ • The MUSCLES of the FOOT are similar in name and number to those of the HAND, (See 6.16) (with the exception of one short intrinsic muscle, the EXTENSOR DIGITORUM BREVIS). (See 6.23)

■ Functions of the FOOT and HAND are different, however:

    ■ FOOT:
      • Provides support
      • Bears heavy body
        weight (no grasping)
    ■ HAND:
      • Complex instrument
        for precise movements,
        such as grasping

★ Topographically, the FOOT MUSCLES are arranged into 3–4 LAYERS, although they are not distinct (even in dissection), and different authorities have differing opinions. Action of the FOOT MUSCLES provide:

    ■ Movement of toes
    ■ Support for arches of
      the foot (by CONTRACTING)

• LUMBRICALES arise from the TENDONS of the FLEXOR DIGITORUM LONGUS (See 6.21)

    Review 6.20, and note the tendons of PERONEUS LONGUS and PERONEUS BREVIS.
    Review 6.21 and note the tendons of FLEXOR HALLUCIS LONGUS and FLEXOR DIGITORUM LONGUS.
    See 6.23 for more details of deep layer plantar muscles.

SUPERFICIAL SHEATH OF THE SOLE
① **PLANTAR APONEUROSIS**

MUSCLES of the SOLE

PLANTAR (BOTTOM) SURFACE:
SOLE of the FOOT

SUPERFICIAL LAYER
2 **ABDUCTOR HALLUCIS**

3 **ABDUCTOR DIGITI MINIMI**

4 **FLEXOR DIGITORUM BREVIS**

MIDDLE LAYER
5 **QUADRATUS PLANTAE**   PRONATOR PEDIS
FLEXOR ACCESSORIUS

6 **LUMBRICALES**   (4)

DEEP LAYER ★ See 6.23:
7 **FLEXOR HALLUCIS BREVIS**

9 **FLEXOR DIGITI MINIMI BREVIS**

11 **PLANTAR INTEROSSEI**

12 **DORSAL INTEROSSEI**

(Medial)   (Lateral)

**MOST SUPERFICIAL
BOTTOM OF THE SOLE
(Right Foot)**

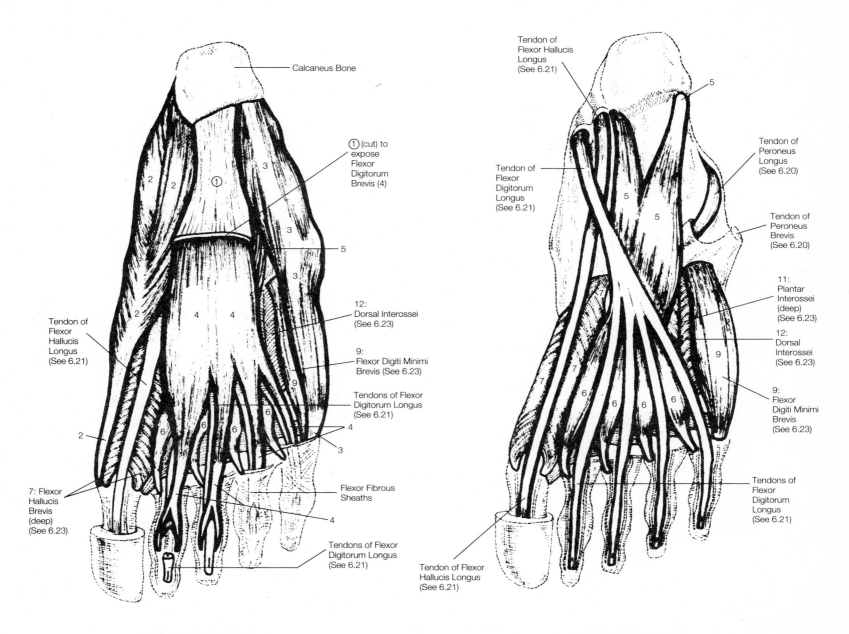

Calcaneus Bone

(1) (cut) to expose Flexor Digitorum Brevis (4)

12: Dorsal Interossei (See 6.23)

9: Flexor Digiti Minimi Brevis (See 6.23)

Tendons of Flexor Digitorum Longus (See 6.21)

Flexor Fibrous Sheaths

Tendons of Flexor Digitorum Longus (See 6.21)

Tendon of Flexor Hallucis Longus (See 6.21)

7: Flexor Hallucis Brevis (deep) (See 6.23)

Superficial Layers

Tendon of Flexor Hallucis Longus (See 6.21)

Tendon of Flexor Digitorum Longus (See 6.21)

Tendon of Peroneus Longus (See 6.20)

Tendon of Peroneus Brevis (See 6.20)

11: Plantar Interossei (deep) (See 6.23)

12: Dorsal Interossei (See 6.23)

9: Flexor Digiti Minimi Brevis (See 6.23)

Tendons of Flexor Digitorum Longus (See 6.21)

Tendon of Flexor Hallucis Longus (See 6.21)

Middle Layer
(1, 2, 3, and 4 removed)

OUTER PLANTAR (BOTTOM) LAYERS OF THE RIGHT FOOT (SOLE)
(Inferior Views)

MUSCULAR SYSTEM: MUSCLES OF THE APPENDICULAR SKELETON
Muscles of the Lower Limb: Muscles of the Foot:
Deep Plantar Layers and Dorsal Surface Muscles

Deep Layer = blues and greens
Dorsal Surface = oranges and reds
★ See 4.15, 6.22

★ In the previous page, we studied the SUPERFICIAL and MIDDLE LAYERS of the BOTTOM (or SOLE) of the FOOT.

Here, we look at the DEEP LAYERS of the SOLE, directly underneath the muscles of the DORSAL (or TOP) surface of the FOOT.

★ Note the SUPERIOR and INFERIOR EXTENSOR RETINACULUM
- Thickenings of the deep fascia in the distal portion of the lower legs that hold tendons in position when muscles contract.

■ SUPERIOR EXTENSOR RETINACULUM
- Crosses the extensor tendons of the foot
- Attached to the lower portions of the TIBIA and FIBULA

■ INFERIOR EXTENSOR RETINACULUM
- Consists of 2 LIMBS
- Both have common origin on lateral surface of CALCANEUS
- Upper limb attached to medial malleolus
- Lower limb curves around instep; attached to the fascia of the ABDUCTOR HALLUCIS on the medial side of the foot.

† TRANSVERSUS PEDIS = Transverse head of ADDUCTOR HALLUCIS

OBLIQUUS HALLUCIS = Oblique head of ADDUCTOR HALLUCIS

★ EXTENSOR HALLUCIS BREVIS is that portion of EXTENSOR DIGITORUM BREVIS that inserts on the Great Toe

Review 6.20
- Tendons of 2-6 shown here
Review 6.21
- Tendons of 5-7 shown here

MUSCLES of the FOOT

PLANTAR (BOTTOM) SURFACE
SOLE of the FOOT
DEEP LAYERS
7 **FLEXOR HALLUCIS BREVIS**

8 **ADDUCTOR HALLUCIS** † (TRANSVERSUS or OBLIQUUS) HALLUCIS

9 **FLEXOR DIGITI MINIMI BREVIS**

10 **OPPONENS DIGITI MINIMI** OPPONENS DIGITI QUINTI

11 **PLANTAR INTEROSSEI** (3) INTEROSSEUS PLANTARIS

DORSAL (TOP) SURFACE
12 **DORSAL INTEROSSEI** (4) INTEROSSEUS DORSALIS PEDIS

13 **EXTENSOR DIGITORUM BREVIS**

14 **EXTENSOR HALLUCIS BREVIS** *

FASCIA RETINACULUM
15 **SUPERIOR EXTENSOR RETINACULUM**

16 **INFERIOR EXTENSOR RETINACULUM**

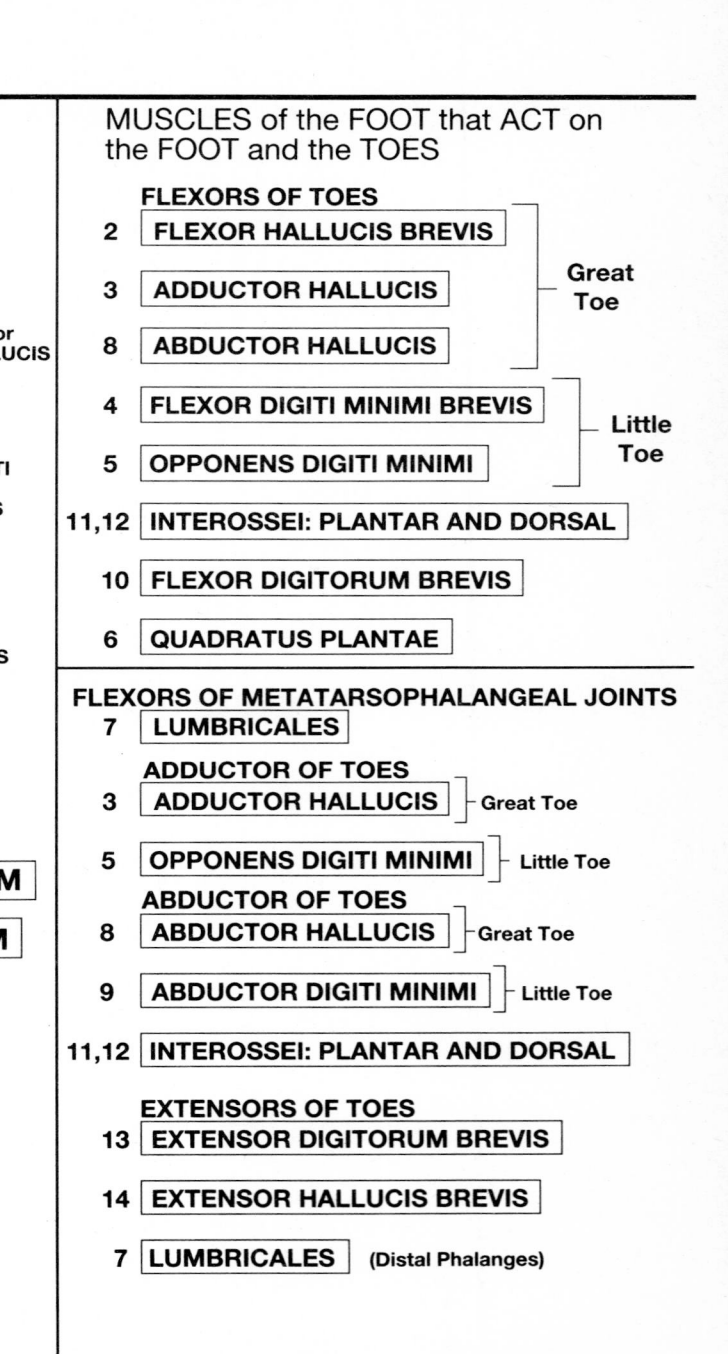

MUSCLES of the FOOT that ACT on the FOOT and the TOES

FLEXORS OF TOES
2 **FLEXOR HALLUCIS BREVIS**
3 **ADDUCTOR HALLUCIS** — Great Toe
8 **ABDUCTOR HALLUCIS**
4 **FLEXOR DIGITI MINIMI BREVIS** — Little Toe
5 **OPPONENS DIGITI MINIMI**

11,12 **INTEROSSEI: PLANTAR AND DORSAL**
10 **FLEXOR DIGITORUM BREVIS**
6 **QUADRATUS PLANTAE**

FLEXORS OF METATARSOPHALANGEAL JOINTS
7 **LUMBRICALES**

ADDUCTOR OF TOES
3 **ADDUCTOR HALLUCIS** — Great Toe

5 **OPPONENS DIGITI MINIMI** — Little Toe

ABDUCTOR OF TOES
8 **ABDUCTOR HALLUCIS** — Great Toe

9 **ABDUCTOR DIGITI MINIMI** — Little Toe

11,12 **INTEROSSEI: PLANTAR AND DORSAL**

EXTENSORS OF TOES
13 **EXTENSOR DIGITORUM BREVIS**

14 **EXTENSOR HALLUCIS BREVIS**

7 **LUMBRICALES** (Distal Phalanges)

DEEP PLANTAR (BOTTOM) LAYERS OF THE RIGHT FOOT (SOLE)

DORSAL (TOP) SURFACE OF THE RIGHT FOOT

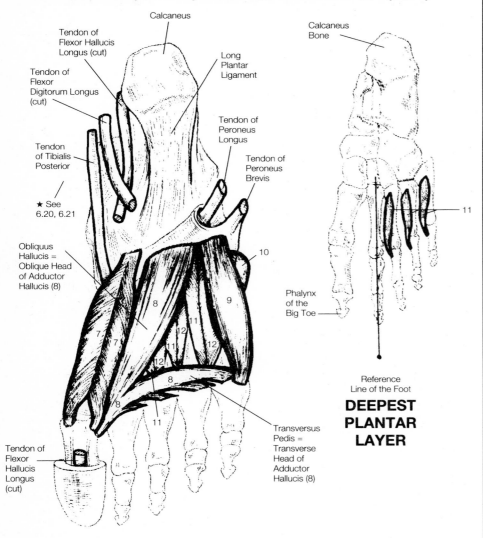

Calcaneus

Tendon of Flexor Hallucis Longus (cut)

Tendon of Flexor Digitorum Longus (cut)

Long Plantar Ligament

Tendon of Tibialis Posterior

Tendon of Peroneus Longus

Tendon of Peroneus Brevis

★ See 6.20, 6.21

Obliquus Hallucis = Oblique Head of Adductor Hallucis (8)

10

Tendon of Flexor Hallucis Longus (cut)

Transversus Pedis = Transverse Head of Adductor Hallucis (8)

Calcaneus Bone

11

Phalynx of the Big Toe

Reference Line of the Foot

DEEPEST PLANTAR LAYER

Inferior Views

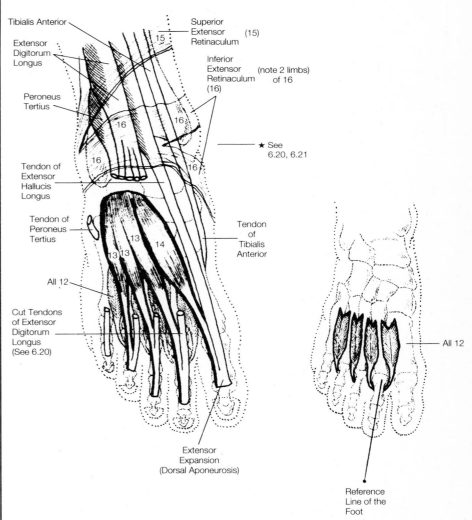

Tibialis Anterior

Extensor Digitorum Longus

Peroneus Tertius

Superior Extensor Retinaculum (15)

15

Inferior Extensor Retinaculum (16)

(note 2 limbs) of 16

16

16

16

16

16

★ See 6.20, 6.21

Tendon of Extensor Hallucis Longus

Tendon of Peroneus Tertius

13 13 13 14

Tendon of Tibialis Anterior

All 12

Cut Tendons of Extensor Digitorum Longus (See 6.20)

Extensor Expansion (Dorsal Aponeurosis)

All 12

Reference Line of the Foot

Superior Views

REVIEW: CROSS-SECTIONAL SUMMARY OF RIGHT ARM, FOREARM, THIGH, AND LEG

- Cross-sectional views of the upper and lower extremities are excellent devices to review the location and relationships of the major skeletal muscles of these areas. (Note that the ANTERIOR POSITION is at the top of each illustration.)

- Reference numbers do not correspond to previous reference numbers. Go back to the previous pages indicated for specific muscles and use the colors chosen from the appropriate structures. Any muscles assigned the same color will be easily distinguished.

- Note the position and location of the major blood vessels, bones, nerves, and interosseous membranes in each cross-section.

★ UPPER EXTREMITY: ARM and FOREARM
- Note the ANTERIOR MUSCLES of the forearm are mostly FLEXORS. (FLEXORS are muscles that bend a limb at a joint.) Note the size of the FLEXOR DIGITORUM MUSCLES. (9 and 10)
- Note the POSTERIOR MUSCLES of the forearm are mostly EXTENSORS. (EXTENSORS are muscles that straighten a limb at a joint.)

★ LOWER EXTREMITY: THIGH and LEG
- Note the muscles of the THIGH that move the leg are surrounded and compartmentalized by tough, fascial sheets. The thigh muscles are generally divided into 2 groups according to function and position:
  ■ ANTERIOR EXTENSOR MUSCLES
  ■ POSTERIOR FLEXOR MUSCLES
The most powerful muscles in the body are found in the thigh and legs, especially in those muscles acting on the hip joint.

★ The chapter ends with a review of the major superficial muscles of the 5 major regions of the body: HEAD, NECK, TRUNK, UPPER EXTREMITY, LOWER EXTREMITY.
(Note that the reference, numbers do not correspond to reference numbers in the cross-section. Use the same colors for identical muscles.)

ARM (Distal Third)

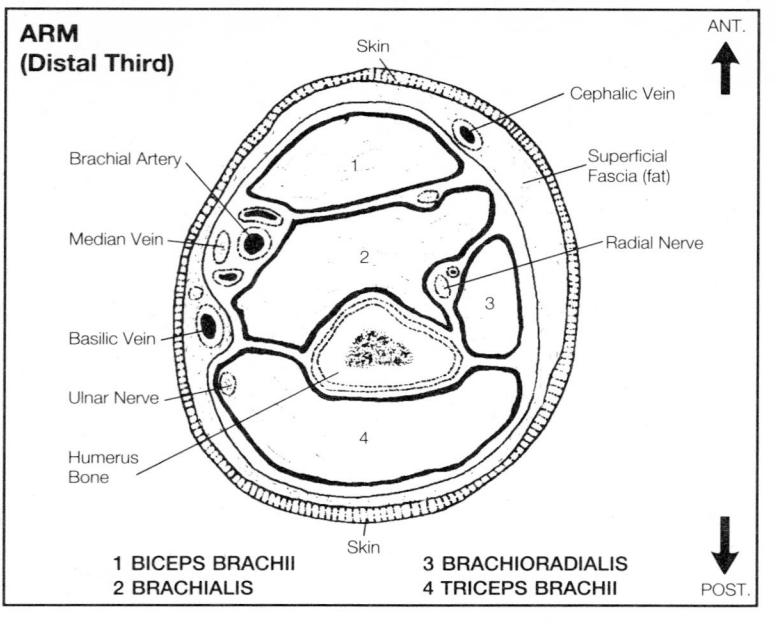

Skin
Cephalic Vein
Brachial Artery
Superficial Fascia (fat)
Median Vein
Radial Nerve
Basilic Vein
Ulnar Nerve
Humerus Bone
Skin

1 BICEPS BRACHII	3 BRACHIORADIALIS
2 BRACHIALIS	4 TRICEPS BRACHII

THIGH (Middle Portion)

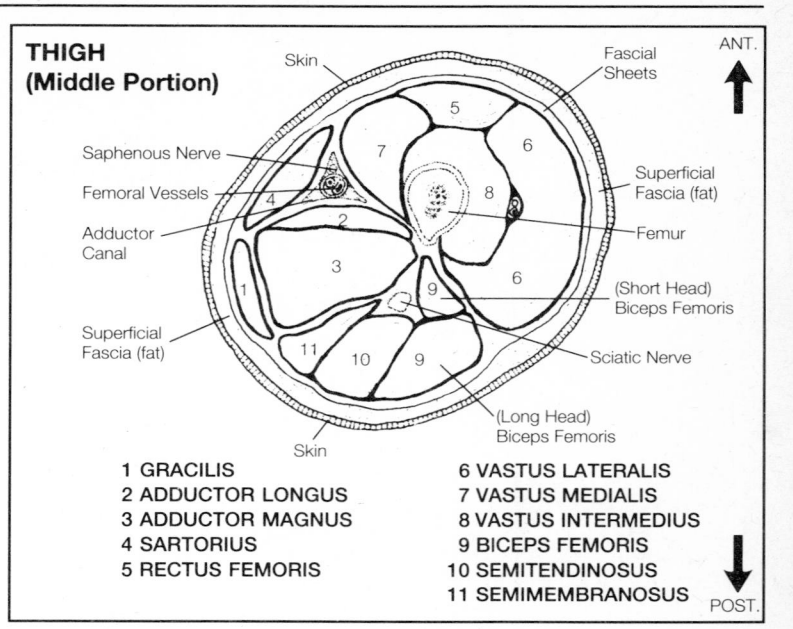

Skin
Fascial Sheets
Saphenous Nerve
Superficial Fascia (fat)
Femoral Vessels
Femur
Adductor Canal
(Short Head) Biceps Femoris
Superficial Fascia (fat)
Sciatic Nerve
(Long Head) Biceps Femoris
Skin

1 GRACILIS	6 VASTUS LATERALIS
2 ADDUCTOR LONGUS	7 VASTUS MEDIALIS
3 ADDUCTOR MAGNUS	8 VASTUS INTERMEDIUS
4 SARTORIUS	9 BICEPS FEMORIS
5 RECTUS FEMORIS	10 SEMITENDINOSUS
	11 SEMIMEMBRANOSUS

FOREARM (Distal Portion)

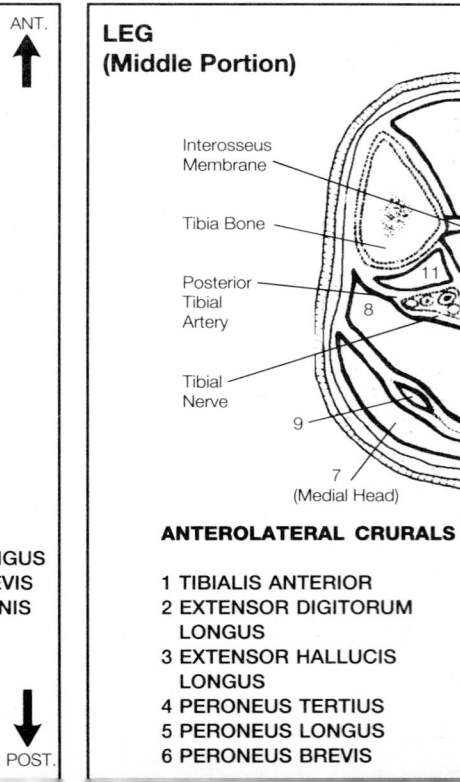

Median Nerve
Radial Artery
Ulnar Nerve
Superficial Fascial (fat)
Ulna Bone
Radius Bone
Interosseus Membrane
Skin

ANTERIOR FLEXORS	POSTERIOR EXTENSORS
5 PRONATOR TERES	12 EXTENSOR CARPI RADIALIS LONGUS
6 FLEXOR CARPI RADIALIS	13 EXTENSOR CARPI RADIALIS BREVIS
7 PALMARIS LONGUS	14 EXTENSOR DIGITORUM COMMUNIS
8 FLEXOR CARPI ULNARIS	15 EXTENSOR DIGITI MINIMI
9 FLEXOR DIGITORUM SUPERFICIALIS	16 EXTENSOR CARPI ULNARIS
10 FLEXOR DIGITORUM PROFUNDUS	17 ABDUCTOR POLLICIS LONGUS
11 FLEXOR POLLICIS LONGUS	18 EXTENSOR POLLICIS BREVIS
	19 EXTENSOR POLLICIS LONGUS
	20 EXTENSOR INDICIS

LEG (Middle Portion)

Interosseus Membrane
Tibia Bone
Posterior Tibial Artery
Fibula Bone
Tibial Nerve
Peroneal Artery
Lateral Head
Medial Head
Superficial Fascia (fat)
Skin

ANTEROLATERAL CRURALS	POSTERIOR CRURALS
1 TIBIALIS ANTERIOR	7 GASTROCNEMIUS
2 EXTENSOR DIGITORUM LONGUS	8 SOLEUS
3 EXTENSOR HALLUCIS LONGUS	9 PLANTARIS
4 PERONEUS TERTIUS	10 FLEXOR HALLUCIS LONGUS
5 PERONEUS LONGUS	11 FLEXOR DIGITORUM LONGUS
6 PERONEUS BREVIS	12 TIBIALIS POSTERIOR

REVIEW: MAJOR SUPERFICIAL MUSCLES OF THE HEAD, NECK, AND TRUNK

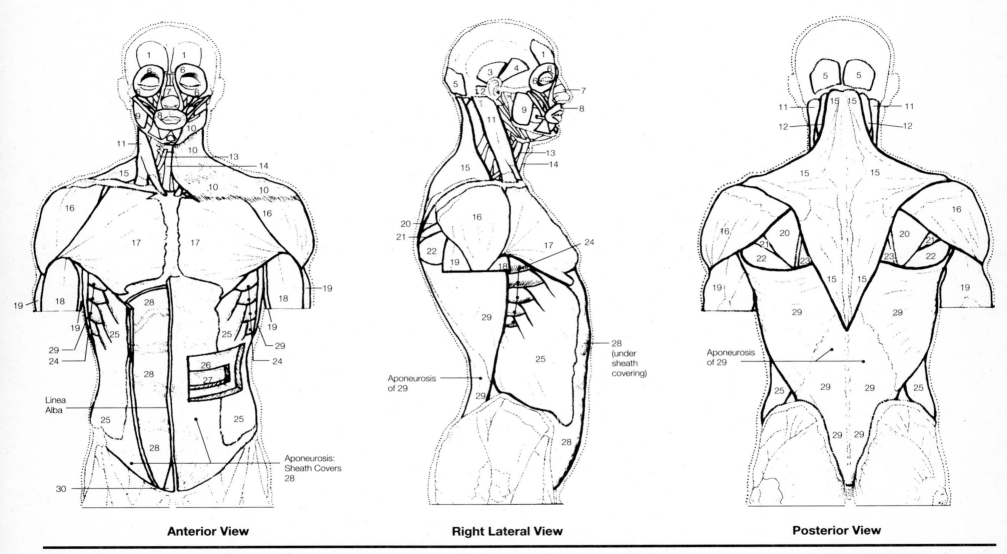

Anterior View

Right Lateral View

Posterior View

1 FRONTALIS	9 MASSETER	17 PECTORALIS MAJOR	25 EXTERNAL OBLIQUE
2 AURICULARIS POSTERIOR	10 PLATYSMA	18 BICEPS BRACHII	26 INTERNAL OBLIQUE
3 AURICULARIS SUPERIOR	11 STERNOCLEIDOMASTOID	19 TRICEPS BRACHII	27 TRANSVERSUS ABDOMINIS
4 AURICULARIS ANTERIOR	12 SPLENIUS CAPITIS	20 INFRASPINATUS	28 RECTUS ABDOMINIS
5 OCCIPITALIS	13 OMOHYOIDEUS	21 TERES MINOR	29 LATISSIMUS DORSI
6 ORBICULARIS OCULI	14 STERNOHYOIDEUS	22 TERES MAJOR	30 PYRAMIDALIS
7 TRANSVERSE NASALIS	15 TRAPEZIUS	23 RHOMBOIDEUS	
8 ORBICULARIS ORIS	16 DELTOIDEUS	24 SERRATUS ANTERIOR	

REVIEW: MAJOR SUPERFICIAL MUSCLES OF THE UPPER AND LOWER EXTREMITIES (RIGHT)

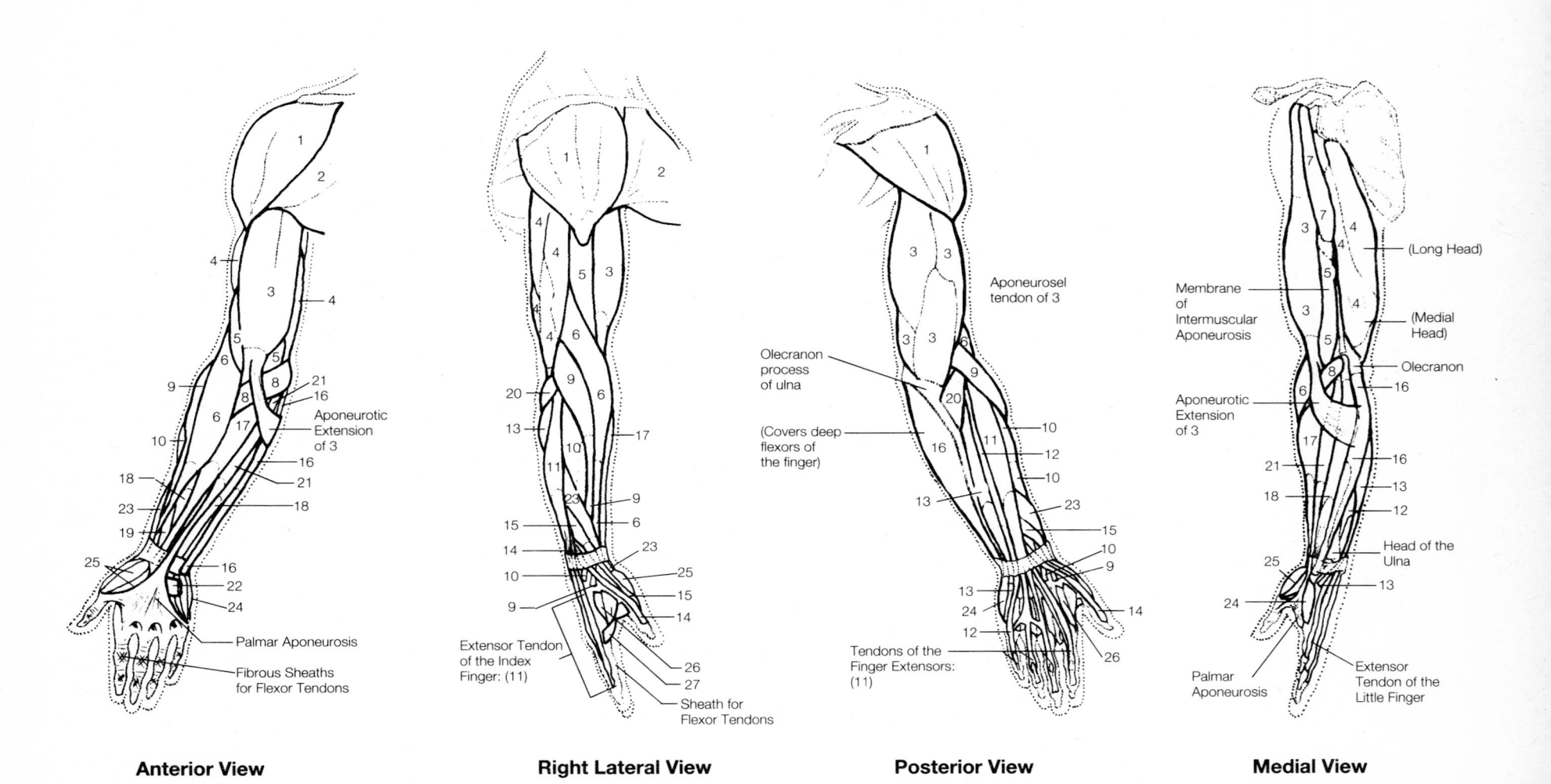

Anterior View   **Right Lateral View**   **Posterior View**   **Medial View**

Aponeurotic Extension of 3

Palmar Aponeurosis

Fibrous Sheaths for Flexor Tendons

Extensor Tendon of the Index Finger: (11)

Sheath for Flexor Tendons

Aponeurosel tendon of 3

Olecranon process of ulna

(Covers deep flexors of the finger)

Tendons of the Finger Extensors: (11)

Membrane of Intermuscular Aponeurosis

Aponeurotic Extension of 3

(Long Head)

(Medial Head)

Olecranon

Head of the Ulna

Palmar Aponeurosis

Extensor Tendon of the Little Finger

1 DELTOIDEUS	8 PRONATOR TERES	15 EXTENSOR POLLICIS BREVIS	22 PALMARIS BREVIS
2 PECTORALIS MAJOR	9 EXTENSOR CARPI RADIALIS LONGUS	16 FLEXOR CARPI ULNARIS	23 ABDUCTOR POLLICIS LONGUS
3 BICEPS BRACHII	10 EXTENSOR CARPI RADIALIS BREVIS	17 FLEXOR CARPI RADIALIS	24 MUSCLES OF THE HYPOTHENAR EMINENCE
4 TRICEPS BRACHII	11 EXTENSOR DIGITORUM COMMUNIS	18 FLEXOR DIGITORUM SUPERFICIALIS	25 MUSCLES OF THE THENAR EMINENCE
5 BRACHIALIS	12 EXTENSOR DIGITI MINIMI	19 FLEXOR POLLICIS LONGUS	26 FIRST DORSAL INTEROSSEUS
6 BRACHIORADIALIS	13 EXTENSOR CARPI ULNARIS	20 ANCONEUS	27 FIRST LUMBRICAL
7 CORACOBRACHIALIS	14 EXTENSOR POLLICIS LONGUS	21 PALMARIS LONGUS	

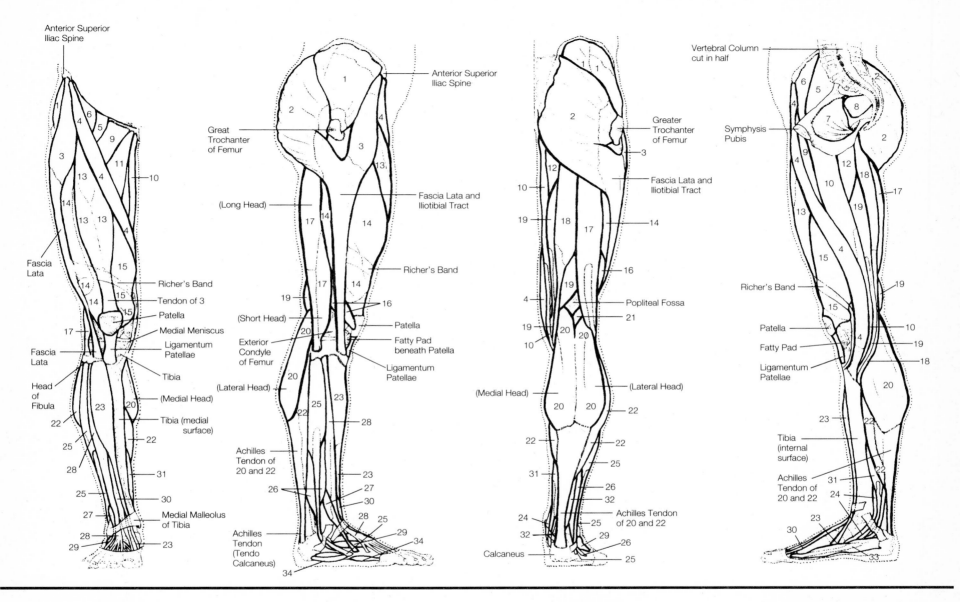

Anterior Superior
Iliac Spine

Fascia
Lata

Richer's Band
Tendon of 3
Patella
Medial Meniscus
Ligamentum
Patellae

Fascia
Lata

Tibia

Head
of
Fibula

Tibia (medial
surface)

Medial Malleolus
of Tibia

Anterior Superior
Iliac Spine

Great
Trochanter
of Femur

Fascia Lata and
Iliotibial Tract

(Long Head)

Richer's Band

(Short Head)

Patella

Exterior
Condyle
of Femur

Fatty Pad
beneath Patella

Ligamentum
Patellae

(Lateral Head)

Achilles
Tendon of
20 and 22

Achilles
Tendon
(Tendo
Calcaneus)

Greater
Trochanter
of Femur

Fascia Lata and
Iliotibial Tract

(Medial Head)

Popliteal Fossa

(Lateral Head)

Achilles Tendon
of 20 and 22

Calcaneus

Vertebral Column
cut in half

Symphysis
Pubis

Richer's Band

Patella

Fatty Pad

Ligamentum
Patellae

Tibia
(internal
surface)

Achilles
Tendon of
20 and 22

1 GLUTEUS MEDIUS	10 GRACILIS	19 SEMIMEMBRANOSUS	28 EXTENSOR DIGITORUM LONGUS
2 GLUTEUS MAXIMUS	11 ADDUCTOR LONGUS	20 GASTROCNEMIUS	29 EXTENSOR DIGITORUM BREVIS
3 TENSOR FASCIAE LATAE	12 ADDUCTOR MAGNUS	21 PLANTARIS	30 EXTENSOR HALLUCIS LONGUS
4 SARTORIUS	13 RECTUS FEMORIS	22 SOLEUS	31 FLEXOR DIGITORUM LONGUS
5 PSOAS MAJOR ⎤ ILIOPSOAS	14 VASTUS LATERALIS	23 TIBIALIS ANTERIOR	32 FLEXOR HALLUCIS LONGUS
6 ILIACUS ⎦	15 VASTUS MEDIALIS	24 TIBIALIS POSTERIOR	33 ABDUCTOR HALLUCIS
7 OBTURATOR INTERNUS	16 VASTUS INTERMEDIUS	25 PERONEUS LONGUS	34 ABDUCTOR DIGITI MINIMI
8 PIRIFORMIS	17 BICEPS FEMORIS	26 PERONEUS BREVIS	35 PLANTAR APONEUROSIS
9 PECTINEUS	18 SEMITENDINOSUS	27 PERONEUS TERTIUS	

Chapter 6: *MUSCULAR SYSTEM* : CHARTS

MUSCLES OF THE AXIAL SKELETON: 6.4–6.11

6.4 MUSCLES OF THE HEAD

FACIAL EXPRESSION	Name	Origin	Insertion	Action	Innervation
SCALP	**FRONTALIS**	Tissue above the supraorbital margin (no bony attachment)	Galea Aponeurotica*	Raises eyebrows; produces transverse furrows in skin of forehead; pulls scalp forward (as in surprise)	Temporal branch of facial nerve (VII)
	OCCIPITALIS	Superior nuchal line and mastoid portion of temporal bone	Galea Aponeurotica*	Pulls scalp backward	Posterior auricular branches of facial nerve (VII)
EYELIDS	**CORRUGATOR**	Inner end of superciliary ridge	Medial half of eyebrow	Frowning lines on forehead	Facial nerve (VII), temporal and zygomatic branches
	ORBICULARIS OCULI	Medial surface of orbit	Skin of eyelids, circularly around orbit	Closes and winks eye(s); Permits blinking, winking, and squinting	Facial nerve (VII), temporal and zygomatic branches
FACE AND NOSE	**PROCERUS**	Skin over nose	Skin of forehead	Draws eyebrows downward; produces transverse wrinkles at root of nose	Buccal branch of facial nerve (VII)
	NASALIS (TRANSVERSE and ALAR NASALIS)	Maxilla	Skin over bridge of nose	Narrows nostrils (depresses cartilaginous part of nose, draws alae medially)	Buccal branches of facial nerve (VII)
	DEPRESSOR SEPTI	Incisor fossa of superior maxillary bone (maxilla)	Ala and septum of nose	Contracts nostril and depresses ala (draws outer wall of nostril downward)	Facial Nerve (VII)

Scalp *GALEA APONEUROTICA The aponeurosis connecting the two bellies of the occipitofrontalis muscle

	Name	Origin	Insertion	Action	Innervation
MOUTH	**LEVATOR LABII SUPERIORIS**	Frontal process of maxilla lower margin of orbit	Skin and cartilage of nose and upper lip	Raises and extends upper lip; dilates nostrils	Infraorbital branch of facial nerve (VII)
	LEVATOR LABII SUPERIORIS ALAEQUE NASI	Nasal process of maxilla	Skin and cartilage of ala of nose and upper lip	Raises upper lip; dilates nostrils	Infraorbital branch of facial nerve (VII)
	LEVATOR ANGULI ORIS	Canine fossa of maxilla	Orbicularis oris muscle of mouth and skin at angle of mouth	Raises angle of mouth	Facial nerve (VII)
	ZYGOMATICUS MAJOR	Lateral aspect of zygomatic bone, zygomatic arch	Angle of the mouth	Draws angle of mouth upward and backward (smiling and laughing)	Buccal branches of facial nerve (VII)
	ZYGOMATICUS MINOR	Zygomatic bone behind the maxillary arch	Orbicularis oris muscle of mouth and levator muscle of upper lip	Draws lip upward and laterally (smiling and laughing)	Buccal branches of facial nerve (VII)
	RISORIUS	Fascia over masseter	Skin at angle of mouth	Draws angle of mouth laterally (tenseness) – ("laughing muscle")	Buccal branches of facial nerve (VII)
	DEPRESSOR LABII INFERIORIS	Anterior surface of lower border of mandible (external oblique line)	Orbicularis oris muscle of mouth and skin of lower lip	Depresses lower lip (expression of irony)	Facial nerve (VII)
	DEPRESSOR ANGULI ORIS	Lateral border of mandible (external oblique line)	Angle of mouth	Pulls down angle of mouth	Facial nerve (VII)
	MENTALIS	Mental symphysis and incisive fossa of mandible	Skin of chin and lower lip	Elevates and protrudes lower lip; wrinkles chin (pouting)	Facial nerve (VII), buccal and mandibular branches
	TRANSVERSUS MENTI	Refers to superficial fibers of depressor muscle of angle of mouth, which turn medially and cross to the opposite side.			
	ORBICULARIS ORIS	Sphincter muscle of mouth; nasal septum and canine fossa of mandible by accessory fibers	1. Labial—fibers restricted to lips 2. Marginal—fibers blending with those of adjacent muscles	Closes, protrudes lips (Forms words, puckers or purses mouth, as in kissing)	Buccal branches of facial nerve (VII)
	BUCCINATOR	Buccinator ridge of mandible; alveolar processes of maxilla; pterygomandibular ligament	Orbicularis oris muscle of mouth at angle of mouth	Compresses cheek (blowing air out of mouth) and retracts angle of mouth Cheeks cave in (as in sucking)	Buccal branches of facial nerve (VII)
MUSCLES OF MASTICATION	**TEMPORALIS**	Temporal fossa and fascia	Coronoid process of mandible	Closes jaw	Mandibular nerve [A branch of the trigeminal nerve (V)]
	MASSETER (consists of superficial and deep portions)	Superficial—zygomatic process of maxilla; lower border of zygomatic arch Deep—lower border and medial surface of zygomatic arch	Superficial—angle and ramus of mandible Deep—upper half of ramus and lateral surface of coronoid process of mandible	Elevates mandible; closes jaw	Masseter nerve, from mandibular division of trigeminal nerve (V)
	MEDIAL PTERYGOID	Medial surface of lateral pterygoid plate; tuber of maxilla	Medial surface of ramus and angle of mandible	Closes jaw by raising and advancing it	Medial pterygoid from mandibular division of trigeminal nerve (V)
	LATERAL PTERYGOID (consists of 2 heads)	Upper head—infratemporal surface of greater wing of sphenoid; infratemporal crest Lower head—lateral surface of lateral pterygoid plate	Neck of condyle of mandible; capsule of temporomandibular joint	Protrudes mandible; opens jaw; moves mandible from side-to-side	Lateral pterygoid from mandibular division of trigeminal nerve (V)

(See Platysma, 6.6)

6.4 MUSCLES OF THE HEAD: THE EYES AND THE EARS (See also 7.19)

THE EYES (ORBITS)	Name	Origin	Insertion	Action	Innervation
EYELIDS	**LEVATOR PALPEBRAE SUPERIORIS**	Upper border of optic foramen	Upper tarsal plate	Raises upper lid	Oculomotor nerve (III)
EYES	**SUPERIOR RECTUS**	Upper border of optic foramen	Upper aspect of sclera	Adducts; rotates eyeball upward and inward	Oculomotor nerve (III)
	INFERIOR RECTUS	Circumference of optic foramen (lower margin)	Underside of sclera	Adducts; rotates eyeball downward and inward	Oculomotor nerve (III)
	LATERAL RECTUS	Outer margin of optic foramen	Outer side of sclera	Abducts eyeball	Abducens nerve (VI)
	MEDIAL RECTUS	Circumference of optic foramen (lower margin)	Inner side of sclera	Adducts eyeball, rotates eyeball inward	Oculomotor nerve (III)
	SUPERIOR OBLIQUE	Lesser wing of sphenoid above optic foramen	By a tendon through trochleapulley to the sclera	Rotates eyeball downward and outward	Trochlear nerve (IV)
	INFERIOR OBLIQUE	Orbital plate of superior maxillary bone	Sclera at right angles to insertion of LATERAL RECTUS just below it	Rotates eyeball upward and outward	Oculomotor nerve (III)
EARS/AURICLES **EXTRINSIC EAR MUSCLES**	**AURICULARIS POSTERIOR**	Mastoid process	Cartilage of ear	Draws pinna of ear backward	Posterior auricular branch of facial nerve (VII)
	AURICULARIS SUPERIOR	Galea aponeurotica	Cartilage of ear	Raises pinna of ear	Temporal branch of facial nerve (VII)
	AURICULARIS ANTERIOR	Superficial temporal fascia	Cartilage of ear	Draws pinna of ear forward	Temporal branch of facial nerve (VII)
INTRINSIC EAR MUSCLES	**HELICUS MAJOR**	Spine of the helix	Anterior border of helix	Tenses skin of acoustic meatus	Auriculotemporal and posterior auricular nerves
	HELICUS MINOR	Anterior rim of the helix	Concha	Tenses skin of acoustic meatus	Temporal and posterior auricular nerves
	TRAGICUS	Refers to a short, flattened vertical band on lateral surface of tragus, innervated by auriculotemporal and posterior auricular nerves			
	ANTITRAGICUS	Outer part of antitragus	Caudate process of helix and antihelix		Temporal and posterior auricular branches of facial nerve (VII)
	TRANSVERSUS AURICULAE	Cranial surface of auricle	Circumference of auricle	Retracts helix	Posterior auricular branch of facial nerve (VII)
	OBLIQUUS AURICULAE	Cranial surface of concha	Cranial surface of auricle above concha		Posterior auricular and temporal nerve
MUSCLES OF THE AUDITORY OSSICLES	**TENSOR TYMPANI**	Cartilaginous portion of auditory tube and canal, temporal tube	Handle of malleus (hammer)	Tenses tympanic membrane	Branch of mandibular nerve through Otic ganglion
	STAPEDIUS	Interior of pyramidal eminence of tympanic cavity	Neck of stapes (stirrup)	Dampens movement of stapes (by depressing its base)	Tympanic branch of facial nerve (VII)

6.5 MUSCLES OF THE PHARYNGEAL AND LARYNGEAL REGIONS

	Name	Origin	Insertion	Action	Innervation
PHARYNX	**STYLOPHARYNGEUS**	Styloid process	Thyroid cartilage and side of pharynx	Raises and dilates pharynx	Pharyngeal plexus; glossopharyngeal nerve
	SALPINGOPHARYNGEUS	Eustachian tube near its orifice at the nasopharynx	Posterior part of palatopharyngeus	Raises nasopharynx	Pharyngeal plexus
	INFERIOR CONSTRICTOR	Sides and under surfaces of cricoid and thyroid cartilages	Medial raphe of posterior wall of pharynx	Constricts pharynx (as in swallowing)	Glossopharyngeal; glossopharyngeal plexus; external and recurrent laryngeal nerves
	MEDIAL CONSTRICTOR	Both cornu of hyoid and styloid ligament	Medial raphe of posterior wall of pharynx	Constricts pharynx (as in swallowing)	Pharyngeal plexus and glossopharyngeal nerve
	SUPERIOR CONSTRICTOR	Medial pterygoid plate; pterygomandibular ligament; mylohyoid ridge of mandible; and mucous membrane of floor of mouth	Medial raphe of posterior wall of pharynx	Constricts pharynx (as in swallowing)	Pharyngeal plexus
LARYNX	**ARYEPIGLOTTIC**	Apex of the arytenoid cartilage	Lateral margin of the epiglottis	Closes larynx inlet (glottis opening)	Recurrent, laryngeal
	THYROEPIGLOTTIC	Lamina of thyroid cartilage	Epiglottis and SACCULIS laryngis	Closes larynx inlet (glottis opening), depresses epiglottis	Recurrent, laryngeal
	ARYTENOID (OBLIQUE and TRANSVERSE)	Dorsal aspect of muscular process of arytenoid cartilage	OBLIQUE portion inserts on aryepiglottic fold; TRANSVERSE portion crosses between the 2 cartilages of the OBLIQUE portion	Closes larynx inlet and approximates arytenoid cartilage	Recurrent, laryngeal
	THYROARYTENOID	Lamina of thyroid cartilage	Muscular process of arytenoid cartilage	Relaxes and shortens vocal cords	Recurrent, laryngeal
	CRICOARYTENOID (LATERAL and POSTERIOR)	Lateral surface (upper border of arch) and back of cricoid cartilage	Muscular process of arytenoid cartilage	Approximates and separates vocal folds (narrows and opens glottis)	Recurrent, laryngeal
	CRICOTHYROID	Front and side of cricoid cartilage	Lamina of thyroid cartilage	Tenses vocal cords	Superior, laryngeal

(6.5, cont'd)

TONGUE	Name	Origin	Insertion	Action	Innervation
INTRINSIC MUSCLES	**SUPERIOR LONGITUDINAL**	Submucosa and septum of tongue	Margins of tongue	Changes shape of tongue in mastication and deglutition	Hypoglossal nerve (XII)
	INFERIOR LONGITUDINAL	Undersurface of tongue at base	Tip of tongue	Changes shape of tongue in mastication and deglutition	Hypoglossal nerve (XII)
	TRANSVERSUS LINGUAE	Median septum of tongue	Dorsum and margins of tongue	Changes shape of tongue in mastication and swallowing	Hypoglossal nerve (XII)
	VERTICAL LINGUAE	Dorsal fascia of tongue	Sides and base of tongue	Causes local variations in tension of vocal fold	Hypoglossal nerve (XII)
EXTRINSIC MUSCLES: MUSCLES THAT MOVE THE TONGUE	**GENIOGLOSSUS**	Superior genial tubercle (inner surface of symphysis of mandible)	Hyoid bone, under surface of tongue	Protrudes and retracts tongue, elevates hyoid	Hypoglossal nerve (XII)
	HYOGLOSSUS	Body and greater horn of hyoid bone	Side of tongue	Depresses side of tongue and retracts tongue	Hypoglossal nerve (XII)
	STYLOGLOSSUS	Styloid process	Margin of tongue	Raises and retracts tongue	Hypoglossal nerve (XII)
	STYLOHYOID	Styloid process	Body of hyoid bone	Draws hyoid and tongue upward (and hyoid back)	Facial nerve (VII)
THE SOFT PALATE:	**PALATOGLOSSUS**	Undersurface of soft palate	Side of tongue	Elevates back of tongue; narrows opening of mouth (fauces) into pharynx	Pharyngeal plexus nerve
	PALATOPHARYNGEUS	Posterior border of bony palate; palatine aponeurosis, soft palate	Posterior border of thyroid cartilage; side of pharynx and esophagus	Aids swallowing (narrows fauces and shuts off nasopharynx)	Pharyngeal plexus nerve
	TENSOR VELI PALATINI	Scaphoid fossa of internal pterygoid process, eustachian tube, and spine of sphenoid	Aponeurosis of soft palate; wall of auditory tube; posterior border of hard palate	Stretches soft palate; opens auditory tube	Mandibular nerve
	LEVATOR VELI PALATINI	Apex of pars petrosa of temporal bone and cartilage of auditory tube	Aponeurosis of soft palate	Raises and draws back soft palate	Pharyngeal plexus nerve
	MUSCULUS UVULAE	Posterior nasal spine of palatine bone and aponeurosis of soft palate	Uvula	Raises uvula	Pharyngeal plexus nerve

6.6 MUSCLES OF THE NECK: CERVICALS, HYOIDS, AND PREVERTEBRALS

CERVICALS: Name	Origin	Insertion	Action	Innervation
SUPERIOR CERVICAL				
PLATYSMA	Clavicle, acromion, fascia of chest wall below clavicle and fascia of deltoid muscle, and pectoralis major	Lower border of mandible and skin around mouth, risorius muscle, and opposite platysma	Wrinkles skin of neck; depresses jaw; draws corners of mouth backward, widens mouth (expressions of horror, sadness, or screaming)	CERVICAL BRANCH of facial nerve (VII)
LATERAL CERVICAL STERNOCLEIDOMASTOID	By 2 heads: manubrium of sternum; medial one-third of clavicle	Mastoid process of temporal bone and superior nuchal line (outer part) of occipital bone	Bilateral action flexes neck; unilateral action rotates neck and head	SPINAL ACCESSORY (XI) and CERVICAL PLEXUS
HYOIDS:				
SUPRA HYOIDS DIGASTRIC (ANTERIOR BELLY)	Digastric fossa of mandible (lower border of lower jaw)	Body of hyoid bone (Intermediate tendon between both bellies)	Draws hyoid bone forward; lowers jaw	Mylohyoid branch of mandibular nerve
DIGASTRIC (POSTERIOR BELLY)	Mastoid notch of temporal bone	Body of hyoid bone	Draws hyoid bone backward; lowers jaw	Facial nerve (VII)
MYLOHYOID	Mylohyoid line of mandible	Median raphe in center of muscle and body of hyoid bone	Elevates hyoid bone; elevates and supports mouth floor, depresses jaw	Mylohyoid branch of the inferior alveolar nerve
GENIOHYOID	Genial tubercles on the internal surface of mandible (mental spine of inferior maxilla)	Body of hyoid bone	Elevates and advances hyoid bone, helps to depress jaw	Branch of the first cervical thru hypoglossal nerve (XII)
STYLOHYOID	Styloid process	Greater cornu of hyoid bone	Fixes hyoid, drawing it up and back	Facial nerve (VII)
HYOGLOSSUS	Body and greater horn of hyoid bone	Side of tongue	Depresses side of tongue and retracts tongue	Hypoglossal nerve (XII)
INFRA HYOIDS STERNOHYOID	Manubrium, medial end of clavicle, and first costal cartilage	Body of hyoid bone	Depresses hyoid bone and larynx	Cervical spinal nerves 1-3 via ansa cervicalis
STERNOTHYROID	Sternum and first costal cartilage	Side of thyroid cartilage	Depresses thyroid cartilage (larynx)	Cervical spinal nerves 1-3 via ansa cervicalis
THYROHYOID	Side of thyroid cartilage	Greater cornu and body of hyoid bone	Depresses hyoid bone; elevates thyroid cartilage if hyoid bone is fixed	Cervical spinal nerves 1-3 via hypoglossal nerve (XII)
OMOHYOID	Superior border of the scapula	Body of hyoid bone	Depresses hyoid bone	Cervical spinal nerves 1-3 via ansa cervicalis

(6.6, cont'd)

PREVERTEBRALS : POSTERIOR NECK REGION

	Name	Origin	Insertion	Action	Innervation
ANTERIOR VERTEBRALS	**RECTUS CAPITIS ANTERIOR**	Lateral mass of atlas	Basilar process of occipital bone	Flexes (inclines), turns, and supports head	Cervical spinal nerves 1–2
	RECTUS CAPITIS LATERALIS	Upper surface of transverse process of atlas	Inferior surface of jugular process of occipital bone	Flexes (inclines) head laterally to same side of contraction	Cervical spinal nerves 1–2
	LONGUS CAPITIS	Transverse processes of C3–C6	Inferior surface of basilar portion of occipital bone	Flexes head	Cervical spinal nerves 1–3
	LONGUS COLLI (3 parts: **1. SUPERIOR OBLIQUE,** **2. INFERIOR OBLIQUE, and** **3. VERTICAL**	1. Transverse processes of C3–C5 2. Bodies of T1-T3 3. Bodies of C5-C7, T1-T3	1. Tubercle of the anterior arch of the atlas 2. Transverse process of C5-C6 3. Bodies of C2-C4	Twists and bends neck forward (flexes and supports cervical vertebrae); supports head	Ventral rami of cervical nerves 2–7
LATERAL VERTEBRALS	**SCALENUS ANTERIOR**	Transverse processes of cervical vertebrae C3–C6	Tubercle of first rib	Elevates first rib; flexes neck; laterally bends neck	Cervical spinal nerves C4–C6
	SCALENUS MEDIUS	Transverse processes of cervical vertebrae C2–C6	Superior surface of first rib	Elevates first rib; bends neck laterally;	Cervical spinal nerves C2–C7
	SCALENUS POSTERIOR	Transverse processes of cervical vertebrae C4–C6	Second rib (outer surfaces)	Elevates second rib; laterally bends lower half of cervical bones; flexes neck	Cervical spinal nerves C6–C8

6.7 DEEP POSTVERTEBRAL MUSCLES OF THE VERTEBRAL COLUMN

SUPERFICIAL STRATUM (columnar)	Name	Origin	Insertion	Action	Innervation
TRUNK EXTENSORS OF THE SPINE :	**SPLENIUS CAPITIS**	Lower half of nuchal ligament; spinous processes of 7th cervical and T1-T3	Mastoid part of temporal bone; occipital bone (superior curved line)	Extends and rotates head	Branches of dorsal divisions of cervical nerve
	SPLENIUS CERVICIS	Spinous processes of C7 and T3-T6	Transverse processes of atlas and axis (C1-C2)	Rotates and flexes head and neck; extends vertebral column	Branches of dorsal divisions of cervical nerve
INTERMEDIATE COLUMN	**ERECTOR SPINAE:**	Refers to superficial fibers of deep muscles of the back, originating from sacrum, spines of lumbar, and 11th and 12th thoracic vertebrae and iliac crest, which split and insert as longissimus, spinalis, and iliocostal muscles			
	LONGISSIMUS CAPITIS	Transverse processes of 4 or 5 upper thoracic vertebrae; articular processes of 3 or 4 lower cervical vertebrae	Mastoid process of temporal bone	Rotates head; draws head backward, or to one side; keeps head erect	Branches of cervical nerve
	LONGISSIMUS THORACIS	Transverse and articular processes of lumbar vertebrae and thoracolumbar fascia	Transverse processes of all thoracic vertebrae and 9 or 10 lower ribs	Extends thoracic vertebrae	Lumbar and thoracic nerves
	LONGISSIMUS CERVICIS	Transverse process of 4 or 5 upper thoracic vertebrae	Transverse processes of 2nd or 3rd to 6th cervical vertebrae	Extends cervical vertebrae	Lower cervical and upper thoracic spinal nerves
MEDIAL COLUMN	**SPINALIS THORACIS**	Spinous processes of L1-L2, and T11-T12	Spinous processes of middle and upper thoracic vertebrae	Extends (erects) vertebral column	Thoracic and lumbar spinal nerves
	SPINALIS CERVICIS	Spinous processes of C5-C7 and nuchal ligament	Spinous processes of the axis (and occasionally the two cervical vertebrae below)	Extends cervical spine	Branches of cervical spinal nerves
	SPINALIS CAPITIS	Inconstant: Lower half of nuchal ligament; spinous processes of 7th cervical and upper thoracic vertebrae	Mastoid part of temporal bone and occipital bone (blends with semispinalis capitus and spinalis cervicis)	Extends and rotates head	Cervical spinal nerves
LATERAL COLUMN	**ILIOCOSTALIS LUMBORUM**	Iliac crest (with sacrospinalis) [ERECTOR SPINAE]	Angles of the lower 6 or 7 ribs (sometimes extends from 5th-12th ribs)	Extends lumbar spine	Thoracic and lumbar nerves
	ILIOCOSTALIS THORACIS	Upper border of angles of 6 lower ribs (7th-12th ribs)	Angles of upper ribs (6th to 1st) and transverse process of 7th cervical vertebra	Thoracic extension; keeps dorsal spine erect	Thoracic spinal nerves
	ILIOCOSTALIS CERVICIS	Angles of 3rd to 6th ribs	Transverse processes of 4th to 6th cervical vertebrae	Extends cervical spine	Branches of cervical nerves

(6.7 cont'd)

	Name	Origin	Insertion	Action	Innervation
DEEPER STRATUM (TRANSVER- SOSPINAL)	SEMISPINALIS CAPITIS	Transverse processes of upper (6-7) thoracic and lower 4 cervical vertebrae	Occipital bone, between inferior and superior curved line	Extends head, (rotates and draws head backward)	Suboccipital, greater occipital, and cervical spinal nerves
	SEMISPINALIS THORACIS	Transverse processes of lower (6-10) thoracic vertebrae	Spinous processes of lower 2 cervical and upper 4 thoracic vertebrae	Extends and rotates vertebral column	Thoracic and cervical spinal nerves
	SEMISPINALIS CERVICIS	Transverse processes of upper (5-6) thoracic vertebrae	Spinous processes of 2nd axis to 5th cervical vertebrae	Extends and rotates vertebral column (to the opposite side)	Cervical spinal nerves
	MULTIFIDUS	Sacrum; sacroiliac ligament; mammillary processes of lumbar; transverse processes of thoracic; and articular processes of cervical vertebrae	Spines of next 4 contiguous vertebrae above	Extends and rotates vertebral column	Spinal nerves throughout column
	ROTATORES	Deep in groove between spinous and transverse processes of vertebrae (Transverse processes of 2nd - 12th dorsal vertebrae)	(Lamina of next vertebra above)	Extends and rotates vertebral column to opposite side	Spinal nerves throughout column
DEEPEST STRATUM (CONTIGUOUS	INTERSPINALES (A SERIES)	Muscle fibers extending on each side between spinous processes of contiguous vertebrae (undersurface of spine of one vertebra)	(Spine of vertebra above)	Supports and extends vertebral column	Spinal nerves throughout column
	INTERTRANSVERSARII	Small muscles passing between transverse processes of adjacent vertebrae		Bends (flex) vertebrae laterally	Spinal nerves throughout column
SUBOCCIPITAL MUSCLES	RECTUS CAPITIS POSTERIOR MAJOR	Spinous process of axis	Inferior curved line of occipital bone	Extends head (rotates and draws head backward)	Suboccipital and greater occipital nerves
	RECTUS CAPITIS POSTERIOR MINOR	Posterior tubercle of atlas	Inferior curved line of occipital bone	Extends head (rotates and draws head backward)	Suboccipital and greater occipital nerves
	OBLIQUUS CAPITIS INFERIOR	Spinous process of axis	Transverse process of atlas	Rotates atlas and head	Suboccipital and cervical spinal nerves
	OBLIQUUS CAPITIS SUPERIOR	Transverse processes of atlas	Occipital bone	Rotates head, extends and moves head laterally	Suboccipital and cervical spinal nerves

POSTERIOR MUSCLES OF THE THORAX: COSTOVERTEBRALS

	Name	Origin	Insertion	Action	Innervation
MUSCLES OF RESPIRATION	LEVATORES COSTARUM	Transverse processes of 7th cervical and upper 11 thoracic vertebrae (T1-T11)	Medial to angle of rib below	Aids in elevation of ribs in respiration; flex vertebral column	Branches of intercostal nerve
	SERRATUS POSTERIOR SUPERIOR	Nuchal ligament; spinous processes of upper 2 thoracic vertebrae and C7	Angles of 2nd through 5th ribs	Elevates ribs in inspiration	Upper 4 thoracic spinal nerves
	SERRATUS POSTERIOR INFERIOR	Spines of lower 2 thoracic and upper 2 lumbar vertebrae	Lower 4 ribs	Depresses ribs in expiration (draws ribs back and downward)	9th through 12th thoracic nerves

6.8 TRUNK CONNECTORS TO THE SHOULDER

Name	Origin	Insertion	Action	Innervation
EXTENSORS OF THE SHOULDER JOINT				
TRAPEZIUS	Superior curved line of occipital bone; nuchal ligament; spinous processes of 7th cervical and all thoracic vertebrae	Clavicle; acromion; base of spine of scapula	Elevates scapula; rotates scapula to raise shoulder in abduction of arm; adducts scapula; draws head back and to the side	Spinal accessory and cervical plexus
LATISSIMUS DORSI	Spines of lower thoracic vertebrae; spines of lumbar and sacral vertebrae through attachment to thoracolumbar fascia; tip of iliac crest; lower ribs; inferior angle of scapula	Floor of intertubercular groove of humerus	Adducts, extends, and rotates humerus medially	Thoracodorsal nerve (brachial plexus)
RHOMBOIDEUS MAJOR	Spinous processes of 2nd through 5th thoracic vertebrae	Vertebral margin of scapula below spine	Adducts and elevates scapula	Dorsal scapular nerve from brachial plexus
RHOMBOIDEUS MINOR	Spinous processes of 7th cervical and 1st thoracic vertebrae; lower part of nuchal ligament	Vertebral margin of scapula at root of (above) spine	Retracts and fixes scapula (adducts and elevates scapula)	Dorsal scapular nerve from brachial plexus
LEVATOR SCAPULAE	Transverse processes of 4 upper cervical vertebrae	Vertebral border (superior edge) of scapula	Elevates posterior angle of scapula	Branches of 3rd and 4th cervical nerves; dorsal scapular from 5th cervical
FLEXORS OF THE SHOULDER JOINT				
PECTORALIS MAJOR	Clavicle; sternum; 6 upper costal cartilages; aponeurosis of external oblique muscle of abdomen	Crest of greater tubercle (bicipital ridge) of humerus	Adducts, flexes, and rotates arm medially	Anterior thoracic, lateral and medial pectoral nerves from brachial plexus
PECTORALIS MINOR	3rd - 5th ribs	Coracoid process of scapula	Draws scapula forward and downward and point to shoulder; elevates ribs	Anterior thoracic nerves from brachial plexus
SUBCLAVIUS	First rib and its cartilage	Lower surface of clavicle	Depresses lateral end of clavicle or elevates the first rib	Special subclavius nerve with fibers from 5th - 6th cervical
SERRATUS ANTERIOR	Upper 8 or 9 ribs	Angles and vertebral border of scapula	Abducts scapula; rotates scapula to raise shoulder in abduction of arm; elevates ribs	Long thoracic nerve from brachial plexus

6.9 MUSCLES OF THE TRUNK: THE THORAX AND THE ABDOMINAL WALL (POSTERIOR)

THORAX:

MUSCLES OF RESPIRATION

Name	Origin	Insertion	Action	Innervation
EXTERNAL INTERCOSTALS	Inferior border of rib	Superior border of rib below	Elevates ribs in inspiration (draws ribs together)	Intercostal nerve
INTERNAL INTERCOSTALS	Inferior border of rib and costal cartilage	Superior border of rib and costal cartilage below	Depresses ribs in expiration; (draws ribs together)	Intercostal nerve
SUBCOSTALS	Inconstant: lower border (inner surface) of ribs	Upper border (inner surface) of 2nd or 3rd rib below	Elevates and depresses ribs in respiration; draws ribs together	Intercostal nerve
TRANSVERSUS THORACIS	Posterior surface of body of sternum and xiphoid process	2nd to 6th costal cartilages	Narrows chest in expiration	Branches of intercostal nerve
DIAPHRAGM	Back of xiphoid process; inner surfaces of lower 6 costal cartilages and lower 4 ribs; medial and lateral arcuate ligaments; bodies of upper lumbar vertebrae	Central tendon of diaphragm	Increases volume of thorax in inspiration (decreases pressure within the thoracic cavity)	Phrenic nerve

THE ABDOMINAL WALL (POSTERIOR)

MUSCLES OF THE TRUNK

Name	Origin	Insertion	Action	Innervation
QUADRATUS LUMBORUM	Iliac crest, thoracolumbar fascia, iliolumbar ligament, lower lumbar vertebrae	12th rib; transverse processes of upper lumbar vertebrae	Extends lumbar region; flexes trunk and vertebral column laterally and forward	Branches of 1st and 2nd lumbar, and 12th thoracic spinal nerves
PSOAS MAJOR	Last thoracic and all of the lumbar vertebrae	Lesser trochanter of femur	Flexes thigh and trunk adducts and rotates it medially	2nd and 3rd lumbar spinal nerves
PSOAS MINOR	Last thoracic and 1st lumbar vertebrae	Arcuate line of hip bone, iliac fascia and iliopectineal tuberosity	Assists psoas major and tenses iliac fascia	First lumbar spinal nerve
ILIACUS	Margin of iliac fossa, ala of sacrum	Greater psoas tendon; lesser trochanter of femur	Flexes thigh and trunk, rotates thigh	Branches of femoral nerve

6.10 MUSCLES OF THE TRUNK: THE ABDOMINAL WALL (ANTEROLATERAL)

Name	Origin	Insertion	Action	Innervation
MUSCLES OF THE TRUNK				
EXTERNAL OBLIQUE	Lower 8 ribs at costal cartilage	Crest of ilium; linea alba through rectus sheath, pubic crest, Poupart's ligament	Flexes and rotates vertebral column; contracts abdominal viscera	Lower thoracic spinal nerves (Iliohypogastric, ilioinguinal, and branches of intercostal nerve)
INTERNAL OBLIQUE	Inguinal ligament; iliac crest; lumbar aponeurosis	Lower 3 or 4 costal cartilages; linea alba; tendon of pubis, crest of ilium, inguinal ligament, lumbar fascia	Flexes and rotates vertebral column; compresses abdominal viscera, flexes thorax forward	Iliohypogastric, ilioinguinal, and branches of lower intercostal nerve
TRANSVERSUS ABDOMINIS	Lumbar fascia; lower 6 costal cartilages; thoracolumbar fascia; iliac crest; inguinal ligament	Linea alba through rectus sheath; conjoined tendon to pubis, pubic crest, iliopectinal line, xiphoid cartilage	Compresses abdominal viscera, flexes thorax	Lower thoracic spinal nerves, iliohypogastric, ilioinguinal, and branches of intercostal nerve
RECTUS ABDOMINIS	Pubic crest and symphysis	Xiphoid process; 5th–7th costal cartilages	Flexes lumbar vertebrae; supports and compresses abdomen	Lower thoracic spinal nerves (branches of 7th - 12th intercostal) nerves
PYRAMIDALIS	Body of the pubis, pubic crest	Linea alba	Tenses abdominal wall, tightens linea alba	Branch of last thoracic spinal nerve
MUSCLE OF THE INGUINAL REGION				
CREMASTER	Inferior margin of internal oblique muscle, midportion of inguinal ligament	Cremasteric-fascia and pubic tubercle	Elevates testis (testicle)	Genital branch of genitofemoral nerve

6.11 MUSCLES OF THE TRUNK: PELVIC OUTLET AND PERINEUM

THE PELVIC DIAPHRAGM AND PELVIC WALL

	Name	Origin	Insertion	Action	Innervation
PELVIC DIAPHRAGM **PELVIC OUTLET: POSTERIOR GROUP**	**LEVATOR ANI:**	Pelvic fascia, pelvic surface of ischial spine; internal obturator fascia, pubis	Central point of perineum; anococcygeal raphe; coccyx, rectum	Draws anus upward and forward, supports viscera (supports rectum and pelvic floor), aids in defecation	Perineal branch of pudendal nerve and 4th sacral nerve
	PUBOCOCCYGEUS	Anterior portion of levator ani, originating in front of the obturator canal	Fibers of the pubo-coccygeus proper, insert in the anococcygeal body and side of the coccyx	(Same as above)	(Same as above)
	ILIOCOCCYGEUS	Posterior portion of the levator ani, as far upward as the obturator canal	Side of coccyx and anococcygeal body	(Same as above)	(Same as above)
	COCCYGEUS	Ischial spine; lesser sacrosciatic ligament	Lateral border of lower sacrum; upper coccyx	Supports and raises coccyx, closes pelvic outlet	3rd and 4th sacral nerves
PELVIC WALL **(See also 6.17)**	**OBTURATOR INTERNUS**	Inner surface of anterolateral wall of pelvis, surrounding most of obturator foramen, attached to inferior rami of pubis and ischium; inner surface of hip bone; obturator membrane and foramen	Inner surface of greater trochanter of femur	Rotates thigh laterally (outward)	1st, 2nd, and 3rd sacral, branch of the obsturator
	PIRIFORMIS	2nd to 4th sacral vertebrae; margins of anterior sacral foramina and sacrosciatic notch of ilium	Upper border of greater trochanter of femur	Abducts and rotates thigh outward	First and second sacral
UROGENITAL DIAPHRAGM **PELVIC OUTLET: ANTERIOR GROUP** ★ See 6.11 for details of ♂ and ♀ action	**SUPERFICIAL TRANSVERSUS PERINEI**	Ramus (tuberosity) of ischium	Central tendon of perineum	Tensor of central point (central tendon) of perineum	Perineal branch of the pudendal nerve
	BULBOSPONGIOSUS	Central point of perineum; medium raphe of bulb	♂ undersurface of bulb, spongy and cavernous part of penis/♀ root of clitoris	Constricts bulbous urethra	Perineal branch of the pudendal nerve
	ISCHIOCAVERNOSUS	Ramus (tuberosity) of ischium and great sacrosciatic ligament	Crus penis/crus clitoris	Maintains erection of penis/clitoris	Perineal branch of the pudendal nerve
	PROFUNDUS TRANSVERSUS PERINEI	Inferior ramus of ischium	Median raphe (central tendon) of perineum	Draws back central point of perineum	Perineal branch of the pudendal nerve
	SPHINCTER URETHRAE	Ramus of pubis	Median raphe behind and in front of urethra	Constricts membranous urethra	Perineal branch of the pudendal nerve
ANAL REGION	**SPHINCTER ANI EXTERNUS**	Tip of coccyx and surrounding fascia, ring of fibers surrounding anus	Tendinous center of perineum and coccyx	Closes anus	Inferior hemorrhoidal branch of the pudendal and 4th sacral nerve
	SPHINCTER ANI INTERNUS	Circular, muscular unstriated rectal fibers of intestine one inch above anal canal	—	Closes anus	Hemorrhoidal branch of the pudenal

6.12 SCAPULAR CONNECTORS TO THE SHOULDER

Name	Origin	Insertion	Action	Innervation
ABDUCTORS OF THE SHOULDER JOINT				
DELTOID	Clavicle; acromion; spine of scapula	Deltoid tuberosity (and shaft) of humerus	Abducts, flexes, and extends arm (raises and rotates it)	Axillary (circumflex) nerve from brachial plexus
SUPRASPINATUS	Supraspinatus fossa of scapula	Greater tubercle of humerus	Abducts arm; rotates arm	Branches of suprascapular nerve from brachial plexus
ROTATORS OF THE SHOULDER				
INFRASPINATUS	Infraspinous fossa of scapula	Greater tubercle of humerus	Rotates arm laterally (back and out)	Branches of suprascapular nerve from brachial plexus
SUBSCAPULARIS	Subscapular fossa of scapula	Lesser tubercle of humerus	Rotates arm medially and lowers it	Branches of subscapular nerve from brachial plexus
TERES MINOR	Lateral margin (axillary border) of scapula	Greater tubercle of humerus	Rotates arm laterally (outward)	Axillary (circumflex) nerve from brachial plexus
TERES MAJOR	Inferior angle (axillary border) of the scapula	Crest of lesser tubercle of humerus	Adducts, extends, and rotates arm medially (draws arm down and back)	Lower subscapular nerve
FLEXOR/ADDUCTOR OF THE SHOULDER				
CORACOBRACHIALIS	Coracoid process of scapula	Middle of the medial surface of shaft of humerus	Flexes (raises) and adducts arm	Musculocutaneous nerve

6.13 MUSCLES OF THE UPPER LIMB: THE BRACHIUM (HUMERUS): THE ARM

Name	Origin	Insertion	Action	Innervation
FLEXOR/ADDUCTOR OF THE SHOULDER JOINT				
CORACOBRACHIALIS	Coracoid process of scapula	Middle of medial surface of shaft of humerus	Flexes (raises) and adducts arm	Musculocutaneous nerve
FLEXORS OF THE FOREARM				
BICEPS BRACHII (consists of 2 heads)	Long head—from the supraglenoid tubercle of scapula (at margin of glenoid cavity) Short head—from the apex of coracoid process	Bicipital tuberosity of radius; antebrachial fascia; ulna	Flexes arm and forearm; supinates hand	Musculocutaneous nerve
BRACHIALIS	Lower half of anterior surface of humerus	Coronoid process of ulna	Flexes forearm	Musculocutaneous and radial nerves
BRACHIORADIALIS (See also 6.15)	Lateral supracondylar ridge of humerus	Lateral surface (styloid process) of lower end of radius	Flexes forearm; supinates forearm	Branch of radial nerve
EXTENSORS OF THE FOREARM				
TRICEPS BRACHII (consists of 3 heads)	Long head—infraglenoid tubercle of scapula Lateral head—posterior surface of humerus below great tubercle Medial head—posterior surface of humerus below groove of radial nerve	Olecranon process of ulna	Extends forearm and arm; long head adducts and extends arm	Branches of radial nerve
ANCONEUS (See also 6.15)	Back of lateral epicondyle of humerus	Olecranon process and posterior surface of ulna	Extends forearm	Branch of radial nerve

6.14 MUSCLES OF THE UPPER LIMB: FOREARM: ANTERIOR (PALMAR) MUSCLES

Name	Origin	Insertion	Action	Innervation
FLEXORS OF THE WRIST				
FLEXOR CARPI RADIALIS	Medial epicondyle of humerus	Bases of 2nd and 3rd metacarpal bones	Flexes and abducts wrist (and hand)	Branch of median nerve
PALMARIS LONGUS	Medial epicondyle of humerus	Flexor retinaculum; palmar aponeurosis; transverse carpal ligament	Tenses palmar aponeurosis flexes wrist (and hand)	Branch of median nerve
FLEXOR CARPI ULNARIS (consists of 2 heads)	Humeral head—medial epicondyle of humerus Ulnar head—olecranon process and posterior border of ulna	Pisiform bone; hook of hamate bone; base of 5th metacarpal bone	Flexes and adducts wrist (and hand)	Branch of ulnar nerve
PRONATORS OF THE FOREARM				
PRONATOR TERES (consists of 2 heads)	Humeral head—medial epicondyle of humerus Ulnar head—coronoid process of ulna	Lateral surface of shaft of radius	Pronates the hand; pronates and flexes forearm	Branch of median nerve
PRONATOR QUADRATUS	Anterior surface and border of distal 3rd or 4th portion of shaft of ulna	Anterior surface of border of distal 4th portion of shaft of radius	Pronates forearm	Anterior volar interosseous nerve (branch of median)
FLEXORS OF THE FINGERS				
FLEXOR DIGITORUM SUPERFICIALIS (consists of 3 heads)	Humeral head—medial epicondyle of humerus Ulnar head—medial side of coronoid process of ulna Radial head—anterior (outer) border of radius	Sides of middle (second) phalanges of each finger (4 medial fingers)	Flexes middle phalanges and hand	Branches of median nerve
FLEXOR DIGITORUM PROFUNDUS	Upper 3/4 of shaft of ulna; coronoid process; interosseous membrane	Bases of distal (terminal) phalanges of each finger (4 medial fingers)	Flexes distal phalanges	Anterior interosseous (branch of median), and ulnar nerves
FLEXORS OF THE THUMB				
FLEXOR POLLICIS LONGUS	Anterior surface of middle 1/3 of radius; medial epicondyle of humerus; coronoid process of ulna	Base of distal (terminal) phalanx of thumb	Flexes thumb	Anterior interosseous nerve (branch of median)

6.15 MUSCLES OF THE UPPER LIMB: FOREARM POSTERIOR (DORSAL) MUSCLES

Name	Origin	Insertion	Action	Innervation
ROTATOR OF THE FOREARM				
SUPINATOR	Lateral epicondyle of humerus; ligaments of elbow; oblique line of ulna	Outer surface of radius	Supinates forearm and hand; rotates forearm	Deep branch of the radial nerve
EXTENSORS OF THE WRIST				
EXTENSOR CARPI RADIALIS LONGUS	Lateral supracondylar ridge of humerus	Back of base of 2nd metacarpal bone	Extends and abducts wrist (hand)	Branch of radial nerve
EXTENSOR CARPI RADIALIS BREVIS	Lateral epicondyle of humerus	Back of bases of 2nd and 3rd metacarpal bones	Extends and abducts wrist (hand)	Deep branch of the radial nerve
EXTENSOR CARPI ULNARIS (consists of 2 heads)	Humeral head—lateral epicondyle of humerus Ulnar head—posterior border of ulna	Base of 5th metacarpal bone	Extends and abducts wrist (hand)	Deep branch of the radial nerve
EXTENSORS OF THE FINGERS				
EXTENSOR DIGITORUM COMMUNIS	Lateral epicondyle of humerus	Extensor expansion of 4 medial fingers; 2nd and 3rd phalanges	Extends wrist (hand) and phalanges	Deep branch of the radial nerve
EXTENSOR DIGITI MINIMI	Lateral epicondyle of humerus	Extensor aponeurosis of small finger; dorsum of 1st phalanx of small finger	Extends small finger	Deep branch of the radial nerve
EXTENSOR INDICIS	Posterior (dorsal) surface of ulna; interosseous membrane	Extensor expansion of index finger; 1st tendon of extensor digitorum communis	Extends index finger	Posterior interosseous nerve (branch of radial)
EXTENSORS OF THE THUMB				
EXTENSOR POLLICIS BREVIS	Posterior (dorsal) surface of radius	Back of proximal (1st) phalanx of thumb	Extends thumb and abducts 1st metacarpal	Posterior interosseous nerve (branch of radial)
EXTENSOR POLLICIS LONGUS	Posterior (dorsal) surface of ulna and interosseous membrane	Back of distal (2nd) phalanx of thumb	Extends distal (2nd) phalanx of thumb and abducts hand	Posterior interosseous nerve (branch of radial)
ABDUCTOR OF THE THUMB				
ABDUCTOR POLLICIS LONGUS	Posterior (dorsal) surface of radius and ulna	Lateral side of base of first metacarpal bone and trapezium	Abducts and assists in extending thumb	Posterior interosseous nerve (branch of radial)

★ BRACHIORADIALIS and ANCONEUS, see 6.13

6.16 MUSCLES OF THE UPPER LIMB: MUSCLES OF THE HAND (MANUS)

Name	Origin	Insertion	Action	Innervation
ABDUCTOR OF THE THUMB				
ABDUCTOR POLLICIS BREVIS	Navicular (scaphoid), ridge of greater multangular (trapezium), and transverse carpal ligament	Lateral surface of base of proximal (1st) phalanx of thumb	Adducts thumb	Branch of median nerve
ADDUCTOR OF THE THUMB				
ADDUCTOR POLLICIS (consists of 2 heads)	Oblique head—sheath of flexor carpi radialis; anterior carpal ligament; capitate bone; bases of 2nd and 3rd metacarpals Transverse head—lower 2/3 of anterior surface of 3rd metacarpal	Medial surface of base of proximal (1st) phalanx of thumb	Adducts and opposes thumb	Ulnar nerve
FLEXORS OF THE THUMB				
OPPONENS POLLICIS	Ridge of greater multangular (trapezium); transverse carpal ligament (flexor retinaculum)	Radial side of 1st metacarpal	Flexes, adducts and opposes thumb	6th and 7th cervical spinal nerves through the median nerve
FLEXOR POLLICIS BREVIS	Transverse carpal ligament; ridge of greater multangular	Base of proximal (1st) phalanx of thumb	Flexes 1st phalanx of thumb	Branch of median and ulnar nerves
ABDUCTORS OF THE FINGERS				
ABDUCTOR DIGITI MINIMI	Pisiform bone; flexor carpi ulnaris tendon	Medial surface of base of proximal (1st) phalanx of small finger	Abducts small finger	Palmar branch of ulnar nerve
DORSAL INTEROSSEI	By two heads from adjacent sides of metacarpal bones	Extensor tendons of 2nd, 3rd, and 4th fingers	Abducts and flexes proximal phalanges (also adducts fingers)	Branch of ulnar nerve
ADDUCTORS OF THE FINGERS				
PALMER INTEROSSEI	Lateral sides of 2nd, 4th, and 5th metacarpal bones	Ulnar side of index finger, radial sides of ring and small fingers (Extensor tendons of 2nd, 3rd, and 4th fingers)	Adduct index finger, adducts and flexes ring and small fingers	Branch of ulnar nerve
FLEXORS OF THE FINGERS				
FLEXOR DIGITI MINIMI BREVIS	Hamulus of hamate (unciform) bone; transverse carpal ligament	Medial aspect of proximal (1st) phalanx of small finger	Flexes 1st phalanx of small finger	Branch of ulnar nerve
LUMBRICALES	Tendons of flexor digitorum profundus	Extensor tendons and 1st phalanx of four lateral fingers	Flexes 1st and extends 2nd and 3rd phalanges	Median and ulnar nerves
MUSCLES THAT CHANGE THE SHAPE OF THE PALM				
OPPONENS DIGITI MINIMI	Hamulus of hamate (unciform) bone; transverse carpal ligament	Medial aspect of 5th metacarpal	Rotates and abducts 5th metacarpal (flexes and adducts small finger)	8th cervical nerve through ulnar nerve
PALMARIS BREVIS	Central portion of palmar aponeurosis and transverse carpal ligament	Skin of medial border (ulnar side) of hand	Tenses palm of hand (wrinkles skin on inner side of hand)	Branch of ulnar nerve

6.17 MUSCLES OF THE GLUTEAL REGION: SUPERIOR POSTERIOR BUTTOCK MUSCLES

Name	Origin	Insertion	Action	Innervation
EXTENSOR OF THE HIP AND/OR FLEXOR OF THE KNEE				
GLUTEUS MAXIMUS	Superior curved iliac line and crest of ilium; dorsal surfaces of sacrum, coccyx, and sacrotuberous ligament	Iliotibial tract of fascia lata; gluteal tuberosity of femur below greater trocanter	Extends, abducts, and rotates thigh laterally	Inferior gluteal nerve
ABDUCTORS OF THE HIP				
GLUTEUS MEDIUS	Dorsal aspect and lateral surface of ilium between anterior and posterior gluteal lines	Greater trochanter of femur	Abducts and rotates thigh medially	Branches of superior gluteal nerve
GLUTEUS MINIMUS	Dorsal aspect and lateral surface of ilium between anterior and posterior gluteal lines	Greater trochanter of femur	Abducts and rotates thigh medially (also extends thigh)	Branches of superior gluteal nerve
TENSOR FASCIAE LATAE	Iliac crest, iliac spine, fascia lata	Iliotibial tract of fascia lata	Abducts, flexes, and rotates thigh medially	Branches of superior gluteal nerve
LATERAL ROTATORS OF THE HIP				
PIRIFORMIS	Margins of anterior sacral foramina and sacrosciatic notch of ilium; 2nd to 4th sacral vertebrae	Upper margin of greater trochanter of femur	Rotates thigh laterally (outward) (also abducts thigh)	1st and 2nd sacral nerves
OBTURATOR INTERNUS	Pelvic surface of hip bone and obturator membrane margin of obturator foramen	Inner surface of greater trochanter of femur	Rotates thigh laterally (outward)	5th lumbar nerves, 1st and 2nd sacral nerves, (branches of obturator nerve)
GEMELLUS SUPERIOR	Spine of ischium	Internal obturator tendon, medial surface of greater trochanter	Rotates thigh laterally (outward)	Nerve to internal obturator, sacral plexus
GEMELLUS INFERIOR	Tuberosity of ischium	Internal obturator tendon, medial surface of greater trochanter	Rotates thigh laterally (outward)	Sacral nerve to quadrate m. of thigh
QUADRATUS FEMORIS	Tuberosity of ischium	Intertrochanteric crest and quadrate tubercle of femur	Rotates thigh laterally (outward)	4th and 5th lumbar nerves; 1st sacral nerve (branches of femoral)
OBTURATOR EXTERNUS	Pubis; ischium; external surface of obturator membrane	Trochanteric fossa of femur	Rotates thigh laterally (outward)	Branch of obturator nerve
GLUTEUS MAXIMUS				

6.18 MUSCLES OF THE LOWER LIMB: MUSCLES OF THE THIGH: MEDIAL AND POSTERIOR FEMORAL

MEDIAL FEMORAL

Name	Origin	Insertion	Action	Innervation
ADDUCTORS OF THE HIP				
GRACILIS	Body and inferior ramus of pubis (symphysis pubis and pubic arch)	Medial surface of shaft of tibia	Adducts thigh; flexes and adducts leg	Branch of obturator nerve
PECTINEUS	Ilio pectineal line of pubis; pubic spine	Pectineal line of femur	Adducts and flexes thigh	Branches of femoral and obturator nerves
ADDUCTOR LONGUS	Pubic crest, pubic symphysis and body of pubis	Middle of linea aspera of femur	Adducts, rotates, and flexes thigh	Branch of obturator nerve
ADDUCTOR BREVIS	Body and inferior ramus of pubis	Upper part of linea aspera of femur	Adducts, flexes, and tends to rotate thigh medially (inward)	Branch of obturator nerve
ADDUCTOR MAGNUS (consists of superficial and deep portions)	Deep—inferior ramus of pubis ramus of ischium Superficial—ischial tuberosity	Deep—linea aspera of femur Superficial—adductor tubercle of femur	Deep—adducts thigh and rotates it laterally (outward) Superficial—extends leg	Branches: Deep—obturator nerve Superficial—sciatic nerve

POSTERIOR FEMORAL: "THE 3 HAMSTRINGS"

Name	Origin	Insertion	Action	Innervation
EXTENSORS OF THE HIP/FLEXORS OF THE KNEE				
BICEPS FEMORIS (consists of 2 heads)	Long head—ischial tuberosity Short head—linea aspera of femur	Head of fibula; lateral condyle of tibia	Extends thigh; flexes knee and rotates knee and leg laterally (outward)	Long head—tibial portions of sciatic; Short head—peroneal and popliteal nerves
SEMITENDINOSUS	Tuberosity of ischium	Upper part of medial surface of tibia below internal tuberosity	Extends thigh; flexes and rotates leg medially (inward)	Tibial portion of sciatic nerve
SEMIMEMBRANOSUS	Tuberosity of ischium	Lateral condyle of femur; medial condyle and border of tibia	Extends thigh; flexes and rotates leg	Tibial portion of sciatic nerve

6.19 MUSCLES OF THE LOWER LIMB: MUSCLES OF THE THIGH: ANTERIOR FEMORAL

Name	Origin	Insertion	Action	Innervation
FLEXORS OF THE HIP/EXTENSORS OF THE KNEE				
SARTORIUS	Anterior superior iliac spine	Upper part of medial surface of tibia near tibial tuberosity	Flexes and rotates thigh and leg	Branches of femoral nerve
QUADRICEPS FEMORIS				
RECTUS FEMORIS	Anterior inferior iliac spine; upper rim of acetabulum	Base of patella; tuberosity of tibia	Flexes thigh; extends leg	Branches of femoral nerve
VASTUS LATERALIS	Lateral aspect of femur (linea aspera to greater trochanter)	Patella; common tendon of quadriceps femoris of thigh	Extends knee and leg	Branches of femoral nerve
VASTUS MEDIALIS	Medial aspect of femur (linea aspera)	Patella; common tendon of quadriceps femoris of thigh	Extends leg; draws patella in	Branches of femoral nerve
VASTUS INTERMEDIUS	Upper part of anterior and lateral surface of shaft of femur	Patella; common tendon of quadriceps femoris of thigh	Extends leg	Branches of femoral nerve

6.20 MUSCLES OF THE LOWER LIMB: MUSCLES OF THE LEG: ANTERIOR AND LATERAL

Name	Origin	Insertion	Action	Innervation
MUSCLES THAT MOVE THE FOOT AND TOES				
ANTERIOR CRURAL MUSCLES (DORSIFLEXORS)				
TIBIALIS ANTERIOR	Lateral condyle and surface of upper tibia; interosseous membrane, intermuscular septum	Internal (medial) cuneiform; base of 1st metatarsal	Dorsiflexes and inverts foot	Deep peroneal nerve
EXTENSOR DIGITORUM LONGUS	Anterior surface of fibula; lateral condyle of tibia; interosseous membrane	Extensor expansion of 2nd and 3rd phalanges of 4 lateral toes	Extends toes, dorsi flexes foot	Deep peroneal nerve
EXTENSOR HALLUCIS LONGUS	Front of fibula; interosseous membrane	Base of distal (terminal) phalanx of great toe	Extends great toe; dorsiflexes ankle (foot)	Deep peroneal nerve
PERONEUS TERTIUS	Anterior surface of lower part of fibula; interosseous membrane	Fascia or base of 5th metatarsal	Everts and dorsiflexes foot	Deep peroneal nerve
LATERAL CRURAL MUSCLES (EVERTORS)				
PERONEUS LONGUS	Lateral condyle of tibia; head of fibula; lateral surface of fibula	By tendon to internal (medial) cuneiform and 1st metatarsal	Plantar flexes, everts, and adducts foot (also extends and abducts foot)	Superficial peroneal nerve
PERONEUS BREVIS	Lateral surface of midportion of shaft (actually an extension of extensor digitorum longus)	Base tuberosity of 5th metatarsal bone	Everts, abducts, plantar flexes (and extends) foot	Superficial peroneal nerve

6.21 MUSCLES OF THE LOWER LIMB: MUSCLES OF THE LEG: POSTERIOR

Name	Origin	Insertion	Action	Innervation
SUPERFICIAL GROUP: TRICEPS SURAE				
GASTROCNEMIUS (consists of 2 heads)	Medial head—popliteal surface of femur; upper part of medial condyle; capsule of knee. Lateral head—lateral condyle; capsule of knee	Aponeurosis unites with tendon of soleus to form Achilles tendon (tendon calcaneus); (Achilles tendon inserts on calcaneus)	Plantar flexes foot and flexes knee (leg)	Branches of tibial nerve
SOLEUS	Upper shaft of fibula; tendinous arch; oblique line of tibia	Calcaneus by Achilles tendon	Plantar flexes foot, extends and rotates foot	Branches of tibial nerve
PLANTARIS	Popliteal surface of femur (External supracondyloid ridge)	Inner border of Achilles tendon or back of calcaneus	Plantar flexes (and extends) foot and flexes leg	Branch of tibial nerve
DEEP GROUP				
POPLITEUS	Lateral condyle of femur; lateral meniscus	Posterior surface of tibia	Flexes leg; rotates leg medially (inward)	Branch of tibial nerve
FLEXOR HALLUCIS LONGUS	Lower portion of shaft of front of fibula; interosseous membrane	Base of distal phalanx of great toe	Extends great toe; dorsiflexes ankle (foot)	Deep peroneal nerve (posterior tibial)
FLEXOR DIGITORUM LONGUS	Posterior surface of shaft of tibia	Distal (terminal) phalanges of 4 lateral (lesser) toes	Flexes phalanges; extends toes	Branch of tibial nerve
TIBIALIS POSTERIOR	Shafts of tibia; and fibula; interosseous membrane	Bases of 2nd to 4th metatarsal bones, internal (medial) cuneiform, tuberosity of scaphoid	Plantar flexes and inverts foot (also extends tarsus and flexes foot)	Branch of tibial nerve

6.22 & 6.23 MUSCLES OF THE LOWER LIMB: MUSCLES OF THE FOOT

Name	Origin	Insertion	Action	Innervation
MUSCLES THAT MOVE THE TOES				
Superficial Plantar Layer:				
ABDUCTOR HALLUCIS	Medial (inner) tubercle of calcaneus; plantar fascia	Medial (inner) side of base of proximal (1st) phalanx of great toe	Abducts and flexes great toe	Medial plantar nerve
FLEXOR DIGITORUM BREVIS	Medial (inner) tuberosity of calcaneus; plantar fascia	Middle (2nd) phalanges of 4 lateral (lesser) toes	Flexes toes	Medial plantar nerve
ABDUCTOR DIGITI MINIMI	Medial and lateral tubercle of calcaneus; plantar fascia, and intermuscular septum	Lateral (external) side of base of proximal (1st) phalanx of small toe	Abducts small toe	Lateral plantar nerve
Middle Plantar Layer:				
LUMBRICALES	Tendons of flexor digitorum longus	Medial side of base of proximal (1st) phalanges of 4 lateral (lesser) toes	Flexes metatarsophalangeal joints (proximal (1st) phalange); extends distal (2nd and 3rd) phalanges	Medial and lateral plantar nerves
QUADRATUS PLANTAE	Inferior surface of calcaneus by 2 heads from outer and inner borders	Tendons of flexor digitorum longus	Aids in flexing toes	Lateral plantar nerves
Deep Plantar Layer:				
FLEXOR HALLUCIS BREVIS	Undersurface of cuboid; middle and lateral cuneiform bones	Both sides of base of proximal (1st) phalanx of great toe	Flexes great toe	Medial plantar nerve
ADDUCTOR HALLUCIS	Oblique head—long plantar ligament, sheath of peroneus longus Transverse head—plantar ligaments from 3 middle metatarsal bones	Lateral side of base of proximal (1st) phalanx of small toe	Flexes and adducts great toe	Lateral plantar nerve
FLEXOR DIGITI MINIMI BREVIS	Sheath of peroneus longus, base of metatarsal of small toe	Lateral surface of base of proximal (1st) phalanx of small toe	Flexes small toe	Lateral plantar nerve
	Hamulus of hamate (unciform) bone; transverse carpal ligament	Medial aspect of 5th metacarpal	Rotates and abducts 5th metacarpal	Eighth cervical through ulnar nerves
PLANTAR INTEROSSEI	Medial side of 3, 4, and 5 metatarsal bones	Medial side of base of proximal (1st) phalanx of corresponding toe	Flexes and abducts 3 outer toes	Lateral plantar nerve
Dorsal Surface:				
DORSAL INTEROSSEI	Sides shafts of adjacent metatarsal bones	Base of proximal (1st) phalanges of 2, 3, and 4 toes	Flexes and abducts 2nd, 3rd and 4th toes; adducts 2nd toe	Lateral plantar nerve
EXTENSOR DIGITORUM BREVIS	Dorsal surface of calcaneus	To 1st phalanx of great toe and to the tendons of the extensor digitorum longis (2nd, 3rd, and 4th toes)	Extends toes	Deep peroneal nerve
EXTENSOR HALLUCIS BREVIS	Refers to the portion of the extensor digitorum brevis that inserts on the great toe			

UNIT 3: INTEGRATION AND CONTROL

Chapter # 7  # Nervous System

---

PERIPHERAL NERVOUS SYSTEM (PNS)

• All the nerve processes that connect the BRAIN and SPINAL CORD with receptors, muscles and glands

AFFERENT SYSTEM of the PNS

Sensory nerve receptors in the periphery of the body pick up informational changes (stimuli) *from* the EXTERNAL and INTERNAL ENVIRONMENT. The stimuli are converted into nerve impulses that travel along nerve processes conveying the information *to* the CENTRAL NERVOUS SYSTEM (CNS).

CENTRAL NERVOUS SYSTEM (CNS)

BRAIN

BONY SKULL ENCASEMENT

BONY VERTEBRAL COLUMN ENCASEMENT

SPINAL CORD

Protected from bone by meninges membrane and suspended in cerebrospinal fluid

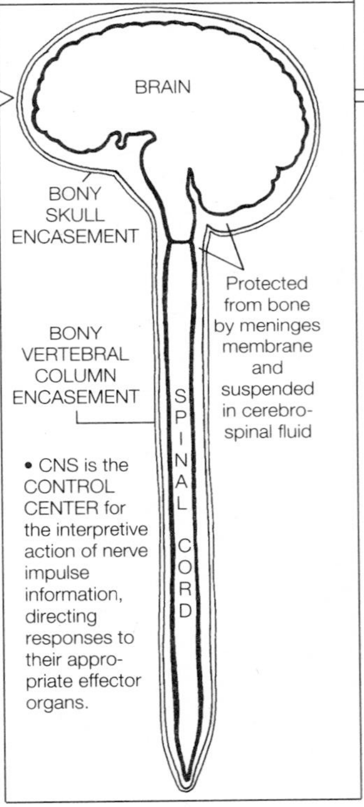

• CNS is the CONTROL CENTER for the interpretive action of nerve impulse information, directing responses to their appropriate effector organs.

EFFERENT SYSTEM of the PNS

Motor nerve cells convey interpreted information *from* the CENTRAL NERVOUS SYSTEM (CNS) along their nerve processes *to* MUSCLES and GLANDS in the body periphery.

SOMATIC NERVOUS SYSTEM (SNS)

Efferent motor nerve cells conduct impulses *from* the CNS *to* SKELETAL MUSCLE TISSUE. Under CONSCIOUS CONTROL and VOLUNTARY

AUTONOMIC NERVOUS SYSTEM (ANS)

Efferent motor nerve cells conduct impulses *from* the CNS, *to* SMOOTH MUSCLE TISSUE (of viscera and blood vessels) CARDIACMUSCLE TISSUE (of the HEART) and GLANDS, (SALIVARY, GASTRIC, SWEAT). Operates without CONSCIOUS CONTROL BELOW the CONSCIOUS LEVEL. INVOLUNTARY

NERVE PORTIONS of ANS

SYMPATHETIC NERVOUS SYSTEM
ADJUSTS ORGAN SYSTEMS for an INCREASE in SKELETAL MUSCLE METABOLISM

PARASYMPATHETIC NERVOUS SYSTEM
ACTIVE in a VARIETY of BODY FUNCTIONS (i.e., GI Function, Defecation, Urination, Rest and Sleep, Sexual Response)

★ See Chart #1 for differences between the CNS and PNS, and anatomical terms of the nervous system

(SENSORY NEURONS of VISCERAL ORGANS and the AUTONOMIC NERVOUS SYSTEM constitute a VISCERAL NERVOUS SYSTEM.)

The NERVOUS SYSTEM is concerned with the INTEGRATION and CONTROL and REGULATION and COORDINATION of all bodily functions through a vast COMMUNICATIONS NETWORK. It is comprised of billions of extremely delicate NERVE CELLS, interlaced with each other in an elaborate system that maintains the vital function of RECEPTION and RESPONSE TO STIMULI. The nervous system is specialized in IRRITABILITY and CONDUCTION. Through IRRITABILITY it is able to receive and bring about responses by which the body adjusts to changes from EXTERNAL and INTERNAL ENVIRONMENTS. These changes constitute STIMULI that initiate electrochemical NERVE IMPULSES in RECEPTORS or SENSE ORGANS. Through CONDUCTION, the nervous system is able to transmit these message-impulses to and from COORDINATING CENTERS.

Two broad functions of the human nervous system are:
    1. **STIMULATION of MOVEMENT** Until stimulated by a NERVE IMPULSE, all skeletal muscle and most smooth muscle cells cannot contract, nor can exocrine glands (salivary, gastric, and sweat) release their secretions.

    2. **SHARING MAINTENANCE of BODY HOMEOSTASIS** by harmoniously interlocking with the ENDOCRINE SYSTEM to coordinate the activities of the other body systems (see Chapter 8). The nervous system can inhibit or stimulate the release of hormones from ENDOCRINE GLANDS while the ENDOCRINE SYSTEM can inhibit or stimulate the flow of nerve impulses.

The nervous system is divided into two divisions:
The **CENTRAL NERVOUS SYSTEM (CNS)** and the **PERIPHERAL NERVOUS SYSTEM (PNS)**.
    ■ **CNS:** consists of the BRAIN and SPINAL CORD, organs that are (1) surrounded by the bony structures of the skull and vertebral column; (2) protected by a meninges membrane; and (3) suspended in cerebrospinal fluid. These areas form central regions of control for: • INTEGRATION • INTERPRETATION • THOUGHT • ASSIMILATION of EXPERIENCES REQUIRED IN MEMORY, LEARNING, and INTELLIGENCE and • TRANSMISSION of MESSAGES TO and FROM the PERIPHERY.

    ■ **PNS:** consists of all the other neural elements lying outside the CNS:
• 12 PAIRS of CRANIAL NERVES and their branches arising from the BRAIN • 31 PAIRS of SPINAL NERVES and their branches arising from the SPINAL CORD • their associated AUTONOMIC GANGLIA (peripheral nerve cell bodies), GANGLIONATED TRUNKS, NERVES, and NERVE PLEXUSES • specialized SENSORY RECEPTORS within the SENSORY ORGANS and • specialized ENDINGS ON MUSCLE. Functionally speaking, the PNS can be subdivided into an **AFFERENT SYSTEM** and an **EFFERENT SYSTEM** (see above chart):
■ **AFFERENT SYSTEM:** consists of nerve cells called AFFERENT (SENSORY) NEURONS that convey impulses toward the CNS. These nerve cells are the first cells to pick up informational changes (stimuli) from the external and internal environment.
■ **EFFERENT SYSTEM:** consists of nerve cells called EFFERENT (MOTOR) NEURONS that convey information from the CNS to MUSCLES and GLANDS. Note the functional subdivision of the EFFERENT SYSTEM of PNS into two systems: a SOMATIC NERVOUS SYSTEM (SNS) conducting impulses from the CNS to SKELETAL MUSCLE TISSUE and an AUTONOMIC NERVOUS SYSTEM (ANS) conducting impulses from the CNS to SMOOTH MUSCLE TISSUE, CARDIAC MUSCLE TISSUE, and GLANDS. (The controlling centers of the SNS and the ANS, which are the nerve cell bodies, are considered part of and within the CNS.)/The peripheral nerve processes of the ANS cell bodies that travel outside the CNS are subdivided into the SYMPATHETIC (STIMULATING) and PARASYMPATHETIC (INHIBITING) SYSTEMS. These complementary divisions carry opposite, or antagonistic, messages to the viscera of the body systems, regulating their activities. Their impulses are adapted and balanced by the CNS BRAIN CENTERS to integrate the actions of the body systems toward the vital maintenance of body homeostasis.

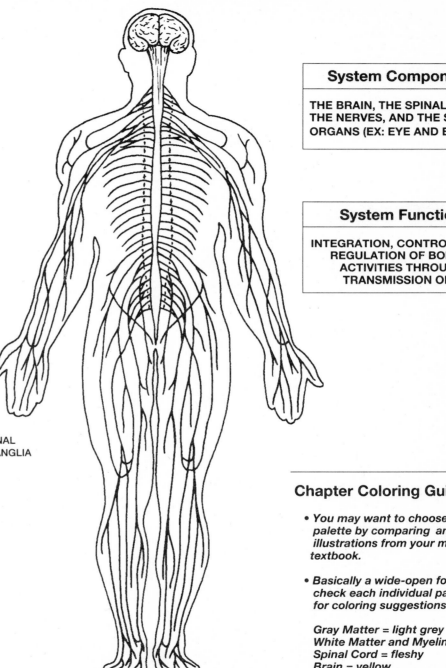

Anterior View

Chapter Coloring Guidelines

• *You may want to choose your color palette by comparing any colored illustrations from your main textbook.*

• *Basically a wide-open format: check each individual page for coloring suggestions*

Gray Matter = light grey
White Matter and Myelin = cream
Spinal Cord = fleshy
Brain = yellow

NERVOUS SYSTEM: GENERAL ORGANIZATION
Histology of Nervous Tissue

Coloring: 1 = green, 2 = blue, 3 = red
4 = pink, 5 = orange, 9 = yellow, 11 = creamy,
C_1 = yellow–orange, C_2 = brown,
7 = red-orange, 8 = purple
★ See Chart #1

★ Despite the elaborate complexity and high specialization of the nervous system, it contains only 2 principal types of cells: NEURONS and NEUROGLIA.

■ NEUROGLIA
  • Offer structural and functional SUPPORT/AID for the NEURONS
    • 5 times more abundant than NEURONS
    • 50% of all brain cells
    • Limited mitotic ability
    • *Not* RECEPTIVE
    • *Not* CONDUCTING
■ NEURONS
  • Structural and functional units of the nervous system
    • Specialized in
■ IRRITABILITY—Response to physical and chemical stimuli
    ■ CONDUCTION—Impulse conduction from one part of the body to another
    ■ INTEGRATION—THINKING, STORAGE OF MEMORY, CONTROLLING MUSCLE ACTIVITY, REGULATING ORGANS and GLANDS.
  • Number of neurons established shortly after birth. Thereafter, they are incapable of mitosis.

■ NEURON CELL BODY
  • Controls metabolism of neuron
  • Contains NEUROFIBRILS—Delicate thread-like strands of protein
  • Contains NISSL BODIES—specialized layers of ROUGH ENDOPLASMIC RETICULUM (See 2.1) involved in protein synthesis of neurofibrils and microtubules; metabolism and transportation of cellular material.
■ 2 TYPES of CYTOPLASMIC EXTENSIONS of the CELL BODY
  1. DENDRITIES—Short ramified branched processes providing a large surface area (DENDRITIC ZONE) to RECEIVE a STIMULUS and CONDUCT IMPULSES to the CELL BODY.
  2. AXON—A single, typically long cylindrical process that conducts nerve impulses AWAY FROM the CELL BODY to another neuron or tissue.
    • AXON HILLOCK—Conical tapering region where the AXON originates from the cell body. Contains *no* NISSL BODIES and *not* enclosed by any sheath coverings.
    • AXOPLASM—(cytoplasm of axon) contains many MITOCHONDRIA, MICROTUBLES and NEUROFIBRILS. Newly synthesized proteins pass into the axoplasm from the cell body at the rate of 1 mm (0.4 inch)/day. They replace proteins lost during metabolism, and are involved in neuron growth and regeneration of severed portions of peripheral nerve fibers.
    • AXOLEMMA—Plasma membrane surrounding the axoplasm (continuation of cell membrane of the cell body).
    • TELODENDRIA—Fine terminal branching filaments of axon.
    • SYNAPTIC KNOBS—Bulblike distal endings of telodendria, important in conduction of the nerve impulse at the SYNAPSE. (See 7.3)
    • NERVE FIBER—refers to an AXON and its SHEATH COVERINGS.

The 2 Principal Kinds of CELLS in the NERVOUS SYSTEM: NEURONS and NEUROGLIA
The Structural and Functional Units
NEURONS

■ 3 Principal Components of a NEURON

1 | 1 CELL BODY (PERIKARYON) | SOMA

2 | RAMIFIED DENDRITES | AFFERENT PROCESSES

3 | 1 AXON (AXIS CYLINDER) | EFFERENT PROCESS

■ Components of the CELL BODY

a | NUCLEUS | b | NUCLEOLUS |

c | CYTOPLASM |

  c_1 | NISSL BODIES |

  c_2 | NEUROFIBRILS | extend into the AXON and DENDRITES for support

■ Components of the AXON

4 | AXON HILLOCK |

5 | AXOPLASM | 6 | AXOLEMMA |

7 | TELODENDRIA |

8 | SYNAPTIC KNOBS (END FEET) |

■ AXON SHEATH COVERINGS
PNS NEUROGLIAL COVERING: SCHWANN

9 | SHEATH of SCHWANN/OUTER NEURILEMMA and CYTOPLASM | ONLY in FIBERS of PNS

10 | NUCLEUS OF SCHWANN CELL |

11 | INNER MYELIN SHEATH | GIVES COLOR to WHITE MATTER

12 | NODES of RANVIER | GAPS BETWEEN SEGMENTS of MYELIN SHEATH

  • Large PERIPHERAL AXONS are surrounded by a MYELIN SHEATH, a multilayered, white, phospholipid segmented covering. MYELIN SHEATH insulates the axon and increases the speed of the nerve impulse conduction. Axons covered by a myelin sheath are called *myelinated*, without it, *unmyelinated*.
  • The MYELIN SHEATH is formed by flattened neuroglial SCHWANN CELLS, which encircle and wind around the axon. Much like toothpaste is rolled from the bottom of a tube, the cytoplasm and nucleus of the Schwann cell are squeezed into an outer layer called the NEURILEMMA (SHEATH OF SCHWANN). The inner, fatty myelin sheath consists of many wrapped layers of the Schwann cell membrane.
  • Unmyelinated axons are enclosed by Schwann cells forming a living NEURILEMMA SHEATH, but without the multiple wrappings characteristic of the myelin sheath.
  • NEURILEMMA aids the regeneration of peripheral nerve fibers.

The Supporting and Protecting Units
NEUROGLIA (GLIAL CELLS): "NERVE GLUE"

■ PNS NEUROGLIA:

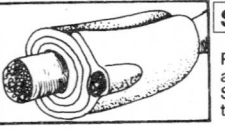

SCHWANN CELLS
  • Wrap around PERIPHERAL AXONS of the PNS, forming an inner MYELIN SHEATH and an outer NEURILEMMA SHEATH. These Schwann cell coverings plus the axon form the PERIPHERAL NERVE FIBER.
  • Flattened cells arranged in series

SATELLITE CELLS
  • Also called CAPSULE CELLS, form supportive capsulelike coverings around the neuron cell bodies within the PERIPHERAL GANGLIA of the PNS. May transfer nutrients to neurons.
  • Small, flattened cells

■ CNS NEUROGLIA:

in GRAY MATTER   in WHITE MATTER

ASTROCYTES
  • Star-shaped with numerous processes, which surround most of the outer surface of BRAIN CAPILLARIES. Contributes to BLOOD-BRAIN BARRIER. Provide structural support by attaching neurons to their blood vessels.
  • Twine around nerve cells of CNS in transversely oriented supporting network
Protoplasmic types are in CNS GRAY MATTER
Fibrous types are in CNS WHITE MATTER

OLIGODENDROCYTES
  • Resemble ASTROCYTES with fewer and shorter processes. Look like "plump pillows."
  • Produce the MYELIN SHEATHS around axons of CNS fibers
  • Form semi-rigid longitudinal supporting rows between neurons of BRAIN and SPINAL CORD.
  • Guide development of CNS neurons

EPENDYMA
  • Wedge-shaped, ciliated columnar cells that line the cavities (VENTRICLES and CENTRAL CANAL) of CNS. Produce small amounts of CEREBROSPINAL FLUID and help circulate it with ciliary motion.

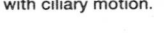

MICROGLIA
  • Also called MESOGLIA, the only NEUROGLIA of MESODERMAL (MESENCHYMAL) origin. (All others are of ECTODERMAL ORIGIN as is all neuron tissue.) Invade the CNS and respond to injury or infection. They become ameboid and phagocytic, engulfing and destroying microbes and cellular debris.
  • Small, flattened, studded with spinelike GEMMULES.

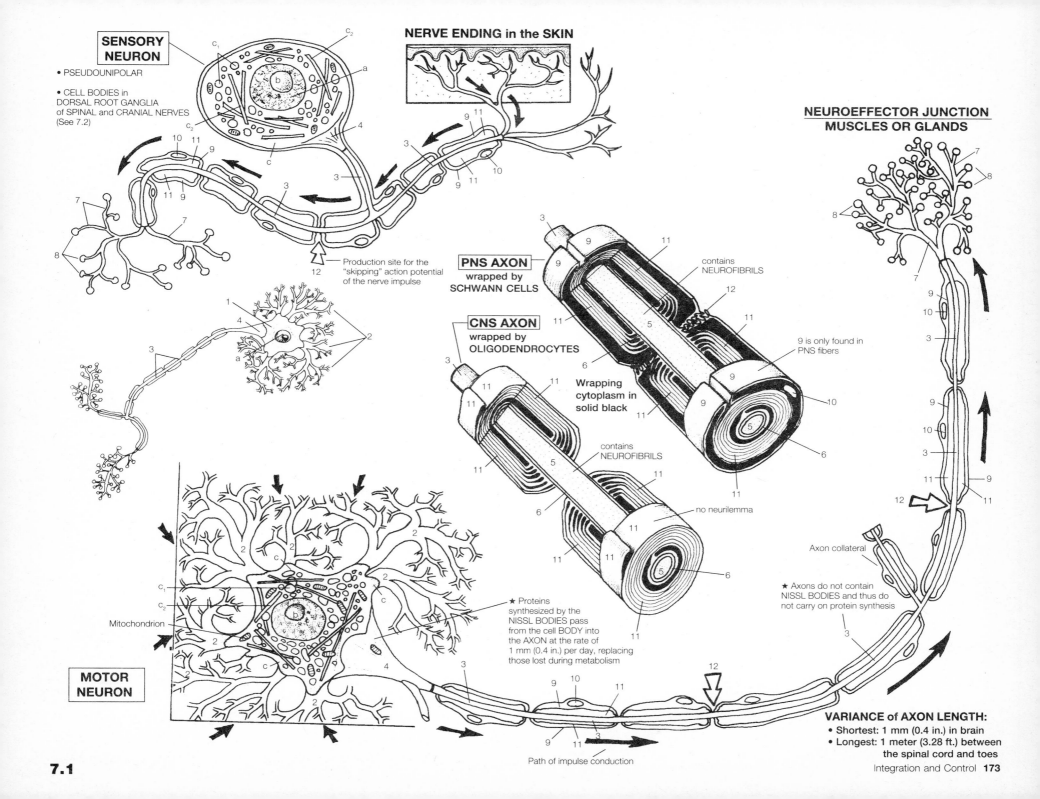

SENSORY NEURON

- PSEUDOUNIPOLAR

- CELL BODIES in DORSAL ROOT GANGLIA of SPINAL and CRANIAL NERVES (See 7.2)

NERVE ENDING in the SKIN

Production site for the "skipping" action potential of the nerve impulse

NEUROEFFECTOR JUNCTION MUSCLES OR GLANDS

PNS AXON wrapped by **SCHWANN CELLS**

CNS AXON wrapped by **OLIGODENDROCYTES**

Wrapping cytoplasm in solid black

contains NEUROFIBRILS

9 is only found in PNS fibers

contains NEUROFIBRILS

no neurilemma

Mitochondrion

★ Proteins synthesized by the NISSL BODIES pass from the cell BODY into the AXON at the rate of 1 mm (0.4 in.) per day, replacing those lost during metabolism

★ Axons do not contain NISSL BODIES and thus do not carry on protein synthesis

Axon collateral

MOTOR NEURON

Path of impulse conduction

VARIANCE of AXON LENGTH:
- Shortest: 1 mm (0.4 in.) in brain
- Longest: 1 meter (3.28 ft.) between the spinal cord and toes

7.1

NERVOUS SYSTEM: GENERAL ORGANIZATION
Structural and Functional Classification
of Nervous Tissue

d = blue 2 = yellow-orange 3 = green
4 = red-orange 5 = grey dot 6 = yellow
7 = yellow-green 8 = red 9 = light brown
10 = fleshy 11 = pink 12 = light blue
★ See Chart #1

★ NEURONS may be classified according to STRUCTURE and FUNCTION
■ STRUCTURAL CLASSIFICATION of NEURONS
• Based on the number of POLES or PROCESSES associated with each cell body.
■ UNIPOLAR
• One process leaves cell body
• Originate embryologically as BIPOLAR NEURONS. Axon and dendrite fuse to form a single process, thus the term PSEUDOUNIPOLAR.
• All SENSORY (AFFERENT) in function
• Found in PNS, transmit sensory information to the CNS (ex: POSTERIOR SENSORY ROOT GANGLIA of the SPINAL and CRANIAL NERVES)
■ BIPOLAR
• Two processes at either end of an elongated cell body, with one process highly modified to respond to particular forms of energy.
• All SENSORY (AFFERENT) in function. Found in organs of special sense (exclusively in retina of eye, auditory nerve in inner ear, taste buds, olfactory epithelium (See 7.18-7.23)
■ MULTIPOLAR
• Many cell processes extend from the cell body. (many dendrites, one axon)
• Mostly MOTOR (EFFERENT) or ASSOCIATION in function
• Concerned with processing or transmitting MOTOR IMPULSES that lead to muscular contraction or gland secretion.

★ ■ FUNCTIONAL CLASSIFICATION of NEURONS
• Based on 3 main modes of direction of impulse conduction
■ SENSORY (AFFERENT) NEURONS
• Transmit impulses from the body periphery (skin, sense organs, muscle and viscera) to the CNS (brain and spinal cord); UNIPOLAR
■ MOTOR (EFFERENT) NEURONS
• Convey impulses from CNS to body periphery (to effector organs, muscles (skeletal or smooth) and glands; MULTIPOLAR
■ ASSOCIATION (INTERNUNCIAL) NEURONS
• Form a network of interconnecting neurons carrying impulses from sensory neurons to motor neurons. They are located in the CNS (brain and spinal cord).

★ The processes (nerve fibers) of afferent and efferent neurons are arranged into bundles called NERVES, which lie outside the CNS, belonging to the PNS. Neurons and their nerve fibers within nerves may be grouped according to a general scheme:
■ SOMATIC AFFERENT
Impulses from sensory receptors in skin, sense organs, joints and skeletal muscle conducted to CNS for interpretation (UNIPOLAR).
■ VISCERAL AFFERENT
• Impulses from sensory receptors in structures with hollow cavities (visceral organs and blood vessels) conducted to the CNS for interpretation (UNIPOLAR).
■ SOMATIC EFFERENT
• Impulses from CNS conducted to motor end plates on skeletal muscle, causing contraction of skeletal muscle (effector organ); (MULTIPOLAR).
■ VISCERAL (AUTONOMIC) EFFERENT
• Belong to the ANS
• Impulses from CNS conducted to smooth and cardiac muscle, causing contraction, and to glands, causing secretion. Two NEURONS are involved through GANGLIONATED TRUNKS outside the CNS.

Structural Classification by Number of PROCESSES:

a | UNIPOLAR or PSEUDOUNIPOLAR |

b | BIPOLAR |

c | MULTIPOLAR |

Types of PROCESSES

in UNIPOLAR and BIPOLAR NEURONS

2 | PERIPHERAL PROCESS | → • FUNCTIONS LIKE a DENDRITE (AFFERENT IMPULSES to the CELL BODY)
• STRUCTURED LIKE an AXON

4 | CENTRAL PROCESS | → FUNCTION and STRUCTURE of AXON (EFFERENT IMPULSES FROM CELL BODY to CNS)

d | DENDRITE | 8 | AXON |

Functional Classification by Direction of IMPULSE CONDUCTION:
NEURONS of the PERIPHERAL NERVOUS SYSTEM (PNS)

■ The AFFERENT SYSTEM of the PNS: UNIPOLAR NEURONS

SENSORY NEURON (AFFERENT):
(Both SOMATIC AFFERENT and VISCERAL AFFERENT)

1 | RECEPTOR |

2 | PERIPHERAL PROCESS (DENDRITIC "AXON") |

3 | CELL BODY in DORSAL ROOT GANGLIA of CRANIAL or SPINAL NERVES |

4 | CENTRAL PROCESS (AXON) |

5 | SYNAPSES | JUNCTIONS BETWEEN NEURONS (See 7.3)

6 | CENTRAL NERVOUS SYSTEM (CNS) |

■ The EFFERENT SYSTEM of the PNS: MULTIPOLAR NEURONS
• SOMATIC NERVOUS SYSTEM (SNS):
SOMATIC MOTOR NEURON (SOMATIC EFFERENT)

7 | CELL BODY in CNS |

8 | AXON |

9 | MOTOR END PLATE (SYNAPSE) |

10 | SKELETAL MUSCLE (EFFECTOR) |

• AUTONOMIC NERVOUS SYSTEM (ANS)
AUTONOMIC MOTOR NEURON (VISCERAL EFFERENT)

11 | PREGANGLIONIC NEURON | (CELL BODY in CNS)

5 | SYNAPSE |

12 | POSTGANGLIONIC NEURON | (CELL BODY outside CNS)

13 | SMOOTH MUSCLE, CARDIAC MUSCLE, or GLANDS |

NEURONS of the CENTRAL NERVOUS SYSTEM (CNS)

Connecting INTERNEURONS

14 | ASSOCIATION NEURON/ INTERNUNCIAL NEURON |

• Found mostly in the CNS, making up the bulk of the neurons of the BRAIN and SPINAL CORD
• ASSOCIATION NEURONS in the SPINAL CORD can be directly related to incoming sensory (afferent) impulses and others to out going (efferent) impulses.
• ASSOCIATION NEURONS linked with higher brain centers integrate sensory input to effect an appropriate motor response (output). Cell bodies of these neurons make up the majority of the GRAY MATTER of the CNS, and their MYELINATED AXONS make up most of the WHITE MATTER NERVE TRACTS of the CNS.

Examples of ASSOCIATION NEURON Linkups

(A) 1 → 1 → 1
RECEPTOR → • ASSOCIATION → • EFFECTOR
NEURON   NEURON   NEURON
(REFLEX ARC—very rare type of linkage)

(B) 1 → 1 → • MANY
RECEPTOR → • ASSOCIATION → • EFFECTOR
NEURON   NEURON   → • NEURONS

(C) MANY → 1 → 1
RECEPTOR → ASSOCIATION → • EFFECTOR
NEURONS → NEURON   NEURON

(D) 1 or more → 1 → • 1 or more
RECEPTOR → ASSOCIATION → • EFFECTOR
NEURONS → NEURON   → • NEURONS

Other association neurons synapsing with the effector neuron(s) allow complex linkups with CNS centers at higher and lower levels of the Brain and Spinal cord.
These functional linkups make the establishment of CONDITIONED INVOLUNTARY REFLEXES possible (ex: SWALLOWING REFLEX). These reflexes probably form the basis of all training. Thus, a fine line exists between the end of INVOLUNTARY REFLEX behavior and the beginning of VOLUNTARY behavior.

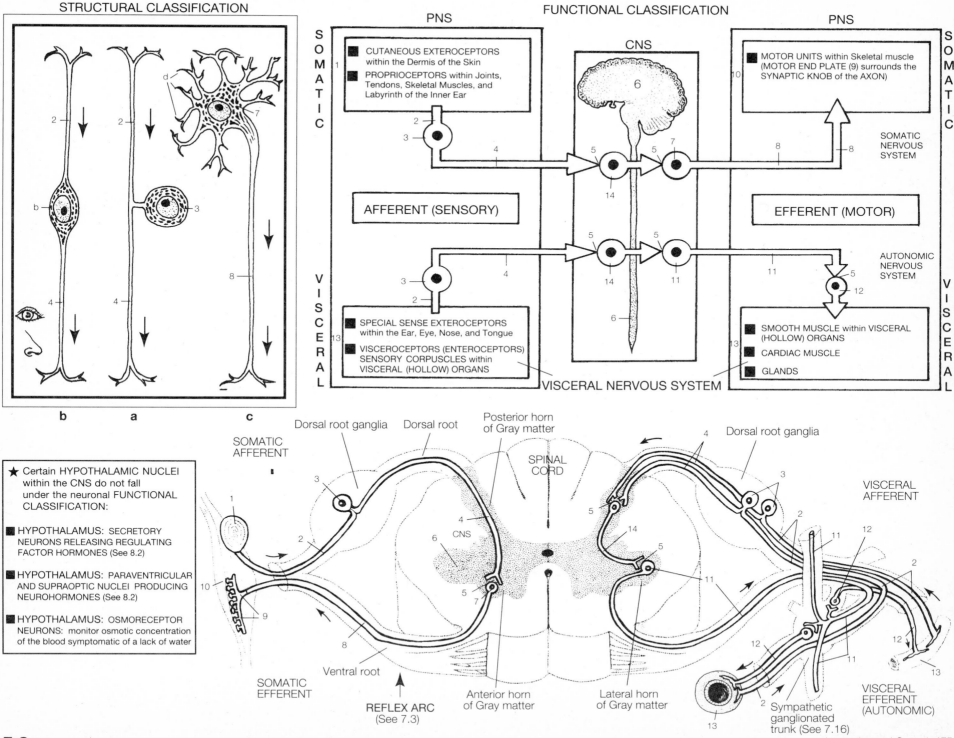

STRUCTURAL CLASSIFICATION

FUNCTIONAL CLASSIFICATION

PNS

CNS

PNS

SOMATIC

- CUTANEOUS EXTEROCEPTORS within the Dermis of the Skin
- PROPRIOCEPTORS within Joints, Tendons, Skeletal Muscles, and Labyrinth of the Inner Ear

- MOTOR UNITS within Skeletal muscle (MOTOR END PLATE (9) surrounds the SYNAPTIC KNOB of the AXON)

SOMATIC

SOMATIC NERVOUS SYSTEM

AFFERENT (SENSORY)

EFFERENT (MOTOR)

AUTONOMIC NERVOUS SYSTEM

VISCERAL

- SPECIAL SENSE EXTEROCEPTORS within the Ear, Eye, Nose, and Tongue
- VISCEROCEPTORS (ENTEROCEPTORS) SENSORY CORPUSCLES within VISCERAL (HOLLOW) ORGANS

- SMOOTH MUSCLE within VISCERAL (HOLLOW) ORGANS
- CARDIAC MUSCLE
- GLANDS

VISCERAL

VISCERAL NERVOUS SYSTEM

b   a   c

★ Certain HYPOTHALAMIC NUCLEI within the CNS do not fall under the neuronal FUNCTIONAL CLASSIFICATION:

- HYPOTHALAMUS: SECRETORY NEURONS RELEASING REGULATING FACTOR HORMONES (See 8.2)

- HYPOTHALAMUS: PARAVENTRICULAR AND SUPRAOPTIC NUCLEI PRODUCING NEUROHORMONES (See 8.2)

- HYPOTHALAMUS: OSMORECEPTOR NEURONS: monitor osmotic concentration of the blood symptomatic of a lack of water

SOMATIC AFFERENT

Dorsal root ganglia   Dorsal root   Posterior horn of Gray matter

SPINAL CORD

Dorsal root ganglia

VISCERAL AFFERENT

CNS

SOMATIC EFFERENT

Ventral root

REFLEX ARC (See 7.3)

Anterior horn of Gray matter

Lateral horn of Gray matter

Sympathetic ganglionated trunk (See 7.16)

VISCERAL EFFERENT (AUTONOMIC)

7.2

Integration and Control   175

NERVOUS SYSTEM: GENERAL ORGANIZATION
Synapses, Neuroeffector Junctions, and Reflex Arcs

★ Simply put, a NERVE IMPULSE is a wave of ELECTRICAL NEGATIVITY that travels along the surface of the MEMBRANE of a NEURON, reversing the electrical state (POSITIVE charge outside-to-inside, and NEGATIVE CHARGE inside-to-outside) "DEPOLARIZING" the MEMBRANE.
■ IRRITABILITY (EXCITABILITY)
 • Ability of a neuron to respond to a stimulus and convert it into an impulse.
■ CONDUCTIVITY
 • Ability of a neuron to transmit an impulse to another neuron or to another tissue (a muscle or a gland).

★ ■ SYNAPSE
 • The junction where one neuron connects to another neuron.
 • PRESYNAPTIC NEURON—the neuron that carries an impulse to the SYNAPSE.
 • POSTSYNAPTIC NEURON—the neuron that carries an impulse AWAY from the SYNAPSE.
 • There is *No* DIRECT PROTOPLASMIC UNION between connecting neurons at the SYNAPSE. (i.e., NO PHYSICAL CONTACT). The two parts of the SYNAPSE are separated by a space called the SYNAPTIC CLEFT.
 • SYNAPSES may occur between the AXON TERMINATIONS of the PRESYNAPTIC NEURON (TELODENDRIA) and 3 possible areas of the POST-SYNAPTIC NEURON:
  • AXON (AXO-AXONIC SYNAPSE)
  • CELL BODY (AXO-SOMATIC SYNAPSE)
  • DENDRITES (AXO-DENDRITIC SYNAPSE)
 • IMPULSE CONDUCTION at a SYNAPSE IS ALWAYS ONE-WAY (from the presynaptic neuron to the postsynaptic neuron). Impulses cannot back up into another presynaptic neuron. This prevents impulse conduction along unsuitable and wrong pathways. ("short-circuits")
 • One neuron usually connects with many other neurons widely scattered throughout the CNS. Thus, intricate complex pathways are built up in the CNS for the integration, coordination, association, and modification of incoming information and memory storage into desired motor responses (muscle contraction). There are approximately 1 x 10^{13} synapses in the CNS, and some neurons are known to have over 5,000 synapses/nerve cell!

■ NEUROEFFECTOR JUNCTIONS in PNS
 • NEUROMUSCULAR JUNCTION—when an impulse is transmitted between a neuron and a muscle cell (fiber).
 • NEUROGLANDULAR JUNCTION—when an impulse is transmitted between a neuron and a glandular cell.

★ Impulse conduction across a synapse or a neuroeffector junction is dependent on CHEMICAL ACTIVITY by NEUROTRANSMITTER SUBSTANCE (produced in the neuron by amino acids, energized for rapid production by MITOCHONDRIA). After production and transportation to the SYNAPTIC KNOB, it is stored in thousands of tiny membrane-enclosed sacs called SYNAPTIC VESICLES. An approaching nerve impulse causes the SYNAPTIC VESICLES to move and BIND to the PRESYNAPTIC MEMBRANE, releasing the neurotransmitter, which diffuses across the SYNAPTIC CLEFT and BINDS to a specific RECEPTOR on the POSTSYNAPTIC MEMBRANE. This BINDING forms the STIMULUS to cause DEPOLARIZATION of the POSTSYNAPTIC MEMBRANE, and the impulse is conducted to the next neuron, muscle fiber, or glandular cell. The neurotransmitter substance is rapidly renewed in the presynaptic neuron.

Types of SYNAPSES

a | AXO- | AXONIC | b   a | AXO- | SOMATIC | c   a | AXO- | DENDRITIC | d

SYNAPSE: INTERNEURONAL FUNCTIONAL JUNCTION

1 PRESYNAPTIC AXON (TELODENDRION)
2 SYNAPTIC KNOB (END FOOT)
3 MITOCHONDRIA
4 SYNAPTIC VESICLES with NEUROTRANSMITTER SUBSTANCE
5 PRESYNAPTIC MEMBRANE

Nonanatomic CONTINUITY JUNCTION:

6 SYNAPTIC CLEFT 150–200 angstrom units (A) wide
7 POSTSYNAPTIC MEMBRANE with RECEPTOR SITES for NEUROTRANSMITTER
8 AXON, CELL BODY, or DENDRITE

SOMATIC NERVOUS SYSTEM (SNS) of the PNS: NEUROMUSCULAR JUNCTION: See 6.2

1a TELODENDRION of AXON BRANCH
2 SYNAPTIC KNOB (END FOOT)
3 MITOCHONDRIA
4 SYNAPTIC VESICLES with NEUROTRANS-MITTERSUBSTANCE
5 PRESYNAPTIC MEMBRANE

Nonanatomic CONTINUITY JUNCTION

6a NEUROMUSCULAR (SUBNEURAL) CLEFT
7a MOTOR END PLATE (FOLDED SARCOLEMMA)
8a MUSCLE FIBER (CYTOPLASM)

REFLEX ARC: Involuntary Functional Unit of the NERVOUS SYSTEM

★ The NEURON is the ANATOMICAL or STRUCTURAL UNIT of the NERVOUS SYSTEM. The NERVOUS REFLEX is the PHYSIOLOGICAL or FUNCTIONAL UNIT.
■ A REFLEX or REFLEX ACTION is an INVOLUNTARY ACTION, AUTOMATIC and UNCONSCIOUS, caused by the STIMULATION of a SENSORY (AFFERENT) RECEPTOR whose response is ALWAYS MUSCULAR CONTRACTION or GLANDULAR SECRETION.
■ Important bodily functions such as DIGESTIVE ACTIVITY, RESPIRATORY MOVEMENTS, HEART BEAT, and POSTURAL ADJUSTMENTS are all controlled through REFLEXES. REFLEXES form the BASIS of all CNS ACTIVITY, occurring at all levels of the BRAIN and SPINAL CORD.
■ The STRUCTURAL BASIS of all REFLEX ACTION is the REFLEX ARC, the simplest type of CONDUCTION NERVE PATHWAY. The ARC leads by a short route from sensory to motor neurons and includes only 2 or 3 neurons. REFLEX ARCS involving only 2 NEURONS are termed MONOSYNAPTIC ARCS (only 1 synapse between the afferent and efferent neurons). In most reflex arcs in humans, AFFERENT and EFFERENT NEURONS are linked by at least one ASSOCIATION NEURON, forming the simplest of the POLYSYNAPTIC ARCS (3 NEURONS, 2 SYNAPSES). In the majority of POLYSYNAPTIC ARCS, a chain of many ASSOCIATION NEURONS forms complex linkups with various levels of the BRAIN and SPINAL CORD.
■ Every SENSORY RECEPTOR NEURON is potentially linked through the CNS with a large number of EFFECTOR ORGNAS all over the body.
■ Every EFFECTOR ORGAN is in similar CNS communication with a large number of RECEPTORS all over the body.
■ REFLEX ACTION in the SPINAL CORD can thus be modified by centers in the BRAIN and BRAINSTEM, which can send inhibiting or facilitating impulses along nerve tracts to cells in the SPINAL CORD.

★ 5 COMPONENTS of a REFLEX ARC
 • RECEPTOR (includes the DENDRITIC "AXON" of SENSORY NEURON)
 • SENSORY NEURON (UNIPOLAR: AXON and CELL BODY in POSTERIOR ROOT GANGLIA)
 • REFLEX CENTER (within CNS: ARC, SYNAPSES, and LINKUPS)
 • MOTOR NEURON
 • EFFECTOR ORGAN

• The simplest form of a REFLEX ARC shows the structural basis of all reflex arcs:
MONOSYNAPTIC ARC (2 NEURONS and 1 SYNAPSE)
Example: STRETCH REFLEX (very rare)
 (IPSILATERAL → RECEPTOR and EFFECTOR on the same side of the SPINAL CORD

e STRETCH RECEPTOR (INCLUDES PERIPHERAL PROCESS "AXON" DENDRITE of SENSORY NEURON)
f AFFERENT (SENSORY) AXON CENTRAL PROCESS OF SENSORY NEURON
g SENSORY NEURON CELL BODY IN DORSAL (POSTERIOR) ROOT GANGLIA
h SYNAPSE
i MOTOR NEURON CELL BODY in CNS
j EFFERENT (MOTOR) AXON
7a MOTOR END PLATE
8a SKELETAL MUSCLE (EFFECTOR)

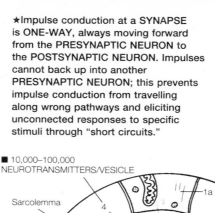

★Impulse conduction at a SYNAPSE is ONE-WAY, always moving forward from the PRESYNAPTIC NEURON to the POSTSYNAPTIC NEURON. Impulses cannot back up into another PRESYNAPTIC NEURON; this prevents impulse conduction from travelling along wrong pathways and eliciting unconnected responses to specific stimuli through "short circuits."

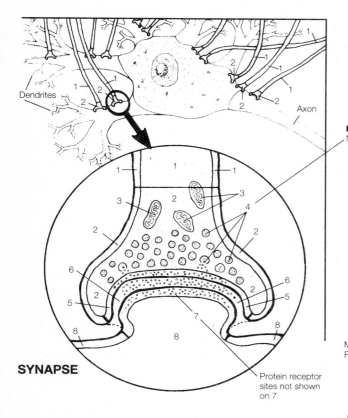

Dendrites

Axon

SYNAPSE

Protein receptor sites not shown on 7

■ 10,000–100,000 NEUROTRANSMITTERS/VESICLE

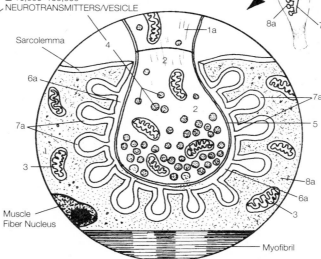

Sarcolemma

Muscle Fiber Nucleus

Myofibril

**MOTOR UNIT: AXON BRANCH inside a
MOTOR END PLATE at NEUROMUSCULAR JUNCTION**

**MONOSYNAPTIC ARC
(STRETCH REFLEX)**

★ Synaptic transmission (0.5–1 msec) is slower than that along a nerve fiber. A certain amount of *control* may be established on the impulse's further progress at the SYNAPTIC POINT.

★ After the neurotransmitter has exerted its effect—whether enhanced (FACILITATION) or reduced (INHIBITION) transmission—a chemical matched to the neurotransmitter is always present to destroy it. This prevents continued stimulation and allows more messages to cross the SYNAPSE.

POLYSYNAPTIC ARCS

Gray Matter

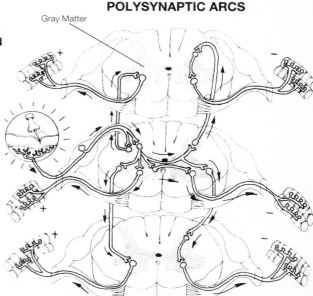

**(FLEXOR REFLEX)
+ EXCITATORY FLEXION**

**(CROSSED EXTENSOR REFLEX)
- INHIBITORY EXTENSION**

TYPES OF SYNAPSES

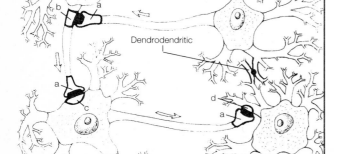

Dendrodendritic

POLYSYNAPTIC ARCS (MULTINEURONAL)

★ Much more common than the rare STRETCH REFLEXES are the POLYSYNAPTIC REFLEXES, which involve many spinal cord segments or centers in the brain. They usually involve at lease 3 neurons and ASSOCIATION or INTERNUNCIAL NEURONS in the gray matter inserted between SENSORY INPUT and MOTOR OUTPUT. (Find them in the drawing and color them.)

■ FLEXOR REFLEX

• PAIN RECEPTOR in the skin responds to a painful stimulus and the FLEXOR MUSCLES are contracted (excited) to remove the body part from the harm.

■ CROSSED EXTENSOR REFLEX

• Combines a FLEXOR REFLEX on one side of the body with muscular EXTENSION on the other side of the body to maintain posture.

• Involves transmission of information across the cord.

a = green a₂ = blue-green b = light green
② = orange ③ = yellow c = yellow-orange
d = pink e = white f = creamy
7 = fleshy 8 = light brown 9 = light gray
★ See 7.5, 7.11, 9.16

NERVOUS SYSTEM: CENTRAL NERVOUS SYSTEM (CNS)
Protective Membranous Coverings
of the Brain and Spinal Cord: Meninges

★ The delicate organs of the CNS are surrounded and protected by a bony encasement—the CRANIUM (SKULL) surrounds the BRAIN, and the VERTEBRAL COLUMN surrounds the SPINAL CORD. Three membranes, collectively referred to as the MENINGES (sing. MENINX), form a protective barrier between the bone and the soft CNS tissue. The 3 connective tissue membranous coverings are the
■ DURA MATTER ■ ARACHNOID MEMBRANE
■ PIA MATER

★① DURA MATER (PACHYMENINX)
■ Outermost and thickest covering
■ Composed of collagen fibers and fibroblasts for tough protection
■ CRANIAL DURA MATER consists of 2 layers, and not only encloses the BRAIN, but also its outer thick layer acts as the PERIOSTEUM for the interior of the CRANIAL CAVITY as it adheres firmly to the CRANIAL BONES of the SKULL. The inner, thin layer follows the general contour of the brain. Around most of the brain these two layers are fused. In specific regions these two layers separate and enclose DURAL SINUSES. (These DURAL SINUSES collect venous blood and drain it to the internal jugular veins of the neck. See 9.16) In four locations, the inner layer turns inward to form SEPTAL PARTITIONS between areas on the surface of the brain and BRAIN ANCHORS to the inside of the CRANIAL CASE. (See box) The aforementioned DURAL SINUSES are found in the MARGINS of these infoldings of the inner meningeal layer of the cranial dura mater.

■ SPINAL DURA MATER
  • Forms a loose, tubular sheath around the spinal cord
  • A single layer corresponding to the inner cranial dura layer
  • Connects to the cranial dura at the FORAMEN MAGNUM, and continues down to the level of the 2nd sacral vertebra, where it is firmly attached. Outer periosteal layer not necessary since vertebrae have their own periosteum.
■ As the SPINAL NERVES exit from the vertebral canal a short portion of their spinal roots are covered by sleeves of the spinal dura.

★② ARACHNOID MEMBRANE
  • The intermediate membrane, thin and delicate
  • Avascular spider web, netlike membrane
  • Lines the inner surface of the DURA spreading over the entire CNS. In the brain it also follows the FALX and TENTORIUM but does not extend into the SULCI or FISSURES of the BRAIN.
  • Delicate strands of connective tissue extend down through the SUBARACHNOID SPACE and connect to the PIA MATER.

★③ PIA MATER
  • Innermost, very thin membrane
  • Directly attached to the surfaces of the CNS, following all irregular contours of the brain and spinal cord. Highly specialized over the ROOFS of the VENTRICLES to form the CHOROID PLEXUSES along with the ARACHNOID (See 7.5)
  • Highly vascular, supporting blood vessels derived from the cerebral and cerebellar arteries that nourish the underlying cells of the brain and spinal cord.

3 PROTECTIVE MENINGES (from outside in)

① | DURA MATER |

② | ARACHNOID MEMBRANE | SPIDER MEMBRANE

③ | PIA MATER |

① External MENINX: DURA MATER

a | DOUBLE-LAYERED CRANIAL DURA MATER | DURA MATER CEREBRI

  a₁ | THICK, OUTER PERIOSTEAL LAYER |

  a₂ | THIN, INNER MENINGEAL LAYER |

b | SINGLE-LAYERED SPINAL DURAL MATER | DURA MATER SPINALIS

  b | THIN, MENINGEAL DURAL SHEATH |

★INFOLDINGS Forming SEPTA from the Thin, Inner MENINGEAL LAYER of the CRANIAL DURA MATER:

4 | FALX CEREBRI |
• Extends downward to lie within the LONGITUDINAL FISSURE
• Separates (partitions) the 2 cerebral hemispheres
• Anterior anchor = CRISTA GALLI of the ETHMOID BONE. Posterior anchor = TENTORIUM

5 | TENTORIUM CEREBELLI |
• Supports the occipital lobes of the cerebrum and separates them from the cerebellum
• Anchors = TENTORIUM, PETROUS BONES, and OCCIPITAL BONE

6 | FALX CEREBELLI (See 7.10) |
• Forms a vertical partition between the 2 cerebellar hemispheres
• Anchor = OCCIPITAL CREST of OCCIPITAL BONE

| DIAPHRAGMA SELLAE |
• Forms the roof of the SELLA TURCICA, the concavity on the superior surface of the SPHENOID BONE that houses the PITUITARY GLAND

② Netlike Middle MENINX: ARACHNOID MEMBRANE
   CRANIAL ARACHNOID (ARACHNOIDEA ENCEPHALI)
  c | ARACHNOID GRANULATIONS (VILLI) | (See 7.5)
• Tiny, puffy outpocketings of the ARACHNOID that invest the SUPERIOR SAGITTAL SINUS. (See 7.5 CSF is drained through these villi.)

     SPINAL ARACHNOID (ARACHNOIDEA SPINALIS)
  d | 3 DORSAL ARACHNOID SEPTA |
• PIAL FIBERS from the PIA MATER that extend to the ARACHNOID at three points on the posterior aspect of the spinal cord

③ Thin, Internal MENINX: PIA MATER
   PIA MATER Extensions of the SPINAL CORD

   Lateral Extensions
  e | LIGAMENTUM DENTICULATUM |
• Extend along the entire spinal cord. Their lateral margins are scalloped and attach to the DURA.

   Inferior Extensions
  f | FILUM TERMINALE |
• Fibrous strands of PIA MATER that extend from the terminal portion of the spinal cord (CONUS MEDULLARIS) through the LUMBAR CISTERN and attach to the DURAL SAC (See 7.11)

★ Spaces Involving the MENINGES:
SEPARATING VERTEBRAL CANAL WALL and DURA MATER

  7 | EPIDURAL SPACE | (POTENTIAL)
• Unlike the adherence of the outer layer of the DURA MATER to the CRANIAL BONES, there is no connection between the SPINAL DURAL SHEATH and the VERTEBRAE forming the VERTEBRAL CANAL; there is a POTENTIAL CAVITY called the EPIDURAL SPACE that forms a *protective pad* around the spinal cord (outside the DURAL SHEATH). This EPIDURAL SPACE consists of AREOLAR and ADIPOSE (FATTY) CONNECTIVE TISSUE, and is HIGHLY VASCULAR. It contains a venous plexus and branches of spinal nerves as they enter or exit the spinal cord.

SEPARATING DURA MATER and ARACHNOID MEMBRANE
  8 | SUBDURAL SPACE |
• A very narrow space found in both portions of the CNS, in which lies a thin film of serous fluid

SEPARATING ARACHNOID MEMBRANE and PIA MATER
  9 | SUBARACHNOID SPACE |
• In most areas throughout the CNS, a considerable space maintained by the web-like strands of the ARACHNOID MEMBRANE
• Contains CEREBROSPINAL FLUID (CSF) (See 7.5)
• In some areas of the BRAIN, this space is hardly recognizable, and the PIA MATER is almost adherent to the ARACHNOID: (THE PIA-ARACHNOID)
At the base of the brain, this space becomes very wide in areas as the SUBARACHNOID CISTERNAE
Another large subarachnoid space, the LUMBAR CISTERN, lies between the end point of the spinal cord (L2) and the end of the DURAL SAC (S2).This space is filled with spinal nerves (roots) as they enter or exit the spinal cord.

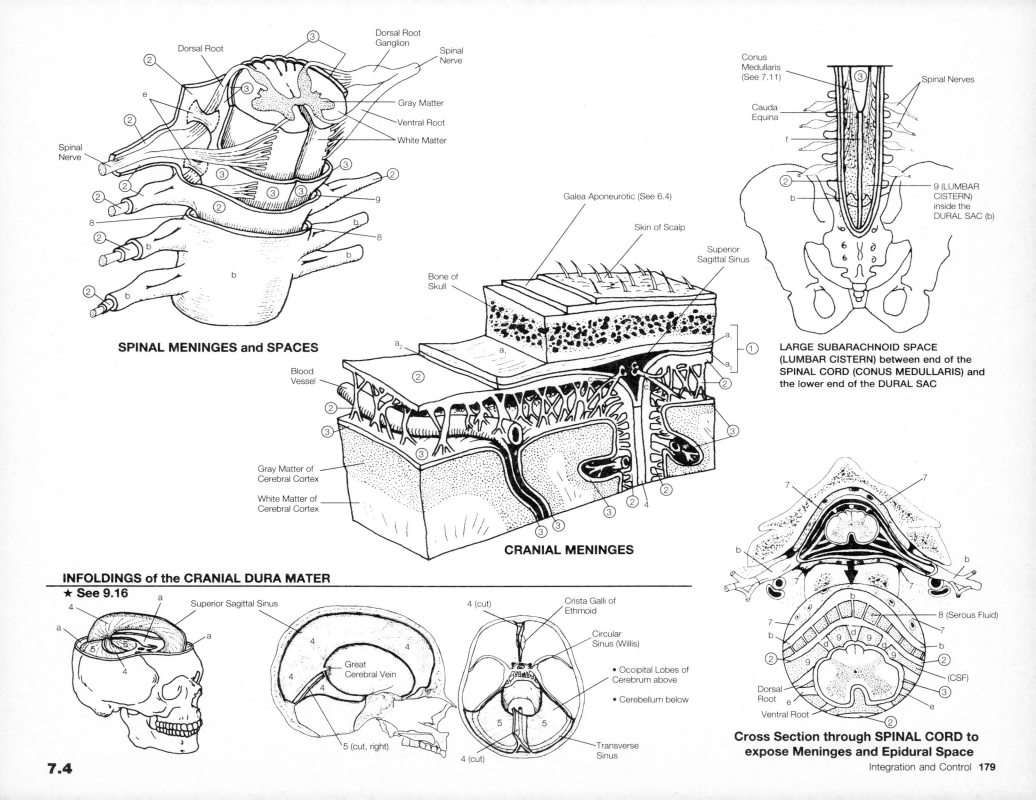

SPINAL MENINGES and SPACES

Dorsal Root
Dorsal Root Ganglion
Spinal Nerve
Gray Matter
Ventral Root
White Matter
Spinal Nerve

CRANIAL MENINGES

Galea Aponeurotic (See 6.4)
Skin of Scalp
Superior Sagittal Sinus
Bone of Skull
Blood Vessel
Gray Matter of Cerebral Cortex
White Matter of Cerebral Cortex

Conus Medullaris (See 7.11)
Cauda Equina
Spinal Nerves
9 (LUMBAR CISTERN) inside the DURAL SAC (b)

LARGE SUBARACHNOID SPACE (LUMBAR CISTERN) between end of the SPINAL CORD (CONUS MEDULLARIS) and the lower end of the DURAL SAC

INFOLDINGS of the CRANIAL DURA MATER
★ See 9.16

Superior Sagittal Sinus
Great Cerebral Vein
5 (cut, right)

4 (cut)
Crista Galli of Ethmoid
Circular Sinus (Willis)
• Occipital Lobes of Cerebrum above
• Cerebellum below
Transverse Sinus
4 (cut)

8 (Serous Fluid)
(CSF)
Dorsal Root
Ventral Root

Cross Section through SPINAL CORD to expose Meninges and Epidural Space

NERVOUS SYSTEM: CENTRAL NERVOUS SYSTEM (CNS)
Protective Buoyant Cushion of the
Brain and Spinal Cord: Cerebrospinal Fluid

*Repeat same colors used for Meninges
from 7.4. Use warm red for 11, light blue
for 5, turquoise for 4.
Ventricles = whatever you have left!*

★ See 7.4, 7.6, 7.9, 9.16

Embryologically, the CNS develops as a neural tube. Structural spaces develop within the brain and spinal cord as remnants of the original tube cavity: The BRAIN contains several VENTRICLES and the CEREBRAL AQUEDUCT. The SPINAL CORD contains the small CENTRAL CANAL. All spaces are lined with EPENDYMAL CELLS.

• Two large LATERAL VENTRICLES lie within the cerebral hemispheres. Each ventricle has a BODY from which arise 3 HORNS. Lying in the BODY and part of the INFERIOR HORN (which extends into the TEMPORAL LOBE) is a CHOROID PLEXUS, a network of capillaries intertwined with the ependymal lining of the ventricles. These 2 CHOROID PLEXUSES in the LATERAL VENTRICLES, secrete the bulk of CSF, or CEREBROSPINAL FLUID.

★ CSF is also formed to a lesser extent by:
• CHOROID PLEXUSES in the roofs of the 3rd and 4th VENTRICLES
• Ciliated ependymal lining covering the plexuses and the central canal of the spinal cord CSF is formed by the ACTIVE TRANSPORT and ULTRAFILTRATION of substances in the BLOOD PLASMA - (800 ml produced each day)

★ CEREBROSPINAL FLUID (CSF)
• Clear, colorless, watery, lymphlike alkaline fluid
• Similar to blood plasma, but does not clot
• Volume: 100–140 ml (23 ml in ventricles, 117 ml in CNS subarachnoid spaces)

★ FUNCTIONS OF CSF
■ *Forms a protective cushion* around and within the CNS, protecting the brain and spinal cord from physical impact.
■ *BUOYS the CNS*, so the brain may function as a heavy organ. Brain weight is about 1500 grams, but buoyed in CSF, it weighs about 50 grams. The specific gravity of CSF is 1.007—a density close to that of brain tissue. Thus, the brain has a NEAR NEUTRAL BUOYANCY, floating in CSF, restricted in its movement by the enclosure of the DURA MATER.
■ *Alteration of volume:* usually shrinking or expanding of cranial contents (due to fluctuations in the amount of blood in the skull) is quickly balanced by increase or decrease of CSF, keeping cranial contents constant.
■ *Exchange of metabolic substances:* CSF receives waste products from nerve cells.
CIRCULATION OF CSF
• Bulk of the CSF formed in the LATERAL VENTRICLES passes through the narrow oval openings of the INTERVENTRICULAR FORAMINA to the single, slitlike cavity of the 3rd VENTRICLE (where more CSF is formed).
• CSF then follows down through the CEREBRAL AQUEDUCT to the 4rd VENTRICLE, (where more CSF is formed). (CSF also seeps down into the CENTRAL CANAL of the SPINAL CORD.)/
• CSF formed by the ependymal cells of the CENTRAL CANAL of the SPINAL CORD drains upward into the 4th VENTRICLE.
• 3 openings in the roof of the 4th VENTRICLE permit CSF to pass from the 4th VENTRICLE into the SUBARACHNOID SPACE (just below the cerebellum) called the CISTERNA MAGNA (GREAT CISTERN).
• CSF circulates over the surface of the BRAIN and SPINAL CORD in the SUBARACHNOID SPACE.
• CSF is reabsorbed into the bloodstream from the cranial space through vascular berrylike tufts of the ARACHNOID (called ARACHNOID GRANULATIONS or VILLI), which project into the upper cranial midline SUPERIOR SAGITTAL SINUS. (See 7.4)

Structures of the CNS through which CEREBROSPINAL FLUID (CSF) Circulates:

| 5 | **VENTRICLES of the BRAIN** |

| 5 | **SUBARACHNOID SPACE of the ENTIRE CNS** |

| 6 | **CENTRAL CANAL of the SPINAL CORD** |

4 VENTRICULAR CAVITIES of the BRAIN:

| 1V, 2V | **2 LATERAL VENTRICLES (1st and 2nd)** | in each Cerebral Hemisphere, inferior to the Corpus Callosum |

| 3V | 3rd VENTRICLE | in the Diencephalon, between the Thalami |

| 4V | 4th VENTRICLE | in the Brainstem, within the Pons, Medulla, and Cerebellum |

FORAMINA Linking VENTRICLES:
Openings Linking LATERAL VENTRICLES to 3rd VENTRICLE:

| 7 | **INTERVENTRICULAR FORAMEN (FORAMEN OF MONRO)** |

Opening Linking 3rd VENTRICLE to 4th VENTRICLE:

| 8 | **CEREBRAL AQUEDUCT (AQUEDUCT OF SYLVIUS)** |

1	**DURA MATER**
2	**ARACHNOID MEMBRANE**
2a	**ARACHNOID GRANULATIONS (VILLI)**
3	**PIA MATER**
4	**SUPERIOR SAGITTAL SINUS** within upper margin of FALX CEREBRI (See 7.4)

Skin of Scalp
Galea Aponeurotica
Bone of Skull
Outer Layer
Inner Layer
Cerebral Artery
Cerebral Vein
Dural Sinus
Falx Cerebri

3 FORAMINA Linking the 4th VENTRICLE to the SUBARACHNOID SPACE (GREAT CISTERN):

| 9 | **MEDIAN APERTURE (FORAMEN) OF MAGENDIE** |

| 10 | **2 LATERAL APERTURES (FORAMINA) OF LUSCHKA** |

Specialized CAPILLARIES in the BRAIN for the Main Production of CSF:

| 11 | **CHOROID PLEXUSES** | LOCATIONS: |

• Medial Walls of LATERAL VENTRICLES
• Roof of the 3rd VENTRICLE
• Roof of the 4th VENTRICLE

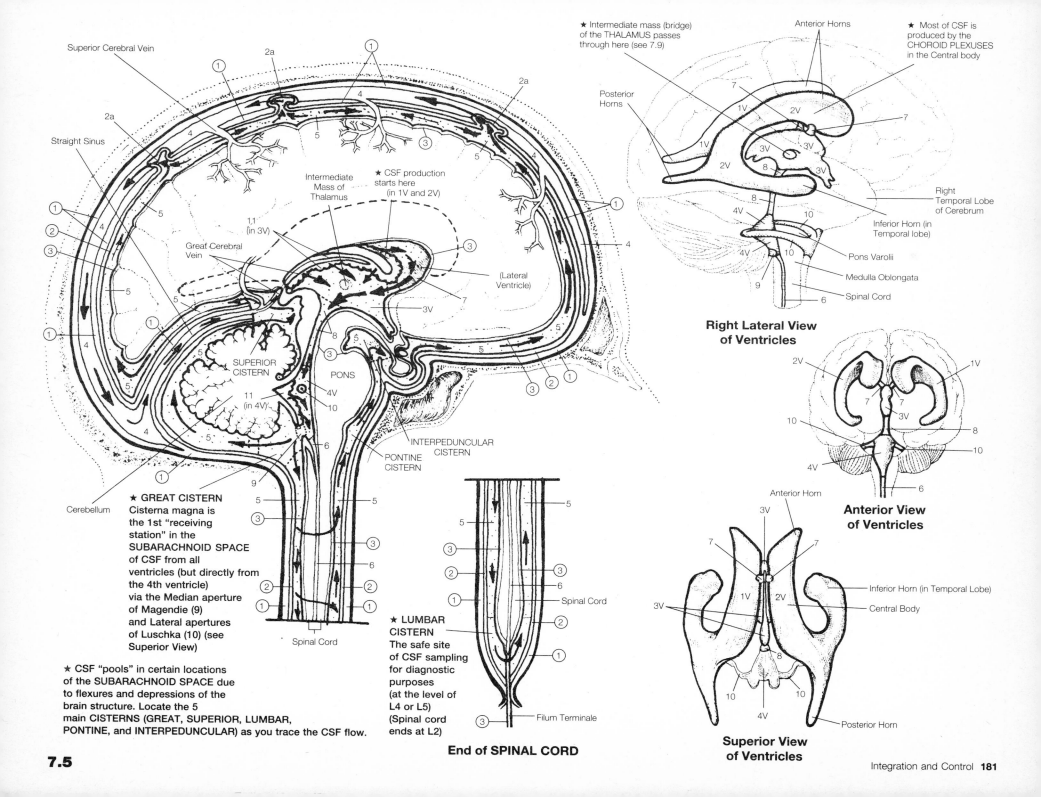

Superior Cerebral Vein

Straight Sinus

2a

★ Intermediate mass (bridge) of the THALAMUS passes through here (see 7.9)

Anterior Horns

★ Most of CSF is produced by the CHOROID PLEXUSES in the Central body

Posterior Horns

2a

4

5

5

3

Intermediate Mass of Thalamus

★ CSF production starts here (in 1V and 2V)

Great Cerebral Vein

1,1 (in 3V)

(Lateral Ventricle)

7

3V

7

SUPERIOR CISTERN

PONS

11 (in 4V)

4V

10

8

3

5

Cerebellum

9

6

Spinal Cord

PONTINE CISTERN

INTERPEDUNCULAR CISTERN

★ GREAT CISTERN
Cisterna magna is the 1st "receiving station" in the SUBARACHNOID SPACE of CSF from all ventricles (but directly from the 4th ventricle) via the Median aperture of Magendie (9) and Lateral apertures of Luschka (10) (see Superior View)

★ CSF "pools" in certain locations of the SUBARACHNOID SPACE due to flexures and depressions of the brain structure. Locate the 5 main CISTERNS (GREAT, SUPERIOR, LUMBAR, PONTINE, and INTERPEDUNCULAR) as you trace the CSF flow.

Right Temporal Lobe of Cerebrum

Inferior Horn (in Temporal lobe)

Pons Varolii

Medulla Oblongata

Spinal Cord

Right Lateral View of Ventricles

2V

1V

7

7

10

3V

8

4V

10

6

Anterior View of Ventricles

Anterior Horn

3V

7

7

1V

2V

3V

3V

8

10

10

4V

Inferior Horn (in Temporal Lobe)

Central Body

Posterior Horn

Superior View of Ventricles

★ LUMBAR CISTERN
The safe site of CSF sampling for diagnostic purposes (at the level of L4 or L5) (Spinal cord ends at L2)

Filum Terminale

End of SPINAL CORD

5

5

6

Spinal Cord

2

7.5

Integration and Control **181**

NERVOUS SYSTEM: CENTRAL NERVOUS SYSTEM (CNS)
Principal Divisions of the Brain

★ **CENTRAL NERVOUS SYSTEM**
■ The **BRAIN**
■ The **SPINAL CORD**

★ The CENTRAL NERVOUS SYSTEM develops in the dorsal (posterior) half of the embryo as a simple, hollow PRIMITIVE NEURAL TUBE:
• The cells lining the tube become the NERVOUS TISSUE of the BRAIN and SPINAL CORD. The inner cavity or NEURAL CANAL undergoes many changes in shape during development of the BRAIN and ultimately forms the VENTRICLES of the BRAIN (See 7.5) and the CENTRAL CANAL of the SPINAL CORD.

★ Rapid differentiation and growth begin in the head (cephalic) portion of the neural tube, and between the end of the 3rd week and middle of the 4th week of embryological development, 3 distinct SWELLINGS are apparent.
■ PROSENCEPHALON (FOREBRAIN)
■ MESENCEPHALON (MIDBRAIN)
■ RHOMBENCEPHALON (HINDBRAIN)
Further development of these 3 swellings form 5 specific regions by the end of the 5th week:
① A massive TELENCEPHALON and ② a central DIENCEPHALON derive from the FOREBRAIN
③ MESENCEPHALON (MIDBRAIN) remains un-changed in its general tubular shape.
④ An upper METENCEPHALON with a large, dorsal outpocketing and ⑤ a lower, thin, narrow MYELENCEPHALON both derive from the HIND-BRAIN. Thus, 5 REGIONS FORM FROM THE ORIGINAL 3 SWELLINGS OF THE NEURAL TUBE.

★ Rapid differentiation occurs in each of these 5 regions shortly after their formation resulting in all of the major structures recognizable later in the adult:
■ Rapid swelling of the TELENCEPHALON forms the 2 EXPANSIVE CEREBRAL HEMISPHERES, which cover the DIENCEPHALON, MIDBRAIN, and portions of the HINDBRAIN.
■ Within the DIENCEPHALON, in the LATERAL WALLS of the 3rd VENTRICLE, 3 swellings form: EPITHALAMUS, THALAMUS, and HYPOTHALAMUS. (The THALAMUS, a large ovoid mass of GRAY MATTER, comprises 4/5 of the DIENCEPHALON.) From the ROOF MIDLINE of the 3rd VENTRICLE the PINEAL GLAND is formed. From the VENTRAL PORTION of the DIEN-CEPHALON, the posterior lobe of the PITUITARY GLAND is formed.
■ The MESENCEPHALON has few changes, the most notable being 4 aggregations of neurons, the paired SUPERIOR and INFERIOR COLLICULI and 2 neural fiber tracts, the CEREBRAL PEDUNCLES.
■ The walls of the METENCEPHALON expand into 2 dorsal midline swellings that join to form the CEREBELLUM and a smaller ventral swelling of nerve fiber bands called the PONS.
■ The adult MYELENCEPHALON forms from an aggregation of nerve fibers, and is called the MEDULLA OBLONGATA. The end (caudal) portion of the MEDULLA is continuous with and structurally similar to the SPINAL CORD.

★ All of the nervous tissue of the CNS (BRAIN and SPINAL CORD) that develops from the walls of the PRIMITIVE NEURAL TUBE specializes into either GRAY or WHITE MATTER (See Chart #1).

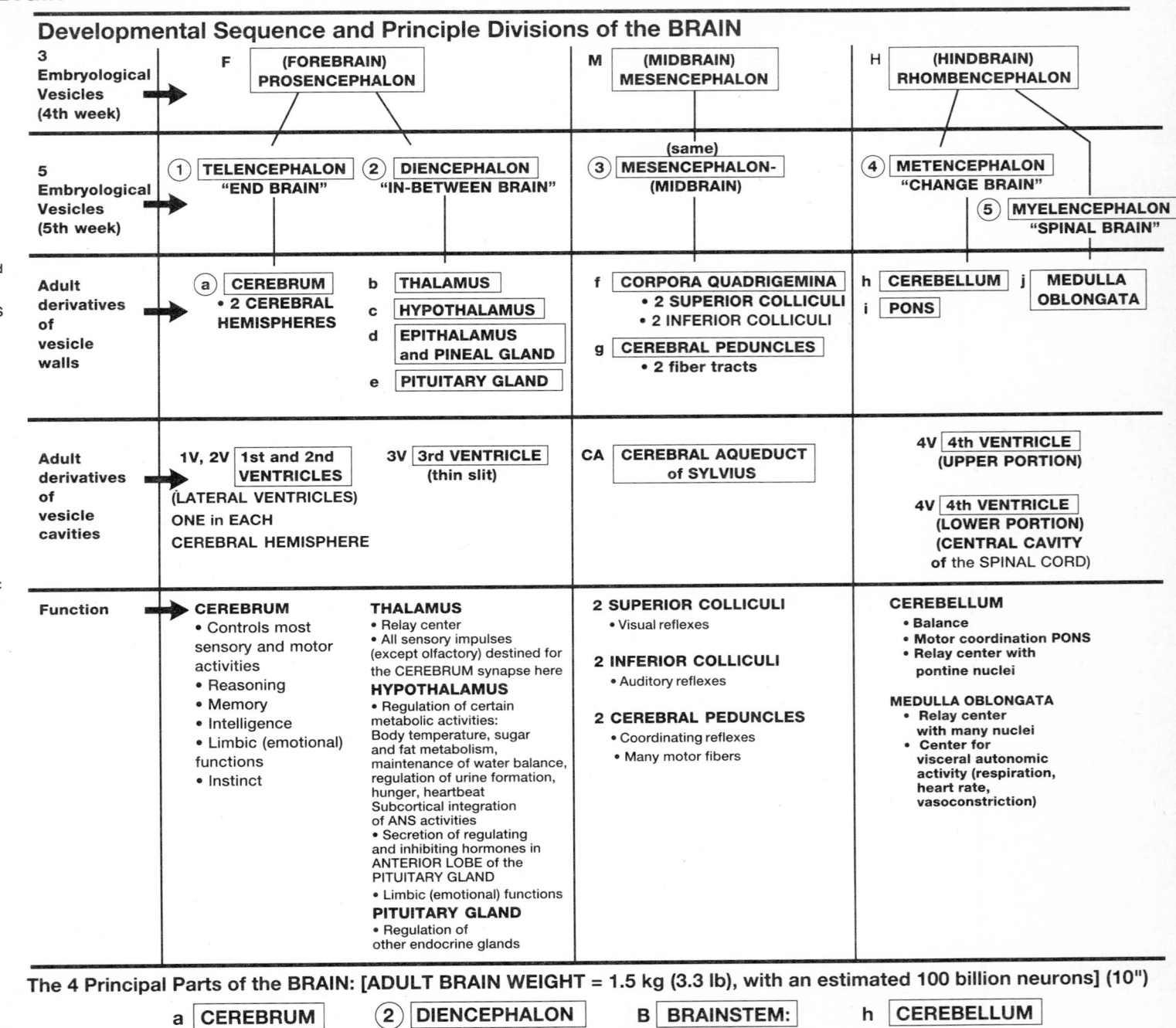

Developmental Sequence and Principle Divisions of the BRAIN

	F	M	H
3 Embryological Vesicles (4th week)	**(FOREBRAIN) PROSENCEPHALON**	**(MIDBRAIN) MESENCEPHALON**	**(HINDBRAIN) RHOMBENCEPHALON**
5 Embryological Vesicles (5th week)	① **TELENCEPHALON** "END BRAIN" ② **DIENCEPHALON** "IN-BETWEEN BRAIN"	(same) ③ **MESENCEPHALON- (MIDBRAIN)**	④ **METENCEPHALON** "CHANGE BRAIN" ⑤ **MYELENCEPHALON** "SPINAL BRAIN"
Adult derivatives of vesicle walls	ⓐ **CEREBRUM** • 2 CEREBRAL HEMISPHERES b **THALAMUS** c **HYPOTHALAMUS** d **EPITHALAMUS and PINEAL GLAND** e **PITUITARY GLAND**	f **CORPORA QUADRIGEMINA** • 2 SUPERIOR COLLICULI • 2 INFERIOR COLLICULI g **CEREBRAL PEDUNCLES** • 2 fiber tracts	h **CEREBELLUM** i **PONS** j **MEDULLA OBLONGATA**
Adult derivatives of vesicle cavities	1V, 2V **1st and 2nd VENTRICLES** (LATERAL VENTRICLES) ONE in EACH CEREBRAL HEMISPHERE 3V **3rd VENTRICLE** (thin slit)	CA **CEREBRAL AQUEDUCT of SYLVIUS**	4V **4th VENTRICLE (UPPER PORTION)** 4V **4th VENTRICLE (LOWER PORTION) (CENTRAL CAVITY of the SPINAL CORD)**
Function	**CEREBRUM** • Controls most sensory and motor activities • Reasoning • Memory • Intelligence • Limbic (emotional) functions • Instinct **THALAMUS** • Relay center • All sensory impulses (except olfactory) destined for the CEREBRUM synapse here **HYPOTHALAMUS** • Regulation of certain metabolic activities: Body temperature, sugar and fat metabolism, maintenance of water balance, regulation of urine formation, hunger, heartbeat Subcortical integration of ANS activities • Secretion of regulating and inhibiting hormones in ANTERIOR LOBE of the PITUITARY GLAND • Limbic (emotional) functions **PITUITARY GLAND** • Regulation of other endocrine glands	**2 SUPERIOR COLLICULI** • Visual reflexes **2 INFERIOR COLLICULI** • Auditory reflexes **2 CEREBRAL PEDUNCLES** • Coordinating reflexes • Many motor fibers	**CEREBELLUM** • Balance • Motor coordination PONS • Relay center with pontine nuclei **MEDULLA OBLONGATA** • Relay center with many nuclei • Center for visceral autonomic activity (respiration, heart rate, vasoconstriction)

The 4 Principal Parts of the BRAIN: [ADULT BRAIN WEIGHT = 1.5 kg (3.3 lb), with an estimated 100 billion neurons] (10")

a **CEREBRUM**   ② **DIENCEPHALON**   B **BRAINSTEM:** (MIDBRAIN, PONS, MEDULLA)   h **CEREBELLUM**

EMBRYOLOGICAL STAGES of BRAIN and VENTRICULAR DEVELOPMENT

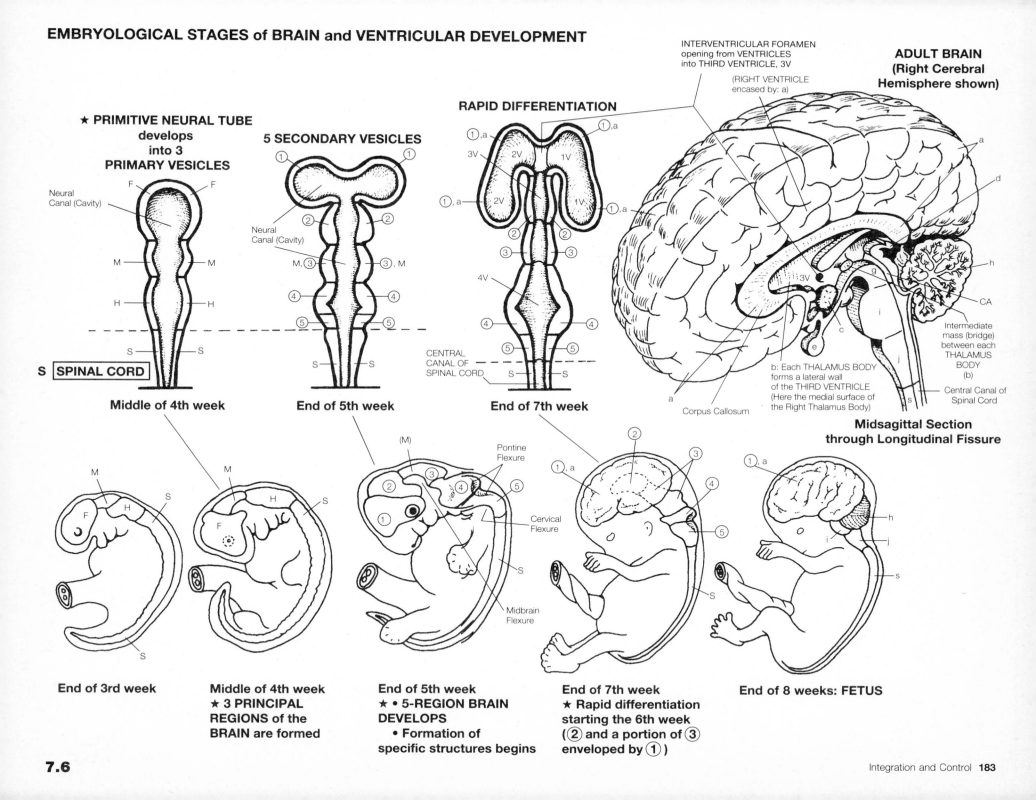

★ **PRIMITIVE NEURAL TUBE**
develops
into 3
PRIMARY VESICLES

Neural
Canal (Cavity)

F   F

M   M

H   H

S   S

S | SPINAL CORD |

Middle of 4th week

5 SECONDARY VESICLES

①   ①

Neural
Canal (Cavity)

②   ②

M, ③   ③, M

④   ④

⑤   ⑤

S   S

End of 5th week

RAPID DIFFERENTIATION

①,a   ①,a

3V   2V   1V

①, a   2V   1V   ①,a

②   ②

③   ③

4V

④   ④

⑤   ⑤

CENTRAL
CANAL OF
SPINAL CORD   S   S

End of 7th week

INTERVENTRICULAR FORAMEN
opening from VENTRICLES
into THIRD VENTRICLE, 3V

(RIGHT VENTRICLE
encased by: a)

ADULT BRAIN
(Right Cerebral
Hemisphere shown)

a

d

3V

h

g

c

i

CA

e

Intermediate
mass (bridge)
between each
THALAMUS
BODY
(b)

b: Each THALAMUS BODY
forms a lateral wall
of the THIRD VENTRICLE
(Here the medial surface of
the Right Thalamus Body)

a

Corpus Callosum

Central Canal of
Spinal Cord

j

s

Midsagittal Section
through Longitudinal Fissure

M   M   (M)   Pontine   ②   ③   ①, a
                 H   Flexure
F   S   F   S   ②   ③   ④   ①, a   ④
                                    ①   ①   ⑤   ⑤
                                          Cervical
                                          Flexure   S   ⑤
                  S   S   S
S   Midbrain
                                                   Flexure

End of 3rd week   **Middle of 4th week**   **End of 5th week**   **End of 7th week**   **End of 8 weeks: FETUS**
                       ★ 3 PRINCIPAL   **★ • 5-REGION BRAIN** **★ Rapid differentiation**
                       REGIONS of the   **DEVELOPS**   **starting the 6th week**
                       BRAIN are formed   **• Formation of**   **(② and a portion of ③**
                                                  specific structures begins **enveloped by ①)**

NERVOUS SYSTEM: CENTRAL NERVOUS SYSTEM (CNS)
The Brain: Lobes and Outer Cerebral Cortex of the Cerebrum

(F) = warm red (P) = yellow (T) = blue
(D) = green (I) = pink (L) = purple

★ See 7.6, 7.8, 7.9

★ The CEREBRUM, derived from the TELENCEPHALON, is the largest portion of the brain, accounting for 80% of the brain mass. It is functionally subdivided into 6 paired lobes comprising 2 large, convoluted CEREBRAL HEMISPHERES. The inferior, medial portions of these hemispheres are connected internally by the CORPUS CALLOSUM, a large tract of white matter (and to a lesser extent by two other white matter tracts, the ANTERIOR and POSTERIOR HIPPOCAMPAL COMMISSURES.) (See 7.8).
 • Where the RIGHT and LEFT HEMISPHERES are not connected by these 3 tracts they are incompletely separated by a large LONGITUDINAL FISSURE. Extending into this midline fissure is the FALX CEREBRI membrane from the cranial dura mater. (See 7.4) There are 4 GENERAL AREAS of the 2 CEREBRAL HEMISPHERES:
 An outer surface layer, called the CEREBRAL CORTEX, and 3 subcortical areas (See 7.8) consisting of:
① CEREBRAL MEDULLARY SUBSTANCE (WHITE MATTER),
② BASAL GANGLIA, and ③ the LATERAL VENTRICLES.

■ CEREBRAL CORTEX: PALLIUM
 • 2–4 mm (0.08–0.16 in.) thick
 • Composed of GRAY MATTER:
nerve cell bodies and their dendrites (and unmyelinated axons and neuroglia)
 • The CONVOLUTED SURFACE AREA is characterized by elevated folds called GYRI (sing. GYRUS), and as these folds tuck away into the brain they form depressed grooves—shallow grooves are called SULCI (sing. SULCUS) and deeper, longer grooves are called FISSURES.
 • Convolutions are extremely important as they *increase the surface area of the gray matter* and thus *increase the total number of nerve cell bodies.*
 • 6 types of cortical gray matter neurons are arranged in 6 interconnecting layers to form an intricate 3-dimensional network (See lower box).
 • The most highly evolved area of the brain concerned with the higher human functions:
  ■ PERCEPTION of SENSORY IMPULSES
  ■ INSTIGATION of VOLUNTARY
MOVEMENT by SKELETAL MUSCLES
  ■ MENTAL PROCESSES
   • INTELLIGENCE
   • MEMORY
   • JUDGMENT
   • IMAGINATION
   • CREATIVE THOUGHT
   • CONSCIOUS THOUGHT
★ • Each cerebral hemisphere is functionally divided into 4 outer lobes and 2 inner lobes. Structurally, the outer, external lobes of each cerebral hemisphere are mirror images of one another, but functionally they are quite different, as one hemisphere dominates the other in specific higher functions.
 • Long tracts to and from the cortex (from both sensory and motor neurons) at some point cross to opposite sides of the CNS.

The Functional Subdivisions of Each CEREBRAL HEMISPHERE: 4 OUTER LOBES and 2 INNER LOBES
4 OUTER LOBES Named in Relation to CRANIAL BONES That Cover Them

(F) FRONTAL LOBE	(P) PARIETAL LOBE	(T) TEMPORAL LOBE	(O) OCCIPITAL LOBE
■ Intellectual Processes • Concentration • Abstract thinking • Decision-making ■ Voluntary Motor Control of Skeletal Muscles ■ Translation of Thought Patterns into Verbal Communication (Speech and Language) ■ Personality • Aggressive and sexual behavior ■ Olfaction (Smell)	■ Somatesthetic Interpretation of Cutaneous and Muscular Sensations ■ Use of Symbols in Understanding ■ Verbal Articulation of Thoughts and Emotions: ■ Ideation and Appropriate Motor Responses for Carrying Ideas through ■ Recollection and Recognition of Specific Objects ■ Interpretation of Textures and Shapes of Handled Objects	■ Interpretation of Auditory Sensations through Sensory Fibers of the Cochlea of the Ear ■ Memory Storage of Both Auditory and Visual Experiences ■ Olfaction ■ Language ■ Emotional Behavior Associated with Preservation of Species and of Self (Anger, Hostility, Sexual	■ Integrates Eye Movements by Directing and Focusing the Eye ■ Conscious Seeing ■ Visual Association: Correlation of Visual Images with Previous Visual Experiences

Major Functional Areas of Each OUTER LOBE

1 PRIMARY MOTOR: (PRECENTRAL GYRUS)	7 PRIMARY SENSORY: (POSTCENTRAL GYRUS)	11 PRIMARY AUDITORY	14 PRIMARY VISUAL
2 PREMOTOR ASSOCIATION	8 SOMATESTHETIC ASSOCIATION	12 AUDITORY ASSOCIATION	15 VISUAL ASSOCIATION
3 SUPPLEMENTARY MOTOR	9 PRIMARY GUSTATORY	13 PRIMARY OLFACTORY (ARCHIPALLIUM)	
4 FRONTAL EYEFIELD	10 GNOSTIC		
5 BROCA'S SPEECH (LEFT LOBE ONLY)	• COMMON INTEGRATIVE AREA • Located between 8, 12, and 15 • Integrates sensory interpretations from 8, 9, 12, 13, and 15, plus the thalamus and lower brainstem, to form a COMMON THOUGHT from these VARIOUS SENSORY INPUTS • Transmits its integrated signal to other parts of the brain to cause the appropriate response		
6 PREFRONTAL CORTEX			

2 INNER LOBES of Each Hemisphere

(I) INSULA:

ISLAND of REIL

 • The central deep lobe, a triangular area of the cerebral cortex, lying in the floor of the lateral fissure
  • Cannot be viewed on the surface

 ■ Involved in the *integration of cerebral activities and memory.*

(L) LIMBIC LOBE:

CINGULATE GYRUS and HIPPOCAMPAL GYRUS

 • Marginal section of each cerebral hemisphere on the medial aspect. Also called the GYRUS FORNICATUS

 • Part of the LIMBIC SYSTEM involved in BASIC EMOTIONAL ACTIVITY (See 7.9)

Major Depressed GROOVES
 ■ FISSURES: (DEEP GROOVES)

a	LONGITUDINAL FISSURE
b	CENTRAL FISSURE of ROLANDO (CENTRAL SULCUS)
c	LATERAL FISSURE of SYLVIUS (LATERAL SULCUS)

 ■ SULCI: (SHALLOW GROOVES, FURROWS) indicated on illustrations

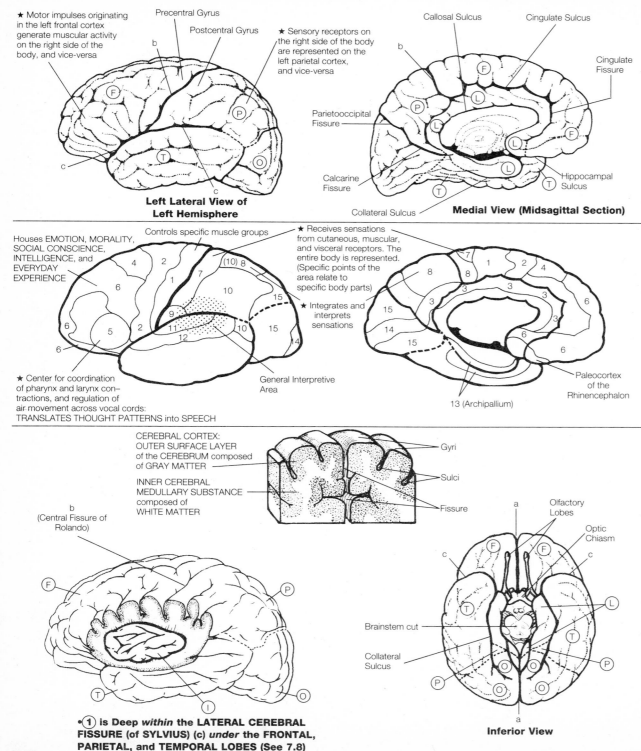

★ Motor impulses originating in the left frontal cortex generate muscular activity on the right side of the body, and vice-versa

Precentral Gyrus

Postcentral Gyrus

★ Sensory receptors on the right side of the body are represented on the left parietal cortex, and vice-versa

Ⓕ

b

ⓒ

ⓒ

Ⓣ

Ⓟ

Ⓞ

Left Lateral View of Left Hemisphere

Callosal Sulcus

Cingulate Sulcus

Cingulate Fissure

Parietooccipital Fissure

b

Ⓕ

Ⓛ

Ⓟ

Ⓛ

Ⓛ

Ⓕ

Ⓣ

Calcarine Fissure

Collateral Sulcus

Hippocampal Sulcus

Medial View (Midsagittal Section)

Houses EMOTION, MORALITY, SOCIAL CONSCIENCE, INTELLIGENCE, and EVERYDAY EXPERIENCE

Controls specific muscle groups

4   2

(10) 8

1   7

6   10

9

2   11   10   15

5   12

6

★ Receives sensations from cutaneous, muscular, and visceral receptors. The entire body is represented. (Specific points of the area relate to specific body parts)

★ Integrates and interprets sensations

7

1   2   4

8   8

3   3   3

15   3

14   3   6   3

15   6

6

General Interpretive Area

13 (Archipallium)

Paleocortex of the Rhinencephalon

★ Center for coordination of pharynx and larynx contractions, and regulation of air movement across vocal cords: TRANSLATES THOUGHT PATTERNS into SPEECH

CEREBRAL CORTEX: OUTER SURFACE LAYER of the CEREBRUM composed of GRAY MATTER

INNER CEREBRAL MEDULLARY SUBSTANCE composed of WHITE MATTER

Gyri

Sulci

Fissure

b (Central Fissure of Rolando)

Ⓕ

Ⓟ

Ⓣ

Ⓘ

Ⓞ

• ① is Deep *within* the LATERAL CEREBRAL FISSURE (of SYLVIUS) (c) *under* the FRONTAL, PARIETAL, and TEMPORAL LOBES (See 7.8)

a   Olfactory Lobes

Optic Chiasm

Ⓕ   Ⓕ

c   c

Ⓣ   Ⓣ

Ⓛ

Brainstem cut

Ⓣ

Collateral Sulcus

Ⓟ   Ⓞ   Ⓞ   Ⓟ

Ⓞ   Ⓞ

a

Inferior View

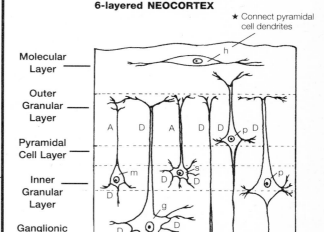

6-layered NEOCORTEX

★ Connect pyramidal cell dendrites

Molecular Layer

Outer Granular Layer

Pyramidal Cell Layer

Inner Granular Layer

Ganglionic Layer

Fusiform Layer

A=AXON   D=DENDRITES

★ **CEREBRAL CORTEX: The PALLIUM**
Outer GRAY MATTER (NERVE CELL BODIES)
NEOCORTEX (NEOPALLIUM) = 90% of the
CEREBRAL CORTEX: 6 Types of CELLS in
the 6–Layered NEOCORTEX:

h   HORIZONTAL (CAJAL'S) CELLS
m   MARTINOTTI'S CELLS
s   STELLATE CELLS
f   FUSIFORM CELLS
p   PYRAMIDAL CELLS
g   GIANT BETZ PYRAMIDAL CELLS (See 7.11)
• Axons, included in PYRAMIDAL PROJECTION TRACTS
• Found mostly in CORTICAL MOTOR AREAS

★ The Other 10% of the CEREBRAL CORTEX:
The **RHINENCEPHALON** (phylogenetically, the oldest portion of the brain. Only 3 layers are present: h, p, g)

■ **PALEOCORTEX (PALEOPALLIUM)**
• Restricted to OLFACTORY REGIONS of the FRONTAL LOBE (includes OLFACTORY BULBS, TRACT and STRIAE, and INTERMEDIATE OLFACTORY AREA)

■ **ARCHICORTEX (ARCHIPALLIUM)-OLDEST**
• Large portions of the old OLFACTORY CORTEX of the TEMPORAL LOBE associated with the LIMBIC SYSTEM (includes FORNIX and HIPPOCAMPUS) (See 7.)

7.7

7.8

Integration and Control **186**

① = light warm colors near yellow
② = light cool colors near purple
★ See 7.6, 7.9, 7.10

NERVOUS SYSTEM: CENTRAL NERVOUS SYSTEM (CNS)
The Brain: Subcortical Areas of the Cerebrum

★ The layer underlying the outer CEREBRAL CORTEX of GRAY MATTER consists mainly of WHITE MATTER (MYELINATED AXONS) arranged into FIBER TRACTS. These tracts offer lines of communication between the CEREBRAL HEMISPHERES involving the CEREBRAL CORTEX in 3 principal directions:

■ ASSOCIATION TRACTS
• Confined to a given hemisphere
• Connect and transmit impulses between the anterior and posterior cortical areas of gyri in various lobes.

■ COMMISSURAL TRACTS
• Connect the cortical areas of gyri of one hemisphere to the corresponding gyri in the opposite hemisphere:

■ CORPUS CALLOSUM
• The largest commissural tract (the GREAT COMMISSURE) composed of a thick layer of transverse fibers arching above the lateral ventricles.
• The major connector between the 2 cerebral hemispheres.

■ Minor connectors are the ANTERIOR and POSTERIOR COMMISSURES

■ FORNIX
• A fibrous, vaulted band connecting the cerebral lobes.

■ HIPPOCAMPAL COMMISSURE
• A thin sheet of transverse fibers under the posterior portion of the corpus callosum, connecting the medial margins of the CRURA of the FORNIX.

■ PROJECTION TRACTS
• Fibers ascending (entering) and descending (leaving) the hemispheres as sensory and motor pathways, respectively. They enable the brain to keep in touch with the peripheral body area.

■ CORONA RADIATA
• Radiating mass of ascending and descending fibers that, above the corpus callosum, extends in all directions to the cerebral cortex.
• Many of the fibers arise in the relay center of the THALAMUS (See 7.8)
• Continuous with the INTERNAL CAPSULE.

■ INTERNAL CAPSULE
• A compact area composed mainly of efferent motor fibers leaving the cerebrum, connecting with the brainstem and spinal cord, and passing through the 2 BASAL GANGLIA.

★ BASAL GANGLIA
• Paired discrete masses of gray matter at the base of the hemispheres on either side of the diencephalon deep within the white matter.
• Interconnected by many fibers to the CEREBRAL CORTEX, INTERNAL CAPSULE, THALAMUS, HYPOTHALAMUS, and GRAY MATTER MASSES of the MIDBRAIN.
• Facilitate or inhibit motor impulses from motor areas of the frontal lobe (except the precentral gyrus).
• CAUDATE NUCLEUS and PUTAMEN assist in the control of large subconscious movements of the skeletal muscles (ex: swinging of arms while walking).
• GLOBUS PALLIDUS regulates muscle tone necessary for specific intentional body movements.

---

3 Major Subcortical Areas:
① **CEREBRAL MEDULLARY SUBSTANCE (BODY)**
② **BASAL GANGLIA**
③ **LATERAL VENTRICLES**

① CEREBRAL MEDULLARY BODY
Inner WHITE MATTER Bundles of DENDRITES and Myelinated AXONS):
Communicating Directional FIBER TRACTS:

• In the Same HEMISPHERE
4 ■ **ASSOCIATION TRACTS**

• Between the Hemispheres
■ **COMMISSURAL TRACTS:** (5–9)
5 **CORPUS CALLOSUM**   8 **FORNIX** (See 7.9)
6 **ANTERIOR COMMISSURE**   9 **HIPPOCAMPAL COMMISSURE OF FORNIX**
7 **POSTERIOR COMMISSURE**
connects the SUPERIOR COLLICULI (See 7.9)

• Ascending and Descending TRACTS Connecting the CORTEX with other Structures of the BRAIN and SPINAL CORD
■ **PROJECTION TRACTS:**
10 **CORONA RADIATA**
11 **INTERNAL CAPSULE**   considered part of the CORPUS STRIATUM and the THALAMUS (See 7.9)

② BASAL GANGLIA (CEREBRAL NUCLEI):
Paired Masses of "Central GRAY MATTER" within WHITE MATTER.
MODIFICATION AND COORDINATION OF VOLUNTARY MUSCLE MOVEMENT THROUGH UNCONSCIOUS, "AUTOMATIC" REFLEXES

CORPUS STRIATUM:
CAUDATE NUCLEUS and the LENTIFORM NUCLEUS
12 **CAUDATE NUCLEUS** UPPER MASS

LENTIFORM NUCLEUS : LOWER MASS

a **PUTAMEN** LATERAL PORTION

b **GLOBUS PALLIDUS** MEDIAL PORTION

★ ③ **LATERAL VENTRICLES**
• Each CEREBRAL HEMISPHERE contains a large cavity called the LATERAL VENTRICLE inferior to (below) the CORPUS CALLOSUM.
• Develop as large cavities of the upper forebrain (TELENCEPHALON) in the embryo. (See 7.5)
• Most of the CSF (cerebrospinal fluid) is secreted within the CHOROID PLEXUSES (specialized capillaries) intertwined with the ciliated ependymal cells (EPENDYMAL VENTRICULORUM CEREBRAL) lining the LATERAL VENTRICLES. (See 7.5, 7.9)

15 ■ **SEPTUM PELLUCIDUM**
• Medial wall of the LATERAL VENTRICLES
• Thin, triangular sheet of nervous tissue
• 2 laminae attached to the CORPUS CALLOSUM above and the FORNIX below (See 7.9)

FORNIX and Related Structures
Left Lateroposterior View

13 **CLAUSTRUM**

14 **AMYGDALOID NUCLEUS**
• Part of LIMBIC SYSTEM (See 7.9)

★ [Some authorities also include SUBSTANTIA NIGRA, SUBTHALAMIC NUCLEI, and the RED NUCLEUS (all of the midbrain) as part of the BASAL GANGLIA.] (See 7.10)

Transverse Section of CEREBRUM through the CORPUS CALLOSUM

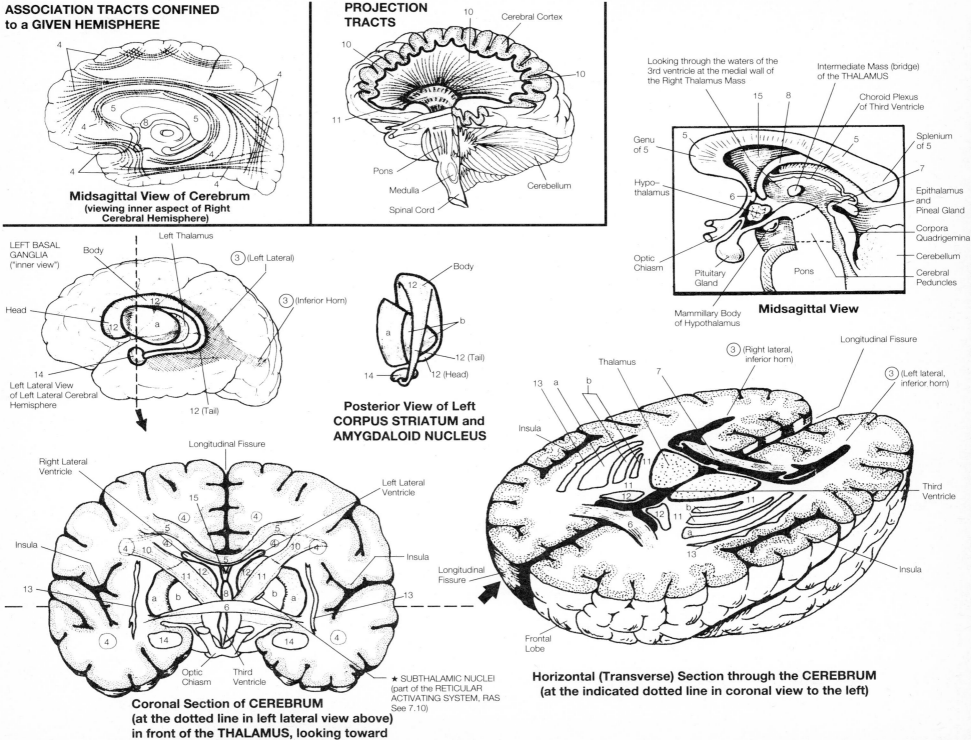

ASSOCIATION TRACTS CONFINED to a GIVEN HEMISPHERE

Midsagittal View of Cerebrum
(viewing inner aspect of Right Cerebral Hemisphere)

PROJECTION TRACTS

10
Cerebral Cortex
10
10
11
Pons
Medulla
Cerebellum
Spinal Cord

Looking through the waters of the 3rd ventricle at the medial wall of the Right Thalamus Mass
Intermediate Mass (bridge) of the THALAMUS
Choroid Plexus of Third Ventricle
15
8
Genu of 5
5
5
Splenium of 5
Hypo-thalamus
6
7
Epithalamus and Pineal Gland
Optic Chiasm
Corpora Quadrigemina
Cerebellum
Pituitary Gland
Pons
Cerebral Peduncles
Mammillary Body of Hypothalamus

Midsagittal View

LEFT BASAL GANGLIA ("inner view")
Left Thalamus
Body
③ (Left Lateral)
Head
12
a
③ (Inferior Horn)
14
12
Left Lateral View of Left Lateral Cerebral Hemisphere
12 (Tail)

Body
12
b
a
12 (Tail)
14
12 (Head)

Posterior View of Left CORPUS STRIATUM and AMYGDALOID NUCLEUS

Thalamus
7
③ (Right lateral, inferior horn)
Longitudinal Fissure
③ (Left lateral, inferior horn)
13
a
b
Insula
11
11
12
11
b
12
11
a
6
13
Longitudinal Fissure
Third Ventricle
Frontal Lobe
Insula

Horizontal (Transverse) Section through the CEREBRUM
(at the indicated dotted line in coronal view to the left)

Longitudinal Fissure
Right Lateral Ventricle
15
Left Lateral Ventricle
Insula
5
4
4
5
Insula
4
10
4
10
4
11
12
12
11
13
a
b
b
a
13
8
6
14
14
4
Optic Chiasm
Third Ventricle
★ SUBTHALAMIC NUCLEI (part of the RETICULAR ACTIVATING SYSTEM, RAS See 7.10)

Coronal Section of CEREBRUM
(at the dotted line in left lateral view above) in front of the THALAMUS, looking toward the posterior aspect of the brain

NERVOUS SYSTEM: CENTRAL NERVOUS SYSTEM (CNS)
The Brain: The Diencephalon

Color all the Hypothalamic Nuclei, (H), and 12 one color.
Carry colors over from 7.6.
Choose your own colors for the rest and have fun!

★ See 7.6, 7.8, 8.1—8.3, 9.9

★ The DIENCEPHALON is the second subdivision of the "old" FOREBRAIN that gave rise to the CEREBRUM.
• A MAJOR AUTONOMIC REGION
• Positioned atop the BRAINSTEM between the inferior portions of the 2 CEREBRAL HEMISPHERES.
• Almost completely engulfed and surrounded by the Hemispheres.
• The 3rd VENTRICLE forms a thin, slitlike midline cavity within the diencephalon.
• Chief components are the EPITHALAMUS, THALAMUS, HYPOTHALAMUS, and PITUITARY GLAND.

★ EPITHALAMUS
• The uppermost portion of the diencephalon. Consists of the PINEAL GLAND, the HABENULAR NUCLEI, and HABENULAR COMMISSURE Includes a thin roof over the 3rd ventricle, whose inside lining forms a CHOROID PLEXUS that produces CSF. (See 7.5)
■ PINEAL GLAND
• Described in 8.1, extends outward from the posterior end of the epithalamus.
• Obscure functions may be concerned with the ONSET OF PUBERTY and ESTABLISHMENT of CIRCADIAN RHYTHMS.
■ HABENULAR NUCLEI
• Receives impulses from the OLFACTORY SYSTEM and projects them to the LIMBIC SYSTEM. The "EMOTIONAL BRAIN." (See box)
• Thus, the nuclei establish a connection between the SENSE of SMELL on the one hand, and SEX DRIVE, RAGE REACTIONS, and ASSOCIATED VISCERAL CHANGES.

★ THALAMUS
• Largest portion of the diencephalon, comprising 4/5th of its mass.
• Many groups of cell bodies or NUCLEI formed into 2 paired ovoid masses.
• Each thalamic mass is positioned immediately below the lateral ventricle in its respective hemisphere, in one of the lateral walls of the 3rd ventricle. The masses are connected by a commissure called the INTERMEDIATE MASS, which passes through a foramen of the 3rd ventricle
• A RELAY CENTER for all SENSORY IMPULSES (except smell) to the CEREBRAL CORTEX. Sensory stimuli are associated, synthesized and then relayed through the CORONA RADIATA to specific cortical areas.
• Impulses are also received from the CORTEX, CORPUS STRIATUM and HYPOTHALAMUS and are relayed to visceral and somatic effectors.
• CENTER for appreciation of primitive, uncritical sensations of CRUDE TOUCH, TEMPERATURE and PAIN, and assists in the initial AUTONOMIC RESPONSE to INTENSE PAIN (SHOCK).

★ HYPOTHALAMUS
• Very small in size, consists of several masses of nuclei interconnected to other vital parts of the brain.
• Forms the floor and lower lateral walls of the 3rd ventricle.
• AUTONOMIC CENTER concerned with HOMEOSTASIS (FACILITATION and INHIBITION) of several VITAL FUNCTIONS NECESSARY FOR ORGANISM SURVIVAL, most of which relate to REGULATION OF VISCERAL ACTIVITIES (but some relate to EMOTION (LIMBIC) and INSTINCT); REGULATION OF:
• HEART RATE, • BODY TEMPERATURE, • WATER and ELECTROLYTE BALANCE (OSMORECEPTOR and THIRST CENTER), HUNGER and GI ACTIVITY (FEEDING and SATIETY CENTER),
• CONSCIOUS ALERTNESS (SLEEP and WAKEFULNESS CENTER),
• SEXUAL RESPONSE (SEXUAL CENTER) HYPOTHALAMUS also produces POLYPEPTIDES that REGULATE the activity of the PITUITARY GLAND in the secretion or suppression of HORMONES. (See 8.2 and 8.3)

4 Major Areas of the DIENCEPHALON plus the 3rd VENTRICLE

(E) **EPITHALAMUS**   (N) **NEUROHYPOPHYSIS (POSTERIOR LOBE) of the PITUITARY GLAND**

(T) **THALAMUS**

(H) **HYPOTHALAMUS**   3V **THIRD VENTRICLE**

★ For details, See 8.1, 8.2, and 8.3 Surrounded by the CIRCLE OF WILLIS, a ring like network of blood vessels (See 9.9)

(E): EPITHALAMUS

1   **PINEAL BODY (GLAND)**
2   **PINEAL STALK (HABENULA)**
(3) **HABENULAR COMMISSURE and HABENULAR NUCLEI**
4   **PIA MATER CAPSULE**

(T): THALAMUS: RELAY CENTER
■ COMMISSURE Connecting the 2 Oval THALAMIC MASSES:

5   **INTERMEDIATE MASS (BRIDGE)**

■ WHITE MATTER Dividing the THALAMIC GRAY MATTER into 3 Groups:

6   **INTERNAL MEDULLARY LAMINA**

■ 3 NUCLEAR Groups of GRAY MATTER MASSES:
ANTERIOR GROUP

7   **ANTERIOR NUCLEUS**

MEDIAL GROUP

8   **MEDIAL NUCLEUS** part of RETICULAR ACTIVATING SYSTEM

LATERAL GROUP

9   **SENSORY RELAY CENTERS**
10 **INTERPRETIVE SENSORY CENTERS**
11 **MOTOR CENTERS**

(H): HYPOTHALAMUS

12 **HYPOTHALAMIC NUCLEI MASSES**
(13) **MAMMILLARY BODIES**

The LIMBIC SYSTEM ("EMOTIONAL BRAIN")

Corpus Callosum
Cingulate Gyrus
16
Corpus Callosum
16
12 Hypothalamus
14
16
15
Olfactory Lobe
15
17
(13)
Pituitary Gland

14 • **FORNIX**
15 • **HIPPOCAMPUS**] (See 7.8)

16 • **CEREBRAL LIMBIC LOBE** (See 7.7)
17 • **AMYGDALOID NUCLEI of BASAL GANGLIA** (See 7.8)
(13) • **MAMMILLARY BODIES of HYPOTHALAMUS**
(7) • **ANTERIOR NUCLEI of THALAMUS**
(3) • **HABENULAR NUCLEI of EPITHALAMUS**

★ The LIMBIC SYSTEM:
• Group of fiber tracts and nuclei that form a neuronal loop inside the brain, roughly in the shape of a doughnut: thalamic regions are in the doughnut hole, and the cerebral cortex is outside.
• Derivative of the old RHINENCEPHALON ("SMELL BRAIN") associated with central processing of olfactory information in vertebrate evolution.
• Presently in humans this system is activated by MOTIVATED BEHAVIOR and AROUSAL. It is a CENTER of BASIC EMOTIONAL DRIVES (ANGER, FEAR, SEX, HUNGER), and also of SHORT-TERM MEMORY VIA the HIPPOCAMPUS.
• Complex circuits of the LIMBIC SYSTEM with the HYPOTHALAMUS influence the ENDOCRINE and AUTONOMIC MOTOR SYSTEMS, and VISCERAL RESPONSES to EMOTIONS.

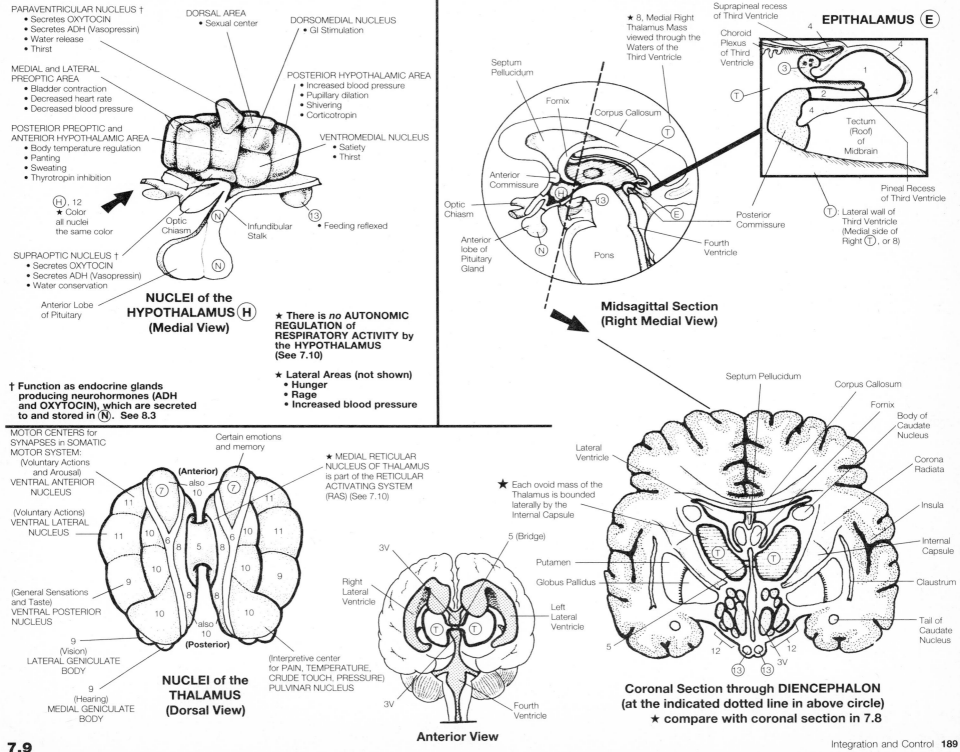

PARAVENTRICULAR NUCLEUS †
• Secretes OXYTOCIN
• Secretes ADH (Vasopressin)
• Water release
• Thirst

DORSAL AREA
• Sexual center

DORSOMEDIAL NUCLEUS
• GI Stimulation

MEDIAL and LATERAL PREOPTIC AREA
• Bladder contraction
• Decreased heart rate
• Decreased blood pressure

POSTERIOR HYPOTHALAMIC AREA
• Increased blood pressure
• Pupillary dilation
• Shivering
• Corticotropin

POSTERIOR PREOPTIC and ANTERIOR HYPOTHALAMIC AREA
• Body temperature regulation
• Panting
• Sweating
• Thyrotropin inhibition

VENTROMEDIAL NUCLEUS
• Satiety
• Thirst

H, 12
★ Color all nuclei the same color

Optic Chiasm

Infundibular Stalk

• Feeding reflexed ⑬

SUPRAOPTIC NUCLEUS †
• Secretes OXYTOCIN
• Secretes ADH (Vasopressin)
• Water conservation

Anterior Lobe of Pituitary

**NUCLEI of the HYPOTHALAMUS (H)
(Medial View)**

★ There is *no* AUTONOMIC REGULATION of RESPIRATORY ACTIVITY by the HYPOTHALAMUS (See 7.10)

★ Lateral Areas (not shown)
• Hunger
• Rage
• Increased blood pressure

† Function as endocrine glands producing neurohormones (ADH and OXYTOCIN), which are secreted to and stored in (N). See 8.3

MOTOR CENTERS for SYNAPSES in SOMATIC MOTOR SYSTEM:
(Voluntary Actions and Arousal)
VENTRAL ANTERIOR NUCLEUS

Certain emotions and memory

(Anterior) also 10

★ MEDIAL RETICULAR NUCLEUS OF THALAMUS is part of the RETICULAR ACTIVATING SYSTEM (RAS) (See 7.10)

(Voluntary Actions)
VENTRAL LATERAL NUCLEUS

(General Sensations and Taste)
VENTRAL POSTERIOR NUCLEUS

also 10
(Posterior)

(Vision)
LATERAL GENICULATE BODY

(Hearing)
MEDIAL GENICULATE BODY

(Interpretive center for PAIN, TEMPERATURE, CRUDE TOUCH, PRESSURE)
PULVINAR NUCLEUS

**NUCLEI of the THALAMUS
(Dorsal View)**

★ 8, Medial Right Thalamus Mass viewed through the Waters of the Third Ventricle

Septum Pellucidum

Fornix

Corpus Callosum

Anterior Commissure

Optic Chiasm

Anterior lobe of Pituitary Gland

Pons

**Midsagittal Section
(Right Medial View)**

Suprapineal recess of Third Ventricle

Choroid Plexus of Third Ventricle

EPITHALAMUS (E)

Tectum (Roof) of Midbrain

Pineal Recess of Third Ventricle

(T): Lateral wall of Third Ventricle (Medial side of Right (T), or 8)

Posterior Commissure

Fourth Ventricle

Septum Pellucidum

Corpus Callosum

Fornix

Body of Caudate Nucleus

Lateral Ventricle

Corona Radiata

★ Each ovoid mass of the Thalamus is bounded laterally by the Internal Capsule

Insula

Internal Capsule

5 (Bridge)

Putamen

Globus Pallidus

Right Lateral Ventricle

Left Lateral Ventricle

Claustrum

Tail of Caudate Nucleus

3V

3V

Fourth Ventricle

Anterior View

**Coronal Section through DIENCEPHALON
(at the indicated dotted line in above circle)
★ compare with coronal section in 7.8**

NERVOUS SYSTEM: CENTRAL NERVOUS SYSTEM (CNS)
The Brain: Mesencephalon and Rhombencephalon
The Brainstem and the Cerebellum

① = orange ② = pink ③ = purple
④ = red h = white g = flesh
Choose your own colors and have fun!
★ See 7.6, 7.15

★ The BRAINSTEM is the stemlike part of the brain that connects the FOREBRAIN (CEREBRAL HEMISPHERES and DIENCEPHALON) with the SPINAL CORD. It consists of:
① The MIDBRAIN (MESENCEPHALON),
② The PONS (ANTERIOR PORTION of the METENCEPHALON), and
③ The MEDULLA OBLONGATA (MYELENCEPHALON)

★ ①MIDBRAIN (MESENCEPHALON)
• A short (1.5 cm), wedge-shaped section of the brainstem between the DIENCEPHALON and the PONS.
• Structures:
■ CORPORA QUADRIGEMINA: 4 rounded elevations on the dorsal portion of the midbrain. The 2 upper "balls," the SUPERIOR COLLICULI, receive fibers from the VISUAL CORTEX and send fibers to eye and neck muscles for VISUAL REFLEXES. The 2 lower "balls," the INFERIOR COLLICULI, receive fibers from the COCHLEA of the ear and send fibers to the MEDIAL GENICULATE BODY of THALAMUS for AUDITORY REFLEXES.
■ CEREBRAL PEDUNCLES
A pair of cylindrical ASCENDING and DESCENDING PROJECTION FIBER TRACTS that support and connect the CEREBRUM to other regions of the brain (bulk of the midbrain).
■ SPECIALIZED NUCLEI considered by some authorities to be part of the BASAL GANGLIA.
■ CEREBRAL AQUEDUCT OF SYLVIUS connects the 3rd and 4th ventricles
★ The "old" HINDBRAIN (RHOMBENCEPHALON) gives rise to the upper METENCEPHALON (PONS and CEREBELLUM) and the lower MYELENCEPHALON (MEDULLA OBLONGATA):
★ ②PONS VAROLII (ANTERIOR METENCEPHALON)
• A rounded eminence of the ventral surface of the brainstem, lying between the cerebral peduncles and the medulla, about 2.5 cm long.
• A bulge on the VENTRAL PONS, called the BASAL PONS, consists of a broad band of surface transverse fiber tracts receiving motor fibers from one cerebral hemisphere and relaying them to the OPPOSITE CEREBELLAR HEMISPHERE via the MIDDLE CEREBELLAR PEDUNCLES. (Pontine nuclei are also found in the BASAL PONS.)
• A deeper DORSAL PONS, called the TEGMENTUM, contain longitudinal motor and sensory tracts that connect the medulla with tracts of the midbrain: (Ascending sensory fibers pass to the THALAMUS and descending fibers pass through the CORTICOSPINAL and RUBROSPINAL tracts.) (The nuclei of cranial nerves II–VII are also found in the TEGMENTUM.)
• The CEREBRAL AQUEDUCT of the MESENCEPHALON enlarges within the entire METENCEPHALON to become the 4th VENTRICLE.
★ ③MEDULLA OBLONGATA (MYELENCEPHALON)
• Lower 3 cm of the BRAINSTEM, the enlarged portion of the spinal cord in the cranium that enters the foramen magnum of the occipital bone.
• Consists mostly of ascending and descending white matter tracts that communicate between the spinal cord and various parts of the brain.
• Most of these fibers cross to the other side at the DECUSSATION AREA of 2 ventral PYRAMIDS
• SPECIALIZED NUCLEI for:
① cranial nerves VIII–XII,
② sensory relay to the thalamus,
③ motor relay from cerebrum to the cerebellum, and
④ autonomic control centers for CARDIAC, VASOMOTOR, and RESPIRATORY activity.

3 Major Areas of the BRAINSTEM plus the 4th VENTRICLE

		① MIDBRAIN (MESENCEPHALON)	② PONS (PONS VAROLII)	③ MEDULLA OBLONGATA (MYELENCEPHALON)
VENTRAL (FRONT, ANTERIOR) PORTION	➡	4 2 CEREBRAL PEDUNCLES	8 BASAL PONS	11 2 PYRAMIDS (CORTEX-to-SPINAL CORD) 12 DECUSSATION AREA of PYRAMIDS 13 2 OLIVES (ONE on EACH LATERAL SURFACE)
ASSOCIATED VENTRICLE	➡	CA CEREBRAL AQUEDUCT of SYLVIUS	4V 4th VENTRICLE (UPPER PORTION)	4V 4th VENTRICLE (LOWER PORTION)
DORSAL (BACK, POSTERIOR) PORTION	➡	TECTUM (ROOF): CORPORA QUADRIGEMINA 5 2 SUPERIOR COLLICULI 6 2 INFERIOR COLLICULI	9 TEGMENTUM (FLOOR of 4th VENTRICLE)	14 PERIPHERAL WHITE MATTER BUNDLE TRACTS (UP and DOWN MEDULLA-SPINAL CORD) (FLOOR of 4th VENTRICLE)
WHITE MATTER STALK PATHWAYS to CEREBELLUM	➡	7 2 SUPERIOR CEREBELLAR PEDUNCLES	10 2 MIDDLE CEREBELLAR PEDUNCLES	15 2 INFERIOR CEREBELLAR PEDUNCLES
MAJOR NUCLEI (CENTRAL DEEP MASSES of NEURON CELL BODIES)	➡	■ RED NUCLEUS • Connects the CEREBRAL HEMISPHERES and the CEREBELLUM • Efferent fibers give rise to descending RUBROSPINAL TRACT • Motor nucleus concerned with MUSCLE TONE, SKILLED MOTOR COORDINATION, and POSTURE ■ SUBSTANTIA NIGRA • Motor nucleus concerned with suppression of purposeless movements	■ PONTINE • Send fibers to the CEREBELLUM • Found in BASAL PONS	■ NUCLEUS GRACILIS and NUCLEUS CUNEATUS • Receive sensory fibers from ascending tracts in dorsal area of SPINAL CORD. Axons cross medulla midline to opposite side of medulla. Information is then relayed to the THALAMUS, and then to the sensory areas of the CEREBRAL CORTEX. ■ OLIVARY NUCLEI (INFERIOR) • Receive fibers from BASAL GANGLIA and CEREBRAL CORTEX, and send fibers to CEREBELLUM • Part of the motor body system and form part of the RETICULAR FORMATION and the RAS
GROUPS of NEURONS FORM NUCLEI of CRANIAL NERVES in BRAINSTEM (See 7.15)	➡	III, IV (in cerebral peduncles)	V, VI, VII cochlear divisions of VIII (in tegmentum)	IX, X, XI, XII vestibular divisions of VIII

★ CEREBELLUM (POSTERIOR METENCEPHALON)
• 2nd largest structure of the BRAIN.
• Occupies the inferior and posterior aspect of the cavity of the skull (CRANIAL CAVITY), overhanging the MEDULLA OBLONGATA.
• Consists of 2 lateral CEREBELLAR HEMISPHERES and a narrow, medial portion, the VERMIS
• Composed of gray matter on the surface, and deep "tree tracts" of white matter collectively called the ARBOR VITAE.
• Cerebellar surface is thrown into numerous parallel ridges called FOLIA that are separated by CEREBELLAR FISSURES.
• Functions totally at the SUBCONSCIOUS (INVOLUNTARY) LEVEL to COORDINATE BODY MOVEMENTS and TO MAINTAIN BALANCE.
• Connected to the BRAINSTEM by 3 pairs of fiber bundles, the SUPERIOR, MIDDLE, and INFERIOR CEREBELLAR PEDUNCLES.
• Incoming afferent impulses from proprioceptors within muscles, tendons, joints, and 2 special sense organs communicate through the peduncles, and are assimilated in the cerebrum. Efferent impulses are discharged through the peduncles to other neurological structures to coordinate voluntary skeletal-muscle contractions.
• Does not serve as a reflex center, but may reinforce some reflexes and inhibit others.

④ The CEREBELLUM
Portions According to Phylogenetic Age:
■ Oldest ARCHICEREBELLUM: FLOCCULONODULAR LOBE
a FLOCCULUS
b NODULUS (NODULE)

■ PALEOCEREBELLUM
c VERMIS (SUP. AND INF.)

■ NEOCEREBELLUM
2 CEREBELLAR HEMISPHERES
d ANTERIOR LOBE
e POSTERIOR LOBE

Extensions of CRANIAL DURA MATER: (See 9.16)
■ Separate the CEREBELLUM from CEREBRUM
f TENTORIUM CEREBELLI

■ Between the HEMISPHERES
FALX CEREBELLI (not shown)

Internal Anatomy
■ Outer GRAY MATTER
g CORTEX

■ Inner Tree of WHITE MATTER TRACTS
h ARBOR VITAE

4 Pairs of Deep NUCLEI within WHITE MATTER
CEREBELLAR NUCLEI (not shown)

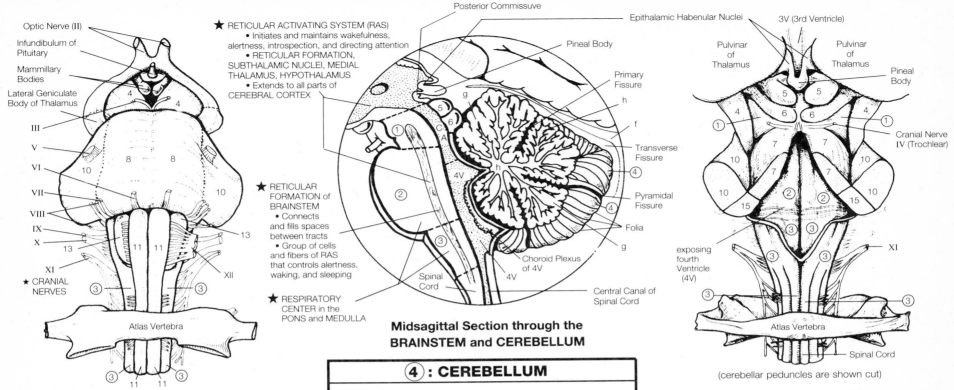

Anterior (Ventral) View

- Optic Nerve (II)
- Infundibulum of Pituitary
- Mammillary Bodies
- Lateral Geniculate Body of Thalamus
- III
- V
- VI
- VII
- VIII
- IX
- X
- XI
- ★ CRANIAL NERVES
- 4 4
- 8 8
- 10
- 10
- 13
- 13
- 11 11
- XII
- 3 3
- 3 3
- 11 11
- Atlas Vertebra

Midsagittal Section through the BRAINSTEM and CEREBELLUM

★ RETICULAR ACTIVATING SYSTEM (RAS)
- Initiates and maintains wakefulness, alertness, introspection, and directing attention
- RETICULAR FORMATION, SUBTHALAMIC NUCLEI, MEDIAL THALAMUS, HYPOTHALAMUS
- Extends to all parts of CEREBRAL CORTEX

★ RETICULAR FORMATION of BRAINSTEM
- Connects and fills spaces between tracts
- Group of cells and fibers of RAS that controls alertness, waking, and sleeping

★ RESPIRATORY CENTER in the PONS and MEDULLA

- Posterior Commissure
- Epithalamic Habenular Nuclei
- Pineal Body
- Primary Fissure
- Transverse Fissure
- Pyramidal Fissure
- Folia
- Choroid Plexus of 4V
- Central Canal of Spinal Cord
- Spinal Cord
- 4V
- 4V
- g
- h
- f
- g
- CA
- 5
- 6
- 1
- 2
- 3

Posterior (Dorsal) View (See 7.15)

- 3V (3rd Ventricle)
- Pulvinar of Thalamus
- Pulvinar of Thalamus
- Pineal Body
- Cranial Nerve IV (Trochlear)
- exposing fourth Ventricle (4V)
- XI
- Spinal Cord
- Atlas Vertebra
- 5 5
- 4 6 6 4
- 1 1
- 7 7
- 10 10
- 7 7
- 2 2
- 10 10
- 15 15
- 3 3
- 3 3
- 3 3

(cerebellar peduncles are shown cut)

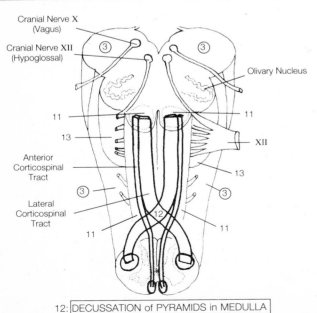

- Cranial Nerve X (Vagus)
- Cranial Nerve XII (Hypoglossal)
- Olivary Nucleus
- 3 3
- 11 11
- 13
- XII
- Anterior Corticospinal Tract
- 3 3
- 13
- Lateral Corticospinal Tract
- 12
- 11 11

12: DECUSSATION of PYRAMIDS in MEDULLA

Anterior View

4 : CEREBELLUM

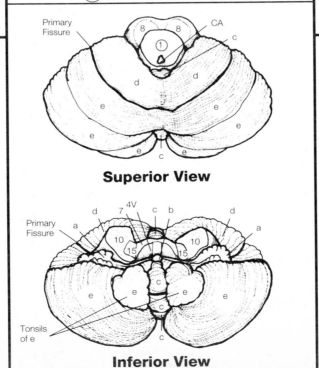

Superior View

- Primary Fissure
- 8 8
- CA
- c
- 1
- d d
- e e
- e e
- e e
- c

Inferior View

- Primary Fissure
- Tonsils of e
- d d
- 7 4V c b
- a a
- 10 10
- 15 15
- e e e e
- c
- c

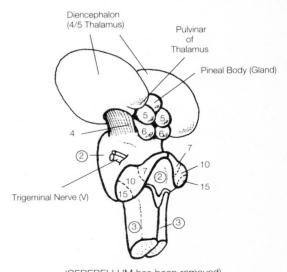

Left Posterolateral View of BRAINSTEM

- Diencephalon (4/5 Thalamus)
- Pulvinar of Thalamus
- Pineal Body (Gland)
- Trigeminal Nerve (V)
- 4
- 5 5
- 6 6
- 2
- 7
- 7
- 2
- 10
- 10
- 15
- 15
- 3 3

(CEREBELLUM has been removed)

NERVOUS SYSTEM: CENTRAL NERVOUS SYSTEM (CNS)
The Spinal Cord: General Structure

c = orange T = yellow L = red a = yellow-orange
b = fleshy c = creamy Ascending Tracts = warms
Descending Tracts - cools

★ See Chart #2, 7.12-7.14

★ **The SPINAL CORD**
- An ovoid column of nervous tissue about 44 cm long, flattened anteroposteriorly
- Continuous with the MEDULLA OBLONGATA, beginning at the FORAMEN MAGNUM of the OCCIPITAL BONE and extends inferiorly through the neural canal of the vertebral column, ending at the level of the SECOND LUMBAR VERTEBRA (L2), terminating in the tapering CONUS MEDULLARIS.
- An average diameter of 1 cm, contains 2 prominent bulges in the lower cervical and lumbar regions (CERVICAL ENLARGEMENT and LUMBAR ENLARGEMENT), expressive of an increased number of neurons that serve the upper and lower extremities, respectively.
- In cross section, it does not fill the vertebral space. It is surrounded immediately by the PIA MATER, then CSF in the SUBARACHNOID SPACE, the ARACHNOID, and the DURA MATER, which fuses with the periosteum of the inner vertebral surfaces.
- Floats in CSF
- Distal to the CONUS MEDULLARIS, the PIA MATER continues as a fibrous strand, the FILUM TERMINALE, that attaches to the coccyx (C1) and joins the extended DURAL SAC of the DURA MATER of the SECOND SACRAL VERTEBRA (S2). This DURAL SAC is filled with CSF as the LUMBAR CISTERN.
- A collection of nerve fibers resembling a horse's tail called the CAUDA EQUINA radiate interiorly from the CONUS MEDULLARIS, floating within the LUMBAR CISTERN as they travel to the SACRAL INTERVERTEBRAL FORAMINA.

★ **The SPINAL SEGMENTS**
- The spinal cord develops as 31 segments, each segment giving rise to a pair of SPINAL NERVES (See 7.12–7.14) that exit posterolaterally from the cord through the intervertebral foramina.

★ **The SPINAL GROOVES**
- 2 GROOVES (ANTERIOR MEDIAN FISSURE and POSTERIOR MEDIAN SULCUS) extend the length of the spinal cord and partially divide the cord into right and left portions.

★ **2 PRINCIPAL FUNCTIONS of the SPINAL CORD:**
① A CENTER FOR SPINAL REFLEXES (CENTRALLY LOCATED GRAY MATTER)
- Specific nerve pathways for reflexive, involuntary movement (See 7.3)
- Gray matter approximates the shape of an "H": the transverse crossbar of the "H" is the GRAY COMMISSURE. The upright projections are the ANTERIOR (VENTRAL) and POSTERIOR (DORSAL) HORNS. Located between these horns to the sides are the short LATERAL HORNS.

② NEURAL COMMUNICATION TO AND FROM THE BRAIN (PERIPHERALLY LOCATED WHITE MATTER TRACTS)
- Gray matter "H" divides the surrounding WHITE MATTER arbitrarily into FUNICULI (COLUMNS). Each FUNICULUS contains a number of specific TRACTS (FASCICULI):
■ SENSORY ASCENDING LONG TRACTS conduct impulses from PERIPHERAL SENSORY RECEPTORS of the BODY to the BRAIN.
■ MOTOR DESCENDING LONG TRACTS conduct motor impulses from the BRAIN to MUSCLES and GLANDS.
■ SHORT INTERSEGMENTAL TRACTS are involved in various spinal reflexes operating on more than one cord level, jumping from segment to segment.

General Structure of the SPINAL CORD

External Anatomy

C	**CERVICAL ENLARGEMENT C3–T2**	
T	**THORACIC SECTION T3–T8**	
L	**LUMBAR ENLARGEMENT T9–T12**	
a	**CONUS MEDULLARIS**	
b	**FILUM TERMINALE**	
c	**CAUDA EQUINA**	

SPINAL GROOVES:
Anterior (Ventral) Surface

d	**ANTERIOR MEDIAN FISSURE**

Posterior (Dorsal) Surface

e	**POSTERIOR MEDIAN SULCUS**

31 SPINAL SEGMENTS Each Give Rise to a Pair of

SN	**SPINAL NERVES**

★ **See Chart #2**

Internal Anatomy

GM	**GRAY MATTER**	• Nerve cell bodies / • Neuroglia / • Unmyelinated association neurons
WM	**WHITE MATTER**	• Tracts (bundles) of myelinated fibers of sensory and motor neurons

GRAY MATTER

1	**2 ANTERIOR (VENTRAL) HORNS**	• Cell bodies of motor neurons (motor axons of bodies exit) / • Association neurons
2	**2 POSTERIOR (DORSAL) HORNS**	• Central processes of sensory neurons enter / • Association neurons
3	**2 LATERAL HORNS**	• Autonomic motor neurons only in T1–L2, S2–4 levels
4	**2 GRAY COMMISSURES**	• Crossing of axons of various neuron / • CENTRAL CANAL in the middle
5	**INTERMEDIATE ZONE**	• Mostly association neurons / • Some motor neurons
6	**RETICULAR FORMATION**	• Diffuse gray column network of cell bodies and fibers. Receives fibers from sensoryLEMNISCUS tract and sends fibers to cerebral cortex, cerebellum, and spinal portion of the RAS (RETICULAR ACTIVATING SYSTEM). (See 7.10)

WHITE MATTER:
SIX FUNICULII (COLUMNS)

7	**2 ANTERIOR (VENTRAL) FUNICULI**	
8	**2 POSTERIOR (DORSAL) FUNICULI**	
9	**2 LATERAL FUNICULI**	
10	**2 WHITE COMMISSURES**	

ASCENDING (SENSORY) WHITE MATTER TRACTS

Pain/Temperature Receptor Pathway:

11	**LATERAL SPINOTHALAMIC TRACT**

THALAMUS (same side) → THALAMOCORTICAL TRACT (INTERNAL CAPSULE and CORONA RADIATA) → POSTCENTRAL GYRUS of CEREBRAL CORTEX

Crude Touch and Pressure Receptor Pathway

12	**ANTERIOR SPINOTHALAMIC TRACT**

Fine Touch, Proprioception and Vibration Receptor Pathway:

13	**FASCICULUS GRACILIS AND FASCICULUS CUNEATUS TRACTS**

→ MEDULLA (DECUSSATION) → MEDIAL LEMNISCUS → THALAMUS → THALAMOCORTICAL TRACT → POSTCENTRAL GYRUS

Muscle/Position Sense Receptor Pathway:

- Tendon stretch impulses **14** | **POSTERIOR SPINOCEREBELLAR TRACTS** |

→ INFERIOR CEREBELLAR PEDUNCLE → CEREBELLUM CORTEX

- Position sense impulses **15** | **ANTERIOR SPINOCEREBELLAR TRACTS** |

→ ASSOCIATION NEURON CROSSES → SUPERIOR CEREBELLAR PEDUNCLE → CEREBELLUM CORTEX

DESCENDING (MOTOR) WHITE MATTER TRACTS
■ All terminate in ANTERIOR (VENTRAL) GRAY HORNS
① CORTICOSPINAL (PYRAMIDAL)TRACTS
- Neuron cell bodies in PRECENTRAL GYRUS of FRONTAL LOBE
- Effect coordinated, precise voluntary movements of skeletal muscle
- 85% DECUSSATE in PYRAMIDS of MEDULLA:

16	**LATERAL CORTICOSPINAL TRACT**

- 25% UNCROSSED in MEDULLA:

17	**ANTERIOR CORTICOSPINAL TRACT**

- Descend directly (no synapses) to lower motor neurons in anterior gray horns

② EXTRAPYRAMIDAL TRACTS and Their (Effector) Activity
- Major extrapyramidal tracts originate in RETICULAR FORMATION

18	**RETICULOSPINAL TRACTS** (Muscle tone, sweat glands)

- No direct descending tracts from the cerebellum—only indirectly through vestibular nuclei, red nuclei, superior colliculi Origins, respectively:

19	**VESTIBULOSPINAL TRACTS** (Muscle tone, equilibrium)
20	**RUBROSPINAL TRACTS** (Muscle tone, posture)
21	**TECTOSPINAL TRACTS** (Head movements)

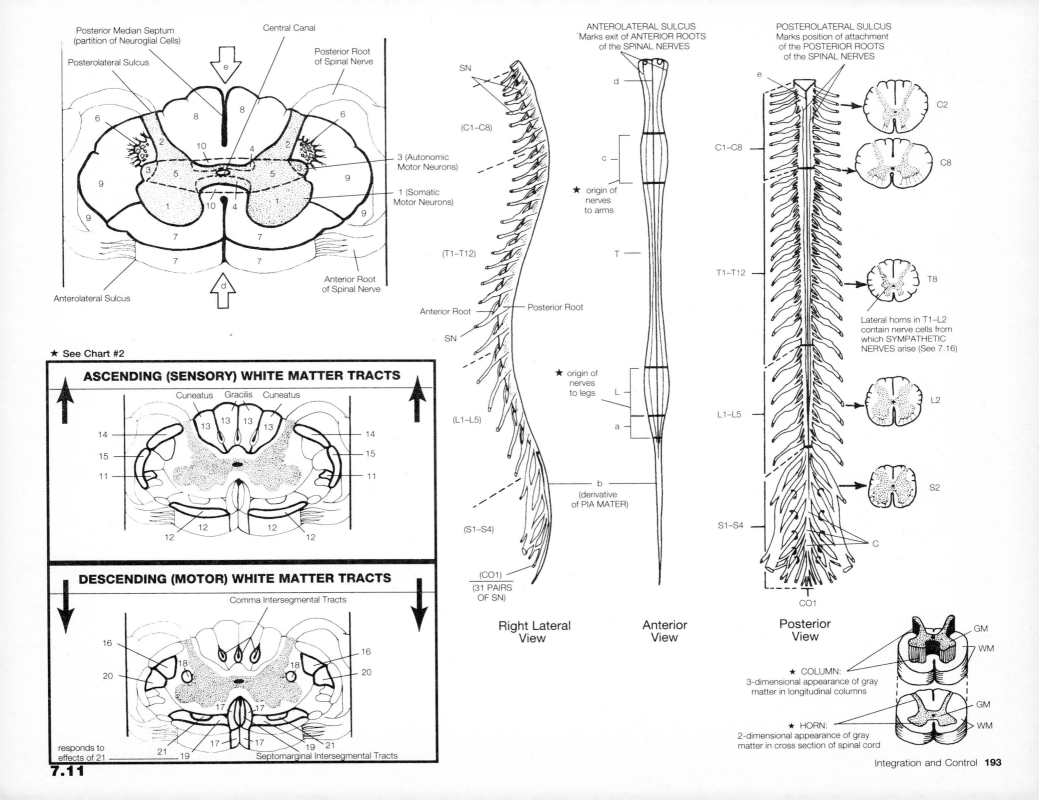

Posterior Median Septum
(partition of Neuroglial Cells)

Central Canal

Posterolateral Sulcus

Posterior Root
of Spinal Nerve

3 (Autonomic
Motor Neurons)

1 (Somatic
Motor Neurons)

Anterior Root
of Spinal Nerve

Anterolateral Sulcus

★ See Chart #2

ASCENDING (SENSORY) WHITE MATTER TRACTS

Cuneatus   Gracilis   Cuneatus

DESCENDING (MOTOR) WHITE MATTER TRACTS

Comma Intersegmental Tracts

Septomarginal Intersegmental Tracts

responds to
effects of 21

7.11

ANTEROLATERAL SULCUS
Marks exit of ANTERIOR ROOTS
of the SPINAL NERVES

SN

(C1–C8)

(T1–T12)

Anterior Root   Posterior Root

SN

(L1–L5)

(S1–S4)

(CO1)
(31 PAIRS
OF SN)

Right Lateral
View

d

c

★ origin of
nerves
to arms

T

★ origin of
nerves
to legs

L

a

b
(derivative
of PIA MATER)

Anterior
View

POSTEROLATERAL SULCUS
Marks position of attachment
of the POSTERIOR ROOTS
of the SPINAL NERVES

e

C1–C8

T1–T12

L1–L5

S1–S4

CO1

Posterior
View

C2

C8

T8

Lateral horns in T1–L2
contain nerve cells from
which SYMPATHETIC
NERVES arise (See 7.16)

L2

S2

C

GM

WM

★ COLUMN:
3-dimensional appearance of gray
matter in longitudinal columns

GM

WM

★ HORN:
2-dimensional appearance of gray
matter in cross section of spinal cord

Integration and Control   **193**

NERVOUS SYSTEM: PERIPHERAL NERVOUS SYSTEM (PNS)
The Spinal Nerves: General Structure and Distribution

★ The PERIPHERAL NERVOUS SYSTEM (PNS) is the portion of the nervous system outside the limits of the CNS. The parts of the PNS convey impulses to and from the brain and spinal cord. The PNS consists of sensory receptors within sensory organs (See 7.18–7.23), NERVE FIBERS, PLEXUSES, GANGLIA, and specialized MOTOR END ORGANS that are derived from the 2 main types of nerves of the PNS: CRANIAL NERVES (arising from the BRAIN) and SPINAL NERVES (arising from the SPINAL CORD).

★ The SPINAL NERVES
•The 31 PAIRS of SPINAL NERVES are named and numbered according to the segment level of the spinal cord from which they emerge
•All spinal nerves (except C1) leave the spinal cord and vertebral canal through the INTERVERTEBRAL FORAMINA to ultimately innervate a DERMATOME. The 1st cervical nerve, C1 emerges between the ATLAS BONE and the OCCIPITAL BONE, and does not innervate a DERMATOME for lack of a DORSAL ROOT. (See box.)
•During fetal development, the spinal cord grows more slowly than the vertebral column. (Recall that the spinal cord terminates at the 2nd LUMBAR VERTEBRA. Thus, the lower lumbar, sacral, and coccygeal nerves that constitute the CAUDA EQUINA must descend to reach their foramina.)
•Spinal nerves are mixed collections of axons of sensory and motor neurons bound in an intricate array of fibrous connective tissue. (See box on bottom page)

★ GENERAL STRUCTURE of a SPINAL NERVE
A spinal nerve attaches at two points to the spinal cord:
① **DORSAL (POSTERIOR) ROOT**
•Sensory axons (central processes) enter the POSTERIOR HORN of the CORD, conveying sensory impulses into the cord
•DORSAL-ROOT GANGLION: A swelling within the root composed of the UNIPOLAR CELL BODIES of the sensory neurons. (See 7.2, 7.3)
② **VENTRAL (ANTERIOR) ROOT**
•Motor axons arise from MULTIPOLAR CELL BODIES of MOTOR NEURONS within the ANTERIOR HORN of the CORD, conveying motor impulses away from the CNS.
•These 2 roots unite a short distance from the cord at the intervertebral foramina to form the SPINAL NERVE.

★ BRANCHES of a SPINAL NERVE
•A typical spinal nerve distribution (thoracic) (T2–T12) divides immediately into several branches after emerging from the intervertebral foramina:
•The SPINAL NERVE splits into 2 large branches, or RAMI, each of which has a number of smaller MUSCULAR and CUTANEOUS branches.
■ **DORSAL RAMUS**
•Innervates the deep back skeletal muscles (and joints) along the vertebral column and the overlying fascia and skin.
■ **VENTRAL RAMUS**
•Innervates the rest of the body (muscles, skin, and fascia on the lateral and anterior side of the torso, coursing between the intercostal muscle layers of the rib cage and muscles of the abdominal wall).
•The SPINAL NERVE also has 2 branches, the RAMI COMMUNICANTES, that connect to a SYMPATHETIC CHAIN GANGLION, part of the SYMPATHETIC DIVISION of the AUTONOMIC NERVOUS SYSTEM. (See 7.16)

■ **GRAY RAMUS** (in ALL SPINAL NERVES)
•Contains UNMYELINATED POSTGANGLIONIC FIBERS
■ **WHITE RAMUS** (ONLY T1–L2)
•Contains MYELINATED PREGANGLIONIC FIBERS

The 31 Pairs of SPINAL NERVES

C	**8 CERVICAL**	C1–C8
T	**12 THORACIC**	T1–T12
L	**5 LUMBAR**	L1–L5
S	**5 SACRAL**	S1–S5
Co	**1 COCCYGEAL**	Co1

General Structure of a SPINAL NERVE:

SN | **SPINAL NERVE**

Posterior SENSORY ROOT

1 | **DORSAL SPINAL ROOT**

1a | **DORSAL-ROOT GANGLION**

Anterior MOTOR ROOT

2 | **VENTRAL SPINAL ROOT**

SPINAL NERVE BRANCHES:
RAMI

3 | **DORSAL RAMUS** POSTERIOR RAMUS

4 | **VENTRAL RAMUS** ANTERIOR RAMUS

RAMI COMMUNICANTES of the
AUTONOMIC NERVOUS SYSTEM

5 | **GRAY RAMUS COMMUNICANS**

6 | **WHITE RAMUS COMMUNICANS**

7 | **SYMPATHETIC CHAIN GANGLION**

8 | **SPLANCHNIC NERVES**

Small Returning BRANCH

9 | **MENINGEAL BRANCH**

•Re-enters spinal canal through the intervertebral foramen to supply the vertebrae, vertebral ligaments, meninges and blood vessels of the cord.

Color spinal nerve groups (C,T,L,S,Co) the same colors as their corresponding DERMATOMES
C = orange T = yellow L = red S = fleshy Co = purple

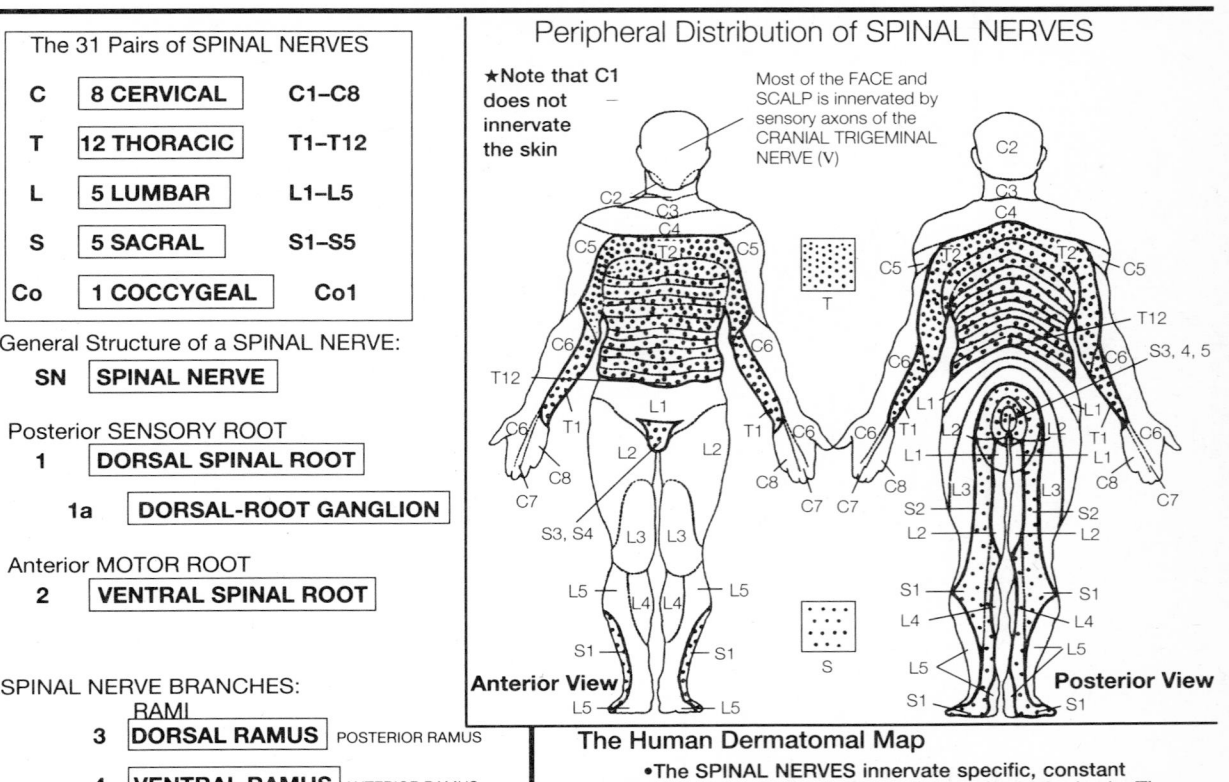

Peripheral Distribution of SPINAL NERVES

★Note that C1 does not innervate the skin

Most of the FACE and SCALP is innervated by sensory axons of the CRANIAL TRIGEMINAL NERVE (V)

Anterior View   Posterior View

The Human Dermatomal Map
•The SPINAL NERVES innervate specific, constant segments of the skin (cutaneous areas) over the entire body. The SKIN SEGMENT whose sensory receptors and axons are supplied by a SINGLE DORSAL (POSTERIOR) SENSORY ROOT of a SPINAL NERVE is called a DERMATOME.
•All spinal nerves supply branches to the skin, except the 1st cervical spinal nerve (C1). (C1 has no dermatome because it lacks a dorsal root.)
•In the neck and trunk, there are consecutive dermatome bands of skin.
•Especially in the trunk area, 3 or 4 dorsal roots may receive inputs from a single dermatome. Thus, overlapping occurs between adjacent dermatomes.
•The human dermatomal map is of clinical importance in determining which segment of the spinal cord or spinal nerve may be malfunctioning, or when a physician desires to anesthetize a particular portion of the body.

Cutaneous Sensory Areas

C	**DORSAL ROOTS of CERVICAL NERVES**	(C2–C8)	L	**DORSAL ROOTS of LUMBAR NERVES**	(L1–L5)
T	**DORSAL ROOTS of THORACIC NERVES**	(T1–T12)	S	**DORSAL ROOTS of SACRAL NERVES**	(S1–S5)

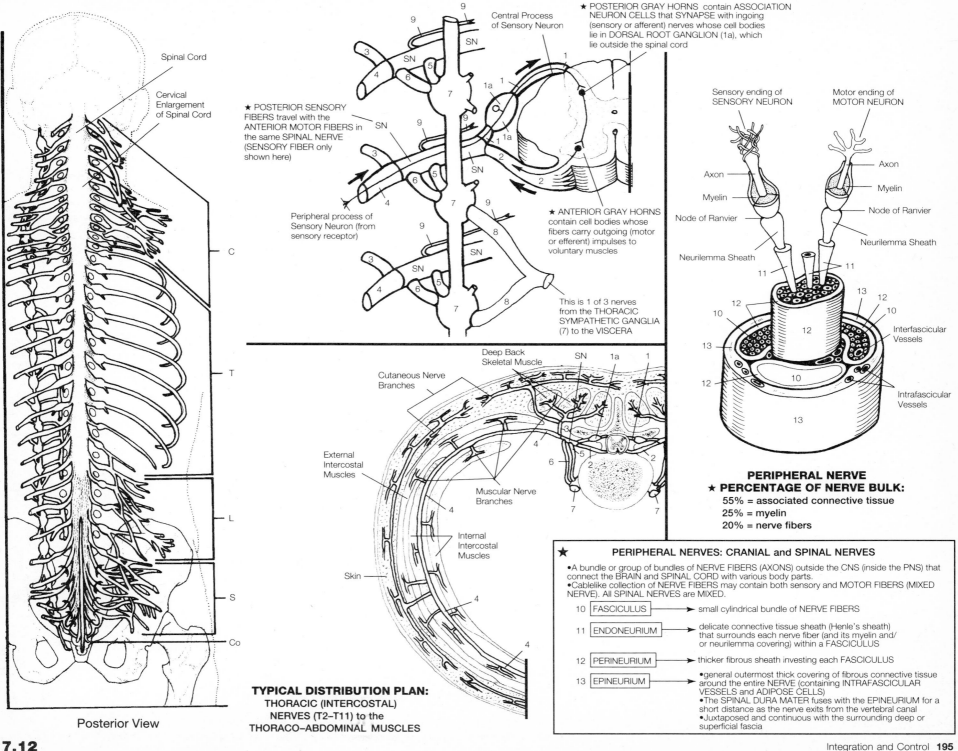

Spinal Cord

Cervical Enlargement of Spinal Cord

C

T

L

S

Co

Posterior View

9 9 9 9 SN

Central Process of Sensory Neuron

★ POSTERIOR GRAY HORNS contain ASSOCIATION NEURON CELLS that SYNAPSE with ingoing (sensory or afferent) nerves whose cell bodies lie in DORSAL ROOT GANGLION (1a), which lie outside the spinal cord

★ POSTERIOR SENSORY FIBERS travel with the ANTERIOR MOTOR FIBERS in the same SPINAL NERVE (SENSORY FIBER only shown here)

Peripheral process of Sensory Neuron (from sensory receptor)

★ ANTERIOR GRAY HORNS contain cell bodies whose fibers carry outgoing (motor or efferent) impulses to voluntary muscles

This is 1 of 3 nerves from the THORACIC SYMPATHETIC GANGLIA (7) to the VISCERA

Sensory ending of SENSORY NEURON

Motor ending of MOTOR NEURON

Axon

Myelin

Node of Ranvier

Neurilemma Sheath

Axon

Myelin

Node of Ranvier

Neurilemma Sheath

Interfascicular Vessels

Intrafascicular Vessels

PERIPHERAL NERVE
★ PERCENTAGE OF NERVE BULK:
55% = associated connective tissue
25% = myelin
20% = nerve fibers

Deep Back Skeletal Muscle

SN 1a 1

Cutaneous Nerve Branches

External Intercostal Muscles

Muscular Nerve Branches

Internal Intercostal Muscles

Skin

TYPICAL DISTRIBUTION PLAN:
THORACIC (INTERCOSTAL)
NERVES (T2–T11) to the
THORACO–ABDOMINAL MUSCLES

★ **PERIPHERAL NERVES: CRANIAL and SPINAL NERVES**
• A bundle or group of bundles of NERVE FIBERS (AXONS) outside the CNS (inside the PNS) that connect the BRAIN and SPINAL CORD with various body parts.
• Cablelike collection of NERVE FIBERS may contain both sensory and MOTOR FIBERS (MIXED NERVE). All SPINAL NERVES are MIXED.

10	FASCICULUS	→ small cylindrical bundle of NERVE FIBERS
11	ENDONEURIUM	→ delicate connective tissue sheath (Henle's sheath) that surrounds each nerve fiber (and its myelin and/ or neurilemma covering) within a FASCICULUS
12	PERINEURIUM	→ thicker fibrous sheath investing each FASCICULUS
13	EPINEURIUM	→ • general outermost thick covering of fibrous connective tissue around the entire NERVE (containing INTRAFASCICULAR VESSELS and ADIPOSE CELLS) • The SPINAL DURA MATER fuses with the EPINEURIUM for a short distance as the nerve exits from the vertebral canal • Juxtaposed and continuous with the surrounding deep or superficial fascia

7.13

Integration and Control **196**

R = yellow T = orange 12 = light blue
13 = light green C = red
★ See Chart #3

NERVOUS SYSTEM: PERIPHERAL NERVOUS SYSTEM (PNS)
The Spinal Nerve Plexuses: The Cervical and Brachial Plexuses

On the previous page, we discussed the branching distribution of the THORACIC SPINAL NERVE GROUP (T2–T12), called the INTERCOSTAL NERVES. This is the only group of spinal nerves not termed a PLEXUS as they go *directly* to the structures they innervate. The VENTRAL RAMI of ALL OTHER 20 PAIRS of SPINAL NERVES (including T1) combine with adjacent nerves on either side of the body and then split again as a network of nerves referred to as a PLEXUS, from which MUSCULAR and CUTANEOUS BRANCHES emerge in a definite pattern. There are 4 PLEXUSES of SPINAL NERVES: CERVICAL, BRACHIAL, LUMBAR and SACRAL PLEXUSES. Names of the nerves indicate structures innervated or general courses taken (nerve numbers in parentheses indicate formation of the plexus from ROOT ORIGINS of ventral rami)

■ indicates major nerves of the plexus.

★ The CERVICAL PLEXUS
• ROOTS (VENTRAL RAMI) = C1–C4 and a portion of C5
• Positioned deep on each side of the neck, alongside the first 4 cervical vertebrae.
• Supplies MUSCLES, FASCIA, and SKIN of the HEAD, NECK, and UPPER PART of the SHOULDERS.
Branches connect with cranial nerves XI (ACCESSORY) and XII (HYPOGLOSSAL) supplying dual innervation to specific neck and pharyngeal muscles.
• MAJOR NERVE = PHRENIC NERVE formed from fibers of C3, C4, and C5 supplying motor fibers to the diaphragm. (These motor impulses cause contraction of the diaphragm, inspiring air into the lungs, essential for normal breathing.)

★ The BRACHIAL PLEXUS
• 5 ROOTS (VENTRAL RAMI) = C5–C8, T1, and portions of C4 and T2
• Positioned on either side of the last 4 cervical and 1st thoracic vertebrae
• Extends downward and laterally, passes over the 1st rib *behind* the clavicle bone and enters the axillary space below the shoulder
• Constitutes the ENTIRE nerve supply for the upper limbs, the MAJORITY of the shoulder region, and some neck muscles.
• Structurally divided into ROOTS, TRUNKS, DIVISIONS, and CORDS:
 ★ 5 ROOTS UNITE TO FORM 3 TRUNKS:
  • Roots C5 and C6 converge to form the UPPER TRUNK
  • Root C7 becomes the MIDDLE TRUNK
  • Roots C8 and T1 converge to form the LOWER TRUNK
 ★ 3 TRUNKS BRANCH into 6 DIVISIONS:
  • Each trunk divides into an ANTERIOR DIVISION and a POSTERIOR DIVISION.
 ★ 6 DIVISIONS CONVERGE INTO 3 CORDS:
  • ANTERIOR DIVISION of the UPPER and MIDDLE TRUNK converge to form the LATERAL CORD.
  • ANTERIOR DIVISION of the LOWER TRUNK continues to become the MEDIAL CORD (thus mostly contains C8 and T1 fibers).
  • POSTERIOR DIVISIONS of all 3 TRUNKS converge to form the POSTERIOR CORD (thus contains C5–C8 fibers). This POSTERIOR CORD is DEEP to the other 2 cords.
• The 3 CORDS are formed in the AXILLA about the AXILLARY ARTERY. These cords give rise to the peripheral nerves, which supply the musculoskeletal structures of the upper limbs.

SPINAL NERVES to the Head and Neck:
The CERVICAL PLEXUS (C1–C4)
(with contributions from C5)

R | ROOTS: (C1–C4)

Superficial Cutaneous BRANCHES

1 | LESSER OCCIPITAL

2 | GREATER AURICULAR   } (C2–C3)

3 | TRANSVERSE CERVICAL

4 | SUPRACLAVICULAR   (C3, C4)

Deep Motor BRANCHES

5 | ANSA CERVICALIS : (C1–C3)

 5a | SUPERIOR ROOT   (C1,C2)

 5b | INFERIOR ROOT   (C2,C3)

6 ■ PHRENIC   (C3–C5)

7 | SEGMENTAL BRANCHES   (C1–C5)

| XI | ACCESSORY CRANIAL NERVE XI |
| XII | HYPOGLOSSAL CRANIAL NERVE XII |

5 MAJOR NERVES (TERMINAL NERVES) from the CORDS of THE BRACHIAL PLEXUS:
LATERAL CORD
 ■ MUSCULOCUTANEOUS NERVE innervates the ANTERIOR UPPER ARM MUSCLES (including BICEPS BRACHII)
POSTERIOR CORD
 ■ AXILLARY NERVE wraps around surgical neck of the humerus bone to innervate the DELTOID and TERES MINOR MUSCLES
 ■ RADIAL NERVE innervates TRICEPS BRACHII, turns halfway about the humerus bone and innervates the POSTERIOR UPPER ARM and FOREARM EXTENSOR MUSCLES (including BRACHIORADIALIS)
MEDIAL CORD
 ■ ULNAR NERVE innervates ANTERIOR FOREARM FLEXOR MUSCLES and the HYPOTHENAR and INTRINSIC FLEXOR MUSCLES of the HAND
 ■ MEDIAN NERVE (LATERAL and MEDIAL HEADS) innervates the ANTERIOR FOREARM FLEXOR MUSCLES and the THENAR MUSCLES of the THUMB

SPINAL NERVES to the Upper Limb:
The BRACHIAL PLEXUS (C5–T1)
(with contributions from C4 and T2)

R | 5 ROOTS : (C5–T1)

8 | DORSAL SCAPULAR BRANCH (C5)

9 | LONG THORACIC BRANCH (C5,C6,C7)

T | 3 TRUNKS (UPPER, MIDDLE, LOWER) (C5–C6)

10 | SUBCLAVIUS   from
11 | SUPRASCAPULAR   } UPPER (C5–C6)
                        TRUNK

D | 6 DIVISIONS (C5–C8)

12 | 3 ANTERIOR DIVISIONS   } (C5–C8)
13 | 3 POSTERIOR DIVISIONS

C | 3 CORDS (C5–T1)

LATERAL CORD: INNERVATES LATERAL ASPECT of the LIMB and SOME SUPERFICIAL BACK MUSCLES

14 ■ MUSCULOCUTANEOUS
15 | MEDIAN (LATERAL HEAD)   } (C5–C7)
16 | LATERAL PECTORAL

POSTERIOR CORD: INNERVATES POSTERIOR ASPECT of the LIMB and 2 SUPERFICIAL BACK MUSCLES

17 | SUBSCAPULAR
18 | THORACODORSAL   } (C5–C6)
19 ■ AXILLARY (CIRCUMFLEX)
20 ■ RADIAL (C5–C8, T1)

MEDIAL CORD: INNERVATES ANTERIOR ASPECT of the LIMB

21 | MEDIAL PECTORAL
22 | MEDIAL BRACHIAL CUTANEOUS   } (C8, T1)
23 | MEDIAL ANTEBRACHIAL CUTANEOUS
24 | MEDIAL (MEDIAL HEAD) (C5–C7, T1)
25 ■ ULNAR (C8, T1)
26 ■ MEDIAN (C6–C8, T1)

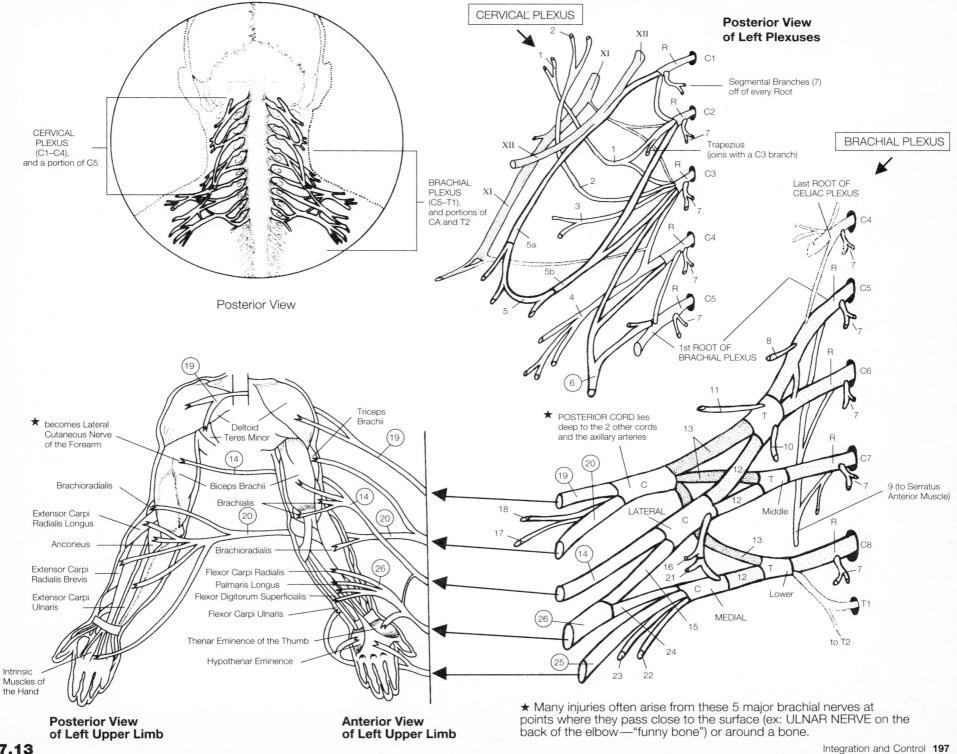

CERVICAL PLEXUS (C1–C4), and a portion of C5

BRACHIAL PLEXUS (C5–T1), and portions of CA and T2

Posterior View

CERVICAL PLEXUS

Posterior View of Left Plexuses

Segmental Branches (7) off of every Root

Trapezius (joins with a C3 branch)

1st ROOT OF BRACHIAL PLEXUS

BRACHIAL PLEXUS

Last ROOT OF CELIAC PLEXUS

9 (to Serratus Anterior Muscle)

★ POSTERIOR CORD lies deep to the 2 other cords and the axillary arteries

★ becomes Lateral Cutaneous Nerve of the Forearm

Deltoid
Teres Minor

Triceps Brachii

Biceps Brachii

Brachialis

Brachioradialis

Brachioradialis

Extensor Carpi Radialis Longus

Flexor Carpi Radialis
Palmaris Longus
Flexor Digitorum Superficialis
Flexor Carpi Ulnaris

Anconeus

Extensor Carpi Radialis Brevis

Extensor Carpi Ulnaris

Thenar Eminence of the Thumb

Hypothenar Eminence

Intrinsic Muscles of the Hand

LATERAL

Middle

MEDIAL

Lower

to T2

Posterior View of Left Upper Limb

Anterior View of Left Upper Limb

★ Many injuries often arise from these 5 major brachial nerves at points where they pass close to the surface (ex: ULNAR NERVE on the back of the elbow —"funny bone") or around a bone.

7.13

Integration and Control **197**

NERVOUS SYSTEM: PERIPHERAL NERVOUS SYSTEM (PNS)
The Spinal Nerve Plexuses: The Lumbar and Sacral Plexuses:
Lumbosacral Plexus

★ **The LUMBAR PLEXUS**
•ROOTS (VENTRAL RAMI) = L1–L4, and a portion of T12
•Positioned on either side of the 1st and 4th lumbar vertebrae
•Innervates lower abdominal wall, scrotum (or labia majora), and anterior and medial portions of the thigh
•Not as complex as the BRACHIAL PLEXUS, with no intricate interlacing of fibers. Structurally, the lumbar plexus has only ROOTS and DIVISIONS (with *no* TRUNKS or CORDS as in the brachial plexus).
  ★ 5 ROOTS BRANCH into 2 DIVISIONS:
•ANTERIOR DIVISION is superficial to the QUADRATUS LUMBORUM MUSCLE
•POSTERIOR DIVISION passes obliquely outward behind the PSOAS MAJOR MUSCLE
•Those divisions give rise to its peripheral nerves. The largest nerve of the lumbar plexus is the FEMORAL NERVE (POSTERIOR DIVISION) which innervates the ANTERIOR THIGH MUSCLES and EXTENSOR LEG MUSCLES.
•OBTURATOR NERVE (ANTERIOR DIVISION) is a large nerve that innervates the ADDUCTOR LEG MUSCLES.

★ **The SACRAL PLEXUS**
•ROOTS (VENTRAL RAMI) = L4–L5, S1–S4
•Positioned immediately caudal to the lumbar, largely *in front of* the sacrum
•Innervates the LOWER BACK, PELVIS, PERINEUM, PUDENDAL GENITALIA, BUTTOCKS, POSTERIOR SURFACE of the THIGH and LEG and DORSAL and VENTRAL SURFACES of the FOOT.
•Like the lumbar plexus, it contains ROOTS that branch into ANTERIOR and POSTERIOR DIVISIONS from which the peripheral nerves arise.
  ★ MAJOR NERVE = SCIATIC NERVE (L4–S3)
SCIATIC NERVE: IS the LARGEST NERVE in the BODY, SUPPLYING the ENTIRE MUSCULATURE of the LEG AND FOOT.
•ACTUALLY COMPOSED of 2 NERVES (the COMMON PERONEAL and the TIBIAL NERVES, wrapped in a common SCIATIC SHEATH).
•Passes from the PELVIS through the GREATER SCIATIC NOTCH of the OS COXA and extends down the posterior aspect of the thigh among the HAMSTRINGS and ADDUCTOR MAGNUS.
•Just above and behind the knee, at the POPLITEAL FOSSA it divides separately into the TIBIAL and COMMON PERONEAL NERVES.
  ■ TIBIAL NERVE (L4–S3)
•Passes down between the heads of the GASTROCNEMIUS MUSCLE (innervating the POSTERIOR LEG), then curves under the medial arch of the foot (innervating the sole of the foot through PLANTAR branches).
  ■ COMMON PERONEAL (L4–S3)
•Rounds the neck of the fibular bone and divides into 2 branches:
     DEEP PERONEAL BRANCH innervates ANTERIOR-LATERAL LEG
     SUPERFICIAL PERONEAL BRANCH innervates PERONEAL MUSCLES of LATERAL LEG

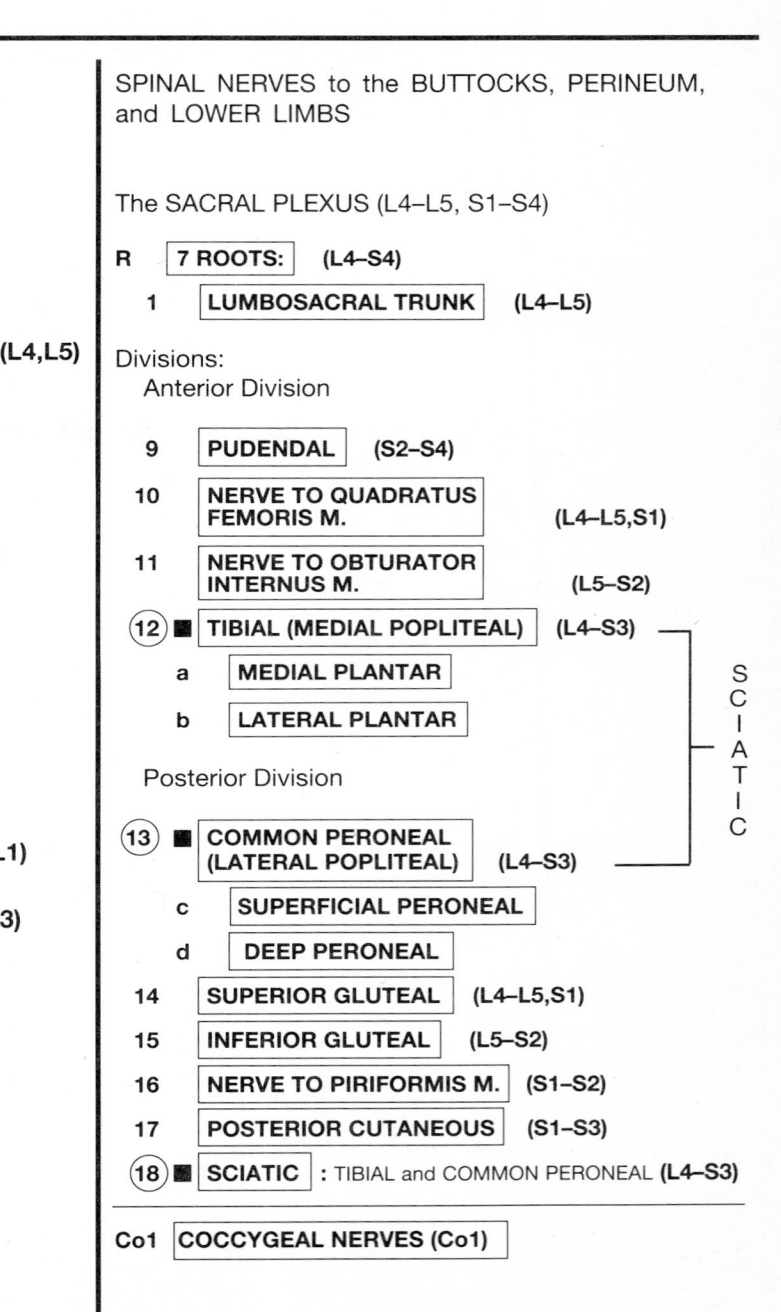

SPINAL NERVES to the GENITALS, ANTEROABDOMINAL WALL, and LOWER LIMBS

The LUMBAR PLEXUS (L1–L4)
(with contributions from T12)

R | 5 ROOTS | : (T12–L4)

1 | LUMBOSACRAL TRUNK | (L4,L5)

Divisions:
Anterior Division

2 | ILIOINGUINAL | (L1)

3 | GENITOFEMORAL | (L1,L2)

4 ■ | OBTURATOR |
                          } (L2–L4)
5 | SAPHENOUS |
       ACCESSORY
       OBTURATOR

Posterior Division

6 | ILIOHYPOGASTRIC | (T12,L1)

7 | LATERAL FEMORAL CUTANEOUS | (L2,L3)

8 ■ | FEMORAL | (L2–L4)

SPINAL NERVES to the BUTTOCKS, PERINEUM, and LOWER LIMBS

The SACRAL PLEXUS (L4–L5, S1–S4)

R | 7 ROOTS: | (L4–S4)

1 | LUMBOSACRAL TRUNK | (L4–L5)

Divisions:
Anterior Division

9 | PUDENDAL | (S2–S4)

10 | NERVE TO QUADRATUS FEMORIS M. | (L4–L5,S1)

11 | NERVE TO OBTURATOR INTERNUS M. | (L5–S2)

12 ■ | TIBIAL (MEDIAL POPLITEAL) | (L4–S3) ┐

a | MEDIAL PLANTAR |

b | LATERAL PLANTAR |

Posterior Division

13 ■ | COMMON PERONEAL (LATERAL POPLITEAL) | (L4–S3) ┘

c | SUPERFICIAL PERONEAL |

d | DEEP PERONEAL |

S C I A T I C

14 | SUPERIOR GLUTEAL | (L4–L5,S1)

15 | INFERIOR GLUTEAL | (L5–S2)

16 | NERVE TO PIRIFORMIS M. | (S1–S2)

17 | POSTERIOR CUTANEOUS | (S1–S3)

18 ■ | SCIATIC | : TIBIAL and COMMON PERONEAL (L4–S3)

Co1 | COCCYGEAL NERVES (Co1) |

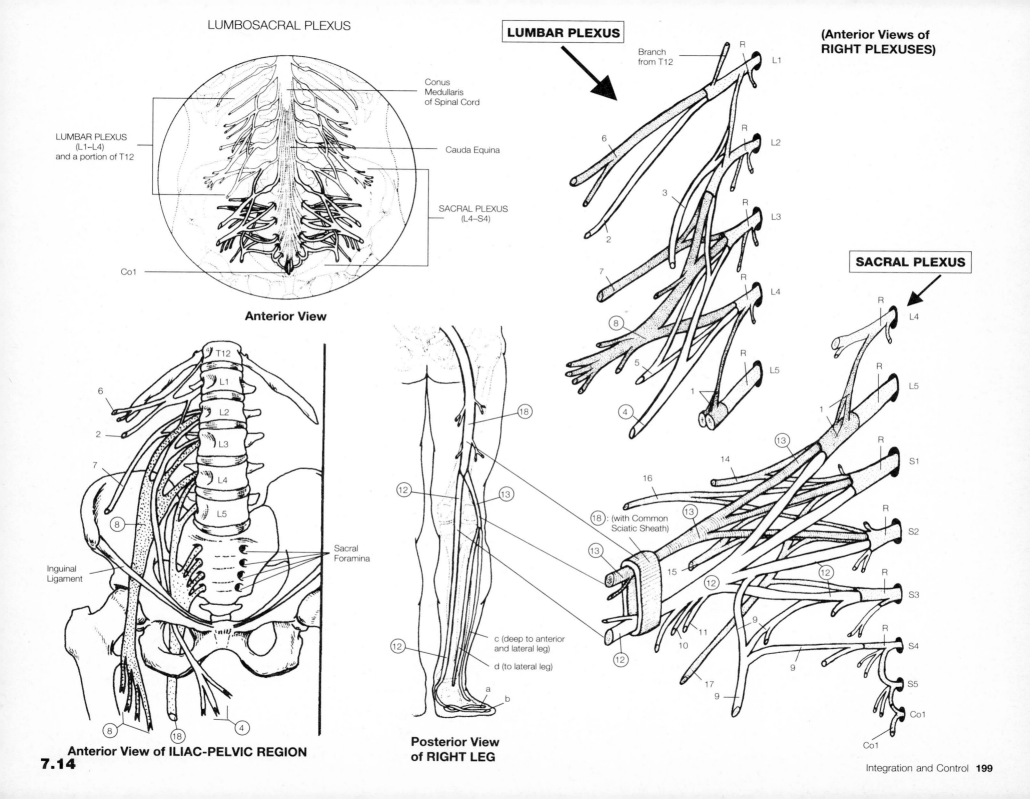

LUMBOSACRAL PLEXUS

Conus
Medullaris
of Spinal Cord

Cauda Equina

LUMBAR PLEXUS
(L1–L4)
and a portion of T12

SACRAL PLEXUS
(L4–S4)

Co1

Anterior View

LUMBAR PLEXUS

Branch
from T12

R

L1

R

L2

6

3

2

R

L3

7

R

L4

8

5

R

L5

1

4

(Anterior Views of RIGHT PLEXUSES)

SACRAL PLEXUS

R

L4

R

L5

13

14

16

1

13

R

S1

18 : (with Common
Sciatic Sheath)

13

15

R

S2

13

12

12

R

S3

11

9

10

9

R

S4

9

17

S5

9

Co1

Co1

T12

L1

6

L2

2

L3

7

L4

8

L5

Inguinal
Ligament

Sacral
Foramina

8

18

4

Anterior View of ILIAC-PELVIC REGION

18

12

13

12

c (deep to anterior
and lateral leg)

d (to lateral leg)

a

b

**Posterior View
of RIGHT LEG**

NERVOUS SYSTEM: PERIPHERAL NERVOUS SYSTEM (PNS)
The Cranial Nerves

*Use a color wheel palette, starting **OLFACTORY (I)** in yellow then continue through yellow-orange, orange, reds, reaching **(VI)** as purple, then continuing through the blues and greens, ending with **HYPOGLOSSAL (XII)** as yellow-green.*

★ See Chart #5, 7.10

See Chart # 5 for Roman Numeral and Name, Foramen Transmitting, Composition of Impulses, Origin (location of Cell Bodies) and Function of All 12 Pairs of Cranial Nerves.

The CRANIAL NERVES
•Of the 12 PAIRS, 2 PAIRS arise from the FOREBRAIN (OLFACTORY I and OPTIC II, both of which contain sensory fibers exclusively). The other 10 pairs arise from the MIDBRAIN and BRAINSTEM.

•All 12 PAIRS leave the skull through foramina of the skull. (See also Chapter 4, Chart #2.)

•All are designated with ROMAN NUMERALS, indicating the order in which the nerves are positioned and arise from the front of the brain (I) to the back (XII). (See cranial origins)

•All are designated with names indicating distribution to innervated structures or function of the nerves.

•Only 3 CRANIAL NERVES contain SENSORY FIBERS ONLY: OLFACTORY I, OPTIC II and VESTIBULOCOCHLEAR VIII, and are associated with special sense.

•The TRIGEMINAL V is primarily sensory, but for a small motor function in the mandibular division.

•The cell bodies of these sensory fibers are located in GANGLIA outside the brain.

•All the other cranial nerves contain both sensory and motor fibers (to varying degrees), and are termed MIXED NERVES. (Recall that ALL SPINAL NERVES are mixed nerves.)

★ Thus, all nerves in the body, both cranial and spinal, are mixed, except for cranials I, II, and VIII.

•Although the SOMATIC NERVOUS SYSTEM has been defined as a CONSCIOUS, VOLUNTARY SYSTEM, some motor fibers control SUBCONSCIOUS, INVOLUNTARY MOVEMENTS. This is because the somatic fibers of the cranial nerves (which carry conscious functions) leave the brain bundled together with some fibers of the AUTONOMIC NERVOUS SYSTEM (which carry subconscious functions).
(See 7.16, 7.17) (The same case may be made for the spinal nerves.)

COMPOSITION OF CRANIAL NERVES

| ENTIRELY SENSORY | | I, II,* VIII |
| ENTIRELY MOTOR | | NONE |

MIXED:

PRIMARILY SENSORY		V
PRIMARILY MOTOR		III, IV, VI, XI, XII
EQUALLY MIXED		VII, IX, X

* Except Edinger-Westphal nucleus of II, motor function (intrinsic eye muscles)

The 12 Pairs of CRANIAL NERVES

COMPOSITION of IMPULSES (SENSORY: ◀ MOTOR: ⟹)

I	OLFACTORY I	
II	OPTIC II	
III	OCULOMOTOR III	
IV	TROCHLEAR IV	SMALLEST
V	TRIGEMINAL V	LARGEST
VI	ABDUCENS VI	(ABDUCENT)
VII	FACIAL VII	
VIII	VESTIBULOCOCHLEAR VIII	
IX	GLOSSOPHARYNGEAL IX	
X	VAGUS X	
XI	ACCESSORY XI	ARISES from BOTH BRAIN and SPINAL CORD
XII	HYPOGLOSSAL XII	

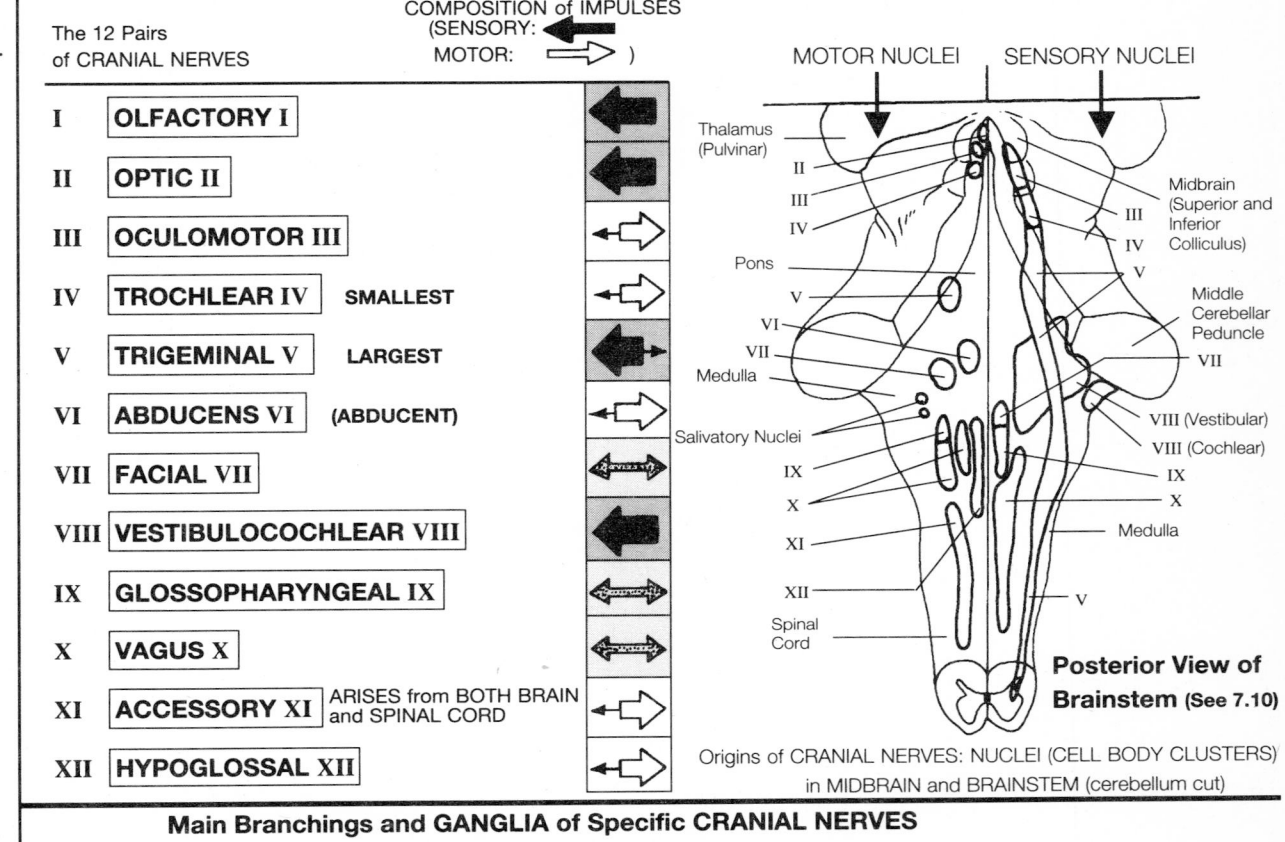

MOTOR NUCLEI   SENSORY NUCLEI

Thalamus (Pulvinar)
Pons
Medulla
Salivatory Nuclei
Spinal Cord

Midbrain (Superior and Inferior Colliculus)
Middle Cerebellar Peduncle
VIII (Vestibular)
VIII (Cochlear)
Medulla

Posterior View of Brainstem (See 7.10)

Origins of CRANIAL NERVES: NUCLEI (CELL BODY CLUSTERS) in MIDBRAIN and BRAINSTEM (cerebellum cut)

Main Branchings and GANGLIA of Specific CRANIAL NERVES

V TRIGEMINAL : 2 roots: A small MOTOR ROOT and a large SENSORY ROOT. The large sensory root enlarges into a swelling called the SEMILUNAR (GASSERIAN) GANGLION, from which arise 3 large branches:
① the OPHTHALMIC BRANCH, ② the MAXILLARY BRANCH, and ③ the MANDIBULAR BRANCH.

VII FACIAL : 2 BRANCHES: INTERMEDIUS SENSORY BRANCH (WITH GENICULATE GANGLION) and a SOMATIC MOTOR BRANCH

VIII VESTIBULOCOCHLEAR : 2 SENSORY BRANCHES: VESTIBULAR BRANCH (with VESTIBULAR GANGLION) and COCHLEAR (AUDITORY) BRANCH

IX GLOSSOPHARYNGEAL : SUPERIOR and INFERIOR GANGLIA and OTIC GANGLION

X VAGUS : SUPERIOR and INFERIOR GANGLIA, and PETROSAL GANGLION

XII HYPOGLOSSAL : 2 MOTOR COMPONENTS: SPINAL MOTOR COMPONENT and CRANIAL MOTOR COMPONENT [BULBAR (MEDULLARY) PORTION]

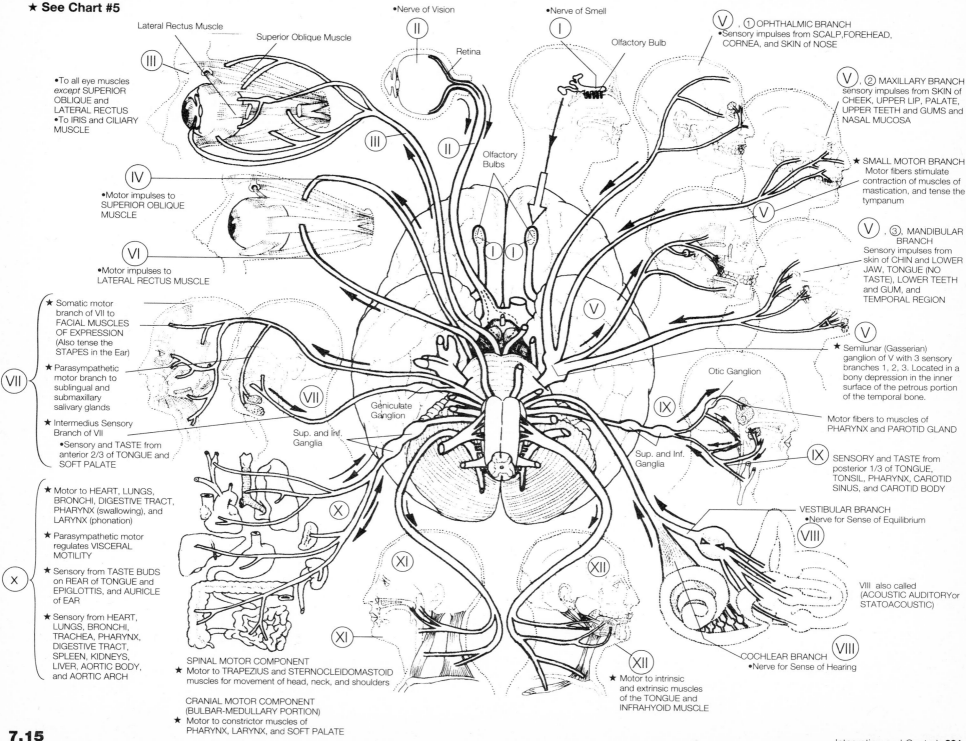

Lateral Rectus Muscle

Superior Oblique Muscle

•Nerve of Vision

•Nerve of Smell

Olfactory Bulb

Retina

Ⓥ , ① OPHTHALMIC BRANCH
•Sensory impulses from SCALP, FOREHEAD,
CORNEA, and SKIN of NOSE

Ⓥ , ② MAXILLARY BRANCH
sensory impulses from SKIN of
CHEEK, UPPER LIP, PALATE,
UPPER TEETH and GUMS and
NASAL MUCOSA

Ⓘ

Ⓘ

Ⓘⓘ

Ⓘⓘⓘ

•To all eye muscles
except SUPERIOR
OBLIQUE and
LATERAL RECTUS
•To IRIS and CILIARY
MUSCLE

Olfactory
Bulbs

★ SMALL MOTOR BRANCH
Motor fibers stimulate
contraction of muscles of
mastication, and tense the
tympanum

Ⓥ

Ⓥ , ③, MANDIBULAR
BRANCH
Sensory impulses from
skin of CHIN and LOWER
JAW, TONGUE (NO
TASTE), LOWER TEETH
and GUM, and
TEMPORAL REGION

•Motor impulses to
SUPERIOR OBLIQUE
MUSCLE

Ⓥ

Ⓥ

Ⓥ

•Motor impulses to
LATERAL RECTUS MUSCLE

★ Semilunar (Gasserian)
ganglion of V with 3 sensory
branches 1, 2, 3. Located in a
bony depression in the inner
surface of the petrous portion
of the temporal bone.

★ Somatic motor
branch of VII to
FACIAL MUSCLES
OF EXPRESSION
(Also tense the
STAPES in the Ear)

★ Parasympathetic
motor branch to
sublingual and
submaxillary
salivary glands

Otic Ganglion

Ⓥ

Ⓘⓧ

Motor fibers to muscles of
PHARYNX and PAROTID GLAND

Ⓥⓘⓘ

Ⓥⓘⓘ

★ Intermedius Sensory
Branch of VII
•Sensory and TASTE from
anterior 2/3 of TONGUE and
SOFT PALATE

Geniculate
Ganglion

Sup. and Inf.
Ganglia

Sup. and Inf.
Ganglia

Ⓘⓧ SENSORY and TASTE from
posterior 1/3 of TONGUE,
TONSIL, PHARYNX, CAROTID
SINUS, and CAROTID BODY

★ Motor to HEART, LUNGS,
BRONCHI, DIGESTIVE TRACT,
PHARYNX (swallowing), and
LARYNX (phonation)

★ Parasympathetic motor
regulates VISCERAL
MOTILITY

Ⓧ

VESTIBULAR BRANCH
•Nerve for Sense of Equilibrium

Ⓥⓘⓘⓘ

★ Sensory from TASTE BUDS
on REAR of TONGUE and
EPIGLOTTIS, and AURICLE
of EAR

Ⓧ

Ⓧⓘ

Ⓧⓘⓘ

VIII also called
(ACOUSTIC AUDITORY or
STATOACOUSTIC)

★ Sensory from HEART,
LUNGS, BRONCHI,
TRACHEA, PHARYNX,
DIGESTIVE TRACT,
SPLEEN, KIDNEYS,
LIVER, AORTIC BODY,
and AORTIC ARCH

Ⓧⓘ

Ⓧⓘⓘ

COCHLEAR BRANCH
•Nerve for Sense of Hearing

Ⓥⓘⓘⓘ

SPINAL MOTOR COMPONENT
★ Motor to TRAPEZIUS and STERNOCLEIDOMASTOID
muscles for movement of head, neck, and shoulders

CRANIAL MOTOR COMPONENT
(BULBAR-MEDULLARY PORTION)
★ Motor to constrictor muscles of
PHARYNX, LARYNX, and SOFT PALATE

★ Motor to intrinsic
and extrinsic muscles
of the TONGUE and
INFRAHYOID MUSCLE

A1 = light grey Spinal Cord = fleshy 1 = yellow 2 = light pink
3 = light green 4 = yellow-green 5 = orange 6–10 = blues
11 = cream 12 = grey 13 = light brown

★ See Chart #6, 7.0, 7.2, 7.17

NERVOUS SYSTEM: AUTONOMIC NERVOUS SYSTEM (ANS)
Sympathetic (Thoracolumbar) Division

★ The AUTONOMIC NERVOUS SYSTEM (ANS)

• The portion of the EFFERENT SYSTEM of the PNS that innervates and regulates SMOOTH MUSCLE (of VISCERAL ORGANS and BLOOD VESSELS), CARDIAC MUSCLE (of the HEART) and GLANDS (SALIVARY, GASTRIC, SWEAT, and the ADRENAL MEDULLA).

• The ANS is entirely MOTOR, often referred to as the VISCERAL EFFERENT NERVOUS SYSTEM.

All of its axons are VISCERAL EFFERENT FIBERS, conveying impulses from the CNS to VISCERAL EFFECTORS [not to SKELETAL MUSCLES, as in the SOMATIC NERVOUS SYSTEM (SNS)]. (See Box)

• Concerned with the control of INVOLUNTARY BODILY FUNCTIONS, but is not self-governing or independent of the CNS. The ANS is itself regulated by BRAIN CENTERS of the CNS, constituting the VISCERAL NERVOUS SYSTEM:

1 ➞ Sensory impulses from VISCERAL EFFECTORS pass over VISCERAL AFFERENT NEURONS (whose cell bodies are in the DORSAL ROOT GANGLIA of the spinal nerves.

2 ➞ These afferent impulses influence mainly the HYPOTHALA-MUS, but also the MEDULLA OBLONGATA and SPINAL CORD. These centers integrate the sensory visceral input with input from higher CNS centers (CEREBRAL CORTEX and LIMBIC SYSTEM).

3 ➞ Appropriate responses are then sent back to the VISCERAL EFFECTOR ORGANS via the ANS.

• The ANS consists of 2 principal divisions:
(organs that receive impulses from both divisions are said to have DUAL INNERVATION)

■ PARA SYMPATHETIC DIVISION (See 7.17)
• Associated with decrease in organ activity, "vegetative reactions" (ex: after a large meal), which conserve and restore body energy: **REST-REPOSE SYSTEM.**

■ SYMPATHETIC DIVISION
• Associated with stimulation to start or increase organ activity. **"FIGHT or FLIGHT RESPONSES"**; expenditure of body energy in response to the need to flee, fight, or be frightened.

★ • Autonomic visceral efferent pathways *always* consist of 2 neurons. The first of the visceral efferent neurons is called a **PREGANGLIONIC NEURON.**

In the SYMPATHETIC DIVISION, the cell bodies of these neurons are exclusive to the LATERAL GRAY HORNS of the SPINAL CORD in all the THORACIC (T1–T12) and the 1st two lumbar segments (L1–L2). (Thus, the SYMPATHETIC DIVISION is also called the THORACOLUMBAR DIVISION.) Their lightly myelinated axons, the PREGANGLIONIC FIBERS, leave the spinal cord through the ventral roots along with the SOMATIC EFFERENT FIBERS of SPINAL NERVES T1–L2. The PREGANGLIONIC FIBERS branch off from the SPINAL NERVE through the WHITE RAMI, and in doing so are collectively referred to as the WHITE RAMI COMMUNICATES.

• In one of 3 routes, each PREGANGLIONIC FIBER courses to an AUTO-NOMIC GANGLION, where it SYNAPSES with the dendrites and cell body of the POSTGANGLIONIC NEURON, the second neuron of the visceral efferent pathway. The POSTGANGLIONIC NEURON lies entirely outside the CNS. Its axon, the POSTGANGLIONIC FIBER, is UNMYELINATED and terminates in a visceral effector organ.

• Each PREGANGLIONIC FIBER via the WHITE RAMI enters the SYMPA-THETIC TRUNK or VERTEBRAL CHAIN GANGLIA, a double series of ganglia that lie in a vertical row on either side of the vertebral column from the base of the skull to the coccyx. (Typically, there are 22 ganglia arranged more or less segmentally along the trunk chain:
3 CERVICAL, 11 THORACIC, 4 LUMBAR, and 4 SACRAL (Most of the preganglionic fibers enter the 11 THORACIC GANGLIA, just ventral to the necks of the corresponding ribs.)

VISCERAL EFFERENT (Motor) Pathways of the ANS
SYMPATHETIC DIVISION: Expends Body Energy

A	SPINAL CORD

A₁	LATERAL GRAY HORNS

■ PREGANGLIONIC NEURONS:

convey efferent impulses from CNS to autonomic ganglia

1	PREGANGLIONIC CELL BODIES

2	PREGANGLIONIC AXONS

■ Types of AUTONOMIC GANGLIA:

3	SYMPATHETIC DOUBLE CHAIN of 22 PARAVERTEBRAL GANGLIA

4	3 COLLATERAL (PREVERTEBRAL) GANGLIA

★ UPON ENTERING the SYMPATHETIC TRUNK:

A PREGANGLIONIC FIBER MAY SYNAPSE in 3 WAYS:
① with the POSTGANGLIONIC CELL BODIES in the FIRST GANGLIA at the LEVEL of ENTRY
② PASS UP or DOWN the trunk, forming the SYMPATHETIC CHAIN of fibers on which the ganglia are strung, SYNAPSING with the POSTGAN-GLIONIC CELL BODIES of the UPPER or LOWER GANGLIA
③ PASS STRAIGHT through the CHAIN at the LEVEL of ENTRY without SYNAPSING in the 1st GANGLION, forming part of the SPLANCH-NIC NERVES. These nerves are PREGANGLIONIC FIBERS from T5–L2, which SYNAPSE with POSTGANGLIONIC CELL BODIES in 3 PREVERTE-BRAL (COLLATERAL) GANGLIA, which are located in front of the vertebral column in the abdominal cavity.
(The POSTGANGLIONIC FIBERS from the Prevertebral ganglia leave them traveling with neighboring blood vessels to the common sites of the ABDOMINAL, PELVIC, and PERINEAL VISCERA)

■ POSTGANGLIONIC NEURONS:
relay impulses from autonomic ganglia to visceral effectors

5	POSTGANGLIONIC CELL BODIES

POSTGANGLIONIC AXONS to 5 Basic Regions:

6	the HEAD AND NECK

7	the THORACIC VISCERA

8	the ABDOMINAL VISCERA

9	the PELVIC/PERINEAL VISCERA

10	the SKIN

COMPARISONS OF DIVISIONS OF EFFERENT SYSTEM OF THE PNS

Feature	SNS (SOMATIC MOTOR SYSTEM)	ANS (AUTONOMIC MOTOR SYSTEM)
Function Mode	Conscious, Voluntary Regulation	Involuntary Regulation (Below Conscious Level)
Effector Organs	Only Skeletal Muscles	Smooth Muscle, Cardiac Muscle, and Glands
Ganglia	None outside CNS	Cell Bodies of Postganglionic Autonomic Fibers in 3 Types of Ganglia (Paravertebral, Prevertebral and Terminal) Outside the CNS
Number of Neurons from CNS to Effector Organ	1	2
Number of Synapses	Fibers Do Not Synapse After They Leave CNS (Go Directly to Skeletal Muscle)	Fibers Synapse Once (at a Ganglion) After They Leave the CNS
Type of Neuromuscular Junction	Specialized Motor End Plate	Nonspecialized Protein Receptor Sites
Nerve Impulse Effect on Effector Cells	Skeletal Muscle: Excitatory Only (All-or-None Principle)	Either Excitatory (+) or Inhibitory (-)
Type of Nerve Fibers	• Thick (9–13 μm) • Myelinated • Fast-conducting	PREGANGLIONIC FIBERS • Thin (3 μm) • Lightly Myelinate • Slow-conducting POSTGANGLIONIC FIBERS • Very Thin (about 1.0 μm) • Unmyelinated • Slow-conducting

★ See Chart #6 for Actions of Parts of the Body Innervated by Postganglionic Fibers (Axons)

Most of the POSTGANGLIONIC FIBERS leaving the SYMPATHETIC TRUNK join the spinal nerves through the GRAY (UNMYELINATED) RAMUS COMMUNICANTES before supplying visceral effectors (However, some pass directly from the trunk to visceral effectors.)

POSTGANGLIONIC AXONS leave the SYMPATHETIC CHAIN at the SUPERIOR CERVICAL GANGLION (the uppermost of the 3 CERVICAL GANGLIA in the cervical portion of the sympathetic trunk, located in the neck anterior to the prevertebral muscles). These axons follow the CAROTID ARTERIES en route to the HEAD and NECK.

POSTGANGLIONIC AXONS to the THORACIC VISCERA leave the upper portions of the SYMPATHETIC CHAIN (the 3 cervical ganglia and T1–T4) and travel directly to the HEART, BRONCHI, LUNGS, and ESOPHAGUS, forming CARDIAC, PULMONARY, and ESOPHAG-EAL PLEXUSES along the way.

POSTGANGLIONIC AXONS to the ABDOMINAL VISCERA have their cell bodies in the 3 sets of PREVERTEBRAL (COLLATERAL) GANGLIA arranged around the major branches of the ABDOMINAL AORTA. PREGANGLIONIC AXONS (T5–L2) pass through the SYMPATHETIC TRUNK without terminating in the trunk, passing through it and forming SPLANCHNIC NERVES, which terminate in the CELIAC (SOLAR) PLEXUS, forming the 3 COLLATERAL GANGLIA. The GREATER SPLANCHNIC NERVE passes to the large, upper CELIAC GANGLION.

POSTGANGLIONIC AXONS to the PELVIS and PERINEUM leave the lower of the 3 COLLAT-ERAL GANGLIA, the INFERIOR MESENTERIC GANGLIA, downward toward the HYPOGAS-TRIC PLEXUS, and from there to the lower intestine, and the pelvic and perineal organs.

POSTGANGLIONIC AXONS to the SKIN leave the SYMPATHETIC CHAIN via the GRAY RAMI COMMUNICANTES at each and every level of the spinal cord to reach BLOOD VESSELS associated with skeletal muscle, arrector pili hair muscles, and sweat glands.

★ Start by coloring all the LATERAL GRAY HORN segments on th SPINAL CORD: These segments are labeled A1: (T1-L2)

★ Then color all the little circles in these horizontal segments (labeled 1): these are the CELL BODIES of the PREGANGLIONIC NEURONS

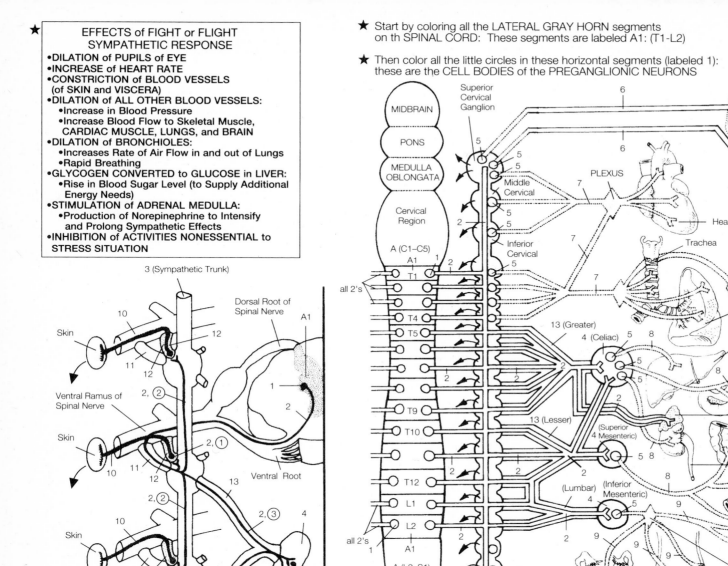
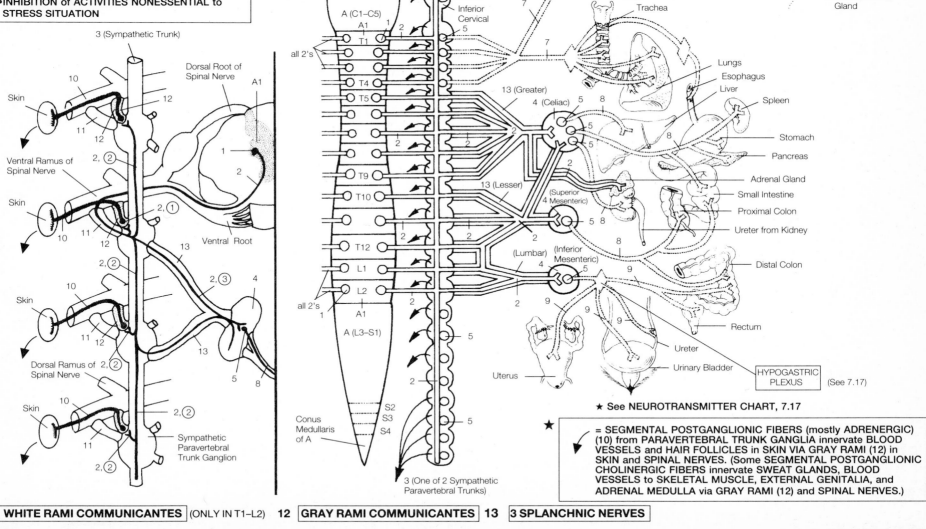

★
EFFECTS of FIGHT or FLIGHT SYMPATHETIC RESPONSE
- DILATION of PUPILS of EYE
- INCREASE of HEART RATE
- CONSTRICTION of BLOOD VESSELS (of SKIN and VISCERA)
- DILATION of ALL OTHER BLOOD VESSELS:
  - Increase in Blood Pressure
  - Increase Blood Flow to Skeletal Muscle, CARDIAC MUSCLE, LUNGS, and BRAIN
- DILATION of BRONCHIOLES:
  - Increases Rate of Air Flow in and out of Lungs
  - Rapid Breathing
- GLYCOGEN CONVERTED to GLUCOSE in LIVER:
  - Rise in Blood Sugar Level (to Supply Additional Energy Needs)
- STIMULATION of ADRENAL MEDULLA:
  - Production of Norepinephrine to Intensify and Prolong Sympathetic Effects
- INHIBITION of ACTIVITIES NONESSENTIAL to STRESS SITUATION

3 (Sympathetic Trunk)

Dorsal Root of Spinal Nerve

A1

Skin

10

12

12

Ventral Ramus of Spinal Nerve

2, ②

11

12

1

2

Skin

2, ①

10

11

12

Ventral Root

13

Skin

2, ③

4

10

11

12

13

Dorsal Ramus of Spinal Nerve

2, ②

5

8

Skin

10

Sympathetic Paravertebral Trunk Ganglion

11

2, ②

2, ②

MIDBRAIN

PONS

MEDULLA OBLONGATA

Cervical Region

A (C1–C5)

A1

Superior Cervical Ganglion

5

5

5

Middle Cervical

5

5

Inferior Cervical

5

A1

1

2

T1

all 2's

T4

T5

T9

T10

T12

L1

L2

all 2's

1

A1

A (L3–S1)

2

2

2

2

2

2

2

2

5

5

5

5

Conus Medullaris of A

S2

S3

S4

3 (One of 2 Sympathetic Paravertebral Trunks)

6

6

6

6

6

Eye

Nasal Mucosa

Parotid Gland

Sublingual Gland

Submandibular Gland

PLEXUS

7

7

7

Heart

Trachea

13 (Greater)

4 (Celiac)

5

8

5

5

5

2

Lungs

Esophagus

Liver

Spleen

Stomach

Pancreas

Adrenal Gland

Small Intestine

Proximal Colon

Ureter from Kidney

13 (Lesser)

(Superior 4 Mesenteric)

5

8

8

2

(Lumbar)

4

(Inferior Mesenteric)

5

2

2

9

9

9

9

9

9

Distal Colon

Rectum

Ureter

Urinary Bladder

Uterus

HYPOGASTRIC PLEXUS (See 7.17)

★ See NEUROTRANSMITTER CHART, 7.17

★ ➤ = SEGMENTAL POSTGANGLIONIC FIBERS (mostly ADRENERGIC) (10) from PARAVERTEBRAL TRUNK GANGLIA innervate BLOOD VESSELS and HAIR FOLLICLES in SKIN VIA GRAY RAMI (12) in SKIN and SPINAL NERVES. (Some SEGMENTAL POSTGANGLIONIC CHOLINERGIC FIBERS innervate SWEAT GLANDS, BLOOD VESSELS to SKELETAL MUSCLE, EXTERNAL GENITALIA, and ADRENAL MEDULLA via GRAY RAMI (12) and SPINAL NERVES.)

11 WHITE RAMI COMMUNICANTES (ONLY IN T1–L2) **12 GRAY RAMI COMMUNICANTES** **13 3 SPLANCHNIC NERVES**

NERVOUS SYSTEM: AUTONOMIC NERVOUS SYSTEM (ANS)
Parasympathetic (Craniosacral) Division

A1 = light grey Spinal Cord = fleshy 1 = yellow 2 = orange
III,VII,IX,X = warm reds 3 = light pink Brainstem = light creams
4–7 = greens 9–13 = blues 8 = light purple

★ See Chart #6, 7.0, 7.2, 7.16

•It is noteworthy that in the SYMPATHETIC DIVISION there are *no* PREGANGLIONIC FIBERS that arise from the BRAINSTEM or SACRAL AREA of the spinal cord. However, this is what occurs in the PARASYM-PATHETIC DIVISION of the ANS: the PREGANGLIONIC CELL BODIES are found in NUCLEI of the BRAINSTEM (MIDBRAIN, PONS and MEDULLA OBLONGATA) and the LATERAL GRAY HORNS of the 2nd–4th SACRAL SEGMENTS of the SPINAL CORD. (Thus, the PARASYMPATHETIC DIVISION is also referred to as the CRANIOSACRAL DIVISION of the ANS.)

•The PREGANGLIONIC AXONS emerge as either a part of a CRANIAL NERVE (limited to III, VII, IX,X) or as part of the VENTRAL ROOT of a SPINAL SACRAL NERVE.

•These PREGANGLIONIC AXONS synapse in GANGLIA that are located next to or *actually within*, the visceral effector organs. These GANGLIA are referred to as TERMINAL GANGLIA and are exclusive to the PARASYMPATHETIC DIVISION.

•Recall that SYMPATHETIC innervation of the SKIN originates from each and every level of the spinal cord. Note that, except for 3 sacral spinal nerves, PARASYMPATHETIC FIBERS do *not* travel within SPINAL NERVES. Thus, CUTANEOUS EFFECTORS in the SKIN (BLOOD VESSELS, ARRECTOR PILI HAIR MUSCLES, and SWEAT GLANDS) do *not* receive PARASYMPATHETIC POSTGANGLIONIC INNERVATION and are *not* found in the long peripheral nerves as are sympathetic postganglionic fibers.

CRANIAL PARASYMPATHETIC OUTFLOW
■ 5 COMPONENTS: 4 PAIRS OF HEAD, or CRANIAL, GANGLIA, plus the PLEXUSES associated with the VAGUS NERVE (X).

•PREGANGLIONIC AXONS of CRANIAL NERVES III, VII, and IX synapse in the CRANIAL GANGLIA. *Short* POSTGANGLIONIC AXONS arise from these ganglia to supply specific head structures. (See box)

•OCULOMOTOR III ⟶ CILIARY GANGLIA (located near the back of the orbit lateral to each optic nerve)

•FACIAL VII: innervates 2 ganglia:
(1) ⟶ PTERYGOPALATINE GANGLIA (lateral to a sphenopalatine foramen in the lateral wall of the nasal cavity near the nasopharynx)
(2) ⟶ SUBMANDIBULAR GANGLIA (deep to the submandibular salivary gland under the mandible near its duct)

•GLOSSOPHARYNGEAL (IX)
⟶ OTIC GANGLIA (just below the foramen ovale in posterior margin of great sphenoidal wing of the sphenoid bone)

■ LONG PREGANGLIONIC AXONS from the VAGUS NERVE (X) pass down the neck through the thorax to the abdomen (to the level of the descending colon), branching off into many plexuses along the way. (Within the plexuses they are joined with POSTGANGLIONIC SYMPA-THETIC AXONS destined for the same visceral effector.)
In the THORAX: Vagal fibers join with sympathetic fibers in the CARDIAC and PULMONARY PLEXUSES. From here, postganglionic axons travel to the HEART, LUNGS, and TRACHEA.
In the ABDOMEN: Continuing vagal fibers mix with sympathetics around the 3 prevertebral ganglia of the CELIAC (SOLAR) PLEXUS. (See 7.16) From here, postganglionic axons follow arteries to the abdominal organs they supply.

SACRAL PARASYMPATHETIC OUTFLOW
•Preganglionic fibers arise from cell bodies in the lateral gray horns of S2-S4 and leave through the ventral roots, joining the spinal nerves. These fibers then bend quickly away toward the pelvic area collectively forming the PELVIC SPLANCHNIC NERVES. The nerves form a HYPOGASTRIC PLEXUS on the rectum with the INFERIOR MESENTERIC GANGLIA (See 7.16). These axons then follow the arteries to the viscera

they supply.
VISCERAL EFFERENT (MOTOR) PATHWAYS of the ANS

PARASYMPATHETIC DIVISION: Conserves Body Energy

A SPINAL CORD	BRAINSTEM		
A₁ LATERAL GRAY HORNS	B MIDBRAIN	C PONS	D MEDULLA

■ PREGANGLIONIC NEURONS

1 PREGANGLIONIC CELL BODIES

Long PREGANGLIONIC AXONS of CRANIAL NERVES:

III   **OCULOMOTOR III**

VII   **FACIAL VII**

IX   **GLOSSOPHARYNGEAL IX**

X   **VAGUS X** : 80% OF CRANIOSACRAL FLOW

Long PREGANGLIONIC SACRAL AXONS

3 S2, S3, S4

3a PELVIC SPLANCHNIC NERVES

■ AUTONOMIC (TERMINAL) GANGLIA: NEAR or WITHIN VISCERAL EFFECTORS

FOUR CRANIAL GANGLIA to the HEAD

4 CILIARY GANGLION

5 PTERYGOPALATINE GANGLION

6 SUBMANDIBULAR GANGLION

7 OTIC GANGLION

Microscopic GANGLIA in Multiple PLEXUSES Associate with the VAGUS NERVE (X) and SACRAL PREGANGLIONIC AXONS

8 INTRAMURAL GANGLIA

STRUCTURAL FEATURES of SYMPATHETIC and PARASYMPATHETIC DIVISIONS

SYMPATHETIC	PARASYMPATHETIC
THORACOLUMBAR nerve outflow (T1–/L2)	CRANIOSACRAL nerve outflow (III, VII, IX, X, S2–S4)
GANGLIA: •Paravertebral (Double Chain Sympathetic Trunks) •Prevertebral (Collateral)	GANGLIA •Terminal
•Far from Visceral Effector Organs (Close to the CNS)	•Near or within Visceral Effector Organs
1 Preganglionic Fiber Synapses with *20 or more* Postganglionic Neurons That Pass to *Many* Visceral Effectors	1 Preganglionic Fiber Synapses with *4–5* Postganglionic Neurons That Pass to *1* Visceral Effector
Comprehensive Distribution throughout the Entire Body, Including the Skin	Limited Distribution Primarily to the Head, and to Thoracic, Abdominal, and Pelvic Viscera
Postganglionic Fibers Relatively Long	Postganglionic Fibers Relatively Short

■ SHORT POSTGANGLIONIC NEURONS

2 POSTGANGLIONIC CELL BODIES

Short POSTGANGLIONIC AXONS to: (*NONE* to the SKIN!)

9 The EYE (CILIARY and IRIS MUSCLES)

10 The LACRIMAL EYE GLANDS, PALATE, NASAL MUCOSA, and PHARYNX

11 The SALIVARY GLANDS (SUBLINGUAL and SUBMANDIBULAR)

12 The PAROTID SALIVARY GLAND

13 The THORACIC and ABDOMINAL PELVIC, and PERINEAL VISCERA

★ **DUAL INNERVATION of VISCERAL EFFECTORS**

•Most visceral effector organs receive fibers (axons) from *both* the SYMPATHETIC and PARASYMPATHETIC DIVISIONS of the ANS. Impulses from one division *stimulate* the organ's activities, and impulses from the other division *inhibit* the organ's activities.

•The STIMULATING DIVISION is *not always the SYMPATHETIC,* as might be assumed. In some instances, (for example in the abdominal viscera) the PARASYMPATHETIC impulses INCREASE DIGESTIVE ACTIVITIES, while SYMPATHETIC impulses INHIBIT them. Under normal conditions, PARASYMPATHETIC IMPULSES to DIGESTIVE GLANDS and SMOOTH MUSCLE of DIGESTIVE ORGANS *dominate* SYMPATHETIC IMPULSES. (DIGESTION and ABSORPTION of NUTRIENTS, DRAINAGE of the URINARY BLADDER, REDUCTION in HEART and RESPIRATORY RATES, and RUSHING of BLOOD to DIGESTIVE VISCERA are all activities under the control of PARASYMPATHETIC STIMULATION, which RESTORES and CONSERVES BODY ENERGY in a REST-REPOSE SYSTEM.

•Integration of the activities of the 2 ANS divisions helps to *maintain homeostasis.* During maintenance of homeostasis, the *primary function* of the SYMPATHETIC DIVISION is to *counteract* the PARASYMPATHETIC activity just enough to carry out normal energy-requiring processes.

•During STRESSFUL SITUATIONS, the SYMPATHETIC activity (which EXPENDS BODY ENERGY) DOMINATES the PARASYMPATHETIC ACTIVITY. (See 7.16, lower page)

★ **CLASSIFICATION OF AUTONOMIC AXONS BASED ON PRODUCTION OF NEUROTRANSMITTER SUBSTANCE:**

■ CHOLINERGIC AXONS release ACETYLCHOLINE (Ach). Effects are short-lived and local.

■ ADRENERGIC AXONS release NOREPINEPHRINE (NE), also called (NOR) ADRENALINE or SYMPATHIN. Effects are long-lasting and widespread.

SYMPATHETIC	PARASYMPATHETIC
PREGANGLIONIC AXONS at SYNAPSES in GANGLIA	
CHOLINERGIC	CHOLINERGIC
POSTGANGLIONIC AXONS at NEUROEFFECTOR JUNCTIONS	
MOSTLY ADRENERGIC (CHOLINERGIC EXCEPTIONS) •Skin •Sweat Glands •External Genitalia •Adrenal Medulla •Smooth Muscles of Blood Vessels to Skeletal Muscle	CHOLINERGIC • None to skin

★ In the head region, preganglionic parasympathetic axons join with preganglionic somatic axons in 3 pairs of cranial nerves (III, IV, IX).

★ The PINHEAD-SIZED CRANIAL GANGLIA innervated by CRANIAL NERVES III, IV, and VII are located just *outside* the VISCERAL ORGANS they supply.

HEAD

Lacrimal Gland

Palate

Pharynx

★ The MICROSCOPIC INTRAMURAL GANGLIA (8) associated with VAGUS X and S2–S4 PREGANGLIONIC AXONS are located *within* the VISCERAL ORGANS they supply (in the SUBMUCOSAL and MUSCULAR TUNIC LAYERS).

(8) Ganglia in Superior Cardiac and Deep Cardiac Plexuses

THORACIC VISCERA

(8) Ganglia in the Pulmonary Plexus

★ The very short PARASYMPATHETIC POST GANGLIONIC AXONS from INTRAMURAL GANGLIA do *not* supply sweat glands and arrector pilli muscles in the skin, or nearby blood vessel musculature. They supply smooth muscle and glands of (13); active in decreased heart rate, bronchoconstriction, and increased secretion/muscular activity.

ABDOMINAL VISCERA

Cervical Region

A (C1–C5)

T1

T12

L1

L2

A (L3–S1)

Inferior Mesenteric Ganglion

Conus Medullaris (of A)

A1

S2

S3

S4

Hypogastric Plexus

(Pelvic Splanchnic Nerves)

PELVIC / PERINEAL VISCERA

NERVOUS SYSTEM: SPECIAL SENSORY ORGANS
Gustatory Sense: Taste
Olfactory Sense: Smell } Chemical Sense

A = fleshy B = yellow 1 = pink 2 = red
3 = creamy 4 = red-orange Modalities = cool colors
D = light blue E = creamy F = pink G = yellow
H = orange f = green 7 = purple e = red
★ See Chart #7

★ **One way of classifying SENSATIONS** is according to the SIMPLICITY or COMPLEXITY of both the RECEPTOR and NEURAL PATHWAYS involved.
   ■ **GENERAL SENSES**—RECEPTORS and NEURAL PATHWAYS are SIMPLE in structure, and WIDESPREAD throughout the BODY in the SKIN, SUBCUTANEOUS TISSUE, MUSCLES, TENDONS, JOINTS, and VISCERA. (See 3.3 for CUTANEOUS SENSATIONS)
   ■ **SPECIAL SENSES**—RECEPTORS are COMPLEX and NEURAL; PATHWAYS are EXTENSIVE in structure, and LOCALIZED in one or two specific areas of the body. We will discuss the special senses of TASTE, SMELL, SIGHT, HEARING, and EQUILIBRIUM in the following six sections.

★ **GUSTATORY SENSATIONS of TASTE**
   •TASTE is a CHEMICAL SENSE (i.e., GUSTATORY RECEPTOR CELLS respond to CHEMICAL STIMULI, which must be in solution, usually SALIVA). The essential organ of taste is the TONGUE and the specialized epithelial receptor cells are most numerous on the DORSUM of the TONGUE (but are also present on the PALATE, OROPHARYNX, EPIGLOTTIS, and LARYNX)
   •A TASTE BUD is a cluster of 40–68 GUSTATORY RECEPTOR CELLS encapsuled by SUPPORTING CELLS. The taste buds are found on connective tissue projections on the upper surface of the tongue called PAPILLAE.
   •Each GUSTATORY CELL contains a hairlike dendritic ending (GUSTATORY HAIR) that projects to the external surface through an opening in the bud called a TASTE PORE CANAL.
   •There are 4 types of TASTE BUDS (all structurally similar), but each type responds more strongly to one of only 4 MODALITIES of TASTE (SWEET, SOUR, BITTER, SALT) on specific areas on the tongue.
   •Each GUSTATORY CELL is innervated by an afferent neuron. The afferent pathway to the brain (via the medulla and thalamus) involves mainly 2 cranial nerves (VII,IX).
Impulses from one side of the tongue pass to the TASTE CENTER of the POSTCENTRAL GYRUS in the PARIETAL LOBE on the *opposite* side of the CEREBRAL CORTEX.
   •TASTE DISCRIMINATION is actually a combination of TASTE SENSATIONS, FOOD TEXTURE, TEMPERATURE and to a significant degree the sensation of *smell*.

★ **OLFACTORY SENSATIONS OF SMELL**
   •SMELL is a CHEMICAL SENSE (i.e., OLFACTORY RECEPTOR CELLS respond to CHEMICAL STIMULI IN SOLUTION). However, the SUBSTANCE OF SMELL must originally be airborne (in a GASEOUS STATE), and then dissolve in solution.
   •The exclusive organ of SMELL is the NOSE (also serves as the main air passage to the RESPIRATORY SYSTEM).
   •The OLFACTORY RECEPTORS are located in the OLFACTORY NASAL EPITHELIUM within the roof of the NASAL CAVITY in the NASAL CLEFT an upper space on both sides of the medial NASAL SEPTUM.
   •The OLFACTORY RECEPTORS are BIPOLAR SENSORY NEURONS whose cell bodies lie between supporting SUSTENTACULAR CELLS and regular PSEUDOSTRATIFIED COLUMNAR EPITHELIUM. The rounded tips of the receptor cell bodies project into the mucus of the nasal cavity. The mucus is secreted by BOWMAN'S GOBLET CELLS and lines the epithelium. Several dendritic hairs (CILIA) from each tip project into the cavity.
   •Gaseous substances drawn upwards dissolve into solution in the mucus, and through the CILIA, stimulate the OLFACTORY RECEPTOR CELLS.
Unmyelinated OLFACTORY AXONS unite to form the OLFACTORY CRANIAL NERVES (I), which pass through the FORAMINA of the CRIBRIFORM PLATE OF THE ETHMOID BONE, and synapse in the paired OLFACTORY BULBS with the dendrites of MITRAL NEURONS. Axons of these MITRAL NEURONS form the OLFACTORY TRACT, which conveys impulses to the brain for odor translation.

GUSTATORY SENSE: TASTE

A | TONGUE |

   1 | CIRCUMVALLATE PAPILLAE | ⎫ contain
                                  ⎬ the
   2 | FUNGIFORM PAPILLAE | ⎭ TASTE BUDS

B | TASTE BUDS |

   3 | ENCAPSULATING SUPPORTING CELLS |

   4 | GUSTATORY RECEPTOR CELLS |

      a | DENDRITIC GUSTATORY HAIRS |

      b | TASTE PORE CANAL |

C | SENSORY NERVE FIBERS |

V | BRANCH OF TRIGEMINAL NERVE |

(VII) | FACIAL CRANIAL NERVE (VII) (CHORDA TYMPANI BRANCH) |

(IX) | GLOSSOPHARYNGEAL CRANIAL NERVE (IX) |

(X) | VAGUS CRANIAL NERVE (X) |

★ See Chart #7: CLASSIFICATION OF SENSATIONS
   •Survival of the human organism depends on the ability to sense the environment and make the necessary homeostatic adjustments:
   •A SENSATION is the arrival of a sensory impulse to the brain.
   •A PERCEPTION is the INTERPRETATION (CONSCIOUS REGISTRATION) of a SENSATION occurring in the CEREBRAL CORTEX. (Thus, you TASTE, SMELL, SEE and HEAR with your brain!)
4 PREREQUISITES for PERCEPTION of a SENSATION:
   ① PRESENCE of a STIMULUS (capable of initiating a response by the nervous system)
   ② A RECEPTOR or SENSE ORGAN that PICKS UP THE STIMULUS and CONVERTS it into a NERVE IMPULSE. (Acts as ENERGY FILTER designed to respond to a limited, narrow range of energy.)/
   ③ CONDUCTION of a NERVE IMPULSE from the receptor or sense organ to the brain along a nerve pathway.
   ④ TRANSLATION of the IMPULSE in the form of a PERCEPTION in a SPECIFIC REGION OF THE BRAIN.

OLFACTORY SENSE: SMELL

D | NASAL CLEFT in NASAL CAVITY |

E | CRIBRIFORM PLATE of ETHMOID BONE |

F | OLFACTORY NASAL EPITHELIUM | MUCOSA

   5 | SUSTENTACULAR/ SUPPORTING CELLS |

   6 | BOWMAN'S GLANDULAR GOBLET CELLS |

   7 | OLFACTORY RECEPTOR CELLS/ BIPOLAR SENSORY NEURONS |

      c | DENDRITIC OLFACTORY HAIRS (CILIA) |

      d | OLFACTORY CELL BODIES |

      e | UNMYELINATED OLFACTORY AXONS |

(I) | OLFACTORY CRANIAL NERVE FIBERS (I) |

G | OLFACTORY BULB |

   f | MITRAL CELL BODIES |

H | OLFACTORY TRACT | AXONS of MITRAL NEURONS

OLFACTORY CENTER of CEREBRAL CORTEX

   8 | LATERAL and MEDIAL OLFACTORY STRIAE |

   9 | OLFACTORY CENTER |

   10 | THALAMIC CENTERS |

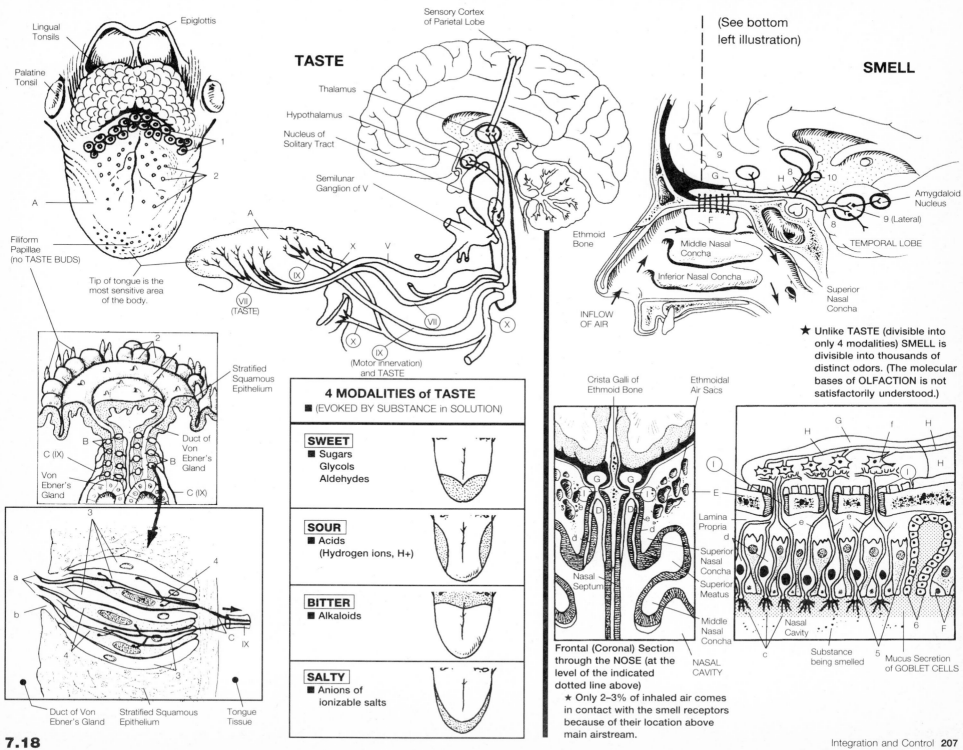

NERVOUS SYSTEM: SPECIAL SENSORY ORGANS
Visual Sense: Sight
Associated Structures of the Eye

★ 80% of all knowledge that the human being acquires is assimilated in through the eyes.

★ The 2 EYES are the ORGANS OF SIGHT or VISION that bend and focus incoming LIGHT WAVES onto SPECIALIZED PHOTORECEPTOR CELLS (RODS and CONES) at the BACK of EACH EYE. The photoreceptor cell layer (RETINA) is encased within a fibrous globe (SCLERA), which becomes transparent in front (CORNER). The EYEBALLS thus formed are subserved by many ASSOCIATED STRUCTURES that either PROTECT the eyeball or provide EYE MOVEMENT.

★ PROTECTION OF THE EYEBALL:
■ BONY ORBIT ■ EYEBROW ■ EYELIDS
■ FACIAL MUSCLES ■ EYELASHES ■ CONJUNCTIVA
■ LACRIMAL APPARATUS

■ **BONY ORBIT**—The hidden posterior 4/5 of the eyeball is encased in a bony socket, the ORBITAL CAVITY (ORBIT), in the skull. Seven bones of the skull form the orbit (See 4.4). A thick layer of AREOLAR and ADIPOSE TISSUE forms a compact cushion between the bone and the eyeball.
The anterior 1/5 of the eyeball is exposed and protected from injury by the following structures:
■ **EYEBROWS**—Short, thick hairs transversely oriented along the superior orbital ridges of the skull directed laterally. They *shade* the eye from the sun, and prevent *perspiration* and falling objects from entering the eye. Eyebrow movement is caused by upper portion of ORBICULARIS OCULI MUSCLE (and a portion of the CORRUGATOR MUSCLE).
■ **EYELIDS (PALPEBRAE)**—Upper and lower eyelids shade eyes during sleep, protect the eyes from excessive light (sun) and foreign objects, and protect the eye from desiccation by reflexively blinking every 7 seconds and moving fluid over the anterior eyeball surface. The upper eyelid is more movable than the lower. CONTRACTION of ORBICULARIS OCULI MUSCLE closes the eyelids over the eye. CONTRACTION of the special LEVATOR PALPEBRAE SUPERIORIS MUSCLE elevates the upper eyelid exclusively to expose the eye. The space that exposes the eyeball between the two eyelids is the PALPEBRAL FISSURE.
★ The EYELIDS develop as reinforced folds of skin. From superficial to deep, each eyelid consists of: EPIDERMIS, DERMIS, SUBCUTANEOUS AREOLAR CONNECTIVE TISSUE, ORBICULARIS OCULI MUSCLE FIBERS, a TARSAL PLATE, TARSAL GLANDS, and a CONJUNCTIVA:
•**TARSAL PLATE**—Thick fold of dense connective tissue that forms most of the inner wall of the eyelids, maintains their shape, and gives them support.
•**TARSAL MEIBOMIAN GLANDS**—Specialized elongated glands embedded in grooves on the deep surface of each tarsal plate in a row. The ducts of these glands open onto the edge of the eyelids. Their oily secretion keeps the eyelids from adhering to each other.
•**CONJUNCTIVA**—A thin, mucus-secreting epithelial membrane. Lines the interior surface of each eyelid as the PALPEBRAL CONJUNCTIVA and reflects from the eyelids onto the eyeball (to the periphery of the CORNEA) as the BULBAR, or OCULAR, CONJUNCTIVA. When the eyelids are closed, a CONJUNCTIVAL SAC forms to protect the eyeball as a barrier against foreign objects (ex: contact lens).
■ **EYELASHES**—A row of short, thick hairs projecting from the border of each lid protects the eye from airborne objects. GLANDS OF ZEIS pour a lubricating fluid into the hair follicles at their base.

1 **BONY ORBIT (ORBITAL CAVITY)**

2 **EYEBROWS**

EYELIDS and EYELASHES

3 **EYELIDS (PALPEBRAE)**

4 **EYELASHES**

a **SEBACEOUS GLANDS of ZEIS**

5 **TARSAL PLATES**

b **TARSAL MEIBOMIAN GLANDS**

CONJUNCTIVA

6 **PALPEBRAL CONJUNCTIVA**

7 **BULBAR (OCULAR) CONJUNCTIVA**

8 **CONJUNCTIVAL SAC**

9 **PALPEBRAL FISSURE:**
SPACE between EYELIDS

COMMISSURES (CANTHI):
ANGLES OF EYELIDS

10 **LATERAL CANTHUS** (NARROW)

11 **MEDIAL CANTHUS** (BROAD)

12 **LACRIMAL CARUNCLE**
:SMALL, REDDISH ELEVATION in the MEDIAL CANTHUS

FACIAL MUSCLES of the EYELIDS

13 **ORBICULARIS OCULI MUSCLE**

14 **LEVATOR PALPEBRAE SUPERIORIS MUSCLE**

EXTRINSIC OCULAR MUSCLES of the EYEBALL:
EYE MOVEMENT

4 RECTUS MUSCLES

SR **SUPERIOR RECTUS M.** MR **MEDIAL RECTUS M.**

LR **LATERAL RECTUS M.** IR **INFERIOR RECTUS M.**

2 OBLIQUE MUSCLES

SO **SUPERIOR OBLIQUE M.** IO **INFERIOR OBLIQUE M.**

LACRIMAL APPARATUS

15 **LACRIMAL GLAND** 18 **LACRIMAL CANALS**

16 **LACRIMAL DUCTS** 19 **LACRIMAL SAC**

17 **PUNCTA LACRIMALIA** 20 **NASOLACRIMAL DUCT**

★ The LACRIMAL APPARATUS refers to the group of structures that MANUFACTURE and SECRETE LACRIMAL FLUID (TEARS) and DRAIN them away (through a SERIES OF DUCTS) into the NASAL CAVITY.
■ **LACRIMAL GLAND**
•A compound, tubuloacinar gland, the size and shape of an almond.
•Located in the SUPERIOR LATERAL portion of each bony ORBIT.
•Produces 1 milliliter of LACRIMAL FLUID (TEARS) per gland per day.
■ **LACRIMAL DUCTS**—6–12 EXCRETORY ducts empty the tears into the CONJUNCTIVAL SAC of the UPPER EYELID. The TEARS are an aqueous, mucus secretion that moistens and lubricates the conjunctival sac, and also protects the sac from infection with a bactericidal enzyme, LYSOZYME.
■ **PUNCTA LACRIMALIA**—With each blink of the eyelids, the tears spread medially and downward over the BULBAR CONJUNCTIVA to two small pore openings (on both sides of the LACRIMAL CARUNCLE) in each PAPILLA of the eyelid, at the MEDIAL CANTHUS. These 2 pores are the PUNCTA LACRIMALIA.
■ **LACRIMAL CANALS**—Tears are conveyed through 2 ducts, the LACRIMAL CANALS, that are located in the LACRIMAL GROOVES of the LACRIMAL BONES of the ORBIT.
■ **LACRIMAL SAC**—From the 2 lacrimal canals, the tears are conveyed into the LACRIMAL SAC, a superior expanded portion of the NASOLACRIMAL DUCT.
■ **NASOLACRIMAL DUCT**—The final canal that empties the tears into the INFERIOR MEATUS of the NASAL CAVITY. (Thus, you can see why you have to blow your nose when you cry!)

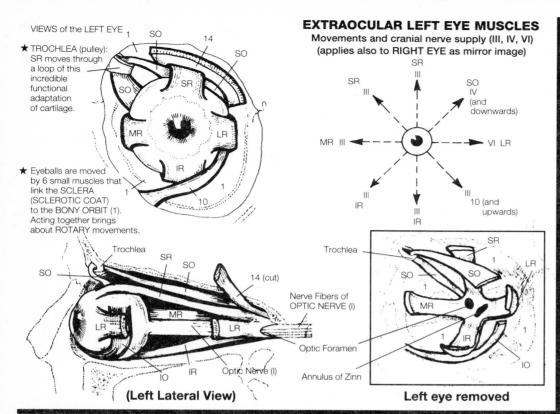

EXTRAOCULAR LEFT EYE MUSCLES
Movements and cranial nerve supply (III, IV, VI)
(applies also to RIGHT EYE as mirror image)

VIEWS of the LEFT EYE

★ TROCHLEA (pulley): SR moves through a loop of this incredible functional adaptation of cartilage.

★ Eyeballs are moved by 6 small muscles that link the SCLERA (SCLEROTIC COAT) (1) to the BONY ORBIT (1). Acting together brings about ROTARY movements.

SO
SR
SO
MR
LR
IR
14

(Left Lateral View)

Trochlea
SR
SO
SO
MR
LR
IO
IR
14 (cut)
Nerve Fibers of OPTIC NERVE (I)
Optic Nerve (I)

Movement diagram:
SR III
SR III — SO IV (and downwards)
MR III ← → VI LR
IR III — III 10 (and upwards)
III IR

Left eye removed

Trochlea
SR
SO
SO
MR
LR
IR
IO
1
Optic Foramen
Annulus of Zinn

LACRIMAL APPARATUS

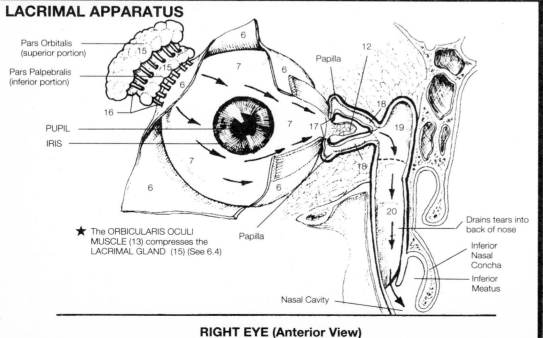

Pars Orbitalis (superior portion)
Pars Palpebralis (inferior portion)
15
15
16
PUPIL
IRIS
6
7
6
7
7
6
6
12
Papilla
18
17
18
19
20
Papilla

★ The ORBICULARIS OCULI MUSCLE (13) compresses the LACRIMAL GLAND (15) (See 6.4)

Drains tears into back of nose
Inferior Nasal Concha
Inferior Meatus
Nasal Cavity

RIGHT EYE (Anterior View)

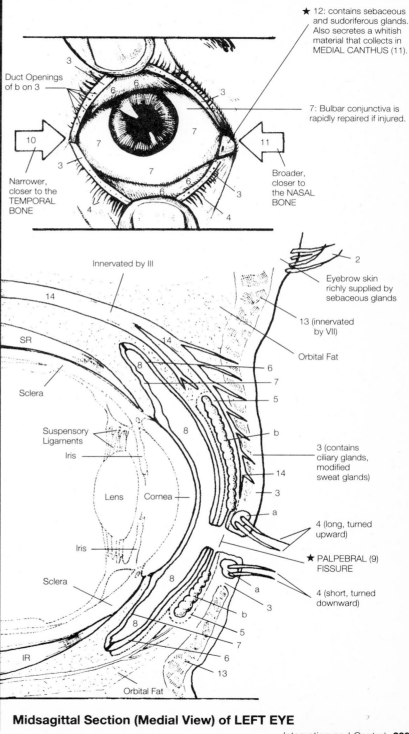

★ 12: contains sebaceous and sudoriferous glands. Also secretes a whitish material that collects in MEDIAL CANTHUS (11).

Duct Openings of b on 3
3
6
6
3
7
7
7
3
6
4
4
6
10
11

7: Bulbar conjunctiva is rapidly repaired if injured.

Narrower, closer to the TEMPORAL BONE

Broader, closer to the NASAL BONE

Innervated by III
14
SR
8
Sclera
Suspensory Ligaments
Iris
Lens
Cornea
Iris
Sclera
IR
Orbital Fat
8
8
8
8
14
6
7
5
b
a
5
3
6
7
13

2
Eyebrow skin richly supplied by sebaceous glands
13 (innervated by VII)
Orbital Fat
3 (contains ciliary glands, modified sweat glands)
3
4 (long, turned upward)
★ PALPEBRAL (9) FISSURE
4 (short, turned downward)

Midsagittal Section (Medial View) of LEFT EYE

NERVOUS SYSTEM: SPECIAL SENSORY ORGANS
Visual Sense: Sight
Structure of the Eyeball

4 = creamy 5 = light gray 6 = pink
7 = orange 8 = light brown 9 = yellow
Chambers = light blues and greens
23 = red 24 = blue

★ The globelike structure of the eyeball is approximately 25 mm (1 in.) in diameter. Only the anterior 1/5 is exposed, the remainder being recessed and protected by the bony orbit in which it snugly fits. The eyeball is composed of 3 BASIC COATS, or TUNICS, from the outside in:
- ■ OUTER FIBROUS TUNIC
- ■ MIDDLE VASCULAR and MUSCULAR TUNIC: UVEA
- ■ INNER NERVOUS TUNIC: RETINA

These 3 tunic layers enclose 2 liquid-filled cavities, basically separated by the LENS, which in itself is not part of any tunic.

★ **OUTER FIBROUS TUNIC** has 2 parts: SCLERA and CORNEA
■ **SCLERA**: Opaque, avascular, outer coat, the 4/5 posterior portion, a white coat of tough fibrous tissue (composed of tightly bound elastic and collagenous fibers). It gives shape to the eyeball and protects its inner parts. [Its posterior surface is pierced by the OPTIC NERVE (II).]
■ **CORNEA**: Clear, transparent (due to tightly packed avascular, dense connective tissue), and CONVEX (permits passage and bending of incoming light waves). Its curvature is greater than the remainder of the bulb, enabling it to function as an important refractive medium. It is composed of 5 layers arranged in unusually regular patterns. The outer layer, the BULBAR (CORNEAL) EPITHELIUM, is continuous with the outer conjunctiva of the SCLERA.

★ **MIDDLE VASCULAR TUNIC**: UVEA is composed of 3 parts: CHOROID, CILIARY BODY, and IRIS.
■ **CHOROID**—Posterior portion, a thin, dark brown vascular membrane that lines most of the internal surface of the SCLERA. Contains numerous blood vessels united by connective tissue containing pigmented cells, and is composed of 5 layers. Its blood supply nourishes the inner nervous tunic, the RETINA. The deep choroid layer, the LAMINA VITREA, is placed next to the pigmentary layer of the RETINA. Like the SCLERA, it is pierced by the OPTIC NERVE at the back of the eyeball, and extends from the OPTIC NERVE to the anterior, jagged margin of the RETINA, called the ORA SERRATA. The choroid absorbs light rays so they are not reflected back out of the eyeball.
■ **CILIARY BODY**—Beyond the ORA SERRATA in the anterior portion of the vascular tunic, the CHOROID becomes the CILIARY BODY, a very thick internal muscular ring composed of 3 distinct planes of smooth muscle fibers (CILIARY MUSCLES). Extensions of the CILIARY BODY, called CILIARY PROCESSES, attach to the ZONULAR FIBERS of ZINN, collectively called the SUSPENSORY LIGAMENT. This circular ligament attaches to the LENS (FOCAL APPARATUS) at the margins of the thin, clear LENS CAPSULE which encloses the LENS. (The CILIARY MUSCLES alter the shape of the pliable LENS for near or far vision.)
■ **IRIS**— A contractile pigmented membrane consisting of circular and radial smooth muscle fibers arranged in the shape of a doughnut. The black doughnut hole is the PUPIL, through which light enters the eyeball. Contraction of the smooth muscle fibers regulates the diameter of the PUPIL. The IRIS is attached at its outer margin to the CILIARY BODY, and is suspended between the LENS and the CORNEA in the fluid-filled AQUEOUS HUMOR of the ANTERIOR CAVITY, separating this cavity into an ANTERIOR and POSTERIOR CHAMBER. These 2 chambers are connected through the PUPIL.

(The larger, posterior cavity of the eye, the VITREOUS CHAMBER, is filled with a transparent, jellylike substance called VITREOUS HUMOR, and fills the inner space of the eyeball between the LENS and the RETINA.)

★ **INNER NERVOUS TUNIC: RETINA (See 7.21)**

3 Basic Layers of the EYEBALL

(1) **FIBROUS TUNIC** (2) **VASCULAR TUNIC** (3) **INTERNAL TUNIC (RETINA) and LENS** (not part of any tunic)

(1) OUTER AVASCULAR FIBROUS TUNIC (2 PARTS)
Posterior 4/5: OPAQUE REGION

4 | SCLERA | "WHITE of the EYE"

  a | VENOUS SINUS/ CANAL of SCHLEMM | At the junction of the SCLERA and CORNEA

Anterior 1/5: TRANSPARENT REGION

5 | CORNEA and MUSCULAR |

(2) MIDDLE VASCULAR TUNIC: UVEA (3 PARTS)
Posterior Portion

6 | CHOROID |

Anterior Portion

7 | CILIARY BODY |

  b | 3 CILIARY SMOOTH MUSCLE PLANES |

  c | CILIARY PROCESSES |

8 | IRIS | continuous with CHOROID through CILIARY BODY

  d | PUPIL | BLACK HOLE OPENING in CENTER of IRIS

  e | SPHINCTER PUPILLAE | CIRCULARLY ARRANGED

  f | DILATOR PUPILLAE | RADIALLY ARRANGED

(3) INNERMOST NERVOUS TUNIC: RETINA

9 | RETINA | (See 7.21)

10 | •ORA SERRATA |

11 | •MACULA LUTEA |

12 | •FOVEA CENTRALIS |

13 | •OPTIC DISC (BLIND SPOT) |

14 | •PARS CILIARIS RETINAE | : Thin portion of the RETINA, situated in front of the ORA SERRATA, which lines the inner surface of the CILIARY BODY.

15 | •PARS IRIDICA | : Covers posterior surface of the IRIS

INTERIOR CAVITIES

Anterior Cavity

16 | ANTERIOR CHAMBER |

17 | POSTERIOR CHAMBER |

18 | FLUID AQUEOUS HUMOR of the ANTERIOR CAVITY |

Posterior Cavity

19 | VITREOUS CHAMBER FILLED with VITREOUS HUMOR |

20 | HYALOID CANAL |
Lymph Channel in vitreous chamber (contains the HYALOID ARTERY in the FETUS)

FOCAL APPARATUS

21 | LENS WITH SURROUNDING LENS CAPSULE |

22 | SUSPENSORY LIGAMENT (ZONULAR FIBERS OF ZINN) |

NERVE TRACT

(II) | OPTIC CRANIAL NERVE II |

BLOOD SUPPLY

23 | CENTRAL RETINAL ARTERY |

24 | CENTRAL VEIN |

25 | CILIARY ARTERIES | : (not shown)
- •Branches of OPHTHALMIC ARTERY that supply the CHOROID LAYER
- •Pierce SCLERA posteriorly and traverse choroid layer to the CILIARY BODY, with extensive anastomoses throughout the choroid

26 | CILIARY VEINS |

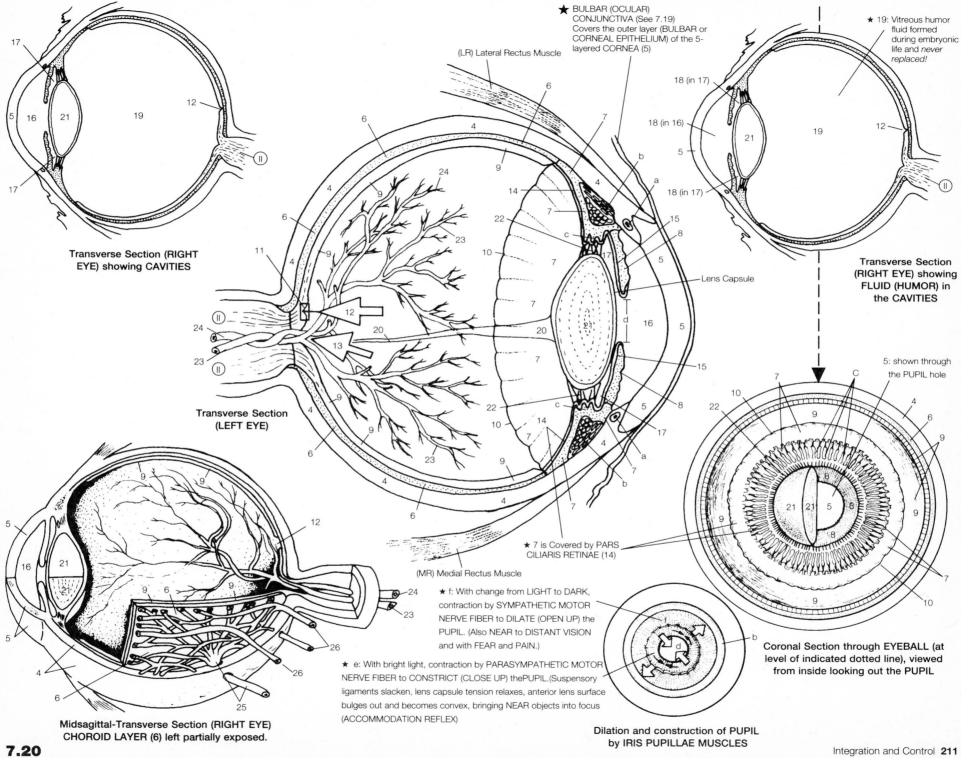

Transverse Section (RIGHT EYE) showing CAVITIES

★ BULBAR (OCULAR) CONJUNCTIVA (See 7.19) Covers the outer layer (BULBAR or CORNEAL EPITHELIUM) of the 5-layered CORNEA (5)

(LR) Lateral Rectus Muscle

★ 19: Vitreous humor fluid formed during embryonic life and *never replaced!*

Lens Capsule

Transverse Section (RIGHT EYE) showing FLUID (HUMOR) in the CAVITIES

Transverse Section (LEFT EYE)

★ 7 is Covered by PARS CILIARIS RETINAE (14)

(MR) Medial Rectus Muscle

5: shown through the PUPIL hole

Coronal Section through EYEBALL (at level of indicated dotted line), viewed from inside looking out the PUPIL

Midsagittal-Transverse Section (RIGHT EYE) CHOROID LAYER (6) left partially exposed.

★ f: With change from LIGHT to DARK, contraction by SYMPATHETIC MOTOR NERVE FIBER to DILATE (OPEN UP) the PUPIL. (Also NEAR to DISTANT VISION and with FEAR and PAIN.)

★ e: With bright light, contraction by PARASYMPATHETIC MOTOR NERVE FIBER to CONSTRICT (CLOSE UP) thePUPIL.(Suspensory ligaments slacken, lens capsule tension relaxes, anterior lens surface bulges out and becomes convex, bringing NEAR objects into focus (ACCOMMODATION REFLEX)

Dilation and construction of PUPIL by IRIS PUPILLAE MUSCLES

7.20

Use same colors for Sclera, Choroid, and Retina
from 7.20. Also use same colors from blow-up
of transverse section from 7.20.

NERVOUS SYSTEM: SPECIAL SENSORY ORGANS
Visual Sense: Sight
Structure of the Retina

★ RETINA:

• The 3rd and innermost tunic of the eye

• Receives inverted light images through the refractive media of the CORNEA, (principal) INNER CAVITY FLUIDS (minimal), and LENS (refining and alteration of refraction).

• The light-sensitive portion of the eye upon which light rays come to a focus.

• The immediate instrument of VISION, whose primary function is IMAGE FORMATION.

• Extends from the point of entrance of the optic nerve posteriorly towards the margin of the pupil anteriorly, completely lining the interior of the eye.

• Consists of 3 parts:

■ *PARS OPTICA*—The nervous or sensory portion of the retina is only in the posterior portion of the eye. It extends from the OPTIC DISC forward to the ORA SERRATA, the ending scalloped border or jagged margin of the retina, where the CHOROID layer ends and the CILIARY BODY begins. The 10-layered PARS OPTICA consists of an OUTER PIGMENTED LAYER (covering the CHOROID'S BOTTOM LAYER) and a 9-layered nervous tissue portion. This nervous tissue layer ends at the ORA SERRATA. (See box)

■ *PARS CILIARIS*—The thin part of the retina lining the inner surface of the CILIARY PROCESS. The outer pigmented layer (along with the internal limiting membrane) extends anteriorly over the back of the ciliary body and processes.

■ *PARS IRIDICA*—These 2 layers continue to form the posterior surface of the IRIS.

★ RODS and CONES (dendrites of the photoreceptor neurons) are the VISUAL RECEPTORS highly specialized for stimulation by LIGHT RAYS.

■ RODS—Specialized for vision in *dim light* for BLACK AND WHITE VISION (*no* discrimination between colors).
Rods are positioned on the peripheral (EXTRAFOVEAL) parts of the PARS OPTICA (over 100 million per eye), and provide poor visual acuity.

■ CONES—specialized for COLOR VISION and VISUAL ACUITY (sharpness of vision), stimulated only by bright light. (Thus, we cannot see color by moonlight.) There are an estimated 7 million cones in each eye.

■ MACULA LUTEA—The special "YELLOW SPOT" in the exact center of the retina, approximately 2mm lateral to the optic nerves exit. Contains an abundance of cones. Contains also a pit, the FOVEA CENTRALIS, where the retina is reduced to a layer of closely-packed cones, the area of the most acute and keenest vision (central vision). (When we look at an object our eyes are directed so that the image will fall on the FOVEA of each eye.) There are *no* RODS present here.

■ OPTIC DISC (PAPILLA)—The "BLIND SPOT," the area where the nerve fibers from all parts of the retina converge to leave the eyeball as the OPTIC NERVE (II). There are *no* RODS or CONES present here, and thus the "blind spot" is not sensitive to light. (RETINAL ARTERIES and VEINS enter or leave the eyeball here.)

★ ■ VISUAL AXIS—A line drawn between the FOVEA and the apparent center of the PUPIL. It establishes the path of light rays when a person is looking at a distant object.

■ EQUATOR—A line drawn perpendicular to the VISUAL AXIS through the greatest lateral expansion of the eyeball.

■ OPTICAL (ANATOMICAL) AXIS—A line drawn between the ANTERIOR POLE (CENTER of the CORNER) and the POSTERIOR POLE (a point between the FOVEA and OPTIC DISC on the posterior retina at the greatest diameter in the anteroposterior dimension).

INNER THIRD TUNIC of the EYEBALL: RETINA

3 Parts of the RETINA:

A │ **PARS OPTICA** │ PARS OPTICA RETINAE

B │ **PARS CILIARIS** │ PARS CILIARIS RETINAE

C │ **PARS IRIDICA** │ PARS IRIDICA RETINAE

Special Points on PARS OPTICA RETINAE:

D │ ORA SERRATA │

E │ MACULA LUTEA │ "YELLOW SPOT"

F │ FOVEA CENTRALIS │

G │ OPTIC DISC ("BLIND SPOT") │ OPTIC PAPILLA

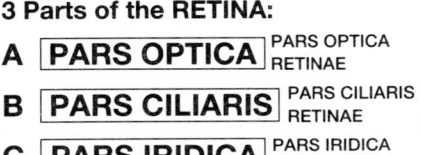

Choroid   Retina   ★ LEFT FUNDUS OCULI

Sclera   • Part of the RETINA seen by

Central Artery   means of an OPHTHALA-
and branches   MOSCOPE, which
shines a beam of

G   light through the
PUPIL onto
the RETINA

Central   E
Vein and
branches

F

Extrafoveal Retina

10 Layers of the RETINA Composed of 3 Types of NEURONS:
PHOTORECEPTOR, BIPOLAR, and GANGLION

1 │ **PIGMENTED EPITHELIUM** │ { Next to the LAMINA VITREA of the CHOROID COAT. (In bright light, PIGMENT GRANULES migrate into cell processes between ROD and CONES, preventing spread of light from one receptor to another.)

2 │ **LAYER of RODS and CONES** │ BACILLAR LAYER   a │ RODS │ b │ CONES │ The 1st RECEPTIVE "NEURONES" DENDRITES of the PHOTORECEPTOR NEURONS

3 │ **EXTERNAL LIMITING MEMBRANE** │ SEPARATES RODS and CONES from THEIR PHOTORECEPTOR CELL BODIES (SUPPORTS THE ROD AND CONE SEGMENTS)

4 │ **OUTER NUCLEAR LAYER** │ c │ CELL BODIES and NUCLEI OF PHOTORECEPTOR NEURONS │

5 │ **OUTER PLEXIFORM LAYER** │ Nerve process and synapses between photoreceptor and bipolar neurons

6 │ **INNER NUCLEAR LAYER** │ d │ CELL BODIES of 2nd OR BIPOLAR NEURONS │

7 │ **INNER PLEXIFORM LAYER** │ Nerve processes and synapses between bipolar and ganglion neurons

8 │ **LAYER of GANGLION CELLS** │ e │ CELL BODIES of 3rd, INTEGRATING GANGLION NEURONS │

9 │ **LAYER of OPTIC NERVE FIBERS** │ Axons of GANGLION NEURONS converge on the OPTIC PAPILLA to leave the eyeball as the OPTIC NERVE (II)

10 │ **INTERNAL LIMITING MEMBRANE** │ Glial membrane of the RETINA, which continues as the PARS CILIARIS RETINAE and the PARS IRIDICA.

◁ │ **DIRECTION of LIGHT RAYS:** │ Must pass through layers 10–2 (not 1) to reach and stimulate the ROD and CONE RECEPTORS

◀ │ **DIRECTION of NERVE IMPULSES:** │ When LIGHT strikes the ROD and CONE RECEPTORS for VISION, nerve impulses are set up that are transmitted via the BIPOLAR NEURONS→GANGLION NEURONS→OPTIC NERVE→VISUAL CEREBRAL CORTEX AREA of OCCIPITAL LOBE

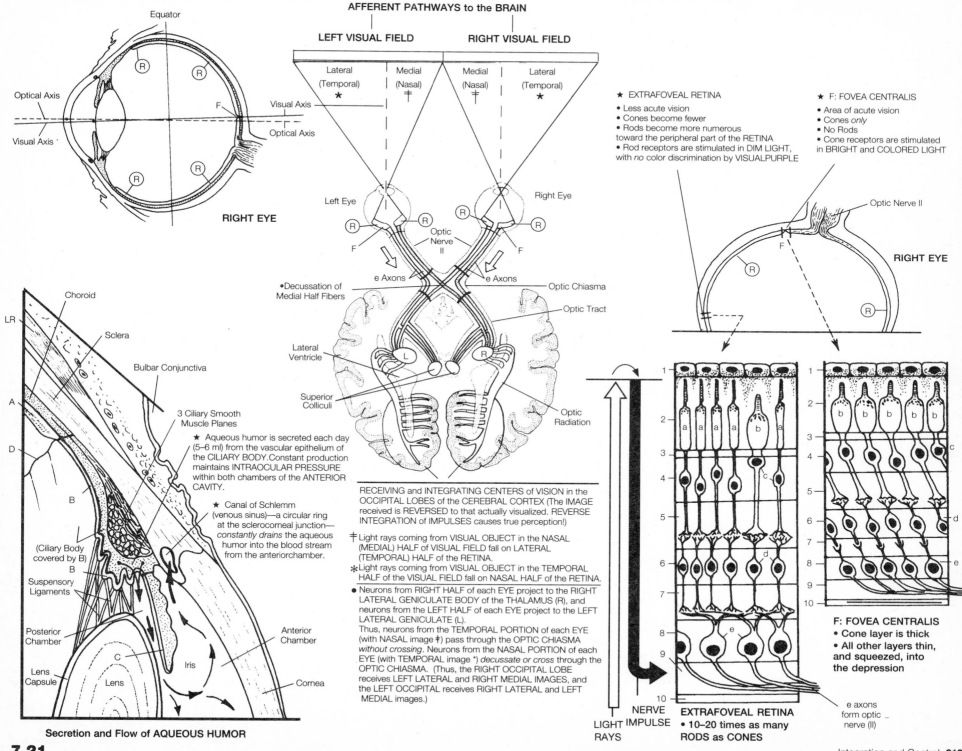

Equator

Optical Axis
Visual Axis
Visual Axis
Optical Axis
F
R R R R

RIGHT EYE

LEFT VISUAL FIELD   RIGHT VISUAL FIELD

Lateral (Temporal) ★ | Medial (Nasal) † | Medial (Nasal) † | Lateral (Temporal) ★

Left Eye   Right Eye
R R R R
F   F
e Axons   e Axons

• Decussation of Medial Half Fibers
Optic Nerve II
Optic Chiasma
Optic Tract

Lateral Ventricle
L   R
Superior Colliculi
Optic Radiation

★ EXTRAFOVEAL RETINA
• Less acute vision
• Cones become fewer
• Rods become more numerous toward the peripheral part of the RETINA
• Rod receptors are stimulated in DIM LIGHT, with *no* color discrimination by VISUALPURPLE

★ F: FOVEA CENTRALIS
• Area of acute vision
• Cones *only*
• No Rods
• Cone receptors are stimulated in BRIGHT and COLORED LIGHT

Optic Nerve II
RIGHT EYE
F
R   R
R

RECEIVING and INTEGRATING CENTERS of VISION in the OCCIPITAL LOBES of the CEREBRAL CORTEX (The IMAGE received is REVERSED to that actually visualized. REVERSE INTEGRATION of IMPULSES causes true perception!)

† Light rays coming from VISUAL OBJECT in the NASAL (MEDIAL) HALF of VISUAL FIELD fall on LATERAL (TEMPORAL) HALF of the RETINA.

✱ Light rays coming from VISUAL OBJECT in the TEMPORAL HALF of the VISUAL FIELD fall on NASAL HALF of the RETINA.

• Neurons from RIGHT HALF of each EYE project to the RIGHT LATERAL GENICULATE BODY of the THALAMUS (R), and neurons from the LEFT HALF of each EYE project to the LEFT LATERAL GENICULATE (L).
Thus, neurons from the TEMPORAL PORTION of each EYE (with NASAL image †) pass through the OPTIC CHIASMA *without crossing*. Neurons from the NASAL PORTION of each EYE (with TEMPORAL image ✱) *decussate or cross* through the OPTIC CHIASMA. (Thus, the RIGHT OCCIPITAL LOBE receives LEFT LATERAL and RIGHT MEDIAL IMAGES, and the LEFT OCCIPITAL receives RIGHT LATERAL and LEFT MEDIAL images.)

Choroid
LR
Sclera
Bulbar Conjunctiva
A
D
B
3 Ciliary Smooth Muscle Planes
★ Aqueous humor is secreted each day (5–6 ml) from the vascular epithelium of the CILIARY BODY. Constant production maintains INTRAOCULAR PRESSURE within both chambers of the ANTERIOR CAVITY.
★ Canal of Schlemm (venous sinus)—a circular ring at the sclerocorneal junction—*constantly drains* the aqueous humor into the blood stream from the anterior chamber.
(Ciliary Body covered by B) B
Suspensory Ligaments
Posterior Chamber
Lens Capsule
Lens
C
Iris
Anterior Chamber
Cornea

Secretion and Flow of AQUEOUS HUMOR

NERVE IMPULSE
LIGHT RAYS

1 2 3 4 5 6 7 8 9 10
a a a a b a
c
d
e

1 2 3 4 5 6 7 8 9 10
b b b b b
c
d
e

F: FOVEA CENTRALIS
• Cone layer is thick
• All other layers thin, and squeezed, into the depression

e axons form optic nerve (II)

EXTRAFOVEAL RETINA
• 10–20 times as many RODS as CONES

NERVOUS SYSTEM: SPECIAL SENSORY ORGANS
Auditory Sense: Sound
Structure of the Ear

4 = fleshy 5 = light orange 6 = yellow 7 = light brown
8 = light grey BL = cream ML = green
10 = light blue 11 = light green Muscles - warms
★ See 7.23

★ The EAR is the organ of both HEARING and EQUILIBRIUM. It contains receptors that convert SOUND WAVES into NERVE IMPULSES (for HEARING) and receptors that respond to movements of head (for EQUILIBRIUM). The VESTIBULOCOCHLEAR CRANIAL NERVE (VIII) transmits impulses from both types of receptors to the brain for interpretation.

★ Anatomically, the EAR is divided into 3 principal regions:
① EXTERNAL (OUTER) EAR ② MIDDLE EAR
③ INNER EAR (LABYRINTH), which contain organs for both HEARING and EQUILIBRIUM

★ ① EXTERNAL or OUTER EAR consists of 3 structures: AURICLE (PINNA), EXTERNAL AUDITORY MEATUS, and TYMPANIC MEMBRANE (EARDRUM).
■ AURICLE (PINNA)—COLLECTS and DIRECTS SOUND WAVES into the EXTERNAL AUDITORY MEATUS. A funnel-like, trumpet-shaped flap of elastic cartilage covered by thick skin, the pinna is attached to the skull by ligaments and poorly developed AURICULAR MUSCLES.
■ EXTERNAL AUDITORY MEATUS—A 2.5 cm (1 in.) slightly S-shaped canal, running slightly upward, leading from the PINNA to the EARDRUM. This small tube lies in the EXTERNAL AUDITORY MEATUS of the TEMPORAL BONE (See 4.5). The walls are lined with cartilage continuous with the cartilage of the PINNA, and are covered with thin, highly sensitive skin, containing fine hairs and sebaceous glands at the entrance.
■ TYMPANIC MEMBRANE—SOUND WAVES SET the TYMPANIC MEMBRANE in VIBRATION. This EARDRUM is a thin, double-layered, semitransparent epithelial partition between the EXTERNAL AUDITORY MEATUS and the MIDDLE EAR. External surface: CONCAVE, covered with skin. Internal surface: CONVEX, covered with a mucous membrane.

★ ② MIDDLE EAR—A small air-filled chamber deep within the PETROUS PORTION of the TEMPORAL BONE. Extending across this TYMPANIC CAVITY are 3 small bones—the AUDITORY OSSICLES—connected by SYNOVIAL JOINTS to form a small LEVER. The "HANDLE" OF THE MALLEUS bone is attached to the internal surface of the EARDRUM. Its head articulates with the INCUS, which is linked to the STAPES, which fits into the OVAL WINDOW, a small opening between the MIDDLE and INNER EAR. These ossicles are attached to the TYMPANIC CAVITY by means of ligaments. VIBRATION of AIR MOVES the LEVER of OSSICLES. This AMPLIFIES and TRANSMITS VIBRATIONS across the MIDDLE EAR so that the FOOTPLATE of the STAPES moves backwards and forwards in the OVAL WINDOW. (There are 2 openings into the TYMPANIC CAVITY, see illustration.)

★ ③ INNER EAR (LABYRINTH)—A complicated series of canals, consisting of 2 main divisions: a BONY LABYRINTH and a MEMBRANOUS LABYRINTH that fits inside the BONY LABYRINTH.
■ BONY LABYRINTH is a series of bony canals in the petrous portion of the temporal bone, lined with periosteum whose cells secrete a pale, limpid fluid, PERILYMPH that fills the space between the 2 labyrinths.
■ MEMBRANOUS LABYRINTH is a series of sacs and tubes lying inside the BONY LABYRINTH, and having the same basic shape as the BONY LABYRINTH. The membranous labyrinth is lined with epithelium and contains an ENDOLYMPH fluid. These 2 fluids provide a LIQUID-CONDUCTING MEDIUM for the vibrations set up by the movement of the STAPES in the OVAL WINDOW. These vibrational movements through fluid stimulate RECEPTORS in the ORGAN of CORTI in the INNER EAR, where nerve impulses are set up in the AUDITORY NERVE (VIII) and pass to AUDITORY CENTERS in the brain. (See 7.23)

3 BASIC REGIONS of the EAR (AUDITORY APPARATUS);

① **EXTERNAL (OUTER) EAR**   ② **MIDDLE EAR**   ③ **INTERNAL EAR** (LABYRINTH)

① EXTERNAL (OUTER) EAR:
4 **AURICLE (PINNA)**
5 **EXTERNAL AUDITORY MEATUS**
6 **TYMPANIC MEMBRANE (EARDRUM)**

② MIDDLE EAR:
7 **TYMPANIC CAVITY**
8 **EUSTACHIAN (AUDITORY) TUBE** ⎤ 2
9 **EPITYMPANIC RECESS** ⎦ OPENINGS into TYMPANIC CAVITY

③ INNER EAR: LABYRINTH—2 Parts:
BL **OUTER BONY LABYRINTH**
ML **INNER MEMBRANOUS LABYRINTH**

Fluids of the LABYRINTH:
10 **PERILYMPH** between the LABYRINTHS
11 **ENDOLYMPH** within MEMBRANOUS LABYRINTH

Innervation of the LABYRINTH:
VII **FACIAL CRANIAL NERVE VII**
VIII **VESTIBULOCOCHLEAR CRANIAL NERVE (VIII): AUDITORY NERVE**
 vb **VESTIBULAR BRANCH**
 cb **COCHLEAR BRANCH** (not shown)

AUDITORY OSSICLES of the MIDDLE EAR
12 **MALLEUS** HAMMER
13 **INCUS** ANVIL
14 **STAPES** STIRRUP

Skeletal Muscles Connected to OSSICLES and Bony Wall
15 **TENSOR TYMPANI** attaches to MALLEUS
16 **STAPEDIUS** attaches to STAPES

Within the MEMBRANOUS LABYRINTH (See also 7.23)
MECHANISMS OF EQUILIBRIUM
■ STATIC EQUILIBRIUM: ORIENTATION of the BODY (HEAD) RELATIVE to the GROUND (GRAVITY)
■ MACULA: MU **MACULA UTRICULI**
    MS **MACULA SACCULI**
a **GELATINOUS MATRIX** PROTEIN-SUGAR LAYERS
b **OTOLITHS**
c **HAIR CELLS**
d **SUPPORTING CELLS**

• Within the UTRICLE and SACCULE of the VESTIBULE of the MEMBRANOUS LABYRINTH, the MACULA contains OTOLITHS (crystals of calcium carbonate) that move back and forth in a GELATINOUS MATRIX with changes in head positions. These moving OTOLITHS mechanically stimulate the HAIR PROCESSES of HAIR CELLS. The information is then carried in fibers of the VESTIBULAR BRANCH of VIII to CENTERS in the BRAIN.

■ DYNAMIC EQUILIBRIUM: MAINTENANCE of BODY POSITION (HEAD) in RESPONSE to SUDDEN MOVEMENTS
CA **CRISTA AMPULLARIS**
e **CUPULA** PROTEIN-SUGAR LAYERS
c **HAIR CELLS**
d **SUPPORTING CELLS**

• Within the AMPULLA of each SEMICIRCULAR CANAL, the CRISTA AMPULLARIS contains a GELATINOUS MASS, the CUPULA, that stretches across each AMPULLA. The APEX swings free in reaction to the starting or stopping of rotary movements of the head in space, when the endolymph moves. The swinging of the CUPULA APEX causes the HAIR PROCESSES of the HAIR CELLS to bend, stimulating the HAIR CELLS. The information is then carried in fibers of the VESTIBULAR BRANCH of VIII to CENTERS in the BRAIN.

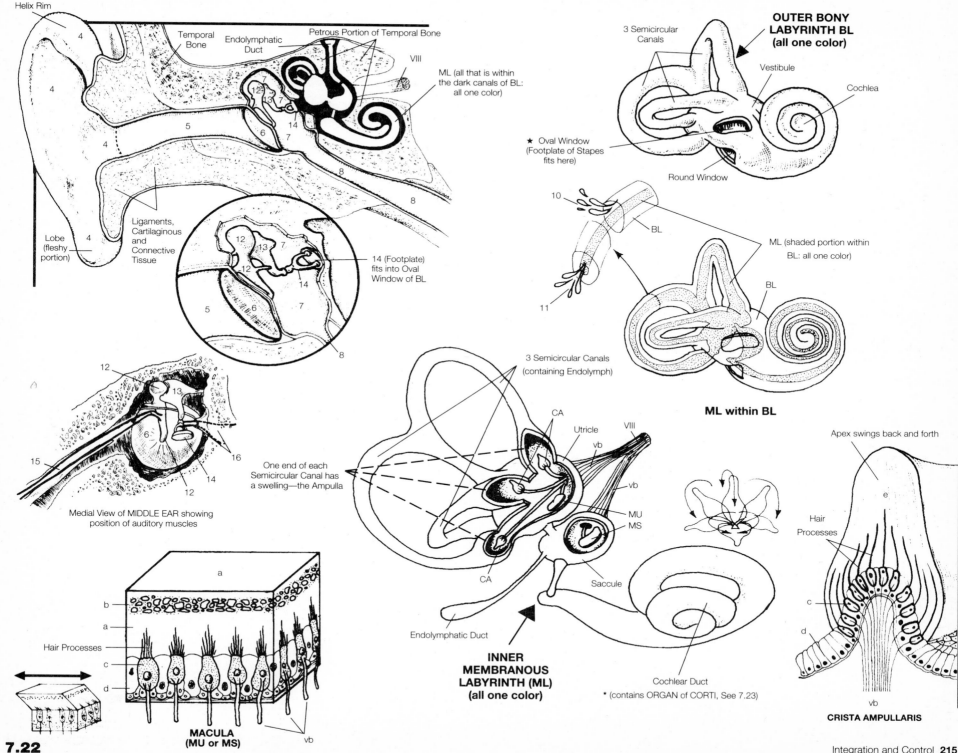

Helix Rim

Temporal Bone

Endolymphatic Duct

Petrous Portion of Temporal Bone

VIII

ML (all that is within the dark canals of BL: all one color)

Ligaments, Cartilaginous and Connective Tissue

Lobe (fleshy portion)

14 (Footplate) fits into Oval Window of BL

Medial View of MIDDLE EAR showing position of auditory muscles

OUTER BONY LABYRINTH BL (all one color)

3 Semicircular Canals

Vestibule

Cochlea

★ Oval Window (Footplate of Stapes fits here)

Round Window

BL

ML (shaded portion within BL: all one color)

BL

ML within BL

3 Semicircular Canals (containing Endolymph)

One end of each Semicircular Canal has a swelling—the Ampulla

CA

Utricle

VIII

vb

vb

MU

MS

CA

Saccule

Endolymphatic Duct

INNER MEMBRANOUS LABYRINTH (ML) (all one color)

Cochlear Duct

* (contains ORGAN of CORTI, See 7.23)

Apex swings back and forth

e

Hair Processes

c

d

vb

CRISTA AMPULLARIS

a

b

a

Hair Processes

c

d

vb

MACULA (MU or MS)

NERVOUS SYSTEM: SPECIAL SENSORY ORGANS
Auditory Sense: Sound
Structure of the Labyrinth of the Ear

★ The INNER EAR (LABYRINTH) contains the ORGANS OF HEARING and EQUILIBRIUM.

★ The BONY LABYRINTH is structurally and functionally divided in 3 areas:
■ VESTIBULE ■ SEMICIRCULAR CANALS ■ COCHLEA
  ■ VESTIBULE—The CENTRAL OVAL PORTION of the BONY LABYRINTH
    • Contains the OVAL WINDOW (into which the STAPES fits) and the ROUND WINDOW directly below it on the opposite end.
    • MEMBRANOUS LABYRINTH within the VESTIBULE contains 2 sacs connected by a small duct: A larger UTRICLE in the upper portion and a smaller SACCULUS. These 2 sacs contain special proprioceptors (MACULA UTRICULI and MACULA SACCULI, respectively) involved in the mechanism of STATIC EQUILIBRIUM.
  ■ SEMICIRCULAR CANALS—3 bony canals of each ear at right angles to each other, projecting upward and posteriorly from the vestibule, with both ends of each canal opening into the vestibule. (One end of each canal contains a swelling called the AMPULLA.)
    • The SEMICIRCULAR DUCTS of the MEMBRANOUS LABYRINTH are almost identical in shape to their bony counterparts and communicate with the UTRICLE. (Each of the 3 DUCTS has a MEMBRANOUS AMPULLA that connects with the upper back of the UTRICLE.) The positioning of the 3 ducts permits detection of an imbalance in 3 planes. Special "proprioceptors" (CRISTAE) in each AMPULLA help to maintain DYNAMIC EQUILIBRIUM in the NONAUDITORY PART of the INNER EAR.
  ■ COCHLEA—Lying in front of the VESTIBULE is the BONY COCHLEA, which spirals 2 3/4 times around a central bony axis (pillar) called the MODIOLUS. View from a cross section (see illustration) shows the spiral canal is divided into 3 separate tunnels resembling the letter Y on its side. The stem of the Y is a bony spiral lamina that projects part way into the interior of the SPIRAL CANAL of the COCHLEA and divides it into 2 SCALAE CHAMBERS: SCALA VESTIBULI (which begins at the OVAL WINDOW and is continuous with the VESTIBULE), and a lower SCALA TYMPANI (terminates at the ROUND WINDOW). The wings of the Y are composed of MEMBRANOUS LABYRINTH, and the third chamber produced between the wings is the SCALA MEDIA, or COCHLEAR DUCT. The upper wing of the Y, the root of the COCHLEAR DUCT, which separates the COCHLEAR DUCT from the SCALA VESTIBULI, is the VESTIBULAR MEMBRANE (of REISSNER). The lower wing of the Y, the floor of the COCHLEAR DUCT, which separates the COCHLEAR DUCT from the SCALA TYMPANI, is the BASILAR MEMBRANE. The BASILAR MEMBRANE joins the tip of the SPIRAL LAMINA and attaches to the outer bony wall by the EXTERNAL SPIRAL LIGAMENT.

★ Resting on the BASILAR MEMBRANE (within the COCHLEAR DUCT) is the ORGAN of CORTI, the functional unit of hearing. The tips of the hairlike processes of the hearing receptor cells of the organ of Corti are imbedded in a flexible gelatinous membrane called the TECTORIAL MEMBRANE. Impulses are passed on from the HAIR CELLS OF CORTI to the branches of the COCHLEAR NERVE (VIII). The cell bodies of these 1st neurons lie within the bone of the SPIRAL LAMINA.

★ The COCHLEAR DUCT is filled with ENDOLYMPH (secreted by the STRIA VASCULARIS, pigmented granular cells with a profuse blood supply), and ends at the ROUND WINDOW. The SCALA VESTIBULI and SCALA TYMPANI contain PERILYMPH, and are completely separated, except where the COCHLEAR SPIRAL ends at a narrow apex, the HELICOTREMA, where they are continuous.
The PERILYMPH of the SCALA VESTIBULI is continuous with that of the VESTIBULE.

■ The BONY LABYRINTH:
3 Structural and Functional Divisional Areas:

A   VESTIBULE

B   3 SEMICIRCULAR CANALS   (90° ANGLES to EACH OTHER)

C   COCHLEA

Central Portion: VESTIBULE

1   OVAL WINDOW (FENESTRAL VESTIBULI)

2   ROUND WINDOW (FENESTRAL COCHLEA)   Enclosed by the SECONDARY TYMPANIC MEMBRANE

Snail-Shaped COCHLEA:
Central BONY AXIS

3   MODIOLUS (CENTRAL PILLAR)

Projecting BONY SHELF

4   OSSEOUS SPIRAL LAMINA

2 Chambers Formed by the SPIRAL LAMINA

5   SCALA VESTIBULI   UPPER CHAMBER

6   SCALA TYMPANI   LOWER CHAMBER

INNERVATION of INNER EAR:
VESTIBULOCOCHLEAR NERVE (VIII):

cb   COCHLEAR BRANCH of VIII

vb   VESTIBULAR BRANCH of VIII

■ The MEMBRANOUS LABYRINTH:
Within the VESTIBULE:

7   UTRICULUS (UTRICLE) ⎤
8   SACCULUS (SACCULE) ⎦ CONTAINS STRUCTURES INVOLVED in EQUILIBRIUM

9   ENDOLYMPHATIC DUCT

10   DUCTUS REUNIENS (HENSEN'S CANAL)   CONNECTS VESTIBULE TO COCHLEAR DUCT

Within the 3 SEMICIRCULAR CANALS:

B1   3 SEMICIRCULAR DUCTS

B2   MEMBRANOUS AMPULLAE of the SEMICIRCULAR DUCTS

Within the COCHLEA   (THE ESSENTIAL ORGAN OF HEARING)

MIDDLE CHAMBER

C1   COCHLEAR DUCT   (SCALA MEDIA)

11   VESTIBULAR MEMBRANE (of REISSNER)   ROOF of COCHLEAR DUCT

12   BASILAR MEMBRANE   FLOOR of COCHLEAR DUCT

OC   ORGAN of CORTI   ORGAN of HEARING

13   TECTORIAL MEMBRANE

14   EXTERNAL SPIRAL LIGAMENT

15   STRIA VASCULARIS

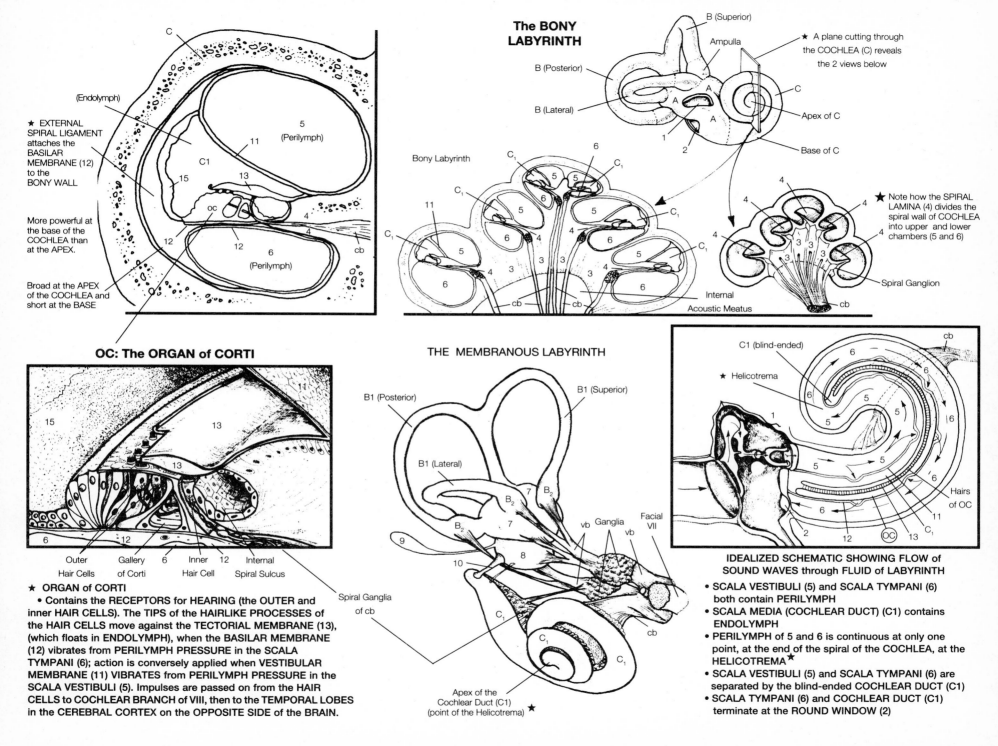

The BONY LABYRINTH

B (Superior)

Ampulla

★ A plane cutting through the COCHLEA (C) reveals the 2 views below

B (Posterior)

B (Lateral)

Apex of C

Base of C

C — Apex of C

Bony Labyrinth

★ Note how the SPIRAL LAMINA (4) divides the spiral wall of COCHLEA into upper and lower chambers (5 and 6)

Spiral Ganglion

Internal Acoustic Meatus

★ EXTERNAL SPIRAL LIGAMENT attaches the BASILAR MEMBRANE (12) to the BONY WALL

More powerful at the base of the COCHLEA than at the APEX.

Broad at the APEX of the COCHLEA and short at the BASE

(Endolymph)

5 (Perilymph)

6 (Perilymph)

OC: The ORGAN of CORTI

Outer Hair Cells | Gallery of Corti | 6 | Inner Hair Cell | 12 | Internal Spiral Sulcus

★ ORGAN of CORTI

• Contains the RECEPTORS for HEARING (the OUTER and inner HAIR CELLS). The TIPS of the HAIRLIKE PROCESSES of the HAIR CELLS move against the TECTORIAL MEMBRANE (13), (which floats in ENDOLYMPH), when the BASILAR MEMBRANE (12) vibrates from PERILYMPH PRESSURE in the SCALA TYMPANI (6); action is conversely applied when VESTIBULAR MEMBRANE (11) VIBRATES from PERILYMPH PRESSURE in the SCALA VESTIBULI (5). Impulses are passed on from the HAIR CELLS to COCHLEAR BRANCH of VIII, then to the TEMPORAL LOBES in the CEREBRAL CORTEX on the OPPOSITE SIDE of the BRAIN.

THE MEMBRANOUS LABYRINTH

B1 (Posterior)

B1 (Superior)

B1 (Lateral)

Ganglia

Facial VII

Spiral Ganglia of cb

Apex of the Cochlear Duct (C1) (point of the Helicotrema) ★

C1 (blind-ended)

★ Helicotrema

Hairs of OC

IDEALIZED SCHEMATIC SHOWING FLOW of SOUND WAVES through FLUID of LABYRINTH

• SCALA VESTIBULI (5) and SCALA TYMPANI (6) both contain PERILYMPH

• SCALA MEDIA (COCHLEAR DUCT) (C1) contains ENDOLYMPH

• PERILYMPH of 5 and 6 is continuous at only one point, at the end of the spiral of the COCHLEA, at the HELICOTREMA★

• SCALA VESTIBULI (5) and SCALA TYMPANI (6) are separated by the blind-ended COCHLEAR DUCT (C1)

• SCALA TYMPANI (6) and COCHLEAR DUCT (C1) terminate at the ROUND WINDOW (2)

Chapter 7: *NERVOUS SYSTEM*: CHARTS

Chart #1 : Structures of the Nervous System (See 7.1)

	THE HUMAN NERVOUS SYSTEM	
	CENTRAL NERVOUS SYSTEM (CNS)	**PERIPHERAL NERVOUS SYSTEM (PNS)** (OUTSIDE THE CNS)
Structures within the System	BRAIN and SPINAL CORD	• 12 Pairs of CRANIAL NERVES and branches • 31 Pairs of SPINAL NERVES and branches • GANGLIA and GANGLIONATED TRUNKS • NERVE PLEXUSES and NERVES arising from them SENSORY RECEPTORS with SENSORY ORGANS • Specialized Endings on MUSCLE
A Cluster of NERVE CELL BODIES (the CELL BODY controls the metabolism of the whole cell unit)	A NUCLEUS (pl. NUCLEI)	A GANGLION (pl. GANGLIA)
A Group of NERVE FIBERS with a Common Origin and Destination (Each NERVE FIBER is an elongated process of a NERVE CELL, or NEURON, usually an AXON)	NERVE TRACT (interconnects regions of CNS) (major portion of the WHITE MATTER of the BRAIN and SPINAL CORD)	NERVE (Cablelike collection of nerve fibers may be exclusively SENSORY or MOTOR, or "MIXED" —carrying impulses from both sensory and motor fibers)
Gross division of system into "MATTER"	■ WHITE MATTER • Nerve fibers covered by a noncellular sheath of fatty substance—MYELIN. Fibers are termed MYELINATED or MEDULLATED. Color of myelin gives name to WHITE MATTER. CNS white matter does not have an outer NEURILEMMA SHEATH. ■ GRAY MATTER • Nerve tissue of grayish color not predominated by myelinated nerve fibers. Contains large numbers of cell bodies of neurons. IN BRAIN: CEREBRAL CORTEX, BASAL GANGLIA, NUCLEI IN SPINAL CORD: H-shaped gray columns	• No GRAY MATTER • Most fibers in PERIPHERAL NERVES are① MYELINATED, and ② have an outer NEURILEMMA SHEATH. (NEURILEMMA found *only* around fibers of PNS.)
The NEUROGLIAL cells forming the MYELIN SHEATH	OLIGODENDROCYTES	SCHWANN CELLS

Chart #2 : Principle Ascending and Descending Tracts of Spinal Cord

	Tract	Funiculus	Origin	Termination	Function
	ASCENDING TRACTS Anterior spinothalamic	Anterior	Posterior horn on one side of cord but crosses to opposite side	Thalamus, then cerebral cortex	Conducts sensory impulses for crude touch and pressure
	Lateral spinothalamic	Lateral	Posterior horn on one side of cord but crosses to opposite side	Thalamus, then cerebral cortex	Conducts pain and temperature impulses that are interpreted as senations of fine touch, precise pressures, and body movements
	Fasciculus gracilis and fasciculus cuneatus	Posterior	Peripheral afferent neurons; does not cross over	Nucleus gracilis and nucleus cuneatus of medulla; eventually thalamus, then cerebral cortex	Conducts sensory impulses from both sides of body to cerebellum for subconscious proprioception necessary for coordinated muscular contractions
	Posterior spinocerebellar	Lateral	Posterior horn; does not cross over	Cerebellum	Conducts sensory impulses from both sides of body to cerebellum for subconscious proprioception necessary for coordinated muscular contractions
	Anterior spinocerebellar	Lateral	Posterior horn; some fibers cross, others do not	Cerebellum	Conducts sensory impulses from both sides of body to cerebellum for subconscious proprioception necessary for coordinated muscular contractions
P Y R A M I D A L	**DESCENDING TRACTS** Anterior corticospinal	Anterior	Cerebral cortex on one side of brain; crosses to opposite side of cord	Anterior horn	Conducts motor impulses from cerebrum to spinal nerves and outward to cells of anterior horns for coordinated, precise voluntary movements of skeletal muscle
	Lateral corticospinal	Lateral	Cerebral cortex on one side of brain; crosses in base of medulla to opposite side of cord	Anterior horn	Conducts motor impulses from cerebrum to spinal nerves and outward to cells of anterior horns for coordinated, precise voluntary movements
E X T R A P Y R A M I D A L	Tectospinal	Anterior	Midbrain; crosses to opposite side of cord	Anterior horn	Conducts motor impulses to cells of anterior horns and eventually to muscles that move the head in response to visual, auditory, or cutaneous stimuli
	Rubrospinal	Lateral	Midbrain (red nucleus); crosses to opposite side of cord	Anterior horn	Conducts motor impulses concerned with muscle tone and posture
	Vestibulospinal	Anterior	Medulla; does not cross over	Anterior horn	Conducts motor impulses that regulate body tone and posture (equilibrium) in response to movements of head
	Anterior and medial reticulospinal	Anterior	Reticular formation of brainstem	Anterior horn	Conducts motor impulses that control muscle tone and sweat gland activity
	Lateral reticulospinal	Lateral	Reticular formation of brainstem	Anterior horn	Conducts motor impulses that control muscle tone and sweat gland activity

From Kent M. Van De Graaff, *Human Anatomy*. 2d ed. Copyright ©1988 Wm. C. Brown Publishers, Dubuque, Iowa. All Rights Reserved. Reprinted by permission.

Chart #3 : The Cervical and Brachial Plexuses (See 7.13)

CERVICAL PLEXUS

Nerve Branches	Origin	Distribution
LESSER OCCIPITAL	C2–C3	Skin of scalp behind and superior to ear
GREATER AURICULAR	C2–C3	Skin in front, below, and over ear and over parotid gland
TRANSVERSE CERVICAL	C2–C3	Skin over anterior aspect of neck
SUPRACLAVICULAR	C3–C4	Skin over upper portion of chest and shoulder
ANSA CERVICALIS		
SUPERIOR ROOT	C1–C2	Infrahyoid, thyrohyoid, and geniohyoid muscles of neck
INFERIOR ROOT	C3–C4	Omohyoid and sternohyoid muscles of the neck
PHRENIC	C3–C5	Diaphragm between thorax and abdomen
SEGMENTAL BRANCHES	C1–C5	Prevertebral (deep) muscles of neck, levator scapulae, and middle scalene muscles

BRACHIAL PLEXUS

Nerve	Origin	Distribution
ROOTS Dorsal Scapular	C5	Levator scapulae, rhomboideus major, and rhomboideus minor muscles
Long thoracic	C5-C7	Serratus anterior muscle
TRUNKS Subclavius	C5-C6	Subclavius muscle
Suprascapular	C5-C6	Supraspinatus and infraspinatus muscles
LATERAL CORD Musculotaneous	C5-C7	Coracobrachialis, biceps brachii, and brachialis muscles
Median (lateral head)	C5-C7	Pronator teres, flexor carpiradialis, flexor digitorum superficialis: skin and lateral two-thirds of palm and fingers
Lateral pectoral	C5-C7	Pectoralis major muscle
POSTERIOR CORD Subscapular	C5-C6	Subscapularis and teres major muscles
Thoracodorsal	C5-C6	Latissimus dorsi muscle
Axillary (circumflex)	C5-C6	Deltoid, teres minor muscles; skin over deltoid and upper posterior aspect of arm
Radial	C5-C6, T1	Triceps brachii, brachoradialis, extensor carpi radialis longus, extensor digitorum, extensor carpi ulnaris, extensor indicis muscles; skin of posterior arm and forearm, lateral two-thirds of dorsum of hand, and fingers over proximal and middle phalanges
MEDIAL CORD		
Medial Pectoral	C8-T1	Pectoralis major and minor muscles
Medial Brachial Cutaneous	C8-T1	Skin of medial and posterior aspects of lower third of arm
Medial Antebrachial Cutaneous	C8-T1	Skin of medial and posterior aspects of forearm
Median (medial head)	C5-C7	Pronator teres, flexor carpi radialis, flexor digitorum superficialis; skin and lateral two-thirds of palm and finger
Ulnar	C8-T1	Flexor carpi ulnaris and flexor digitorum profundus muscles; skin of medial side of hand, little finger, and medial half of ring finger

Chart #4 : The Lumbar and Sacral Plexuses (See 7.14)

LUMBAR PLEXUS

Nerve	Origin	Distribution
Anterior Division		
ILIOINGUINAL	L1	External oblique, internal oblique, transversus abdominis muscles; skin of upper medial aspect of thigh, root of penis and scrotum in male, and labia majora and mons pubis in female
GENITOFEMORAL	L1–L2	Cremaster muscle; skin over middle anterior surface of thigh, scrotum in male, and labia majora in female
OBTURATOR	L2–L4	Obturator externus, pectineus, adductor longus, adductor brevis adductor magnus, gracilis muscles; skin over medial aspect of thigh
Saphenous (Accessory Obturator)	L2–L4	Skin of medial aspect of leg
Posterior Division		
ILIOHYPOGASTRIC	T12–L1	External oblique, internal oblique, transversus abdominis muscles; skin of lower abdomen and buttock
LAT. FEMORAL CUTANEOUS	L2–L3	Skin over lateral, anterior, and posterior aspect of thigh
FEMORAL	L2–L4	Iliacus, psoas major, pectineus, rectus femoris, sartorius muscles; rectus femoris, vastus lateralis, vastus medialis, vastus intermedius muscles; skin on front and medial aspect of thigh and medial side of leg and foot

SACRAL PLEXUS

Nerve	Origin	Distribution
PUDENDAL	S2–/S4	Muscles of perineum; skin of penis and scrotum in male, clitoris, labia majora, and lower vagina in female
QUADRATUS FEMORIS	L4–L5, S1	Inferior gemellus and quadratus femoris muscles
OBTURATOR INTERNUS	L5–S2	Superior gemellus and obturator internus muscles
TIBIAL (MED. POPLITEAL)	L4–S3	Gastrocnemius, plantaris, soleus, popliteus, tibialis posterior, flexor digitorum, and hallicus muscles
MEDIAL PLANTAR	L4–S3	Abductor hallicus, flexor digitorum brevis, and flexor hallicus muscles; skin over medial two-thirds of plantar surface of foot
LATERAL PLANTER	L4–S3	Abductor digiti minimi, quadratus plantae, lumbricales, adductor hallicus, flexor digiti minimi brevis, interossei muscles; skin over lateral third of plantar surface of foot
COMMON PERONEAL (LAT. POPLITEAL)	L4–S2	
SUPERFICIAL PERONEAL	L4–S2	Peroneus longus and peroneus brevis muscles; skin over distal third of anterior aspect of leg and dorsum of foot
DEEP PERONEAL	L4–S2	Tibialis anterior, extensor hallicus longus, peroneus tertius, and extensor digitorum brevis muscles; skin over great and second toes
SUP. GLUTEAL	L4–L5, S1	Gluteus minimus, and gluteus medius muscles and tensor fasciae latae
INF. GLUTEAL	L5–S2	Gluteus maximus muscle
PIRIFORMIS	S1–S2	Piriformis muscle
POST. CUTANEOUS	S1–S3	Skin over anal region, lower aspect of buttock, upper posterior aspect of thigh, upper part of calf, scrotum in male, and labia majora in female
SCIATIC	L4–S3	See Tibial and Common Peroneal

Chart #5 : The 12 Cranial Nerves (See 7.15)

Number and name	Foramen transmitting	Composition	Location of cell bodies	Function
I Olfactory	Foramen in ethmoidal cribriform plate	Sensory	Bipolar cells in nasal mucosa	Olfaction
II Optic	Optic canal	Sensory	Ganglion cells of retina	Vision
III Oculomotor	Superior orbital fissure	Motor	Oculomotor nucleus	Motor impulses to levator palpebrae superioris and extrinsic eye muscles except superior oblique and lateral rectus; innervation to muscles that regulate amount of light entering eye and that focus the lens
		Sensory: proprioception		Proprioception from muscles innervated with motor fibers
IV Trochlear	Superior orbital fissure	Motor	Trochlear nucleus	Motor impulses to superior oblique muscle of eyeball
		Sensory: proprioception		Proprioception from superior oblique muscle of eyeball
V Trigeminal				
Ophthalmic division	Superior orbital fissure	Sensory	Semilunar ganglion	Sensory impulses from cornea, skin of nose, forehead, and scalp
Maxillary division	Foramen rotundum	Sensory	Semilunar ganglion	Sensory impulses from nasal mucosa, upper teeth and gums, palate, upper lip, and skin of cheek
		Sensory	Semilunar ganglion	Sensory impulses from temporal region, tongue, lower teeth and gum, and skin of chin and lower jaw
Mandibular division	Foramen ovale	Sensory: proprioception		Proprioception from muscles of mastication
		Motor	Motor trigeminal nucleus of the PONS	Motor impulses to muscles of mastication and muscle that tenses tympanum
VI Abducens	Superior orbital fissure	Motor	Abducens nucleus	Motor impulses to lateral rectus muscle of eyeball
		Sensory: proprioception		Proprioception from lateral rectus muscle of eyeball
VII Facial	Stylomastoid foramen	Motor	Motor facial nucleus	Motor impulses to muscles of facial expression and muscle that tenses the stapes
		Motor: parasympathetic	Superior salivatory nucleus	Secretion of tears from lacrimal gland and salivation from sublingual and submandibular salivary glands
		Sensory	Geniculate ganglion	Sensory impulses from taste buds on anterior two-thirds of tongue; nasal and palatal sensation
		Sensory: proprioception		Proprioception from muscles of facial expression

Chart #5 Cont'd

Number and name	Foramen transmitting	Composition	Location of cell bodies	Function
VIII Vestibulocochlear	Internal acoustic meatus	Sensory	Vestibular ganglion	Sensory impulses associated with equilibrium
			Spiral ganglion	Sensory impulses associated with hearing
IX Glossopharyngeal	Jugular foramen	Motor	Nucleus ambiguus	Motor impulses to muscles of pharynx used in swallowing
		Sensory: proprioception	Petrosal ganglion	Proprioception from muscles of pharynx
		Sensory	Petrosal ganglion	Sensory impulses from taste buds on posterior one-third of tongue; pharynx, middle ear cavity, carotid sinus
		Parasympathetic	Interior salivatory nucleus	Salivation from parotid salivary gland
X Vagus	Jugular foramen	Motor	Nucleus ambiguus	Contraction of muscles of pharynx (swallowing) and larynx (phonation)
		Sensory: proprioception		Proprioception from muscles of pharynx and larynx
		Sensory	Nodose ganglion	Sensory impulses from taste buds on rear of tongue; sensations from auricle of ear; general visceral sensations
		Motor: parasympathetic	Dorsal motor nucleus	Regulate visceral motility
XI Accessory	Jugular foramen	Motor	Nucleus ambiguus	Laryngeal movement
			Spinal accessory nucleus	Motor impulses to trapezius and sternocleidomastoid muscles for movement of head, neck, and shoulders
		Sensory: proprioception		Proprioception from muscles that move head, neck, and shoulders
XII Hypoglossal	Hypoglossal canal	Motor	Hypoglossal nucleus	Motor impulses to intrinsic and extrinsic muscles of tongue and infrahyoid muscle
		Sensory: proprioception		Proprioception from muscles of tongue

 **: The Autonomic Nervous System (See 7.16)
(The Visceral Efferent Nervous System)**

Activities of Autonomic Nervous System	Sympathetic Division	Parasympathetic Division
Structure	**Effect of Stimulation**	**Effect of Stimulation**
EYE		
IRIS	Dilation of pupil, contraction of dilator muscle of iris	Constriction of pupil, contracts pupillary sphincter muscle of iris
CILIARY MUSCLE	(No innervation)	Contracts, accommodates lens for near vision
GLANDS		
SWEAT	Increased secretion rate	(No innervation)
LACRIMAL	Vasoconstriction, which inhibits secretion	Increased secretion (Normal or excessive)
SALIVARY	Vasoconstriction, which decreases secretion rate	Increased secretion, vasodilation
GASTRIC	Vasoconstriction, which inhibits secretion	Increased secretion
INTESTINAL	Vasoconstriction, which inhibits secretion	Increased secretion
PANCREAS	Vasoconstriction, which inhibits secretion	Promotes secretion
PANCREATIC ISLETS of LANGERHANS	Decreases secretion of INSULIN	Increases secretion of INSULIN
ADRENAL MEDULLA	Increases release of epinephrine and norepinephrine	(No innervation)
ADRENAL CORTEX	Increases glucocorticoid release	(No innervation)
LUNGS		
BRONCHIAL AIRWAYS	Dilation	Constriction
HEART		
RATE and STRENGTH of CONTRACTION	Increases	Decreases
CORONARY VESSELS	Dilation	Constriction
BLOOD VESSELS	Vasoconstriction**	Mostly no innervation
LIVER	Release of stored (glycogen) glucose: GLYCOGENOLYSIS Decreased bile secretion	Promotes storage of glucose (glycogen): GLYCOGENESIS Increased bile secretion
STOMACH	Decreased motility	Increased motility
INTESTINES	Decreased motility	Increased motility
SPLEEN	Contraction and release of stored blood to circulating volume	(No innervation)
URINARY BLADDER	Increases tone in internal sphincter and relaxes muscular wall	Contracts muscular wall, relaxes internal sphincter
HAIR FOLLICLES	Contraction of arrector pili muscles ("goose flesh")	(No innervation)
GENITALS	Contraction of transport structures during orgasm. In male, vasoconstriction of ductus deferens, seminal vesicle, prostate gland, in female, reverse uterine peristalsis	Vasodilation and erection in both sexes; vaginal secretion increase
KIDNEY	Decreased urine volume (due to constriction of blood vessels)	(No innervation)

**The dilatation of the blood vessels of the heart, lung, and skeletal muscle during exercise
is a well-documented fact. However, the role the sympathetic nervous system plays in this dilation
as compared to local control has been a continuous controversy in physiology. Please consult
current literature and your instructor for the current status of this controversy.

Chart #7 : **Classification of the Senses (See 7.18)**

3 Types of Classification of the Senses: ① GENERAL or SPECIAL (COMPLEXITY of RECEPTORS)
② SOMATIC or VISCERAL (LOCATION of RECEPTORS)
③ EXTERO-, VISCERO-, and PROPRIO-, CEPTORS (LOCATION of RECEPTORS and TYPES of STIMULI TO WHICH THEY RESPOND)

① **GENERAL SENSES** • Widespread throughout the body • Simple in structure • Simple neural pathways	**SPECIAL SENSES** • Localized in 1 or 2 specific regions • Complex in structure • Extensive neural pathways

② **SOMATIC (SOMASTHETIC) SENSES**
• Receptors localized within body wall
• Includes: (a) EXTEROCEPTORS: (1) (in skin) and (2) in Special Sense Organs
    (b) PROPRIOCEPTORS (in joints, tendons, muscles,
        LABYRINTH of INNER EAR)
	VISCERAL SENSES
	• Receptors located in visceral organs
	• Includes VISCEROCEPTERS

③ **(a) EXTEROCEPTORS** • Located near the surface of the body • Respond to stimuli from *external environment* • Connected with: (1) SKIN, (2) SPECIAL SENSE ORGANS	**(b) PROPRIOCEPTORS** • Stimulated by changes in the *locomotor system* of the BODY • Sense of EQUILIBRIUM or BALANCE • Awareness of POSITION • MOVEMENT of BODY in SPACE	**VISCEROCEPTORS (ENTEROCEPTORS)** • Located in viscera (hollow organs) • Respond to stimuli changes from the *internal environment* (e.g., distension in hollow organs):

(1) CUTANEOUS (SKIN) RECEPTORS within the DERMIS:
• TACTILE RECEPTORS → TOUCH
• MECHANORECEPTORS → PRESSURE
• THERMORECEPTORS → TEMPERATURE
• NOCIORECEPTORS ·
  (CHEMORECEPTOR) → PAIN
 	
(2) SPECIALIZED RECEPTORS in SPECIAL SENSE ORGANS
• In EAR:
  PRESSURE RECEPTORS in the hair cells of the organ of Corti in the COCHLEA of the inner ear
• In EYE:
  PHOTORECEPTORS (RODS and CONES) in the retina of the eye
• In NOSE:
  OLFACTORY CHEMORECEPTORS* in the nasal epithelium of the nasal cavity
• In TONGUE
  GUSTATORY (TASTE) CHEMORECEPTORS* in the dorsum of the tongue
 	
*CHEMORECEPTORS are stimulated by CHEMICALS
	• In JOINTS
	  Stretch and Pressure
	• In TENDONS
	  Tension and stretch
	• In MUSCLES
	  Stretch
	• In LABYRINTH of the INNER EAR
	  Movements and position of head
		• PAIN
		• HUNGER
		• THIRST
		• FATIGUE
		• NAUSEA
		
		Specialized BARORECEPTORS in the blood vessels of the circulatory system are sensitive to CHANGES in BLOOD PRESSURE

EXTEROCEPTORS • May convey information to CONSCIOUSNESS, with awareness of sensation, leading to responses planned in the cerebral cortex • May serve as afferent pathways for reflex (involuntary) action with or without rising to consciousness.	**VISCEROCEPTOR and PROPRIOCEPTOR** • Most information does not rise to consciousness

OVERSTIMULATION *of any* RECEPTOR CAN GIVE RISE to the SENSATION of *pain.*

UNIT 3: INTEGRATION AND CONTROL

Chapter 8 Endocrine System

The body could not function without the well-coordinated integrating and control mechanisms of both the ENDOCRINE SYSTEM and THE NERVOUS SYSTEM. Both systems work together harmoniously in an interlocking manner to maintain BODY HOMEOSTASIS by INHIBITING or STIMULATING cell activity and organ function.

THE NERVOUS SYSTEM can inhibit or stimulate the release of hormones (biologically active chemicals—REGULATORY MOLECULES) and the ENDOCRINE SYSTEM can inhibit or stimulate the flow of nerve impulses.

The NERVOUS SYSTEM controls and integrates body activities by transmission of electrical impulses over neurons; the ENDOCRINE SYSTEM does so by secretion of HORMONES into the blood stream where they are transported throughout the body to all body tissues.

ENDOCRINE GLANDS secrete their hormones into the extra-cellular spaces around the secretory cells and directly into the blood via capillaries. (They do *not* secrete their products via ducts into body cavities or organ lumens, or onto a free surface as do EXOCRINE GLANDS.) Thus, ENDOCRINE GLANDS are also called DUCTLESS GLANDS.

Hormones are delivered to every cell in the body, but each hormone only affects SPECIFIC TARGET CELLS and ORGANS. (The effect a hormone has on a specific site varies with its concentration and the concentration of other hormones in the blood.)

A TARGET CELL must have SPECIFIC RECEPTOR PROTEINS in order to respond to a specific hormone. Hormones can be functionally categorized into 3 groups based on the *location of receptor proteins:*

RECEPTOR PROTEIN SITE

■ THYROID HORMONES → within the nucleus of the target cells (Thyroid Gland)

■ STEROID HORMONES → within the cytoplasm of the target cells (Adrenal Cortex, Gonads)

■ CATECHOLAMINES, POLYPEPTIDES, AND GLYCOPROTEINS → in the outer surface of the target cell membrane (All endocrine glands except the above)

Hormones do not generally accumulate in the blood because their half-life is very short (from under 2 minutes to 2 hours). They are removed quickly by the target organs (and by the liver, where they are converted to less active products by enzyme reactions).

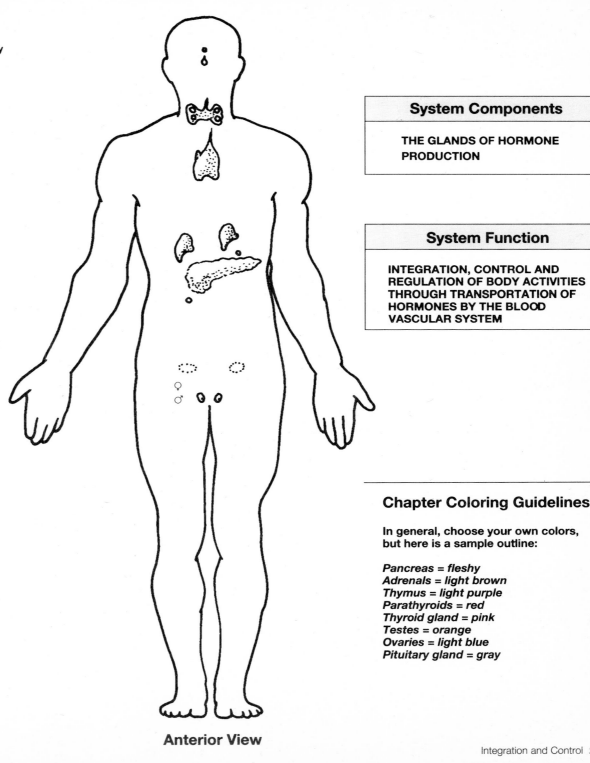

Anterior View

System Components

THE GLANDS OF HORMONE PRODUCTION

System Function

INTEGRATION, CONTROL AND REGULATION OF BODY ACTIVITIES THROUGH TRANSPORTATION OF HORMONES BY THE BLOOD VASCULAR SYSTEM

Chapter Coloring Guidelines

In general, choose your own colors, but here is a sample outline:

Pancreas = fleshy
Adrenals = light brown
Thymus = light purple
Parathyroids = red
Thyroid gland = pink
Testes = orange
Ovaries = light blue
Pituitary gland = gray

C = pink D = red E = light purple
F = light brown G = flesh
I = light blue J= orange
A–J will be used throughout the chapter.

★ See Charts #1 and #2

ENDOCRINE SYSTEM: GENERAL ORGANIZATION
Location of Major Endocrine Glands

★ There are 2 main types of ENDOCRINE (DUCTLESS) GLANDS:

■ EXCLUSIVELY ENDOCRINE
(located generally in the head and neck regions, close to the CNS control centers)

■ ENDOCRINE GLANDS WITHIN ORGANS THAT SERVE OTHER FUNCTIONS—they occur in "islands" of endocrine tissue within the larger structures (located generally in the lower abdominal areas).

★ The ENDOCRINE STRUCTURES do *not* form a usual SYSTEM where several organs work together to accomplish a single process. They are *diverse* organs that produce a wide variety of HORMONES that help regulate:
■ Total body metabolism
■ Many aspects of homeostasis
■ Growth
■ Reproduction

In many cases, the effects of certain hormones are:
■ Vital to the survival of the organism
■ Necessary for life

★ CRITERIA FOR DETERMINING IF A STRUCTURE IS AN ENDOCRINE GLAND

■ It is DUCTLESS, passing products directly into BLOOD VESSELS within the gland itself

■ It is extremely VASCULAR

■ It consists of specific cells (circumscribed in groups) that are obviously different in both structure and function from other body cells—they produce specific hormones.

■ Its hormones have specific and well-defined effects on body function; their removal or injection have clear-cut alterations in the body.

Regions of the Body

CRANIAL REGION
A [PITUITARY GLAND (HYPOPHYSIS)]

B [PINEAL GLAND (EPIPHYSIS CEREBRI)]

CERVICAL REGION
C [THYROID GLAND]

D [PARATHYROID GLANDS (4)] BEHIND AND EMBEDDED IN THE THYROID GLAND

THORACIC REGION
E [THYMUS GLAND]

ABDOMINAL REGION
F [ADRENAL (SUPRARENAL) GLANDS (2)]

G [PANCREAS]

H [GASTROINTESTINAL SECRETORY CELLS OF STOMACH AND DUODENUM]

PELVIC-PERINEAL REGION
I [OVARIES ♀]

J [TESTES ♂]

**ORGANS EXCLUSIVELY ENDOCRINE
(ONLY FUNCTION: SECRETION OF HORMONES)**

A   PITUITARY GLAND
B   PINEAL GLAND
C   THYROID GLAND
D   PARATHYROID GLANDS
F   ADRENAL GLANDS

**ORGANS NOT EXCLUSIVELY ENDOCRINE
(SERVING OTHER FUNCTIONS IN ADDITION TO PRODUCTION AND SECRETION OF HORMONES)**

ENDOCRINE GLANDS FOUND IN SMALL ISLANDS

E   THYMUS GLAND
G   PANCREAS
H   STOMACH
H   DUODENUM OF SMALL INTESTINE
I, J   OVARIES ♀ AND TESTES ♂

• SKIN
• LIVER
• KIDNEYS

• [HYPOTHALAMUS] (X)

• PLACENTA DURING PREGNANCY

★ See Chart #1
comparing EXOCRINE GLANDS and ENDOCRINE GLANDS

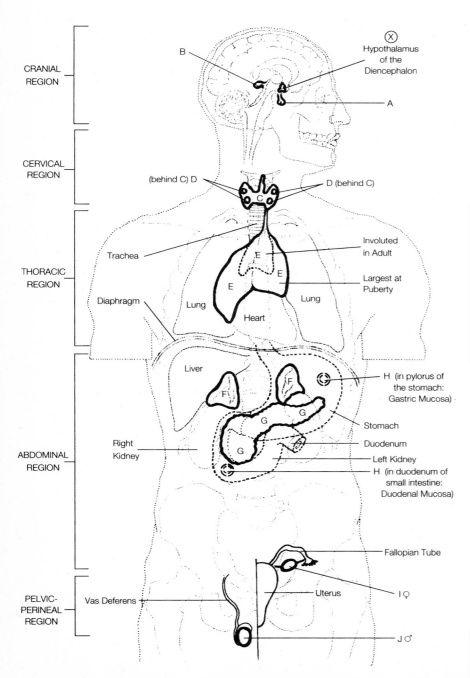

CRANIAL
REGION

B

Ⓧ
Hypothalamus
of the
Diencephalon

A

CERVICAL
REGION

(behind C) D

D (behind C)

Trachea

C

THORACIC
REGION

Involuted
in Adult

E

E

Largest at
Puberty

Diaphragm

E

Lung

Lung

Heart

Liver

H (in pylorus of
the stomach:
Gastric Mucosa)

F

F

G

G

Stomach

Duodenum

Right
Kidney

G

Left Kidney

H (in duodenum of
small intestine:
Duodenal Mucosa)

ABDOMINAL
REGION

Fallopian Tube

PELVIC-
PERINEAL
REGION

Vas Deferens

Uterus

I ♀

J ♂

GLANDS REGRESSIVE in the ADULT: PINEAL and THYMUS GLANDS

B PINEAL GLAND (See 7.9)

The PINEAL GLAND is very
small (5–8 mm long, 5 mm wide).
■ Cone-shaped (flattened)
■ Attached to the roof of the 3rd
VENTRICLE, just above the tectum
(roof) of the MIDBRAIN. Located in
a pocket near the splenium of the
corpus callosum (see illustration 8.2).
■ Covered by a capsule formed
from the PIA MATER (part of the
MENINGES covering the BRAIN).

The PINEAL GLAND is larger in
the child, and begins to harden
at puberty. (Calcification
deposits form BRAIN SAND)

Around the secretory cells (PINEALOCYTES
and NEUROGLIAL CELLS) is a high
innervation by the SYMPATHETIC
NERVOUS SYSTEM from the SUPERIOR
CERVICAL GANGLION. These nerves are
connected to the RETINAL CELLS
and appear to be an evolutionary remnant.

★ The PINEAL GLAND has *no* direct
connections to the rest of the BRAIN.
The physiology of the PINEAL GLAND
is obscure, but its main activity appears
to be the INHIBITION OF THE PITUITARY-
GONAD AXIS (specifically the OVARIES,
thus affecting MENSTRUATION).
MELATONIN is the only chemical
produced *solely* by the PINEAL GLAND.

HORMONES SECRETED by PINEAL	SECRETION EFFECTS
MELATONIN	• Active in the onset of puberty (?) • *Inhibits* PITUITARY-GONAD AXIS (?) • Entraining circadian rhythms (?)
ADRENOGLOMERULOTROPIN	• *Stimulates* ADRENAL CORTEX to secrete ALDOSTERONE (?)
SEROTONIN	• Normal BRAIN PHYSIOLOGY
GROWTH-INHIBITING FACTOR	• Inhibits growth

E THYMUS GLAND (See 10.3)

The THYMUS GLAND is a bilobed
lymphatic gland (see illustration 8.4)
■ Located in the ANTERIOR (and
SUPERIOR) MEDIASTINUM
  • Posterior to the STERNUM
  • In front of the AORTA
  • Between the LUNGS
■ Size and structure varies with age:
  • Very large in the infant
  • Maximum size at puberty
    (about 40 grams)
  • After puberty, the thymic
    tissue is replaced by fat and
    connective tissue
  • At maturity, the gland
    has atrophied

• A FIBROUS CAPSULE envelops both
lobes; invaginations of the capsule
separate the lobes into many LOBULES.
■ The OUTER CORTEX of each lobule
is the site for the production and
maturation of densely-packed
THYMUS-DEPENDENT LYMPHOCYTES
of the immune system, called
THYMOCYTES or T-CELLS, which
are packed in EPITHELIORETICULAR
FIBROUS TISSUE.
■ In the INNER MEDULLA of each
lobule the T-CELLS are widely
scattered amid large EPITHELIO-
RETICULAR CELLS and huge THYMIC
(HASSALL'S) CORPUSCLES.
• A number of hormones secreted
by the THYMUS GLAND (including
THYMOSIN) may help regulate the
immune system.

MAIN HORMONE SECRETED by THYMUS	SECRETION EFFECTS
THYMOSIN	• Possibly influences B-CELL LYMPHOCYTES (which may be pro- cessed in the fetal liver and spleen) to develop into PLASMA CELLS • PLASMA CELLS produce ANTIBODIES against ANTIGENS

8.2

Integration and Control **230**

Arteries = red
Capillary Plexuses = purple
Veins = blue

★ See also 7.9, 8.3

ENDOCRINE SYSTEM: CRANIAL REGION
The Pituitary Gland (Hypophysis) and the Hypothalamus: Structure and Circulation

★ The PITUITARY GLAND (HYPOPHYSIS) is small and round, about 1/2 in. wide.
■ Shaped like a pea

■ Attached to the inferior aspect of the brain (region of the diencephalon) by a stalk: the INFUNDIBULUM

■ Covered by DURA MATER (part of the meninges covering the brain)

■ Supported by the SELLA TURCICA (concave space) of the SPHENOID BONE

The PITUITARY GLAND is an ENDOCRINE GLAND that secretes a number of hormones that regulate many bodily processes:
■ Growth ■ Reproduction
■ Various metabolic activities such as:
• Maintenance of water balance
• Sugar and fat metabolism
• Regulation of body temperature

The secretions of the PITUITARY GLAND are controlled by:
■ The HYPOTHALAMUS (See 7.9)
■ NEGATIVE FEEDBACK INHIBITION from the TARGET GLANDS (See 8.3)

★ The PITUITARY GLAND is structurally and functionally divided into 2 lobes:
■ ANTERIOR LOBE (ADENOHYPOPHYSIS)
■ POSTERIOR LOBE (NEUROHYPOPHYSIS)
Both lobes of the pituitary gland receive hormones from the HYPOTHALAMUS via the INFUNDIBULAR PORTION of the HYPOPHYSEAL STALK. Each lobe has its own particular circulatory communicating pathway with specific regions of the HYPOTHALAMUS.

■ THE ANTERIOR LOBE (ADENOHYPOPHYSIS)
• Secretes its own hormones
• Hormone secretion is a response from HYPOTHALAMIC REGULATING HORMONES (REGULATING FACTORS) released from NEUROSECRETORY CELLS in the HYPOTHALAMUS
• Circulatory pattern involves a HYPOTHALAMIC-HYPOPHYSEAL PORTAL SYSTEM

■ THE POSTERIOR LOBE (NEUROHYPOPHYSIS)
• Does not secrete its own hormones
• Receives hormones from specific HYPOTHALAMIC NUCLEI, which function as endocrine glands. These hormones pass through fibers of the HYPOTHALAMIC-HYPOPHYSEAL TRACT and are stored in the posterior lobe.

P PITUITARY GLAND (HYPOPHYSIS)
① ADENOHYPOPHYSIS (ANTERIOR LOBE)
② NEUROHYPOPHYSIS (POSTERIOR LOBE)
③ HYPOPHYSEAL STALK

① ADENOHYPOPHYSIS: NON-NEURAL PORTION HAS SECRETORY CELLS
a ANTERIOR LOBE PARS DISTALIS
b INTERMEDIATE LOBE PARS INTERMEDIA
c PARS TUBERALIS

② NEUROHYPOPHYSIS: NEURAL PORTION HAS NO SECRETORY CELLS
d POSTERIOR LOBE PARS NERVOSA
e INFUNDIBULUM A DOWNWARD EXTENSION OF THE FLOOR OF THE 3rd VENTRICLE

③ HYPOPHYSEAL STALK: THE PATHWAY BETWEEN HYPOTHALAMUS AND HYPOPHYSIS (cte)
c PARS TUBERALIS OF ANTERIOR LOBE
e INFUNDIBULUM OF POSTERIOR LOBE CONTAINS NERVE FIBERS and NEUROGLIA-LIKE CELLS CALLED PITUICYTES

PITUITARY GLAND LOCATION IN BONE CONCAVITY
4 SELLA TURCICA OF SPHENOID BONE

Ⓧ HYPOTHALAMUS (HYPOTHALAMIC SECRETIONS TO PITUITARY)

5 PARAVENTRICULAR NUCLEUS (ENDOCRINE GLANDS) PRODUCE HORMONES SUPRAOPTIC NUCLEUS 6

10 SECRETORY NEURONS (NEUROSECRETORY CELLS)

AXONS OF THE HYPOTHALAMO-HYPOPHYSEAL NERVE TRACK

RELEASES REGULATING FACTORS (STIMULATING AND INHIBITING HORMONES) TO
X₁ MEDIAN EMINENCE (BASAL PORTION OF HYPOTHALAMUS)

PITUITARY (HYPOPHYSIS)

CIRCULATORY PATHWAYS OF PITUITARY

TO NEUROHYPOPHYSIS (POSTERIOR LOBE)
• Secretes hormones from HYPOTHALAMUS into the blood stream
• Only a storage bin
• Not an endocrine gland

7 CAPILLARY PLEXUS OF THE INFUNDIBULAR PROCESS

8 INFERIOR HYPOPHYSEAL ARTERY

9 POSTERIOR HYPOPHYSEAL VEIN

TO ADENOHYPOPHYSIS (ANTERIOR LOBE)

11 SUPERIOR HYPOPHYSEAL ARTERY

STALK HYPOTHALAMIC-HYPOPHYSEAL PORTAL SYSTEM:
12 PRIMARY CAPILLARY PLEXUS
13 HYPOPHYSEAL PORTAL VEINS
14 VENULE
15 SINUSOIDS (VENOUS) (SECONDARY CAPILLARY PLEXUS)
16 ANTERIOR HYPOPHYSEAL VEIN

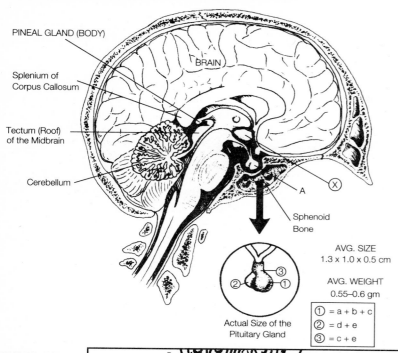

PINEAL GLAND (BODY)

BRAIN

Splenium of
Corpus Callosum

Tectum (Roof)
of the Midbrain

Cerebellum

A

Ⓧ

Sphenoid
Bone

AVG. SIZE
1.3 x 1.0 x 0.5 cm

AVG. WEIGHT
0.55–0.6 gm

③
②   ①

Actual Size of the
Pituitary Gland

| ① = a + b + c |
| ② = d + e |
| ③ = c + e |

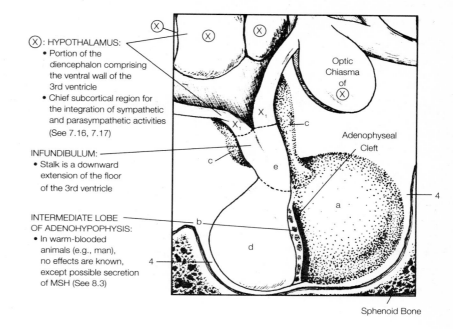

Ⓧ : HYPOTHALAMUS:
- Portion of the diencephalon comprising the ventral wall of the 3rd ventricle
- Chief subcortical region for the integration of sympathetic and parasympathetic activities (See 7.16, 7.17)

INFUNDIBULUM:
- Stalk is a downward extension of the floor of the 3rd ventricle

INTERMEDIATE LOBE OF ADENOHYPOPHYSIS:
- In warm-blooded animals (e.g., man), no effects are known, except possible secretion of MSH (See 8.3)

Optic Chiasma of Ⓧ

Adenohypyseal Cleft

Sphenoid Bone

★ HORMONAL AND CIRCULATOR PATHWAYS BETWEEN THE PITUITARY GLAND (HYPOPHYSIS) AND THE HYPOTHALAMUS

★ COMPARISON OF ADENOHYPOPHYSIS AND NEUROHYPOPHYSIS

- 2 DISTINCT TYPES OF TISSUES

- 2 DIFFERENT EMBRYONIC ORIGINS

- SECRETE DIFFERENT HORMONES REGU- LATED BY DIFFERENT CONTROL SYSTEMS

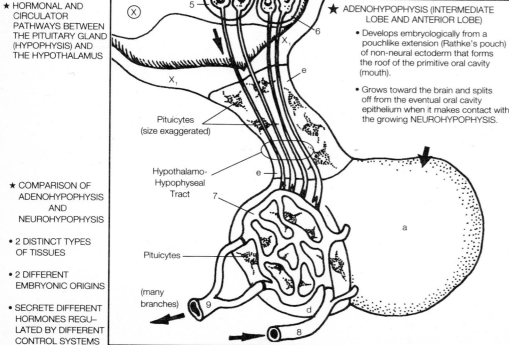

Pituicytes (size exaggerated)

Hypothalamo- Hypophyseal Tract

Pituicytes

(many branches)

★ ADENOHYPOPHYSIS (INTERMEDIATE LOBE AND ANTERIOR LOBE)

- Develops embryologically from a pouchlike extension (Rathke's pouch) of non-neural ectoderm that forms the roof of the primitive oral cavity (mouth).

- Grows toward the brain and splits off from the eventual oral cavity epithelium when it makes contact with the growing NEUROHYPOPHYSIS.

NEUROHYPOPHYSIS (Posterior Lobe)

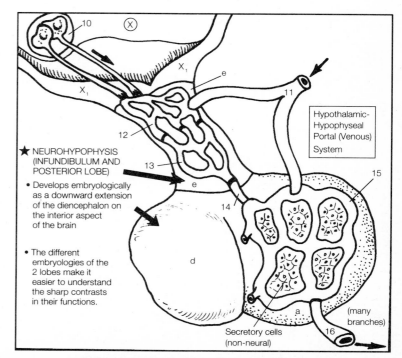

★ NEUROHYPOPHYSIS (INFUNDIBULUM AND POSTERIOR LOBE)

- Develops embryologically as a downward extension of the diencephalon on the interior aspect of the brain

- The different embryologies of the 2 lobes make it easier to understand the sharp contrasts in their functions.

Hypothalamic- Hypophyseal Portal (Venous) System

Secretory cells (non-neural)

(many branches)

ADENOHYPOPHYSIS (Anterior Lobe)

ENDOCRINE SYSTEM: CRANIAL REGION

The Pituitary Gland (Hypophysis) and the Hypothalamus:
Hormonal Secretions, Target Organs, and Their Hormones

Pick 9 of your favorite colors and 1st do the 9 hormones: 1–9. Then color in the target organ hormone secretions f–I (7 of them).

★ See Chart #3

★ The HYPOPHYSIS and HYPOTHALAMUS are strongly coordinated in the regulation of:
- ■ SPECIFIC ENDOCRINE GLANDS
- ■ A NUMBER OF METABOLIC FUNCTIONS

There are 9 major hormones secreted by the PITUITARY GLAND: The ADENOHYPOPHYSIS secretes 7 hormones and the NEUROHYPOPHYSIS secretes 2 hormones

ADENOHYPOPHYSIS: secretes 7 hormones
- ■ The secretory cells are ACIDOPHILS, BASOPHILS, and CHROMOPHOBES.

- ■ Most of the hormones secreted are called TROPHIC HORMONES: they make their target organs HYPERTROPHY (increase in size).

- ■ The secretory cells are stimulated (or inhibited) by REGULATING FACTORS formed from SECRETORY NEURONS in the HYPOTHALAMUS.

- ■ Their axon endings do *not* enter the ANTERIOR LOBE, but end in the MEDIAN EMINENCE OF THE HYPOTHALAMUS.

- ■ From here, the hormones are drained by VENULES IN THE INFUNDIBULUM, then to a SECONDARY CAPILLARY PLEXUS (SINUSOIDS) IN THE LOBE, receiving venous blood.

- ■ This vascular link with the MEDIAN EMINENCE forms a HYPOTHALAMIC-HYPOPHYSEAL PORTAL SYSTEM (similar to the HEPATIC PORTAL SYSTEM in the DIGESTIVE SYSTEM).

- ■ The REGULATING FACTORS (and thus the trophic hormones) are controlled by HORMONES produced by the TARGET ORGANS!! Thus, the ANTERIOR PITUITARY and HYPOTHALAMUS are *not* "MASTER GLANDS."

A chain of specificity exists for each TROPHIC HORMONE ⟶ TARGET ORGAN(S) ⟶ TARGET ORGAN HORMONE ⟶ REGULATING FACTORS (HYPOTHALAMIC HORMONES) ⟶ back to the specific TROPHIC HORMONE

★ NEUROHYPOPHYSIS: secretes 2 hormones
- ■ Contains no secretory cells, and receives its hormones from the HYPOTHALAMUS along axon fibers into the lobe.
- ■ Secretion of these hormones is controlled by NEUROENDOCRINE REFLEXES.

★ The BLOOD SUPPLY OF THE PITUITARY GLAND is extremely rich and is furnished by the CIRCLE OF WILLIS. (See 9.9)

HORMONES of the HYPOTHALAMUS

① (ADH) ANTIDIURETIC HORMONE — VASOPRESSIN

② OXYTOCIN — "HORMONE OF LABOR"

HYPOTHALAMIC HORMONES = RF
(REGULATING FACTORS)

HORMONES of the HYPOPHYSIS (PITUITARY GLAND)

NEUROHYPOPHYSIS (POSTERIOR LOBE)

ADH and OXYTOCIN are *produced* in the HYPOTHALAMUS (by the PARAVENTRICULAR and SUPRAOPTIC NUCLEI). They are trans-ported along the AXONS of the *HYPOTHALAMO-HYPOPHYSEAL NERVE TRACT* and then *stored* in the POSTERIOR LOBE. The POSTERIOR LOBE has *no* SECRETORY CELLS and IS NOT AN ENDOCRINE GLAND, BUT A STORAGE WAREHOUSE.

① (ADH) ANTIDIURETIC H. — VASOPRESSIN

② OXYTOCIN — "HORMONE OF LABOR"

ADENOHYPOPHYSIS (ANTERIOR LOBE): TROPHIC HORMONES

③ (TSH) THYROTROPHIN — THYROID-STIMULATING HORMONE

④ (ACTH) ADRENOCORTICOTROPHIN — CORTICOTROPHIN

⑤ (FSH) FOLLICULOTROPHIN — FOLLICLE-STIMULATING HORMONE

⑥ (LH) ♀ LUTEOTROPHIN — LUTEINIZING HORMONE*

⑦ PROLACTIN — LUTEOTROPIC HORMONE

⑧ (GH)/SOMATOTROPHIN (STH) — GROWTH HORMONE

⑨ (MSH) MELANOCYTE-STIMULATING HORMONE

* In the male ♂, LH is known as ICSH: INTERSTITIAL CELL-STIMULATING HORMONE

HYPOPHYSEAL HORMONE	TARGET ORGANS	TARGET ORGAN HORMONE SECRETIONS/EFFECTS
③ →	THYROID GLAND	→ f THYROID HORMONES — THYROXINE (T₄)
④ →	ADRENAL GLANDS (CORTEX)	→ g ADRENAL CORTICAL HORMONES
⑤⑥⑦ →	OVARIES ♀	→ h ESTROGEN — i PROGESTERONE
⑤⑥ →	TESTES ♂	→ j TESTOSTERONE
① →	KIDNEYS	→ k WATER CONSERVATION
②⑦ →	BREAST	→ l MILK EJECTION/SUCKLING
⑧ →	ALL BODY ORGANS (BONE, MUSCLE, ADIPOSE TISSUE)	
⑨ →	SKIN	

SELF-REGULATION BY NEGATIVE FEEDBACK INHIBITION

★ Chart #3 includes specific HYPOTHALAMIC REGULATING FACTORS that stimulate release of PITUITARY HORMONES

★ NOTE: Reference # 1–16, A, X, X₁, a–e are carried over from previous pages (8.2) ∞∞∞∞∞∞ indicates secretion of hormone by a secretory cell.

ADENOHYPOPHYSIS
3 TYPES OF SECRETORY CELLS:

HORMONES SECRETED

- ■ ACIDOPHILS ⟶ ⑦⑧
- ■ BASOPHILS ⟶ ③④⑤⑥⑨
- ■ CHROMOPHOBES ⟶ ?

Ⓧ HYPOTHALAMIC NUCLEI CENTERS:

5

ADH

①

②

6

10

RF

X₁
Median Eminence of Hypothalamus

Optic Chiasma of Hypothalamus

11

X₁
Median Eminence of Hypothalamus

12

13

14

15

KIDNEYS

① ADH

②

7

9

8

b

d

a

16

size of PITUICYTES exaggerated

BREAST

UTERUS

SKIN

⑨

HORMONES SECRETED into BLOOD CIRCULATION

NEGATIVE FEEDBACK INHIBITION

SELF-REGULATION of HORMONAL SECRETION

TARGET ORGANS

RF

⑧ GH

⑦

⑥ LH ICSH

⑤ FSH

④ ACTH

③ TSH

?

SECRETORY CELLS

GROWTH:
BONE
MUSCLE
FATTY TISSUE

BREAST

OVARY ♀

TESTES ♂

ADRENAL GLANDS (CORTEX)

THYROID GLAND

l

h   i

j

g

f

TARGET ORGAN HORMONES

8.3

Integration and Control **233**

ENDOCRINE SYSTEM: NECK (CERVICAL) REGION
The Thyroid Gland

Color (C) Thyroid Gland = pink, and the individual divisions, follicles, and secretions any colors you want.
Arteries = warm Veins = cool

★ The THYROID GLAND is the largest of the endocrine glands, and weighs about 1 oz. (25 g.)
- Located just below the LARYNX in front of the TRACHEA (at 2nd–4th TRACHEAL RINGS)
- Covered by a thin connective tissue capsule

★ Microscopically, the THYROID GLAND consists of many spherical hollow sacs, or THYROID FOLLICLES.
- The interior of each FOLLICLE is filled with a protein-rich fluid, COLLOID. In this fluid is THYROGLOBULIN: A STORED FORM OF THE PRINCIPAL THYROID HORMONES: T_4 and T_3

★ ■ The PRINCIPAL FOLLICULAR CELLS form a simple epithelial wall around the THYROID FOLLICLES, and reach the lumen of their hollow sacs.
- They synthesize T_4 and T_3.
- T_4 and T_3 are hormones extremely important in CONTROLLING OVERALL BODY METABOLIC RATE and ARE ESSENTIAL FOR HEALTHY GROWTH and DEVELOPMENT.

- The wall cells between the FOLLICULAR CELLS that do *not* reach the inner lumen are the C CELLS, or PARAFOLLICULAR CELLS.
- C CELLS secrete a hormone called TCT (THYROCALCITONIN) important in BONE and CALCIUM metabolism.

★ The THYROID GLAND has a rich supply of BLOOD. It receives 80–120 ml of BLOOD per MINUTE

NERVE SUPPLY:
- Postganglionic fibers from the superior and middle CERVICAL SYMPATHETIC GANGLIA.

© **THE THYROID GLAND**

Structure

a **CONNECTIVE TISSUE CAPSULE**   c **ISTHMUS**

b **R. AND L. LATERAL LOBES**   d **PYRAMIDAL LOBE OF THE ISTHMUS**

Histology:∞➤ =Hormones Synthesized/Secreted)

SPHERICAL HOLLOW SACS

e **THYROID FOLLICLES**

(1) **COLLOID** **THYROGLOBULIN** (in COLLOID, T_4 and T_3 are attached to THYROGLOBULIN PROTEIN)

SIMPLE CUBOIDAL EPITHELIAL WALL around FOLLICLES

f **PRINCIPAL FOLLICULAR CELLS**

(2) ∞➤ **THYROXINE (T_4)**
(TEYRAIODOTHYRONINE)

(3) ∞➤ **TRIIODOTHYRONINE (T_3)**

SECRETION EFFECTS
- Increases rate of protein synthesis
- Increases rate of energy release from carbohydrates
- Increases oxygen consumption (all tissues)
- Regulates rate of growth and development
- Stimulates maturity of nervous system

HORMONAL REGULATION SOURCE
- HYPOTHALAMUS
▽
- Release of TSH from Anterior Lobe of the PITUITARY GLAND

g **PARAFOLLICULAR CELL [C CELLS]**

(4) ∞➤ **THYROCALCITONIN (TCT)**
(CALCITONIN)

- Regulates calcium homeostasis
- Inhibits release of calcium from bone tissue, thus lowering blood calcium levels
- Antagonizes action of the PARATHYROID HORMONE and vitamin D_3

- Calcium levels in blood

Blood Circulation

1 **SUPERIOR THYROID ARTERY** BRANCH OF EXTERNAL CAROTID ARTERY

2 **INFERIOR THYROID ARTERY** BRANCH OF SUBCLAVIAN ARTERY

3 **SUPERIOR THYROID VEIN**
4 **MIDDLE THYROID VEIN** PASS INTO THE INTERNAL JUGULAR VEINS

5 **INFERIOR THYROID VEIN** JOINS THE BRACHIOCEPHALIC VEIN

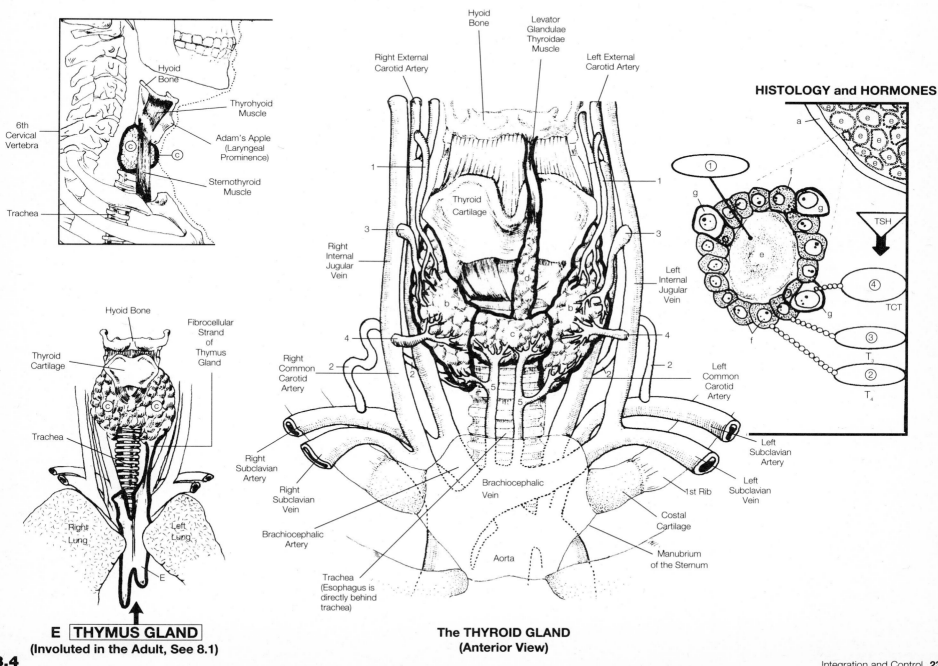

6th Cervical Vertebra

Hyoid Bone

Thyrohyoid Muscle

Adam's Apple (Laryngeal Prominence)

Sternothyroid Muscle

Trachea

Hyoid Bone

Thyroid Cartilage

Fibrocellular Strand of Thymus Gland

Trachea

Right Lung

Left Lung

E

E **THYMUS GLAND**
(Involuted in the Adult, See 8.1)

Hyoid Bone

Right External Carotid Artery

Levator Glandulae Thyroidae Muscle

Left External Carotid Artery

Thyroid Cartilage

Right Internal Jugular Vein

Left Internal Jugular Vein

Right Common Carotid Artery

Left Common Carotid Artery

Right Subclavian Artery

Right Subclavian Vein

Left Subclavian Artery

Left Subclavian Vein

Brachiocephalic Artery

Brachiocephalic Vein

1st Rib

Costal Cartilage

Trachea (Esophagus is directly behind trachea)

Aorta

Manubrium of the Sternum

The THYROID GLAND
(Anterior View)

HISTOLOGY and HORMONES

TSH

④ TCT

③ T_3

② T_4

ENDOCRINE SYSTEM: NECK (CERVICAL) REGION
The Parathyroid Glands

★ The 4 PARATHYROID GLANDS are small, rounded masses of tissue:
■ Embedded on posterior surface of each lateral lobe of the THYROID GLAND

■ 0.1–0.3 in. long, 0.07–0.2 in. wide 0.02–0.07 in. thick

Histologically, 2 kinds of epithelial cells are immersed in CONNECTIVE TISSUE SEPTA:

■ PRINCIPAL (CHIEF) CELLS synthesize the PARATHYROID HORMONE (PTH), which acts on the bones, kidneys, and intestines promoting a rise in blood calcium levels.

■ OXYPHIL CELLS support the CHIEF CELLS, and *may* synthesize a reserve capacity of PTH.

NERVE SUPPLY
Receives fibers arising from:
■ Pharyngeal branches of the VAGUS NERVE

■ CERVICAL SYMPATHETIC GANGLIA

★ BLOOD SUPPLY and DRAINAGE
■ Similar to that of the THYROID GLAND

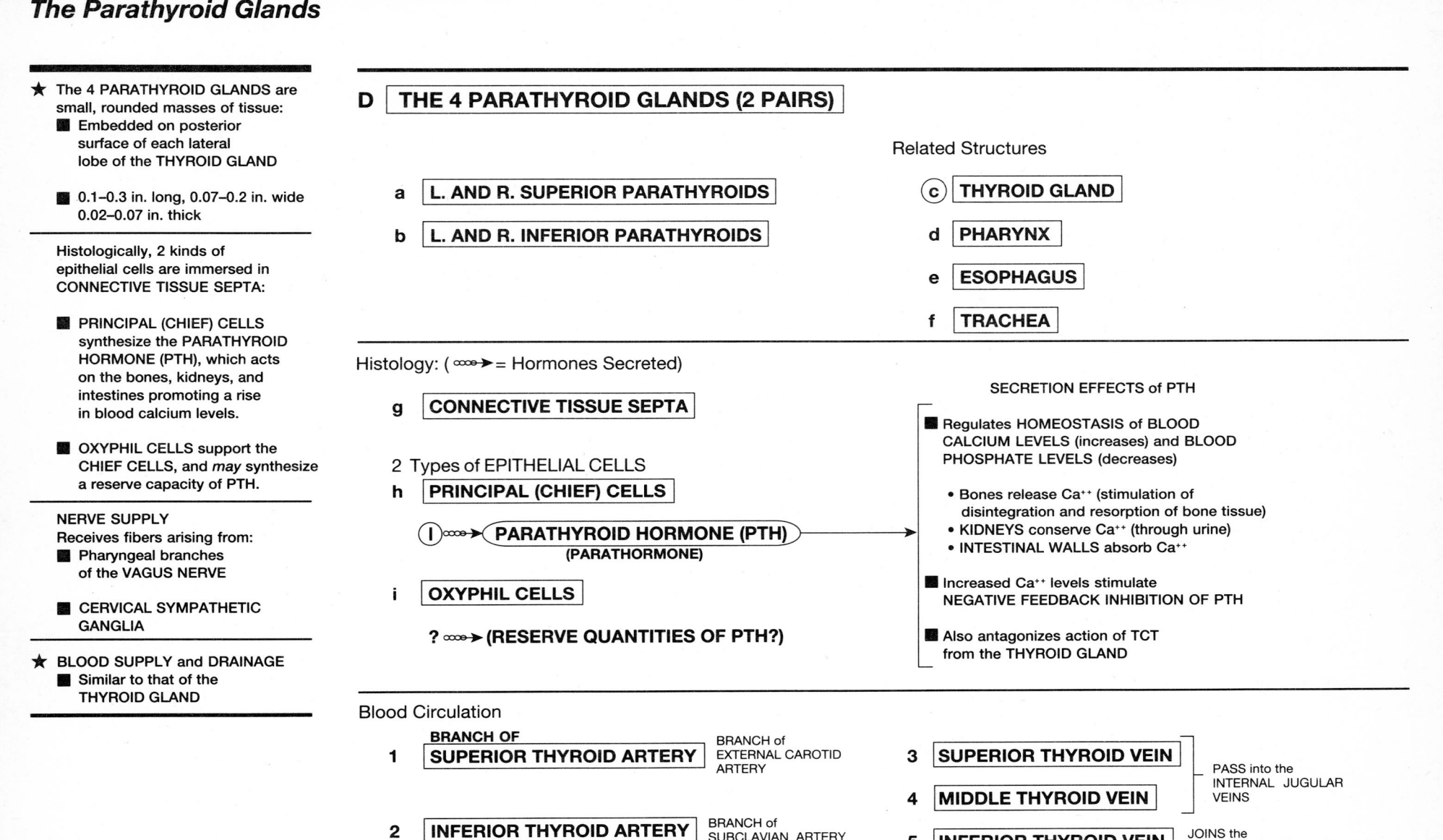

D | THE 4 PARATHYROID GLANDS (2 PAIRS)

Related Structures

a | L. AND R. SUPERIOR PARATHYROIDS

b | L. AND R. INFERIOR PARATHYROIDS

c | THYROID GLAND

d | PHARYNX

e | ESOPHAGUS

f | TRACHEA

Histology: (∞→ = Hormones Secreted)

SECRETION EFFECTS of PTH

g | CONNECTIVE TISSUE SEPTA

■ Regulates HOMEOSTASIS of BLOOD CALCIUM LEVELS (increases) and BLOOD PHOSPHATE LEVELS (decreases)

2 Types of EPITHELIAL CELLS
h | PRINCIPAL (CHIEF) CELLS

l ∞→ **PARATHYROID HORMONE (PTH)**
(PARATHORMONE)

• Bones release Ca^{++} (stimulation of disintegration and resorption of bone tissue)
• KIDNEYS conserve Ca^{++} (through urine)
• INTESTINAL WALLS absorb Ca^{++}

i | OXYPHIL CELLS

■ Increased Ca^{++} levels stimulate NEGATIVE FEEDBACK INHIBITION OF PTH

? ∞→ (RESERVE QUANTITIES OF PTH?)

■ Also antagonizes action of TCT from the THYROID GLAND

Blood Circulation

BRANCH OF
1 | SUPERIOR THYROID ARTERY
BRANCH of EXTERNAL CAROTID ARTERY

3 | SUPERIOR THYROID VEIN

4 | MIDDLE THYROID VEIN
PASS into the INTERNAL JUGULAR VEINS

2 | INFERIOR THYROID ARTERY
BRANCH of SUBCLAVIAN ARTERY

5 | INFERIOR THYROID VEIN
JOINS the BRACHIOCEPHALIC VEIN

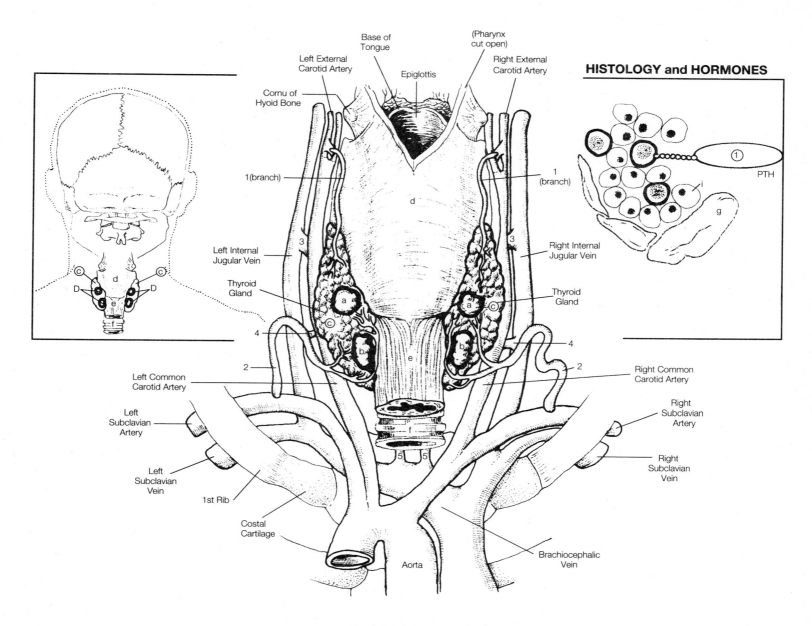

Base of Tongue

(Pharynx cut open)

Left External Carotid Artery

Right External Carotid Artery

Epiglottis

Cornu of Hyoid Bone

HISTOLOGY and HORMONES

1(branch)

1 (branch)

PTH

d

Left Internal Jugular Vein

Right Internal Jugular Vein

Thyroid Gland

Thyroid Gland

Left Common Carotid Artery

Right Common Carotid Artery

Left Subclavian Artery

Right Subclavian Artery

Left Subclavian Vein

Right Subclavian Vein

1st Rib

Costal Cartilage

Brachiocephalic Vein

Aorta

The 4 PARATHYROID GLANDS

(Posterior View)

ENDOCRINE SYSTEM: ABDOMINAL REGION
The Adrenal (Suprarenal) Glands

★ The pair of ADRENAL (SUPRARENAL) GLANDS lie atop the SUPERIOR and MEDIAL BORDERS OF THE KIDNEYS (within the RENAL FASCIA). Like the KIDNEYS, they are RETROPERITONEAL ORGANS.

★ The ADRENAL GLANDS are actually 2 different glands (CORTEX and MEDULLA) in a FIBROUS CAPSULE. They have different and separate:
- Hormonal secretions
- Functions in the body
- Control mechanisms
- Embryological germ layers (see below)

■ ADRENAL CORTEX is divided into 3 zones that secrete different types of corticosteroid hormones, which help regulate:
- Mineral balance (ZONA GLOMERULOSA)
- Energy balance and stress (ZONA FASCICULATA)
- Reproductive function (ZONA RETICULARIS)

■ ADRENAL MEDULLA secretes catecholamine hormones from specialized postganglionic nerve cells called CHROMAFFIN CELLS. Each cluster of CHROMAFFIN CELLS receives direct autonomic innervation, arranged near blood vessels.
- Mimic effects similar to SYMPATHETIC NERVOUS SYSTEM
- Stimulates metabolic rate increasing energy in response to acute stress

★ NERVE SUPPLY (INNERVATION)
■ PREGANGLIONIC FIBERS of the
g GREATER SPLANCHNIC NERVES

■ FIBERS OF THE
h CELIAC PLEXUS
and ASSOCIATED SYMPATHETIC PLEXUSES

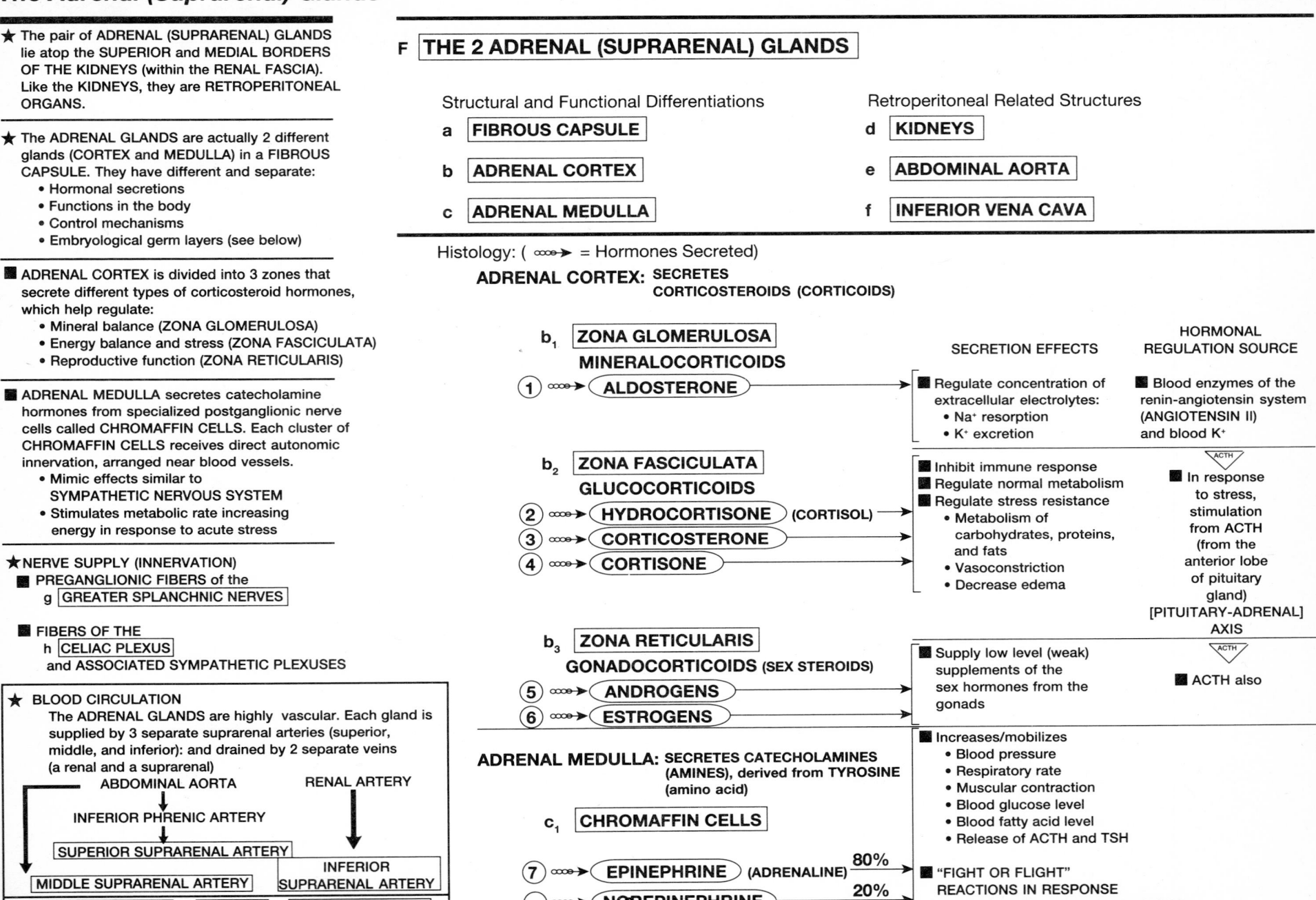

★ BLOOD CIRCULATION
The ADRENAL GLANDS are highly vascular. Each gland is supplied by 3 separate suprarenal arteries (superior, middle, and inferior): and drained by 2 separate veins (a renal and a suprarenal)

ABDOMINAL AORTA → RENAL ARTERY
INFERIOR PHRENIC ARTERY
SUPERIOR SUPRARENAL ARTERY
MIDDLE SUPRARENAL ARTERY
INFERIOR SUPRARENAL ARTERY
RIGHT SUPRARENAL VEIN | RENAL VEIN | LEFT SUPRARENAL VEIN
INFERIOR VENA CAVA ← LEFT RENAL VEIN

F | THE 2 ADRENAL (SUPRARENAL) GLANDS

Structural and Functional Differentiations

a FIBROUS CAPSULE
b ADRENAL CORTEX
c ADRENAL MEDULLA

Retroperitoneal Related Structures

d KIDNEYS
e ABDOMINAL AORTA
f INFERIOR VENA CAVA

Histology: (∞∞➤ = Hormones Secreted)

ADRENAL CORTEX: SECRETES CORTICOSTEROIDS (CORTICOIDS)

SECRETION EFFECTS | HORMONAL REGULATION SOURCE

b₁ ZONA GLOMERULOSA
MINERALOCORTICOIDS
① ∞∞➤ ALDOSTERONE

■ Regulate concentration of extracellular electrolytes:
- Na^+ resorption
- K^+ excretion

■ Blood enzymes of the renin-angiotensin system (ANGIOTENSIN II) and blood K^+

b₂ ZONA FASCICULATA
GLUCOCORTICOIDS
② ∞∞➤ HYDROCORTISONE (CORTISOL)
③ ∞∞➤ CORTICOSTERONE
④ ∞∞➤ CORTISONE

■ Inhibit immune response
■ Regulate normal metabolism
■ Regulate stress resistance
- Metabolism of carbohydrates, proteins, and fats
- Vasoconstriction
- Decrease edema

ACTH
■ In response to stress, stimulation from ACTH (from the anterior lobe of pituitary gland) [PITUITARY-ADRENAL] AXIS

b₃ ZONA RETICULARIS
GONADOCORTICOIDS (SEX STEROIDS)
⑤ ∞∞➤ ANDROGENS
⑥ ∞∞➤ ESTROGENS

■ Supply low level (weak) supplements of the sex hormones from the gonads

ACTH
■ ACTH also

ADRENAL MEDULLA: SECRETES CATECHOLAMINES (AMINES), derived from TYROSINE (amino acid)

c₁ CHROMAFFIN CELLS

⑦ ∞∞➤ EPINEPHRINE (ADRENALINE) 80%
○ ∞∞➤ NOREPINEPHRINE 20%

■ Increases/mobilizes
- Blood pressure
- Respiratory rate
- Muscular contraction
- Blood glucose level
- Blood fatty acid level
- Release of ACTH and TSH

■ "FIGHT OR FLIGHT" REACTIONS IN RESPONSE TO LIFE-THREATENING SITUATIONS

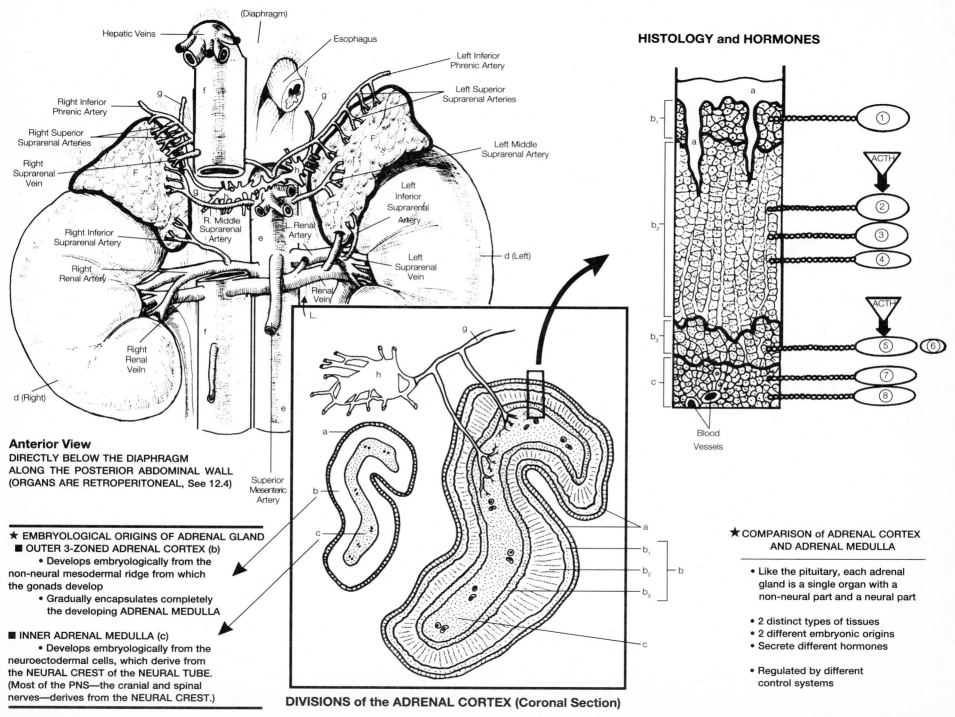

Hepatic Veins

(Diaphragm)

Esophagus

Left Inferior Phrenic Artery

Left Superior Suprarenal Arteries

Right Inferior Phrenic Artery

Right Superior Suprarenal Arteries

Right Suprarenal Vein

Left Middle Suprarenal Artery

Right Inferior Suprarenal Artery

R. Middle Suprarenal Artery

L. Renal Artery

Left Inferior Suprarenal Artery

Right Renal Artery

Left Suprarenal Vein

d (Left)

Right Renal Veiln

Renal Vein

L.

Right Renal Vein

Right Renal Veiln

Superior Mesenteric Artery

d (Right)

Anterior View
DIRECTLY BELOW THE DIAPHRAGM
ALONG THE POSTERIOR ABDOMINAL WALL
(ORGANS ARE RETROPERITONEAL, See 12.4)

★ EMBRYOLOGICAL ORIGINS OF ADRENAL GLAND
■ OUTER 3-ZONED ADRENAL CORTEX (b)
   • Develops embryologically from the non-neural mesodermal ridge from which the gonads develop
      • Gradually encapsulates completely the developing ADRENAL MEDULLA

■ INNER ADRENAL MEDULLA (c)
   • Develops embryologically from the neuroectodermal cells, which derive from the NEURAL CREST of the NEURAL TUBE. (Most of the PNS—the cranial and spinal nerves—derives from the NEURAL CREST.)

DIVISIONS of the ADRENAL CORTEX (Coronal Section)

HISTOLOGY and HORMONES

ACTH

ACTH

Blood Vessels

★COMPARISON of ADRENAL CORTEX AND ADRENAL MEDULLA

• Like the pituitary, each adrenal gland is a single organ with a non-neural part and a neural part

• 2 distinct types of tissues
• 2 different embryonic origins
• Secrete different hormones

• Regulated by different control systems

ENDOCRINE SYSTEM: ABDOMINAL REGION
Gastrointestinal Hormones and Pancreatic Secretions

Color Stomach and Duodenum = light orange
Pancreas = fleshy   Gallbladder and
Common Bile Duct = light green
Liver = reddish brown; (that is, if
you want to color the organs!)

★ See 12.9

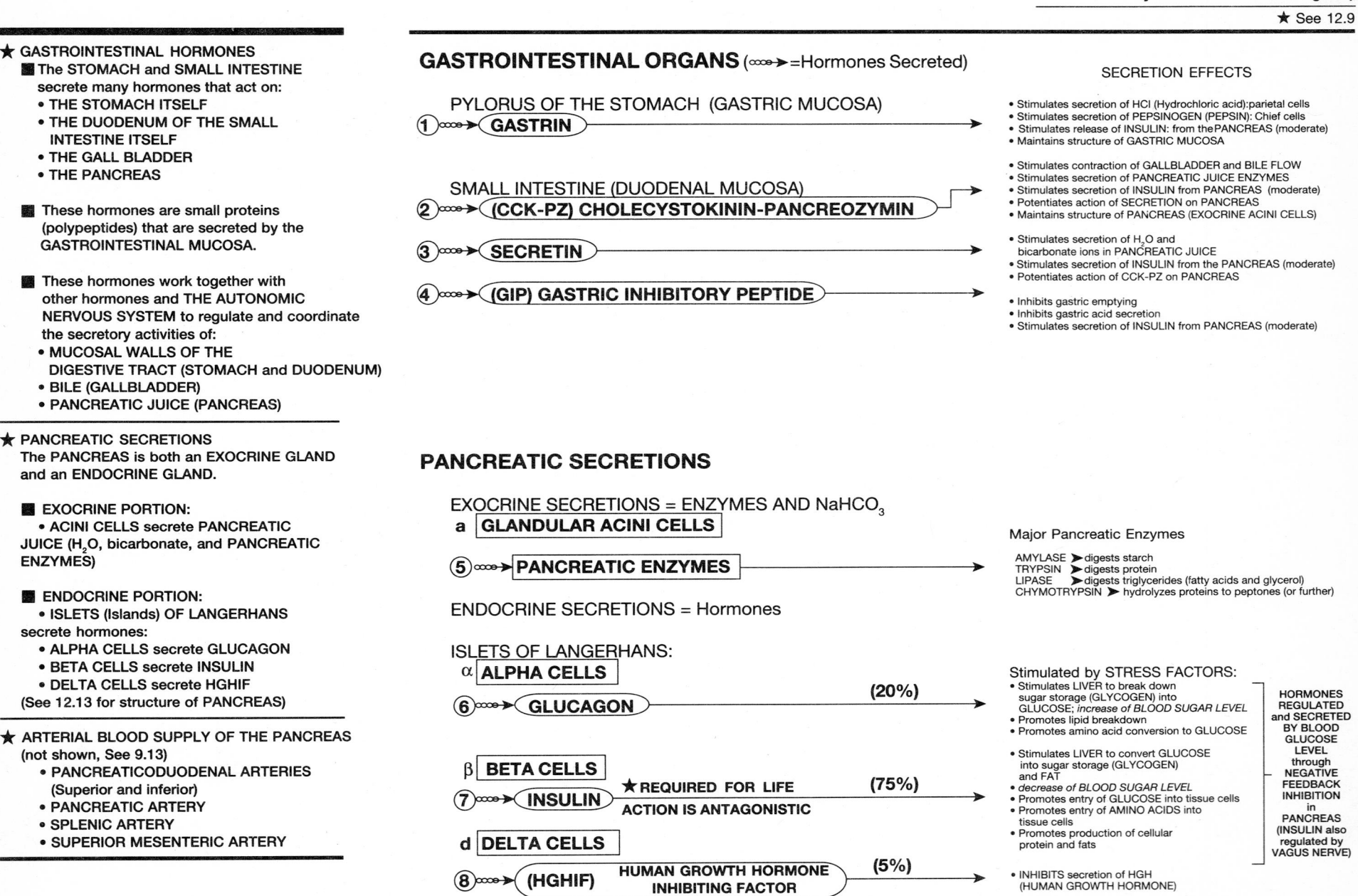

★ **GASTROINTESTINAL HORMONES**
■ The STOMACH and SMALL INTESTINE secrete many hormones that act on:
- THE STOMACH ITSELF
- THE DUODENUM OF THE SMALL INTESTINE ITSELF
- THE GALL BLADDER
- THE PANCREAS

■ These hormones are small proteins (polypeptides) that are secreted by the GASTROINTESTINAL MUCOSA.

■ These hormones work together with other hormones and THE AUTONOMIC NERVOUS SYSTEM to regulate and coordinate the secretory activities of:
- MUCOSAL WALLS OF THE DIGESTIVE TRACT (STOMACH and DUODENUM)
- BILE (GALLBLADDER)
- PANCREATIC JUICE (PANCREAS)

★ **PANCREATIC SECRETIONS**
The PANCREAS is both an EXOCRINE GLAND and an ENDOCRINE GLAND.

■ EXOCRINE PORTION:
- ACINI CELLS secrete PANCREATIC JUICE (H_2O, bicarbonate, and PANCREATIC ENZYMES)

■ ENDOCRINE PORTION:
- ISLETS (Islands) OF LANGERHANS secrete hormones:
- ALPHA CELLS secrete GLUCAGON
- BETA CELLS secrete INSULIN
- DELTA CELLS secrete HGHIF
(See 12.13 for structure of PANCREAS)

★ ARTERIAL BLOOD SUPPLY OF THE PANCREAS (not shown, See 9.13)
- PANCREATICODUODENAL ARTERIES (Superior and inferior)
- PANCREATIC ARTERY
- SPLENIC ARTERY
- SUPERIOR MESENTERIC ARTERY

GASTROINTESTINAL ORGANS (∞→ =Hormones Secreted)

SECRETION EFFECTS

PYLORUS OF THE STOMACH (GASTRIC MUCOSA)
① ∞→ **GASTRIN**
- Stimulates secretion of HCl (Hydrochloric acid):parietal cells
- Stimulates secretion of PEPSINOGEN (PEPSIN): Chief cells
- Stimulates release of INSULIN: from the PANCREAS (moderate)
- Maintains structure of GASTRIC MUCOSA

SMALL INTESTINE (DUODENAL MUCOSA)
② ∞→ **(CCK-PZ) CHOLECYSTOKININ-PANCREOZYMIN**
- Stimulates contraction of GALLBLADDER and BILE FLOW
- Stimulates secretion of PANCREATIC JUICE ENZYMES
- Stimulates secretion of INSULIN from PANCREAS (moderate)
- Potentiates action of SECRETION on PANCREAS
- Maintains structure of PANCREAS (EXOCRINE ACINI CELLS)

③ ∞→ **SECRETIN**
- Stimulates secretion of H_2O and bicarbonate ions in PANCREATIC JUICE
- Stimulates secretion of INSULIN from the PANCREAS (moderate)
- Potentiates action of CCK-PZ on PANCREAS

④ ∞→ **(GIP) GASTRIC INHIBITORY PEPTIDE**
- Inhibits gastric emptying
- Inhibits gastric acid secretion
- Stimulates secretion of INSULIN from PANCREAS (moderate)

PANCREATIC SECRETIONS

EXOCRINE SECRETIONS = ENZYMES AND $NaHCO_3$
a | **GLANDULAR ACINI CELLS** |

⑤ ∞→ **PANCREATIC ENZYMES**

Major Pancreatic Enzymes

AMYLASE ➤digests starch
TRYPSIN ➤digests protein
LIPASE ➤digests triglycerides (fatty acids and glycerol)
CHYMOTRYPSIN ➤ hydrolyzes proteins to peptones (or further)

ENDOCRINE SECRETIONS = Hormones

ISLETS OF LANGERHANS:
α | **ALPHA CELLS** |
⑥ ∞→ **GLUCAGON**   (20%)

Stimulated by STRESS FACTORS:
- Stimulates LIVER to break down sugar storage (GLYCOGEN) into GLUCOSE; *increase of BLOOD SUGAR LEVEL*
- Promotes lipid breakdown
- Promotes amino acid conversion to GLUCOSE

β | **BETA CELLS** |
⑦ ∞→ **INSULIN** ★REQUIRED FOR LIFE   (75%)
                    ACTION IS ANTAGONISTIC
- Stimulates LIVER to convert GLUCOSE into sugar storage (GLYCOGEN) and FAT
- *decrease of BLOOD SUGAR LEVEL*
- Promotes entry of GLUCOSE into tissue cells
- Promotes entry of AMINO ACIDS into tissue cells
- Promotes production of cellular protein and fats

d | **DELTA CELLS** |
⑧ ∞→ **(HGHIF)** HUMAN GROWTH HORMONE   (5%)
                    INHIBITING FACTOR
- INHIBITS secretion of HGH (HUMAN GROWTH HORMONE)

HORMONES REGULATED and SECRETED BY BLOOD GLUCOSE LEVEL through NEGATIVE FEEDBACK INHIBITION in PANCREAS (INSULIN also regulated by VAGUS NERVE)

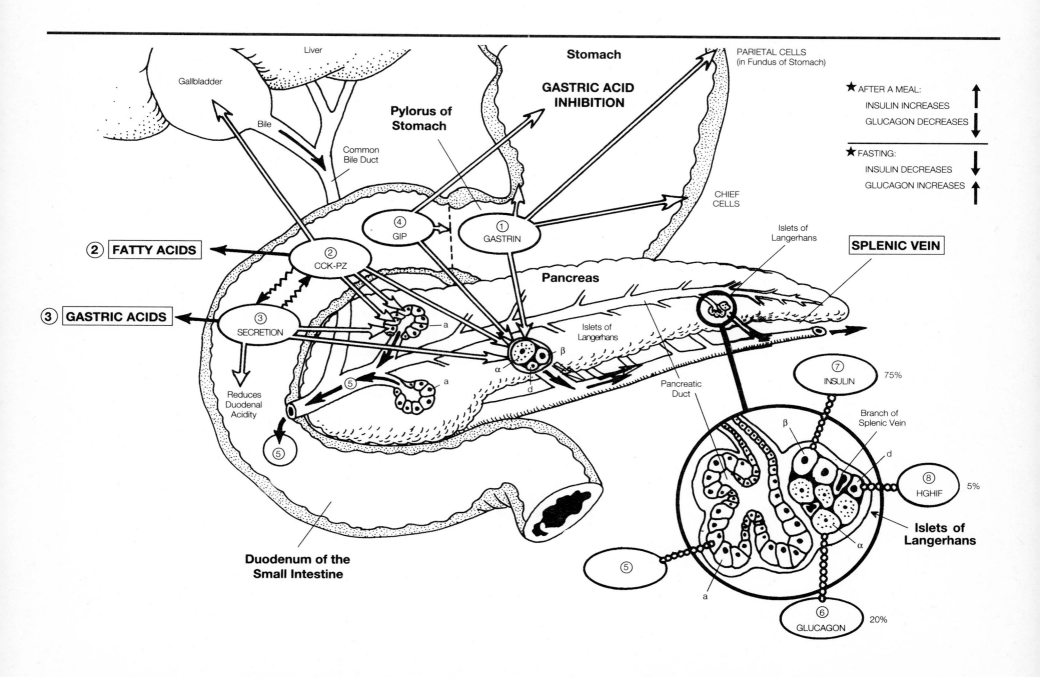

Liver

Gallbladder

Bile

Common Bile Duct

Pylorus of Stomach

Stomach

GASTRIC ACID INHIBITION

PARIETAL CELLS (in Fundus of Stomach)

CHIEF CELLS

④ GIP

① GASTRIN

② **FATTY ACIDS**

② CCK-PZ

③ **GASTRIC ACIDS**

③ SECRETION

Reduces Duodenal Acidity

⑤

⑤

a

a

α

β

d

Pancreas

Islets of Langerhans

Islets of Langerhans

Pancreatic Duct

Duodenum of the Small Intestine

SPLENIC VEIN

⑦ INSULIN 75%

Branch of Splenic Vein

β

d

⑧ HGHIF 5%

Islets of Langerhans

α

⑤

a

⑥ GLUCAGON 20%

8.7

ENDOCRINE SYSTEM: PELVIC-PERINEAL REGION
The Gonads and the Placenta

★ The MALE and FEMALE GONADS

The GONADS secrete SEX STEROIDS:
■ The OVARIES ♀ : (ESTROGENS and PROGESTROGENS)
  • In 1st half of menstrual cycle, the OVARIAN FOLLICLES (which contain the OOCYTE egg cell) along with their GRANULOSA CELLS and THECA INTERNA secrete ESTROGEN. (See 14.8)

  • In midcycle, *one* of the FOLLICLES grows very large, ruptures, and discharges its OVUM from the OVARY. The empty follicle is stimulated by LH (luteinizing hormone) from the ANTERIOR PITUITARY and is transformed into a different endocrine structure: the CORPUS LUTEUM (YELLOW BODY).
      The CORPUS LUTEUM secretes ESTROGEN and PROGESTERONE

■ The TESTES ♂ : (ANDROGENS)
  • In the interstitial tissue between the SEMINIFEROUS TUBULES are the LEYDIG CELLS (in "islands").
  • They secrete TESTOSTERONE, (main androgen), other androgens, and even small amounts of ESTROGEN (ESTRADIOL-17β).

★ The PLACENTA during Pregnancy
■ The PLACENTA is an ovoid, spongy structure in the UTERUS of the MOTHER, which gives the FETUS NOURISHMENT and allows WASTE EXCHANGE.
    The PLACENTA is also an ENDOCRINE GLAND. It secretes large amounts of:
    • ESTROGENS
    • PROGESTERONES
    • POLYPEPTIDE and PROTEIN HORMONES

The GONADS (⚬⚬⚬➤ Hormones Secreted)

SECRETION EFFECTS: STIMULATES DEVELOPMENT and MAINTENANCE of:

REGULATION from ANTERIOR LOBE of the PITUITARY GLAND

The OVARIES ♀

a | OVARIAN FOLLICLES WITH THEIR GRANULOSA AND THECA INTERNA:

① ⚬⚬⚬➤ (ESTROGEN (ESTRADIOL-17 β))
→ • Female secondary sex characteristics
  • Cyclic changes in vaginal epithelium and uterine endothelium

FSH

b | CORPUS LUTEUM

① ⚬⚬⚬➤ (ESTROGEN (ESTRADIOL-17 β))
→ • Same as above (lesser amounts)

FSH

② ⚬⚬⚬➤ (PROGESTERONE)
→ • Changes in uterine endometrium in 2nd (secretory) phase of menstrual cycle
  • Placenta
  • Mammary glands

LH

The TESTES ♂

c | INTERSTITIAL CELLS OF LEYDIG:

③ ⚬⚬⚬➤ (TESTOSTERONE)— (MAIN ANDROGEN)

④ ⚬⚬⚬➤ (OTHER ANDROGENS)
→ • Male genitalia (penis and scrotum)
  • Male sex accessory organs: (seminal vesicles, epididymides, ductus deferens, prostate gland)
  • Male secondary sex characteristics

ICSH (LH)

The PLACENTA during Pregnancy

① ⚬⚬⚬➤ (ESTROGENS)
→ • Same as above (as applied to ①)

② ⚬⚬⚬➤ (PROGESTERONE)
→ • Same as above (as applied to ②)

⑤ ⚬⚬⚬➤ ((HCG) HUMAN CHORIONIC GONADOTROPHIN)
→ • Similar to LH (Luteinizing hormone)

⑥ ⚬⚬⚬➤ (SOMATOMAMMOTROPHIN)
→ • Similar to GH (Growth hormone, SOMATOTROPHIN)
  • Similar to PROLACTIN

Similar to hormones secreted by the anterior lobe of the pituitary gland (See 8.3)

THE OVARIES ♀
(See 14.5, 14.7)

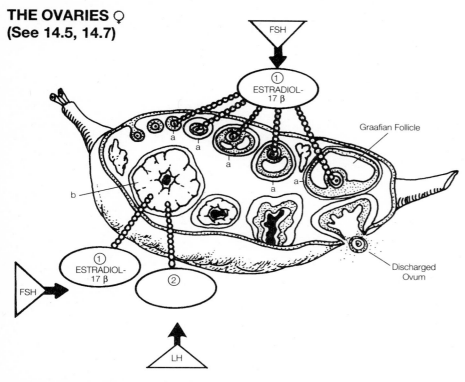

FSH

① ESTRADIOL-17 β

a
a
a
a
a-
a

b

Graafian Follicle

FSH → ① ESTRADIOL-17 β

②

Discharged Ovum

LH

★ Note how each hormone secreted by the anterior lobe of the pituitary gland (FSH, LH, ICSH) regulates and stimulates the secretion of specific sex steroids

The PLACENTA during PREGNANCY

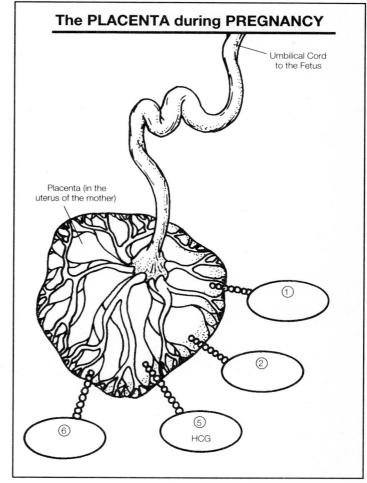

Umbilical Cord to the Fetus

Placenta (in the uterus of the mother)

①

②

⑥

⑤ HCG

THE TESTES ♂
(See 14.1, 14.2)

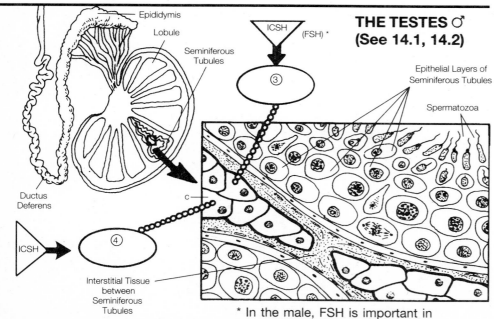

Epididymis
Lobule
Seminiferous Tubules

ICSH (FSH) *

③

Epithelial Layers of Seminiferous Tubules

Spermatozoa

Ductus Deferens

c

ICSH → ④

Interstitial Tissue between Seminiferous Tubules

* In the male, FSH is important in maintaining SPERMATOGENESIS

Chapter 8: *ENDOCRINE SYSTEM*: CHARTS

Chart #1 : Glands of the body (See 8.1)

ALL GLANDS OF THE BODY ARE CLASSIFIED AS EITHER EXOCRINE OR ENDOCRINE

	EXOCRINE GLANDS	**ENDOCRINE GLANDS**
LOCATION of SECRETION	Products secreted into *ducts* or *tubes* that empty at the surface of the covering and lining epithelium (Released either at: (1) the skin surface or(2) into the lumen or cavity of a hollow organ)	*Ductless* Glands: Products are secreted into the extracellular space around the secretory cells, then passed into the capillaries (and into the blood) of the blood vascular system
EXAMPLES of SECRETIONS	■ Enzymes ■ Oils ■ Sweat ■ Mucus	■ Always HORMONES (chemicals that regulate various physiological activities of cells to maintain homeostasis) Hormones may affect activities of either: 1. All cells in the body 2. Specific (target) cells or organs
EXAMPLES of GLANDS	• Sweat Glands (eliminate perspiration to cool skin) • Salivary Glands (secrete a digestive enzyme) • Sebaceous Glands (connected to hair follicles in skin) (See 2.3, 3.2, and 12.7)	■ PITUITARY GLAND ■ PINEAL GLAND ■ THYROID GLAND ■ PARATHYROIDS ■ ADRENAL GLANDS ■ PANCREAS ■ OVARIES, TESTES ■ THYMUS GLAND

Chart #2 : Summary of Endocrine Structures and Functions (See 8.1)

Gland or Organ	Hormone(s) Secreted	Primary Target Organ(s)	Primary Effect(s)
■ ADRENAL CORTEX	Cortisol	Liver, muscles	Glucose metabolism
	Aldosterone	Kidneys	Na⁺ retention, K⁺ excretion
■ ADRENAL MEDULLA	Epinephrine	Heart, bronchioles, blood vessels	Similar to sympathetic (adrenergic) stimulation of the ANS; cardiac stimulant, bronchiole relaxer, vasoconstriction
	Norepinephrine	Blood sugar, sweat production	Vasoconstriction only
DUODENUM OF THE SMALL INTESTINE	Secretin	Pancreas and gallbladder	Increased secretion by pancreas and gall bladder and gallbladder contraction
	Cholecystokinin-Pancreozymin		
	Enterogastrone	Stomach	Decreased gastric emptying
	Enterokinase	Trypsin precursor	Activates trypsin
HYPOTHALAMUS	Releasing and inhibiting hormones	Anterior pituitary	Regulates secretion of anterior pituitary hormones
KIDNEYS LIVER	Erythropoietin (alpha-globulin)	Bone marrow	Stimulates red blood cell production
OVARIES	Estradiol and progesterone	Female genital tract and mammary glands	Maintains structure of genital tract; promotes secondary sex characteristics
PANCREAS (ISLETS OF LANGERHANS)	Insulin	Many organs	Promotes decrease in blood sugar
	Glucagon		Promotes increase in blood sugar
■ PARATHYROID	Parathyroid Hormone	Bone, intestines, kidneys	Increases blood calcium level
■ PINEAL	Melatonin	Hypothalamus and anterior pituitary	Affects secretion of gonadotrophins
■ PITUITARY	See Chart #3		
■ PLACENTA	Human Chorionic Hormone	Corpus luteum	Maintains corpus luteum for first 5 1/2 weeks of pregnancy
	Placental Lactogen	Mammary glands	Promotes mammary gland growth
	Progesterone	Uterus, mammary glands	Maintains uterus during pregnancy and promotes mammary gland growth
	Estrogen	Uterus, mammary glands	Maintains uterus and promotes mammary gland growth
STOMACH	Gastrin	Stomach	Stimulates acid secretion
TESTES	Testosterone	Prostate, seminal vesicles, and other organs	Stimulates secondary sex characteristics
THYMUS	Thymosin	Lymph nodes	Stimulates white blood cell production
■ THYROID	Thyroxine (T_3 and T_4)	Most organs	Growth and development; stimulates basal rate of cell respiration (basal metabolic rate or BMR)
	Thyrocalcitonin	Bone, intestines, kidneys	Decreases blood calcium level

■ = ORGANS (GLANDS) EXCLUSIVELY ENDOCRINE
Adapted from Kent M. Van De Graaff and Stuart Ira Fox, *Concepts of Human Anatomy and Physiology*, 2d ed., p. 578.

Chart #3 : Summary of Pituitary Hormones and Functions (See 8.3)

Hormone	Target Tissue(s)	Effect(s) of Hormone	Regulation of Secretion ‡
ANTERIOR PITUITARY			
ACTH (ADRENOCORTICOTROPIC HORMONE) (CORTICOTROPHIN)	Adrenal cortex	Secretion of glucocorticoids	Stimulated by CRH (corticotrophin-releasing hormone); inhibited by glucocorticoids
TSH (THYROID-STIMULATING HORMONE) (THYROTROPHIN)	Thyroid gland	Secretion of thyroid hormone	Stimulated by TRH (thyrotrophin-releasing hormone); inhibited by thyroid hormones
GH (GROWTH HORMONE) (SOMATOTROPHIN) (STH)	Most tissues	Protein synthesis and growth; lipolysis and increased blood glucose	Stimulated by GHRH (growth hormone–releasing hormone); inhibited by somatostatin
FSH (FOLLICLE-STIMULATING HORMONE) LH (LUTEINIZING HORMONE)	Gonads	Gamete production and sex steroid hormone secretion	Stimulated by GnRH (gonadotrophin-releasing hormone); inhibited by sex steroids
PROLACTIN (LUTEOTROPIC HORMONE)	Mammary glands and other accessory sex organs	Milk production Controversial action in other organs	Inhibited by PIH (prolactin-inhibiting hormone)
LH (LUTEINIZING HORMONE) (LUTEOTROPHIN)	Gonads	Sex hormone secretion; ovulation and corpus luteum formation	Stimulated by GnRH (gonadotrophin-releasing hormone)
POSTERIOR PITUITARY			
ADH (ANTIDIURETIC HORMONE) (made in hypothalamus but stored in posterior pituitary)	Kidneys and blood vessels	Promotes water retention and vasoconstriction	Osmoreceptors of hypothalamus
OXYTOCIN (made in hypothalamus but stored in posterior pituitary)	Uterus Mammary glands	Stimulates uterine contractions Stimulates milk release	Hypothalamus Hypothalamus

‡ Hormones that stimulate release of pituitary secretion (such as CRH, TRH, GHRH, GnRH) are HYPOTHALAMIC REGULATING FACTORS; those hormones that inhibit further release of pituitary secretion are TARGET ORGAN HORMONES (negative–feedback inhibition)

* LH is found in females; in males it is called ICSH (interstitial cell–stimulating hormone)

Adapted from Stuart Ira Fox, *Human Physiology*, 3d ed., p. 298.

UNIT 4: REGULATION AND MAINTENANCE

Chapter **9** 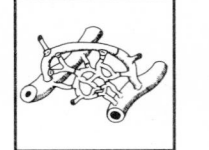 Cardiovascular System

The CARDIOVASCULAR SYSTEM is composed of 3 basic elements: the HEART, BLOOD VESSELS, and BLOOD. It is the chief transport system of the body.

The HEART is the center of the system and pumps BLOOD (TOTAL VOLUME is ≈ 5,000 mL) through the BLOOD VESSELS. The BLOOD VESSELS form a tubular network throughout the body (approximately 60,000 miles) permitting blood to flow from the heart to all living body cells (via 40 BILLION microscopic CAPILLARIES) and then back to the HEART. Under normal conditions, it takes about 1 minute for blood to be circulated from the HEART to the most distal extremity and back to the HEART again.

The BLOOD VASCULAR SYSTEM provides 4 basic functions essential to the life of the human organism:
- ■ TRANSPORTATION OF SUBSTANCES   ■ BODY PROTECTION
- ■ REGULATION OF BODY FUNCTIONS   ■ TEMPERATURE REGULATION

■ TRANSPORTATION OF SUBSTANCES
RESPIRATORY SUBSTANCES:
• OXYGEN—RED BLOOD CELLS (ERYTHROCYTES) transport OXYGEN (O_2) to the tissue cells. (O_2 received from inhaled air in the lungs attaches to the hemoglobin molecules in the RED BLOOD CELLS.)
• CARBON DIOXIDE—produced by cellular respiration, released into the BLOOD and carried to the lungs for eventual release into the exhaled air.

NUTRIENT-RICH MOLECULES:
• Food is mechanically and chemically broken down in the DIGESTIVE TRACT and absorbed through the intestinal wall into the blood stream (mostly CARBOHYDRATES and PROTEINS, with some LIPIDS). These nutritive molecules do not travel straight to the heart for distribution to the body system, but first travel through the HEPATIC PORTAL SYSTEM for filtration through the sinusoids (specialized capillaries) of the liver then enter the heart through the VENOUS SYSTEM. (There is also LIPID ABSORPTION through the LYMPH VASCULAR SYSTEM before entrance into the heart. See Chapter 10.)

EXCRETORY SUBSTANCES:
• Metabolic wastes, excessive water, and ions released from cellular metabolism, plus other molecules from the fluid portion of the blood (plasma) are filtered through the capillaries of the KIDNEYS of the URINARY SYSTEM and excreted to the outside through the urine.

■ REGULATION OF BODY FUNCTIONS
• Released hormones and regulatory molecules from the ENDOCRINE SYSTEM are transported to specific,distant target organs and body tissues.

■ BODY PROTECTION
• The clotting mechanism provided by BLOOD PLATELETS protects against blood loss when blood vessels are damaged.
• WHITE BLOOD CELLS (LEUKOCYTES) protect against foreign microbes or antigens entering the body. The LYMPH VASCULAR SYSTEM also acts as a protective mechanism against disease-causing viruses and toxins.

■ TEMPERATURE REGULATION
• Body heat comes from cellular metabolism (mostly muscle cells), and the CARDIOVASCULAR SYSTEM works in conjunction with the INTEGUMEN-TARY SYSTEM by losing excess heat through radiation from dilated blood vessels.

Thus, to maintain HOMEOSTASIS, the BLOOD VASCULAR SYSTEM works closely with the LYMPH VASCULAR SYSTEM, RESPIRATORY SYSTEM, URINARY SYSTEM,DIGESTIVE SYSTEM, ENDOCRINE SYSTEM, and INTEGUMENTARY SYSTEM.

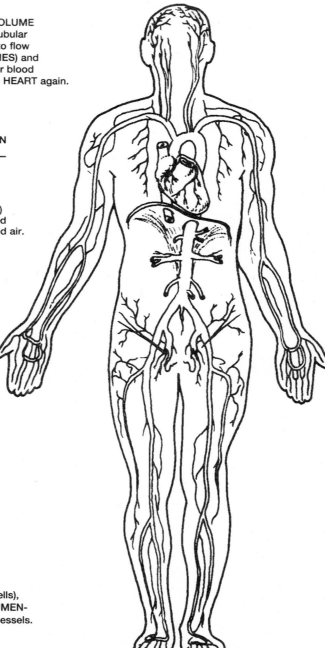

**Anterior View
(Arterial System
Only Shown)**

System Components

THE BLOOD, HEART, AND BLOOD VESSELS

System Function

- **DISTRIBUTION OF OXYGEN AND NUTRIENTS TO THE CELLS**

- **CARRYING OF CARBON DIOXIDE AND WASTES FROM THE CELLS**

- **MAINTENANCE OF BODY ACID-BASE BALANCE**

- **PROTECTION AGAINST DISEASE**

- **PREVENTION OF HEMORRHAGE (BLOOD CLOT FORMATION)**

- **REGULATION OF BODY TEMPERATURE**

Chapter Coloring Guidelines

In general, Arteries = red and warm-related colors; Veins = blue and cool-related colors; (except in the PULMONARY CIRCULATION, where the Arteries are blue and the Veins are red because of the difference in oxygen content). Capillary Networks = purple

There probably will be overlapping of color choices when coloring in the review plates. Any vessels assigned the same colors will be easily distinguished according to position.

All capillary beds = purple   1 = bright red
2,3 = blues   4 = green   5 = light blue
6 = pink   7,8 = blue greens

CARDIOVASCULAR SYSTEM: GENERAL ORGANIZATION
General Scheme of Circulation

★ The CIRCULATORY SYSTEM of the human body can be generally divided into 2 systems:
- ■ CARDIOVASCULAR SYSTEM (BLOOD VASCULAR SYSTEM)

- ■ LYMPHATIC SYSTEM (LYMPH VASCULAR SYSTEM)

The CARDIOVASCULAR SYSTEM (Chapter 9) consists of:
- ■ The HEART   ■ BLOOD
- ■ BLOOD VESSEL

The LYMPHATIC SYSTEM (Chapter 10) consists of:
- ■ LYMPH NODES   ■ LYMPH VESSELS
- ■ LYMPH ORGANS   ■ LYMPH

★ The HEART is a double muscular pump and consists of 4 chambers:
- ■ RIGHT ATRIUM   ■ LEFT ATRIUM
- ■ RIGHT VENTRICLE   ■ LEFT VENTRICLE

It can be considered as 2 separate pumps that pump blood in rhythm simultaneously.

★ The BLOOD VESSELS form the 2 main circulatory routes throughout the entire body, which are:
■ ROUTE 1 PULMONARY CIRCULATION: Blood vessels that transport blood to the LUNGS (for GAS EXCHANGE), and back to the HEART. It consists of:
- • RIGHT VENTRICLE
- • PULMONARY TRUNK
- • PULMONARY ARTERIES (L. and R.)
- • PULMONARY CAPILLARIES (in LUNGS)
- • PULMONARY VEINS (2 L. and 2 R.)
- • LEFT ATRIUM

■ ROUTE 2 SYSTEMIC CIRCULATION: Is composed of all of the blood vessels in the body that are *not* part of the PULMONARY CIRCULATION: Blood leaves the heart and travels through the aorta and its branches to all the body regions through the capillaries and back to the HEART via the venous vessels (veins):
- • LEFT VENTRICLE
- • AORTA and AORTIC BRANCHES
- • ALL ARTERIES (except PULMONARY ARTERIES)
- • All CAPILLARIES (except in LUNGS)
- • All VEINS (except PULMONARY VEINS)
- • RIGHT ATRIUM

★ CORONARY CIRCULATION, and CEREBRAL CIRCULATION, are discussed separately.

General Circulatory Routes
SYSTEMIC CIRCULATION:
(LEFT VENTRICLE → RIGHT ATRIUM)

1   AORTA AND BRANCHES

2   SUPERIOR VENA CAVA AND BRANCHES

3   INFERIOR VENA CAVA AND BRANCHES

PULMONARY CIRCULATION:
(RIGHT VENTRICLE → LEFT ATRIUM)

4   PULMONARY TRUNK

5   PULMONARY ARTERIES (L. AND R.)

6   PULMONARY VEINS (2 L. AND 2 R.)

HEPATIC PORTAL CIRCULATION

7   HEPATIC PORTAL VEIN

8   HEPATIC VEIN

Blood from the STOMACH (E), GASTROINTESTINAL TRACT (H) and SPLEEN (G) travels through the HEPATIC PORTAL SYSTEM: a secondary CAPILLARY NETWORK (SINUSOIDAL FILTER) *inside* the LIVER—before returning to the HEART via the INFERIOR VENA CAVA.

Capillary Beds

RIGHT ATRIUM

RIGHT VENTRICLE

LEFT ATRIUM

LEFT VENTRICLE

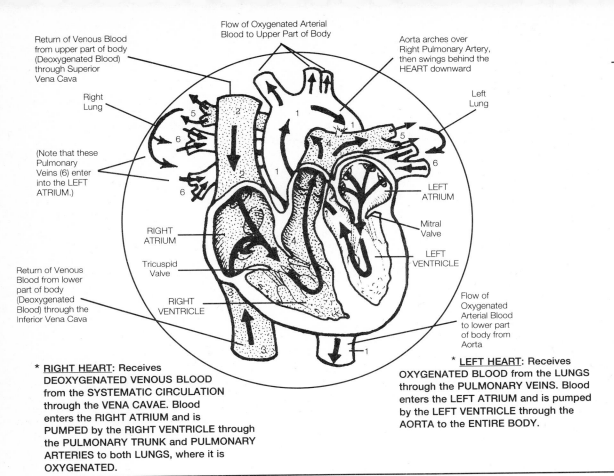

Return of Venous Blood from upper part of body (Deoxygenated Blood) through Superior Vena Cava

Flow of Oxygenated Arterial Blood to Upper Part of Body

Aorta arches over Right Pulmonary Artery, then swings behind the HEART downward

Right Lung

Left Lung

(Note that these Pulmonary Veins (6) enter into the LEFT ATRIUM.)

LEFT ATRIUM

RIGHT ATRIUM

Mitral Valve

Tricuspid Valve

LEFT VENTRICLE

Return of Venous Blood from lower part of body (Deoxygenated Blood) through the Inferior Vena Cava

RIGHT VENTRICLE

Flow of Oxygenated Arterial Blood to lower part of body from Aorta

* **RIGHT HEART:** Receives DEOXYGENATED VENOUS BLOOD from the SYSTEMATIC CIRCULATION through the VENA CAVAE. Blood enters the RIGHT ATRIUM and is PUMPED by the RIGHT VENTRICLE through the PULMONARY TRUNK and PULMONARY ARTERIES to both LUNGS, where it is OXYGENATED.

* **LEFT HEART:** Receives OXYGENATED BLOOD from the LUNGS through the PULMONARY VEINS. Blood enters the LEFT ATRIUM and is pumped by the LEFT VENTRICLE through the AORTA to the ENTIRE BODY.

CAPILLARY AND SINUSOID* STATIONS
(Color the capillary areas on either side of these boxes in the illustration light purple)

PULMONARY CIRCULATION
A ☐ LUNGS

SYSTEMIC CIRCULATION
B ☐ HEAD AND NECK

C ☐ UPPER EXTREMITIES (LIMBS)

D ☐ THORACIC/ABDOMINAL WALLS

E ☐ LIVER *

F ☐ STOMACH

G ☐ SPLEEN *

All exchanges of materials between BLOOD and TISSUE CELLS occur across the walls of the 1-layered, endothelial-tubed CAPILLARIES, the thinnest and most numerous of all blood vessels.

*SINUSOID SPACES (lined with phagocytic Kupffer cells) act as CAPILLARY SUBSTITUTES in the LIVER and SPLEEN. They are a discontinuous type of capillary in contrast to the continuous type found throughout the body, and are larger than capillaries. They are highly permeable and allow direct contact of LIVER and SPLEEN cells with the blood.

H ☐ GASTROINTESTINAL TRACT

I ☐ KIDNEYS

J ☐ PELVIS AND PERINEUM

K ☐ LOWER EXTREMITIES (LIMBS)

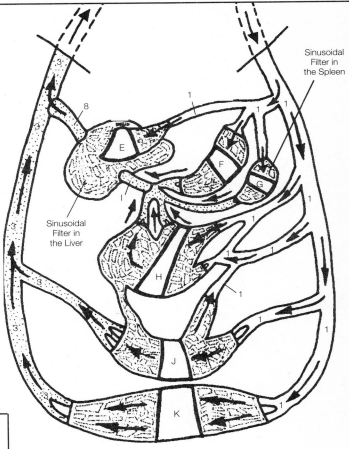

Sinusoidal Filter in the Spleen

Sinusoidal Filter in the Liver

E
F
G
I
H
J
K

SCHEMATIC DIAGRAM of the GENERAL CIRCULATORY ROUTES of the VASCULAR SYSTEM

Oxygen Content in the Blood Vessels

☐ = DEOXYGENATED BLOOD

▦ = OXYGENATED BLOOD

CARDIOVASCULAR SYSTEM: GENERAL ORGANIZATION
Structure of Blood Vessels

- In general, ARTERIES transport BLOOD *away from the heart* to the CAPILLARIES; (in the systemic circulation, arterial blood is oxygenated) VEINS transport BLOOD *toward the heart* from the CAPILLARIES (in the systemic circulation, venous blood is deoxygenated).

★ • The thin-walled structure of the CAPILLARIES enables exchange of: ■ Plasma Fluid
    ■ Dissolved Molecules
    ■ Gases
between BLOOD and the Surrounding Tissues (Tissue cells). CAPILLARIES are only 1-layer thick, consisting of flat squamous epithelium, called ENDOTHELIUM.

■ BLOOD transported away from the HEART travels in vessels of Progressively Decreasing Diameters: ELASTIC ARTERIES ➤ MUSCULAR ARTERIES ➤ ARTERIOLES ➤ CAPILLARIES

■ CAPILLARIES are microscopic blood vessels surrounding tissue cells that connect the arterial flow into them to the venous flow out of them.

■ BLOOD returning to the HEART travels in Vessels of Progressively Increasing Diameters: CAPILLARIES ➤ VENULES ➤ MUSCULAR VEINS ➤ GREAT VEINS

★ ■ ARTERIES have more muscle and smaller diameters than VEINS when vessels of similar size are compared.
■ ARTERIES are rounder, and VEINS are more collapsed in nature. (They are not usually filled and function as reservoirs since they can stretch to receive more blood)
★ ■ Most VEINS have semilunar, 1-way valves; (■ ARTERIES do not have valves.)

Types of BLOOD VESSELS

		Size Transition Ratios
1	ARTERY	3–10
2	ARTERIOLE	100
3	CAPILLARIES	500
4	VENULES	100
5	VEINS	10–3

3a SINUSOIDS Discontinuous capillary spaces lined with phagocytic KUPFFER CELLS (act as a capillary substitution in the LIVER and SPLEEN).

MICROCIRCULATION

6 METARTERIOLES (ARTERIOVENOUS ANASTOMOSES)

7 PRECAPILLARY SPHINCTER MUSCLES

- In some tissues, small arterioles (20 – 30 μm in diameter) by-pass capillary beds and transport blood *directly* to venules via shunts: METARTERIOLES (or ARTERIOVENOUS ANASTOMOSES).

- PRECAPILLARY SPHINCTER MUSCLES regulate the flow of blood from the arterioles through the capillaries.
- Arterioles also have more smooth muscle than elastin, providing a narrow consistent lumina, and the greatest *resistance* to blood flow through the arterial system.

General Wall Structure of BLOOD VESSELS

3 MAIN LAYERS

8 OUTER (longitudinal arrangement)
TUNICA EXTERNA (ADVENTITIA)

9 MIDDLE (circular arrangement)
TUNICA MEDIA

10 INNER (longitudinal arrangement)
(TUNICA INTERNA INTIMA)

8 TUNICA EXTERNA (ADVENTITIA)

A FIBROUS CONNECTIVE TISSUE

B [VASA VASORUM] Tiny blood vessels distributed to large arteries and veins

9 TUNICA MEDIA

C AREOLAR AND EXTERNAL ELASTIC TISSUE

D SMOOTH MUSCLE CELLS

10 TUNICA INTERNA

E AREOLAR AND INTERNAL ELASTIC TISSUE

F ENDOTHELIUM Faces the inner lumen of the vessel

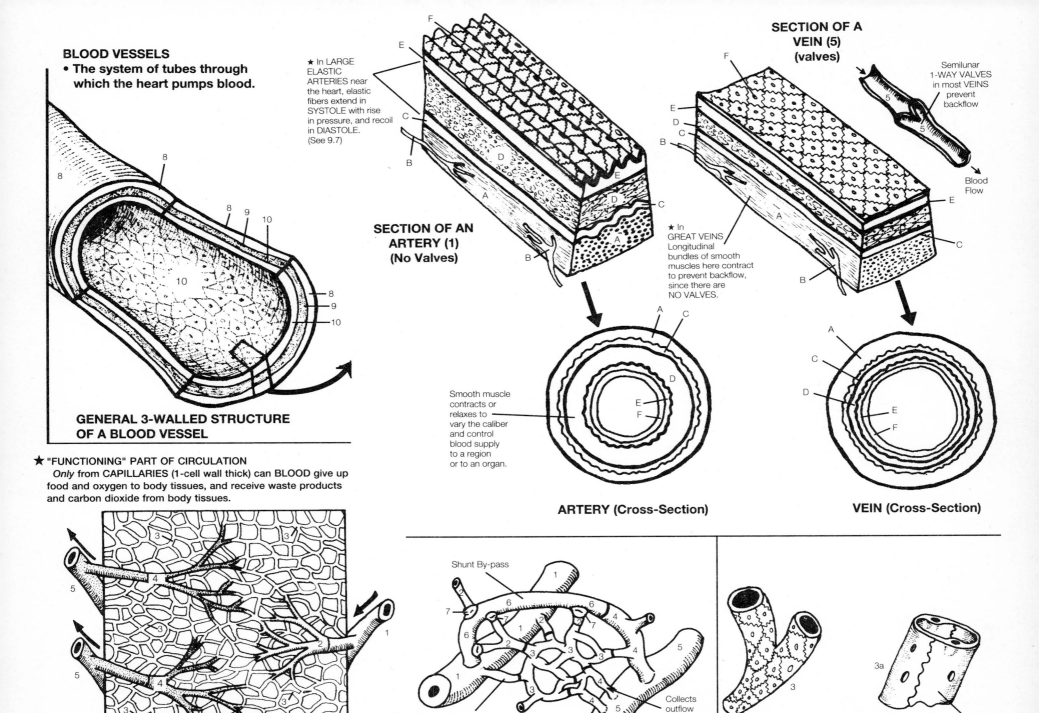

BLOOD VESSELS

- **The system of tubes through which the heart pumps blood.**

GENERAL 3-WALLED STRUCTURE OF A BLOOD VESSEL

★ "FUNCTIONING" PART OF CIRCULATION

Only from CAPILLARIES (1-cell wall thick) can BLOOD give up food and oxygen to body tissues, and receive waste products and carbon dioxide from body tissues.

★ In LARGE ELASTIC ARTERIES near the heart, elastic fibers extend in SYSTOLE with rise in pressure, and recoil in DIASTOLE. (See 9.7)

SECTION OF AN ARTERY (1) (No Valves)

★ In GREAT VEINS Longitudinal bundles of smooth muscles here contract to prevent backflow, since there are NO VALVES.

SECTION OF A VEIN (5) (valves)

Semilunar 1-WAY VALVES in most VEINS prevent backflow

Blood Flow

Smooth muscle contracts or relaxes to vary the caliber and control blood supply to a region or to an organ.

ARTERY (Cross-Section)

VEIN (Cross-Section)

Capillary Bed

Shunt By-pass

Smooth muscle contracts or relaxes to control inflow to CAPILLARY BED

Collects outflow from CAPILLARY BED

METARTERIOLE MICROCIRCULATION

Kupffer Cell

1-Cell Wall Thickness (1 layer)

CARDIOVASCULAR SYSTEM: GENERAL ORGANIZATION
Blood Composition

Human blood is composed of FLUID BLOOD PLASMA in which are suspended RED and WHITE BLOOD CELLS (CORPUSCLES), PLATELETS, FAT GLOBULES, and a great variety of chemical substances (including Carbohydrates, Proteins, Hormones, and Gases [O_2, CO_2 and N])

In the red bone marrow undifferentiated mesenchymal cells—called STEM CELLS are transformed into HEMOCYTOBLASTS. Each in turn undergoes differentiation into 5 types of cells from which all types of blood cells develop:

- ■ ≈7.7% of Body weight
- ■ Avg. amount = 5 liters, varies with body weight
- ■ pH = 7.3–7.4

100% BLOOD=BLOOD CELLS (45%) + BLOOD PLASMA (55%)

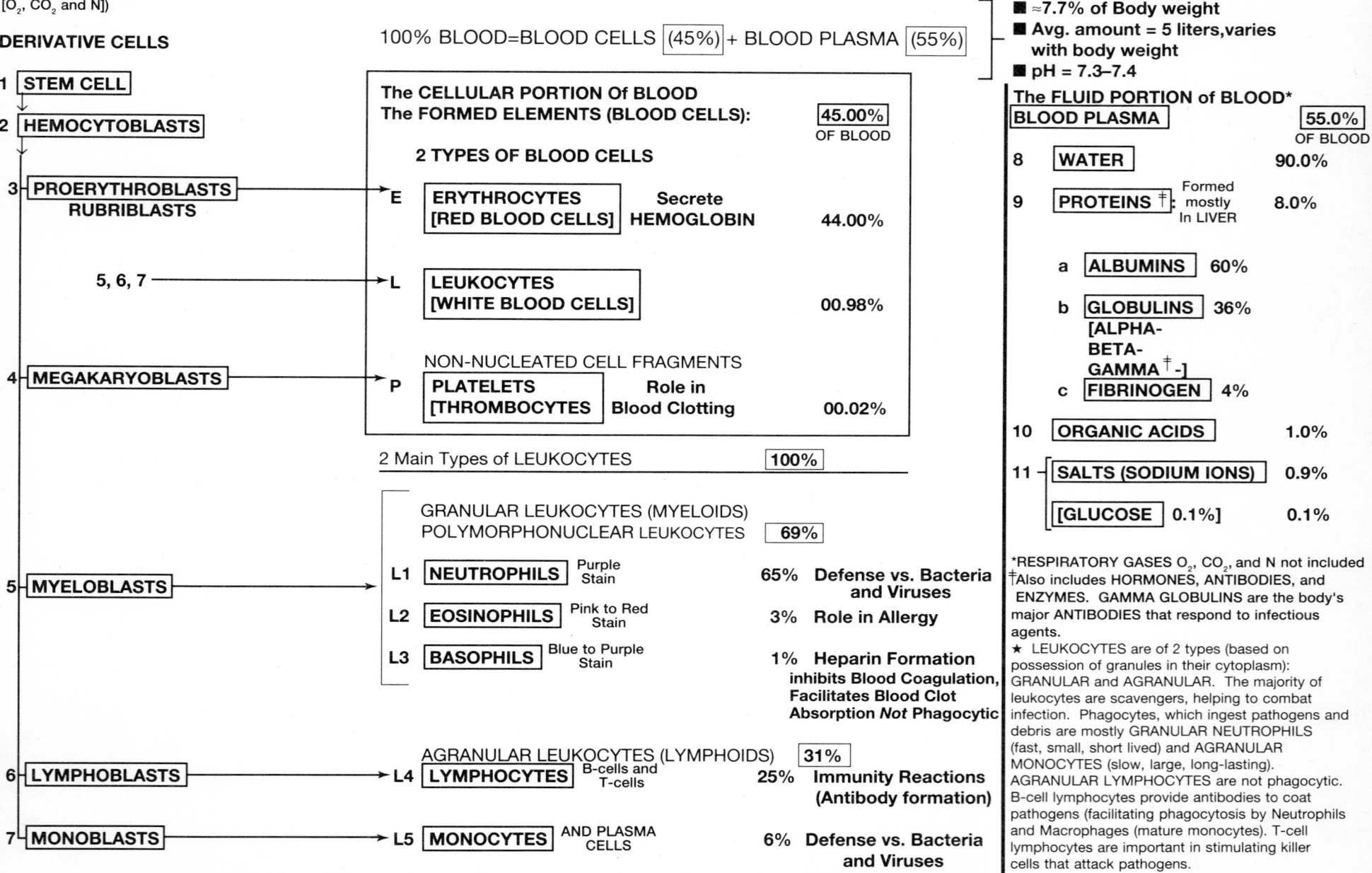

DERIVATIVE CELLS

1 | STEM CELL

2 | HEMOCYTOBLASTS

3 | PROERYTHROBLASTS
RUBRIBLASTS

5, 6, 7

4 | MEGAKARYOBLASTS

The CELLULAR PORTION Of BLOOD
The FORMED ELEMENTS (BLOOD CELLS): **45.00%** OF BLOOD

2 TYPES OF BLOOD CELLS

E | ERYTHROCYTES [RED BLOOD CELLS] | Secrete HEMOGLOBIN | 44.00%

L | LEUKOCYTES [WHITE BLOOD CELLS] | 00.98%

NON-NUCLEATED CELL FRAGMENTS

P | PLATELETS [THROMBOCYTES] | Role in Blood Clotting | 00.02%

2 Main Types of LEUKOCYTES | **100%**

GRANULAR LEUKOCYTES (MYELOIDS)
POLYMORPHONUCLEAR LEUKOCYTES | **69%**

5 | MYELOBLASTS

L1 | NEUTROPHILS | Purple Stain | 65% | Defense vs. Bacteria and Viruses

L2 | EOSINOPHILS | Pink to Red Stain | 3% | Role in Allergy

L3 | BASOPHILS | Blue to Purple Stain | 1% | Heparin Formation inhibits Blood Coagulation, Facilitates Blood Clot Absorption *Not* Phagocytic

AGRANULAR LEUKOCYTES (LYMPHOIDS) | **31%**

6 | LYMPHOBLASTS

L4 | LYMPHOCYTES | B-cells and T-cells | 25% | Immunity Reactions (Antibody formation)

7 | MONOBLASTS

L5 | MONOCYTES | AND PLASMA CELLS | 6% | Defense vs. Bacteria and Viruses

The FLUID PORTION of BLOOD*
BLOOD PLASMA | **55.0%** OF BLOOD

8 | WATER | 90.0%

9 | PROTEINS † | Formed mostly In LIVER | 8.0%

a | ALBUMINS | 60%

b | GLOBULINS | 36% [ALPHA-BETA-GAMMA †-]

c | FIBRINOGEN | 4%

10 | ORGANIC ACIDS | 1.0%

11 | SALTS (SODIUM IONS) | 0.9%

[GLUCOSE | 0.1%] | 0.1%

*RESPIRATORY GASES O_2, CO_2, and N not included
†Also includes HORMONES, ANTIBODIES, and ENZYMES. GAMMA GLOBULINS are the body's major ANTIBODIES that respond to infectious agents.
★ LEUKOCYTES are of 2 types (based on possession of granules in their cytoplasm): GRANULAR and AGRANULAR. The majority of leukocytes are scavengers, helping to combat infection. Phagocytes, which ingest pathogens and debris are mostly GRANULAR NEUTROPHILS (fast, small, short lived) and AGRANULAR MONOCYTES (slow, large, long-lasting). AGRANULAR LYMPHOCYTES are not phagocytic. B-cell lymphocytes provide antibodies to coat pathogens (facilitating phagocytosis by Neutrophils and Macrophages (mature monocytes). T-cell lymphocytes are important in stimulating killer cells that attack pathogens.

HEMOPOIESIS:
The Formation of Blood Cells
(X% = % of LEUKOCYTE Population

★ Red blood cells (Erythrocytes) are produced at the rate of 2.4 million/second! (Each red blood cell lives for ≈ 120 days.)

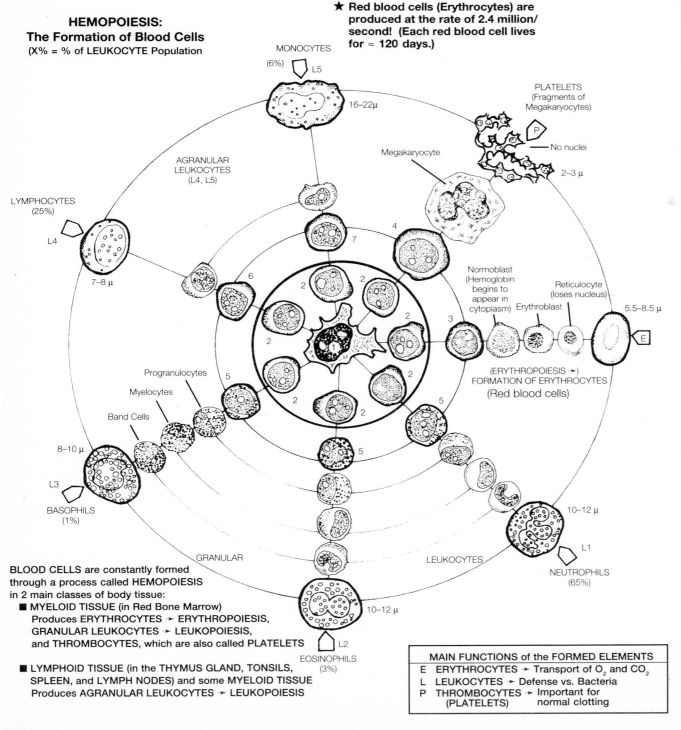

MONOCYTES
(6%)
L5
16–22 μ

PLATELETS
(Fragments of Megakaryocytes)
P
No nuclei
2–3 μ

Megakaryocyte

AGRANULAR LEUKOCYTES
(L4, L5)

LYMPHOCYTES
(25%)
L4
7–8 μ

Normoblast
(Hemoglobin begins to appear in cytoplasm)
Erythroblast
Reticulocyte
(loses nucleus)
5.5–8.5 μ
E

(ERYTHROPOIESIS →)
FORMATION OF ERYTHROCYTES
(Red blood cells)

Progranulocytes
Myelocytes
Band Cells
8–10 μ
L3
BASOPHILS
(1%)

GRANULAR
LEUKOCYTES

10–12 μ
L1
NEUTROPHILS
(65%)

L2
EOSINOPHILS
(3%)
10–12 μ

BLOOD CELLS are constantly formed through a process called HEMOPOIESIS in 2 main classes of body tissue:

■ MYELOID TISSUE (in Red Bone Marrow)
Produces ERYTHROCYTES → ERYTHROPOIESIS, GRANULAR LEUKOCYTES → LEUKOPOIESIS, and THROMBOCYTES, which are also called PLATELETS

■ LYMPHOID TISSUE (in the THYMUS GLAND, TONSILS, SPLEEN, and LYMPH NODES) and some MYELOID TISSUE
Produces AGRANULAR LEUKOCYTES → LEUKOPOIESIS

MAIN FUNCTIONS of the FORMED ELEMENTS	
E ERYTHROCYTES →	Transport of O_2 and CO_2
L LEUKOCYTES →	Defense vs. Bacteria
P THROMBOCYTES → (PLATELETS)	Important for normal clotting

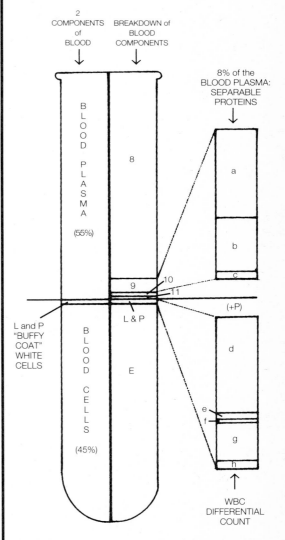

2 COMPONENTS of BLOOD

BREAKDOWN of BLOOD COMPONENTS

8% of the BLOOD PLASMA: SEPARABLE PROTEINS

BLOOD PLASMA (55%)

8
a
b
c

9
10
11
L & P

L and P "BUFFY COAT" WHITE CELLS

BLOOD CELLS (45%)

E

(+P)
d
e
f
g
h

WBC DIFFERENTIAL COUNT

$$(HEMATOCRIT) = \frac{\% \ of \ BLOOD \ CELLS}{TOTAL \ BLOOD \ VOLUME}$$

★ In a test tube, WHITE BLOOD CELLS (LEUKOCYTES) and PLATELETS (THROMBOCYTES) form a thin, "BUFFY COAT" (light-colored) between the PLASMA on top and the heavier RED BLOOD CELLS (ERYTHRO-CYTES), which sink to the bottom.

CARDIOVASCULAR SYSTEM: THE HEART
Location, Pericardium, and Heart Wall

1-4 = neutral colors
5-8 and mediastinum = cool colors
Ⓐ1 = *cream* Ⓐ2 = *yellow* Ⓒ = *flesh*
D = red-orange E = pink
★ See 1.4

★ **GENERAL INFORMATION**
- The HEART is a hollow 4-chambered muscular organ.
- The HEART functions as a double pump: 2 PUMPS in 1, each with a different function, but working together with simultaneous contractions.
  - Right pump collects blood *low in oxygen* from the body (via the SUPERIOR and INFERIOR VENA CAVAE) and pumps it to the LUNGS (via the PULMONARY TRUNK).
  - Left pump collects blood *high in oxygen* from the LUNGS (via the PULMONARY VEINS) and pumps it to the BODY via the AORTA.

★ **LOCATION:**
- In the THORACIC CAVITY
- In the MEDIASTINUM between the LUNGS
- 2/3 is left of the midline body axis

- APEX points downward and to the left and comes in contact with the DIAPHRAGM

- The BASE is the superior aspect of the HEART where the large vessels are attached.

★ **The PERICARDIUM (PARIETAL PERICARDIUM)**
- Also called the PERICARDIAL SAC
- A double membranous, fibroserous sac enclosing the heart and the origin of the great blood vessels, and protecting them from the rest of the thoracic organs.
- Consists of 2 layers:
  ① FIBROUS PERICARDIUM (OUTER)
  - Tough, fibrous connective tissue
  - Prevents overdistension of the heart, provides protection, anchors heart in the MEDIASTINUM.
  ② SEROUS PERICARDIUM (INNER)
  - Thin, delicate
  - Continuous with the EPICARDIUM (VISCERAL PERICARDIUM), the outer layer of the wall of the heart, at the base of the heart and around the great blood vessels.
- Between the SEROUS PERICARDIUM and the EPICARDIUM is the PERICARDIAL CAVITY, containing some watery fluid (secreted by the SEROUS PERICARDIUM). This PERICARDIAL FLUID prevents friction between the membranes during heart movement.

Location of the HEART

Orientation

1 **BASE**

2 **APEX**

3 **AXIS**

4 **MIDLINE BODY AXIS**

Borders

5 **LEFT**

6 **SUPERIOR**

7 **RIGHT**

8 **INFERIOR**

Surface

9 **STERNOCOSTAL (ANTERIOR)**

10 **DIAPHRAGMATIC (INFERIOR)**

MEDIASTINUM (See 1.4, 11.7)

SM **SUPERIOR**

AM **ANTERIOR**

MM **MIDDLE**

PM **POSTERIOR**

A **PERICARDIUM: PERICARDIAL SAC**

A **PARIETAL PERICARDIUM**
   2 LAYERS:

Ⓐ1 **FIBROUS PERICARDIUM LAYER** OUTER

Ⓐ2 **SEROUS PERICARDIUM LAYER** INNER

B **PERICARDIAL CAVITY\:**
   PERICARDIAL FLUID Potential space between the PARIETAL PERICARDIUM AND THE VISCERAL PERICARDIUM with lubricating watery fluid

3 Layers of the HEART WALL

Ⓒ **EPICARDIUM\VISCERAL PERICARDIUM** EXTERNAL

D **MYOCARDIUM** MIDDLE

E **ENDOCARDIUM** INNER

★ EPICARDIUM
- Thin, transparent, outer serous layer
- Continuous with the thin, inner serous layer of the PERICARDIUM

★ MYOCARDIUM
- Cardiac muscle tissue responsible for heart contraction
- Bulk of the heart
- Cardiac muscle fibers are INVOLUNTARY, STRIATED, BRANCHED; arranged in INTERLACING BUNDLES

★ ENDOCARDIUM
- Thin layer of serous ENDOTHELIUM that lines the inner surface and cavities of the heart, lining the inside of the MYOCARDIUM. Covers the valves of the heart and the tendons that hold them.
- Continuous with the endothelium or TUNICA INTERNA of the blood vessels attached to the heart

4 CHAMBERS of the HEART:
   RIGHT HEART   *LEFT HEART*

- **RIGHT ATRIUM (RA)**   ■ **LEFT ATRIUM (LA)**

- **RIGHT VENTRICLE (RV)**   ■ **LEFT VENTRICLE (LV)**

PERICARDIUM and HEART WALL

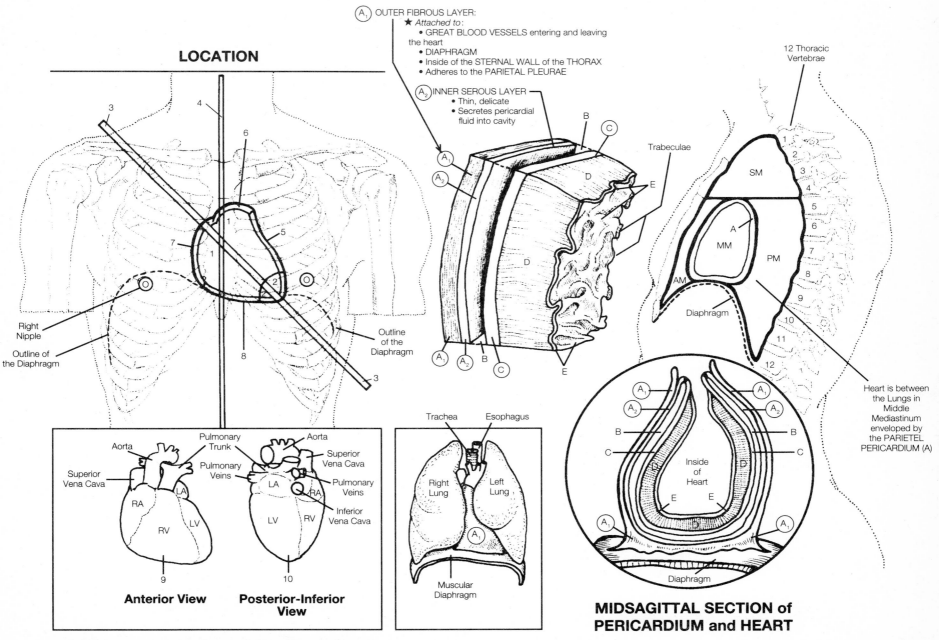

LOCATION

(A₁) OUTER FIBROUS LAYER:
★ *Attached to*:
- GREAT BLOOD VESSELS entering and leaving the heart
- DIAPHRAGM
- Inside of the STERNAL WALL of the THORAX
- Adheres to the PARIETAL PLEURAE

(A₂) INNER SEROUS LAYER
- Thin, delicate
- Secretes pericardial fluid into cavity

12 Thoracic Vertebrae

Trabeculae

B
C
D
E

SM
A
MM
PM
AM
Diaphragm

3
4
6
1
7
5
1
2
8

Right Nipple

Outline of the Diaphragm

Outline of the Diaphragm

3

Heart is between the Lungs in Middle Mediastinum enveloped by the PARIETEL PERICARDIUM (A)

Aorta
Pulmonary Trunk
Aorta
Superior Vena Cava
Superior Vena Cava
Pulmonary Veins
Pulmonary Veins
Inferior Vena Cava
LA
RA
RA
LA
RV
LV
LV
RV

9
10

Anterior View   **Posterior-Inferior View**

Trachea   Esophagus

Right Lung   Left Lung

A₁

Muscular Diaphragm

A₁
A₂
B
C
D
E
Inside of Heart
E
A₁
A₂
B
C
D
E
A₁

Diaphragm

MIDSAGITTAL SECTION of PERICARDIUM and HEART

SURFACE LOCATION of HEART CHAMBERS

9.4

Regulation and Maintenance 257

CARDIOVASCULAR SYSTEM: THE HEART
Gross Anatomy and Great Blood Vessels

Heart Chambers = warm colors (light)
Aorta = reds Vena Cavae = blues
Sulci = yellows, creams
PT = green-blue PA = light blue PV = pink

★ ■ HEART = 4-CHAMBERED DOUBLE PUMP
■ *Simultaneously*, the UPPER ATRIA from each pump contract and empty blood into the LOWER VENTRICLES, then,
■ *Simultaneously*, the LOWER VENTRICLES contract and push blood out of the heart at the upper base:
   • RIGHT VENTRICLE pushes blood (low O_2) through the PULMONARY TRUNK
   • LEFT VENTRICLE pushes blood (high O_2) through the AORTA

■ Partition between the ATRIA is thin and muscular: INTERATRIAL SEPTUM

■ Partition between the VENTRICLES is thick and muscular: INTERVENTRICULAR SEPTUM

■ Valves between the ATRIA and VENTRICLES are ATRIOVENTRICULAR (CUSPID) VALVES
   • MITRAL (BICUSPID) VALVE between LEFT ATRIA and LEFT VENTRICLE

   • TRICUSPID VALVE between RIGHT ATRIA and RIGHT VENTRICLE

■ Valves at the base of the PULMONARY TRUNK and AORTA are SEMILUNAR VALVES.

★ ■ Sulci (or grooves) on the surface of the HEART are indicative of the partitions between the chambers. They contain the CORONARY ARTERIES (See 9.7).
   • ATRIOVENTRICULAR (CORONARY) SULCUS encircles the HEART: indicates the partition between ATRIA and VENTRICLES
   • ANTERIOR and POSTERIOR INTERVENTRICULAR SULCI indicate partitioning between both VENTRICLES

★ ■ The cusps (leaflets) of the ATRIOVENTRICULAR VALVES are each held in position by strong, tendinous cords, or CHORDAE TENDINEAE. These cords in turn are secured to the ventricular walls by cone-shaped PAPILLARY MUSCLES.
   These cords and muscles prevent the atrioventricular valves from INVERTING into the ATRIA when the VENTRICLES contract.

★ ■ LIGAMENTUM ARTERIOSUM is a remnant of the DUCTUS ARTERIOSUS shunt in the FETUS.

Gross Anatomy of the HEART

CHAMBERS
RA   RIGHT ATRIUM
LA   LEFT ATRIUM
RV   RIGHT VENTRICLE
LV   LEFT VENTRICLE

AURICLES
1   RIGHT AURICLE] conical pouches of the atria that project from the upper anterior portions
2   LEFT AURICLE

SEPTA (PARTITIONS)
3   INTERATRIAL SEPTUM
    FO   FOSSA OVALIS   only in Right Atrium
4   INTERVENTRICULAR SEPTUM

VALVES
    ATRIOVENTRICULAR VALVES
5   TRICUSPID VALVE
6   BICUSPID (MITRAL) VALVE

SEMILUNAR VALVES
7   PULMONARY SEMILUNAR VALVE
8   AORTIC SEMILUNAR VALVE

ACCESSORY MUSCLES AND CORDS
9   CHORDAE TENDINEAE
10   PAPILLARY MUSCLES
11   TRABECULAE CARNEAE
12   MUSCULI PECTINATI   Anterior Auricular combed bundles

SULCI (GROOVES)
13   ATRIOVENTRICULAR (CORONARY) SULCUS
14   ANTERIOR INTERVENTRICULAR SULCUS
15   POSTERIOR INTERVENTRICULAR SULCUS

LIGAMENTS
16   LIGAMENTUM ARTERIOSUM

GREAT BLOOD VESSELS:
    VENOUS
SVC   SUPERIOR VENA CAVA
IVC   INFERIOR VENA CAVA
CS   CORONARY SINUS
PV   PULMONARY VEINS   (2 from each lung)

ARTERIAL
PT   PULMONARY TRUNK
PA   PULMONARY ARTERIES (R. AND L.)
CA   CORONARY ARTERIES
A   AORTA

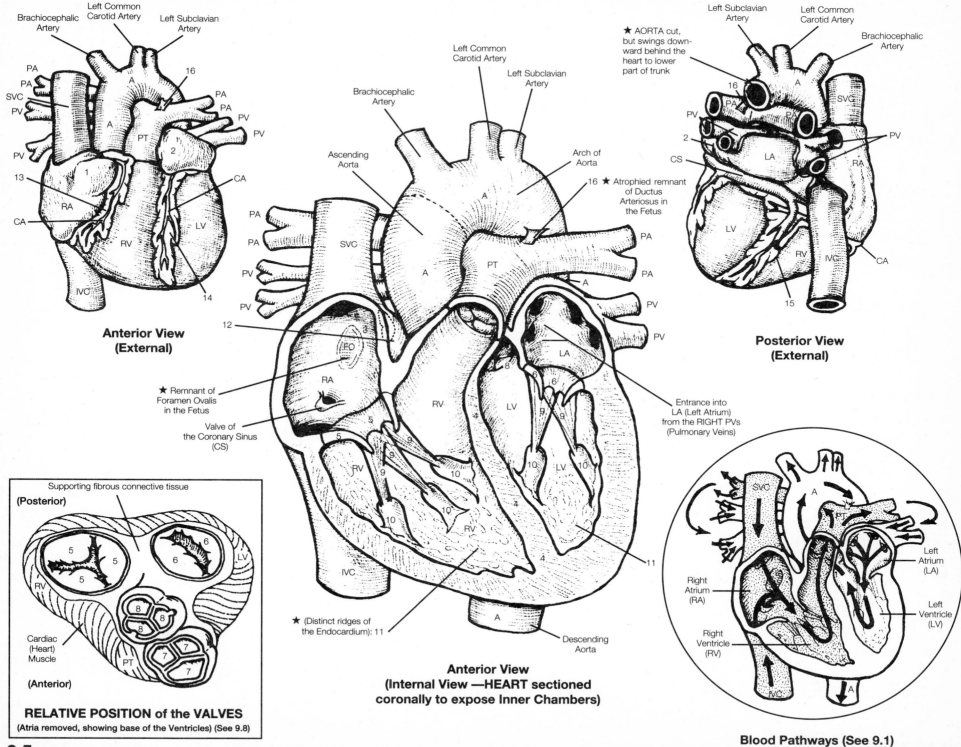

Brachiocephalic Artery

Left Common Carotid Artery

Left Subclavian Artery

PA
PA
SVC
PV

A

16

PA
PA
PV
PV

A

PT

2

PV

13

1

CA

RA

CA

LV

RV

IVC

14

**Anterior View
(External)**

Left Subclavian Artery

Left Common Carotid Artery

Brachiocephalic Artery

★ AORTA cut, but swings downward behind the heart to lower part of trunk

16

A

PA
PA

SVC

PV

2

PV

CS

LA

PV

RA

LV

RV

IVC

CA

15

**Posterior View
(External)**

Brachiocephalic Artery

Left Common Carotid Artery

Left Subclavian Artery

Ascending Aorta

A

Arch of Aorta

16 ★ Atrophied remnant of Ductus Arteriosus in the Fetus

PA

PA

SVC

PV

PV

PV

PT

A

PA

PA

PV

PV

12

FO

RA

8

8

6

6

LA

9

9

Entrance into LA (Left Atrium) from the RIGHT PVs (Pulmonary Veins)

★ Remnant of Foramen Ovalis in the Fetus

Valve of the Coronary Sinus (CS)

5

5

RV

4

LV

10

10

LV

10

9

9

RV

9

10

10

4

RV

4

11

IVC

A

★ (Distinct ridges of the Endocardium): 11

Descending Aorta

**Anterior View
(Internal View —HEART sectioned
coronally to expose Inner Chambers)**

Supporting fibrous connective tissue

(Posterior)

5

5

5

6

6

LV

RV

8

8

8

Cardiac (Heart) Muscle

7

7

PT

7

(Anterior)

RELATIVE POSITION of the VALVES
(Atria removed, showing base of the Ventricles) (See 9.8)

SVC

A

PT

Right Atrium (RA)

Left Atrium (LA)

Right Ventricle (RV)

Left Ventricle (LV)

IVC

A

Blood Pathways (See 9.1)

CARDIOVASCULAR SYSTEM: THE HEART
Conduction System and Contraction

Color the same (1–5) on 9.7
Color 1–5 a progression of colors;
(example yellow-orange-red-purple-blue)
8 = red-orange   9 and 11 = light colors

★ The HEART is INNERVATED by the AUTONOMIC NERVOUS SYSTEM, but these autonomic neurons:
  • *DO NOT INITIATE CONTRACTIONS!*
  • *ONLY REGULATE TIME* IT TAKES TO COMPLETE A SINGLE CARDIAC CYCLE.

★ ■ CONTRACTION and RELAXATION of the HEART WALLS (MYOCARDIUM) is accomplished by an intrinsic regulating system, THE CARDIAC CONDUCTION SYSTEM:
  ■ Specialized muscle tissue cells within the heart, which cannot contract themselves but generate and distribute ELECTRICAL IMPULSES that stimulate CONTRACTION of CARDIAC MUSCLE FIBERS:
    ■ SA NODE
    ■ AV NODE
    ■ AV BUNDLE and BRANCHES
    ■ PURKINJE FIBERS

★ The ATRIA and VENTRICLES act as if they were composed of only one muscle cell: The entire MYOCARDIUM is electrically stimulated as a SINGLE UNIT and CONTRACTS as a SINGLE UNIT! (Actually, VENTRICULAR CONTRACTION starts 0.1–0.2 seconds after ATRIAL CONTRACTION.) This relatively long single contraction (≈300 msec) is equivalent to a twitch produced by a single skeletal muscle fiber (but only lasting 20–100 msec). The heart, unlike skeletal muscle, cannot sustain a contraction of a summation of many twitches. Thus, rhythmic pumping action is insured because the heart cannot be stimulated again until after it has relaxed from its previous long contraction.
  ■ SA NODE (SINOATRIAL NODE/PACEMAKER)
    • INITIATES EACH CARDIAC CYCLE
    • SETS THE *PACE* FOR HEART RATE
    • PACE is regulated by ANS and Thyroid Hormone or EPINEPHRINE hormone
    • Located in right atrial wall inferior to the opening of the SVC
  ■ INTERNODAL TRACTS
    • Spreads impulse to both atria
    • *ATRIA CONTRACT*
    • AV NODE is DEPOLARIZED
  ■ AV BUNDLE OF HIS carries impulse to the top of the INTERVENTRICULAR SEPTUM down the septum via the RIGHT and LEFT BUNDLE BRANCHES
  ■ AV BUNDLE OF HIS distributes the charge over the MEDIAL SURFACE OF THE VENTRICLES
  ■ PURKINJE FIBERS within the ventricular muscle end the conduction chain causing the
    • Actual CONTRACTION of VENTRICLES as the impulses traveling over these fibers penetrate into MYOCARDIUM

CONDUCTILE MUSCLE TISSUE

CARDIAC CONDUCTION SYSTEM

1 SA NODE (SINOATRIAL NODE) PACEMAKER

2 INTERNODAL TRACTS

3 AV NODE (ATRIOVENTRICULAR NODE)

4 AV BUNDLE OF HIS

4a R. AND L. BUNDLE BRANCHES

5 PURKINJE FIBERS

HEART INNERVATION
  SYMPATHETIC (STIMULATORY): INCREASE ♥ RATE
CAC CARDIOACCELETORY CENTER

6 CARDIAC NERVE

  PARASYMPATHETIC (INHIBITORY): DECREASE ♥ RATE
CIC CARDIOINHIBITORY CENTER

7 VAGUS NERVE

PRESSORECEPTOR BLOOD PRESSURE BALANCE
CSR CAROTID SINUS REFLEX

AR AORTIC REFLEX

CONTRACTILE MUSCLE TISSUE

CARDIAC TISSUE COMPONENTS

8 FIBERS

9 INTERCALATED DISKS

10 NUCLEI

11 SARCOPLASMIC RETICULUM

12 MITOCHONDRIA

★ ■ Heart muscle tissue has a striated appearance similar to skeletal muscle; contracts by sliding filament.
  ■ INVOLUNTARY MUSCLE (Skeletal is voluntary)
  ■ QUADRANGULAR in shape (Skeletal is spindle-shaped)
  ■ CENTRALLY-LOCATED NUCLEI (Skeletal muscle has peripherally-located nuclei)
  ■ Each fiber is separated from the next fiber by electrical synapses called INTERCALATED DISKS (GAP JUNCTIONS)
    • Transverse thickening of SARCOLEMMA
    • Strengthen cardiac muscle tissue
    • Aid in impulse conduction
  ■ Short, bifurcated and interconnected (Skeletal muscle is long, fibrous, single, and separated)
  ■ Contracts without nervous stimulation although regulated by it. (Skeletal muscle contracts *only* by nervous stimulation.)
  ■ While a person is at rest, the CARDIAC CONDUCTION SYSTEM contracts and relaxes cardiac muscle tissue 72 times a minute *without stopping*:
    • RAPIDLY
    • CONTINUOUSLY
    • RHYTHMICALLY
(Skeletal muscle does not contract in a person at rest.)

• AUTONOMIC CONTROL
INNERVATION OF THE HEART

• SA NODE PACEMAKER would set an unvarying heart rate by itself.

• Many factors (chemicals, blood pressure, body temperature, emotions and stress, age) stimulate regulatory mechanisms from the ANS (AUTONOMIC NERVOUS SYSTEM) to adjust heart rate.

AUTONOMIC HEART CONTROL is a result of COMPLEMENTARY ACTION OF:

■ SYMPATHETIC STIMULATION from CARDIOACCELETORY CENTER in the MEDULLA.

■ PARASYMPATHETIC INHIBITION from CARDIOINHIBITORY CENTER in the MEDULLA.

SENSORY NERVE CELL PRESSORECEPTORS (BARORECEPTORS) act on these 2 centers to maintain a balance between stimulation and inhibition:

★ CSR • CAROTID SINUS REFLEX
→ maintains normal brain blood pressure

★ AR • AORTIC REFLEX
→ maintains general systemic blood pressure

• RIGHT ATRIAL REFLEX
→ responds to venous blood pressure (not shown)

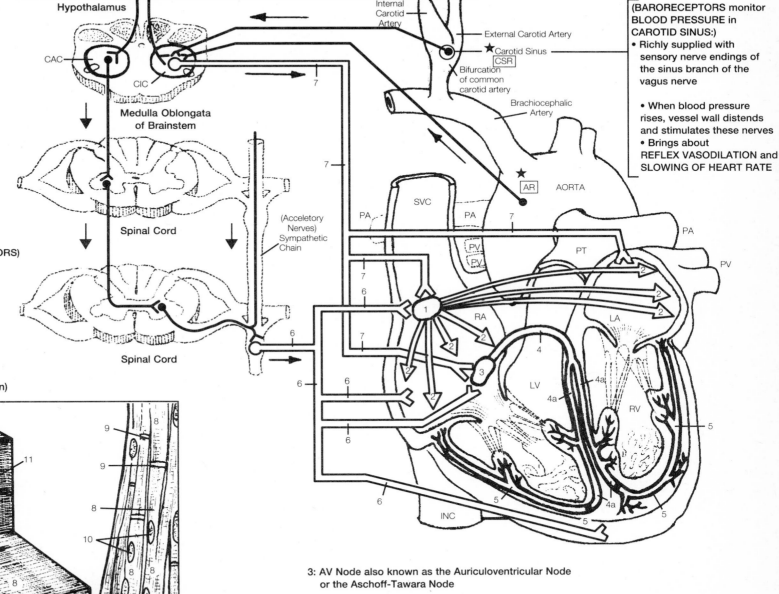

(BARORECEPTORS monitor BLOOD PRESSURE in CAROTID SINUS:)
• Richly supplied with sensory nerve endings of the sinus branch of the vagus nerve

• When blood pressure rises, vessel wall distends and stimulates these nerves
• Brings about REFLEX VASODILATION and SLOWING OF HEART RATE

3: AV Node also known as the Auriculoventricular Node or the Aschoff-Tawara Node

CARDIAC CONDUCTION SYSTEM

CONTRACTILE (CARDIAC) MUSCLE TISSUE

9.6

CARDIOVASCULAR SYSTEM: THE HEART
Coronary (Cardiac) Circulation

Arteries = reds and warm colors
Veins = blues and cool colors
★ See 9.5

★ The AORTA is the major systemic vessel arising from the HEART, and begins as the ASCENDING AORTA, ascending from the LEFT VENTRICLE. Just beyond the segment of the AORTIC SEMILUNAR VALVE (See 9.5) are the LEFT and RIGHT AORTIC SINUSES. These aortic sinuses, also known as the SINUS OF VALSALVA, are small dilatations of the AORTA that precede the only branches of the Ascending Aorta—the LEFT and RIGHT CORONARY ARTERIES.

★ ■ CORONARY ARTERIES supply the HEART itself, serving the MYOCARDIUM. They lie within a groove bedding that encircles the HEART, called the ATRIOVENTRICULAR SULCUS (indicative of the inner partition between the ATRIA and VENTRICLES).

  ■ Branches of LEFT CORONARY ARTERY
    • ANTERIOR INTERVENTRICULAR A.
      ➤ Serves both VENTRICLES
    • CIRCUMFLEX A.
      ➤ Serves LEFT CHAMBERS
       (LEFT ATRIUM and LEFT VENTRICLE)
      • Branch of the CIRCUMFLEX supplies
       the LEFT VENTRICLE
  ■ Branches of RIGHT CORONARY ARTERY
    • POSTERIOR INTERVENTRICULAR A.
      ➤ Serves both VENTRICLES
    • MARGINAL ARTERY
      ➤ Serves RIGHT CHAMBERS
       (mainly the RIGHT VENTRICLE)

★ ■ CORONARY VEINS
    • Parallel coronary arteries
    • Blood enters from capillaries
     in the MYOCARDIUM
    • More superficial than arteries
    • Walls thinner than arteries
    PRINCIPAL VEINS:
    ■ GREAT CARDIAC VEIN
      ➤ Drains anterior aspect
       of the HEART
    ■ MIDDLE CARDIAC VEIN
      ➤ Drains posterior aspect
       of the HEART
    ■ SMALL CARDIAC VEIN
      ➤ Drains anterior-inferior
       aspect of the heart
  ■ CORONARY SINUS
    • LARGE VENOUS CHANNEL on
  posterior aspect of the HEART receives
  venous blood from cardiac (coronary) veins and
  then empties into the RIGHT ATRIUM (See 9.5).

CORONARY CIRCULATION of the HEART

Vessels Supplying the HEART

    THE AORTIC SINUSES AND THE CORONARY ARTERIES

1 | LEFT AORTIC SINUS |

    ② | LEFT CORONARY ARTERY |

      3 | ANTERIOR INTERVENTRICULAR BRANCH |

      4 | CIRCUMFLEX ARTERY (BRANCH) |

      5 | POSTERIOR ARTERY OF THE LEFT VENTRICLE |

6 | RIGHT AORTIC SINUS |

    ⑦ | RIGHT CORONARY ARTERY |

      8 | POSTERIOR INTERVENTRICULAR BRANCH |

      9 | MARGINAL BRANCH |

Vessels Draining the HEART (13 drains into 12; 10, 12, 14-17 drain into 11)

THE CARDIAC (CORONARY) VEINS

10 | ANTERIOR CARDIAC VEINS |

11 | CORONARY SINUS |

    12 | GREAT CARDIAC VEIN | 15 | MIDDLE CARDIAC VEIN |

    13 | LEFT MARGINAL VEIN | 16 | POSTERIOR VEIN OF THE LEFT VENTRICLE |

    14 | SMALL CARDIAC VEIN | 17 | OBLIQUE VEIN OF THE LEFT ATRIUM |

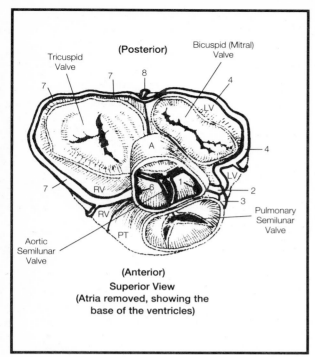

Tricuspid Valve
(Posterior)
Bicuspid (Mitral) Valve
7
7
8
4
LV
A
4
7
LV
RV
6
1
2
3
RV
3
Pulmonary Semilunar Valve
Aortic Semilunar Valve
PT

(Anterior)
Superior View
(Atria removed, showing the base of the ventricles)

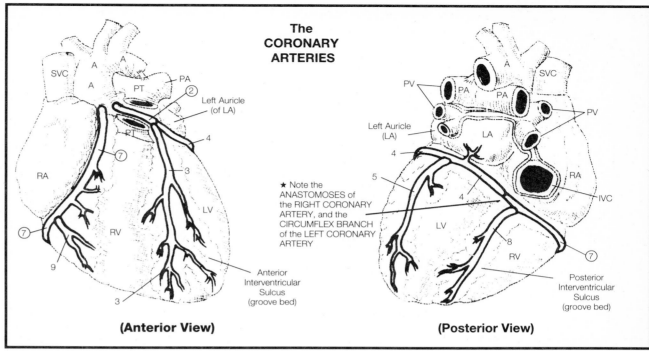

The CORONARY ARTERIES

SVC
A
A
PA
PT
②
Left Auricle (of LA)
PT
4
RA
⑦
3
LV
⑦
RV
9
3
Anterior Interventricular Sulcus (groove bed)

(Anterior View)

PV
A
SVC
PA
PA
Left Auricle (LA)
LA
4
5
4
RA
IVC
LV
8
RV
⑦
Posterior Interventricular Sulcus (groove bed)

★ Note the ANASTOMOSES of the RIGHT CORONARY ARTERY, and the CIRCUMFLEX BRANCH of the LEFT CORONARY ARTERY

(Posterior View)

RA Right Atrium
LA Left Atrium
RV Right Ventricle
LV Left Ventricle
A   Aorta
SVC Superior Vena Cava
IVC Inferior Vena Cava
PT Pulmonary Trunk
PA Pulmonary Arteries
PV Pulmonary Veins

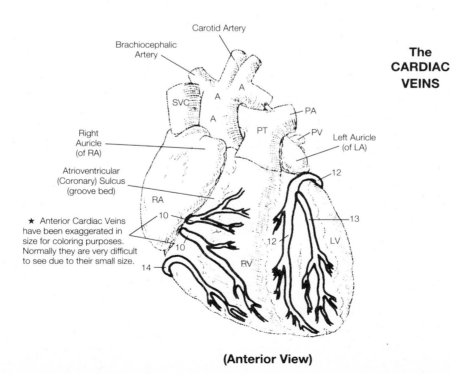

Carotid Artery
Brachiocephalic Artery
A
SVC
A
PA
PT
PV
Left Auricle (of LA)
Right Auricle (of RA)
Atrioventricular (Coronary) Sulcus (groove bed)
RA
12
10
13
10
12
LV
14
RV

★ Anterior Cardiac Veins have been exaggerated in size for coloring purposes. Normally they are very difficult to see due to their small size.

The CARDIAC VEINS

(Anterior View)

Carotid Artery
Left Subclavian Artery
Brachiocephalic Artery
A
PA
SVC
PA
PV
PV
Left Auricle (LA)
LA
12
17
RA
12
11
IVC
16
LV
15
RV
14

(Posterior View)

CARDIOVASCULAR SYSTEM: PRINCIPAL ARTERIES OF THE BODY
Arteries of the Head and Neck (Right Lateral View)

*Use strong, bright warm colors
for major arteries and major branches,
and lighter, warm colors and neutral
colors (or no colors) for minor branches.*

★ Compare with 9.17

The 2 COMMON CAROTID ARTERIES pass upward along either side of the trachea in the NECK (the RIGHT COMMON CAROTID ARTERY is shown here). Small branches of the common CAROTID ARTERY supply various neck structures:
- Larynx
- Thyroid Gland
- Anterior Neck Muscles
- Lymph Glands

The COMMON CAROTID ARTERY bifurcates into:
- ■ INTERNAL CAROTID ARTERY
  Enters the base of the skull and supplies the brain (See 9.10).
- ■ EXTERNAL CAROTID ARTERY
  Mainly supplies structures in the neck and head area *external to the skull.*
  Main branches of the EXTERNAL CAROTID ARTERY include:
- ■ SUPERIOR THYROID ARTERY
  - ➤ • Hyoid Muscles
  - • Larynx and Vocal Cords
  - • Thyroid Gland
- ■ ASCENDING PHARYNGEAL ARTERY
  - ➤ • Pharyngeal Area
  - • Lymph Nodes
- ■ LINGUAL ARTERY
  - ➤ • Tongue (extensive)
  - • Sublingual Salivary Gland
- ■ FACIAL ARTERY passes through a notch on the inferior side of the mandible:
  - ➤ • Pharyngeal Area
  - • Palate
  - • Chin
  - • Lips
  - • Nose
- ■ OCCIPITAL ARTERY
  - ➤ • Posterior Scalp
  - • Meninges covering the Brain
  - • Mastoid Process of Temporal Bone
  - • Posterior Neck Muscles
- ■ POSTERIOR AURICULAR ARTERY
  - ➤ • Ear
  - • Scalp over Ear

The EXTERNAL CAROTID ARTERY divides into 2 large arteries near the mandibular condyle:
- ■ SUPERFICIAL TEMPORAL ARTERY
  - ➤ • Parotid Salivary Gland
  - • Superficial Structures on the sides of the Head
- ■ MAXILLARY ARTERY
  - ➤ • Teeth and Gums
  - • Muscles of Mastication
  - • Nasal Cavity
  - • Eyelids
  - • Meninges

1ST VESSEL to Branch from
the AORTIC ARCH (veering RIGHT):

B | BRACHIOCEPHALIC (INNOMINATE) A.

2 BIFURCATIONS OF THE BRACHIOCEPHALIC

S | RIGHT SUBCLAVIAN ARTERY

CC | RIGHT COMMON CAROTID ARTERY

Branches of the RIGHT SUBCLAVIAN A.

IT | INTERNAL THORACIC A. | (INTERNAL MAMMARY A.)

V | VERTEBRAL A. | (See 9.9)

TT | THYROCERVICAL TRUNK

- 1 INFERIOR THYROID A.
- 2 SUPRASCAPULAR A.
- 3 TRANSVERSE CERVICAL A.

CT | COSTOCERVICAL TRUNK

- 4 DEEP CERVICAL A.
- 5 HIGHEST INTERCOSTAL A.

Branches of the RIGHT COMMON CAROTID A.

IC | INTERNAL CAROTID A. | (See 9.9)

EC | EXTERNAL CAROTID A.

Branches of the EXTERNAL CAROTID A.

St | SUPERIOR THYROID A.

AP | ASCENDING PHARYNGEAL A.

L | LINGUAL A.

F | FACIAL A.

- 6 SUPERIOR AND INFERIOR LABIAL BR.
- 7 LATERAL NASAL BRANCH
- 8 ANGULAR BRANCH
- 9 SUBMENTAL BRANCH

O | OCCIPITAL A.

PA | POSTERIOR AURICULAR A.

EXTERNAL CAROTID ARTERY Terminates by Dividing into 2 Separate ARTERIES:

ST | SUPERFICIAL TEMPORAL A.

- 10 TRANSVERSE FACIAL BRANCH
- 11 ZYGOMATIC-ORBITAL BRANCH

M | MAXILLARY A.

- 12 SUPERIOR AND INFERIOR ALVEOLAR BR.
- 13 INFRAORBITAL BRANCH
- 14 MIDDLE MENINGEAL BRANCH
- 15 BUCCAL BRANCH

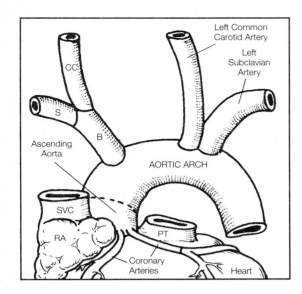

★ Three (3) major vessels arise
from the AORTIC ARCH:
  ◼ BRACHIOCEPHALIC ARTERY (TRUNK)
  ◼ LEFT COMMON CAROTID ARTERY
  ◼ LEFT SUBCLAVIAN ARTERY

  ◼ The BRACHIOCEPHALIC ARTERY
    bifurcates into 2 vessels:

    • RIGHT COMMON CAROTID ARTERY
      (supplying the neck and head
      on the right side)
    • RIGHT SUBCLAVIAN ARTERY
      (supplying the right arm)

★ As you recall, the ASCENDING AORTA
is the SYSTEMIC VESSEL that arises
from the LEFT VENTRICLE of the HEART.
(Its *only* branches, the CORONARY ARTERIES,
supply the HEART). It continues as the
AORTIC ARCH, which curves up and to the
left posteriorly over the RIGHT PULMONARY ARTERY. (See 9.5)
(The RIGHT PULMONARY ARTERY branches
off to the right from the PULMONARY TRUNK (PT),
shown cut in the illustration.)

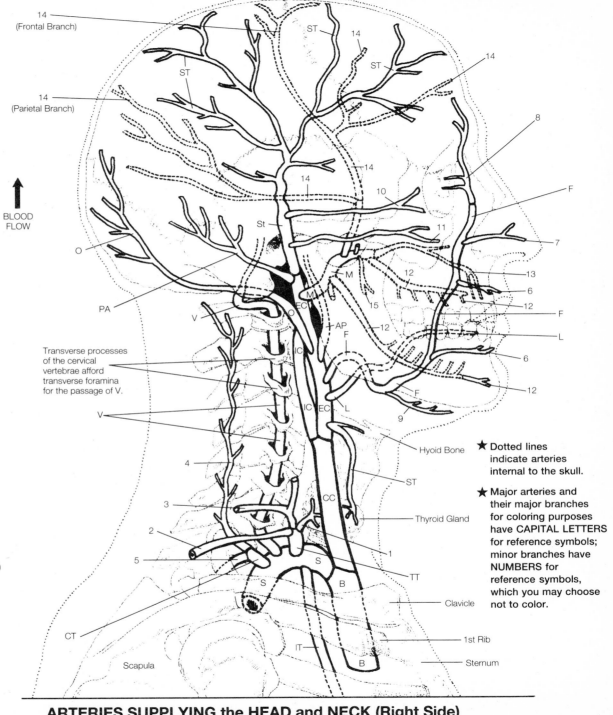

Transverse processes
of the cervical
vertebrae afford
transverse foramina
for the passage of V.

★ Dotted lines
indicate arteries
internal to the skull.

★ Major arteries and
their major branches
for coloring purposes
have CAPITAL LETTERS
for reference symbols;
minor branches have
NUMBERS for
reference symbols,
which you may choose
not to color.

ARTERIES SUPPLYING the HEAD and NECK (Right Side)
(Right Lateral View)

CARDIOVASCULAR SYSTEM: PRINCIPAL ARTERIES OF THE BODY
Arteries of the Brain

Color in the CIRCLE OF WILLIS first, then transfer those colors to the Inferior view of theBrain. Use bright reds and warm colors.

★ Compare with 9.16

★ Blood to the BRAIN is supplied by:
2 INTERNAL CAROTID ARTERIES and
2 VERTEBRAL ARTERIES. (4 Vessels)

■ They all eventually unite in a unique circular arrangement of vessels on the inferior aspect of the BRAIN surrounding the PITUITARY GLAND (HYPOPHYSIS) called the CIRCLE OF WILLIS. (see below)

■ The PAIRED VERTEBRAL ARTERIES enter the SKULL through the FORAMEN MAGNUM, and once inside the brain case, unite to form the BASILAR ARTERY at the level of the PONS. The main branches of the VERTEBRAL ARTERIES are the:

• 2 POSTERIOR CEREBRAL ARTERIES, which supply the posterior portion of the CEREBRUM.

• 2 POSTERIOR COMMUNICATING ARTERIES, which branch from the POSTERIOR CEREBRAL ARTERIES and form part of the ARTERIAL CIRCLE OF WILLIS by joining with the INTERNAL CAROTID ARTERIES.

■ The INTERNAL CAROTID ARTERIES enter the base of the SKULL through the CAROTID CANAL of the TEMPORAL BONE.

★ CIRCLE OF WILLIS:
Arterial circle surrounding the PITUITARY GLAND (HYPOPHYSIS)

• ANASTOMOSES CENTER: if one of the 4 main vessels becomes occluded, the other vessels may offer limited alternate routes for passage of blood to the brain (although, practically and functionally speaking, they are not very effective).

• Circle of Willis may relieve pressure from larger vessels to the smaller branches of the circle (mainly those to the PITUITARY GLAND.)

Main Arterial Supply to the BRAIN:

(IC) | INTERNAL CAROTID ARTERIES | (PAIRED)

(V) | VERTEBRAL ARTERIES | (PAIRED)

Branches of the 2 INTERNAL CAROTID ARTERIES

1 | ANTERIOR CHOROID A. | (2)

2 | ANTERIOR CEREBRAL A. | (2)

3 | ANTERIOR COMMUNICATING A. | (1)

4 | MIDDLE CEREBRAL A. | (2)

BRANCHES OF THE 2 VERTEBRAL ARTERIES

5 | ANTERIOR SPINAL A. | (1)

6 | POSTERIOR INFERIOR CEREBELLAR A. | (2)

7 | BASILAR ARTERY | (1)

Within the BRAIN CASE, the 2 VERTEBRAL ARTERIES Unite to Form
1 Large BASILAR ARTERY. The Following Are Branches of the BASILAR ARTERY:

8 | ANTERIOR INFERIOR CEREBELLAR A. | (2)

9 | LABYRINTHINE A. | (2) INTERNAL AUDITORY A.

10 | PONTINE BRANCHES | (MANY)

11 | SUPERIOR CEREBELLAR A. | (2)

12 | POSTERIOR CEREBRAL A. | (2)

13 | POSTERIOR COMMUNICATING ARTERIES | (2)

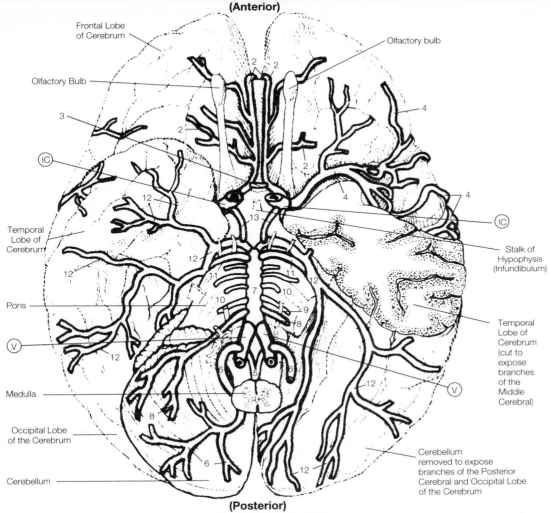

(Anterior)

Frontal Lobe of Cerebrum

Olfactory bulb

Olfactory Bulb

Temporal Lobe of Cerebrum

Pons

Medulla

Occipital Lobe of the Cerebrum

Cerebellum

Stalk of Hypophysis (Infundibulum)

Temporal Lobe of Cerebrum (cut to expose branches of the Middle Cerebral)

Cerebellum removed to expose branches of the Posterior Cerebral and Occipital Lobe of the Cerebrum

(Posterior)
INFERIOR VIEW OF THE BRAIN
SHOWING CEREBRAL ARTERIAL CIRCULATION

CIRCLE OF WILLIS (arrows indicate BLOOD FLOW)

Hypophysis

(INFERIOR VIEW of BRAIN)

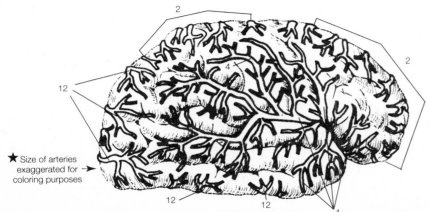

★ Size of arteries exaggerated for coloring purposes

(Right Lateral View)
RIGHT CEREBRAL HEMISPHERE

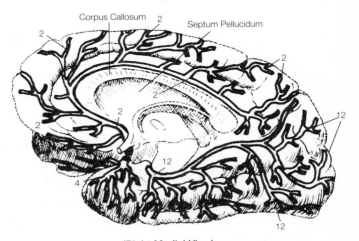

Corpus Callosum

Septum Pellucidum

(Right Medial View)
RIGHT CEREBRAL HEMISPHERE

CARDIOVASCULAR SYSTEM: PRINCIPAL ARTERIES OF THE BODY
Arteries of the Upper Extremity (Right Limb)

Color the major arteries and branches first.
Use bright, warm colors for major
arteries and branches, and lighter warm
color and neutral colors for minor branches.
★ Compare with 9.18

★ Note: On the RIGHT SIDE of the BODY, the RIGHT SUBCLAVIAN ARTERY branches from the BRACHIOCEPHALIC ARTERY, whereas on the LEFT SIDE of the BODY, the LEFT SUBCLAVIAN ARTERY branches DIRECTLY from the AORTIC ARCH.

★ The SUBCLAVIAN ARTERIES pass laterally deep to the CLAVICLE, and as they enter the Axillary region, they become the AXILLARY ARTERIES (from the outer border of the 1st rib to the lower border of the TERES MAJOR MUSCLE). Small branches supply:
- Tissues of the Upper Thorax
- Tissues of the Shoulder Region

★ As the AXILLARY ARTERIES enter the Brachial Region, they become the BRACHIAL ARTERIES as they continue along the medial side of the HUMERUS. The major branch of the BRACHIAL ARTERY is:
- DEEP BRACHIAL ARTERY → curves posteriorly (near the RADIAL NERVE) and supplies the TRICEPS MUSCLE.

★ Just below the CUBITAL FOSSA on the anterior side of the arm, the BRACHIAL ARTERY bifurcates into a LATERAL RADIAL ARTERY and a MEDIAL ULNAR ARTERY.
Supply:
- Forearm
- All of the Hands and Digits
■ First and largest branch of the RADIAL ARTERY is the RADIAL RECURRENT ARTERY → supplies elbow region

■ First and largest branches of the ULNAR ARTERY are the ANTERIOR and POSTERIOR ULNAR RECURRENT ARTERIES

★ At the wrist, the RADIAL and ULNAR ARTERIES anastomose to form the PALMAR ARCHES, which branch into the PALMAR DIGITAL ARTERIES to the fingers.

★ BLOOD PRESSURE POINTS
- BRACHIAL ARTERY (on the medial side of the HUMERUS)
★ PULSATION POINTS
- SUBCLAVIAN ARTERY (just above medial portion of CLAVICLE)
- RADIAL ARTERY (lateral side of wrist at base of thumb)
- ULNAR ARTERY (medial side of wrist)

1ST Branch of the AORTIC ARCH:
B | BRACHIOCEPHALIC (INNOMINATE) A.

2 Branches of the BRACHIOCEPHALIC ARTERY:
CC | RIGHT COMMON CAROTID ARTERY

S | RIGHT SUBCLAVIAN ARTERY

As the SUBCLAVIAN ARTERY Passes through the AXILLARY REGION, It Becomes the:
AX | AXILLARY ARTERY

As the AXILLARY ARTERY Continues through the BRACHIAL REGION, It Becomes the:
BR | BRACHIAL ARTERY

The BRACHIAL ARTERY Bifurcates at the Elbow into 2 Main ARTERIES:
R | RADIAL ARTERY LATERAL
U | ULNAR ARTERY MEDIAL

The RADIAL and ULNAR ARTERIES Anastomose in the Palm to Form 2 PALMAR ARCHES:
SP | SUPERFICIAL PALMAR ARCH
DP | DEEP PALMAR ARCH

ANASTOMOSES:
PC | PALMAR CARPAL ARCH
DC | DORSAL CARPAL ARCH

19 DORSAL METACARPAL ARTERIES
20 DORSAL DIGITAL ARTERIES

Branches of the AXILLARY ARTERY
1 SUPERIOR THORACIC A.
2 THORACOACROMIAL A.
3 LATERAL THORACIC A.
4 SUBSCAPULAR A.
5 ANTERIOR AND POSTERIOR HUMERAL CIRCUMFLEX ARTERIES

Branches of the BRACHIAL ARTERY
6 PROFUNDA BRACHII DEEP BRACHIAL
7 SUPERIOR ULNAR COLLATERAL A.
8 INFERIOR ULNAR COLLATERAL A.

Branches of the ULNAR ARTERY
9 ANTERIOR ULNAR RECURRENT A.
10 POSTERIOR ULNAR RECURRENT A.
11 COMMON INTEROSSEOUS A.
 a ANTERIOR INTEROSSEOUS A.
 b POSTERIOR INTEROSSEOUS A.
 c INTEROSSEOUS RECURRENT A.
12 DORSAL CARPAL BRANCH

Branches of the RADIAL ARTERY
13 RADIAL RECURRENT A.
14 DORSAL CARPAL BRANCH
 d DORSALIS POLLICIS
 e DORSALIS INDICIS

Branches of the DEEP PALMAR ARCH
15 PRINCEPS POLLICUS PRINCIPAL ARTERY OF THE THUMB
16 RADIALIS INDICIS
17 PALMAR METACARPAL ARTERIES (4)

Branches of the SUPERFICIAL PALMAR ARCH
18 PALMAR DIGITAL ARTERIES

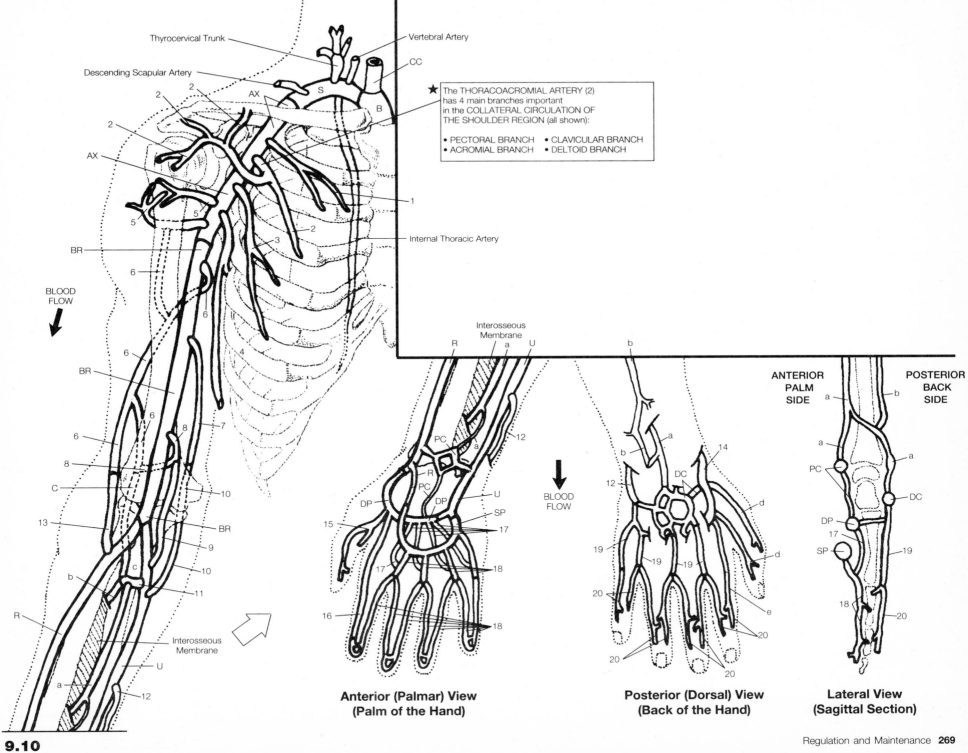

Thyrocervical Trunk

Descending Scapular Artery

Vertebral Artery

CC

★ The THORACOACROMIAL ARTERY (2)
has 4 main branches important
in the COLLATERAL CIRCULATION OF
THE SHOULDER REGION (all shown):

- PECTORAL BRANCH   • CLAVICULAR BRANCH
- ACROMIAL BRANCH   • DELTOID BRANCH

AX

S

B

2
2
2

AX

1

5
5

Internal Thoracic Artery

BR

6

3

6

2

4

6

BLOOD
FLOW

6

BR

6

Interosseous
Membrane

R   a   U

ANTERIOR
PALM
SIDE

POSTERIOR
BACK
SIDE

b

a   a

8

7

PC

a

12

b

PC

BR

8

R

PC

b

DC

DP

6

C

DP   DP

U

BLOOD
FLOW

14

DP

10

DC

13

15

SP

12

a

DC

9

17

19

17

DP

BR

b

c

10

17

18

19   19

SP

19

11

16

18

20

18

R

Interosseous
Membrane

16

18

20

20

U

20

a

12

20

20

**Anterior (Palmar) View
(Palm of the Hand)**

**Posterior (Dorsal) View
(Back of the Hand)**

**Lateral View
(Sagittal Section)**

CARDIOVASCULAR SYSTEM: PRINCIPAL ARTERIES OF THE BODY
Aortic Branches Supplying the Thorax and Abdomen

*Use bright, warm colors for Aorta and major branches;
light warms and neutral colors for minor branches.
Color A₃ as TA and AA separately.*

★ See Chart #5 for organs supplied; compare with 9.19

The AORTA Is Generally Divided
into 3 Main Sections:

A₁ | ASCENDING AORTA |
A₂ | AORTIC ARCH |
A₃ | DESCENDING AORTA | (TA and AA)

The DESCENDING AORTA Is
Divided into 2 Main Sections:

TA | THORACIC AORTA |
AA | ABDOMINAL AORTA |

Branches of the AORTIC ARCH
Supplying the THORAX:

5 | SUBCLAVIAN ARTERIES (L. AND R.) |
1 | COSTOCERVICAL TRUNKS |
2 | HIGHEST INTERCOSTAL A. |
3 | INTERNAL THORACIC ARTERIES |
4 | ANTERIOR INTERCOSTAL ARTERIES |
5 | MUSCULOPHRENIC A. |
6 | SUPERIOR EPIGASTRIC A. |

Branches of the DESCENDING AORTA
Supplying the THORAX: THORACIC AORTA

• **Thoracic aorta passes downward from the
4th –12th thoracic vertebrae and ends at
the diaphragm.**

VISCERAL BRANCHES: supply viscera

7 | BRONCHIAL ARTERIES | (2 left, 1 right)
8 | ESOPHAGEAL ARTERIES | (4–5)
**(Not shown: PERICARDIAL, MEDIASTINAL
and MUSCULAR VISCERAL BRANCHES)**

PARIETAL BRANCHES: supply body wall structures

9 | SUPERIOR PHRENIC ARTERIES |
10 | POSTERIOR INTERCOSTAL ARTERIES |
                              (9 PAIRS)

**(Not shown: L. AND R. SUBCOSTAL ARTERIES,
similar to the POSTERIOR INTERCOSTAL ARTERIES)**

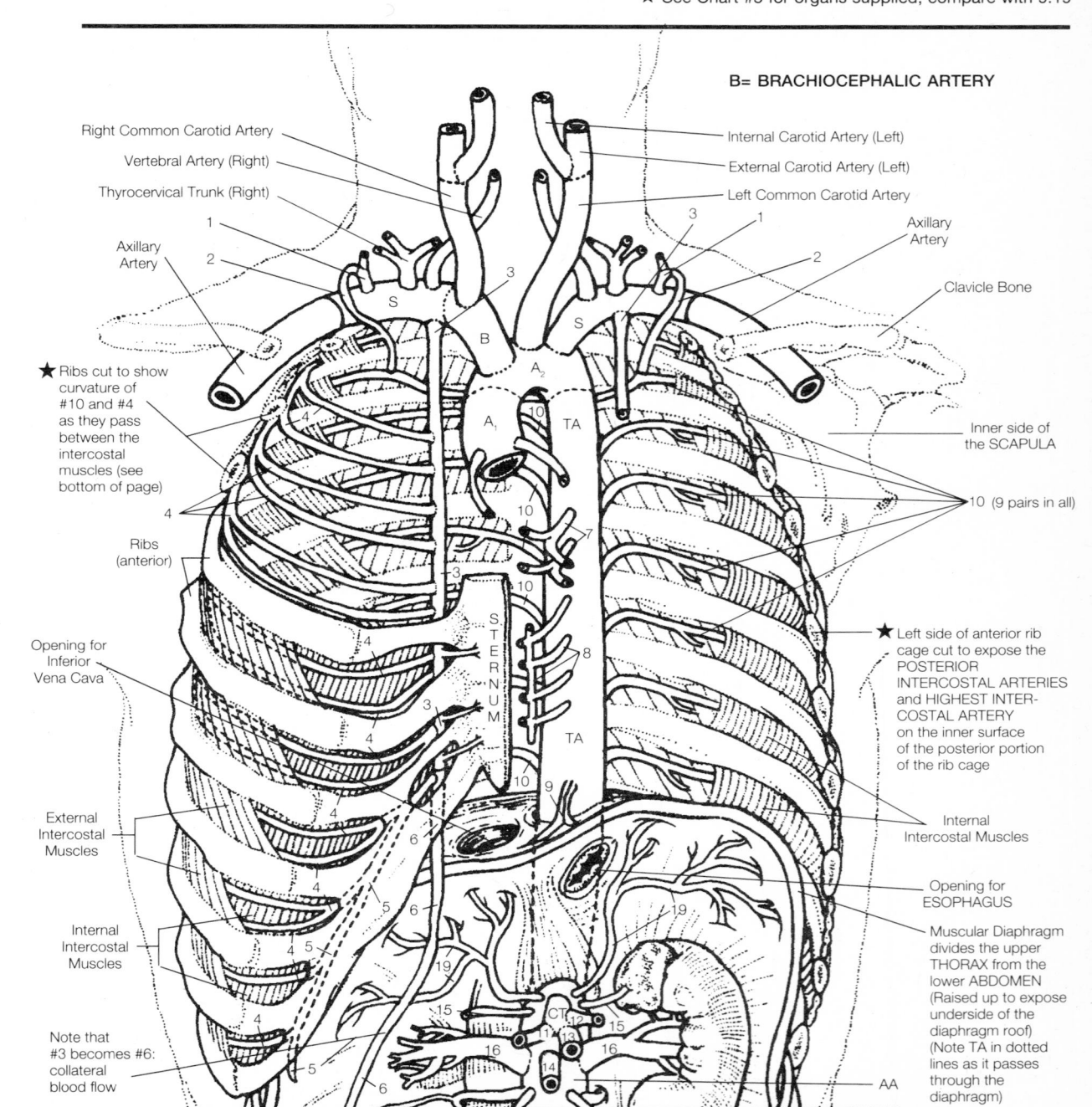

B= BRACHIOCEPHALIC ARTERY

Right Common Carotid Artery
Vertebral Artery (Right)
Thyrocervical Trunk (Right)
Axillary Artery
Internal Carotid Artery (Left)
External Carotid Artery (Left)
Left Common Carotid Artery
Axillary Artery
Clavicle Bone
Inner side of the SCAPULA

★ Ribs cut to show curvature of #10 and #4 as they pass between the intercostal muscles (see bottom of page)

Ribs (anterior)

Opening for Inferior Vena Cava

External Intercostal Muscles

Internal Intercostal Muscles

Note that #3 becomes #6: collateral blood flow

10 (9 pairs in all)

★ Left side of anterior rib cage cut to expose the POSTERIOR INTERCOSTAL ARTERIES and HIGHEST INTER-COSTAL ARTERY on the inner surface of the posterior portion of the rib cage

Internal Intercostal Muscles

Opening for ESOPHAGUS

Muscular Diaphragm divides the upper THORAX from the lower ABDOMEN (Raised up to expose underside of the diaphragm roof) (Note TA in dotted lines as it passes through the diaphragm)

AA

As the DESCENDING AORTA enters the ABDOMEN, it passes through the muscular diaphragm and becomes the ABDOMINAL AORTA. The ABDOMINAL AORTA descends along the posterior wall of the abdomen and ends at the level of the 4th Lumbar Vertebra.

Branches of the DESCENDING AORTA
Supplying the ABDOMEN: ABDOMINAL AORTA

VISCERAL BRANCHES

CT | **CELIAC TRUNK** | COELIAC TRUNK

11 | **COMMON HEPATIC A.**

12 | **LEFT GASTRIC A.**

13 | **SPLENIC A.**

14 | **SUPERIOR MESENTERIC A.**

15 | **SUPRARENAL (ADRENAL) ARTERIES (L. AND R.)**

16 | **RENAL ARTERIES (L. AND R.)**

GONADAL ARTERIES (L. AND R.)

17 — ♂ **TESTICULAR ARTERIES**

♀ **OVARIAN ARTERIES**

18 | **INFERIOR MESENTERIC A.**

PARIETAL BRANCHES

19 | **INFERIOR PHRENIC ARTERIES**

20 | **LUMBAR ARTERIES**

21 | **MIDDLE SACRAL (MIDSACRAL) A.**

• At the level of the 4th Lumbar Vertebra, the ABDOMINAL AORTA bifurcates into 2 large COMMON ILIAC ARTERIES: (See 9.13)

CI | **COMMON ILIAC ARTERIES (L. AND R.)**

★ VERTEBRAL COLUMN and SPINAL CORD supplied by branches of the:
- POSTERIOR INTERCOSTAL ARTERIES
- VERTEBRAL ARTERY
- LUMBAR ARTERIES — not shown
- LATERAL SACRAL ARTERIES

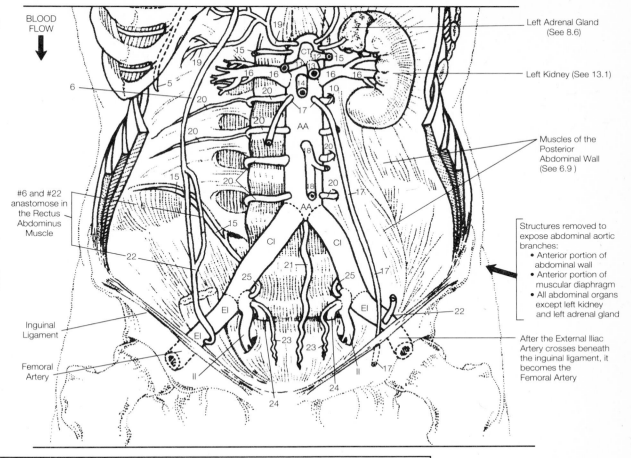

BLOOD FLOW

Left Adrenal Gland (See 8.6)

Left Kidney (See 13.1)

Muscles of the Posterior Abdominal Wall (See 6.9)

Structures removed to expose abdominal aortic branches:
- Anterior portion of abdominal wall
- Anterior portion of muscular diaphragm
- All abdominal organs except left kidney and left adrenal gland

#6 and #22 anastomose in the Rectus Abdominus Muscle

Inguinal Ligament

Femoral Artery

After the External Iliac Artery crosses beneath the inguinal ligament, it becomes the Femoral Artery

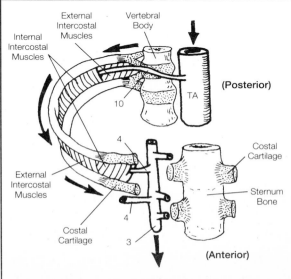

External Intercostal Muscles

Vertebral Body

Internal Intercostal Muscles

External Intercostal Muscles

(Posterior)

Costal Cartilage

Costal Cartilage

Sternum Bone

(Anterior)

ARTERIES That Serve the Thoracic Wall

• Segmental Posterior and Anterior Intercostal Arteries serve the external and intercostal muscles and structures of the thoracic wall.

• Note the blood flow from the Thoracic Aorta (TA) to the Posterior Intercostal Arteries (10) around the rib cage in front to the anterior intercostal arteries (4) and then into the Internal Thoracic Arteries (3), which become the Superior Epigastric Arteries (6).

Branches of the COMMON ILIAC ARTERIES (See 9.13)
EI EXTERNAL ILIAC ARTERIES
22 Inferior Epigastric A.

II INTERNAL ILIAC ARTERIES (HYPOGASTRIC ARTERIES)

23 Lateral Sacral A.

24 Superior Gluteal A.

25 Iliolumbar A.

CARDIOVASCULAR SYSTEM: PRINCIPAL ARTERIES OF THE BODY
Arteries of the Organs of Digestion

Use bright, warm colors for major arteries (and repeat colors from vessels in 9.11). Use lighter warm and neutral colors for minor branches.

★ See chart #5; compare with 9.20; See also 12.9-12.14

There are 3 principal ABDOMINAL
★ ARTERIES supplying the digestive
organs as they branch off from the
ABDOMINAL AORTA (See 9.11):

CT | CELIAC TRUNK

SM | SUPERIOR MESENTERIC A.

IM | INFERIOR MESENTERIC A.

■ CELIAC TRUNK
★ • Thick, short artery immediately
divides into 3 arteries:
  ■ COMMON HEPATIC ARTERY
  ■ SPLENIC ARTERY
  ■ LEFT GASTRIC ARTERY
■ COMMON HEPATIC ARTERY has 3 main branches:
  • HEPATIC ARTERIES (L. AND R.)─[Liver and Gallbladder
  ┌Stomach
  • GASTRODUODENAL A.─┤Body of Pancreas
  └Duodenum
  • RIGHT GASTRIC A.─[Stomach

■ SPLENIC ARTERY has 3 main branches
  • LEFT GASTROEPIPLOIC A.─[Stomach
  • PANCREATIC A.─[Tail of Pancreas
  • POLARS (SUP. and INF.)─[Spleen
(SPLENIC ARTERY also ends in numerous
direct branches to the SPLEEN and some offshoots
to the STOMACH)

■ LEFT GASTRIC ARTERY ─┤• Lesser curvature
                          of stomach
                        • Esophagus

■ SUPERIOR MESENTERIC ARTERY
An unpaired vessel arising
anteriorly from the ABDOMINAL AORTA,
just *below* the CELIAC TRUNK with numerous
branches supplying abdominal organs
  • PANCREAS
  • DUODENUM
  • SMALL INTESTINE
  • CECUM & APPENDIX
  • ASCENDING & TRANSVERSE COLONS
    OF THE LARGE INTESTINE
■ INFERIOR MESENTERIC ARTERY
An unpaired vessel arising anteriorly from the
ABDOMINAL AORTA, just *above* the BIFURCATION
of the AORTA, supplying lower abdominal organs:
  • DESCENDING AND SIGMOID COLONS
    OF THE LARGE INTESTINE
  • RECTUM

3 Unpaired Vessels Supplying the ORGANS OF
DIGESTION in the Abdomen: CELIAC TRUNK,
INFERIOR MESENTERIC ARTERY, SUPERIOR
MESENTERIC ARTERY

CT | CELIAC TRUNK
▼

CH | COMMON HEPATIC ARTERY
  H | HEPATIC ARTERIES (L. AND R.)
    1 | CYSTIC ARTERY FROM RIGHT HEPATIC A.
  G | GASTRODUODENAL A.
    2 | ANTERIOR SUPERIOR PANCREATICODUODENAL A.
    3 | RIGHT GASTROEPIPLOIC A.
    3a | OMENTAL BRANCHES
    4 | POSTERIOR SUPERIOR PANCREATICODUODENAL A. [RETRODUODENAL]
  R-G | RIGHT GASTRIC A.
S | SPLENIC ARTERY
  5 | LEFT GASTROEPIPLOIC A.
  6 | PANCREATIC A.
  7 | SUPERIOR AND INFERIOR POLAR ARTERIES
L-G | LEFT GASTRIC ARTERY

IM | INFERIOR MESENTERIC ARTERY
▼
Branches to Rectum
  15 | SUPERIOR RECTAL A.

Branches to Large Intestine
  16 | LEFT COLIC A.
  17 | SIGMOID A.

SM | SUPERIOR MESENTERIC ARTERY
▼
Branches to Pancreas and Duodenum
  9 | INFERIOR PANCREATICODUODENAL A.

Branches to Large Intestine
  10 | ILEOCOLIC A.
    a | ANTERIOR CECAL BRANCH
    b | POSTERIOR CECAL BRANCH
    c | ILEAL BRANCH
    d | APPENDICULAR BRANCH
  11 | RIGHT COLIC A.
  12 | MIDDLE COLIC A.

Branches to Small Intestine
  13 | ILEAL INTESTINAL A.
  14 | JEJUNAL INTESTINAL A.

Ascending Colon

Ending of Small Intestine (ILEUM)

Vermiform Appendix

Cecum

**Posterior View
VERMIFORM APPENDIX and the
CECUM of the LARGE INTESTINE**

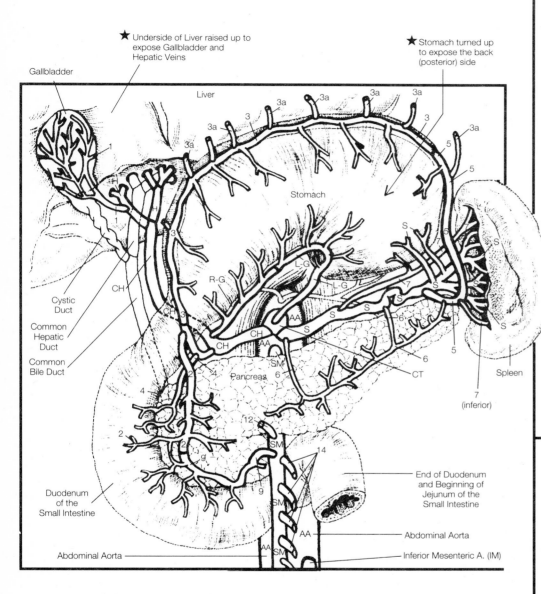

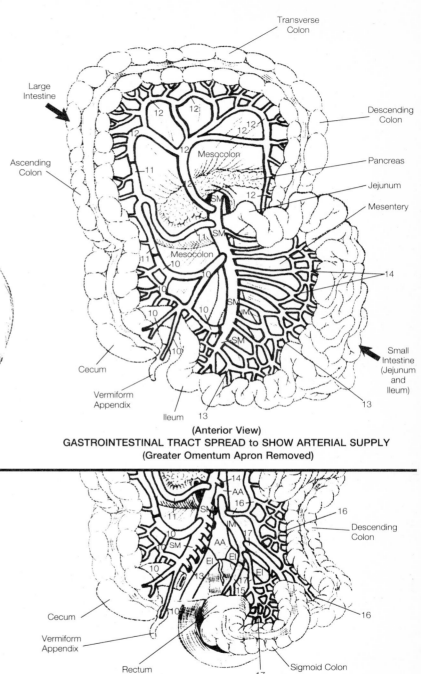

ARTERIAL SUPPLY OF MAJOR ABDOMINAL ORGANS. The 3 principal abdominal branches of the ABDOMINAL AORTA: the CELIAC TRUNK (CT), the SUPERIOR MESENTERIC ARTERY (SM), and the INFERIOR MESENTERIC ARTERY (IM).

AA = Abdominal Aorta
EI = External Iliac Arteries

GASTROINTESTINAL TRACT SPREAD to SHOW ARTERIAL SUPPLY
(Greater Omentum Apron Removed)
(Anterior View)

SMALL INTESTINE REMOVED TO EXPOSE ARTERIAL SUPPLY TO THE SIGMOID COLON and RECTUM. (Note cut branches from the JEJUNAL (14) and ILEAL (13) INTESTINAL ARTERIES of the SUPERIOR MESENTERIC ARTERY.)

CARDIOVASCULAR SYSTEM: PRINCIPAL ARTERIES OF THE BODY
Arteries of the Pelvis and Perineum

Repeat colors used for Aorta and Common Iliac and Femoral Arteries from previous page. Use light warm and neutral colors for minor branches.

★ The ABDOMINAL AORTA bifurcates into the L. and R. COMMON ILIAC ARTERIES at the level of 4th LUMBAR VERTEBRA. The COMMON ILIAC ARTERIES divide 2 – 3 in. inferiorly into 2 main branches:

- ■ EXTERNAL ILIAC ARTERIES (L. and R.)
- ■ INTERNAL ILIAC ARTERIES (L. and R.)

★ ■ INTERNAL ILIAC ARTERIES
- Principal supply of the PELVIS and PERINEUM

Branches:
- ILIOLUMBAR and LATERAL SACRAL ARTERIES → PELVIC WALL and MUSCLES (Psoas Major, Quadratus Lumborum)
- MIDDLE RECTAL ARTERY → the Internal Pelvic Organs
- VESICULAR ARTERIES (SUP., MID., INF.) → the Urinary Bladder
- SUPERIOR and INFERIOR GLUTEAL ARTERIES → the Buttocks
- OBTURATOR ARTERY → Upper Medial Thigh Muscles

★ INTERNAL PUDENDAL ARTERY → external Genitalia in the PERINEAL area:

■ INTERNAL PUDENDAL ARTERY is an IMPORTANT VESSEL IN SEXUAL ACTIVITY: Penile erection and female genital engorgement are vascular expressions (controlled by the AUTONOMIC NERVOUS SYSTEM).

★ ■ THE EXTERNAL ILIAC ARTERIES become the FEMORAL ARTERIES as they exit the Pelvic cavity and cross the INGUINAL LIGAMENT. 2 Branches arise from the EXTERNAL ILIAC ARTERIES just before this change.
- INFERIOR EPIGASTRIC ARTERY → skin and Abdominal Wall Muscles
- DEEP ILIAC CIRCUMFLEX ARTERY → muscles of the Iliac Fossa

■ THE FEMORAL ARTERIES Send branches back into the pelvic region to supply the Genitals and Abdominal Wall.

★ The FEMORAL ARTERIES and THE INFERIOR MESENTERIC ARTERY assist the INTERNAL ILIAC ARTERIES in supplying the PELVIS and PERINEUM.

Primary Blood Supply of the PELVIS and PERINEUM: **INTERNAL ILIACS**

| II | **INTERNAL ILIAC ARTERIES (L. AND R.) (HYPOGASTRIC ARTERIES)** |

Posterior Division of Internal Iliacs

1	**LATERAL SACRAL A.**
2	**SUPERIOR GLUTEAL A.**
3	**ILIOLUMBAR A.**

Secondary Blood Supply of the PELVIS and PERINEUM

AA	**ABDOMINAL AORTA**
11	**MIDSACRAL A.**
CI	**LEFT AND RIGHT COMMON ILIAC ARTERIES**
EI	**EXTERNAL ILIAC ARTERIES (L. AND R.)**
12	**INFERIOR EPIGASTRIC A.**
13	**DEEP ILIAC CIRCUMFLEX A.**
F	**FEMORAL ARTERIES (L. AND R.)**
14	**EXTERNAL PUDENDAL ARTERIES**
15	**DEEP FEMORAL ARTERIES** PROFUNDA FEMORIS
15a	**LATERAL FEMORAL CIRCUMFLEX A.**
15b	**MEDIAL FEMORAL CIRCUMFLEX A.**

Anterior Division of Internal Iliacs

PARIETAL BRANCHES

4	**INFERIOR GLUTEAL A.**
5	**OBTURATOR A.**
6	**INTERNAL PUDENDAL A.** : ♂ ♀
a	**POSTERIOR SCROTAL [LABIAL] A.**
b	**ARTERY OF THE BULB OF THE PENIS [VESTIBULE]**
c	**ARTERIES OF PENIS [CLITORIS]**
d	**PERINEAL A.**
e	**INFERIOR RECTAL A.**

VISCERAL BRANCHES

7	**MIDDLE RECTAL A.**
8	**INFERIOR VESICLE A.**
9	**UMBILICAL A.**
9a	**SUPERIOR VESICLE A.**
10	**[VAGINAL/UTERINE] A.**

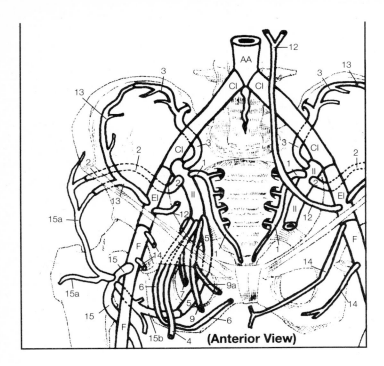

(Anterior View)

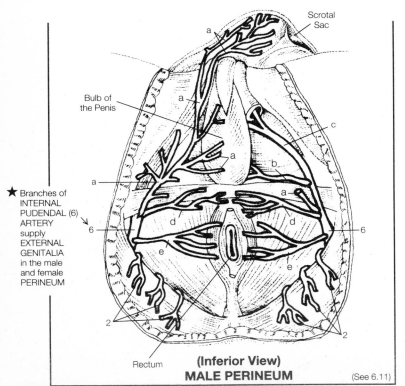

Scrotal Sac

Bulb of the Penis

★ Branches of INTERNAL PUDENDAL (6) ARTERY supply EXTERNAL GENITALIA in the male and female PERINEUM

Rectum

(Inferior View)
MALE PERINEUM

(See 6.11)

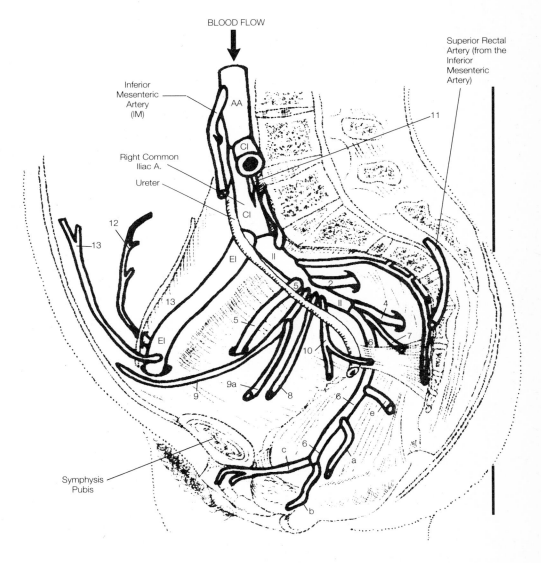

BLOOD FLOW

Inferior Mesenteric Artery (IM)

Right Common Iliac A.

Ureter

Superior Rectal Artery (from the Inferior Mesenteric Artery)

Symphysis Pubis

(Sagittal View)
Viewing branches of the Right Internal Iliac Artery
FEMALE PELVIC REGION
★★ SAME NOMENCLATURE IS IDENTICAL TO JUXTAPOSED VEINS: All these vessels and their names can generally be applied to the venous system as well, where COMMON ILIAC VEIN drains into the INFERIOR VENA CAVA. (BLOOD FLOW ↑) Therefore, there are no pages in the VENOUS section of this chapter covering veins of the Pelvis and Perineum.

CARDIOVASCULAR SYSTEM: PRINCIPAL ARTERIES OF THE BODY
Arteries of the Lower Extremity (Right Limb)

Check last page for same vessels with different reference numbers and use same colors for consistency.
Use light, warm and neutral colors for minor branches.
★ Compare with 9.21

★ Reviewing, the EXTERNAL ILIAC ARTERIES diverge through the pelvis and enter the thigh region, becoming the FEMORAL ARTERIES.

★ ■ FEMORAL TRIANGLE
Superiorly, the FEMORAL ARTERIES are close to the surface. A triangular area around this point serves as an important pressure point in diagnosis and palpation. This area is formed by the:
- Inguinal Ligament
- Sartorius Muscle
- Adductor Longus Muscle (See 6.18-6.19)

★ Branches of FEMORAL ARTERY:
■ DEEP FEMORAL ARTERY (PROFUNDIS FEMORIS)
- Largest branch
- Supplies hamstring muscles (passes posteriorly)
■ MEDIAL and LATERAL FEMORAL CIRCUMFLEX ARTERIES
Supplies muscles in the proximal thigh via branches encircling the FEMUR bone.

■ POPLITEAL ARTERY
- The FEMORAL ARTERY passes down the medial and posterior side of the thigh at the back of the knee joint, where it becomes the POPLITEAL ARTERY.
- Serves small branches to the knee joint (GENICULARS), then divides below the knee into 2 branches:
■ ANTERIOR TIBIAL ARTERY
- Traverses over the INTEROSSEOUS MEMBRANE and serves anterior aspect of leg (and leg muscles).
- At the ankle, it becomes the DORSALIS PEDIS, serving the ankle and dorsum of the foot.
■ POSTERIOR TIBIAL ARTERY
- Continues down the back of the leg between the knee and the ankle
- Sends off a large PERONEAL ARTERY serving the Peroneal Leg muscles
- At the ankle, it bifurcates into the LATERAL and MEDIAL PLANTAR ARTERIES, which serve the sole (or bottom of the foot).
★ The LATERAL PLANTAR ARTERY anastomoses with the DORSAL PEDIS ARTERY (not shown) to form the PLANTAR ARCH!

| AA | ABDOMINAL AORTA |

| CI | RIGHT COMMON ILIAC A. |

| II | INTERNAL ILIAC A. |

| 1 | SUPERIOR GLUTEAL A. |

| 2 | INFERIOR GLUTEAL A. |

| 3 | OBTURATOR A. |

| EI | EXTERNAL ILIAC A. |

As It Enters the Thigh, the EXTERNAL ILIAC Becomes the FEMORAL ARTERY:

| F | FEMORAL ARTERY |

| 4 | DEEP FEMORAL A. (PROFUNDA FEMORIS) |

| a | PERFORATING BRANCHES |

| b | MEDIAL FEMORAL CIRCUMFLEX A. |

| C | LATERAL FEMORAL CIRCUMFLEX A. |

| C_1 | DESCENDING BRANCH |

| 5 | DESCENDING GENICULAR A. |

As It Crosses the Posterior Aspect of the Knee, The FEMORAL ARTERY Becomes the POPLITEAL ARTERY:

| P | POPLITEAL ARTERY |

| 6 | GENICULAR ARTERIES (SUPERIOR, MIDDLE, AND INFERIOR) |

Below the Knee, the POPLITEAL ARTERY Divides into the ANTERIOR TIBIAL and POSTERIOR TIBIAL:

| AT | ANTERIOR TIBIAL ARTERY |

At the Ankle, the ANTERIOR TIBIAL Becomes the DORSALIS PEDIS ARTERY:

| DP | DORSALIS PEDIS A. |

| 7 | ARCUATE A. |

| 8 | DORSAL METATARSAL ARTERIES |

| 8a | DORSAL DIGITAL ARTERIES |

| 9 | LATERAL TARSAL ARTERIES |

| PT | POSTERIOR TIBIAL ARTERY |

| 10 | PERONEAL A. |

| 11 | MEDIAL PLANTAR A. |

| 12 | LATERAL PLANTAR A. |

| PA | PLANTAR ARCH |

| 13 | PLANTAR METATARSAL ARTERIES |

| 13a | PLANTAR DIGITAL ARTERIES |

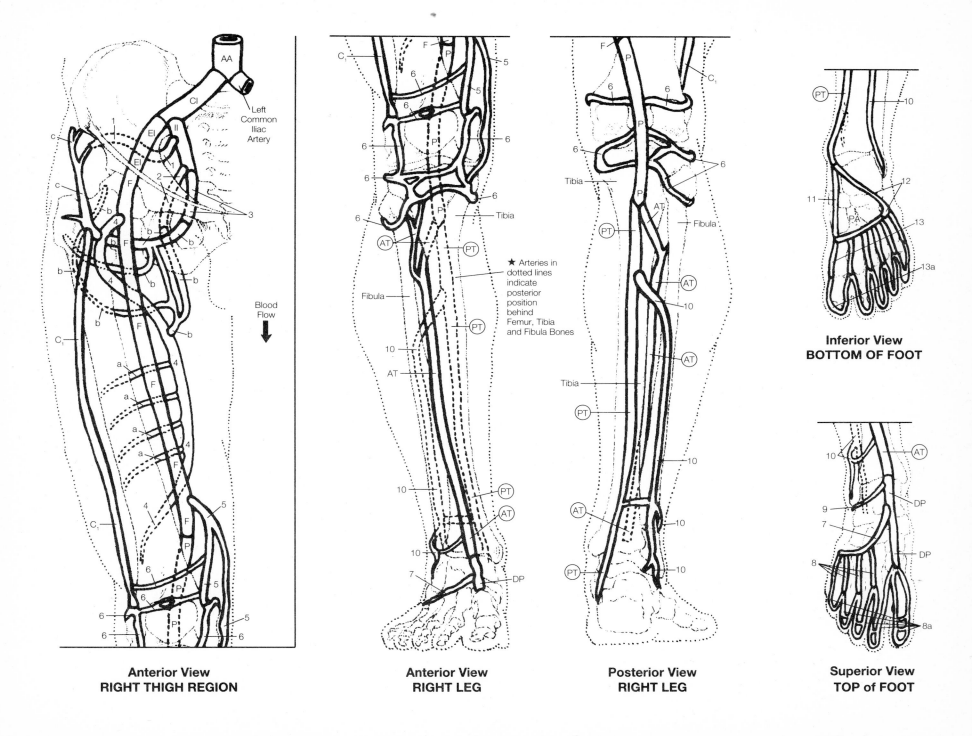

★ Arteries in dotted lines indicate posterior position behind Femur, Tibia and Fibula Bones

Blood Flow

AA

Left Common Iliac Artery

Anterior View RIGHT THIGH REGION

Anterior View RIGHT LEG

Posterior View RIGHT LEG

Inferior View BOTTOM OF FOOT

Superior View TOP of FOOT

9.14

CARDIOVASCULAR SYSTEM: REVIEW OF THE MAJOR ARTERIES

Use color choices from previous pages. Any vessels assigned the same color will be easily distinguished according to position.

★ See Chart #4 for organs supplied; compare with 9.22

Note: Vessels drawn in dotted lines indicate a position *behind* bone

CO	**CORONARY ARTERIES**
Ⓐ	**AORTA**
B	**BRACHIOCEPHALIC A.**
CC	**COMMON CAROTID A.**
EC	**EXTERNAL CAROTID A.**
IC	**INTERNAL CAROTID A.**
S	**SUBCLAVIAN A.**
AX	**AXILLARY A.**
BR	**BRACHIAL A.**
R	**RADIAL A.**
U	**ULNAR A.**
SP	**SUPERFICIAL PALMAR ARCH**
DP	**DEEP PALMAR ARCH**
CI	**COMMON ILIAC A.**
II	**INTERNAL ILIAC A.**
EI	**EXTERNAL ILIAC A.**
F	**FEMORAL A.**

EC
IC
CC
S
Clavicle Bone
AX
CO
Humerus Bone
BR

Sup. Ulnar Collateral A.
Inf. Ulnar Collateral A.
Profunda Brachii A.
Ant. Ulnar Recurrent A.
Post. Ulnar Recurrent A.
Interosseous Recurrent A.
Common Interosseous A.
R
U
Ant. Interosseous A.
Palmar Carpal Arch
DP
SP

B
S
A
CT
A
A
CI CI
EI
II
F
EI
II
F

Deep Femoral Artery (Profunda Femoris A.)
Midsacral Artery
Palmar Digital Arteries

Vertebral Artery
Thyrocervical Trunk
Descending Scapular Artery
Internal Thoracic Artery
Superior Thoracic Artery
Thoracoacromial Artery
Lateral Thoracic Artery
Anterior and Posterior Humeral Circumflex Arteries
Subscapular Artery
Profunda Brachii Artery
Inferior Phrenic Artery
Suprarenal (Adrenal) Artery
Left Gastric Artery
Common Hepatic Artery
Splenic Artery
Renal Artery
Superior Mesenteric Artery
Gonadal Artery
Lumbar Arteries
Inferior Epigastric Artery
Inferior Mesenteric Artery
Iliolumbar Artery
Deep Circumflex Iliac Artery
Princeps Pollicis
Radialis Indicis
Palmar Metacarpal Arteries

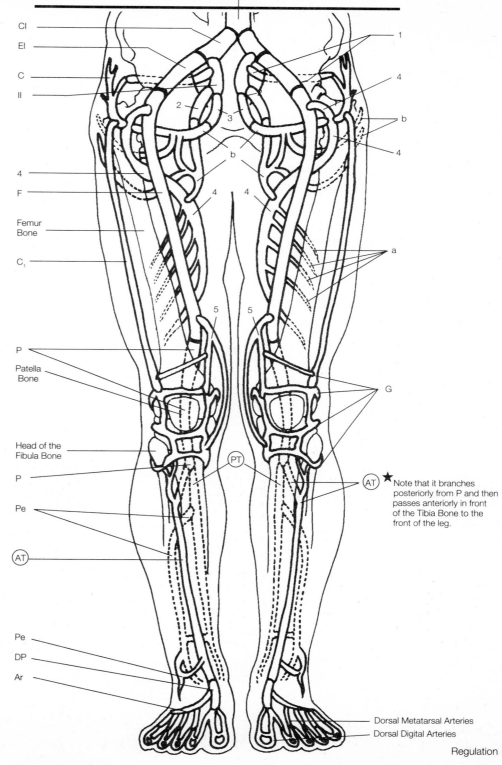

II INTERNAL ILIAC A.

1 SUPERIOR GLUTEAL A.

2 INFERIOR GLUTEAL A.

3 OBTURATOR A.

F FEMORAL A.

4 DEEP FEMORAL A. PROFUNDA FEMORIS

a PERFORATING BRANCHES OF PROFUNDA FEMORIS A.

b MEDIAL FEMORAL CIRCUMFLEX A.

c LATERAL FEMORAL CIRCUMFLEX A.

C₁ DESCENDING BRANCH

5 DESCENDING GENICULAR A.

P POPLITEAL A.

G GENICULAR ARTERIES

AT ANTERIOR TIBIAL A.

DP DORSALIS PEDIS A.

Ar ARCUATE A.

PT POSTERIOR TIBIAL A.

Pe PERONEAL A.

Labels on figure: CI, EI, C, II, 4, F, Femur Bone, C₁, P, Patella Bone, Head of the Fibula Bone, P, Pe, AT, Pe, DP, Ar

1, 4, b, 4, b, 2, 3, 4, 4, 5, 5, a, G, PT, AT

★ Note that it branches posteriorly from P and then passes anteriorly in front of the Tibia Bone to the front of the leg.

Dorsal Metatarsal Arteries
Dorsal Digital Arteries

9.15

CARDIOVASCULAR SYSTEM: PRINCIPAL VEINS OF THE BODY
The Cranium and Venous Sinuses

★There are 2 MAIN VESSELS that drain blood from the HEAD and NECK REGION
- ■ EXTERNAL JUGULAR VEINS (See 9.17)
- ■ INTERNAL JUGULAR VEINS

■ INTERNAL JUGULAR VEINS are larger and deeper than the EXTERNAL JUGULAR VEINS.
- INTERNAL JUGULAR VEINS arise from numerous VENOUS SINUSES in the cranial area. These vessels (sinuses) are actually spaces between 2 LAYERS OF DURA MATER, which act as conduits (dilated channels) for the BLOOD. (See 7.4)

- DURAL VENOUS SINUSES are a series of bulbous PAIRED and UNPAIRED BLOOD CHANNELS that receive blood from the BRAIN through the:
  - CEREBRAL VEINS
  - OPHTHALMIC VEINS
  - CEREBELLAR VEINS
  - MENINGEAL VEINS

■ SIGMOID SINUSES become the INTERNAL JUGULAR VEINS at the JUGULAR FORAMEN

■ INTERNAL JUGULAR VEINS drain:
- BRAIN
- MENINGES
- DEEP REGIONS OF FACE and NECK

As each INTERNAL JUGULAR VEIN passes down the neck beneath the STERNOCLEIDOMASTOID MUSCLE, it is surrounded in common with the COMMON CAROTID ARTERY and VAGUS NERVE by a protective sheath—the CAROTID SHEATH.

The INTERNAL JUGULAR VEINS merge with the SUBCLAVIAN VEINS to form the BRACHIOCEPHALIC VEINS.

■ The BRACHIOCEPHALIC VEINS then merge into a single large SUPERIOR VENA CAVA, which enters the RIGHT ATRIUM of the HEART.

VEINS of the BRAIN

1 | SUPERIOR CEREBRAL V. |

(2) | INFERIOR AND MEDIAL CEREBRAL VEINS | (See 9.19)

3 | GREAT CEREBRAL VEIN OF GALEN | VENA CEREBRI MAGNA

(4) | SUPERIOR AND INFERIOR OPHTHALMIC VEINS | (See 9.19)

[Not Shown: CEREBELLAR VEINS, MENINGEAL VEINS]

VEINS of CRANIAL (SPONGY) BONE

5 | DIPLOË VEINS | END IN THE DURAL SINUSES

External-Internal Connectors

(6) | EMISSARY VEINS | (See 9.19)

- Small, valveless veins that establish communication between SINUSES inside the SKULL and VEINS outside the SKULL.

- Receive blood from the scalp, pass through the cranial bones, and empty into DURAL SINUSES

SINUSES of the DURA MATER

COS | CONFLUENCE OF THE SINUSES | Located at the INTERNAL OCCIPITAL PROTUBERANCE

UNPAIRED SINUSES

7 | SUPERIOR SAGITTAL SINUS | (See 7.4, 7.5)

8 | INFERIOR SAGITTAL SINUS |

9 | STRAIGHT SINUS |

10 | BASILAR PLEXUS | : [Between the CAVERNOUS and INFERIOR PETROSAL SINUSES]

PAIRED SINUSES

11 | CAVERNOUS SINUSES |

12 | SUPERIOR PETROSAL SINUSES |

13 | INFERIOR PETROSAL SINUSES |

14 | OCCIPITAL SINUSES |

15 | TRANSVERSE/LATERAL SINUSES |

16 | SIGMOID SINUSES | : [Become the INTERNAL JUGULAR VEINS]

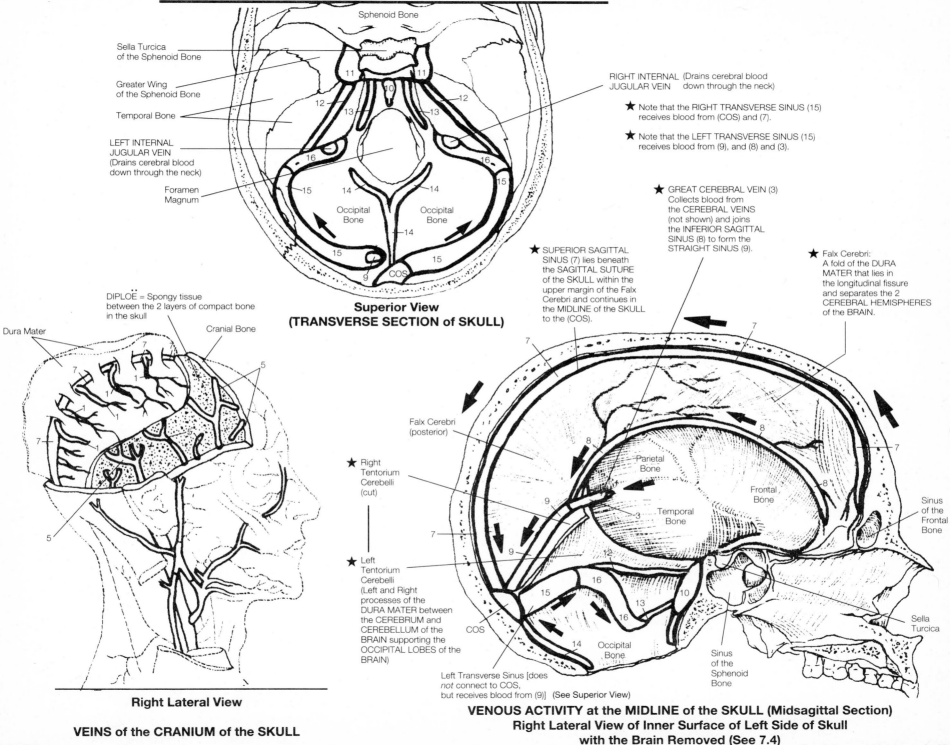

Sphenoid Bone

Sella Turcica
of the Sphenoid Bone

Greater Wing
of the Sphenoid Bone

Temporal Bone

LEFT INTERNAL
JUGULAR VEIN
(Drains cerebral blood
down through the neck)

Foramen
Magnum

Occipital
Bone

Occipital
Bone

COS

**Superior View
(TRANSVERSE SECTION of SKULL)**

RIGHT INTERNAL (Drains cerebral blood
JUGULAR VEIN   down through the neck)

★ Note that the RIGHT TRANSVERSE SINUS (15)
  receives blood from (COS) and (7).

★ Note that the LEFT TRANSVERSE SINUS (15)
  receives blood from (9), and (8) and (3).

★ GREAT CEREBRAL VEIN (3)
  Collects blood from
  the CEREBRAL VEINS
  (not shown) and joins
  the INFERIOR SAGITTAL
  SINUS (8) to form the
  STRAIGHT SINUS (9).

★ SUPERIOR SAGITTAL
  SINUS (7) lies beneath
  the SAGITTAL SUTURE
  of the SKULL within the
  upper margin of the Falx
  Cerebri and continues in
  the MIDLINE of the SKULL
  to the (COS).

★ Falx Cerebri:
  A fold of the DURA
  MATER that lies in
  the longitudinal fissure
  and separates the 2
  CEREBRAL HEMISPHERES
  of the BRAIN.

DIPLOË = Spongy tissue
between the 2 layers of compact bone
in the skull

Dura Mater

Cranial Bone

Falx Cerebri
(posterior)

★ Right
  Tentorium
  Cerebelli
  (cut)

★ Left
  Tentorium
  Cerebelli
  (Left and Right
  processes of the
  DURA MATER between
  the CEREBRUM and
  CEREBELLUM of the
  BRAIN supporting the
  OCCIPITAL LOBES of the
  BRAIN)

Parietal
Bone

Frontal
Bone

Temporal
Bone

Sinus
of the
Frontal
Bone

COS

Occipital
Bone

Sinus
of the
Sphenoid
Bone

Sella
Turcica

Left Transverse Sinus [does
not connect to COS,
but receives blood from (9)] (See Superior View)

Right Lateral View

VEINS of the CRANIUM of the SKULL

**VENOUS ACTIVITY at the MIDLINE of the SKULL (Midsagittal Section)
Right Lateral View of Inner Surface of Left Side of Skull
with the Brain Removed (See 7.4)**

CARDIOVASCULAR SYSTEM: PRINCIPAL VEINS OF THE BODY
Veins of the Head and Neck (Right Lateral View)

Use bright, cool colors for the major veins and major tributaries. Use light, cool colors and neutral colors (or no colors) for minor tributaries.
★ Compare with 9.8; see 9.16

★ In general, VEINS are more abundant than ARTERIES, and are both DEEP *and* SUPERFICIAL, whereas ARTERIES are basically DEEP. VEINS *receive* smaller TRIBUTARIES, while ARTERIES *give off* smaller BRANCHES.

• The INTERNAL JUGULAR VEINS drain the deeper regions of the head and neck, and the EXTERNAL JUGULAR VEINS drain the SUPERFICIAL REGIONS.
• The INTERNAL JUGULAR VEINS pass *beneath* the STERNOCLEIDOMASTOID MUSCLE, and the EXTERNAL JUGULAR VEINS pass *superficially* to it, but deep to the SUPERFICIAL PLATYSMA MUSCLE. (See 6.6)

■ EXTERNAL JUGULAR VEINS
• Descend laterally along the INTERNAL JUGULAR VEINS

• Drain into the right and left SUBCLAVIAN VEINS just posterior to the clavicle bones

★ ■ RETROMANDIBULAR VEIN connects the EXTERNAL JUGULAR VEIN to the FACIAL VEIN (a major tributary of the INTERNAL JUGULAR VEIN)
• Retromandibular vein drains blood from the scalp through the branches of the SUPERFICIAL TEMPORAL VEIN.

★ ■ The VERTEBRAL VEIN descends (after forming a plexus posteriorly with the OCCIPITAL VEIN) through the TRANSVERSE FORAMINA OF THE CERVICAL VERTEBRAE, and empties directly into the SUBCLAVIAN VEIN just medial to the entrance of the EXTERNAL JUGULAR VEIN.

★ The ANTERIOR JUGULAR VEINS connect the FACIAL VEINS and the EXTERNAL JUGULAR VEINS through a unique JUGULAR VENOUS ARCH.

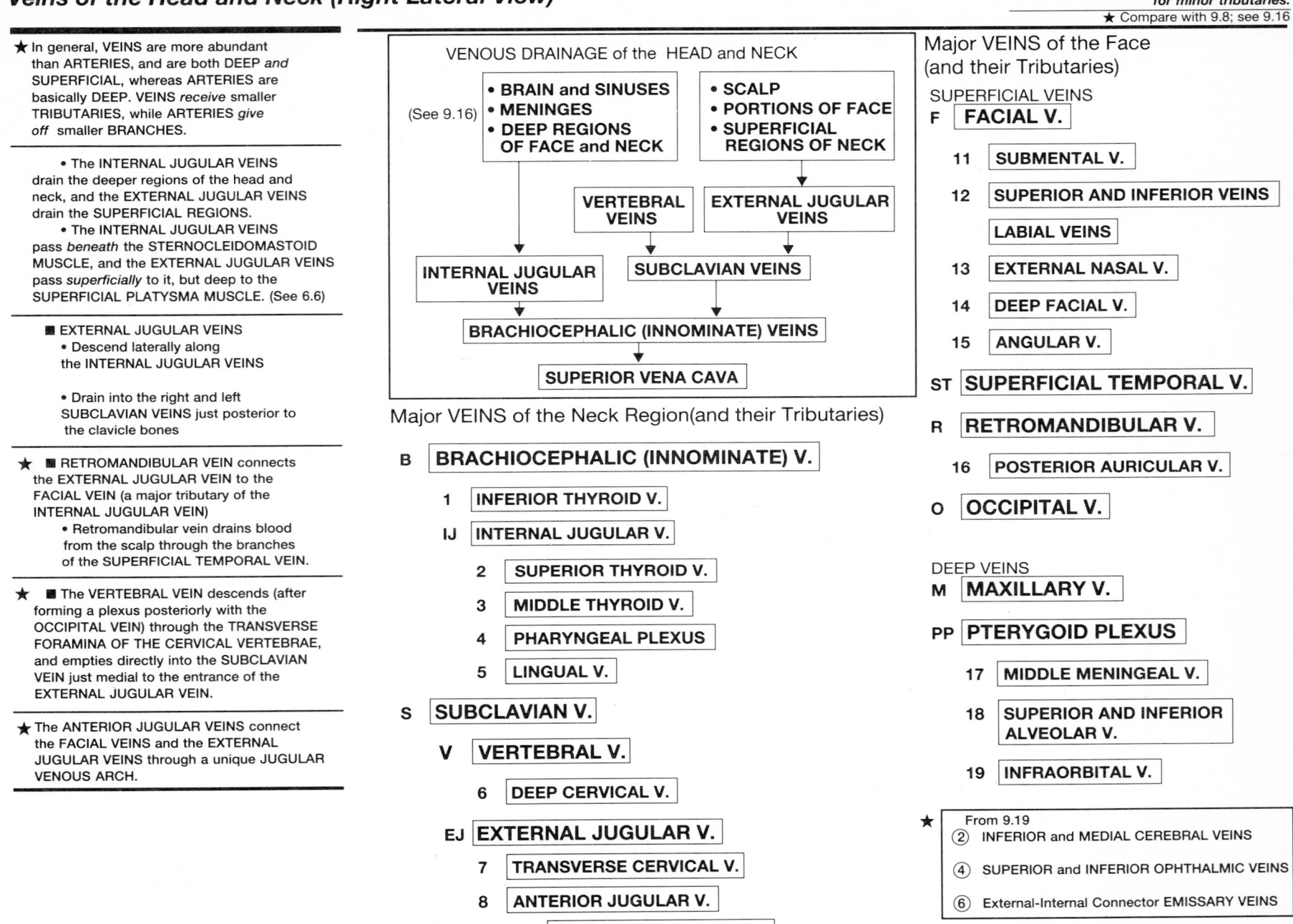

VENOUS DRAINAGE of the HEAD and NECK

(See 9.16)

• BRAIN and SINUSES
• MENINGES
• DEEP REGIONS OF FACE and NECK

• SCALP
• PORTIONS OF FACE
• SUPERFICIAL REGIONS OF NECK

VERTEBRAL VEINS

EXTERNAL JUGULAR VEINS

INTERNAL JUGULAR VEINS

SUBCLAVIAN VEINS

BRACHIOCEPHALIC (INNOMINATE) VEINS

SUPERIOR VENA CAVA

Major VEINS of the Neck Region (and their Tributaries)

B **BRACHIOCEPHALIC (INNOMINATE) V.**

1 INFERIOR THYROID V.

IJ INTERNAL JUGULAR V.

2 SUPERIOR THYROID V.

3 MIDDLE THYROID V.

4 PHARYNGEAL PLEXUS

5 LINGUAL V.

S **SUBCLAVIAN V.**

V **VERTEBRAL V.**

6 DEEP CERVICAL V.

EJ **EXTERNAL JUGULAR V.**

7 TRANSVERSE CERVICAL V.

8 ANTERIOR JUGULAR V.

8a JUGULAR VENOUS ARCH

9 POSTERIOR EXTERNAL JUGULAR V.

Major VEINS of the Face
(and their Tributaries)

SUPERFICIAL VEINS

F **FACIAL V.**

11 SUBMENTAL V.

12 SUPERIOR AND INFERIOR VEINS

LABIAL VEINS

13 EXTERNAL NASAL V.

14 DEEP FACIAL V.

15 ANGULAR V.

ST **SUPERFICIAL TEMPORAL V.**

R **RETROMANDIBULAR V.**

16 POSTERIOR AURICULAR V.

O **OCCIPITAL V.**

DEEP VEINS

M **MAXILLARY V.**

PP **PTERYGOID PLEXUS**

17 MIDDLE MENINGEAL V.

18 SUPERIOR AND INFERIOR ALVEOLAR V.

19 INFRAORBITAL V.

★ From 9.19
② INFERIOR and MEDIAL CEREBRAL VEINS
④ SUPERIOR and INFERIOR OPHTHALMIC VEINS
⑥ External-Internal Connector EMISSARY VEINS

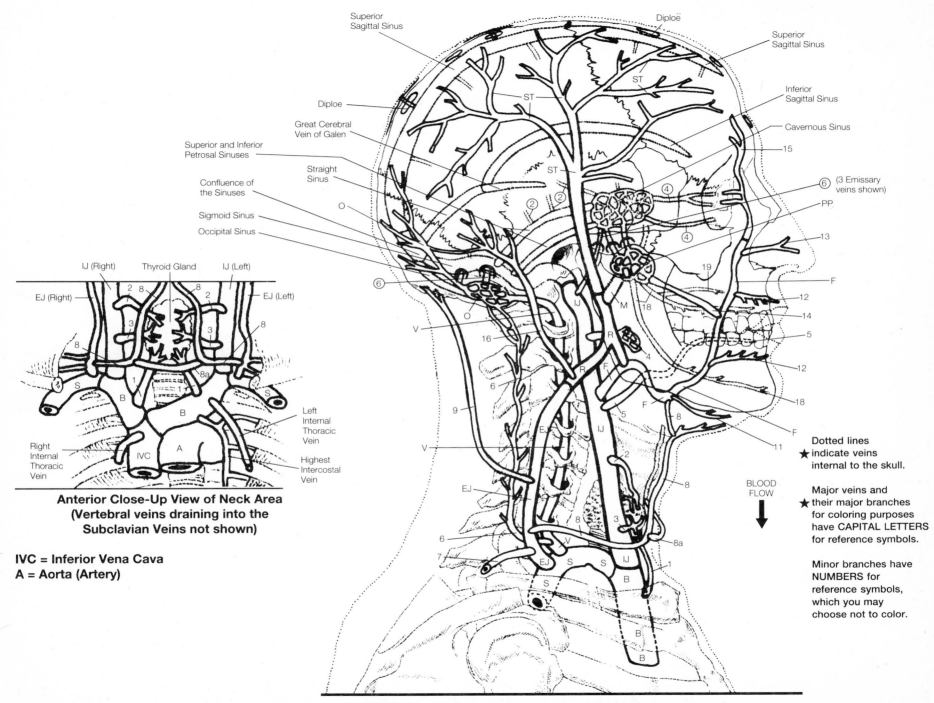

Superior Sagittal Sinus

Diploë

Superior Sagittal Sinus

Inferior Sagittal Sinus

Cavernous Sinus

Diploë

Great Cerebral Vein of Galen

Superior and Inferior Petrosal Sinuses

Straight Sinus

Confluence of the Sinuses

Sigmoid Sinus

Occipital Sinus

ST

ST

ST

O

15

(6) (3 Emissary veins shown)

PP

13

F

12

14

5

12

18

F

11

19

IJ

M

18

V

O

16

R

R

F

4

9

5

8

6

V

EJ

IJ

F

8

EJ

8

2

EJ

S

S

IJ

8

3

B

V

8a

6

7

EJ

S

1

S

B

B

B

BLOOD FLOW

Dotted lines
★ indicate veins internal to the skull.

Major veins and
★ their major branches for coloring purposes have CAPITAL LETTERS for reference symbols.

Minor branches have NUMBERS for reference symbols, which you may choose not to color.

Anterior Close-Up View of Neck Area (Vertebral veins draining into the Subclavian Veins not shown)

IJ (Right) Thyroid Gland IJ (Left)

EJ (Right)

EJ (Left)

Right Internal Thoracic Vein

Left Internal Thoracic Vein

Highest Intercostal Vein

2 8 8 2 8

3 3 8

8 8a

S S

B

V 1

B

A

IVC

IVC = Inferior Vena Cava
A = Aorta (Artery)

VEINS DRAINING the HEAD and NECK (Right Lateral View)

CARDIOVASCULAR SYSTEM: PRINCIPAL VEINS OF THE BODY
Veins of the Upper Extremity (Right Limb)

Use bright, cool colors for the Vena Cava and major tributaries. Use light, cool colors and neutral colors for minor tributaries.
★ Compare with 9.10

SUPERFICIAL VEINS
- Located just beneath skin
- Extensive anastomoses between and with DEEP VEINS

■ **CEPHALIC VEIN**
Origin at middle of the DORSAL PALMAR ARCH, winds upward around radial border of forearm.
- Joins the BASILIC VEIN via the MEDIA CUBITAL VEIN, in the cubital fossa in front of the elbow
- Then unites with ACCESSORY CEPHALIC VEIN to form the CEPHALIC VEIN OF THE UPPER EXTREMITY
- Empties into the AXILLARY VEIN just above the clavicle, where it pierces the fascia.

■ **BASILIC VEIN**
- Origin at ulnar end of the DORSAL ARCH, passes upward along the medial side of the arm (ulnar side of forearm)
- Merges with the BRACHIAL VEIN just below the head of the HUMERUS to form the AXILLARY VEIN.

DEEP VEINS
■ **RADIAL VEINS** (lateral side of forearm) and **ULNAR VEINS** (medial side of forearm) drain the DEEP and SUPERFICIAL PALMAR (VOLAR) ARCHES of the hand.
- Join in the cubital fossa to form the BRACHIAL VEIN

■ **BRACHIAL VEIN** continues up the medial side of the BRACHIUM, and merges with the BASILIC VEIN near the head of the Humerus to form the AXILLARY VEIN.

■ As the AXILLARY VEIN passes the first rib, it becomes the SUBCLAVIAN VEIN.

■ **SUBCLAVIAN VEINS** (L. and R.) unite with INTERNAL JUGULAR VEINS from the neck to form the BRACHIOCEPHALIC VEINS. (Compare with illustration 9.8)

★THORACIC DUCT of LYMPHATIC SYSTEM enters the LEFT BRACHIOCEPHALIC VEIN at junction of the LEFT INTERNAL JUGULAR VEIN and SUBCLAVIAN VEINS. (The RIGHT LYMPHATIC DUCT enters the RIGHT BRACHIOCEPHALIC VEIN at junction of the RIGHT INTERNAL VEIN and SUBCLAVIAN VEIN.) (See 10.2)

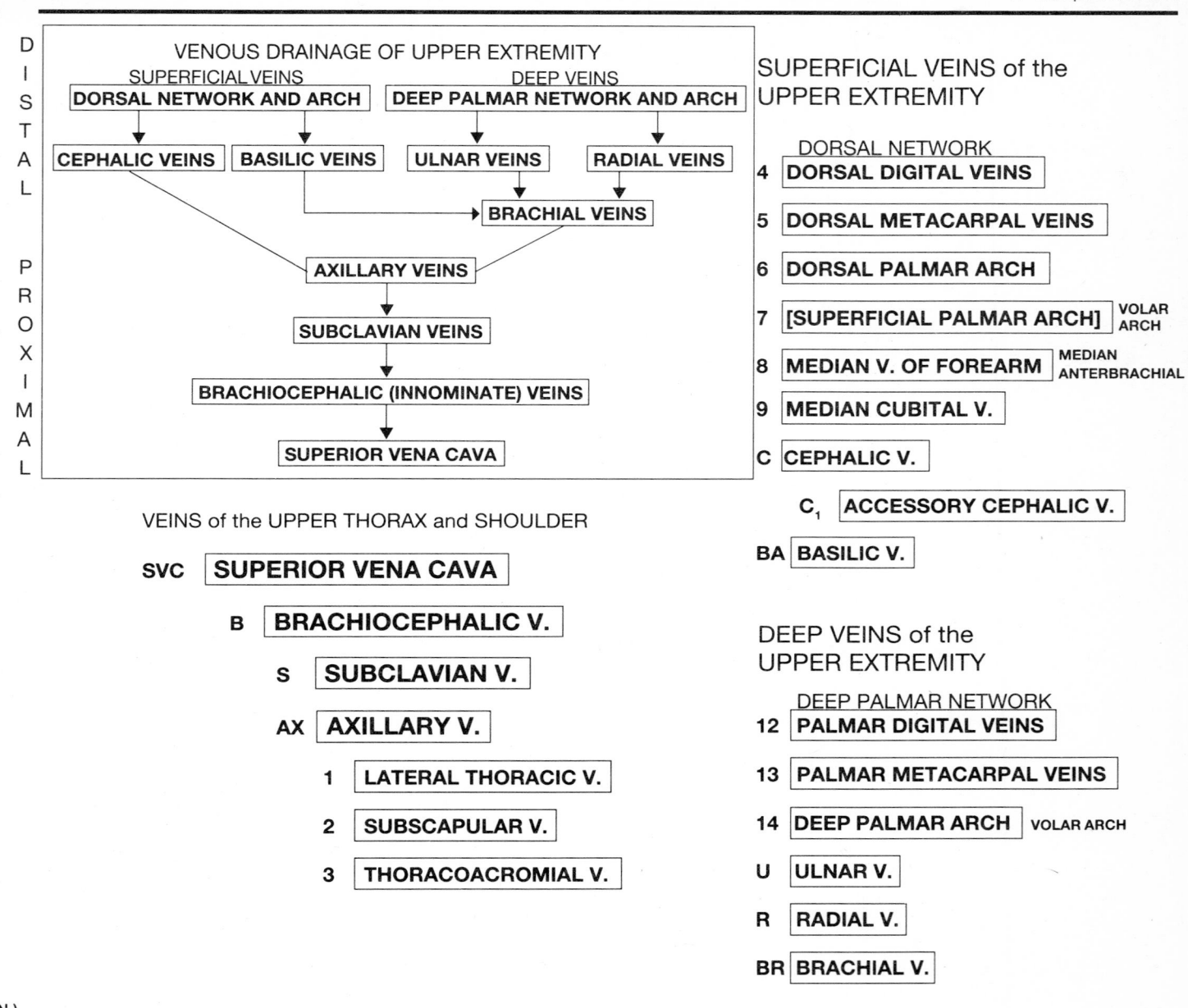

D I S T A L

P R O X I M A L

VENOUS DRAINAGE OF UPPER EXTREMITY

SUPERFICIAL VEINS — DEEP VEINS

DORSAL NETWORK AND ARCH — DEEP PALMAR NETWORK AND ARCH

CEPHALIC VEINS — BASILIC VEINS — ULNAR VEINS — RADIAL VEINS

BRACHIAL VEINS

AXILLARY VEINS

SUBCLAVIAN VEINS

BRACHIOCEPHALIC (INNOMINATE) VEINS

SUPERIOR VENA CAVA

VEINS of the UPPER THORAX and SHOULDER

SVC **SUPERIOR VENA CAVA**

B **BRACHIOCEPHALIC V.**

S **SUBCLAVIAN V.**

AX **AXILLARY V.**

1 **LATERAL THORACIC V.**

2 **SUBSCAPULAR V.**

3 **THORACOACROMIAL V.**

SUPERFICIAL VEINS of the UPPER EXTREMITY

DORSAL NETWORK

4 **DORSAL DIGITAL VEINS**

5 **DORSAL METACARPAL VEINS**

6 **DORSAL PALMAR ARCH**

7 **[SUPERFICIAL PALMAR ARCH]** VOLAR ARCH

8 **MEDIAN V. OF FOREARM** MEDIAN ANTERBRACHIAL

9 **MEDIAN CUBITAL V.**

C **CEPHALIC V.**

C_1 **ACCESSORY CEPHALIC V.**

BA **BASILIC V.**

DEEP VEINS of the UPPER EXTREMITY

DEEP PALMAR NETWORK

12 **PALMAR DIGITAL VEINS**

13 **PALMAR METACARPAL VEINS**

14 **DEEP PALMAR ARCH** VOLAR ARCH

U **ULNAR V.**

R **RADIAL V.**

BR **BRACHIAL V.**

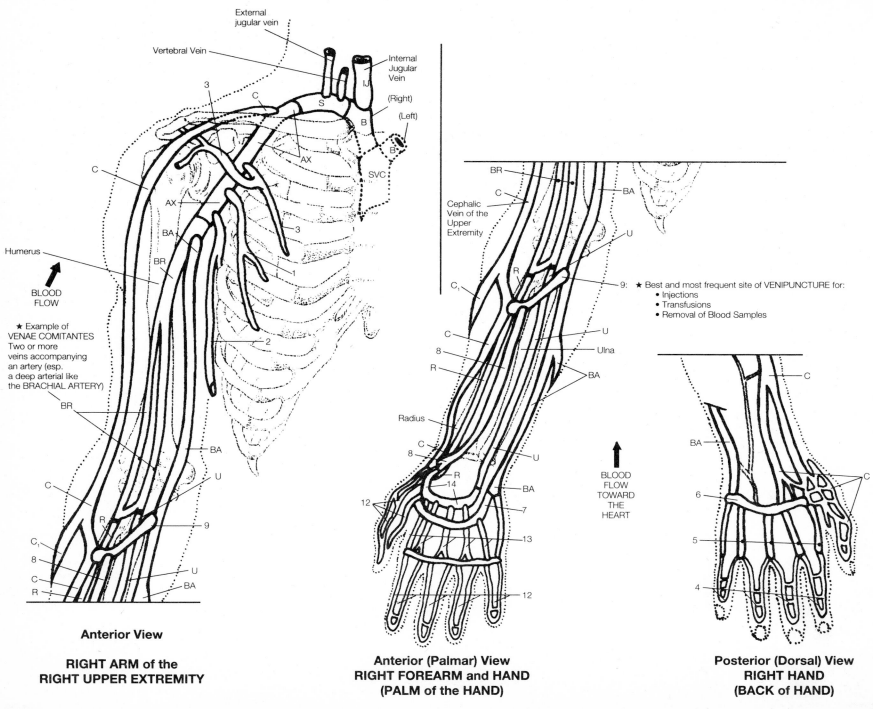

External
jugular vein

Vertebral Vein

Internal
Jugular
Vein

(Right)

(Left)

3

C

S

AX

B

B'

SVC

C

AX

BA

BR

Humerus

1

3

**BLOOD
FLOW**

★ Example of
VENAE COMITANTES
Two or more
veins accompanying
an artery (esp.
a deep arterial like
the BRACHIAL ARTERY)

2

BR

BA

C

U

R

9

C₁

8

C

R

U

BA

BR

C

Cephalic
Vein of the
Upper
Extremity

BA

U

U

Ulna

BA

Radius

9:

★ Best and most frequent site of VENIPUNCTURE for:
• Injections
• Transfusions
• Removal of Blood Samples

C₁

R

C

8

R

C

8

R

N

14

U

BA

12

BA

7

13

12

**BLOOD
FLOW
TOWARD
THE
HEART**

C

BA

6

C

5

4

C

Anterior View

**RIGHT ARM of the
RIGHT UPPER EXTREMITY**

**Anterior (Palmar) View
RIGHT FOREARM and HAND
(PALM of the HAND)**

**Posterior (Dorsal) View
RIGHT HAND
(BACK of HAND)**

CARDIOVASCULAR SYSTEM: PRINCIPAL VEINS OF THE BODY
Venous Tributaries of the Superior and Inferior Vena Cavae

Use bright, cool colors for vena cavae and major tributaries.
Use light, cool colors and neutral colors for minor tributaries.
★ See Chart #5 for organs drained; compare with 9.11; see 9.20

THORACIC REGION:

• Blood flow in the veins is from smaller vessels (tributaries) that drain organs and structures in the thoracic region and ultimately emptying into the large BRACHIOCEPHALIC VEINS (L. and R.)

★ • The L. and R. BRACHIOCEPHALIC VEINS merge slightly to the right at the top of the STERNUM (MANUBRIUM) to form the SUPERIOR VENA CAVA.

■ SUPERIOR VENA CAVA and the BRACHIOCEPHALIC VEINS *do not* have valves and are large vessels.

SVC | SUPERIOR VENA CAVA

★ **BRACHIOCEPHALIC TRIBUTARIES** of the **SUPERIOR VENA CAVA:**

B | BRACHIOCEPHALIC V.

IT | INTERNAL THORACIC V.

1 | ANTERIOR INTERCOSTAL VEINS

2 | MUSCULOPHRENIC V.

3 | SUPERIOR EPIGASTRIC V. ‡

4 | SUPERIOR PHRENIC V.

S | SUBCLAVIAN V.

AX | AXILLARY V.

5 | THORACOEPIGASTRIC V. ‡

(‡ Collateral venous flow via #3 and #5)
[Not shown: PERICARDIAL, MEDIASTINAL and MUSCULAR tributaries]

★ ■ The SUPERIOR VENA CAVA, in addition to collecting blood from the BRACHIOCEPHALIC VEINS and their tributaries, also collects blood from vessels along the POSTERIOR THORACIC WALL: AZYGOS VENOUS SYSTEM.

• AZYGOS VEIN is a single vein arising in the abdomen as a branch of the RIGHT ASCENDING LUMBAR VEIN. It passes upward through the AORTIC HIATUS of the DIAPHRAGM (through which the AORTA passes into the ABDOMEN) into the *THORAX*, then along the *right side* of the vertebral column to the level of the 4th thoracic vertebra, where it turns and enters the SUPERIOR VENA CAVA in the MEDIASTINUM.

• If the INFERIOR VENA CAVA becomes obstructed, the AZYGOS VEIN is the principal vein by which blood can return to the heart from the abdomen and lower extremities.

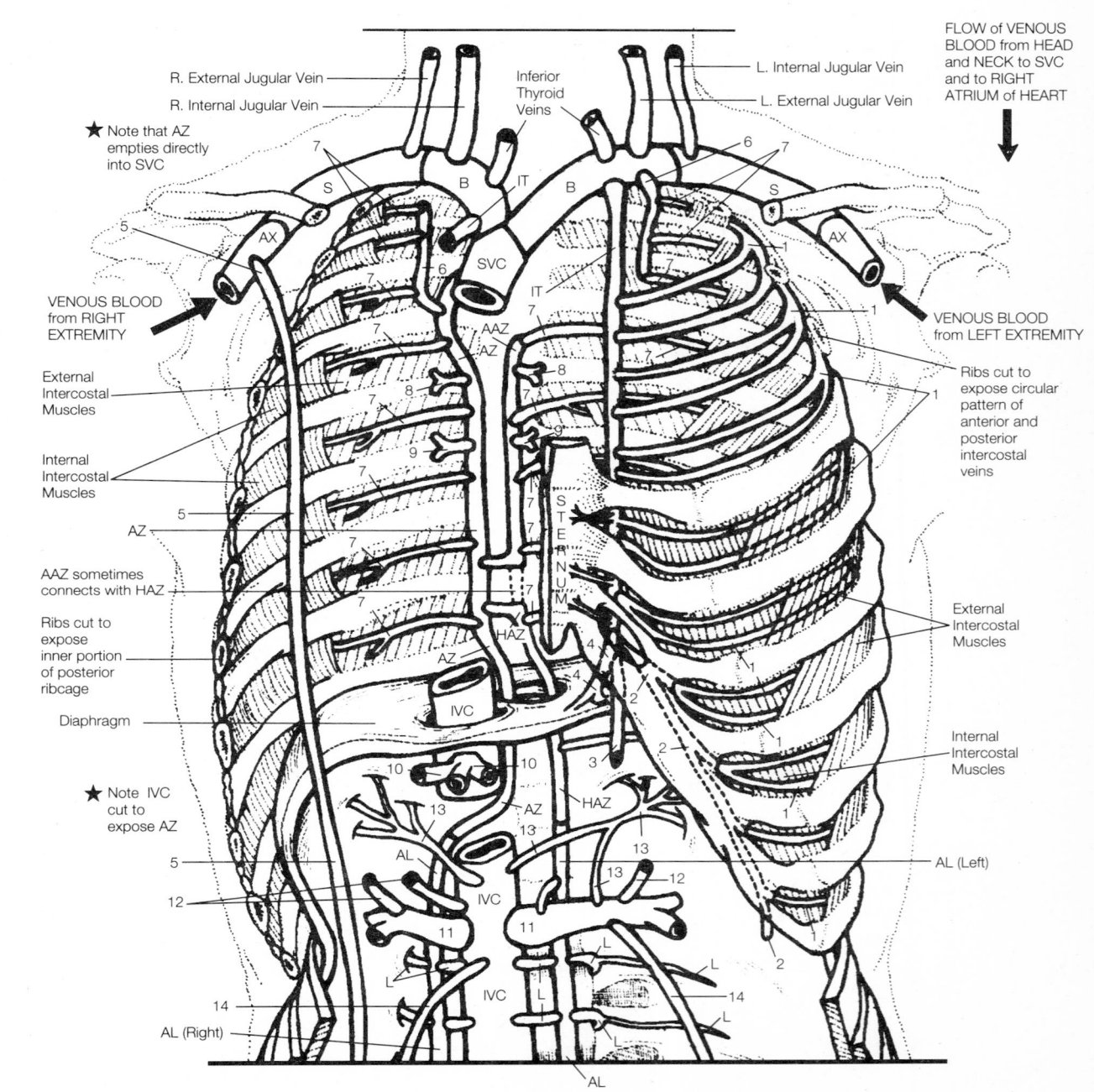

FLOW of VENOUS BLOOD from HEAD and NECK to SVC and to RIGHT ATRIUM of HEART

R. External Jugular Vein
R. Internal Jugular Vein
Inferior Thyroid Veins
L. Internal Jugular Vein
L. External Jugular Vein

★ Note that AZ empties directly into SVC

VENOUS BLOOD from RIGHT EXTREMITY

VENOUS BLOOD from LEFT EXTREMITY

Ribs cut to expose circular pattern of anterior and posterior intercostal veins

External Intercostal Muscles
Internal Intercostal Muscles

AZ

AAZ sometimes connects with HAZ

Ribs cut to expose inner portion of posterior ribcage

Diaphragm

External Intercostal Muscles

Internal Intercostal Muscles

★ Note IVC cut to expose AZ

AL (Left)

AL (Right)

AL

In the THORAX, the AZYGOS VEIN receives the HEMIAZYGOS, ACCESSORY AZYGOS and BRONCHIAL VEINS (plus the RIGHT POSTERIOR and HIGHEST INTERCOSTAL VEINS, and RIGHT SUBCOSTAL VEINS)

PRINCIPAL AZYGOS VEINS

AZ | AZYGOS V. | AZYGOUS V.

AL | ASCENDING LUMBAR VEINS

HAZ | HEMIAZYGOS V. | Drains *left* side of VERTEBRAL COLUMN

AAZ | ACCESSORY AZYGOS V.

TRIBUTARIES of the PRINCIPAL AZYGOS VEINS

L | LUMBAR VEINS | Drain • POSTERIOR ABDOMINAL
• VERTEBRAL COLUMN
• SPINAL CORD

6 | HIGHEST INTERCOSTAL VEINS

7 | POSTERIOR INTERCOSTAL VEINS

8 | BRONCHIAL VEINS

9 | ESOPHAGEAL VEINS

L & R subcostals not shown, similar to (7)

ABDOMINAL REGION:

★ The L. and R. COMMON ILIAC VEINS merge at the level of the 5th vertebra and form the INTERIOR VENA CAVA.

■ INFERIOR VENA CAVA
• Vessel with the largest diameter in the body
• Ascends through the abdominal cavity, parallel and to the right of the ABDOMINAL AORTA. Receives an abundance of *tributaries* (similar in name and position to corresponding arteries).

IVC | INFERIOR VENA CAVA

ABDOMINAL TRIBUTARIES of the INFERIOR VENA CAVA:

10 | HEPATIC V. (L. AND R.) | ORIGINATE IN CAPILLARY SINUSOIDS OF THE LIVER

11 | RENAL V. (L. AND R.)

12 | SUPRARENAL (ADRENAL) VEINS

13 | INFERIOR PHRENIC VEINS

14 | TESTICULAR-OVARIAN [GONADAL] VEINS

9.19

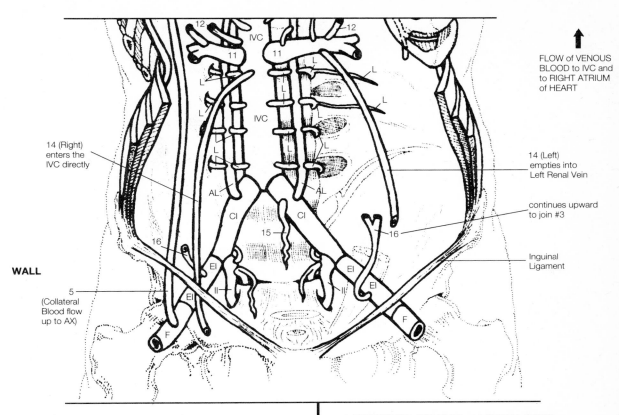

12   12
IVC
11   11
L   L
L   L
L
IVC
L   L
L
AL   AL
CI   CI
15   16
16
EI   EI
II   EI   II   EI
F   F

FLOW of VENOUS BLOOD to IVC and to RIGHT ATRIUM of HEART

14 (Right) enters the IVC directly

14 (Left) empties into Left Renal Vein

continues upward to join #3

Inguinal Ligament

WALL

5 (Collateral Blood flow up to AX)

---

2 LARGE ABDOMINAL TRIBUTARIES Join to Form the INFERIOR VENA CAVA:

CI | COMMON ILIAC VEINS (L. AND R.)

ABDOMINAL TRIBUTARIES of the COMMON ILIAC VEINS:

6 | MIDSACRAL V.

EI | EXTERNAL ILIAC VEINS

7 | INFERIOR EPIGASTRIC VEINS

II | INTERNAL ILIAC VEINS (HYPOGASTRIC VEINS) | (See 9.11) for branches

F | FEMORAL VEINS

★ VERTEBRAL COLUMN and SPINAL CORD drained by tributaries of the:
• POSTERIOR INTERCOSTAL VEINS (7)
• VERTEBRAL VEIN
• LUMBAR VEINS | not shown
• LATERAL SACRAL VEINS

★ NOTE: The INFERIOR VENA CAVA *does not* receive blood directly from the GASTROINTESTINAL TRACT, SPLEEN OR THE PANCREAS!! The venous blood from *these* organs passes through a filtering system in the LIVER before entering the IVC. See HEPATIC PORTAL SYSTEM, 9.20!

★★ The veins draining the PELVIS and PERINEUM are parallel to and carry the same names as arteries supplying those regions.
    The main difference lies in the fact that the COMMON ILIAC ARTERIES are branches of the ABDOMINAL AORTA, but the COMMON ILIAC VEINS are tributaries that join to form the INFERIOR VENA CAVA. . . . See 9.13!!

CARDIOVASCULAR SYSTEM: PRINCIPAL VEINS OF THE BODY
Venous Tributaries of the Vena Cavae:
Inferior Vena Cava: Hepatic Portal System

Repeat color choices for collateral
anastomotic back-up vessels from previous
pages, but now use lighter tones for them
(or don't color them at all!).
Use strong, bright, cool colors for
Hepatic Portal System and tributaries
★ See 9.19

★ A PORTAL SYSTEM is specialized venous flow consisting of two capillary beds joined by a vein.
Under normal conditions, veins draining a capillary network drain blood directly into systemic veins. In a VENOUS PORTAL SYSTEM, veins draining one capillary network deliver blood to another capillary network. Blood then continues to the usual systemic veins and enters the Vena Cavae (and ultimately the right atrium of the heart).

★ In the HEPATIC PORTAL SYSTEM, the HEPATIC PORTAL VEIN collects blood from the capillaries of the ABDOMINAL VISCERA (containing the absorbed nutrients of digestion) and conveys it to the SINUSOIDS in the LIVER (discontinuous spaces lined with specialized capillary cells called KUPFFER CELLS) for filtration through the HEPATOCYTE LIVER CELLS (See 12.16). From the LIVER, the blood then passes through the HEPATIC VEINS to the INFERIOR VENA CAVA, (then to the heart).

★ THE HEPATIC PORTAL VEIN receives blood from the digestive organs, and is formed by the union of 2 main vessels:
• The SUPERIOR MESENTERIC VEIN and the SPLENIC VEIN
■ SUPERIOR MESENTERIC VEIN (Drains blood from small intestine)
■ SPLENIC (LIENAL) VEIN (Drains blood from spleen, and drains blood from 4 tributaries):
• INFERIOR MESENTERIC VEIN (from large intestine)
• PANCREATIC VEIN (from pancreas)
• LEFT and RIGHT GASTROEPIPLOIC VEINS (from stomach)

The HEPATIC PORTAL VEIN also drains blood from the R. and L. GASTRIC veins (from stomach) and the CYSTIC VEIN (from gallbladder through its right branch).

NOTE: LIVER SINUSOIDS receive blood from 2 sources. It is a mixture of oxygenated and deoxygenated blood.
■ OXYGEN-RICH "RED" BLOOD from the HEPATIC ARTERY
■ NUTRIENT-RICH "BLUE" BLOOD from the HEPATIC PORTAL VEIN
The SINUSOIDS have a unique ability to modify the chemical nature of the venous blood from the digestive tract.

Drainage of the Organs of Digestion: The HEPATIC PORTAL SYSTEM

PORTAL SYSTEM OF VEINS

H   HEPATIC VEINS R. AND L.

S   SINUSOIDS

HP   HEPATIC PORTAL VEIN

HP_1 RIGHT BRANCH OF THE HEPATIC PORTAL VEIN

HP_2 LEFT BRANCH OF THE HEPATIC PORTAL VEIN

COLLATERAL ANASTOMOTIC BACK-UP: SYSTEMIC CAVAL SYSTEM CONNECTIONS

IVC   INFERIOR VENA CAVA

E   ESOPHAGEAL V.

CI   COMMON ILIAC V.

1   ASCENDING LUMBAR V.

2   LUMBAR V.

EI   EXTERNAL ILIAC V.

II   INTERNAL ILIAC V.

3   MIDDLE RECTAL V.

4   INFERIOR RECTAL V.

P-U   PARA-UMBILICALS

TRIBUTARIES of the HEPATIC PORTAL VEIN

HP1   RIGHT BRANCH OF THE HEPATIC PORTAL VEIN

5   CYSTIC V.

HP2   LEFT BRANCH OF THE HEPATIC PORTAL VEIN

LT   LIGAMENTUM TERES   ROUND LIGAMENT

LV   [LIGAMENTUM VENOSUM]

SM   SUPERIOR MESENTERIC V.

6   MIDDLE COLIC V.

7   RIGHT COLIC V.

8   ILEOCOLIC V.

9   SMALL INTESTINAL V.   JEJUNAL AND ILEAL

10   RIGHT GASTROEPIPLOIC V.

11   PANCREATICODUODENAL V.

SP   SPLENIC (LIENAL) V.

12   SHORT GASTRIC V.

13   LEFT GASTROEPIPLOIC V.

14   PANCREATIC V.

IM   INFERIOR MESENTERIC V.

15   SUPERIOR RECTAL V.

16   SIGMOID V.

17   LEFT COLIC V.

G_2 LEFT GASTRIC (CORONARY) V.

G_1 RIGHT GASTRIC (PYLORIC) V.

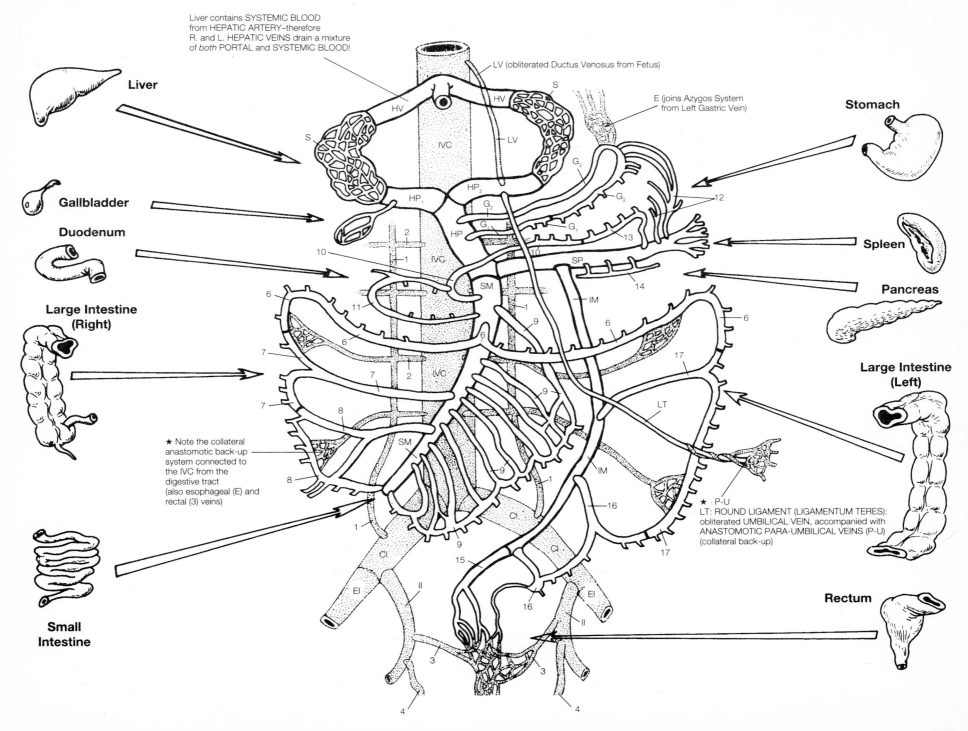

Liver contains SYSTEMIC BLOOD
from HEPATIC ARTERY–therefore
R. and L. HEPATIC VEINS drain a mixture
of *both* PORTAL and SYSTEMIC BLOOD!

LV (obliterated Ductus Venosus from Fetus)

E (joins Azygos System
from Left Gastric Vein)

Liver

Gallbladder

Duodenum

**Large Intestine
(Right)**

★ Note the collateral
anastomotic back-up
system connected to
the IVC from the
digestive tract
(also esophageal (E) and
rectal (3) veins)

**Small
Intestine**

Stomach

Spleen

Pancreas

**Large Intestine
(Left)**

★ P-U
LT: ROUND LIGAMENT (LIGAMENTUM TERES):
obliterated UMBILICAL VEIN, accompanied with
ANASTOMOTIC PARA-UMBILICAL VEINS (P-U)
(collateral back-up)

Rectum

CARDIOVASCULAR SYSTEM: PRINCIPAL VEINS OF THE BODY
Veins of the Lower Extremity (Right Limb)

Use bright, cool colors for major tributaries. (Repeat colors for IVC, CI, II, EI)
Use light, cool colors and neutral colors for minor tributaries.
★ Compare with 9.14

★ The veins of the lower extremities are generally divided into 2 groups:
- ■ SUPERFICIAL GROUP
- ■ DEEP GROUP

DEEP VEINS have more valves than the SUPERFICIAL VEINS:

★ **DEEP VEINS**
- ■ ANTERIOR and POSTERIOR TIBIAL VEINS drain blood from the deep veins of the foot and join below the knee to form the POPLITEAL VEIN.

- ■ The POPLITEAL VEIN becomes the FEMORAL VEIN just above the knee. The FEMORAL VEIN drains blood from the DEEP FEMORAL VEIN and LATERAL-MEDIAL CIRCUMFLEX VEINS in the upper thigh.

- ■ As the FEMORAL VEIN approaches the inguinal ligament, it drains blood from the GREAT SAPHENOUS VEIN, and then becomes the EXTERNAL ILIAC VEIN after entering the abdominal cavity. The EXTERNAL ILIAC VEIN merges with the INTERNAL ILIAC VEIN at the level of the sacroiliac joint in the pelvic region to form the COMMON ILIAC VEIN.

- ■ The L. and R. COMMON ILIAC VEINS merge at the level of the 5th vertebra to form the large INFERIOR VENA CAVA.

SUPERFICIAL VEINS
The DORSAL DIGITAL and METATARSAL VEINS enter the DORSAL VENOUS ARCH, which gives rise to two large superficial veins:
- • The GREAT and SMALL SAPHENOUS VEINS
- ■ GREAT SAPHENOUS VEIN:
  - • The longest vessel in the body
  - • Ascends along the MEDIAL aspect of the leg and enters the FEMORAL VEIN.

- ■ SMALL SAPHENOUS VEIN
  - • Ascends along the LATERAL aspect of the foot, POSTERIORLY up the calf of the leg, and passes deep to enter the POPLITEAL VEIN in the POSTERIOR FOSSA behind the knee.
- ■ The GENICULARS (superior, medial, and inferior) drain the knee region and enter the POPLITEAL VEIN in the POSTERIOR KNEE REGION.

SUPERFICIAL VEINS of the LOWER EXTREMITY

1 **SMALL SAPHENOUS V.** enters POPLITEAL VEIN

2 **GREAT SAPHENOUS V.** enters FEMORAL VEIN

3 **DORSAL VENOUS ARCH**

4 **DORSAL METATARSAL VEINS**

5 **DORSAL DIGITAL VEINS**

6 **GENICULAR VEINS**

DEEP VEINS of the LOWER EXTREMITY

IVC **INFERIOR VENA CAVA**

CI **COMMON ILIAC VEINS (L. AND R.)**

II **INTERNAL ILIAC V.**

EI **EXTERNAL ILIAC V.**

As It Approaches the Thigh, the FEMORAL VEIN Becomes the EXTERNAL ILIAC VEIN

F **FEMORAL V.**

F1 **LATERAL FEMORAL CIRCUMFLEX VEIN**

F2 **MEDIAL FEMORAL CIRCUMFLEX VEIN**

F3 **DEEP FEMORAL V.**

Just above the knee, the POPLITEAL VEIN becomes the FEMORAL VEIN. Just below the knee, the POPLITEAL VEIN is formed by the junction of the Anterior and Posterior Tibial Veins.

P **POPLITEAL V.**

(AT) **ANTERIOR TIBIAL V.**

7 **DORSALIS PEDIS VEIN**

(PT) **POSTERIOR TIBIAL V.**

8 **PERONEAL V.**

9 **MEDIAL PLANTAR V.**

10 **LATERAL PLANTAR V.**

11 **PLANTAR VENOUS ARCH**

12 **PLANTAR METATARSAL VEINS**

13 **PLANTAR DIGITAL VEINS**

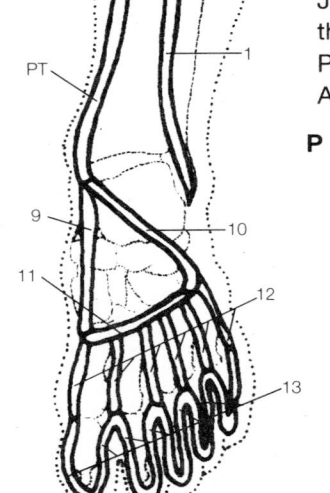

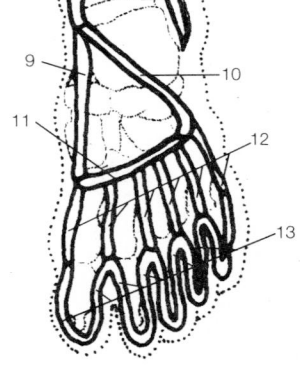

BLOOD FLOW

(Superior View)
DORSAL–TOP SURFACE of the FOOT

(Inferior View)
VENTRAL–BOTTOM SURFACE of the FOOT

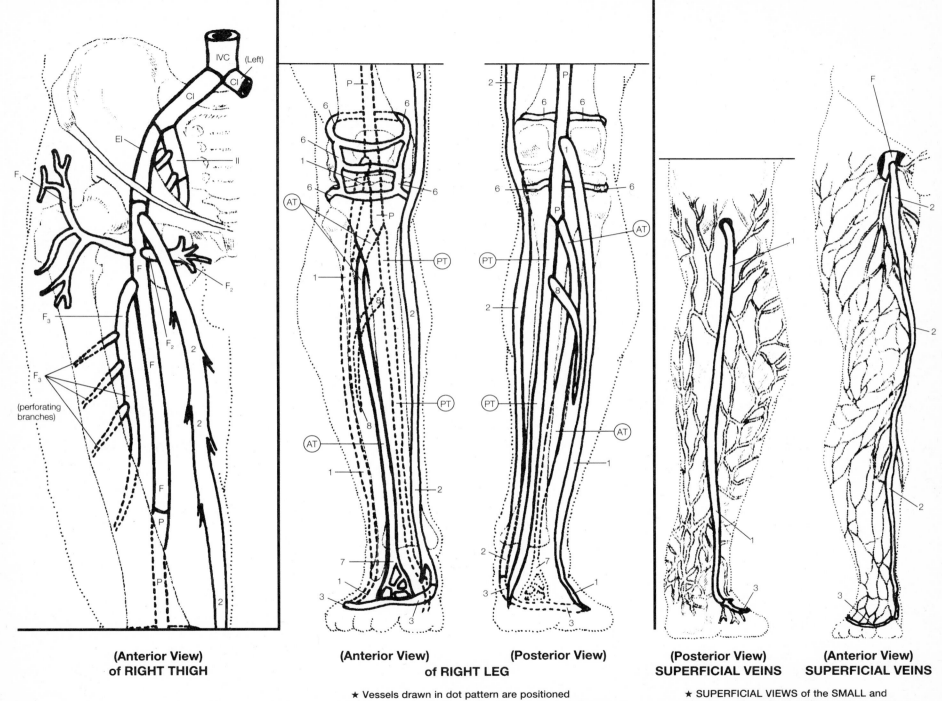

(Anterior View) of RIGHT THIGH

(Anterior View) of RIGHT LEG

(Posterior View)

(Posterior View) SUPERFICIAL VEINS

(Anterior View) SUPERFICIAL VEINS

★ Vessels drawn in dot pattern are positioned behind bone on the opposite side of indicated view. You may choose not to color these.

★ SUPERFICIAL VIEWS of the SMALL and GREAT SAPHENOUS VEINS include numerous venous tributaries

9.21

CARDIOVASCULAR SYSTEM: REVIEW OF THE MAJOR VEINS

Use color choices from previous pages.
Any vessels assigned the same color
will be easily distinguished according
to position.

★ See Chart #5 for organs drained; compare with 9.15

Note: Vessels drawn in dotted lines
indicate a position *behind* bone.

CO	CORONARY VEINS
SVC	SUPERIOR VENA CAVA
B	BRACHIOCEPHALIC V.
IJ	INTERNAL JUGULAR V.
EJ	EXTERNAL JUGULAR V.
S	SUBCLAVIAN V.
AX	AXILLARY V.
BR	BRACHIAL V.
BA	BASILIC V.
R	RADIAL V.
U	ULNAR V.
MC	MEDIAN CUBITAL V.
MV	MEDIAN V. OF THE FOREARM
SP	SUPERFICIAL PALMAR ARCH
DP	DEEP PALMAR ARCH
IVC	INFERIOR VENA CAVA

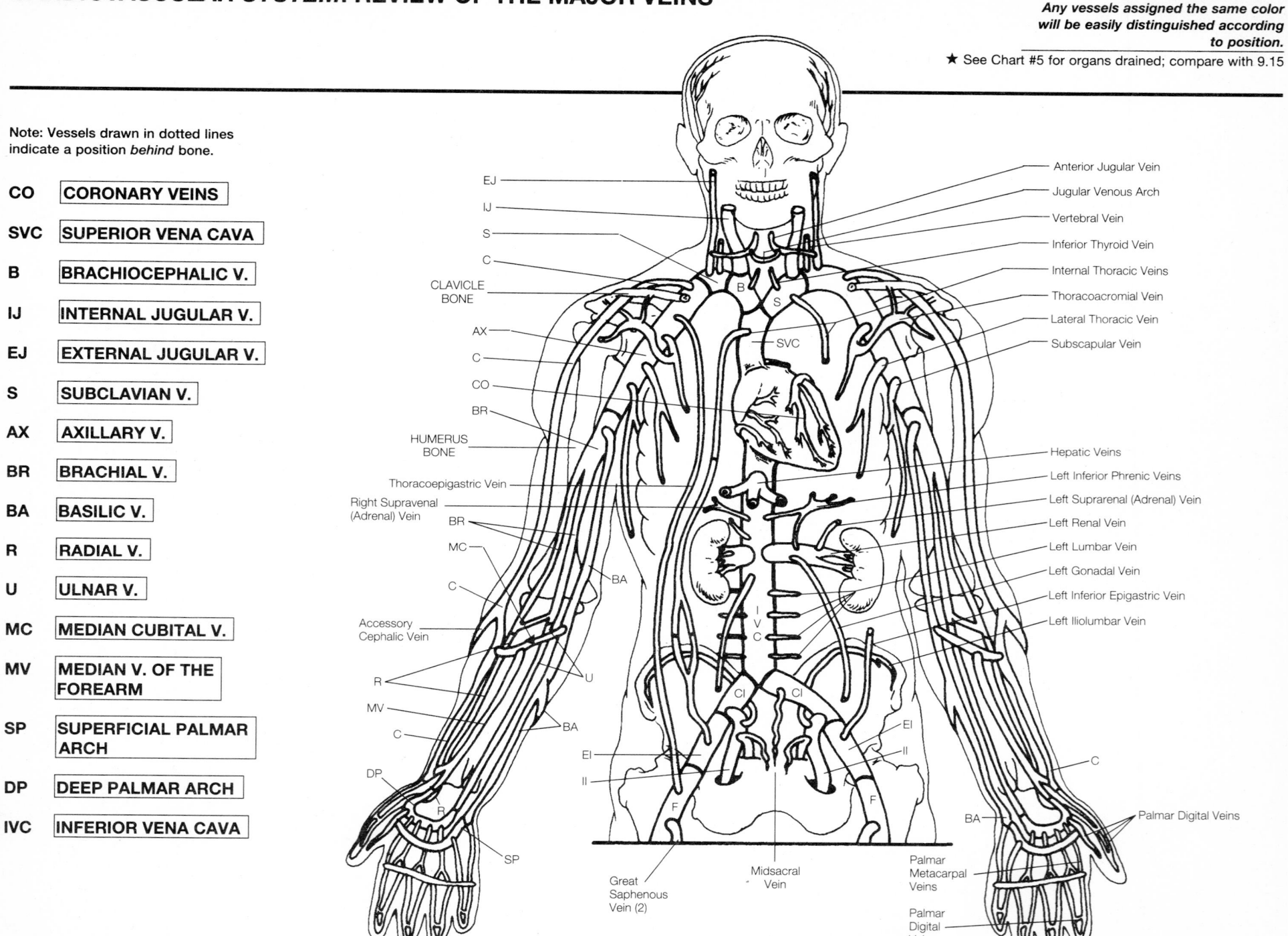

EJ
IJ
S
C
CLAVICLE BONE
AX
C
CO
BR
HUMERUS BONE
Thoracoepigastric Vein
Right Supravenal (Adrenal) Vein
BR
MC
C
Accessory Cephalic Vein
R
MV
C
DP
R
SP

Anterior Jugular Vein
Jugular Venous Arch
Vertebral Vein
Inferior Thyroid Vein
Internal Thoracic Veins
Thoracoacromial Vein
Lateral Thoracic Vein
Subscapular Vein
Hepatic Veins
Left Inferior Phrenic Veins
Left Suprarenal (Adrenal) Vein
Left Renal Vein
Left Lumbar Vein
Left Gonadal Vein
Left Inferior Epigastric Vein
Left Iliolumbar Vein

B
S
SVC
BA
U
IVC
CI
CI
EI
II
F
F
EI
II
BA
C

Great Saphenous Vein (2)
Midsacral Vein
Palmar Metacarpal Veins
Palmar Digital Veins
Palmar Digital Veins

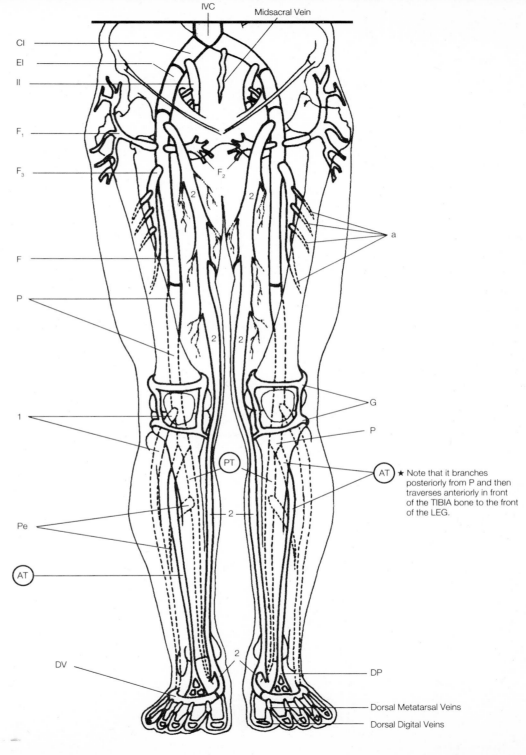

CI | COMMON ILIAC V.
II | INTERNAL ILIAC V.
EI | EXTERNAL ILIAC V.
F | FEMORAL V.
2 | GREAT SAPHENOUS V.
F₁ | LATERAL FEMORAL CIRCUMFLEX V.
F₂ | MEDIAL FEMORAL CIRCUMFLEX V.
F₃ | DEEP FEMORAL V.
P | POPLITEAL V.
G | GENICULAR VEINS
(AT) | ANTERIOR TIBIAL V.
DP | DORSALIS PEDIS V.
(PT) | POSTERIOR TIBIAL V.
1 | SMALL SAPHENOUS V.
DV | DORSAL VENOUS ARCH

IVC

Midsacral Vein

★ Note that it branches posteriorly from P and then traverses anteriorly in front of the TIBIA bone to the front of the LEG.

Dorsal Metatarsal Veins

Dorsal Digital Veins

Chapter 9: *CARDIOVASCULAR SYSTEM*: Charts

Chart #1 : Elements of the Vascular System (See 9.2)

Vessel Type	Size	Wall Structure	Function	Example
ELASTIC ARTERY	1–2 cm	Dominated by elastin; also contains collagen, smooth muscle, and an endothelial lining	Conducts blood away from the heart; expands in order to accommodate each stroke volume; and recoils to supply additional propulsive force	Aorta
MUSCULAR ARTERY	.1–1 cm	Even amounts of elastin, collagen, and muscle with an endothelial lining	Conducts blood from the aorta to the organs with as little drop in mean arterial pressure as possible	Renal Artery
ARTERIOLE	20–200 microns	Dominated by smooth muscle, with smaller amounts of elastin and collagen and an endothelial lining	Regulates the amount of blood flow to the capillary bed in each region	Found in every tissue
CAPILLARY	5–7 microns	Single layer of endothelium (flat, squamous epithelium)	Primary exchange vessel	Found in every tissue
VENULE–SMALL VEIN	20–500 microns	Dominated by smooth muscle with smaller amounts of elastin and collagen and an endothelial lining	Conducts blood from capillary bed to veins with larger capacity	Found in every tissue
MEDIUM-SIZED VEIN	.15–1.5 cm	Even amounts of smooth muscle, elastin, and collagen with an endothelial lining	Acts as reservoirs of blood that can alter the "circulating" blood volume through venoconstriction; also conducts blood to the heart	Renal Vein
LARGE (GREAT) VEIN	1.5–3 cm	Dominated by collagen, with moderate amounts of elastin and smooth muscle with an endothelial lining	Returns blood to the heart	Vena Cava

Chart #2 : Formed Elements of the Blood: 45% of Blood (See 9.3)
(Blood Plasma = 55% of Blood)

Component	Description	Number Present	Function
ERYTHROCYTE (RED BLOOD CELL)	Biconcave disk lacking nucleus that contains hemoglobin survives 120 hours	2.4 million produced/second 4 million–6 million/mm³ (♂: 5.5 million/mm³) (♀: 4.8 million/mm³)	Transports oxygen and carbon dioxide

LEUKOCYTE (WHITE BLOOD CELL)

Component	Description	Number Present	Function
Granular Leukocyte **GRANULOCYTE**	About twice the size of red blood cell; cytoplasmic granules present; lobed nuclei; develop from red bone marrow survives 12 hours to 3 days		
1. EOSINOPHIL	Nucleus bilobed; cytoplasmic granules stain red in eosin stain	1%–3% of white cells present	Antigen/antibody reactions; defends against allergic reactions. Helps to detoxify foreign substances; secretes enzymes that break down clots
2. NEUTROPHIL	Nucleus with 2–5 lobes; cytoplasmic granules stain purple	54%–62% of white cells present	Phagocytic; defends against acute infections small, highly mobile, short-lived
3. BASOPHIL	Nucleus lobed; cytoplasmic granules stain blue in hematoxylin stain	Less than 1% of white cells present	Releases anticoagulant heparin and stimulates blood clot absorbtion
Agranular Leukocyte **AGRANULOCYTE**	Cytoplasmic granules absent; spherical nuclei; develop from lymphoid and myeloid tissue survives 100–300 days		
1. MONOCYTE	2–3 times larger than red blood cell; nuclear shape varies from round to lobed (grow into large, phagocytic, macrophages)	3%–9% of white cells present	Phagocytic; defends against chronic infections large, slow moving, long-lasting
2. LYMPHOCYTE	Only slightly larger than red blood cell; nucleus nearly fills cell	25%–33% of white cells present	Provides specific immune response to pathogens B lymphocytes produce antibodies T lymphocytes stimulate T attack cells

Component	Description	Number Present	Function
THROMBOCYTE (PLATELET)	Cytoplasmic fragment; survives 5–9 days	130,000–360,000/mm³	Clotting

From Kent M. Van De Graaff, *Human Anatomy*, 2d ed. Copyright © 1988 Wm. C. Brown Publishers, Dubuque, Iowa. All Rights Reserved. Reprinted by permission.

Chart #3 : **Blood Plasma: The Fluid Portion of Blood: 55% of Blood (See 9.3)**
(Formed Elements = 45% of Blood)

■ WATER (H_2O): 90%

■ SOLUTES: 10% | PLASMA PROTEINS = 8%
INORGANIC and ORGANIC SUBSTANCES = 2%

• PLASMA PROTEINS: 8% formed mainly in the LIVER

SERUM ALBUMIN: (60%) and SERUM GLOBULIN (α,β,γ): (36%) : 96%

• Exert OSMOTIC PRESSURE (25–30 mm Hg)
• Help regulate BLOOD VOLUME and BODY'S FLUID BALANCE
• Responsible for BLOOD'S VISCOSITY (which is important
in maintaining Blood Pressure)

FIBRINOGEN: (4%)
• Precursor of Fibrin, which forms the FRAMEWORK OF BLOOD CLOT

• REGULATORY AND PROTECTIVE PROTEINS

HORMONES
• Chemical messengers from Endocrine Glands (important
in coordination and control of body's activities)

ANTIBODIES (γ GAMMA GLOBULINS)
• Important in Immunity reactions
• Contribute to BLOOD'S VISCOSITY

ENZYMES
• Organic complex protein catalysts capable of inducing chemical
changes in specific substances without being changed themselves

INORGANIC SUBSTANCES: 0.9%

SODIUM, CHLORIDE, CALCIUM, POTASSIUM, BICARBONATE, IODINE, IRON

ORGANIC SUBSTANCES: 1.1%

WASTE MATERIALS: (Urea, Uric Acid, Xanthine, Creatine, Ammonia)
• Products of Tissue Activity being transported from tissues
to the KIDNEYS and THE SKIN for EXCRETION

NUTRITIVE MATERIALS (Amino Acids, Fats, Cholesterol, GLUCOSE 0.1%)
• Foodstuffs in solution ABSORBED from the gut going to
tissues for UTILIZATION and STORAGE

■ RESPIRATORY GASES
OXYGEN
• Small amounts of inspired O_2 in solution
CARBON DIOXIDE
• Small amounts of CO_2 in solution, and as bicarbonate being carried to
the LUNGS for EXPIRATION

Adapted from *Illustrated Physiology* McNaught and Callander, ©1975 Churchill Livingstone

Chart #4

Order of Branching	Organ or Region Supplied
9.7	
ASCENDING AORTA	Initial upward section of main trunk of Arterial System
LEFT AORTIC SINUS	
LEFT CORONARY ARTERY	Left atrium and ventricle
ANTERIOR INTERVENTRICULAR BRANCH	(Myocardium of heart)
CIRCUMFLEX BRANCH	
POSTERIOR ARTERY	
RIGHT AORTIC SINUS	Right atrium and ventricle
RIGHT CORONARY ARTERY	(Myocardium of heart)
POSTERIOR INTERVENTRICULAR BRANCH	
MARGINAL BRANCH	
9.8	
BRACHIOCEPHALIC (INNOMINATE) ARTERY	(Right side of head, neck, and upper extremity)
RIGHT SUBCLAVIAN ARTERY	(Brain, meninges, spinal cord, neck, thoracic walls, upper extremity)
INTERNAL THORACIC ARTERY	Anterior thoracic wall, mediastinal structures and diaphragm
VERTEBRAL ARTERY	Muscles of the neck, spinal cord, cerebellum, and internal cerebrum
THYROCERVICAL TRUNK	(Larynx, trachea, thyroid gland)
INFERIOR THYROID ARTERY	Larynx, esophagus, trachea, muscles of the neck and thyroid gland
SUPRASCAPULAR ARTERY	Bone and periosteum of clavicle and scapula, shoulder joint and muscles
TRANSVERSE CERVICAL ARTERY	Muscles of the neck
COSTOCERVICAL TRUNK	(Neck, rib cage and back muscles)
DEEP CERVICAL ARTERY	Deep muscles of the neck
HIGHEST INTERCOSTAL ARTERY	Intercostal muscles, vertebral column and back muscles
RIGHT COMMON CAROTID ARTERY	(Neck and thyroid gland)
INTERNAL CAROTID ARTERY	Middle ear, brain, pituitary gland, orbit, and choroid plexus of lateral ventricle
EXTERNAL CAROTID ARTERY	Neck, face, scalp and skull
SUPERIOR THYROID ARTERY	Hyoid muscles, larynx, thyroid gland and pharynx
ASCENDING PHARYNGEAL ARTERY	Pharynx, soft palate, ear, meninges, cranial nerves and capitis muscles
LINGUAL ARTERY	Tongue, sublingual gland, lingual tonsils and epiglottis
FACIAL ARTERY	Face, tonsils, palate and submandibular gland
SUPERIOR AND INFERIOR LABIAL BRANCH	Inferior–lower lip; superior–upper lip and nose
LATERAL NASAL BRANCH	Structures bordering nasal cavity, nasal septum and adjacent sinuses
ANGULAR BRANCH	Lacrimal sac, inferior portion of orbicularis oculi muscle and nose
SUBMENTAL BRANCH	Tissue under chin
OCCIPITAL ARTERY	Muscles of the neck, scalp, meninges and mastoid cells
POSTERIOR AURICULAR ARTERY	Middle ear, mastoid cells, auricle, parotid gland, digastric and other neck muscles
SUPERFICIAL TEMPORAL ARTERY	Parotid gland, auricle, scalp, skin of face and masseter muscle
TRANSVERSE FACIAL ARTERY	Parotid gland, masseter muscle and skin of face
ZYGOMATIC-ORBITAL ARTERY	Orbicularis oculi muscle
MAXILLARY ARTERY	Both jaws, teeth, muscles of mastication, ear, meninges, nose, nasal sinus and palate
INFERIOR ALVEOLAR ARTERY	Lower teeth, gums, mandible, lower lip and chin;
SUPERIOR ALVEOLAR ARTERY	Incisors and canine teeth of upper jaw, maxillary sinus and buccinator muscle

Order of Branching	Organ or Region Supplied

MAXILLARY ARTERY
  INFRAORBITAL BRANCH — Maxilla, maxillary sinus, upper teeth, lower lip, cheek, and side of nose
  MIDDLE MENINGEAL BRANCH — Cranial bones and dura mater
  BUCCAL BRANCH — Buccinator muscle and mucous membranes of the mouth

9.9

INTERNAL CAROTID ARTERY — Middle ear, brain, pituitary gland, orbit, and choroid plexus of lateral ventricle
  ANTERIOR CHOROID ARTERY — Choroid plexus of lateral ventricle, hippocampus, and fimbria
  ANTERIOR CEREBRAL ARTERY — Occipital, frontal and parietal cortex, and corpus collosum
  ANTERIOR COMMUNICATING ARTERY — Establishes connection between the anterior cerebral arteries
  MIDDLE CEREBRAL ARTERY — Occipital, frontal, parietal and temporal cortex, and basal ganglia
VERTEBRAL ARTERY — Muscles of the neck, vertebrae, spinal cord, cerebellum, and internal cerebrum
  ANTERIOR SPINAL ARTERY — Anterior spinal cord
  POSTERIOR INFERIOR CEREBELLAR ARTERY — Lower cerebellum, medulla, and choroid plexus of the fourth ventricle
  BASILAR ARTERY — Brain stem, internal ear, cerebellum, and posterior cerebrum
    ANTERIOR INFERIOR CEREBELLAR ARTERY — Lower anterior cerebellar cortex and inner ear
    INTERNAL AUDITORY ARTERY — Internal ear
    PONTINE BRANCHES — Pons
    SUPERIOR CEREBELLAR ARTERIES — Upper cerebellum, midbrain, pineal body, and choroid plexus of third ventricle
    POSTERIOR CEREBRAL ARTERY — Occipital and temporal lobes, basal ganglia, choroid plexus of lateral ventricle, thalamus, and midbrain

      POSTERIOR COMMUNICATING ARTERY — Hippocampus and thalamus

9.10

BRACHIOCEPHALIC (INNOMINATE) ARTERY — (Right side of head, neck, and upper extremity)
  RIGHT COMMON CAROTID ARTERY — (Neck and thyroid gland)
  SUBCLAVIAN ARTERY — (Brain, meninges, spinal cord, neck, thoracic walls, and upper extremities)
    AXILLARY ARTERY — (Forms the brachial artery and seven branches)
      SUPERIOR THORACIC ARTERY — Intercostal, serratus anterior, and pectoralis major and minor muscles
      THORACOACROMIAL ARTERY — Pectoral, deltoid, subclavian muscles, and acromion process
      LATERAL THORACIC ARTERY — Pectoral muscles and mammary gland
      SUBSCAPULAR ARTERY — Bone and periosteum of clavicle and scapula, shoulder joint and muscles
      ANTERIOR HUMERAL CIRCUMFLEX ARTERY — Shoulder joint and head of humerus, long tendon of biceps and tendon of pectoralis major muscle
      POSTERIOR HUMERAL CIRCUMFLEX ARTERY — Deltoid muscle, shoulder joint, teres minor and triceps muscle
    BRACHIAL ARTERY — Shoulder, arm, forearm, and hand
      PROFUNDA BRACHII ARTERY — Humerus, muscles, and skin of arm
      SUPERIOR ULNAR COLLATERAL ARTERY — Forearm, wrist, and hand
      INFERIOR ULNAR COLLATERAL ARTERY — Forearm, wrist, and hand

(continued)

9.10

ULNAR ARTERY	Forearm, wrist, and hand
ANTERIOR ULNAR RECURRENT ARTERY	Forearm, wrist, and hand
POSTERIOR ULNAR RECURRENT ARTERY	Forearm, wrist, and hand
COMMON INTEROSSEOUS ARTERY	Deep structures of the forearm
ANTERIOR INTEROSSEUS ARTERY	Flexor digitorum profundus muscle, flexor pollicis longus muscle, radius, and ulna
POSTERIOR INTEROSSEOUS ARTERY	Superficial and deep muscles on posterior forearm
INTEROSSEOUS RECURRENT ARTERY	Forearm
RADIAL ARTERY	Forearm, wrist, and hand
RADIAL RECURRENT ARTERY	Brachioradialis muscle, brachialis muscle, and elbow joint
DORSAL CARPAL BRANCH ARTERY	Wrist, hand, and fingers
DORSALIS POLLICIS ARTERY	Thumb
DORSALIS INDICIS ARTERY	Fingers
SUPERFICIAL PALMAR ARCH	
PALMAR DIGITAL ARTERY	Fingers
DEEP PALMAR ARCH	
PRINCEPS POLLICIS ARTERY	Sides and palmar aspect of thumb
RADIALIS INDICIS ARTERY	Index finger
PALMER METACARPAL ARTERIES	Hand and fingers
PALMAR CARPAL ARCH	Wrist, forearm, and hand
DORSAL CARPAL ARCH	Wrist, forearm, and hand
DORSAL METACARPAL ARTERIES	Fingers and hand
DORSAL DIGITAL ARTERY	Fingers

9.11

AORTIC ARCH	Arch of main trunk of arterial system
SUBCLAVIAN ARTERY	(Brain, meninges, spinal cord; neck, thoracic walls, upper extremity)
COSTOCERVICAL TRUNK	(Branch of subclavian supplies neck and posterior intercostal muscles)
HIGHEST INTERCOSTAL ARTERY	First and second intercostal muscles, upper back muscles, and vertebral column
INTERNAL THORACIC ARTERY	Anterior thoracic wall, mediastinal structures, and diaphragm
ANTERIOR INTERCOSTAL ARTERIES	Thoracic wall
MUSCULOPHRENIC ARTERY	Diaphragm, and abdominal and thoracic walls
SUPERIOR EPIGASTRIC ARTERY	Abdominal muscles, diaphragm, skin, and peritoneum
THORACIC AORTA	(Thorax)
BRONCHIAL ARTERIES (2 or more)	Bronchi, lower trachea, pulmonary vessels, pericardium, part of esophagus
ESOPHAGEAL ARTERIES (4-5)	Esophagus
PERICARDIAL ARTERIES	Pericardium of heart
MEDIASTINAL ARTERIES	Lymph glands, posterior mediastinum
SUPERIOR PHRENIC ARTERIES	Upper surface of vertebral portion of diaphragm
POSTERIOR INTERCOSTAL ARTERIES	Thoracic wall, intercostal spaces, back muscles, vertebral column
SUBCOSTAL ARTERIES	Upper abdominal wall

(continued)

9.11

ABDOMINAL AORTA	(Abdomen)
CELIAC TRUNK	(Stomach, liver, pancreas, duodenum, spleen)
COMMON HEPATIC ARTERY	Stomach, body of pancreas, gallbladder, liver, duodenum, and greater omentum
LEFT GASTRIC ARTERY	Esophagus and lesser curvature of stomach
SPLENIC ARTERY	Spleen, stomach, pancreas, and greater omentum
SUPERIOR MESENTERIC ARTERY	Small intestine and proximal two-thirds of colon
SUPRARENAL (ADRENAL) ARTERY	Suprarenal gland (adrenal)
RENAL ARTERY	Kidney, adrenal gland, and ureter
TESTICULAR/OVARIAN ARTERIES	Testicular–ureter, epididymis, and testis; ovarian–ureter, ovary, and oviducts
INFERIOR MESENTERIC ARTERY	Inferior surface of descending colon, sigmoid colon, and rectum
INFERIOR PHRENIC ARTERY	Diaphragm and adrenal gland
LUMBAR ARTERIES	Abdominal wall, vertebrae, lumbar muscles, and renal capsule
MIDDLE SACRAL (MIDSACRAL) ARTERY	Sacrum, coccyx, and rectum
COMMON ILIAC ARTERY	Pelvis, abdominal wall, and lower extremity
EXTERNAL ILIAC ARTERY	Abdominal wall, external genitalia, and lower limb
INTERNAL ILIAC (HYPOGASTRIC) ARTERY	Wall and viscera of pelvis, gluteal region, reproductive organs, and medial aspect of thigh
LATERAL SACRAL ARTERIES	Coccyx, sacrum

9.12

CELIAC TRUNK	(Stomach, liver, pancreas, duodenum, spleen)
COMMON HEPATIC ARTERY	Stomach, body of pancreas, gallbladder, liver, duodenum, and greater omentum
LEFT HEPATIC ARTERY	Left portion of liver
RIGHT HEPATIC ARTERY	Right portion of liver and gallbladder
CYSTIC ARTERY	Gallbladder
GASTRODUODENAL ARTERY	Stomach, duodenum, pancreas, and greater omentum
ANTERIOR SUPERIOR PANCREATICODUODENAL ARTERY	Pancreas and duodenum
RIGHT GASTROEPIPLOIC ARTERY	Stomach and greater omentum
POSTERIOR SUPERIOR PANCREATICODUODENAL ARTERY (RETRODUODENAL ARTERY)	Pancreas and duodenum
RIGHT GASTRIC ARTERY	Lesser curvature of the stomach
SPLENIC ARTERY	Spleen, stomach, pancreas, and greater omentum
LEFT GASTROEPIPLOIC ARTERY	Stomach and greater omentum
PANCREATIC ARTERY	Pancreas
SUPERIOR AND INFERIOR POLAR ARTERIES	Stomach
LEFT GASTRIC ARTERY	Esophagus and lesser curvature of stomach
OMENTAL BRANCHES	Omentum

(continued)

9.12

SUPERIOR MESENTERIC ARTERY	Small intestine and proximal two-thirds of colon
INFERIOR PANCREATICODUODENAL ARTERY	Pancreas and duodenum
ILEOCOLIC ARTERY	Ileum, cecum, vermiform appendix, and ascending colon
ANTERIOR CECAL BRANCH	Cecum
POSTERIOR CECAL BRANCH	Cecum
ILEAL BRANCH	Ileum
APPENDICULAR BRANCH	Vermiform appendix
RIGHT COLIC ARTERY	Ascending colon
MIDDLE COLIC ARTERY	Transverse colon
ILEAL INTESTINAL ARTERY	Small intestine and ileum
JEJUNAL INTESTINAL ARTERY	Small intestine and jejunum
INFERIOR MESENTERIC ARTERY	Descending colon, sigmoid colon, and rectum
SUPERIOR RECTAL ARTERY	Rectum
LEFT COLIC ARTERY	Descending colon
SIGMOID ARTERY	Sigmoid colon

9.13

COMMON ILIAC ARTERY	Pelvis, abdominal wall, and lower extremity
EXTERNAL ILIAC ARTERY	Abdominal wall, external genitalia, and lower extremity
INFERIOR EPIGASTRIC ARTERY	Cremaster and abdominal muscles and peritoneum
DEEP CIRCUMFLEX ILIAC ARTERY	Psoas, sartorius, tensor fasciae latae, obliques and transverse abdominal muscles and adjacent skin
FEMORAL ARTERY	Lower extremity, pelvis, and genitalia
EXTERNAL PUDENDAL ARTERY	External genitalia and medial thigh muscles
DEEP FEMORAL (PROFUNDA FEMORIS) ARTERY	Hip joints, thigh muscles, gluteal muscles, and femur
LATERAL FEMORAL CIRCUMFLEX ARTERY	Hip joints and thigh muscles
MEDIAL FEMORAL CIRCUMFLEX ARTERY	Hip joints and thigh muscles
INTERNAL ILIAC ARTERY	Wall and viscera of pelvis, gluteal region, reproductive organs, and medial aspect of thigh
LATERAL SACRAL ARTERY	Structures near the coccyx and sacrum
SUPERIOR GLUTEAL ARTERY	Upper portions of gluteal muscles and overlying skin, obturator internus, piriformis, levator ani, and coccygeus muscles and hip joint
ILIOLUMBAR ARTERY	Pelvic muscles and bones, fifth lumbar segment, and sacrum
INFERIOR GLUTEAL ARTERY	Gluteal region and posterior thigh
OBTURATOR ARTERY	Pelvic muscles
INTERNAL PUDENDAL ARTERY	External genitalia and anal canal
POSTERIOR SCROTAL (LABIAL) ARTERIES	Male–scrotum; (female–labia)
ARTERY OF THE BULB OF THE PENIS (VESTIBULE)	Male–penis; (female–lower vagina)
ARTERIES OF THE PENIS (CLITORIS)	Male–penis; (female–clitoris)
PERINEAL ARTERY	Perineum and skin of external genitalia
INFERIOR RECTAL ARTERY	Rectum, levator ani, and external sphincter muscles and overlying skin
MIDDLE RECTAL ARTERY	Rectum and vagina
INFERIOR VESICLE ARTERY	Bladder, prostate gland, and seminal vesicles

Order of Branching	Organ or Region Supplied
(continued)	
9.13	
UMBILICAL ARTERY	Vas deferens, seminal vesicles, testis, urinary bladder, and ureter
SUPERIOR VESICLE ARTERY	Bladder, urachus (ligamentum umbilicale medium), and ureter
VAGINAL/UTERINE ARTERIES	Vaginal–vagina, fundus of bladder; uterine–uterus, vagina, round ligament of uterus, oviducts, and ovary
ABDOMINAL AORTA	(Abdomen)
MIDSACRAL ARTERY	Sacrum, coccyx, and rectum
9.14	
ABDOMINAL AORTA	(Abdomen)
COMMON ILIAC ARTERY	Pelvis, abdominal wall, and lower extremity
INTERNAL ILIAC ARTERY	Wall and viscera of pelvis, gluteal region, reproductive organs, and medial aspect of thigh
SUPERIOR GLUTEAL ARTERY	Upper gluteal muscles and overlying skin, obturator internus, piriformis, levator ani, and coccygeus muscles and hip joint
INFERIOR GLUTEAL ARTERY	Gluteal region and back of thigh
OBTURATOR ARTERY	Pelvic muscles and hip joint
FEMORAL ARTERY	Lower abdominal wall, external genitalia, and lower extremity
DEEP FEMORAL (PROFUNDA FEMORIS) ARTERY	Thigh muscles, hip joint, gluteal muscles, and femur
PERFORATING BRANCHES	Adductor, hamstring, and gluteal muscles, and femur
MEDIAL FEMORAL CIRCUMFLEX ARTERY	Hip joint and thigh muscles
LATERAL FEMORAL CIRCUMFLEX ARTERY	Hip joint and thigh muscles
DESCENDING BRANCH	Thigh muscles
DESCENDING GENICULATE ARTERY	Knee and calf
POPLITEAL ARTERY	Knee and calf
GENICULARIS (SUPERIOR, MIDDLE AND INFERIOR) ARTERY	Knee
ANTERIOR TIBIAL ARTERY	Leg, ankle, and foot
DORSALIS PEDIS ARTERY	Foot and toes
ARCUATE ARTERY	Foot and toes
DORSAL METATARSAL DORSAL ARTERY	Toes
DORSAL DIGITAL ARTERY	Toes
LATERAL TARSAL ARTERY	Muscles and joints of tarsus
POSTERIOR TIBIAL ARTERY	Leg, foot, and heel
PERONEAL ARTERY	Posterior and lateral aspect of ankle and deep calf muscles
MEDIAL PLANTAR ARTERY	Muscles, articulations, and skin of the medial aspect of the sole of the foot and toes
PLANTAR ARCH	
PLANTAR METATARSAL ARTERIES	Foot and toes
PLANTAR DIGITAL ARTERIES	Toes

Chart #5 : Drainage of Blood through the Venous Tree (9.17–9.22)
(Blood Flow from the Smaller Tributaries to the Major Veins)

Order of Confluence	Organ or Region Drained
9.17	
BRACHIOCEPHALIC (INNOMINATE) VEIN	(Head, neck, and upper extremity) paired veins unite to form superior vena cava
INFERIOR THYROID VEIN	Thyroid plexus
INTERNAL JUGULAR VEIN	Brain and cranial sinuses, superficial parts of face and neck
SUPERIOR THYROID VEIN	Upper portion of thyroid
MIDDLE THYROID VEIN	Thyroid gland
PHARYNGEAL PLEXUS VEIN	Deep structures of the neck
LINGUAL VEIN	Lower jaw
SUBCLAVIAN VEIN	(Continuation of axillary vein, main venous trunk of upper extremity) Joins internal jugular vein to form the brachiocephalic vein
VERTEBRAL VEIN	Suboccipital venous plexus
DEEP CERVICAL VEIN	Suboccipital triangle
EXTERNAL JUGULAR VEIN	Lower jaw, deep parts of face, exterior of cranium
TRANSVERSE CERVICAL VEIN	Posterior neck
ANTERIOR JUGULAR VEIN	Anterior neck
JUGULAR VENOUS ARCH	(Connects both anterior jugular veins)
POSTERIOR EXTERNAL JUGULAR VEIN	Posterior neck and base of skull
FACIAL VEIN	Deep structures of the face
SUBMENTAL VEIN	Chin and lower mouth
SUPERIOR AND INFERIOR LABIAL VEINS	Upper and lower lip
EXTERNAL NASAL VEIN	Nose
DEEP FACIAL VEIN	Pterygoid plexus
ANGULAR VEIN	Eye and base of nose
SUPERFICIAL TEMPORAL VEIN	Lateral portion of scalp in the frontal and parietal regions
RETROMANDIBULAR VEIN	Parotid gland and deep mandible
POSTERIOR AURICULAR VEIN	Lateral portion of scalp above ear
OCCIPITAL VEIN	Occipital region of scalp
MAXILLARY VEIN	Pterygoid plexus
PTERYGOID PLEXUS	(Deep vein plexus drains cavernous sinuses of skull, connects with maxillary vein, deep facial vein, and alveolar veins)
MIDDLE MENINGEAL VEIN	Multiple veins of dura mater and anterior cranium
SUPERIOR AND INFERIOR ALVEOLAR VEINS	Teeth, and upper and lower jaw
INFRAORBITAL VEIN	Maxilla, maxillary sinus, upper teeth, cheek, and side of nose
9.18	
SUPERIOR VENA CAVA	(Drains blood from upper half of body)
BRACHIOCEPHALIC (INNOMINATE) VEIN	(Paired veins draw blood from head, neck, and upper extremities. They unite to form the superior vena cava.)
SUBCLAVIAN VEIN	(Main venous trunk of upper extremity)
AXILLARY VEIN	Portions of venous trunk of upper extremity
LATERAL THORACIC VEIN	Mammary plexus
SUBSCAPULAR VEIN	Posterior of axilla, shoulder, and scapular muscles
THORACOACROMIAL VEIN	Pectoral, deltoid, subclavian muscles, and acromion process
DORSAL DIGITALS VEINS	Fingers
DORSAL METACARPAL VEINS	Fingers
DORSAL ARCH	Middle and distal phalanges, and dorsal surface of hand
SUPERFICIAL PALMAR ARCH	Hand
MEDIAN VEIN OF FOREARM	Ventral forearm and thumb

Order of Confluence	Organ or Region Drained

9.18 (continued)

MEDIAN CUBITAL VEIN	Ventral forearm and hand (connects with the cephalic, median, and basilic veins)
CEPHALIC VEIN	Lateral forearm and hand
ACCESSORY CEPHALIC VEIN	Back of hand and fingers
BASILIC VEIN	Medial forearm and hand
PALMAR DIGITAL VEINS	Fingers
PALMAR METACARPAL VEINS	Fingers
DEEP PALMAR ARCH VEIN	Fingers and palm
ULNAR VEIN	Medial forearm and hand
RADIAL VEIN	Lateral forearm and hand
BRACHIAL VEIN	Forearm and hand

9.19

SUPERIOR VENA CAVA	(Drains blood from upper half of body and returns it to the heart, right atrium)
BRACHIOCEPHALIC VEIN	(Drains blood from head, neck, and upper extremities)
INTERNAL THORACIC VEIN	(Intercostal spaces)
ANTERIOR INTERCOSTAL VEINS	Lower nine intercostal muscles, vertebral column, contents of vertebral column, and muscles and skin of the back
MUSCULOPHRENIC VEIN	Diaphragm and abdominal and thoracic walls
SUPERIOR EPIGASTRIC VEIN	Skin of abdomen and superficial fascia
SUPERIOR PHRENIC VEIN	Upper surface of vertebral portion of diaphragm
SUBCLAVIAN VEIN	(Main venous trunk of upper extremity)
AXILLARY VEIN	(Portions of venous trunk of upper extremity)
THORACOEPIGASTRIC VEIN	Anterior and lateral subcutaneous layer of torso
PERICARDIAL VEIN	Pericardium of heart
MEDIASTINAL VEIN	Lymph glands and posterior mediastinum

AZYGOS VEIN	(Connects superior and inferior vena cava)
ASCENDING LUMBAR VEIN	(Connects lumbar veins)
HEMIAZYGOS VEIN	(Single vein of lower left thoracic wall)
ACCESSORY HEMIAZYGOUS VEIN	(Left side of vertebral column, from intercostal spaces above level of 6th–7th vertebrae)
LUMBAR VEINS	(Posterior abdominal wall in region of lumbar vertebrae)
HIGHEST INTERCOSTAL VEINS	(Upper two intercostal spaces)
POSTERIOR INTERCOSTAL VEINS	(Intercostal spaces, back muscles, vertebral column)
BRONCHIAL VEINS	(Larger bronchi and roots of lungs)
ESOPHAGEAL VEINS	(Esophagus)
SUBCOSTAL VEINS	(Upper abdominal wall)

INFERIOR VENA CAVA	(Drains blood from lower half of body and returns it to the heart, right atrium)
HEPATIC VEIN	Liver
RENAL VEIN	Kidney
SUPRARENAL (ADRENAL) VEIN	Suprarenal gland (adrenal)
INFERIOR PHRENIC VEIN	Inferior surface of diaphragm and adrenal gland
TESTICULAR/OVARIAN (GONADAL) VEIN	Testis/Ovary
COMMON ILIAC VEIN	(Drains blood from pelvis and leg. One on each side meets to form inferior vena cava.)
MIDSACRAL VEIN	Sacrum and coccyx, posterior pelvic wall
EXTERNAL ILIAC VEIN	Abdominal wall, external genitalia, and lower limb
INFERIOR EPIGASTRIC VEIN	Abdominal muscles and superficial fascia
FEMORAL VEIN	Lower extremity
INTERNAL ILIAC VEIN	Walls and viscera of pelvis; gluteal region; reproductive organs, and medial side of thigh
LATERAL SACRAL VEIN	Sacrum and coccyx, posterior pelvic wall

(continued)

9.20

Order of Confluence	Organ or Region Drained
HEPATIC VEIN	(Drains liver)
SINUSOIDS	Minute spaces resembling a sinus, slightly larger than a capillary; drains liver tissue
HEPATIC PORTAL VEIN	(Subdivision of systemic venous system. Collects blood from digestive tract and conveys it to the liver.)
RIGHT BRANCH	
CYSTIC VEIN	Gallbladder
LEFT BRANCH	
LIGAMENTUM TERES (LIGAMENTUM VENOSUM)	
SUPERIOR MESENTERIC VEIN	(Large vein from small intestine) unites with splenic vein behind pancreas to form the hepatic portal vein
MIDDLE COLIC VEIN	Transverse colon
RIGHT COLIC VEIN	Ascending colon
ILEOCOLIC VEIN	Cecum, appendix, and ascending colon
RIGHT GASTROEPIPLOIC VEIN	Stomach and greater omentum
PANCREATICODUODENAL VEIN	Pancreas and duodenum
SPLENIC (LIENAL) VEIN	Spleen, pancreas, stomach, and greater omentum
SHORT GASTRIC VEIN	Left portion greater curvature of stomach
PANCREATIC VEIN	Pancreas
INFERIOR MESENTERIC VEIN	(Drains blood from rectum, and sigmoid and descending parts of colon)
SUPERIOR RECTAL VEIN	Upper part of rectal plexus
SIGMOID VEIN	Sigmoid colon
LEFT COLIC VEIN	Descending colon
LEFT GASTRIC (CORONARY) VEIN	Esophagus and lesser curvature of stomach
RIGHT GASTRIC (PYLORIC) VEIN	Greater curvature of stomach

9.21

Order of Confluence	Organ or Region Drained
INFERIOR VENA CAVA	(Returns blood from lower half of body)
COMMON ILIAC VEIN	Thighs and legs
INTERNAL ILIAC VEIN	Lower pelvic organs
EXTERNAL ILIAC VEIN	Lower extremity
FEMORAL VEIN	Leg and thigh
LATERAL FEMORAL CIRCUMFLEX VEIN	Hip joint and thigh muscles
MEDIAL FEMORAL CIRCUMFLEX VEIN	Hip joint and thigh muscles
DEEP FEMORAL VEIN	Hip joint and thigh muscles
POPLITEAL VEIN	Knee and distal portion of lower limb
ANTERIOR TIBIAL VEIN	Leg, ankle, and foot
DORSALIS PEDIS VEIN	Dorsum of foot and toes
POSTERIOR TIBIAL VEIN	Leg, foot, and heel
PERONEAL VEIN	Posterior and lateral aspect of ankle and deep calf muscles
MEDIAL PLANTAR VEIN	Abductor hallicus and flexor digitorum brevis muscles
LATERAL PLANTAR VEIN	Soles and toes
PLANTAR VENOUS ARCH	Soles and toes
PLANTAR METATARSAL VEIN	Skin, joints, flexor tendons, and sheaths of the toes
PLANTAR DIGITAL VEIN	Toes
SMALL SAPHENOUS VEIN	Foot
GREAT SAPHENOUS VEIN	Lower extremity
DORSAL VENOUS ARCH	
DORSAL METATARSAL VEINS	Toes
DORSAL DIGITAL VEINS	Toes
GENICULATE VEINS	Knee joint

UNIT 4: REGULATION AND MAINTENANCE

Chapter 10  Lymph Vascular System

---

GENERAL ORGANIZATION   10.1

General Scheme of
Drainage and Circulation

BODY AREAS AND PRINCIPAL CHANNELS
OF LYMPH DRAINAGE   10.2

THE 4 LYMPHOID ORGANS   10.3

Lymph Nodes, Spleen,
Tonsils, and Thymus Gland

Charts #1–2

★ The **LYMPH VASCULAR SYSTEM** is made
up of a series of small masses of lymphoid
tissue called **LYMPH NODES**, connected by
LYMPHATIC VESSELS through which flows
LYMPH FLUID, and 3 organs (glands)—
SPLEEN, TONSILS, and **THYMUS.**

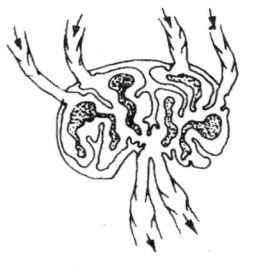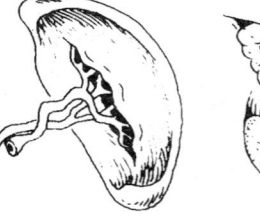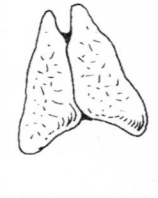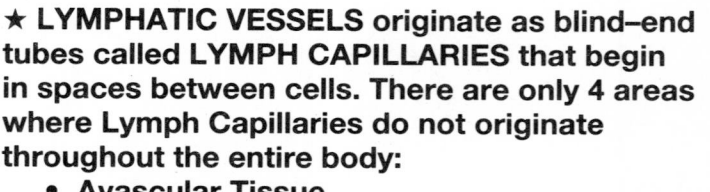

★ **LYMPHATIC VESSELS** originate as blind–end
tubes called **LYMPH CAPILLARIES** that begin
in spaces between cells. There are only 4 areas
where Lymph Capillaries do not originate
throughout the entire body:
  • **Avascular Tissue**
  • **Central Nervous System**
  • **Splenic Pulp**
  • **Bone Marrow**
Lymph Capillaries are slightly larger and more
permeable than Blood Capillaries

■ THE LYMPH VASCULAR SYSTEM

• Prevents the accumulation of tissue fluid around the body tissue cells (a condition called EDEMA).

• Aids in the necessary return of escaped extravascular fluids to the CARDIOVASCULAR SYSTEM, increasing the amount of fluid returned to the heart for requisite constant fluid circulation.

• Microscopic, unconnected lymph capillaries (ending blindly in swollen or rounded ends) begin draining protein– containing fluid from the intercellular tissue (interstitial spaces).

• This interstitial fluid has previously escaped through the highly permeable blood capillaries of the CARDIOVASCULAR SYSTEM.

■ THE LYMPH VASCULAR SYSTEM

• Collects interstitial fluid as LYMPH, and ultimately returns this fluid to the Venous System of the CARDIOVASCULAR SYSTEM through a series of progressively larger lymphatic vessels (purifying and filtering the LYMPH at periodic filtering stations called LYMPH NODES).

• Arises from VEINS in the developing embryo.
• Associated with VEINS throughout the body and drains lymph into the 2 large BRACHIO–CEPHALIC VEINS at the rate of 2 liters/day.

• Does not have a separate heart to circulate LYMPH, and relies on the alternating contraction and relaxation of nearby skeletal muscles to accomplish lymph flow.

■ THE LYMPH VASCULAR SYSTEM

• Also transports into the CARDIOVASCULAR SYSTEM, large aggregations of Fat Monomers (Fatty Acids and Glycerol) and other aggregations that are unable to be directly absorbed through the digestive process.

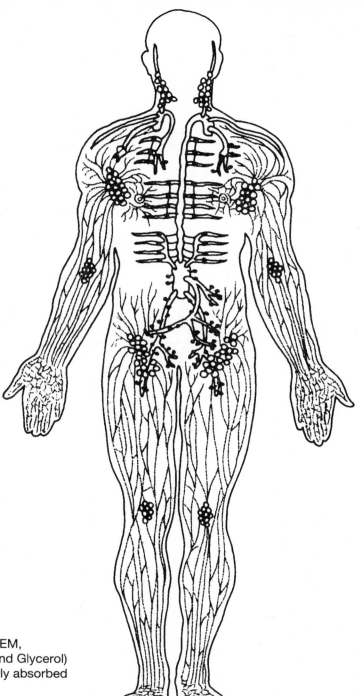

Anterior View

System Components

LYMPH, LYMPH NODES, LYMPHATIC VESSELS, AND 3 GLANDS (EX: SPLEEN, TONSILS, AND THYMUS GLAND)

System Function

• **RETURN OF PROTEIN AND FLUID TO THE BLOOD VASCULAR SYSTEM**

• **TRANSPORTATION OF FATS FROM THE DIGESTIVE SYSTEM TO THE BLOOD VASCULAR SYSTEM**

• **FILTRATION OF BLOOD**

• **PRODUCTION OF WHITE BLOOD CELLS**

• **PROTECTION AGAINST DISEASE**

Chapter Coloring Guidelines

You may want to choose your color palette by comparing any colored illustration from your main textbook.

In general, lymph vessels, nodes, trunk, and ducts are colored GREEN. Use different types of greens (light, lime-green, blue-green, etc.) for all lymphatic structures.

LYMPH VASCULAR SYSTEM: GENERAL ORGANIZATION
General Scheme of Drainage and Circulation

General Scheme of LYMPH DRAINAGE

1 LYMPH CAPILLARIES/PLEXUSES

2 AFFERENT LYMPHATIC VESSELS

3 LYMPH NODES

4 EFFERENT LYMPHATIC VESSELS

5 LYMPH TRUNKS

6 LYMPH DUCTS

7 SUBCLAVIAN VEINS

BLOOD CIRCULATION

S SYSTEMIC CIRCULATION

P PULMONARY CIRCULATION

SCHEME of BLOOD OXYGENATION

OXYGENATED BLOOD

DEOXYGENATED BLOOD

Origination of LYMPH VASCULAR SYSTEM

★ The LYMPH VASCULAR SYSTEM
is *not* a CLOSED-LOOP
SYSTEM (as is the BLOOD
VASCULAR SYSTEM).

The LYMPH VASCULAR SYSTEM
originates in a
vast network (plexus) of
LYMPH CAPILLARIES, which
are BLIND or CLOSED-ENDED
microscopic tubes that *begin* in
the INTERCELLULAR SPACES
between tissue cells.

The LYMPH VASCULAR SYSTEM
ends in the SUBCLAVIAN VEINS
of the Venous System of the
BLOOD VASCULAR SYSTEM.

★ LYMPH CAPILLARIES are
composed of highly permeable
simple squamous epithelium.
LYMPH fluid, which can easily
enter them is formed as a
filtrate of plasma through
blood capillaries.

• INTERSTITIAL FLUID and LYMPH
are basically the same as BLOOD
PLASMA, but they have a lower
concentration of protein. Large
proteins are held back by filtration.

1 LYMPH CAPILLARIES/PLEXUSES

8 CAPILLARY BEDS

9 ARTERIOLES

10 VENULES

11 TISSUE CELLS

FILTRATION of BLOOD FLUID
From BLOOD CAPILLARIES into
LYMPH CAPILLARIES

IN BLOOD CAPILLARY
12 BLOOD PLASMA

IN INTERCELLULAR SPACES
13 INTERSTITIAL FLUID

IN LYMPH CAPILLARY
14 LYMPH

LYMPH "PUMP"

• FLOW OF LYMPH
is due to:

• Skeletal muscle
contractions

• Respiratory
movements

• Lymph valves

GENERAL FLOW OF LYMPH (GENERAL DIRECTION OF FLOW IN THE BODY IS UPWARD, TOWARD THE NECK REGION)

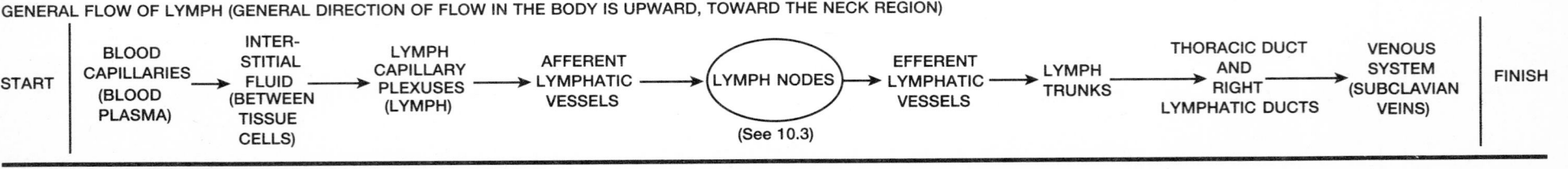

START → BLOOD CAPILLARIES (BLOOD PLASMA) → INTER-STITIAL FLUID (BETWEEN TISSUE CELLS) → LYMPH CAPILLARY PLEXUSES (LYMPH) → AFFERENT LYMPHATIC VESSELS → LYMPH NODES (See 10.3) → EFFERENT LYMPHATIC VESSELS → LYMPH TRUNKS → THORACIC DUCT AND RIGHT LYMPHATIC DUCTS → VENOUS SYSTEM (SUBCLAVIAN VEINS) → FINISH

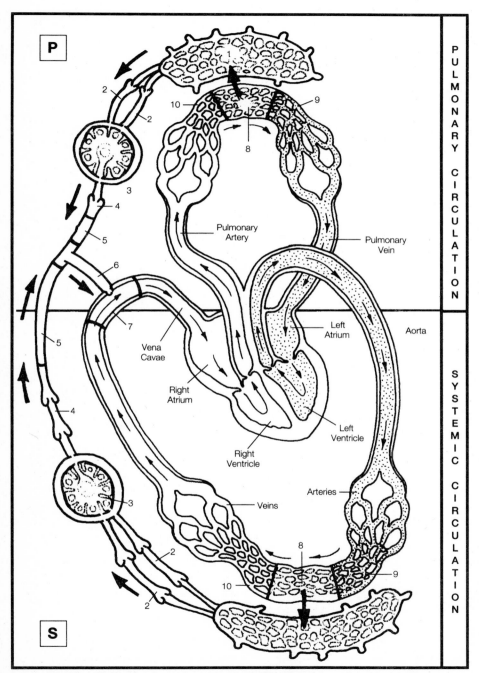

GENERAL SCHEME of DRAINAGE and CIRCULATION
of LYMPH (small arrows → = blood flow)
(large arrows ➡ = lymph flow)

Within the circulation diagram:

PULMONARY CIRCULATION

SYSTEMIC CIRCULATION

P

Pulmonary Artery

Pulmonary Vein

Left Atrium

Aorta

Vena Cavae

Right Atrium

Left Ventricle

Right Ventricle

Arteries

Veins

S

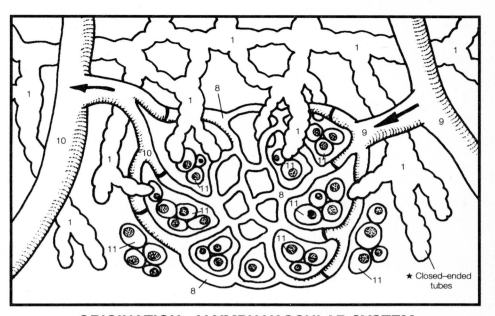

ORIGINATION of LYMPH VASCULAR SYSTEM
(Arrows indicate Blood Flow)
★ Within the villi of the small intestine, lymph capillaries
are called LACTEALS, which transport products of
fat absorption away from the digestive tract (See 12.12).

★ Closed-ended tubes

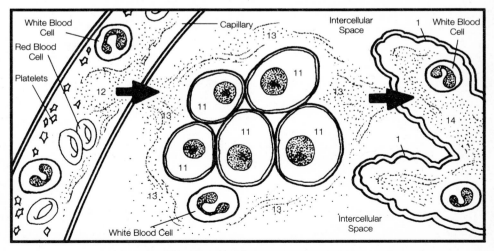

FILTRATION of BLOOD FLUID
(Arrows indicate Blood Plasma Flow)

White Blood Cell

Red Blood Cell

Platelets

Capillary

Intercellular Space

White Blood Cell

White Blood Cell

Intercellular Space

LYMPH VASCULAR SYSTEM: BODY AREAS AND PRINCIPAL CHANNELS OF LYMPH DRAINAGE

Cisterna Chyli = yellow
Large Thoracic Duct = yellow-green
Use colors close to greens
and blues as much as you can. (cool colors)
★ See Chart #1, and 12.12

★ The BASIC LYMPHATIC SYSTEM consists of:
- LYMPHATIC VESSELS • LYMPH NODES
- LYMPH

■ 3 BASIC FUNCTIONS OF LYMPHATIC SYSTEM:
① Returns lost blood filtrate—called INTERSTITIAL (TISSUE) FLUID—back to the blood stream (venous system).
② Transports absorbed fats from the intestine to the blood.
③ Contains LYMPHOCYTE cells, which help protect the body from diseases.

★ LYMPH fluid differs from BLOOD in that red corpuscles are absent and the protein content is lower. (Thus, LYMPH is similar to BLOOD PLASMA except for the low protein content.)

LYMPH contains: • PROTEINS (serum albumin, serum globulin, serum fibrinogen) • ORGANIC SUBSTANCES (urea, glucose, neutral fats, creatinine) • WATER • LYMPHOCYTE CELLS (produced in lymphoid organs, See 10.3) INTESTINAL LYMPH (CHYLE) contains FATS absorbed from the intestine.

★ LYMPH is formed in tissue spaces all over the body (from blood filtrate that escapes through the thin-walled blood capillaries). It is gathered into small vessels that carry it centrally. All lymph eventually enters either the RIGHT LYMPHATIC DUCT or the LARGE (LEFT) THORACIC DUCT.

★ The THORACIC DUCT begins in the abdomen as a dilated sac, the CISTERNA CHYLI, which receives lymph vessels from the PELVIS, LOWER LIMBS, INTESTINE, and DIGESTIVE ORGANS.

★ LYMPH is usually a clear, transparent, colorless fluid. In vessels draining the intestines, however, it appears milky, owing to the presence of absorbed fats, and is called CHYLE. The absorption of fats takes place through the EPITHELIAL CELLS of the INTESTINE and those of the VILLI. These cells carry it to the LACTEALS (the lymph vessels of the small intestine). These LACTEALS take up the CHYLE and pass it to the lymph circulation and, via the thoracic duct to the bloodstream. (See 12.12)

2 Main Body Areas Drained by 2 Principal LYMPH DUCTS

Arrows indicate flow of lymph to the two large ducts

Ⓡ **RIGHT LYMPHATIC DUCT** Ⓛ **LARGE THORACIC DUCT** LEFT LYMPHATIC DUCT

The Principal DEEP CHANNELS [TRUNKS and DUCTS] of LYMPH DRAINAGE

Ⓡ **RIGHT LYMPHATIC DUCT** Ⓛ **LARGE THORACIC DUCT**

1 **RIGHT JUGULAR TRUNK**
2 **RIGHT SUBCLAVIAN TRUNK**
3 **RIGHT BRONCHOMEDIASTINAL TRUNK**

4 **LEFT JUGULAR TRUNK**
5 **LEFT SUBCLAVIAN TRUNK**
6 **LEFT BRONCHOMEDIASTINAL TRUNK**
7 **LOWER INTERCOSTAL TRUNKS**
7a **UPPER INTERCOSTAL LYMPH VESSELS**

8 **CISTERNA CHYLI**
9 **RIGHT AND LEFT LUMBAR TRUNKS**
10 **INTESTINAL TRUNKS**

The RIGHT LYMPHATIC DUCT returns LYMPH to the Venous System by emptying into the RIGHT SUBCLAVIAN VEIN (at the junction of the RIGHT INTERNAL JUGULAR VEIN).

The LARGE THORACIC DUCT returns LYMPH to the Venous System by emptying into the LEFT SUBCLAVIAN VEIN (at the junction of the LEFT INTERNAL JUGULAR VEIN).

The Principal Superficial Groups of NODE CENTER CLUSTERS

★ LYMPH NODES usually appear in NODE CLUSTERS in specific regions of the body. ➡

PEYER'S PATCHES in the submucosa of the small intestine contain:
- Many scattered LYMPHOCYTES
- Large aggregations of lymphatic tissue
- Lymphatic nodules

NECK
11 **CERVICAL NODES**
SUBMAXILLARY

CHEST
12 **THORACIC NODES**

UPPER EXTREMITY
13 **AXILLARY NODES**
14 **CUBITAL NODES**

SMALL INTESTINE
15 **PEYER'S PATCHES**

LOWER EXTREMITY
16 **INGUINAL NODES**
17 **POPLITEAL NODES**

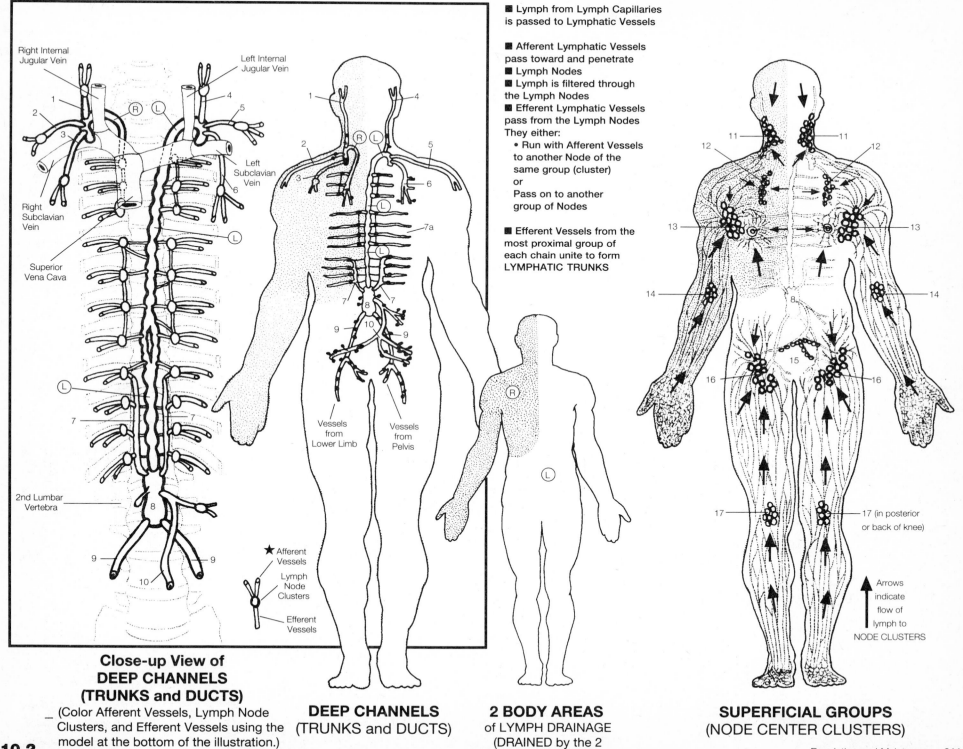

Right Internal Jugular Vein

Left Internal Jugular Vein

Right Subclavian Vein

Left Subclavian Vein

Superior Vena Cava

2nd Lumbar Vertebra

★ Afferent Vessels

Lymph Node Clusters

Efferent Vessels

**Close-up View of
DEEP CHANNELS
(TRUNKS and DUCTS)**

(Color Afferent Vessels, Lymph Node Clusters, and Efferent Vessels using the model at the bottom of the illustration.)

10.2

■ Lymph from Lymph Capillaries is passed to Lymphatic Vessels

■ Afferent Lymphatic Vessels pass toward and penetrate
■ Lymph Nodes
■ Lymph is filtered through the Lymph Nodes
■ Efferent Lymphatic Vessels pass from the Lymph Nodes They either:
  • Run with Afferent Vessels to another Node of the same group (cluster) or
  Pass on to another group of Nodes

■ Efferent Vessels from the most proximal group of each chain unite to form LYMPHATIC TRUNKS

Vessels from Lower Limb

Vessels from Pelvis

**DEEP CHANNELS
(TRUNKS and DUCTS)**

**2 BODY AREAS
of LYMPH DRAINAGE
(DRAINED by the 2
PRINCIPAL DUCTS)**

17 (in posterior or back of knee)

Arrows indicate flow of lymph to NODE CLUSTERS

**SUPERFICIAL GROUPS
(NODE CENTER CLUSTERS)**

Regulation and Maintenance **311**

LYMPH VASCULAR SYSTEM: THE 4 LYMPHOID ORGANS:
Lymph Nodes, Spleen, Tonsils, and Thymus Gland

Spleen = color body dark red
Lymph vessels, lymph capsule, and   trabeculae
= greens   Parenchyma = yellows and blues
Tonsils = warm colors Thymus = light brown
See Chart #2

The FOUR LYMPHOID ORGANS: LYMPH NODES, SPLEEN, TONSILS, and THYMUS GLAND

LYMPH NODES	SPLEEN	THYMUS GLAND (See 8.1, 8.4)	TONSILS
• In passing from any region of the body to the 2 main lymph ducts, lymph must pass through lymph vessels that lead to LYMPH NODES.	• Elongated, dark red ovoid body lying posterior and inferior to the stomach	• An unpaired bilobed organ located in the mediastinal cavity anterior to and above the heart.	• Masses of lymphatic tissue located in depressions of the mucous membrane of the FAUCES and the PHARYNX
• LYMPH travels from the numerous afferent vessels to the CORTICAL SINUSES just under the capsule, then to the MEDULLARY SINUSES, then out through one or two efferent vessels.	• Largest collection of RETICULOENDOTHELIAL (power to ingest or phagocytose bacteria or colloidal particles) cells in the body • *Not* a vital body organ • Functions in various stages of development in BLOOD FORMATION, BLOOD STORAGE, and BLOOD FILTRATION. (See Chart #2)	• Each lobe is divided by the trabeculae into many lobules. • Very large in fetus and children, regresses during puberty, becomes degenerate and involuted in the adult.	• Filtering action protects body from invasion of bacteria
• Filter lymph, freeing it of foreign particulate matter, especially bacteria GERMINAL CENTERS produce LYMPHOCYTES—Agranular LEUKOCYTES or WHITE BLOOD CELLS.	• WHITE PULP produces LYMPHOCYTES (WHITE BLOOD CELLS) • CONTAINS EFFERENT VESSELS *ONLY* VIA THE WHITE PULP. SINCE THERE ARE NO AFFERENT VESSELS, IT DOES NOT FILTER LYMPH.	• Essential for maturation of the thymic lymphoid cells, or the T CELLS (THYMOCYTES). Important in the body's cellular immune response.	• Aids in the formation of LEUKOCYTES
• Nodes lie between afferent vessels and efferent vessels. • 1–25 mm (0.04–1 inch) in length		• Important in development of immune response in children.	

STROMA (FRAMEWORK)	*STROMA*	*STROMA*	
A CAPSULE	A DENSE CAPSULE	A FIBROUS CAPSULE	
B TRABECULAE OF CAPSULE	B TRABECULAE OF CAPSULE	B TRABECULAE OF CAPSULE	**3 PAIRS of TONSILS** form the **"RING of WALDEYER":**
C HILUS	C HILUS	C LOBULES	

PARENCHYMA *OUTER CORTEX*	*PARENCHYMA: SPLENIC PULP* *WHITE PULP:* LYMPHOID TISSUE ARRANGED AROUND ARTERIES	*PARENCHYMA of A LOBULE* *OUTER CORTEX*	14 **PHARYNGEALS (ADENOIDS)**
1 CORTICAL SINUSES	6 MALPIGHIAN CORPUSCLES (SPLENIC NODULES) (DENSELY PACKED LYMPHOCYTES)	10 THYMOCYTES (T CELLS) DENSELY PACKED	15 PALATINES
2 CORTICAL NODULES (DENSELY PACKED LYMPHOCYTES)	7 SPLENIC ARTERY AND BRANCHES	11 EPITHELIORETICULAR FIBROUS TISSUE	16 LINGUALS
3 GERMINAL CENTERS			
INNER MEDULLA	RED PULP: PULP INFILTRATED WITH RED BLOOD CELLS ASSOCIATED WITH VEINS	*INNER MEDULLA*	
4 MEDULLARY SINUSES	8 BILLROTH'S CORDS (SPLENIC CORDS)	10 THYMOCYTES (T CELLS) WIDELY SCATTERED	
5 MEDULLARY CORDS (LYMPHOCYTES arranged in strands)	9 SPLENIC VEIN AND VENOUS SINUSES FILLED WITH BLOOD	12 EPITHELIORETICULAR CELLS 13 THYMIC (HASSALL'S) CORPUSCLES	

LYMPH NODES

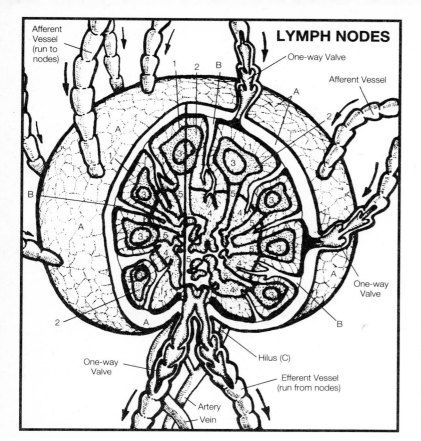

Afferent Vessel (run to nodes)

One-way Valve

Afferent Vessel

One-way Valve

Hilus (C)

One-way Valve

Efferent Vessel (run from nodes)

Artery

Vein

THYMUS GLAND

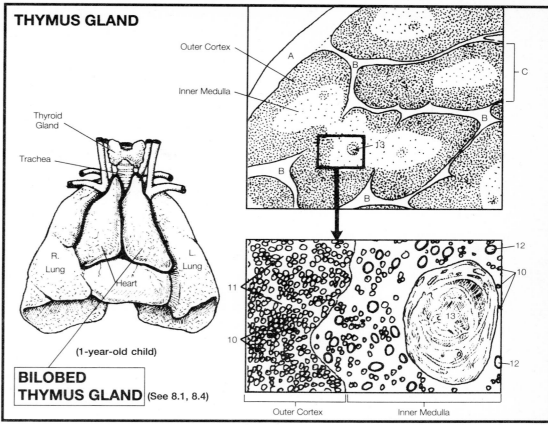

Outer Cortex

Inner Medulla

Thyroid Gland

Trachea

R. Lung

L. Lung

Heart

(1-year-old child)

BILOBED THYMUS GLAND (See 8.1, 8.4)

Outer Cortex

Inner Medulla

TONSILS

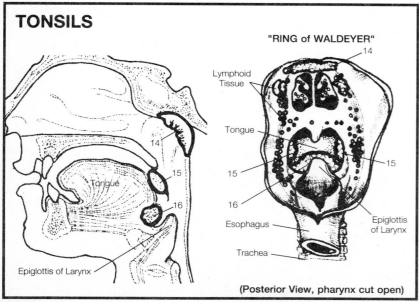

"RING of WALDEYER"

Lymphoid Tissue

Tongue

Tongue

Epiglottis of Larynx

Esophagus

Trachea

Epiglottis of Larynx

(Posterior View, pharynx cut open)

SPLEEN

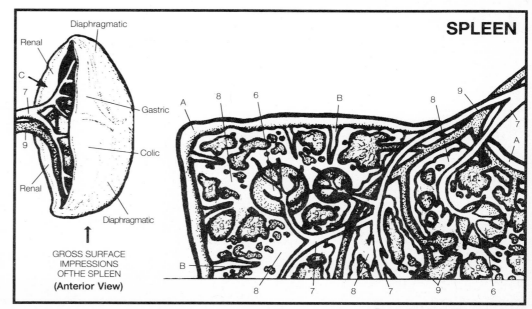

Diaphragmatic

Renal

Gastric

Colic

Renal

Diaphragmatic

GROSS SURFACE IMPRESSIONS OF THE SPLEEN

(Anterior View)

Chart #1 : **Body Regions and Parts that Drain Lymph into Major Lymphatic Trunks and Ducts (See 10.2)**

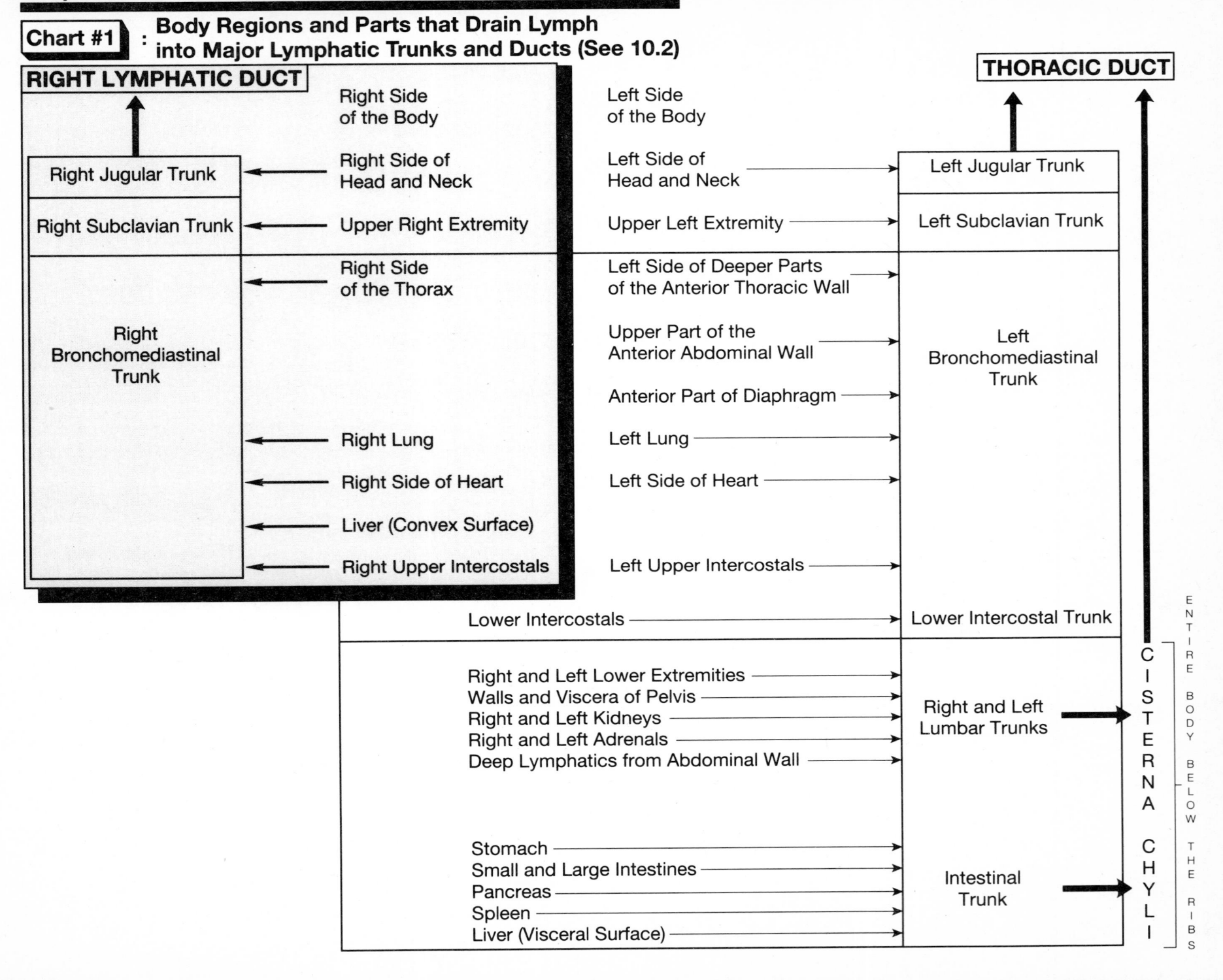

Chart #2 : The 4 Lymphoid Organs (See 10.3)

	LYMPH NODES	SPLEEN	TONSILS	THYMUS GLAND
CHARACTERISTICS →	• Small, ovoid bodies enclosed in fibrous connective tissue capsules, divided internally by trabecular bands	• Largest mass of lymphatic tissue (reticuloendothelial cells) in the body • Not a vital organ in the adult • Since there are no afferent vessels, the spleen does *not* filter lymph	• 3 pairs of tonsils • Masses of lymphoid tissue embedded in mucous membrane	• Very large in the fetus and children • Regression during puberty • Degeneration and involution in adult
LOCATION ——→	• Superficial clusters interspersed along the paths of lymphatic vessels	• Left hypochondriac region of abdomen (upper left portion) between diaphragm and fundus of stomach; beneath the diaphragm and suspended from the stomach	• Pharyngeal region • Forms a circular area around the pharynx, called the "RING of WALDEYER"	• Within the mediastinum between the lungs In the anterior thorax, behind the manubrium of the sternum, along the trachea

FUNCTION ——→	• Production of LYMPHOCYTES (agranular LEUKOCYTES, or WHITE BLOOD CELLS)
	
	• Contains PHAGOCYTES that ingest and destroy bacteria, foreign particles, and worn-out cells from the lymph
	
	• Houses immune-responsive T –LYMPHOCYTES and B –LYMPHOCYTES
		■ BLOOD FORMATION
		  • In embryo → All types of blood cells
		  • In adult → Only LYMPHOCYTES and MONOCYTES
		(If adult bone marrow is damaged, spleen can produce various blood cells.)
		
		■ BLOOD STORAGE
		  Blood reservoir: Can contract and discharge stored blood cells into circulation (through action of smooth muscle and elastic tissue fibers).
		
		■ BLOOD FILTRATION
		  Contains PHAGOCYTES that ingest and destroy elements from circulation, including:
		  • BACTERIA
		  • NON-FUNCTIONING WORN-OUT RED BLOOD CELLS
		  • INCLUSION (HEINZ) BODIES
		  • NUCLEI OF IMMATURE RED BLOOD CELLS
			• Combat infection of the ear, nose, and throat regions
			
			• May become sites of infection due to excess accumulation of infectious material
				• Immunity site in the child (and fetus)
				
				• Produces LYMPHOCYTES that destroy invading microbes
				
				• Produces T –LYMPHOCYTES from undifferentiated lymphocytes that manufacture antibodies against invading microbes

UNIT 4: REGULATION AND MAINTENANCE

Chapter 11 Respiratory System

Cells need a constant supply of OXYGEN (O_2) for the release of energy in carrying out vital cell activities.

These cell activities release CARBON DIOXIDE (CO_2), which becomes toxic to the cell. (It leads to an acid condition by forming carbonic acid, which in turn ionizes yielding hydrogen ions.) Thus, the presence of CO_2 in the cell environment demands its immediate and efficient elimination.

Thus, all living cells require elimination of CARBON DIOXIDE to the fluid around them and obtain OXYGEN from it.

The RESPIRATORY SYSTEM and theCARDIO-VASCULAR SYSTEM are the two systems in the body that eliminate CO_2 and supply O_2. They both share in the function of RESPIRATION.

The RESPIRATORY SYSTEM consists of a series of passageways and two pliable LUNGS that exchange GASES between the external ATMOSPHERE and the BLOOD. The CARDIOVASCULAR SYSTEM transports theGASES in the BLOOD between the LUNGS and the BODY TISSUE CELLS.

RESPIRATION is the *overall* EXCHANGE of GASES between the ATMOSPHERE, the BLOOD, and the CELLS, and involves 3 basic processes:

3 BASIC PROCESSES of RESPIRATION:

■ VENTILATION (BREATHING)
   The movement of air between the ATMOSPHERE and the LUNGS

■ GAS EXCHANGE
   •EXTERNAL RESPIRATION
   Gas exchange of oxygen (O_2) and carbon dioxide (CO_2) between the LUNGS and the BLOOD

   •INTERNAL RESPIRATION
   Gas exchange of oxygen (O_2) and carbon dioxide (CO_2) between the BLOOD and the TISSUES (CELLS)

   It is not possible to have normal respiratory exchange of gases (O_2 and CO_2) in the lungs unless the pulmonary tissue is adequately perfused with blood.

■ OXYGEN UTILIZATION
   Cell respiration releases energy needed for tissue metabolism by reacting with food substances in the cytoplasm and mitochondria of the cell.
   These reactions are called GLYCOLYSIS, the KREB'S CITRIC-ACID CYCLE, and the ELECTRON TRANSPORT SYSTEM.

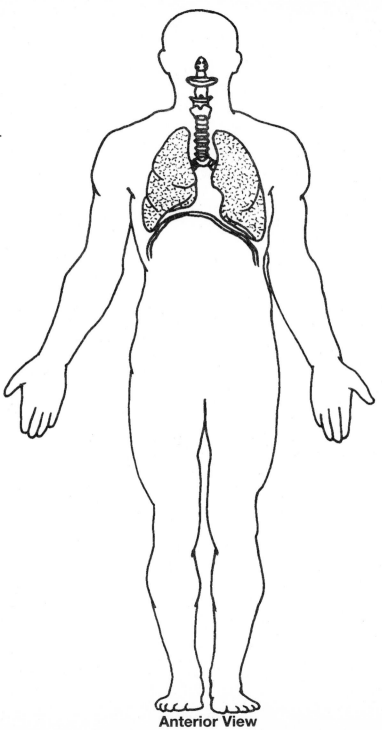

Anterior View

SYSTEM COMPONENTS

THE LUNGS, AND A SERIES OF PASSAGEWAYS (NOSE, MOUTH, PHARYNX, LARYNX, TRACHEA, AND BRONCHI) IN AND OUT OF THE LUNGS

SYSTEM FUNCTION

•SUPPLIES OXYGEN

•ELIMINATION OF CARBON DIOXIDE

•REGULATION OF BODY ACID-BASE BALANCE

Chapter Coloring Guidelines

•*You may want to choose your color palette by comparing any colored illustrations from your main textbook.*

•*In general, color Bronchial Tree in Blues and cool colors.*

•*Lungs = light brown or purple*

•*Remember, Pulmonary Arterioles should be colored Blue (not red) because they carry deoxygenated blood to the alveoli, and Pulmonary Venules are colored Red (not blue) because they carry oxygenated blood to the heart!*

RESPIRATORY SYSTEM: GENERAL ORGANIZATION
General Scheme of the Organs and System

URT = *warm colors (flush, light oranges)*
LRT = *cool colors (blues, purples, etc.)*
Lungs = *light brown or light purple*

A–O will be used throughout the
entire chapter ★ See Chart #1

★ The RESPIRATORY SYSTEM consists of organs that exchange GASES between the ATMOSPHERE and the BLOOD. These organs can be conveniently divided into 2 RESPIRATORY TRACTS:
  ■ UPPER RESPIRATORY TRACT (URT)
  • NOSE
  • NASAL CAVITY
  • PHARYNX
  ■ LOWER RESPIRATORY TRACT (LRT)
  • LARYNX
  • TRACHEA
  • BRONCHIAL TREE (BRONCHI and their increasingly smaller subdivisions)
  • LUNGS

★ The RESPIRATORY SYSTEM can also be viewed as being divided into 2 ZONES (see chart this page):
  ■ CONDUCTION ZONE
  ■ RESPIRATION ZONE

★ THE CARDIOVASCULAR SYSTEM (See Ch. 9) transports the GASES in the BLOOD between the LUNGS and the CELLS. Both the RESPIRATORY SYSTEM and the CARDIOVASCULAR SYSTEM supply OXYGEN and eliminate CARBON DIOXIDE. Both systems are equal partners in RESPIRATION (the OVERALL exchange of GASES between the ATMOSPHERE, BLOOD, and CELLS). If either system fails, there is rapid death of cells from oxygen starvation and disruption of body homeostasis.

★ RESPIRATORY MOVEMENTS
  • There are 2 alternating phases to the RESPIRATORY (BREATHING) CYCLE— the 1st process of RESPIRATION:
  ① INSPIRATION (Inhalation or intake of AIR to LUNGS)
  ② EXPIRATION (Exhalation or expulsion of AIR from LUNGS)
  • Rhythmical bellowlike movements that aid the cycle are achieved through the ALTERNATING VOLUME CHANGES within the THORAX and LUNGS. These volume changes are the result of ALTERNATING MUSCLE CHANGES, principally from the DIAPHRAGM and INTERCOSTAL MUSCLES (See 6.9), and the expansive qualities of the RIB CAGE.
  • The movement of air in and out of the lungs is due to differences in pressure between the air outside of the body and the air inside the lungs (and thorax).
  • EXAMPLE: Diaphragm lowers, rib cage elevates, thorax expands, thoracic volume increases, pulmonary pressure decreases, air enters lungs (INSPIRATION).

Division of the RESPIRATORY SYSTEM into 2 RESPIRATORY TRACTS:

UPPER RESPIRATORY TRACT (URT)

A | NOSE | [†] EXTERNAL NOSE (covered with SKIN and supported by paired NASAL BONES and cartilage) (See 4.7)

B | NASAL CAVITY | INTERNAL NOSE

C | PHARYNX | THROAT, OR GULLET

[†] The MOUTH is considered as a secondary organ of importance for the entrance and exit of air.

LOWER RESPIRATORY TRACT (LRT)

D | LARYNX | VOICE BOX

E | TRACHEA | WINDPIPE

THE BRONCHIAL TREE

F | PRIMARY BRONCHI (2) |

G | SECONDARY (LOBAR) BRONCHI |

H | TERTIARY (SEGMENTAL) BRONCHI |

I | BRONCHIOLES |

J | TERMINAL BRONCHIOLES |

The RESPIRATION ZONE

K | RESPIRATORY BRONCHIOLES |

L | ALVEOLAR DUCTS | ATRIA

M | ALVEOLI |

N | ALVEOLAR SACS |

O | LUNGS |

DIVISION OF THE RESPIRATORY SYSTEM INTO 2 ZONES:

CONDUCTION ZONE: PERMANENTLY OPENED PASSAGEWAYS FOR CONDUCTION OF AIR TO LUNG TISSUE

• Transportation of air to the RESPIRATORY ZONE in the LUNGS

• Various epithelia along passageways warm, filter (cleanse), and humidify the air

• *No* exchange of gases
INCLUDES:
  • NOSE   • PRIMARY BRONCHI
  • NASAL CAVITY   • SECONDARY BRONCHI
  • PHARYNX   • TERTIARY BRONCHI
  • LARYNX   • BRONCHIOLES
  • TRACHEA   • TERMINAL BRONCHIOLES

:(URT, LARYNX, TRACHEA, and the BRONCHIAL TREE)

RESPIRATION ZONE:

• O_2–CO_2 GAS EXCHANGE with the BLOOD

INCLUDES:
  • RESPIRATORY BRONCHIOLES
  • ALVEOLAR DUCTS
  • ALVEOLI, and
  • ALVEOLAR SACS

★ All structures of the respiratory system beyond the PRIMARY BRONCHI (including the BRONCHIAL TREE and ALVEOLI) are contained in the LUNGS.

★ Most of the mass of the lungs is comprised of the ALVEOLAR DUCTS and their terminal ALVEOLI and ALVEOLAR SACS.

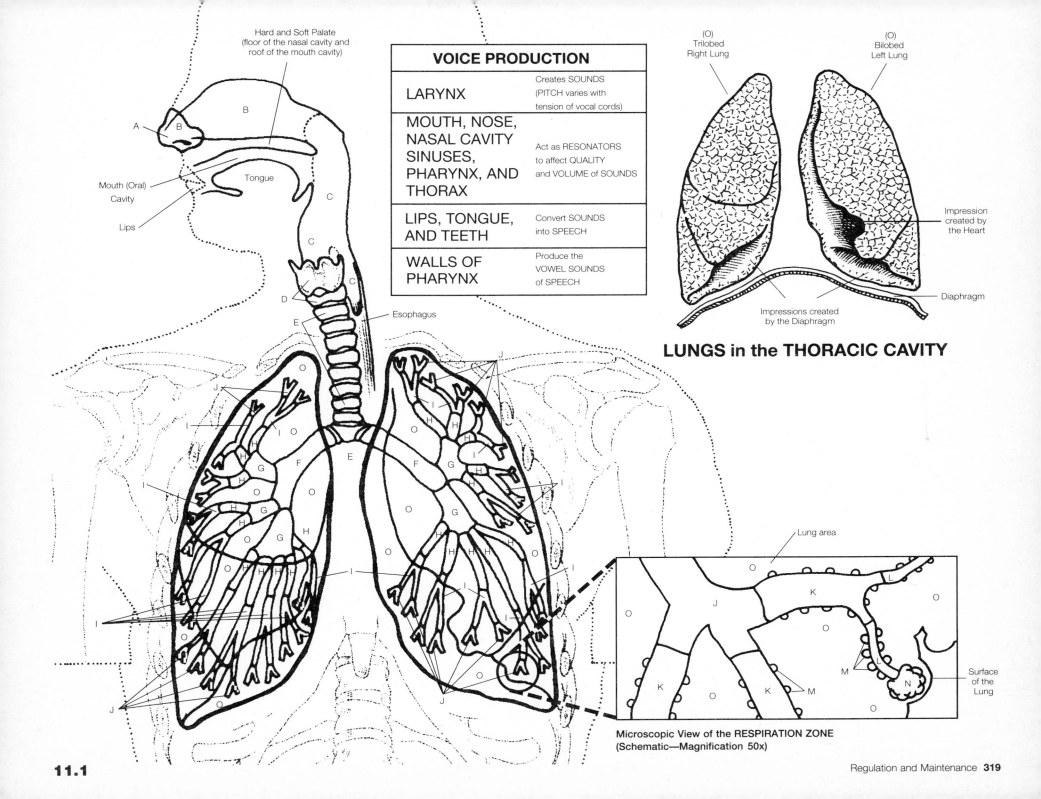

Hard and Soft Palate
(floor of the nasal cavity and
roof of the mouth cavity)

B

A
B

Mouth (Oral)
Cavity

Tongue

Lips

C

C

C

C

D

E

Esophagus

VOICE PRODUCTION	
LARYNX	Creates SOUNDS (PITCH varies with tension of vocal cords)
MOUTH, NOSE, NASAL CAVITY SINUSES, PHARYNX, AND THORAX	Act as RESONATORS to affect QUALITY and VOLUME of SOUNDS
LIPS, TONGUE, AND TEETH	Convert SOUNDS into SPEECH
WALLS OF PHARYNX	Produce the VOWEL SOUNDS of SPEECH

(O)
Trilobed
Right Lung

(O)
Bilobed
Left Lung

Impression
created by
the Heart

Impressions created
by the Diaphragm

Diaphragm

LUNGS in the THORACIC CAVITY

Lung area

Surface
of the
Lung

Microscopic View of the RESPIRATION ZONE
(Schematic—Magnification 50x)

11.1

RESPIRATORY SYSTEM: UPPER RESPIRATORY TRACT
Conduction Zone: The Nasal Cavity and the Pharynx

Use warm colors for the Nasal Cavity and Pharynx. Larynx = turquoise blue
Concha = choose your own bright colors

★ See also 4.6, 4.7, 6.4, 6.5, 7.18, 12.8

★ THE NASAL CAVITY

• Cavity between the floor of the cranium and the roof of the mouth, separated into 2 nasal fossae chambers by a medial nasal septum (See 4.7)

• As inspired air travels from the nose into the cavity, the NASAL EPITHELIUM warms, moistens, and cleanses (filters) the air. (Inhalation of air can also take place through the mouth, or oral cavity, with lesser effect.)

• Cavity acts as a resonating chamber for voice phonetics.

• Olfactory epithelium in the ethmoid bone (in the roof of the cavity) provides a sense of smell (See 7.18)

• 3 scroll-like turbinate bones called CONCHAE project medially from each lateral wall of the nasal cavity. Each of these 3 CONCHAE overlies a MEATUS (tubelike passageway). The SUPERIOR and MIDDLE CONCHAE are projections of the lateral mass of the ethmoid bone (See 4.6, 4.7). The INFERIOR CONCHA is a FACIAL BONE.

★ The NOSTRILS are the 2 EXTERNAL APERTURES (NARES) into the NASAL CAVITY. The CHOANAE are the 2 INTERNAL APERTURES (NARES) providing a funnel-shaped opening for communication between the NASAL FOSSAE and the PHARYNX.

★ THE PHARYNX (See 6.4, 12.8)

• Funnel-shaped passageway for:

■ AIR into the LARYNX

■ Food and Fluid into the ESOPHAGUS

• Connects both the NASAL and ORAL CAVITIES with the LARYNX (at the base of the skull)

• Lumen is lined with mucous membrane

• The supporting pharyngeal walls are composed of skeletal muscles (See 6.5)

• Constriction and relaxation of these wall muscles (constrictors) produce the VOWEL SOUNDS of SPEECH

• Acts as a resonating chamber for certain speech sounds

★ If air from the PHARYNX enters the ESOPHAGUS instead of the LARYNX (or if gas enters the ESOPHAGUS from the STOMACH), a BELCH (BURP) may occur. If FOOD or FLUID enters the LARYNX, COUGHING often results.

Lateral Wall of the Nasal Cavity:

ANTERIOR PORTION (OF NOSTRILS)

1 | VESTIBULE |

POSTERIOR PORTION

2,(2) | SUPERIOR CONCHA (AND MEATUS) |

3,(3) | MIDDLE CONCHA (AND MEATUS) |

4,(4) | INFERIOR CONCHA (AND MEATUS) |

Openings of the Nasal Cavity:
The NASAL APERTURES

FRONT: EXTERNAL (ANTERIOR) NARES

5 | NOSTRILS |

BACK: INTERNAL (POSTERIOR) NARES

6 | CHOANAE |

C: The PHARYNX
3 Divisions of the PHARYNX (See 12.8)

7 | NASOPHARYNX |
• The section above the PALATE
• Contains PHARYNGEAL TONSILS
• Lined with PSEUDOSTRATIFIED CILIATED epithelium with GOBLET CELLS (continuation of the NASAL CAVITY epithelium)

8 | OROPHARYNX |
• The section between the PALATE and HYOID BONE
• Contains PALATINE and LINGUAL TONSILS
• Transition to STRATIFIED COLUMNAR epithelium.

9 | LARYNGOPHARYNX |
• The section below the HYOID BONE and above the beginning of the ESOPHAGUS
• Lined with STRATIFIED SQUAMOUS epithelium

★ The PHARYNX communicates (has openings) with 5 different structures: EUSTACHIAN TUBES, NASAL CAVITY, ORAL CAVITY, LARYNX, and ESOPHAGUS (See 12.8)

OPENINGS INTO THE PHARYNX:

NASOPHARYNX

10 | EUSTACHIAN TUBE OPENING | OPENS FROM LATERAL WALL

6 | CHOANAE | 2 OPENINGS FROM NASAL CAVITY SEPARATED BY THE NASAL SEPTUM

OROPHARYNX

11 | FAUCES | OPENING FROM ORAL (MOUTH) CAVITY

LARYNGOPHARYNX

D₁ | GLOTTIS OPENING OF LARYNX |

12ₐ | BEGINNING OF ESOPHAGUS |

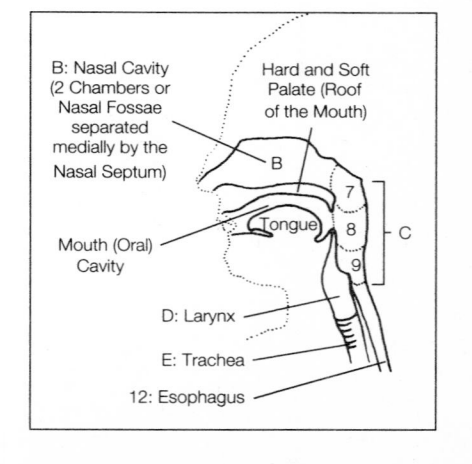

B: Nasal Cavity (2 Chambers or Nasal Fossae separated medially by the Nasal Septum)

Hard and Soft Palate (Roof of the Mouth)

Mouth (Oral) Cavity

Tongue

D: Larynx

E: Trachea

12: Esophagus

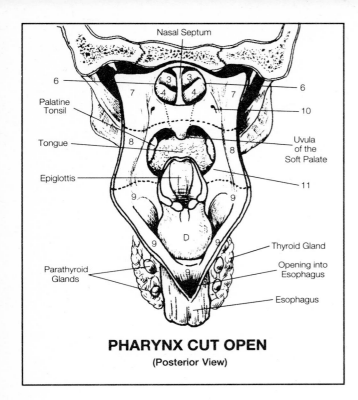

PHARYNX CUT OPEN

(Posterior View)

Nasal Septum

6

Palatine Tonsil

Tongue

Epiglottis

Parathyroid Glands

6

10

Uvula of the Soft Palate

11

Thyroid Gland

Opening into Esophagus

Esophagus

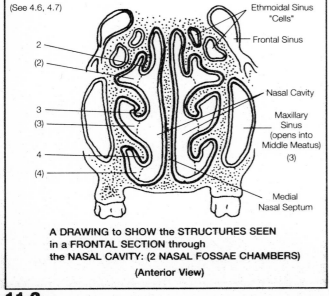

(See 4.6, 4.7)

Ethmoidal Sinus "Cells"

Frontal Sinus

2

(2)

3

(3)

4

(4)

Nasal Cavity

Maxillary Sinus (opens into Middle Meatus)

(3)

Medial Nasal Septum

A DRAWING to SHOW the STRUCTURES SEEN in a FRONTAL SECTION through the NASAL CAVITY: (2 NASAL FOSSAE CHAMBERS)

(Anterior View)

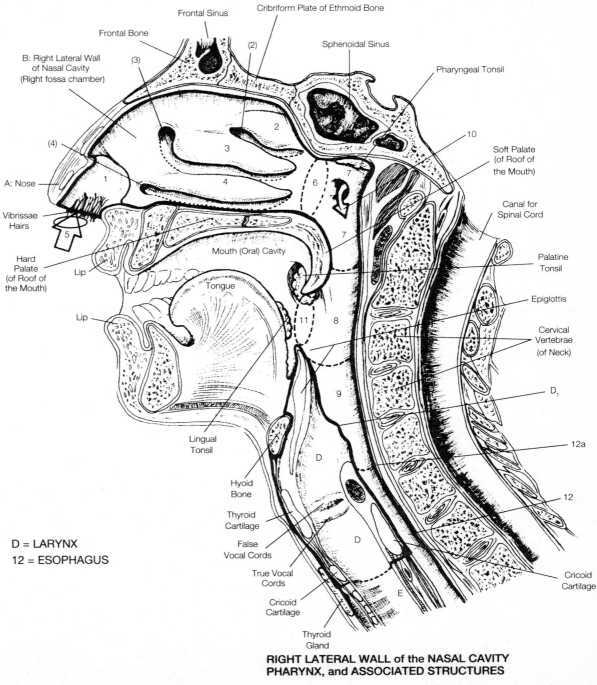

Frontal Sinus

Frontal Bone

Cribriform Plate of Ethmoid Bone

Sphenoidal Sinus

Pharyngeal Tonsil

B: Right Lateral Wall of Nasal Cavity (Right fossa chamber)

(3)

(2)

10

(4)

2

Soft Palate (of Roof of the Mouth)

3

A: Nose

4

Canal for Spinal Cord

Vibrissae Hairs

Mouth (Oral) Cavity

Palatine Tonsil

Hard Palate (of Roof of the Mouth)

Lip

Tongue

Epiglottis

Lip

Cervical Vertebrae (of Neck)

Lingual Tonsil

Hyoid Bone

Thyroid Cartilage

False Vocal Cords

True Vocal Cords

Cricoid Cartilage

Thyroid Gland

D₁

12a

12

Cricoid Cartilage

D = LARYNX

12 = ESOPHAGUS

RIGHT LATERAL WALL of the NASAL CAVITY PHARYNX, and ASSOCIATED STRUCTURES

RESPIRATORY SYSTEM: UPPER RESPIRATORY TRACT
Conduction Zone: Paranasal Air Sinuses and Openings into the Nasal Cavity

Choose a bright color scheme for the 4 sinuses: yellow, blue, green, and red.
- *Use the same colors for the sinuses and their duct openings*

Nasal Septum = light fleshy color.

★ See 11.2 for other reference numbers and use the same colors

★ See Chart #2, 4.7

★ **PARANASAL AIR SINUSES**
Are
PAIRED AIR SPACES
located within 4
BONES of the SKULL:
3 CRANIAL BONES
- FRONTAL
- SPHENOID
- ETHMOID
1 FACIAL BONE
- MAXILLA

PARANASAL AIR SINUSES
★
- Produce mucus
- Decrease weight of skull
- Give structural strength to the skull

- Warm and moisten
- inspired air

- Function as sound resonating chambers

★ **EACH PARANASAL AIR SINUS**
Communicates within
the **NASAL CAVITY**
via drainage ducts
on its own side
(openings are through
the lateral walls of nasal cavity).

★ **PARANASAL AIR SINUSES** are
lined with mucous
membranes that
are *continuous*
with the
NASAL CAVITY.

★ Tears secreted
from the **LACRIMAL GLAND** drain into
the **NASOLACRIMAL DUCT** and out into
the **NASAL CAVITY** through the Inferior
Meatus. One blows
the nose after
crying—see why?

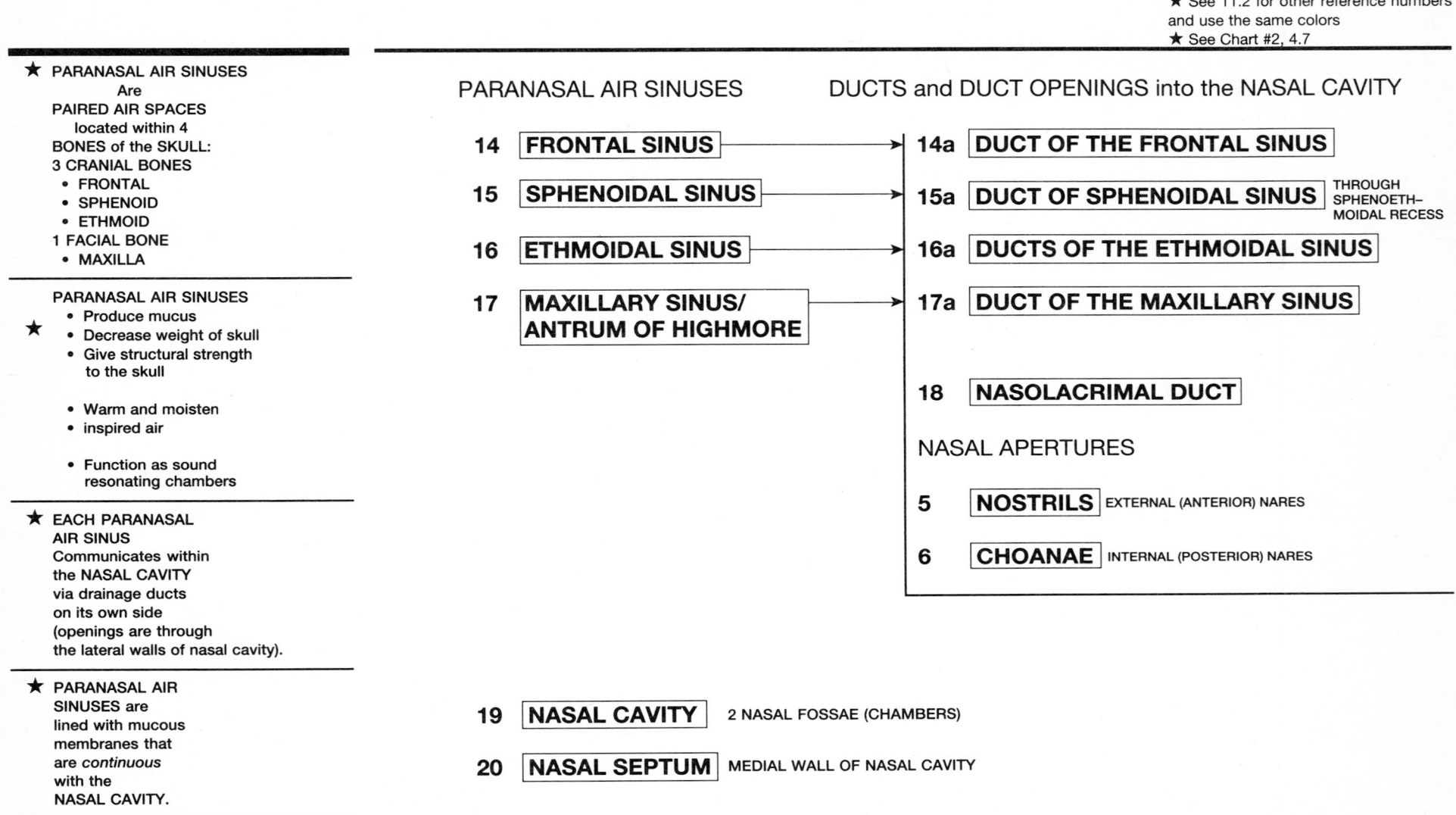

PARANASAL AIR SINUSES   DUCTS and DUCT OPENINGS into the NASAL CAVITY

14   **FRONTAL SINUS** → 14a   **DUCT OF THE FRONTAL SINUS**

15   **SPHENOIDAL SINUS** → 15a   **DUCT OF SPHENOIDAL SINUS**   THROUGH SPHENOETH–MOIDAL RECESS

16   **ETHMOIDAL SINUS** → 16a   **DUCTS OF THE ETHMOIDAL SINUS**

17   **MAXILLARY SINUS/ ANTRUM OF HIGHMORE** → 17a   **DUCT OF THE MAXILLARY SINUS**

18   **NASOLACRIMAL DUCT**

NASAL APERTURES

5   **NOSTRILS**   EXTERNAL (ANTERIOR) NARES

6   **CHOANAE**   INTERNAL (POSTERIOR) NARES

19   **NASAL CAVITY**   2 NASAL FOSSAE (CHAMBERS)

20   **NASAL SEPTUM**   MEDIAL WALL OF NASAL CAVITY

3 FUNCTIONS OF THE PARANASAL SINUSES ASSOCIATED WITH THE NASAL CAVITY

- WARMS, MOISTENS, FILTERS, and CLEANS AIR
- SENSE of SMELL from OLFACTORY EPITHELIUM
- SOUND RESONATING CHAMBER (VOICE PHONETICS)

PARANASAL AIR SINUSES
within 4 SKULL BONES

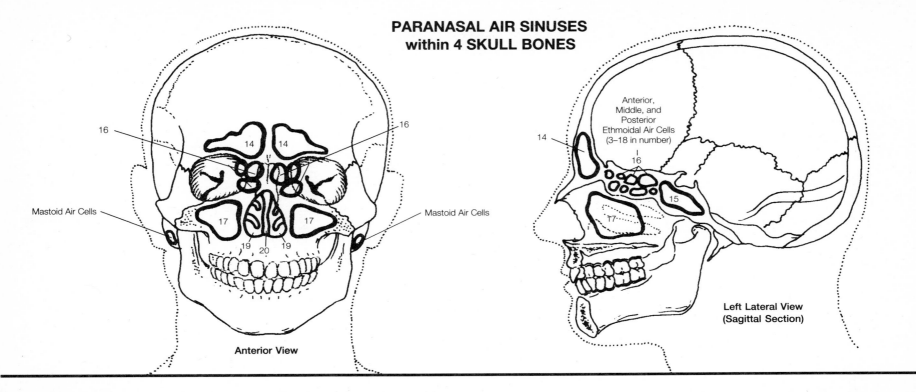

16   14 14   16

Mastoid Air Cells

17   17
19 20 19

Anterior View

Mastoid Air Cells

Anterior,
Middle, and
Posterior
Ethmoidal Air Cells
(3–18 in number)

14
16
15
17

Left Lateral View
(Sagittal Section)

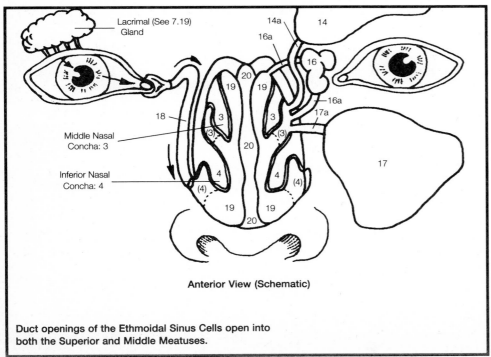

Lacrimal (See 7.19)
Gland

14a   14
16a
16
20
19 19
16a
18   3   3   17a
(3)   (3)
Middle Nasal
Concha: 3
20
17
4   4
(4)   (4)
Inferior Nasal
Concha: 4
19 19
20

Anterior View (Schematic)

Duct openings of the Ethmoidal Sinus Cells open into
both the Superior and Middle Meatuses.

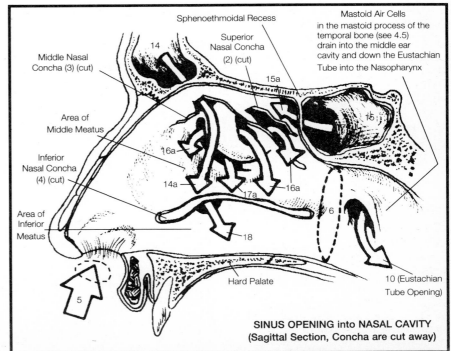

Sphenoethmoidal Recess
Mastoid Air Cells
in the mastoid process of the
temporal bone (see 4.5)
drain into the middle ear
cavity and down the Eustachian
Tube into the Nasopharynx

Superior
Nasal Concha
(2) (cut)

Middle Nasal
Concha (3) (cut)

14
15a

Area of
Middle Meatus

15

16a

Inferior
Nasal Concha
(4) (cut)

14a   16a
17a

Area of
Inferior
Meatus

6

18

Hard Palate

5

10 (Eustachian
Tube Opening)

SINUS OPENING into NASAL CAVITY
(Sagittal Section, Concha are cut away)

RESPIRATORY SYSTEM: LOWER RESPIRATORY TRACT (LRT)
Conduction Zone: The Larynx

All cartilages = blues
Membranes and Ligaments =
light browns and oranges
Hyoid Bone = gray   Vocal cords = fleshy

★ The LARYNX or VOICE BOX:

• The enlarged upper end of the TRACHEA below the ROOT of the TONGUE, shaped like a triangular box

• Forms the entranceway into the LOWER RESPIRATORY TRACT (LRT)

• Connects the PHARYNX (lower portion, LARYNGOPHARYNX) with the TRACHEA

• Musculocartilaginous structure lined with mucous membrane that forms the organ of voice

• Located in the anterior midline of the neck between the 4th–6th cervical vertebrae

STRUCTURE OF THE LARYNX:

• Consists of 9 CARTILAGES (3 large single and 3 small paired) bound together by an elastic membrane and moved by muscles

• Extrinsic muscles include the INFRAHYOIDS (See 6.6) and others

• Intrinsic muscles (See 6.5)

• The CAVITY of the LARYNX is divided into 3 regions: an UPPER VESTIBULE, a lower INFRAGLOTTIC LARYNX, and a middle VENTRICLE, which is the space between the TRUE and FALSE VOCAL CORDS (FOLDS).

• The mucous membrane of the larynx is arranged into 2 pairs of FOLDS:

■ UPPER PAIR: VENTRICULAR FOLDS (FALSE VOCAL CORDS)—*not* used in sound production, help support the:

■ LOWER PAIR: VOCAL FOLDS (TRUE VOCAL CORDS)—*are* used in sound production.

These 2 pairs of strong connective tissue bands are stretched across the upper opening of the larynx from anterior to posterior. Under the mucous membrane of the TRUE VOCAL CORDS lie the CONUS ELASTICUS bands, which stretch between the large THYROID CARTILAGE and the small ARYTENOID CARTILAGES.

• The space (slit) between the TRUE VOCAL CORDS is the GLOTTIS (RIMA GLOTTIDIS), or the LARYNGEAL OPENING.

• The spoon-shaped EPIGLOTTIS lies on top of the LARYNX behind the root of the tongue and aids in closing the GLOTTIS during swallowing. The "stem" portion is hinged to the THYROID CARTILAGE by the THYROEPIGLOTTIC LIGAMENT. The "leaf" portion is free and can move up and down like a trap door. During swallowing, the LARYNX elevates (due to the contraction of the EXTRINSIC MUSCLES), causes the free end of the epiglottis to form a lid over the GLOTTIS, which closes the GLOTTIS. (Thus, it is impossible to breathe and swallow at the same time.)

9 CARTILAGES of the LARYNX

3 LARGE (SINGLES)

1 | EPIGLOTTIS |

2 | THYROID CARTILAGE | "ADAM'S APPLE"

3 | CRICOID CARTILAGE |

6 SMALL (3 PAIRED)

4 | ARYTENOID CARTILAGE |

5 | CORNICULATE CARTILAGE |

6 | CUNEIFORM CARTILAGE |

LIGAMENTS and MEMBRANES
EXTRINSIC

7 | THYROHYOID MEMBRANE |

8 | THYROHYOID LIGAMENT |

9 | CRICOTRACHEAL LIGAMENT |

INTRINSIC

10 | CRICOTHYROID LIGAMENT |

11 | THYROEPIGLOTTIC LIGAMENT |

INNER MEMBRANE OF TRUE VOCAL CORDS

12 | CONUS ELASTICUS |

INNER LIGAMENT OF TRUE VOCAL CORDS

13 | VOCAL LIGAMENT |

TWO PRINCIPAL FUNCTIONS of the LARYNX

① ■ Permits passage of air to the TRACHEA during breathing
■ Prevents food or fluid from entering the TRACHEA during swallowing (if they enter, COUGHING results to expel the material)

② SOUND PRODUCTION:
• Vocal cords controlled by intrinsic muscles produce sound

LARYNGEAL OPENING

18 | GLOTTIS | RIMA GLOTTIDIS

LARYNGEAL CAVITIES

19 | VESTIBULE |

20 | VENTRICLE |

21 | INFRAGLOTTIC LARYNX |

(INFERIOR ENTRANCE TO THE GLOTTIS)

LARYNGEAL FOLDS

14 | INTERARYTENOID FOLD |

15 | ARYEPIGLOTTIC FOLD |

16 | TRUE VOCAL CORDS (VOCAL CORDS) |

17 | FALSE VOCAL CORDS (VENTRICULAR FOLDS) |

★ The INTRINSIC SKELETAL MUSCLES of the LARYNX (See 6.5), when contracted, change the LENGTH, POSITION, AND TENSION of the VOCAL CORDS. These muscles are attached internally to the THYROID CARTILAGE, the pyramid-shaped ARYTENOID CARTILAGES, and to the VOCAL CORDS themselves. When the muscles contract, they pull the CONUS ELASTICUS "strings" tight and stretch the cords out into the air passageways, narrowing the GLOTTIS. If air is directed from the lungs against the folds, they vibrate and create sound waves in the column of air in the PHARYNX, NASAL CAVITY, and ORAL CAVITY. (These 3 latter cavities and the PARANASAL SINUSES act as RESONATING CHAMBERS, giving the voice its individual, human quality.)

During breathing, the TRUE VOCAL CORDS are ABDUCTED; during COUGHING, they are closed, then rapidly released; during PHONATION (speech, sounds) they are ADDUCTED with a thin GLOTTIS opening.

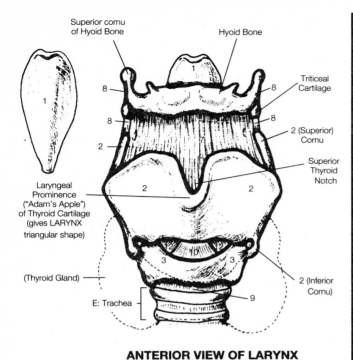

Superior cornu of Hyoid Bone
Hyoid Bone
Triticeal Cartilage
2 (Superior) Cornu
Superior Thyroid Notch
Laryngeal Prominence ("Adam's Apple") of Thyroid Cartilage (gives LARYNX triangular shape)
(Thyroid Gland)
E: Trachea
2 (Inferior Cornu)

ANTERIOR VIEW OF LARYNX
(22 ⬚HYOID BONE⬚)

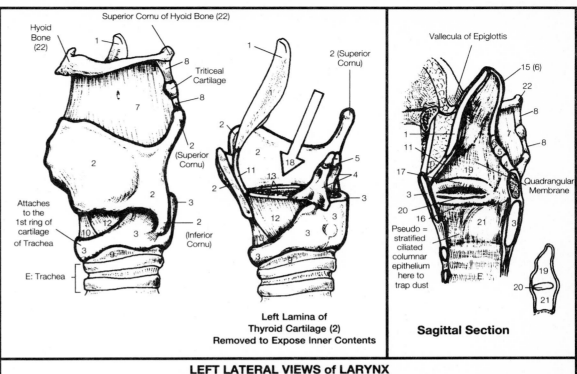

Hyoid Bone (22)
Superior Cornu of Hyoid Bone (22)
Triticeal Cartilage
2 (Superior Cornu)
2 (Superior Cornu)
Attaches to the 1st ring of cartilage of Trachea
2 (Inferior Cornu)
E: Trachea

Left Lamina of Thyroid Cartilage (2) Removed to Expose Inner Contents

Vallecula of Epiglottis
Quadrangular Membrane
Pseudo = stratified ciliated columnar epithelium here to trap dust

Sagittal Section

LEFT LATERAL VIEWS of LARYNX

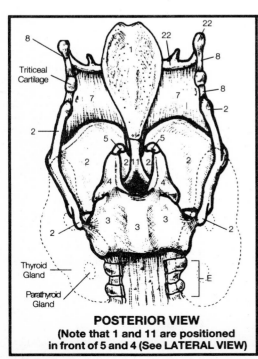

Triticeal Cartilage
Thyroid Gland
Parathyroid Gland

POSTERIOR VIEW
(Note that 1 and 11 are positioned in front of 5 and 4 (See LATERAL VIEW)

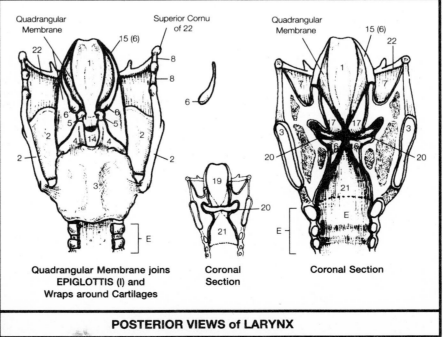

Quadrangular Membrane
Superior Cornu of 22
Quadrangular Membrane

Quadrangular Membrane joins EPIGLOTTIS (I) and Wraps around Cartilages

Coronal Section

Coronal Section

POSTERIOR VIEWS of LARYNX

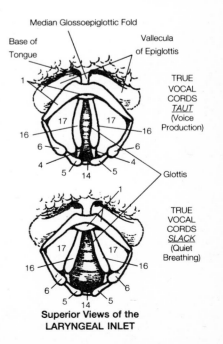

Median Glossoepiglottic Fold
Base of Tongue
Vallecula of Epiglottis
TRUE VOCAL CORDS *TAUT* (Voice Production)
Glottis
TRUE VOCAL CORDS *SLACK* (Quiet Breathing)

Superior Views of the LARYNGEAL INLET

11.4

RESPIRATORY SYSTEM: LOWER RESPIRATORY TRACT
Conduction Zone: The Trachea, the Bronchial Tree, and Bronchopulmonary Segmentation

Trachea, Primary and Secondary Bronchi color in Blues
(Fibrous connective tissue on Trachea = light brown)
Tertiary Bronchi = open up and have fun! REMEMBER TO
COLOR THE BRONCHOPULMONARY SEGMENTS OF THE LUNGS
THE SAME AS THEIR CORRESPONDING BRONCHIAL SEGMENTS.

★ E, F, G, H, I, J are reference numbers from 11.1

★ The CRICOID CARTILAGE of the LARYNX forms a ring and attaches to the 1st ring of cartilage of the TRACHEA.

★ The TRACHEA (WINDPIPE) is a cylindrical, cartilaginous, rigid tube left permanently open for the passage of air, that connects the LARYNX to the BRONCHIAL TUBES.
  • 11.3 cm (4 1/2 in.) in length
  • 2.5 cm (1 in.) in diameter
  • Supported by a stack of C-shaped rings of HYALINE CARTILAGE (interspersed with fibrous connective tissue) that are positioned anteriorly, with the open end of the C's in posterior position
  • Extends from the LARYNX to the STERNAL ANGLE in the THORAX (CHEST), (from the 6th cervical vertebra to the 5th thoracic vertebra), where it divides at a point called the CARINA into the RIGHT and LEFT PRIMARY BRONCHI, one leading to each LUNG. (See 11.6)

★ THE BRONCHIAL TREE
  • The RIGHT PRIMARY BRONCHUS is shorter and more vertical than the LEFT PRIMARY BRONCHUS.
  • After entering the LUNG, each PRIMARY BRONCHUS divides further into increasingly smaller tubes, terminating in microscopic BRONCHIOLES.
  • The continuous branching of the TRACHEA into PRIMARY BRONCHI, SECONDARY BRONCHI, TERTIARY BRONCHI, BRONCHIOLES, and finally into TERMINAL BRONCHIOLES resembles an upside-down tree trunk with its branches; it is commonly referred to as the BRONCHIAL TREE.

★ BRONCHOPULMONARY SEGMENTATION of the LUNGS
  • The lungs are conveniently divided into BRONCHOPULMONARY SEGMENTS according to the bronchial branches that penetrate them. (see ill.)
  • As they enter the lungs, the PRIMARY BRONCHI divide to form smaller bronchi, the SECONDARY (LOBAR) BRONCHI, one for each lobe of the lung. (The RIGHT LUNG has 3 lobes, the LEFT LUNG has 2 lobes). The secondary bronchi branch into smaller TERTIARY (SEGMENTAL) BRONCHI, which branch into still smaller BRONCHIOLES, which in turn branch into even smaller TERMINAL BRONCHIOLES.

★ Structural Changes through the BRONCHIAL TREE:
  ① RINGS OF CARTILAGE (in E and F), replaced by PLATES OF CARTILAGE (in G and H), which disappear in I and J.
  ② As cartilage decreases, smooth muscle content increases.
  ③ Pseudostratified ciliated columnar epithelium (in E, F, G, and H) changes to simple cuboidal epithelium in I and J.

The BRONCHIAL TREE

E **TRACHEA**   "Trunk" of the upside-down BRONCHIAL TREE

(TRILOBED RIGHT LUNG)   **(BILOBED LEFT LUNG)**

2 PRIMARY BRONCHI

F₁ **RIGHT PRIMARY BRONCHUS**   F₂ **LEFT PRIMARY BRONCHUS**

5 SECONDARY (LOBAR) BRONCHI

G₁ **RIGHT SUPERIOR LOBAR BRONCHUS**   G₄ **LEFT SUPERIOR LOBAR BRONCHUS**

G₂ **RIGHT MIDDLE LOBAR BRONCHUS**

G₃ **RIGHT INFERIOR LOBAR BRONCHUS**   G₅ **LEFT INFERIOR LOBAR BRONCHUS**

H:18 TERTIARY (SEGMENTAL) BRONCHI

Number of Tertiary Bronchial Segments	Corresponding Lobar Bronchus and Lung Lobe				Corresponding Lobar Bronchus and Lung Lobe	Number of Tertiary Bronchial Segments
3	G₁	① APICAL ② POSTERIOR ③ ANTERIOR		APICOPOSTERIOR ① and ② ANTERIOR ③ SUPERIOR LINGULAR ④ INFERIOR LINGULAR ⑤	G₄	4
2	G₂	④ MEDIAL ⑤ LATERAL				
5 10	G₃	⑥ SUPERIOR ⑦ MEDIAL BASAL ⑧ ANTERIOR BASAL ⑨ LATERAL BASAL ⑩ POSTERIOR BASAL		SUPERIOR ⑥ ANTERIOR MEDIAL BASAL ⑦ and ⑧ LATERAL BASAL ⑨ POSTERIOR BASAL ⑩	G₅	4 8

★ BRONCHIOLES are analogous in function to the ARTERIOLES in the VASCULAR SYSTEM—they provide the greatest resistance to air flow in the conducting passages as they contain thick smooth muscle for contraction and dilation

I **BRONCHIOLES**

J **TERMINAL BRONCHIOLES**

★ (The CONDUCTION ZONE ends with the TERMINAL BRONCHIOLES.)

★ Similar to the lower cavity of the LARYNX, the inner cavity (lumen) of the TRACHEA and all BRONCHI is lined with a mucosa of pseudostratified ciliated columnar epithelium. It contains many GOBLET CELLS, which secrete mucus. Dust particles cling to the mucus, are swept up by the cilia, and are expelled from the body by a cough reflex.

★ The posterior portion of the TRACHEA consists of connective tissue (imbedded with smooth trachealis muscle) in contact with the esophagus. This soft area of tissue allows the esophagus to expand as swallowed food passes toward the stomach.

BRONCHIAL BRANCHES inserting into the TRILOBED RIGHT LUNG

- 1 PRIMARY BRONCHUS
- 3 SECONDARY (LOBAR) BRONCHI
- 10 TERTIARY (SEGMENTAL) BRONCHI

BRONCHIAL BRANCHES inserting into the BILOBED LEFT LUNG

- 1 PRIMARY BRONCHUS
- 2 SECONDARY (LOBAR) BRONCHI
- 8 TERTIARY (SEGMENTAL) BRONCHI

Larynx : D

Thyroid Gland

(16–20 rings): Hyaline Cartilage

Fibrous Connective Tissue

Foreign objects frequently lodge here

Carina Tracheae (Point of Bifurcation)

E

Lobes of Right Lung

Lobes of Left Lung

RIGHT LUNG

Right Lateral Views

LEFT LUNG

Left Lateral Views

RIGHT LUNG

Right Medial Views

★ See next page for subdivision of TERMINAL BRONCHIOLES into RESPIRATORY BRONCHIOLES

LEFT LUNG

Left Medial Views

RESPIRATORY SYSTEM: LOWER RESPIRATORY TRACT
Respiration Zone: The Lungs: The Alveolus: The Functional Respiratory Unit of Gas Exchange

Color the Pulmonary Venule Red, because it returns oxygenated blood to the heart from the Alveolus.
Color the Pulmonary Arteriole Blue for deoxygenated blood.
Color Lymphatic Vessel lime green.

J, K, L, M, N are reference numbers from 11.1.

★TERMINAL BRONCHIOLES subdivide into microscopic branches called RESPIRATORY BRONCHIOLES. (The simple cuboidal epithelium found in the BRONCHIOLES and TERMINAL BRONCHIOLES changes gradually in the distal portions of the RESPIRATORY BRONCHIOLES to simple squamous epithelium.)

★RESPIRATORY BRONCHIOLES branch into several ALVEOLAR DUCTS (or ATRIA), usually from 2–11 DUCTS per RESPIRATORY BRONCHIOLE. Around each ALVEOLAR DUCT are numerous polyhedral-shaped outpocketings called ALVEOLI. (A TERMINAL BRONCHIOLE is considered a RESPIRATORY BRONCHIOLE when it contains outpocketings of ALVEOLI.)

★ALVEOLI are individually located along the respiratory bronchioles and alveolar ducts.

ALVEOLI are usually clustered together in a honeycomblike pattern in groups called ALVEOLAR SACS, which share a common chamber. ALVEOLI are the FUNCTIONAL UNITS OF THE LUNGS WHERE GAS EXCHANGE OCCURS.

ALVEOLI (AIR CELLS) are composed of simple squamous epithelium for the most efficient gas exchanges and their walls are extremely tiny (0.25–0.50 mm in diameter).

There are over *350 million* alveoli per lung (providing a staggering surface area of over 70 square meters for gas exchange!).

It is not possible to have normal respiratory exchange of gases in the lungs unless the pulmonary tissue is adequately perfused with blood. Each ALVEOLUS in an ALVEOLAR SAC is surrounded by a rich supply of blood from a vast capillary network.

3 basic structures that comprise the RESPIRATION ZONE.
- ◼ ALVEOLAR DUCTS
- ◼ ALVEOLAR SACS
- ◼ ALVEOLI

End of the CONDUCTION ZONE

J | TERMINAL BRONCHIOLE

The RESPIRATION ZONE

★ Each bronchopulmonary segment of a lung is broken up into numerous small compartments called LOBULES. Each LOBULE is wrapped in an ELASTIC CONNECTIVE TISSUE COVERING and contains the following:

CONTENTS of a LOBULE:

K | RESPIRATORY BRONCHIOLE

L | ALVEOLAR DUCTS | ATRIA

M | ALVEOLI | (SINGLE = ALVEOLUS)

N | ALVEOLAR SACS

1 | LYMPHATIC VESSEL

2 | PULMONARY ARTERIOLE

3 | PULMONARY VENULE

4 | PULMONARY NERVE

★ The ALVEOLUS: CELLS of the ALVEOLAR (EPITHELIAL) WALL

- ◼ SQUAMOUS PULMONARY EPITHELIAL CELLS (S.P.E. CELLS)
  - • TYPE I alveolar cells
  - Small alveolar cells
  - • Form a continuous lining of the alveolar wall
- ◼ SEPTAL CELLS
  - • TYPE II alveolar cells
  - • great alveolar cells
  - • Small, cuboidal cells interspersed occasionally among the S.P.E. cells
  - • Produce a phospholipid substance SURFACTANT that coats the inner lining of the Individual alveolus and lowers surface tension, preventing its collapse.
- ◼ ALVEOLAR MACROPHAGES
  - • Phagocytic, free DUST CELLS that remove foreign particulate matter (especially dust)

6 | S.P.E. CELLS [TYPE I]

7 | SEPTAL CELLS [TYPE II]

7 | SURFACTANT LAYER

8 | ALVEOLAR MACROPHAGE

The ALVEOLAR-CAPILLARY (RESPIRATORY) MEMBRANE (#6–11)

- • The thin membrane through which respiratory gases move (by the process of DIFFUSION) between the LUNGS and the BLOOD
- • Its extremely small thickness (0.2–0.5 mm) notwithstanding its many layers, allows for MAXIMAL EFFICIENCY of GAS DIFFUSION
- • In addition to the cells of the ALVEOLAR WALL (above), this membrane consists of:

9 | ALVEOLAR BASEMENT MEMBRANE

10 | CAPILLARY BASEMENT MEMBRANE

11 | CAPILLARY ENDOTHELIUM

The ALVEOLAR BASEMENT MEMBRANE is ELASTIC and permits distension of lung tissue in INSPIRATION, passive recoil in EXPIRATION. It is most often fused to the CAPILLARY BASEMENT MEMBRANE.

12 | RED BLOOD CELL

13 | DIFFUSION OF O_2

14 | DIFFUSION OF CO_2

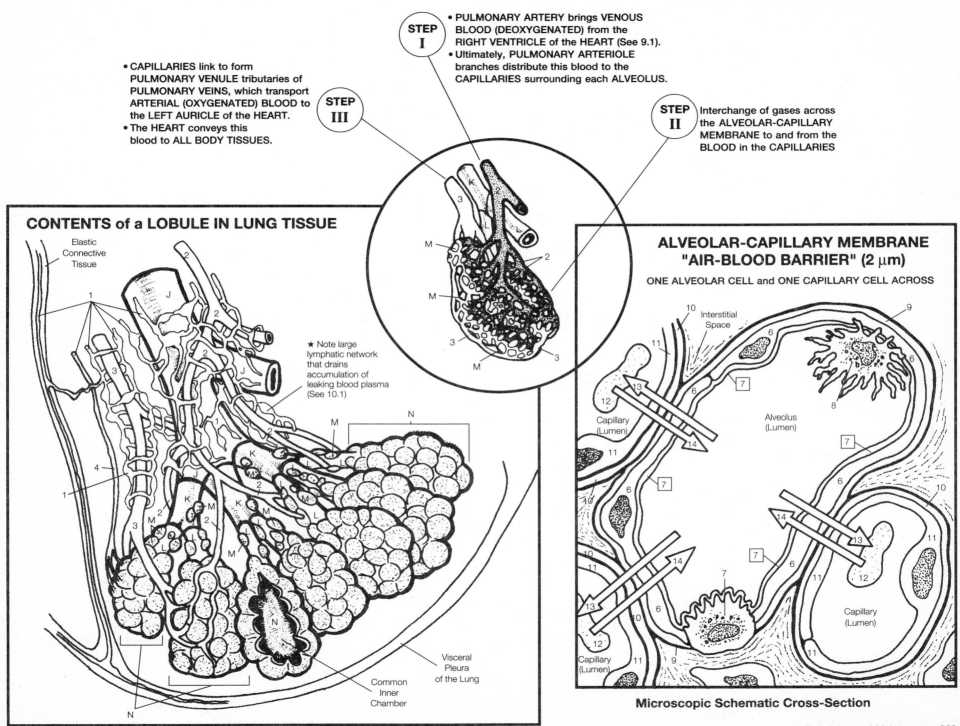

• PULMONARY ARTERY brings VENOUS BLOOD (DEOXYGENATED) from the RIGHT VENTRICLE of the HEART (See 9.1).
• Ultimately, PULMONARY ARTERIOLE branches distribute this blood to the CAPILLARIES surrounding each ALVEOLUS.

STEP I

STEP III

• CAPILLARIES link to form PULMONARY VENULE tributaries of PULMONARY VEINS, which transport ARTERIAL (OXYGENATED) BLOOD to the LEFT AURICLE of the HEART.
• The HEART conveys this blood to ALL BODY TISSUES.

STEP II Interchange of gases across the ALVEOLAR-CAPILLARY MEMBRANE to and from the BLOOD in the CAPILLARIES

CONTENTS of a LOBULE IN LUNG TISSUE

Elastic Connective Tissue

★ Note large lymphatic network that drains accumulation of leaking blood plasma (See 10.1)

Visceral Pleura of the Lung

Common Inner Chamber

ALVEOLAR-CAPILLARY MEMBRANE "AIR-BLOOD BARRIER" (2 μm)

ONE ALVEOLAR CELL and ONE CAPILLARY CELL ACROSS

Interstitial Space

Capillary (Lumen)

Alveolus (Lumen)

Capillary (Lumen)

Capillary (Lumen)

Microscopic Schematic Cross-Section

RESPIRATORY SYSTEM: LOWER RESPIRATORY TRACT
Respiration Zone: The Lungs: Lobes, Pleurae, and Surfaces

Pulmonary Artery = blue Pulmonary Vein = red Heart, Aorta, and Arteries = warms (reds, oranges) Veins, Vena Cavae = Cools (blues, greens) Color the Lung Lobes the same colors used on the Secondary (Lobar) Bronchi and Lung Lobes from 11.5

★ See 1.4, 11.5

★ The RIGHT LUNG and LEFT LUNG are separately contained in PLEURAL MEMBRANES

3 FUNCTIONS of the PLEURAE:

■ Lubricant secreted by the PLEURAE in pleural cavities allows lungs to slide along chest wall

■ Pressure in pleural cavities is less than pressure in lungs, which is required for ventilation

■ Provides effective, tight separation of major thoracic organs

★ VISCERAL PLEURA
Covers the LUNGS and cannot be separated from them. (Also invaginates the interlobular fissures)

★ PARIETAL PLEURA
Lines and adheres to the rib cage, the diaphragm, and the pericardium of the heart
It is an extension of the VISCERAL PLEURA (at the root of the LUNG) as it bends around and out.

★ PLEURAL CAVITY
The potential space between the two pleural layers. Filled with a lubricating fluid secreted by the cells of the PLEURA.

★ Each lung has 4 BORDERS or SURFACES that match the contour of the thoracic cavity:

1. APEX (CUPOLA):

Anterior border that extends above the level of the clavicle bone

2. BASE:

Inferior border whose concave form results from the convex dome of the diaphragm

3. COSTAL SURFACE: The broad, rounded surface of the lungs in contact with the membranes covering the ribs

4. MEDIASTINAL SURFACE:

The medial surface of the lungs as they hug the lateral aspects of the MEDIASTINUM (See 1.4)

★ The MEDIASTINAL SURFACE (or MEDIAL SURFACE) of each LUNG contains a vertical slit called a ROOT (HILUS), through which BRONCHI, PULMONARY VESSELS, and NERVES enter and exit.

LOBES of the LUNGS

TRILOBED RIGHT LUNG

1 | RIGHT SUPERIOR LOBE

2 | RIGHT MIDDLE LOBE

3 | RIGHT INFERIOR LOBE

BILOBED LEFT LUNG

4 | LEFT SUPERIOR LOBE

5 | LEFT INFERIOR LOBE

FISSURES of the LUNGS

6 | HORIZONTAL FISSURE

7 | OBLIQUE FISSURE

PLEURAE of the LUNGS:
SEROUS MEMBRANES THAT ENCLOSE AND PROTECT THE LUNGS

8 | VISCERAL PLEURA
(PULMONARY PLEURA)

9 | PARIETAL PLEURA

10 | PLEURAL CAVITY

Left and Right MEDIASTINAL SURFACES of the LUNGS

★ The LUNGS are very spongy and pliable. Permanent impressions from surrounding organs in the thoracic cavity help mold their form, especially along the mediastinal surface. Studying these impressions helps in understanding the relationships of the organs in the thoracic cavity. (See MEDIASTINUM, 1.4)

(A) = ANTERIOR MEDIASTINUM

11 | ROOT [HILUM]

12 | PRIMARY BRONCHUS

13 | PULMONARY ARTERY

14 | PULMONARY VEIN

15 | PULMONARY LIGAMENT
(The bending of the VISCERAL PLEURAE to become the PARIETAL PLEURAE)

LEFT LUNG (Medial View)

RIGHT LUNG (Medial View)

POSTERIOR ◄──────► ANTERIOR ◄──────► POSTERIOR

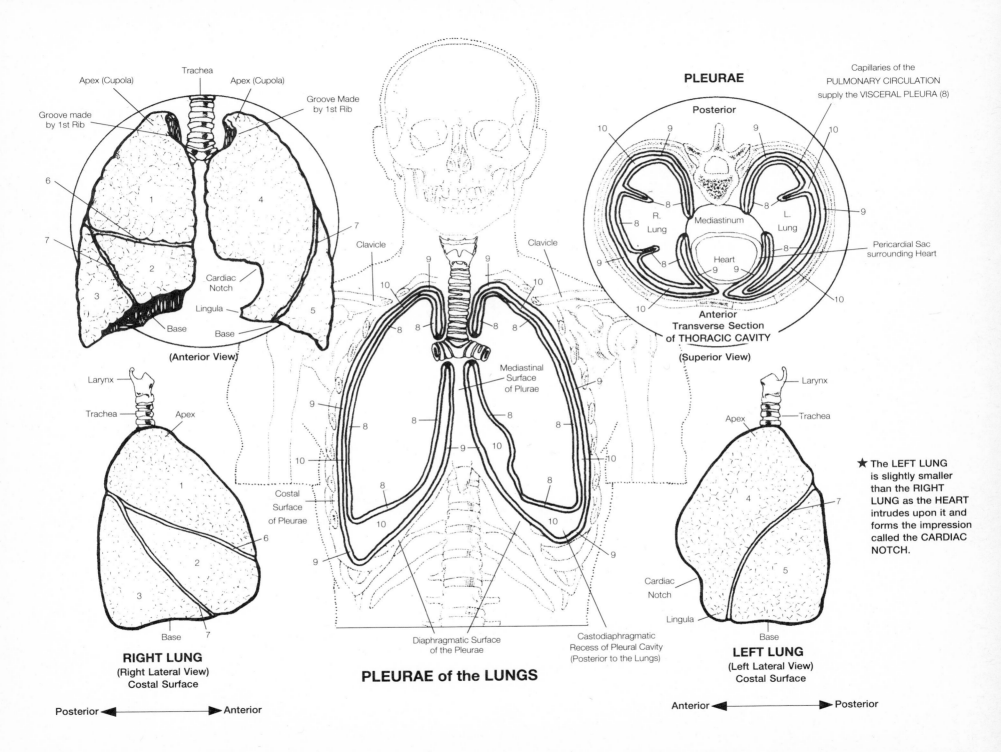

Apex (Cupola)   Trachea   Apex (Cupola)
                                    Groove Made
Groove made   by 1st Rib
by 1st Rib

6
        1   4
7
                                    7
        2
    3   Cardiac
                            Notch
                            Lingula
            Base   Base   5

(Anterior View)

PLEURAE
                            Capillaries of the
                            PULMONARY CIRCULATION
Posterior   supply the VISCERAL PLEURA (8)

10   9   9   10

    8   8
    8   R.   L.   9
        Lung Mediastinum Lung
    8   8   9
9
                Heart   Pericardial Sac
    8   9   9   surrounding Heart
10
                                            10
            Anterior
        Transverse Section
        of THORACIC CAVITY
                                    10
            (Superior View)

Larynx
Trachea   Apex

        1

                6

        2

    3

        Base   7

RIGHT LUNG
(Right Lateral View)
Costal Surface

Posterior ◄────► Anterior

Clavicle   Clavicle

            9   10
10   8
    8   8   8   8
                                    Mediastinal
                                    Surface
                                    of Plurae
        9   9

            9
9
        8   8   8
                        9

    10   10

Costal   8   8
Surface
of Pleurae   9   10

        8   8

    10   10

9   9

Diaphragmatic Surface   Castodiaphragmatic
of the Pleurae   Recess of Pleural Cavity
                            (Posterior to the Lungs)

PLEURAE of the LUNGS

Larynx
Apex   Trachea

        4   7

    Cardiac
    Notch   5

    Lingula

        Base

LEFT LUNG
(Left Lateral View)
Costal Surface

Anterior ◄────► Posterior

★ The LEFT LUNG
is slightly smaller
than the RIGHT
LUNG as the HEART
intrudes upon it and
forms the impression
called the CARDIAC
NOTCH.

Chapter 11 : *RESPIRATORY SYSTEM* : CHARTS

Chart #1 : Major Structures of the Respiratory System (See 11.1)

Structure	Description	Function	Epithelial Tissue Type
NOSE	Jutting exterior portion that is part of the face	Warms, moistens, and filters inhaled air as it is conducted to the pharynx through the NASAL cavity	Stratified squamous epithelium
PARANASAL SINUSES	Air spaces in certain facial bones	Produce mucus; provide sound resonance; lighten the skull	Stratified squamous epithelium
PHARYNX	Chamber connecting oral and nasal cavities to larynx	Passageway for air into larynx and for food into esophagus	Pseudostratified ciliated* columnar epithelium
LARYNX	Voice box; short passageway that connects pharynx to trachea	Passageway for air; sound production; prevents foreign materials from entering trachea	Pseudostratified ciliated columnar epithelium
TRACHEA	Flexible, tubular connection between larynx and bronchial tree	Passageway for air; air filter	Pseudostratified ciliated columnar epithelium
BRONCHIAL TREE	Bronchi and branching bronchioles in the lung; tubular connection between trachea and alveoli	Passageway for air; continued air filtration	Pseudostratified ciliated† columnar epithelium
ALVEOLI	Microscopic, membranous air sacs within the lungs	Functional units of respiration; site of gaseous exchange between the respiratory and circulatory systems	Simple squamous epithelium
LUNGS	Major organs of respiratory system; located in pleural cavities of the thorax	Contain bronchial trees, alveoli, and associated pulmonary vessels	Simple squamous epithelium
PLEURAE	Serous membranes covering the lungs and lining the thoracic cavity forming a closed sac	Compartmentalizes, protects, and lubricates the lungs	Simple squamous epithelium

* NASOPHARYNX = Pseudostratified ciliated columnar
OROPHARYNX = Stratified columnar
LARYNGOPHARYNX = Stratified squamous

† Bronchioles = Simple cuboidal

Adapted from Kent M. Van De Graaff, *Human Anatomy*, 2d ed., p. 561.

Chart #2 : Paranasal Sinuses (See 11.3)

FACIAL BONE HOUSING SINUS	4 PAIRED PARANASAL SINUSES (one on either side of midline axis)	SPACES INTO WHICH THEY DRAIN (Each sinus drains on its own side through the lateral walls of the nasal cavity)
FRONTAL BONE	FRONTAL SINUS	MIDDLE MEATUS
SPHENOID BONE	SPHENOIDAL SINUS	Opens into NASAL CAVITY above the SUPERIOR CONCHA, behind the SPHENOETHMOIDAL RECESS
ETHMOID BONE	ETHMOIDAL SINUS (Anterior, Middle, and Posterior Ethmoidal "Cells")	SUPERIOR MEATUS and MIDDLE MEATUS
MAXILLARY BONE	MAXILLARY SINUS	MIDDLE MEATUS

	4 PAIRED PARANASAL SINUSES	SPACES INTO WHICH THEY DRAIN
	NASOLACRIMAL DUCT (Part of the Lacrimal Apparatus, See 7.19)	INFERIOR MEATUS

UNIT 4: REGULATION AND MAINTENANCE

Chapter 12 Digestive System

INGESTION of organic food substances is a vital requirement for life of the human organism and other animals (unlike plants, which can form them from inorganic compounds within their cells).

■ The 3 basic organic food substances are CARBOHYDRATES, FATS (LIPIDS), and PROTEINS, which come in a wide variety of edible plant and animal forms. They provide the source for VITAL CELL ACTIVITIES in the cells of the body tissues:

▸ *LIFE ENERGY (Calories)*
Cell Respiration releases this energy for the production of ATP and Heat. [All 3 basic organic food substances are involved.]

▸ *GROWTH and REPAIR OF BODY TISSUES*
[especially PROTEIN, which contains Nitrogen]

Initial ingestion of these organic food substances is in the form of aggregations of large molecules called POLYMERS—forms not suitable for ready use by the tissue cells.
■ *THE PROCESSES OF DIGESTION* (both MECHANICAL and CHEMICAL) break down these long chains of POLYMERS into their subunit MONOMERS—forms readily, absorbable across the intestinal wall and accessible for use by the tissue cells. These free MONOMERS are the END PRODUCTS OF DIGESTION.

Simple foods that can be absorbed unchanged are: SALT, the SIMPLEST SUGARS (such as GLUCOSE), CRYSTALLOIDS (in general), and WATER. STARCHES, FATS, and PROTEINS are not absorbable until split into these smaller MONOMER molecules.

The digestion or breaking down of these food molecules occurs as a series of catalytic HYDROLYSIS REACTIONS (water and a wide variety of digestive enzymes). Each enzyme acts in an ACID or ALKALINE or NEUTRAL JUICE according to its peculiar properties.

■ The location of the fully active digestive enzymes is limited primarily to the cavity (lumen) of the GASTRO-INTESTINAL TRACT (GI TRACT) (the STOMACH and INTESTINES).
The process of ABSORPTION transports the formed monomers across the wall of the INTESTINES to the CARDIOVASCULAR and LYMPH VASCULAR SYSTEMS for distribution to the tissue cells.

The monomers of CARBOHYDRATES (MONOSACCHARIDES) and the monomers of PROTEINS (AMINO ACIDS) are absorbed into the CARDIOVASCULAR SYSTEM, as are small aggregations of FAT monomers (FATTY ACIDS and GLYCEROL). Larger aggregations of FAT monomers are absorbed into the LYMPH VASCULAR SYSTEM (See Chapter 10).

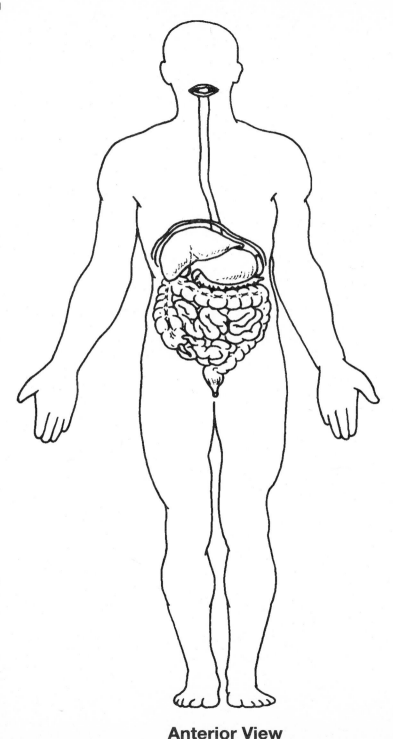

Anterior View

System Components

A TUBULAR PASSAGEWAY AND ASSOCIATED ORGANS (ex: SALIVARY GLANDS, LIVER, GALLBLADDER, AND PANCREAS)

System Function

• PHYSICAL AND CHEMICAL BREAKDOWN OF FOOD FOR CELL USAGE

• ELIMINATION OF SOLID WASTES

Chapter Coloring Guidelines

You may want to choose your color palette by comparing any colored illustrations from your main textbook.

General:
Liver = reddish brown
Esophagus and Gastrointestinal Tract = light oranges, flesh
Gallbladder = yellow-green
Pancreas = yellow-orange

DIGESTIVE SYSTEM: GENERAL ORGANIZATION
General Scheme of the Alimentary Canal:
The Principal Organs of Digestion

B = light brown   C = flesh
D = light orange   E = light yellow
F = light red
★ See Chart #1, Chart #7, 12.2

★ The DIGESTIVE SYSTEM can be anatomically and functionally divided into 2 main divisions:

■ The ALIMENTARY CANAL
■ ACCESSORY ORGANS (See 12.2)

★ The ALIMENTARY CANAL or TRACT is a continuous digestive tube transversing the VENTRAL BODY CAVITY (See 1.4) and extending from the: MOUTH to the ANUS, including the ORAL (BUCCAL) CAVITY, PHARYNX, ESOPHAGUS, STOMACH, SMALL and LARGE INTESTINES, and the RECTUM (considered part of the LARGE INTESTINE).

★ The GASTROINTESTINAL TRACT (or GI TRACT) is a subdivision of the ALIMENTARY CANAL and consists of the STOMACH and INTESTINES (SMALL and LARGE). It is in the GI TRACT where most of the fully active digestive enzymes are located, produced by the "brush-border" enzymes of the intestinal mucosal lining, or entering into the tract within juice secreted by ACCESSORY ORGANS (See 12.2)

The ALIMENTARY CANAL is a *1-way transport system* open at both ends. It is therefore continuous with the outside world, and is not in truth part of the internal environment. Thus, indigestible material that passes from one end (mouth) to the other (anus) without being absorbed across the lining of the tract NEVER ENTERS THE BODY, AND IS CONSIDERED TO OCCUR OUTSIDE THE BODY.

The LENGTH of the ALIMENTARY CANAL:
  •In cadaver = 30 feet (9 meters)
  •In living person = slightly shorter due to muscle tone in tract walls

The ALIMENTARY CANAL: MOUTH to ANUS

A | ORAL (BUCCAL) CAVITY | MAJOR PORTION OF THE MOUTH

B | PHARYNX

C | ESOPHAGUS

GASTROINTESTINAL TRACT (GI TRACT)

D | STOMACH

E | SMALL INTESTINE

F | LARGE INTESTINE

★ The ALIMENTARY CANAL contains the FOOD from the time it is INGESTED until it is DIGESTED and prepared for DEFECATION. As the FOOD progresses along the canal, it is subject to CHEMICAL and MECHANICAL changes that break it down into suitable units for ABSORPTION and ultimate ASSIMILATION (utilization of these simple units by all the cells of the body).

■ CHEMICAL DIGESTION
  •Enzymatic secretions produced by cells along the tract and combined with enzymatic secretions from cells in the accessory organs (See 12.2) break down the large food molecules (PROTEINS, CARBOHY-DRATES, and LIPIDS) into usable units (AMINO ACIDS, MONOSACCHARIDES, GLYCEROL and FATTY ACIDS, respectively).

  •These units are small enough to pass through the walls of the digestive organs (mainly the small intestine) into the blood capillaries (and in the case of larger fat molecules into lymph capillaries), then travel through venous blood entering and exiting the heart, which distributes them to all the body's tissues and then through the plasma membranes of the body's cells.

■ MECHANICAL DIGESTION
  •Various movements that aid chemical digestion

  •Teeth physically break down the food (MASTICATION) before it can be swallowed.

  •Muscular contractions of the smooth muscles in the walls of the STOMACH and SMALL INTESTINE physically break down the food by churning it; the food is thus thoroughly mixed with the enzymes involved in the catabolic reactions of chemical digestion.

■ Different regions of the alimentary canal are specialized for different functions of the digestive process.
■ There are 7 basic functions:
- INGESTION
- MASTICATION
- DEGLUTITION
- PERISTALSIS
- DIGESTION
- ABSORPTION
- DEFECATION

■ See Chart #1
★ Review the relationships of the digestive organs to the nine regions of the abdominopelvic cavity (See 1.3).

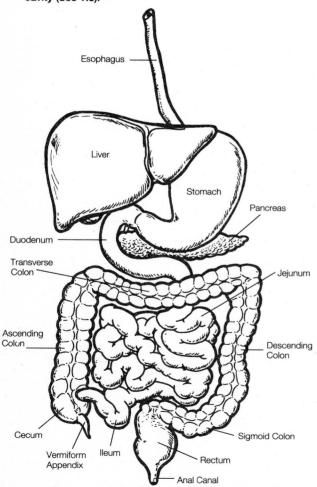

ESOPHAGUS, GASTROINTESTINAL TRACT, and ACCESSORY ORGANS of the ABDOMINAL REGION

12.1

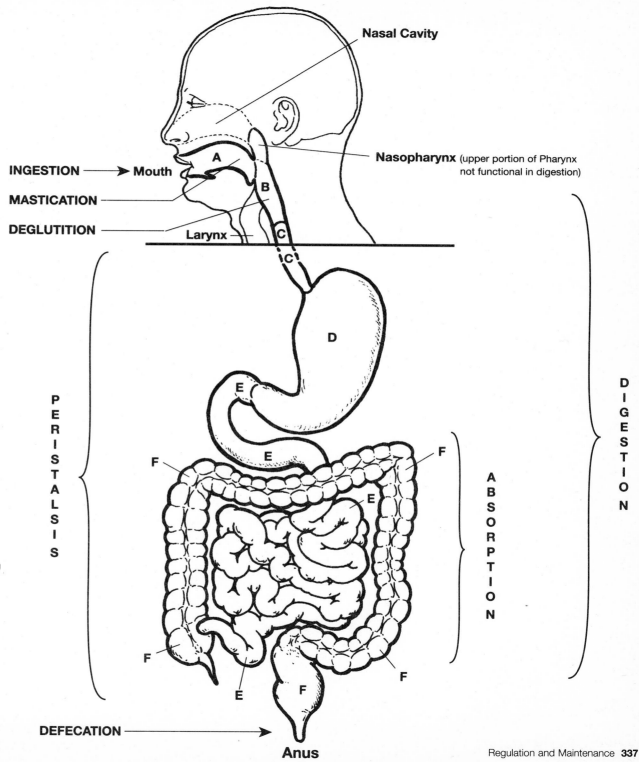

DIGESTIVE SYSTEM: GENERAL ORGANIZATION
The Accessory Structures of Digestion

Salivary Glands = neutral colors
L = reddish brown
M = yellow-green
N = yellow-orange
Duct System = cool colors
★ See 12.1

★The secondary group of digestive organs comprising the other anatomical and functional division of the digestive system are the ACCESSORY ORGANS:
• TEETH, TONGUE, SALIVARY GLANDS, LIVER, GALLBLADDER, PANCREAS (and GASTROINTESTINAL GLANDS)

★ACCESSORY ORGANS INSIDE ALIMENTARY CANAL
■ TEETH protrude into the alimentary canal (oral cavity) and aid *in physical breakdown of food.* (See 12.5, 12.6)

■ TONGUE comprises the floor of the oral cavity. A freely movable muscular organ whose digestive function is manipulation of the FOOD BOLUS in MASTICATION and DEGLUTITION [also involved in speech production and taste (See 6.5, 7.18, 11.1, 12.7)]

■ Numerous BUCCAL GLANDS are located in the mucous membranes lining the oral cavity (in the palatal region).

■ GASTRIC and INTESTINAL GLANDS located mainly in submucosal layers of GI tract with openings into the gastrointestinal lumen.

★ACCESSORY ORGANS OUTSIDE ALIMENTARY CANAL
■ SALIVARY GLANDS (See 12.7)
■ LIVER (See 12.15, 12.16, Chart 9)
■ GALLBLADDER ⎫
■ PANCREAS   ⎬ (See 12.14, Chart 8)
• These accessory organs produce or store secretions that aid in the chemical breakdown of food.

• These secretions are released into the alimentary tract through DUCTS.

ACCESSORY STRUCTURES of the ORAL CAVITY

G | TEETH |

H | TONGUE |

Major SALIVARY GLANDS

PAIRED GLANDS WITH DUCTS

I | PAROTID GLANDS (2) |

J | SUBMANDIBULAR GLANDS (2) | SUB-MAXILLARY

K | SUBLINGUAL GLANDS (2) |

ACCESSORY ORGANS of the ABDOMINAL REGION

L | LIVER |

M | GALLBLADDER |

N | PANCREAS |

GASTRIC AND INTESTINAL GLANDS (not shown)

(See 12.1, bottom page)

DUCT SYSTEM of the ABDOMINAL ACCESSORY ORGANS

Ducts of the abdominal accessory organs carry their secretions into the DUODENUM via the common AMPULLA of VATER, where the common BILE DUCT and MAIN PANCREATIC DUCT meet.

1 | R. AND L. HEPATIC DUCTS |

2 | COMMON HEPATIC DUCT |

3 | CYSTIC DUCT |

4 | COMMON BILE DUCT |

5 | MAIN PANCREATIC DUCT | DUCT OF WIRSUNG

6 | ACCESSORY PANCREATIC DUCT | DUCT OF SAN-TORINI

7 | AMPULLA OF VATER |

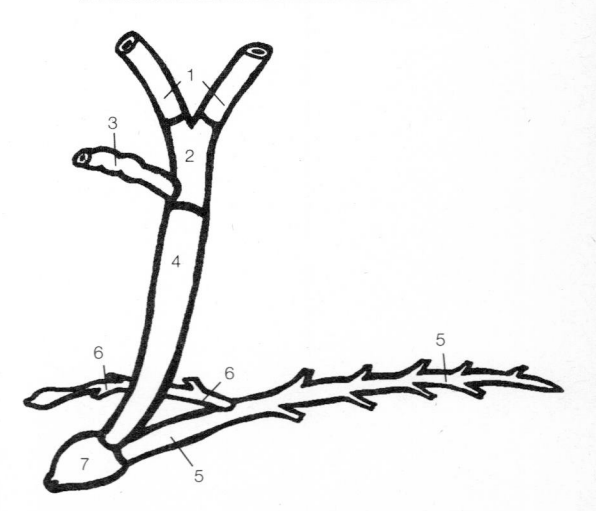

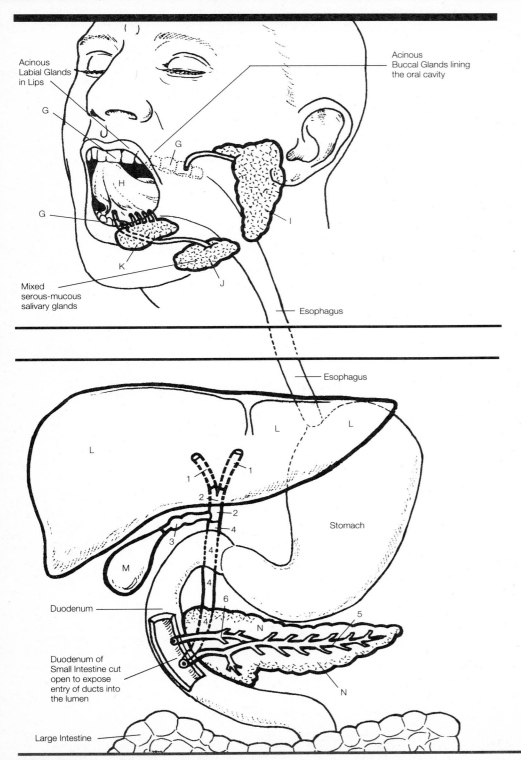

Acinous Labial Glands in Lips

Acinous Buccal Glands lining the oral cavity

G

G

H

G

K

I

J

Mixed serous-mucous salivary glands

Esophagus

Esophagus

L

L

L

1

1

2

2

4

3

4

4

4

M

6

Stomach

Duodenum

Duodenum of Small Intestine cut open to expose entry of ducts into the lumen

N

N

5

Large Intestine

GENERAL SCHEME OF DIGESTION AND ABSORPTION OF CARBOHYDRATERS, PROTEINS, AND FATS (LIPIDS) (See 12.12, 12.14)

The large complex food molecules of CARBOHYDRATES, PROTEINS, and LIPIDS are broken down into usable units by combinations of enzymatic secretions of the GI tract and the accessory organs.
In the STOMACH, the ingested FOOD BOLUS is converted into a pasty CHYME. When CHYME reaches the SMALL INTESTINE, CARBOHYDRATES and PROTEINS have only been PARTLY DIGESTED, while LIPID digestion has not actually begun to any extent. In the SMALL INTESTINE, the INTESTINAL JUICE (SUCCUS ENTERICUS) contains enzymes which complete the digestion of CARBOHYDRATES and PROTEINS; LIPIDS are digested in the SMALL INTESTINE by PANCREATIC LIPASE enzyme from PANCREATIC JUICE.

★ CARBOHYDRATES
■ CARBOHYDRATE DIGESTION begins in the MOUTH with the action of SALIVARY AMYLASE (from Salivary Glands), converting cooked STARCH (a POLYSACCHARIDE) first into DEXTRIN and then into MALTOSE (a DISACCHARIDE). [The action of SALIVARY AMYLASE is stopped in the STOMACH by the low pH of GASTRIC JUICE]. Ingested SUCROSE (Table sugar) and LACTOSE (Milk sugar) - 2 DISACCHARIDES - are not acted upon until they reach the SMALL INTESTINE.

■ Most CARBOHYDRATE DIGESTION occurs in the DUODENUM of the SMALL INTESTINE by the action of PANCREATIC AMYLASE, from PANCREATIC JUICE, which cleaves the STRAIGHT, LARGE POLYSACCHARIDE CHAINS into simpler Glucose Chains of DISACCHARIDES (MALTOSE), TRISACCHARIDES (MAL-TRIOSE), and short–chained OLIGOSACCHARIDES. These 3 simple Glucose Chains, together with SUCROSE and LACTOSE, are then hydrolyzed to MONOSACCHARIDES in the SMALL INTESTINE by "BRUSH–BORDER" ENZYMES located in the MICROVILLI of the lining epithelium.
■ The MONOSACCHARIDES are then absorbed across the microvilli and secreted unaltered from the epithelial cells into BLOOD CAPILLARIES within the villi, and then into the HEPATIC PORTAL VEIN enroute to the LIVER.

★ PROTEINS
■ PROTEIN DIGESTION begins in the STOMACH with the action of acidic PEPSIN secreted in the GASTRIC JUICE from the STOMACH WALL. PEPSIN breaks down ENTWINED LARGE POLYPEPTIDE CHAINS into shorter-chained PEPTONES or PROTEOSES. [The action of PEPSIN is stopped in the DUODENUM by alkaline PANCREATIC JUICE].
■ Most PROTEIN DIGESTION occurs in the DUODENUM and JEJUNUM of the SMALL INTESTINE by PANCREATIC JUICE enzymes, TRYPSIN, CHYMOTRYPSIN and ELASTASE, resulting in simpler DIPEPTIDES, TRIPEPTIDES (and some FREE AMINO ACIDS). These all are absorbed into the DUODENAL EPITHELIAL CELLS, and within these cells, the PEPTIDES eventually hydrolyze into more FREE AMINO ACIDS.
■ The FREE AMINO ACIDS are then absorbed and secreted unaltered from the epithelial cells into the BLOOD CAPILLARIES that carry them to the HEPATIC PORTAL VEIN enroute to the LIVER.

★ LIPIDS
■ FAT DIGESTION begins in the STOMACH with the action of LINGUAL LIPASE (from Lingual Salivary Gland), at a very limited level.

■ Most of the ingested FAT enters the DUODENUM *undigested* as FAT GLOBULES. These Globules are *not* digested (the bonds joining the subunits are *not* hydrolyzed), but *emulsified* into finer droplets by BILE SALTS from the LIVER released into the DUODENUM -- which increases the surface area for subsequent digestion.
■ Almost all FAT DIGESTION occurs in the SMALL INTESTINE by the action of PANCREATIC LIPASE (STEAPSIN) -- the only active fat-digesting enzyme in the adult body. Complex TRIGLYCERIDES are hydrolized into FREE FATTY ACIDS and MONOGLYCERIDES. [PHOSPHOLIPASE A breaks up PHOSPHOLIPIDS into FREE FATTY ACIDS and LYSOLECITHIN]. These simpler components then dissolve into MICELLE PARTICLES (AGGREGATIONS) of BILE SALTS, CHOLESTEROL and LECITHIN secreted by the LIVER, producing "MIXED MICELLES" which move to the "BRUSH BORDER" of the SMALL INTESTINAL EPITHELIUM. (See 12.12)
■ Unlike CARBOHYDRATES and PROTEIN MONOMERS which pass through intestinal epithelial cells unaltered, these LIPID products, after having been absorbed through the membrane of the microvilli, *RESYNTHESIZE* into TRIGLYCERIDES and PHOSPHOLIPIDS within the epithelial cells ! They (and Cholesterol) are then combined with PROTEIN inside the epithelial cells to form tiny CHYLOMICRONS.
■ These CHYLOMICRONS are too large to pass into the HEPATIC PORTAL BLOOD, and bypass the Liver by secreting into the LYMPHATIC CAPILLARIES of the intestinal villi. They later enter Venous Blood by way of the LEFT THORACIC DUCT of the LYMPHATIC SYSTEM (See 10.2), to the HEART, then to the LIVER via the HEPATIC ARTERY. Once in the blood, TRIGLYCERIDES are removed from the CHYLOMICRONS and hydrolyzed by LIPOPROTEIN LIPASE (found in Blood Plasma), producing FREE FATTY ACIDS and GLYCEROL for use by the tissue cells.

DIGESTIVE SYSTEM: GENERAL ORGANIZATION
General Histology of Canal Walls

★ The ALIMENTARY CANAL is the entire digestive tube from the MOUTH to the ANUS.

The MOUTH (ORAL CAVITY) and PHARYNX are 3-layered and do not contain the 4 tunic layers specified for digestion. They are lined with nonkeratinized, stratified squamous epithelium (mucosa) and do not contain a SEROSA LAYER. (See 12.8) Therefore, the 4 tunic layers are only continuous from the beginning of the ESOPHAGUS down to the end of the ANAL CANAL. Each of these 4 layers, called a TUNIC, performs its own specific function in the digestive process, characterized by a dominant, specialized type of tissue.

■ MUCOSA
Inner layer of columnar epithelial cells in close contact with contents of the inner lumen (cavity). Specialized functions include:
  •SECRETION
  •ABSORPTION (of nutrients and hormones to the bloodstream)
  •MOBILITY– Folds of the mucosa form projections into the lumen that can continually change shape and surface area. There are different types of folds in different parts of the system: in the STOMACH, the mucosa folds are temporary and are called RUGAE. In the SMALL INTESTINE, there are 2 types of permanent folds: PLICAE CIRCULARES (also includes submucosa) and mucosal projections called VILLI.

■ SUBMUCOSA
  •Thick, vascular layer of connective tissue serves the MUCOSA
  •STRONG BLOOD SUPPLY–Huge supply of blood vessels remove absorbed materials and supply mucosa
  •COORDINATION of motor and secretory activities of mucosa through MEISSNER'S NERVE PLEXUS

■ MUSCULARIS
  •MOVEMENT
      •Controls diameter of tube
      •Mixes contents
      •Peristaltic action moves contents along the tube
      •COORDINATION of muscular and secretory activities through AUERBACH'S NERVE PLEXUS

■ SEROSA
  •COVERS many digestive and accessory organs in the abdominal cavity
  •REDUCES FRICTION between contacting organs with its smooth, moist texture
  •Carries blood and lymph vessels, and nerves to and from MESENTERY
  •Consists of a layer of simple squamous epithelium (the visceral peritoneum), reinforced with underlying areolar connective tissue.

General Histology
4 Layer (TUNICS) of the ESOPHAGUS and GI TRACT

a | MUCOSA | TUNICA MUCOSA

b | SUBMUCOSA | TUNICA SUBMUCOSA

c | MUSCULARIS | TUNICA MUSCULARIS

d | SEROSA/VISCERAL PERITONEUM | TUNICA SEROSA

Sublayers and Contents of the MUCOSA, SUBMUCOSA, and MUSCULARIS

MUCOSA

1 | LINING COLUMNAR EPITHELIUM | (with glands and Goblet cells)

2 | LAMINA PROPRIA | (loose fibrous connective tissue with capillaries and lymphatic vessels)

3 | MUSCULARIS MUCOSA | (thin layers of smooth muscle)

SUBMUCOSA

4 | SUBMUCOSAL GLANDS | (in esophagus and first part of duodenum)

5 | MEISSNER'S PLEXUS | (SUBMUCOUS NERVE PLEXUS regulates movements of mucosal villi through the muscularis mucosa)

MUSCULARIS

6 | INNER CIRCULAR MUSCULARIS

7 | OUTER LONGITUDINAL MUSCULARIS

8 | AUERBACH'S PLEXUS | (MYENTERIC NERVE PLEXUS regulates movements of inner and outer muscularis)

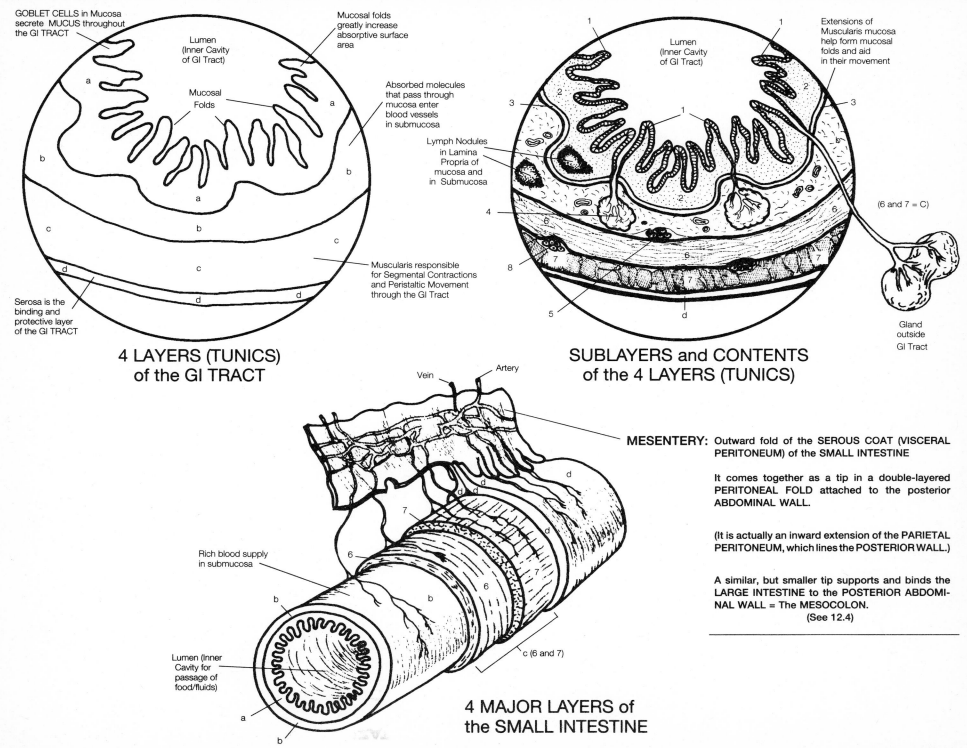

GOBLET CELLS in Mucosa secrete MUCUS throughout the GI TRACT

Lumen (Inner Cavity of GI Tract)

Mucosal Folds

Mucosal folds greatly increase absorptive surface area

Absorbed molecules that pass through mucosa enter blood vessels in submucosa

Muscularis responsible for Segmental Contractions and Peristaltic Movement through the GI Tract

Serosa is the binding and protective layer of the GI TRACT

4 LAYERS (TUNICS) of the GI TRACT

Lumen (Inner Cavity of GI Tract)

Extensions of Muscularis mucosa help form mucosal folds and aid in their movement

Lymph Nodules in Lamina Propria of mucosa and in Submucosa

(6 and 7 = C)

Gland outside GI Tract

SUBLAYERS and CONTENTS of the 4 LAYERS (TUNICS)

Vein   Artery

Rich blood supply in submucosa

Lumen (Inner Cavity for passage of food/fluids)

c (6 and 7)

4 MAJOR LAYERS of the SMALL INTESTINE

MESENTERY: Outward fold of the SEROUS COAT (VISCERAL PERITONEUM) of the SMALL INTESTINE

It comes together as a tip in a double-layered PERITONEAL FOLD attached to the posterior ABDOMINAL WALL.

(It is actually an inward extension of the PARIETAL PERITONEUM, which lines the POSTERIOR WALL.)

A similar, but smaller tip supports and binds the LARGE INTESTINE to the POSTERIOR ABDOMINAL WALL = The MESOCOLON.
(See 12.4)

DIGESTIVE SYSTEM: GENERAL ORGANIZATION
The Peritoneum: Serous Membranes of the Abdominal Cavity

★ The organs of the GI tract (the STOMACH and INTESTINES) and the ABDOMINAL ACCESSORY DIGESTIVE ORGANS, located in the abdominal cavity, are covered and supported by the PERITONEUM—the largest serous membrane in the body. It is composed of SIMPLE SQUAMOUS EPITHELIUM with portions reinforced with connective tissue (This epithelium is also referred to as MESOTHELIUM, as it derives from the MESODERM lining the primitive body cavity in the embryo). Lesser serous membranes are found in the thoracic cavity: the PERICARDIUM surrounds the heart (See 9.4), and the PLEURAE surround the lungs. (See 11.7)

★ Unlike the PERICARDIUM and the PLEURAE, the PERITONEUM contains *large folds* that weave in and out of the abdominal organs, binding the organs together and to the wall of the abdominal cavity. The folds contain:
- •Blood Vessels •Lymph Vessels
- •Nerves that supply the abdominal organs

★ All serous membranes have a PARIETAL PORTION that lines the body wall, and a VISCERAL PORTION that covers the internal organs. The PARIETAL PERITONEUM lines the walls of the abdominal cavity. Along the dorsal aspect of the cavity (back portion) each side joins together to form a double-layered fold, called the MESENTERY, which projects into the cavity and supports the GI tract. The DORSAL MESENTARY acts as a pendulum for the free-moving Small Intestine during peristaltic movements. A specific portion of the mesentery—the MESOCOLON—supports the Large Intestine. The peritoneal covering continues around most of the abdominal organs as the VISCERAL PERITONEUM. (SEROSA TUNIC)

The space between the PARIETAL PERITONEUM and the VISCERAL PERITONEUM is called the PERITONEAL CAVITY, which contains a small amount of lubricating fluid secreted by the peritoneum. This fluid minimizes the friction created as the viscera glide on each other, or against the wall of the abdominal cavity.

★ Extensions of the PARIETAL PERITONEUM include the FALCIFORM LIGAMENT, and the LESSER and GREATER OMENTA.
■ LESSER OMENTUM suspends the STOMACH and DUODENUM from the LIVER
■ GREATER OMENTUM is called the "FATTY APRON," hanging like an apron over the front of the intestines arising from the serosa of the stomach. It stores large quantities of fat and protects against the spread of infections by numerous lymph nodes.

The PERITONEUM:
The VISCERAL and PARIETAL PERITONEUM

1 | PARIETAL PERITONEUM |

2 | VISCERAL PERITONEUM | SEROSA

PERITONEAL CAVITIES

3 | GREATER PERITONEAL CAVITY |

4 | OMENTAL BURSA | LESSER PERITONEAL CAVITY WITHIN THE GREATER OMENTUM

5 | EPIPLOIC FORAMEN OF WINSLOW |
THE OPENING BETWEEN THE GREATER AND LESSER PERITONEAL CAVITIES

Extensions and Folds of the PERITONEUM:

6 | FALCIFORM LIGAMENT | } ATTACHES LIVER TO ANTERIOR ABDOMINAL WALL AND DIAPHRAGM

7 | LESSER OMENTUM | 2 LAYERS } DOUBLE FOLDS BETWEEN ORGANS

8 | GREATER OMENTUM | 4 LAYERS }
"FATTY APRON"

MESENTERY

9 | DORSAL MESENTERY | } DOUBLE FOLD BETWEEN POSTERIOR ABDOMINAL WALL AND SMALL INTESTINE

10 | MESOCOLON | } DOUBLE FOLD BETWEEN POSTERIOR ABDOMINAL WALL AND LARGE INTESTINE

11 | MESOAPPENDIX | } DOUBLE FOLD BETWEEN SMALL INTESTINE AND APPENDIX (See 12.10)

Posterior

9 or 10

Gut — Lumen of Gut

7 or 8

ORGAN

Anterior

Schematic Representation
of the SEROUS MEMBRANES
(Cross-section through abdominal cavity)

★ It is extremely important for the surgeon to understand the relationships within the ABDOMINAL CAVITY of the RETROPERITONEAL, INTRAPERITONEAL, and ANTI-PERITONEAL organs and structures:

RETROPERITONEAL (behind the peritoneum)
- •SUPRARENAL GLANDS (PAIRED)
- •KIDNEYS and URETERS (PAIRED)
- •PANCREAS
- •DUODENUM of SMALL INTESTINE (except the first few centimeters attach to a portion of the mesentery or lesser omentum)
- •ASCENDING and DESCENDING COLONS of the LARGE INTESTINE
- •ABDOMINAL AORTA and INFERIOR VENA CAVA (and their branches)

INTRAPERITONEAL (within the peritoneum)
- •LIVER   •GALLBLADDER
- •STOMACH   •SPLEEN
- •SMALL INTESTINE
- •TRANSVERSE and SIGMOID COLONS of the LARGE INTESTINE
- •MESENTERIES packed with NERVES and GANGLIA, BLOOD VESSELS, LYMPHATIC VESSELS, and NODES

ANTIPERITONEAL (in front of the peritoneum)
- •URINARY BLADDER
- •(UTERUS in the female)

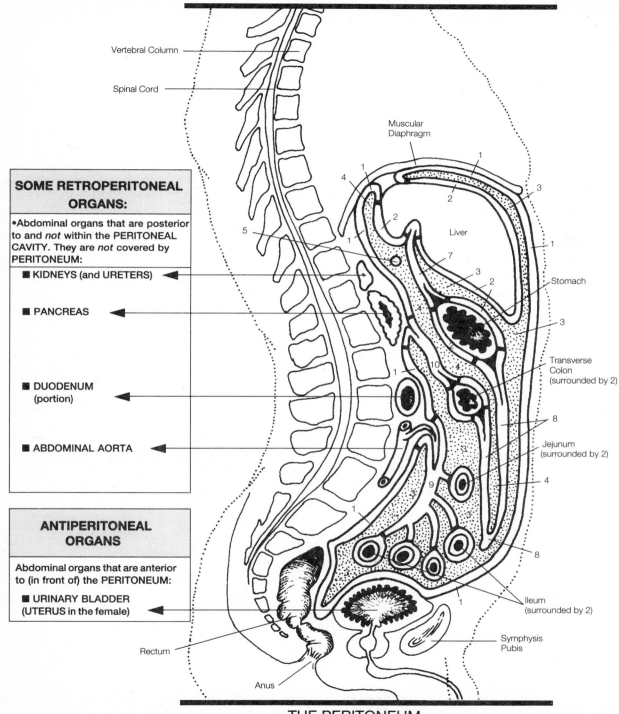

Vertebral Column

Spinal Cord

Muscular Diaphragm

Liver

Stomach

Transverse Colon (surrounded by 2)

Jejunum (surrounded by 2)

Ileum (surrounded by 2)

Symphysis Pubis

Rectum

Anus

THE PERITONEUM
(Schematic Right Lateral View, Sagittal Section)
(Falciform Ligament not Shown)

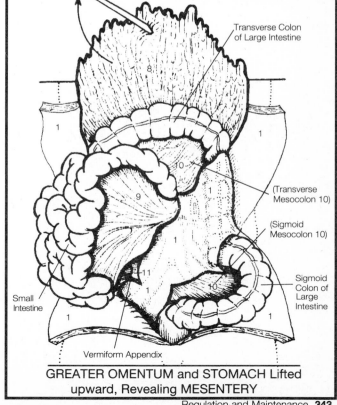

Stomach

Gall-bladder

Liver

"Fatty Apron"

Torso cut open, Liver lifted up, Revealing the LESSER OMENTUM and FALCIFORM LIGAMENT

Transverse Colon of Large Intestine

(Transverse Mesocolon 10)

(Sigmoid Mesocolon 10)

Sigmoid Colon of Large Intestine

Small Intestine

Vermiform Appendix

GREATER OMENTUM and STOMACH Lifted upward, Revealing MESENTERY

DIGESTIVE SYSTEM: THE ALIMENTARY CANAL: HEAD REGION
The Mouth and the Oral Cavity

1 = red 4 = pink 5 = orange
8 = light red 2b = yellow
★ See also 12.6

★The process of digestion begins with the INGESTION of food into the ORAL CAVITY of the MOUTH.

■ Chewing movements of the LIPS, CHEEKS, TONGUE, TEETH, and LOWER JAW break down the food in a destructive process.

■ Then the food is mixed with saliva secreted by the SALIVARY GLANDS and is rolled into a BOLUS, or soft moist mass that becomes suitable for SWALLOWING through the PHARYNX.

★The MOUTH consists of the VESTIBULE and the ORAL CAVITY.

The VESTIBULE is an anterior slitlike cavity (depression) bounded externally by the LIPS and CHEEKS and internally by the GUMS and TEETH. The ORAL CAVITY is that part of the MOUTH enclosed by the TEETH. It is thus a slightly smaller area than the MOUTH itself, excluding the VESTIBULE.

★The mouth is lined with nonkeratinized stratified squamous epithelium (mucosa). Imbedded in this mucosal membrane are thousands of mucous glands that, along with the SALIVARY GLANDS, keep the MOUTH constantly moist.

The LIPS are the anterior terminal portion of the CHEEKS. On the outside they are covered by SKIN, on the inside by a mucous membrane. The transition zone between the two is called the VERMILION BORDER.

FUNCTIONS OF THE LIPS:

- SPEECH   • SUCKLING
- MANIPULATION OF FOOD
- KEEPING FOOD BETWEEN UPPER AND LOWER TEETH

The ORAL ORIFICE = Opening of the MOUTH

The MOUTH
A VESTIBULE

B ORAL CAVITY PROPER EXTENDS FROM THE VESTIBULE TO THE FAUCES

BOUNDARIES of the MOUTH

1 LIPS   Anterior

2 CHEEKS   Lateral Walls

3 HARD PALATE   Superior (roof)

4 SOFT PALATE   Posterior (roof)

5 TONGUE   Inferior (floor)

Associated Structures of the Mouth
6 LABIAL FRENULUM

7 UVULA OF THE SOFT PALATE

8 GINGIVA (GUMS)

9 TEETH

2a BUCCINATOR MUSCLE ⎤
                       ⎬ CHEEK
2b BUCCAL FAT   ⎦

The FAUCES = Opening between the ORAL CAVITY and the PHARYNX

Anterior "PILLAR" of FAUCES
10 PALATOGLOSSAL ARCH

GLOSSOPALATINE ARCH

Posterior "PILLAR" of FAUCES
11 PALATOPHARYNGEAL ARCH

PHARYNGOPALATINE ARCH

12 TONSILLAR FOSSA

13 PALATINE TONSIL BETWEEN THE TWO ARCHES

14 TRIANGULAR FOLD

BOUNDARIES of the FAUCES (ISTHMUS)

SULCUS TERMINALIS
(on DORSUM of TONGUE)   Inferior

SOFT PALATE   Superior

PALATOGLOSSAL ARCH   Lateral
and PALATOPHARYNGEAL ARCH

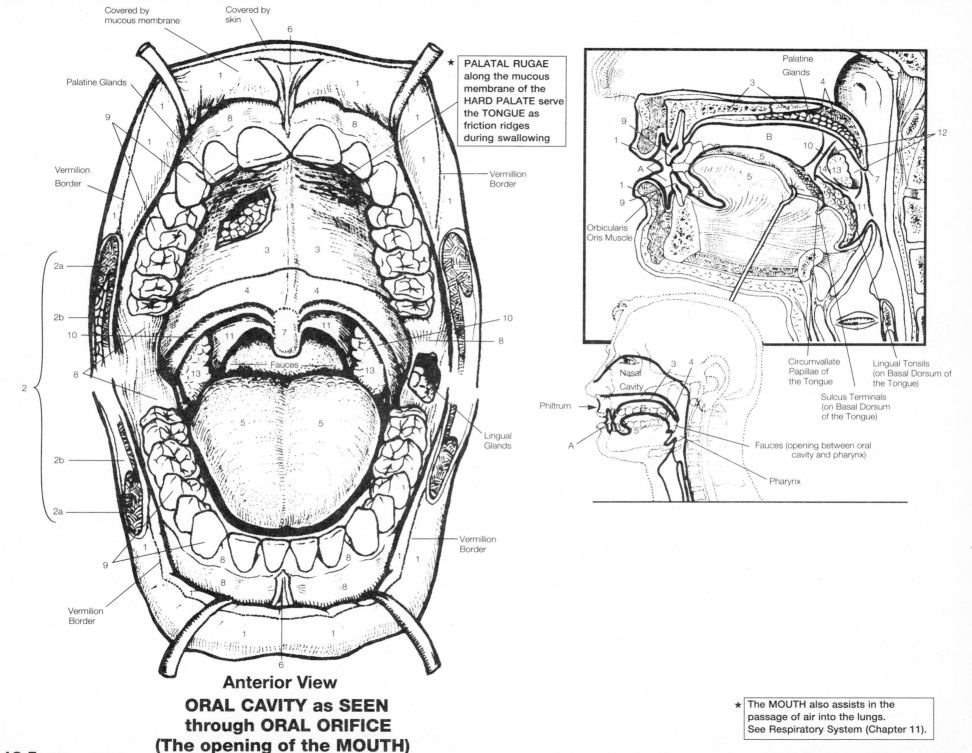

Covered by mucous membrane

Covered by skin

Palatine Glands

Vermilion Border

Orbicularis Oris Muscle

Philtrum

Nasal Cavity

Palatine Glands

Circumvallate Papillae of the Tongue

Sulcus Terminals (on Basal Dorsum of the Tongue)

Lingual Tonsils (on Basal Dorsum of the Tongue)

Fauces (opening between oral cavity and pharynx)

Pharynx

★ **PALATAL RUGAE** along the mucous membrane of the **HARD PALATE** serve the **TONGUE** as friction ridges during swallowing

Vermillion Border

Fauces

Lingual Glands

Vermillion Border

Vermilion Border

Anterior View

ORAL CAVITY as SEEN through ORAL ORIFICE (The opening of the MOUTH)

★ The MOUTH also assists in the passage of air into the lungs. See Respiratory System (Chapter 11).

DIGESTIVE SYSTEM: ACCESSORY ORGANS OF THE ORAL CAVITY
The Teeth (Dentes)

1 = white 5 = light pink 7 = purple
9 = light yellow 11 = light red
Artery = red Vein = blue Nerve = grey

★ See Charts #1 and #4

★ The MECHANICAL ACTION of TEETH begins the DESTRUCTION and BREAKDOWN of INGESTED FOOD: the process of MASTICATION. (See Chart #1)
MASTICATION is under VOLUNTARY CONTROL.

★ A typical tooth consists of three major portions:
 ■CROWN is the portion above the level of the GUMS.
 ■ROOT consists of 1–3 projections embedded in the socket (pockets in the alveolar processes of the mandible and maxillae bones).
 ■NECK or CERVIX is a general constricted area between the CROWN and ROOT (in the area of the GUMS).

★ The primary component of a tooth is a bonelike substance called DENTIN.
 DENTIN encloses a cavity. The enlarged part of the cavity lies in the crown and is called the PULP CAVITY or PULP CHAMBER. It is filled with PULP. Extensions of the PULP CAVITY into the root are the narrow ROOT CANALS, which end at an opening at the base of the root called the APICAL FORAMEN. The tooth receives nourishment from blood vessels, and also receives lymph vessels and nerves, which traverse the APICAL FORAMEN and become PULP.

★ The DENTIN of the CROWN is covered by ENAMEL. ENAMEL is the hardest substance in the body, being 99% mineral (calcium phosphate and calcium carbonate). ENAMEL protects the tooth from the wear of chewing. It cannot be replaced after a tooth erupts.

The TEETH [DENTES]
Major Portions of a Tooth
1 CROWN ENAMEL

2 NECK/CERVIX

3 ROOT ANCHOR

Structure of Tooth
4 ENAMEL COVERS DENTIN

5 DENTIN BULK of TOOTH (HARDER than BONE)

6 ODONTOBLAST LAYER

7 CEMENTUM COVERS ROOT

PULP CAVITY: Central Region
8 PULP CHAMBER HOLLOW CORE inside DENTIN

9 PULP CONNECTIVE TISSUE, BLOOD VESSELS, LYMPH VESSELS, and NERVES

10 ROOT CANAL

Supporting Structures of a Tooth
11 GINGIVAE (GUMS)

12 PERIODONTAL MEMBRANE LINES ALVEOLUS SOCKET (FIBERS INSERT into CEMENTUM)

13 ALVEOLUS SOCKET POCKET for ROOT

DENTITION: TYPE, NUMBER, ARRANGEMENT OF TEETH IN THE DENTAL ARCH

★ Humans are DIPHYODONT, meaning they develop 2 sets of teeth during a lifetime. See Chart #4.

FIRST DENTITION: MILK, or BABY, TEETH
14 DECIDUOUS 20

SECOND DENTITION: REPLACES BABY TEETH

15 PERMANENT 32

TYPES OF TEETH

I { CENTRAL INCISOR

LATERAL INCISOR

C CANINE CUSPID (EYETOOTH)

P { 1st PREMOLAR

2nd PREMOLAR } BICUSPIDS

M { 1st MOLAR

2nd MOLAR

3rd MOLAR WISDOM TEETH

■ BABY TEETH erupt at about 6 months of age, beginning with the INCISORS (eruption ends at 2 1/2 years)

■ PERMANENT TEETH begin to replace the baby teeth at about age 6, in sequence from the INCISORS to the MOLARS, continuing to about age 17

■ WISDOM TEETH (3rd MOLARS) are the last to erupt (between the ages of 17–25)

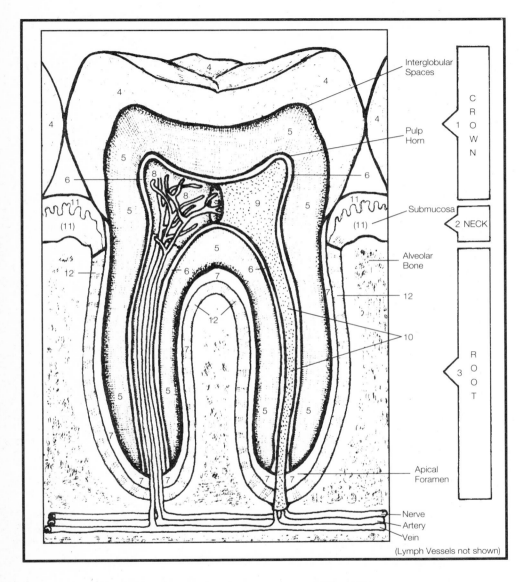

Interglobular
Spaces

Pulp
Horn

6

Submucosa

Alveolar
Bone

12

10

Apical
Foramen

Nerve
Artery
Vein

(Lymph Vessels not shown)

CROWN 1

NECK 2

ROOT 3

CROSS-SECTION of a LOWER MOLAR

FRONT   BACK

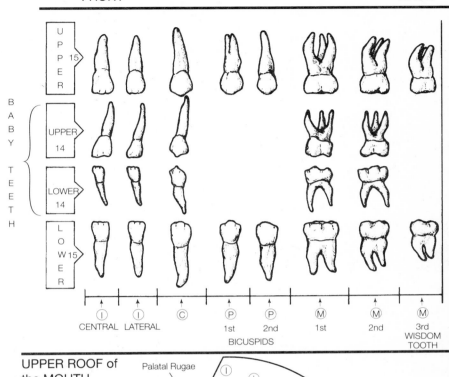

UPPER 15

BABY TEETH

UPPER 14

LOWER 14

LOWER 15

I CENTRAL   I LATERAL   C   P 1st   P 2nd   M 1st   M 2nd   M 3rd WISDOM TOOTH

BICUSPIDS

UPPER ROOF of the MOUTH (Underside of the Upper Jaw)

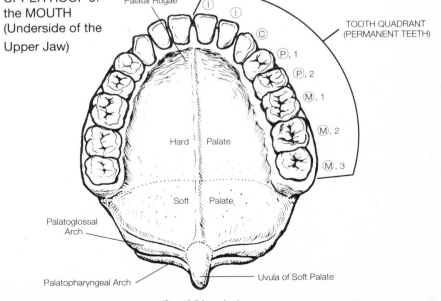

Palatal Rugae

TOOTH QUADRANT (PERMANENT TEETH)

I   I

C

P, 1

P, 2

M, 1

Hard   Palate

M, 2

Soft   Palate

M, 3

Palatoglossal
Arch

Palatopharyngeal Arch

Uvula of Soft Palate

(See 12.5 for colors)

Glands = light browns, greys
Ducts = cool colors
8–10 = warm colors
11 = yellow

★ See Chart #5, 6.4, 7.18

DIGESTIVE SYSTEM: ACCESSORY ORGANS OF THE ORAL CAVITY
The Salivary Glands and the Tongue

★ The chemical activity of SALIVA (through the enzyme SALIVARY AMYLASE or PTYALIN) INITIATES THE BREAKDOWN OF INGESTED CARBOHYDRATES (or POLYSACCHARIDES) ultimately into MONOSACCHARIDES. This is the only chemical digestion that occurs in the MOUTH. (Actually, only 3–5% of CARBOHYDRATES are reduced to DISACCHARIDES in the mouth since food is swallowed so quickly.)*

Most of the SALIVA is produced by 3 pairs of SALIVARY GLANDS outside the ORAL CAVITY and transported to it by SALIVARY DUCTS. SALIVA is a slightly acid solution of SALTS and ORGANIC SUBSTANCES, whose secretion is REFLEX and INVOLUNTARY.

Saliva also functions as a solvent in cleansing teeth. Solid substances must be dissolved in watery SALIVA to stimulate the TASTE BUDS on the TONGUE.

The NERVE ENDINGS (TASTE BUDS) in the PAPILLAE of the TONGUE react to the food placed on them and they activate the secretion of the SALIVARY GLANDS.

★ The TONGUE is a highly specialized, striated muscular organ covered with a mucous membrane. (See 6.4)
■ FUNCTIONS OF THE TONGUE
•MANIPULATION OF FOOD DURING MASTICATION
•SPEAKING
•CLEANSING OF TEETH
•SWALLOWING
•CONTAINS TASTE BUDS THAT SENSE FOOD TASTES (See 7.18)

Only the anterior two-thirds of the TONGUE lie in the ORAL CAVITY. The remaining one-third lies in the PHARYNX and is attached to the HYOID BONE.

The TONGUE is divided into symmetrical lateral halves by a MEDIAN SEPTUM (dorsally seen as the MEDIAN SULCUS). Each lateral half consists of an identical complement of EXTRINSIC and INTRINSIC MUSCLES. Rounded masses of LINGUAL TONSILS can be found on the dorsal surface of the base of the TONGUE. Roughened PAPILLAE aid in the handling of food.

The SALIVARY GLANDS
A PAROTID GLANDS (2) *

B SUBMANDIBULAR GLANDS (2) SUBMAXILLARY

C SUBLINGUAL GLANDS (2)

DUCTS of the SALIVARY GLANDS
PAROTID GLANDS
A₁   PAROTID (STENSEN'S) DUCTS

SUBMANDIBULAR GLANDS
B₁   WHARTON'S DUCTS

SUBLINGUAL GLANDS
C₁   RIVINUS'S DUCTS

C₂   BARTHOLIN'S DUCTS

C₃   SUBLINGUAL CARUNCLES

SECRETORY CELLS of SALIVARY GLANDS
1 MUCOUS CELLS secrete thick, slimy, stringy MUCUS

2 SEROUS CELLS secrete watery fluid containing digestive enzymes

3 SEROUS DEMILUNES crescent-shaped serous cells capping mucous acini cells
DEMILUNES OF GIANNUZZI

*PAROTID GLANDS
secrete the digestive enzyme PTYALIN (or SALIVARY AMYLASE), which starts to split cooked STARCH → DEXTRINS → MALTOSE: MALTOSE is a DISACCHARIDE that is later digested into MONOSACCHARIDES lower in the GI tract,

D The TONGUE (LINGUA)

4 LINGUAL FRENULUM

5 MEDIAN SULCUS

6 FORAMEN CAECUM

7 SULCUS TERMINALIS

PAPILLAE
Containing the NERVE ENDINGS or TASTE BUDS

8 FILIFORM PAPILLAE do not contain many taste buds

9 FUNGIFORM PAPILLAE

10 CIRCUMVALLATE PAPILLAE

TASTE ZONES OF THE TASTE BUDS

a BITTER   c SALT

b SOUR   d SWEET

11 LINGUAL TONSILS

★ See Chart #5

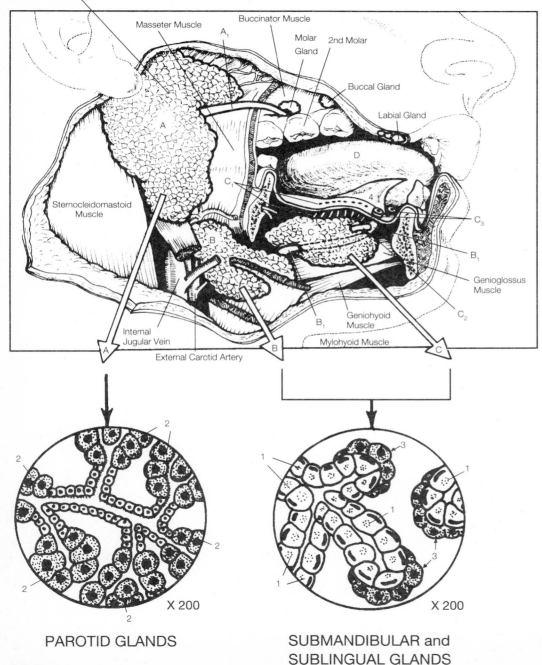

★MUMPS are the result of infected and swollen PAROTID GLANDS and/or DUCTS

Masseter Muscle
Buccinator Muscle
A₁
Molar Gland
2nd Molar
Buccal Gland
Labial Gland
A
D
C₁
4
C₃
Sternocleidomastoid Muscle
B
C
B₁
Genioglossus Muscle
C₂
Internal Jugular Vein
B₁
Geniohyoid Muscle
External Carotid Artery
Mylohyoid Muscle
A
B
C

2
2
2
2
2
2
X 200

PAROTID GLANDS

1
3
1
1
1
1
3
3
X 200

SUBMANDIBULAR and
SUBLINGUAL GLANDS

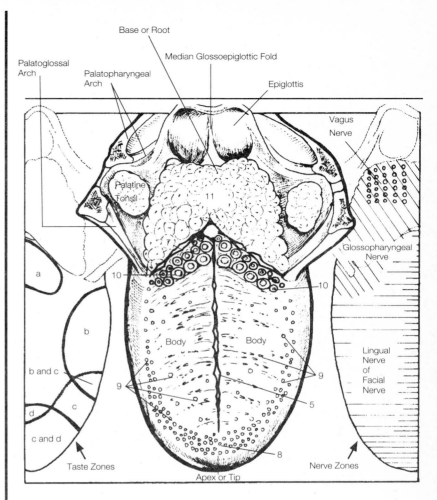

Palatoglossal Arch
Base or Root
Median Glossoepiglottic Fold
Palatopharyngeal Arch
Epiglottis
Vagus Nerve
Palatine Tonsil
11
7
7
a
10
10
Glossopharyngeal Nerve
b
Body
Body
9
b and c
9
5
c
Lingual Nerve of Facial Nerve
d
c
9
8
c and d
Taste Zones
Apex or Tip
Nerve Zones

9
8

FILIFORM and FUNGIFORM PAPILLAE
at the APEX OF THE TONGUE

12.7

DIGESTIVE SYSTEM: THE ALIMENTARY CANAL: NECK AND THORACIC REGION: *The Pharynx and the Esophagus*

1–4 = warm colors Ⓒ *= flesh*
a = yellow b = light orange
c₁, c₂ = red, orange

★ See 6.4, 11.2

★ The PHARYNX is a fibromuscular pouch, posterior to the MOUTH, that connects the ORAL CAVITY and NASAL CAVITY with the ESOPHAGUS and LARYNX. It is therefore a common passageway for both the DIGESTIVE SYSTEM and the RESPIRATORY SYSTEM.

• It extends from the base of the skull (attached by PHARYNGOBASILAR FASCIA) to the level of the 6th cervical vertebra, where it is continuous with the ESOPHAGUS

• Regarding the role of the pharynx in the digestive process, it does *not* produce digestive enzymes and does *not* carry on absorption. It secretes mucus continually from the lining of the mucous membrane, which along with saliva keeps the PHARYNX constantly moist for efficient passage of the food bolus. The PHARYNX aids in the transport of the bolus to the ESOPHAGUS.

★ The PHARYNX is 3-layered (with an inner MUCOSA, SUBMUCOSA, and skeletal, voluntary MUSCULARIS) but does not contain outer SEROSA, as does the rest of the ALIMENTARY CANAL. Around the pharynx and around the thoracic portion of the ESOPHAGUS, in place of a true SEROSA, is the ADVENTITIA. This is a layer of connective tissue blending without a noticeable demarcation into the underlying surrounding tissue of the muscularis.

★ The ESOPHAGUS is a collapsible muscular tube approximately 25 cm long (10 in.) that conveys ingested food and fluid from the MOUTH (and PHARYNX) to the STOMACH.

• The anterior portion of the ESOPHAGUS is connected to the posterior portion of the TRACHEA by connective tissue imbedded with smooth muscularis muscle. This tissue attaches the open ends of the C-rings of the tracheal hyaline cartilage. This soft area of tissue allows the ESOPHAGUS to expand as swallowed food passes toward the STOMACH.

B | **The PHARYNX [THROAT]**

Muscular Wall of PHARYNX
1 | **SUPERIOR CONSTRICTOR MUSCLE**

2 | **MIDDLE CONSTRICTOR MUSCLE**

3 | **INFERIOR CONSTRICTOR MUSCLE**

4 | **CRICOPHARYNGEUS MUSCLE**

Ⓒ | **The ESOPHAGUS (GULLET)**
4 Basic TUNIC COATS
a | **MUCOSA** INNER LINING

b | **SUBMUCOSA**

c MUSCULARIS
c₁ | **INNER CIRCULAR MUSCLE**

c₂ | **OUTER LONGITUDINAL MUSCLE**

d | **ADVENTITIA/"SEROSA"**

Posterior View

Left Lateral View

The ESOPHAGUS does *not* secrete digestive enzymes and does *not* carry on absorption. It is involved in the DEGLUTITION REFLEX (SWALLOWING) and also the downward PERISTALTIC movement of the food bolus. (See diagram)

The ESOPHAGUS is the first digestive organ to contain all 4 TUNIC LAYERS:

■ MUCOSA (Inner lining)
•Nonkeratinized STRATIFIED SQUAMOUS COLUMNAR EPITHELIUM; thick, protective surface layer open to the LUMEN (CANAL)

■ SUBMUCOSA
•Contains mucous glands that secrete MUCUS along DUCTS to the inner mucosa for lubrication

■ MUSCULARIS
•Arranged in 2 layers:
INNER CIRCULAR MUSCLE
OUTER LONGITUDINAL MUSCLE
•In the upper third of the esophagus, the muscle is SKELETAL (STRIATED, or VOLUNTARY)
•In the middle third the muscle is a mixture of SKELETAL and VISCERAL
•In the lower third, the muscle is VISCERAL (SMOOTH, or INVOLUNTARY)

■ SEROSA
An outer serous coat combined with connective tissue blends with the TRACHEA above and the STOMACH below

Except during the passage of food, the ESOPHAGUS is closed and flattened with its mucosa thrown into many longitudinal folds

3 LOWER ESOPHAGEAL CARDIAC SPHINCTER

★ Just above the level of the DIAPHRAGM, the ESOPHAGUS is slightly narrowed due to the CARDIAC (GASTROESOPHAGEAL) SPHINCTER.

■ During swallowing, RELAXATION of the CARDIAC SPHINCTER permits food bolus to enter the STOMACH.

■ After the food bolus passes into the STOMACH, this sphincter CONSTRICTS to prevent reflux of stomach contents back up into the ESOPHAGUS. (This reflux would normally occur due to pressure differentials in the THORACIC and ABDOMINAL CAVITIES during RESPIRATION.) A common problem, "Heart Burn", is caused by esophageal refluxing.

■ Movement of the DIAPHRAGM against the STOMACH during RESPIRATION also helps prevent regurgitation of gastric contents.

PERISTALSIS

■ Progressive, wavelike movement of smooth muscle (longitudinal and circular layers), especially in hollow tubes of alimentary canal

•A function of the Tunica Muscularis
•Controlled by Medulla in the Brain

Just above and around the top of the Bolus, the Inner Circular Muscles Contract, thus Constricting the esophageal wall, squeezing the Bolus Downward.

Just below and around the bottom of the Bolus, the Outer Longitudinal Muscles Contract, thus Shortening this lower section and pushing the esophageal wall Outward to receive the Bolus.

■ These Contractions of the Muscularis are repeated in a wave form that moves down the Esophagus, pushing the Bolus toward the Stomach.

PASSAGE of SUBSTANCES from MOUTH to STOMACH

Solid, or Semisolid Food
4–8 seconds

Very Soft Foods and Liquids
1 second

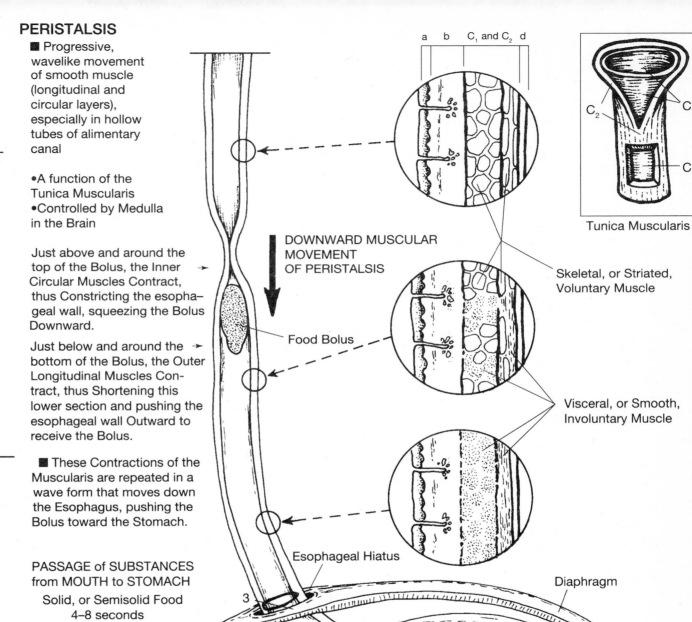

Tunica Muscularis

DOWNWARD MUSCULAR MOVEMENT OF PERISTALSIS

Food Bolus

Skeletal, or Striated, Voluntary Muscle

Visceral, or Smooth, Involuntary Muscle

Esophageal Hiatus

Diaphragm

Cardia

Stomach

The ESOPHAGUS Originates at the LARYNX Posterior to the TRACHEA, as a continuation of the LARYNGOPHARNX. Opens through the DIAPHRAGM at the ESOPHAGEAL HIATUS, and Terminates at the CARDIA of the STOMACH.

DIGESTIVE SYSTEM: THE ALIMENTARY CANAL: THE GI TRACT
The Stomach (Gaster): Gross Anatomy

A–D = oranges, reds, pinks E = flesh
Arrows = cool colors Bolus = ochre
★ See 12.10, 12.14

★ The STOMACH is the most distensible part of the GI TRACT. It has two openings: it is continuous with the ESOPHAGUS superiorly at the UPPER CARDIAC ORIFICE and empties into the DUODENUM of the SMALL INTESTINE through the LOWER PYLORIC ORIFICE inferiorly.

★ FUNCTIONS OF THE STOMACH
■ Acts as a RESERVOIR (STORAGE BASE) for food as the food is mechanically CHURNED and mixed with GASTRIC SECRETIONS (ACID and ENZYMES) made by the STOMACH CELLS. Food BOLUS is changed into a pasty, liquid CHYME, aided by peristaltic MIXING WAVES and Muscular CONTRACTIONS (See 2.10)
■ Delivers CHYME in small quantities (through REGULATORY MOVEMENTS of the PYLORUS) to the DUODENUM. (Pyloric movements also inhibit backflow of CHYME back into STOMACH.)
■ INITIATES DIGESTIONS OF PROTEINS (PARTIAL DIGESTION of PROTEINS by the action of the enzyme PEPSIN.) (See 12.10, 12.14)

★ Note: CARBOHYDRATES ARE NOT DIGESTED AT ALL IN THE STOMACH. FATS are very slightly digested. (The complete digestion and absorption of food molecules occurs when the CHYME enters the SMALL INTESTINE.) Saliva swallowed into the STOMACH continues to act on cooked STARCH changing it into a form of SUGAR.

■ LIMITED ABSORPTION: The STOMACH wall is impermeable to the passage of most substances into the blood. (The only commonly ingested substances are ALCOHOL, ASPIRIN, GLUCOSE, and some water and salts.)

■ Important in ACID-BASE EQUILIBRIUM of body (especially when vomiting removes ELECTROLYTES).
■ Regulates DILUTION/CONCENTRATION of FLUIDS so they are the same concentration as the body's own fluids
■ STOMACH ACID kills a large amount of the MICROBES present in food
■ Secretion of INTRINSIC FACTOR, which is needed for VITAMIN B_{12} (an EXTRINSIC FACTOR) absorption in the intestine

Four Regions of the STOMACH
A CARDIA

B FUNDUS

C BODY

D PYLORUS "GATEKEEPER"

Portions of the PYLORUS
D₁ PYLORIC SPHINCTER PYLORIC VALVE (circular muscle)

D₂ PYLORIC VESTIBULE

D₃ PYLORIC ANTRUM

Related Structures
E ESOPHAGUS

F CARDIAC SPHINCTER

G DUODENUM OF SMALL INTESTINE

H DIAPHRAGM

2 Surfaces (Broadly rounded)
1 ANTERIOR SURFACE

2 POSTERIOR SURFACE

2 CURVATURES (BORDERS)
3 GREATER CURVATURE Lateral Convex

4 LESSER CURVATURE Medial Concave

INDENTATIONS
5 CARDIAC INCISURA

6 ANGULAR INCISURA

7 PYLORIC INTERMEDIATE SULCUS

FOLDS
8 GASTRIC RUGAE FOLDS of the MUCOSA

★ The STOMACH is the beginning of the GASTROINTESTINAL TRACT (GI TRACT). It is a J-shaped, dilated, saclike bulge in the alimentary canal located directly below the DIAPHRAGM to the right of the SPLEEN, partly under the LIVER. The STOMACH is generally found in the EPIGASTRIC and LEFT HYPOCHONDRIAC REGIONS of the abdomen (See 1.3), although its position and size vary continually due to:
• Individuality
• Respiratory movements
• Degree of fullness (STOMACH distends/stretches as swallowed food collects in it)

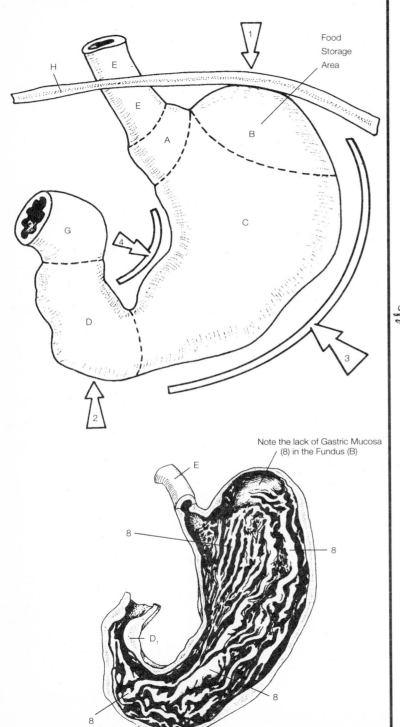

1

Food
Storage
Area

H

E

E

A

B

4

G

C

D

3

2

Note the lack of Gastric Mucosa
(8) in the Fundus (B)

E

8

8

8

D_1

8

8

STOMACH Partly Sectioned at
Both Orifices to Expose SPHINCTER
MUSCLES, Portions of the PYLORUS,
and GASTRIC RUGAE FOLDS of the
MUCOSA

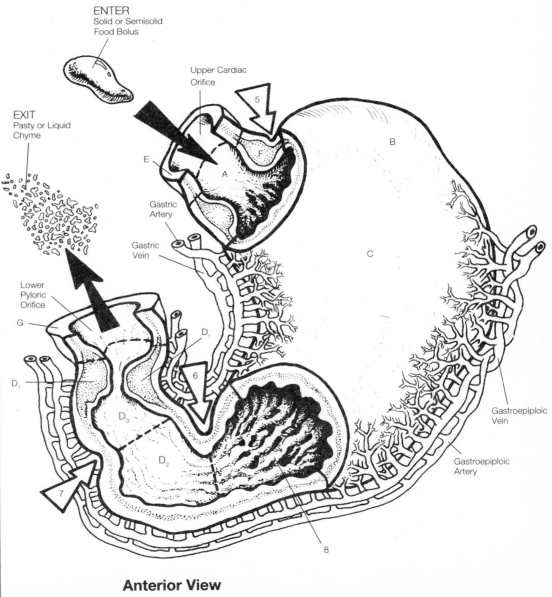

ENTER
Solid or Semisolid
Food Bolus

Upper Cardiac
Orifice

5

EXIT
Pasty or Liquid
Chyme

E

A

F

B

Gastric Artery

Gastric Vein

C

Lower
Pyloric
Orifice

G

D_1

D_1

6

D_3

D_2

8

Gastroepiploic
Vein

Gastroepiploic
Artery

7

Anterior View

DIGESTIVE SYSTEM: THE ALIMENTARY CANAL: THE GI TRACT
The Stomach (Gaster): Microscopic Anatomy

a_1 = light pink
a_2 = light yellow
a = warm colors
Secretory Cells = cool colors

★ The STOMACH WALL is composed of the 4 basic layers as the rest of the alimentary canal (GI tract), but with specific structural and functional modifications: (review 12.3)

■ STOMACH MUCOSA
•Shaped into many large, longitudinal folds, called GASTRIC RUGAE (visible to the naked eye) when the STOMACH is empty. When the STOMACH fills, it distends and stretches out as the RUGAE smooth out and disappear.
•Microscopically, consists of 2 layers of SIMPLE COLUMNAR EPITHELIUM folded downward to form numerous GASTRIC PITS (GASTRIC GLANDS) lined with an assortment of SECRETING CELLS. (See right of page)
•Secretions of the GASTRIC GLANDS together are called GASTRIC JUICE, a strongly acid (PH 0.9—1.6), thin, colorless fluid containing: PEPSIN (initiates PROTEIN DIGESTION), HYDRO–CHLORIC ACID, MUCIN, GASTRIC LIPASE, INTRINSIC FACTOR, and small quantities of inorganic salts (Na, K, Ca, Mg, Chlorides, Phosphates and Sulphates).

■ STOMACH SUBMUCOSA
•Connects MUCOSA to MUSCULARIS, composed of LOOSE AREOLAR CONNECTIVE TISSUE.

■ STOMACH MUSCULARIS
•Composed of 3 smooth muscle layers, compared to the normal 2 layers in the other areas of the alimentary canal, with the addition of an OBLIQUE LAYER inside the CIRCULAR LAYER. This triple arrangement of fibers allows a wide variety of gastric contractions to churn food, break it into smaller particles, mix it with GASTRIC JUICE, and then pass it into the DUODENUM

■ STOMACH SEROSA (See 12.4)
•Part of the VISCERAL PERITONEUM
•At the LESSER CURVATURE, the 2 layers of the VISCERAL PERITONEUM come to-gether to form the LESSER OMENTUM, which extends upward to the LIVER.
•At the GREATER CURVATURE, the 2 layers of the VISCERAL PERITONEUM come together as the GREATER OMENTUM, which continues downward as an "apron" hanging over the intestines.

4 Basic LAYERS

a   **STOMACH MUCOSA**

b   **STOMACH SUBMUCOSA**

c   **STOMACH MUSCULARIS**

d   **STOMACH SEROSA**

STOMACH MUCOSA

a_1 **SIMPLE COLUMNAR LINING EPITHELIUM**

a_2 **LAMINA PROPRIA**

a_3 **MUSCULARIS MUCOSA**

STOMACH MUSCULARIS (SMOOTH MUSCLE)

c_1 **INNER OBLIQUE LAYER**

c_2 **MIDDLE CIRCULAR LAYER**

c_3 **OUTER LONGITUDINAL LAYER**

*It has not been determined with any degree of certainly whether INTRINSIC FACTOR is secreted by GOBLET, CHIEF, or ARGENTAFFEN CELLS

Differentiations of LINING EPITHELIUM

1   **SURFACE EPITHELIAL CELLS**

G   **GASTRIC PIT/GLAND**

SECRETORY CELLS	SECRETIONS (GASTRIC JUICE)
2 **GOBLET CELLS**	• MUCUS (MUCIN) • INTRINSIC FACTOR (of the • ANTIANEMIC PRINCIPLE)*
3 **PARIETAL/ OXYNTIC CELLS**	• HYDROCHLORIC ACID (HCL)
4 **CHIEF/ ZYMOGENIC CELLS**	• PEPSINOGEN (inactive pro–enzyme): Converts into active • PEPSIN when mixed with • HYDROCHLORIC ACID in the GASTRIC JUICE
5 **ARGENTAFFEN CELLS**	• SEROTONIN (vasoconstrictor) • HISTAMINE (pharmacologic action): Increases gastric secretion, among other things
6 **PYLORIC G CELLS**	• GASTRIN (hormone): Stimulates secretion of GASTRIC ACID. Also affects secretory activity of the GALLBLADDER, PANCREAS, and SMALL INTESTINE.

★ Remember, the chief chemical activity of the STOMACH is TO BEGIN THE DIGESTION OF PROTEINS, achieved primarily through the enzyme PEPSIN. PEPSIN cannot digest the proteins in the CHIEF CELLS that produce it because it is secreted in an inactive form called PEPSINOGEN. It is converted into PEPSIN when activated by the HYDROCHLORIC ACID secreted by the PARIETAL CELLS. Once PEPSIN has been activated by HCL, the GOBLET CELLS secrete MUCUS, which lines the MUCOSA and forms a protective barrier between the stomach lining and the acidic gastric juices to prevent self–digestion within the stomach epithelium.

4 BASIC TUNIC LAYERS
of the STOMACH

Narrow openings of the Gastric
Pits open to the inner Lumen

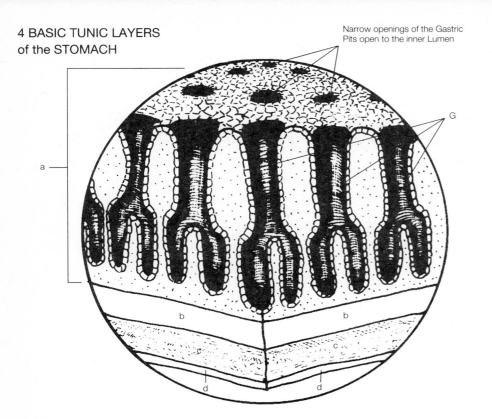

G

a

b   b

c   c

d   d

1 (of a₁)

Narrow openings
of the Gastric Pits

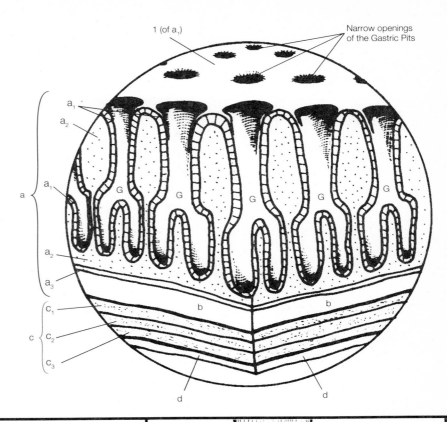

a₁
a₂

a

a₁

G   G   G   G   G

a₂

a₃

b   b

c₁

c   c₂

c₃

d   d

Differentiated Cells of the Lining Epithelium

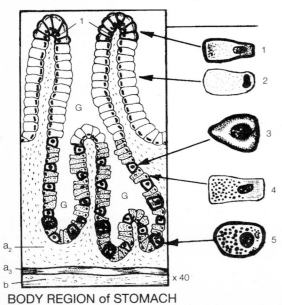

1
1

G

G   G

a₂

a₃

b

x 40

BODY REGION of STOMACH

1

2

3

4

5

Pyloric Glands Secrete
Alkaline Mucus

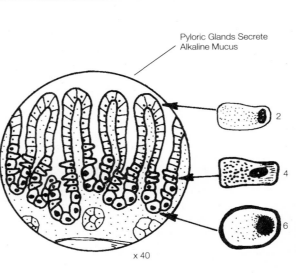

x 40

2

4

6

PYLORIC REGION of STOMACH

C₃

C₁

C₂   C₂

C₂

C₁

C₂

C₃

3 LAYERS of the STOMACH MUSCULARIS

DIGESTIVE SYSTEM: THE ALIMENTARY CANAL: THE GI TRACT
The Small Intestine: Gross Anatomy

★ The INTESTINE is the portion of the alimentary canal extending from the PYLORUS of the STOMACH to the ANUS. It is divided into the SMALL INTESTINE and the COLON (LARGE INTESTINE).

★ The SMALL INTESTINE is a long, muscular tube (during life it is approximately 7 meters, or 23 feet long). It begins at the PYLORIC VALVE of the STOMACH, coils repeatedly through the central and lower parts of the abdominal cavity (See 1.3), and ends at the ILEOCECAL VALVE opening into the LARGE INTESTINE.

★ The SMALL INTESTINE is the MAJOR SITE FOR DIGESTION and ABSORPTION of INGESTED ORGANIC FOOD SUBSTANCES.

★ The SMALL INTESTINE is named so because of its small, 1 in. diameter. It is divided into 3 SEGMENTS, or REGIONS: DUODENUM, JEJUNUM, and ILEUM:

■ DUODENUM
•Widest and most fixed region. Short first part 20–28 cm long (8–11 in.), from the PYLORIC SPHINCTER (PYLORIC VALVE) to the DUODENOJEJUNAL FLEXURE.
•Retroperitoneal, except for a beginning part of the SUPERIOR PORTION near the STOMACH. It has *no* mesentery.
•A crucial section of the ALIMENTARY CANAL: It receives secretions from the LIVER, GALLBLADDER, and PANCREAS through a SYSTEM OF DUCTS opening into the DESCENDING PORTION.
•Mixes ACID CHYME it receives in small quantities from the STOMACH with BILE from the LIVER and GALLBLADDER, PANCREATIC JUICE from the PANCREAS, and intestinal juices secreted by BRUNNER'S GLANDS and CRYPTS OF LIEBERKUHN from the DUODENAL WALL.
•CHYLE is formed here. (See 10.2, 12.12)

■ JEJUNUM
•Extends from DUODENUM to ILEUM
•Second region, 2.4 meters (8 feet) long, with a slightly larger lumen and more internal fixed folds than the ILEUM
•PRIMARY SITE OF ABSORPTION (although some occurs in the DUODENUM and ILEUM) (See 12.2)

■ ILEUM
•Third and last region of small intestine
•Varies in length in adult (9.6—4.7 meters) =(31 1/2—15 1/2 feet)
•Abundance of lymphatic tissue in ILEUM walls
•Joins LARGE INTESTINE at ILEOCECAL VALVE

Three Regions of the SMALL INTESTINE

A DUODENUM

B JEJUNUM (≈ 2/5)

C ILEUM (≈ 3/5)

5 Portions of the DUODENUM

A₁ SUPERIOR PARS SUPERIOR DUODENAL CAP/BULB

A₂ DESCENDING PARS DESCENDENS

A₃ HORIZONTAL PARS INFERIOR

A₄ ASCENDING PARS ASCENDENS

A₅ DUODENOJEJUNAL FLEXURE

SUPPORTIVE STRUCTURE

1 DORSAL MESENTERY

The SMALL INTESTINE is supported, except for the DUODENAL REGIONS, by MESENTERY. Recall that the MESENTERY extends from a TIP from the posterior abdominal wall. (See 12.3, 12.4) Therefore, it functions as a FAN-SHAPED ATTACHMENT allowing MOBILITY and little chance for the coiled organ of the small intestine to become twisted. MESENTERY encloses the structures that feed and supply it:BLOOD VESSELS, LYMPHATIC VESSELS, and NERVES.

Openings into the DESCENDING DUODENUM

2 ACCESSORY PANCREATIC DUCT
See 12.14

3 HEPATOPANCREATIC AMPULLA (AMPULLA OF VATER)
See 12.14

Elevation for Entry of Accessory PANCREATIC DUCT

4 LESSER DUODENAL PAPILLA

Elevation for Entry of HEPATOPANCREATIC AMPULLA

5 GREATER DUODENAL PAPILLA

Entry into Medial Side of the CECUM (OF THE COLON) from the Terminal Portion of the ILEUM

6 ILEOCECAL VALVE COLIC VALVE

The ILEOCECAL VALVE:
Opens and closes during digestion to allow spurts of chyme from the ILEUM to enter the LARGE INTESTINE. Initiated when food enters the STOMACH by the GASTROILEAL REFLEX (controlled by the AUTONOMIC NERVOUS SYSTEM). DIGESTION and ABSORPTION of food are usually complete by the time the CHYME RESIDUE reaches the ILEOCECAL VALVE.

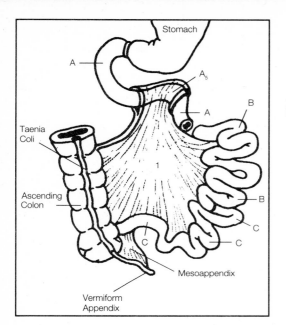

Stomach

A

A₅

A

Taenia
Coli

Ascending
Colon

1

B

B

C

C

C

Mesoappendix

Vermiform
Appendix

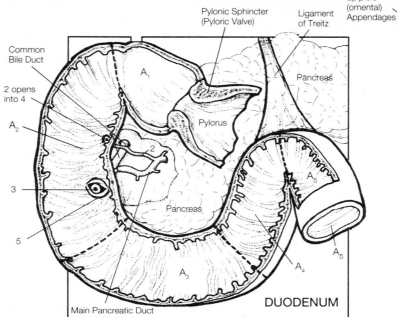

Pylonic Sphincter
(Pyloric Valve)

Ligament
of Treitz

Epiploic
(omental)
Appendages

Common
Bile Duct

A₁

Pancreas

2 opens
into 4

Pylorus

A₂

2

3

Pancreas

A₅

5

A₅

A₃

A₄

Main Pancreatic Duct

DUODENUM

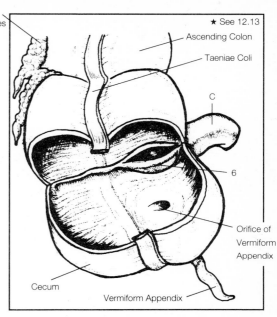

★ See 12.13

Ascending Colon

Taeniae Coli

C

6

Orifice of
Vermiform
Appendix

Cecum

Vermiform Appendix

CECUM Cut Open to Expose
ILEOCECAL VALVE

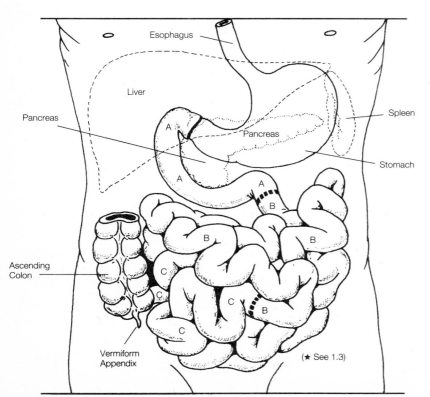

Esophagus

Liver

Pancreas

A

Pancreas

Spleen

A

Stomach

A

A

B

Ascending
Colon

B

C

B

C

C

C

B

Vermiform
Appendix

(★ See 1.3)

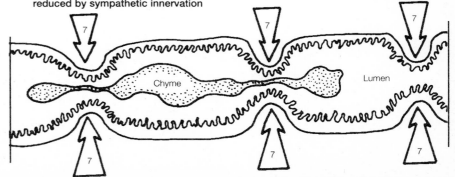

2 MAJOR TYPES of CONTRACTION in the SMALL INTESTINE

PERISTALSIS
- ■ Much weaker than in the ESOPHAGUS and STOMACH
- ■ Movement of chyme through the intestine (INTESTINAL MOTILITY) is very slow and mainly due to greater pressure at the pyloric end of the small intestine than at the distal end

7 SEGMENTATION
- ■ Simultaneous contractions of smooth muscle at different intestinal segments, regulated by the ANS
- ■ Muscular constrictions of the LUMEN mix CHYME more thoroughly with digestive enzymes and mucus
- ■ Contraction stimulated by parasympathetic (vagus nerve) innervation; contraction reduced by sympathetic innervation

Chyme

Lumen

7

7

7

7

7

7

a = yellow b = light orange
c = orange, red 5 = light yellow
6 = light orange 7 = red
8 = green 9 = purple

★ See 12.2

DIGESTIVE SYSTEM: THE ALIMENTARY CANAL: THE GI TRACT
The Small Intestine: Microscopic Anatomy

★ The products of digestion are absorbed at a very fast rate across the epithelial lining of the intestinal mucosa, particularly in the JEJUNUM OF THE SMALL INTESTINE. MUCOSA and SUBMUCOSA are specially adapted for the rapid absorption of nutrients. This high absorptive rate is due to an increased mucosal surface area furnished by a number of folds:
- •PLICAE CIRCULARES
- •VILLI
- •MICROVILLI

★ ■ PLICAE CIRCULARES
•Unlike the GASTRIC RUGAE FOLDS in the STOMACH, which change their shape and surface area continually, the PLICAE CIRCULARES are deep, large, permanent folds projecting into the intestinal lumen. They either extend all the way around the intestinal circumference or just part way. They cause the CHYME to SPIRAL as it passes through the long intestinal tube.

★ ■ VILLI
•4–5 million VILLI, each 0.5–1 mm high, project into the intestinal lumen, giving the mucosa its velvety, carpet-like appearance
•Each VILLUS has a core of LAMINA PROPRIA, the connective tissue layer of the MUCOSA, which is embedded with:①AN ARTERIOLE,②A VENULE,③A CAPILLARY NET-WORK, and④A CENTRAL LACTEAL (LYMPHATIC VESSEL).
•Digested PROTEINS (AMINO ACIDS) and CARBO-HYDRATES (MONOSACCHARIDES), and small aggregations (chylomicrons) of FAT (FATTY ACIDS and GLYCEROL) are absorbed into BLOOD CAPILLARIES.
•Larger fat chylomicrons enter the LACTEAL and mix with the LYMPH, forming a milky, alkaline CHYLE. (See 10.2)
•At the BASE of each VILLUS are vertical pouches or pits lined with glandular epithelium that open through pores to the intestinal lumen. These pits are called CRYPTS OF LIEBERKÜHN or the INTESTINAL GLANDS where the intestinal juice (succus entericus) is secreted. New epithelial cells are formed here and are pushed up to the top of each VILLUS as the surface epithelial cells are being continually sloughed off. CELLS OF PANETH deep in the crypts secrete intestinal digestive enzymes.
•Absorption of most digested food occurs through the BORDER EPITHELIUM covering the VILLI.

★ ■ MICROVILLI
•Cell membranes of each lining epithelial cell have folded projections of their own called MICROVILLI (collectively, they are called the BRUSH–BORDER, open to the intestinal lumen)
•Provide an extremely large surface area for absorption
•BRUSH–BORDER ENZYMES that stay attached to the cell membrane further hydrolyze CHYME to increase efficiency of rapid absorption.

4 Basic Layers
a SMALL INT. MUCOSA

b SMALL INT. SUBMUCOSA

c SMALL INT. MUSCULARIS

d SMALL INT. SEROSA

FOLDS of SMALL INTESTINE
LARGE, PERMANENT, DEEP FOLDS in MUCOSA and SUBMUCOSA: (seen with naked eye)
P PLICAE CIRCULARES KERCKRING'S FOLDS

MICROSCOPIC FINGERLIKE FOLDS of the MUCOSA:
V VILLI

FOLDINGS of the APICAL PLASMA CELL MEMBRANE of EPITHELIAL CELLS: (only seen with electron microscope)
M MICROVILLI

BORDER FORMED by COLLECTIVE MICROVILLI on EDGES of the COLUMNAR EPITHELIAL CELLS:
B BRUSH BORDER
Contains permanently attached BRUSH–BORDER ENZYMES which hydrolize DISACCHARIDES, POLYPEPTIDES, and other substrates in preparation for absorption into the columnar epithelium

SMALL INTESTINAL MUCOSA
a_1 SIMPLE COLUMNAR LINING EPITHELIUM

1 SURFACE EPITHELIAL CELLS
CONTINUOUSLY EXFOLIATED at TIP of EACH VILLUS

2 ABSORPTIVE CELLS

3 GOBLET CELLS SECRETE MUCUS

4 CRYPTS of LIEBERKÜHN
SECRETE ALKALINE INTESTINAL JUICE, (SUCCUS ENTERICUS) containing digestive enzymes (mainly from the secretory cells of Paneth)

a_2 LAMINA PROPRIA (CORE OF EACH VILLUS)
5 LYMPHATIC VESSEL/LACTEAL
6 VENULE   7 ARTERIOLE

a_3 MUSCULARIS MUCOSA
8 MESENTERIC (PEYER'S) PATCHES

SMALL INTESTINAL SUBMUCOSA
9 DUODENAL BRUNNER'S GLANDS
SECRETE ALKALINE MUCUS to PROTECT WALLS from ENZYME ACTION

SMALL INTESTINAL MUSCULARIS
C_1 INNER CIRCULAR LAYER

C_2 OUTER LONGITUDINAL LAYER

SMALL INTESTINAL SEROSA
(VISCERAL PERITONEUM)
•Completely covers the SMALL INTESTINE, except for a MAJOR PORTION of the DUODENUM

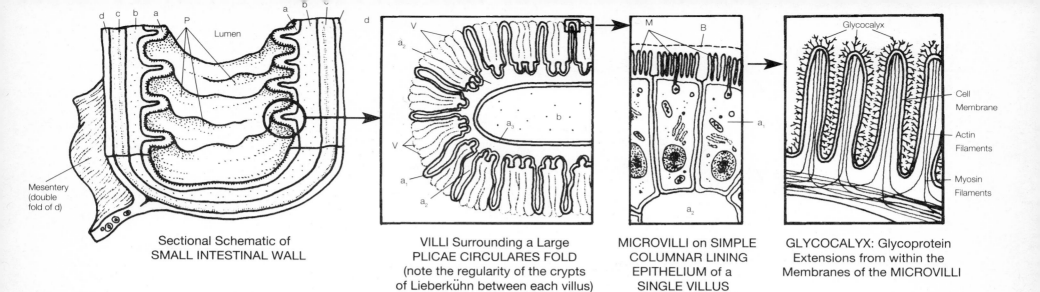

Sectional Schematic of SMALL INTESTINAL WALL

VILLI Surrounding a Large PLICAE CIRCULARES FOLD (note the regularity of the crypts of Lieberkühn between each villus)

MICROVILLI on SIMPLE COLUMNAR LINING EPITHELIUM of a SINGLE VILLUS

GLYCOCALYX: Glycoprotein Extensions from within the Membranes of the MICROVILLI

Labels in figure: d c b a, P, Lumen, a, b, c, d, Mesentery (double fold of d), V, a_2, a_3, b, V, a_1, a_2, M, B, a_1, a_2, Glycocalyx, Cell Membrane, Actin Filaments, Myosin Filaments

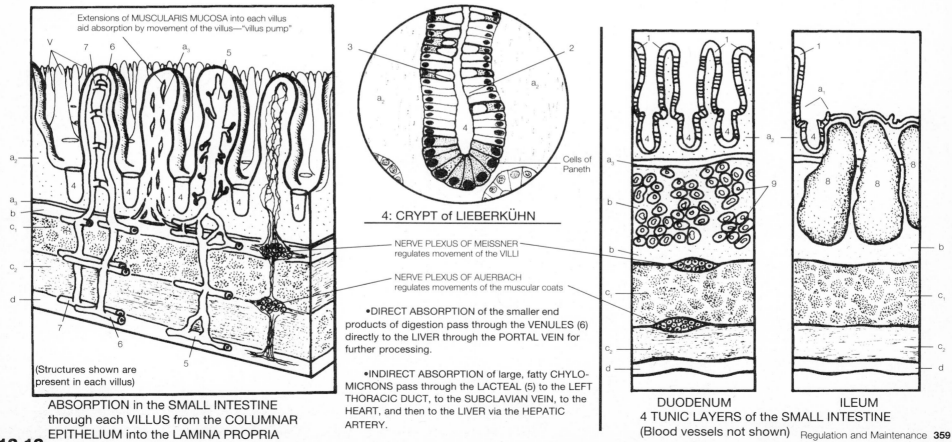

ABSORPTION in the SMALL INTESTINE through each VILLUS from the COLUMNAR EPITHELIUM into the LAMINA PROPRIA

Extensions of MUSCULARIS MUCOSA into each villus aid absorption by movement of the villus—"villus pump"

(Structures shown are present in each villus)

4: CRYPT of LIEBERKÜHN

Cells of Paneth

NERVE PLEXUS OF MEISSNER regulates movement of the VILLI

NERVE PLEXUS OF AUERBACH regulates movements of the muscular coats

•DIRECT ABSORPTION of the smaller end products of digestion pass through the VENULES (6) directly to the LIVER through the PORTAL VEIN for further processing.

•INDIRECT ABSORPTION of large, fatty CHYLO-MICRONS pass through the LACTEAL (5) to the LEFT THORACIC DUCT, to the SUBCLAVIAN VEIN, to the HEART, and then to the LIVER via the HEPATIC ARTERY.

DUODENUM

ILEUM

4 TUNIC LAYERS of the SMALL INTESTINE (Blood vessels not shown)

DIGESTIVE SYSTEM: THE ALIMENTARY CANAL: THE GI TRACT
The Large Intestine: Gross and Microscopic Anatomy

a = yellow b = light orange
c = orange, red d = grey, neutral
12 = yellow orange A–D = cool colors

★ See Charts #6, #7

★ The LARGE INTESTINE is about 1.5 meters (5 feet) long and about 6.5 cm (2 1/2 in.) in diameter. It is named the large intestine because its diameter is larger than the small intestine. The LARGE INTESTINE extends from the ILEOCECAL VALVE to the ANUS. It is attached to the POSTERIOR ABDOMINAL WALL by the MESOCOLON, a supportive specialized portion of the DORSAL MESENTERY. It is larger but shorter than the small intestine and is regularly arranged (fixated) within the abdominal cavity; it is shaped like an upside-down *U.*

★ FUNCTIONS OF THE LARGE INTESTINE

■ There is *LITTLE* or *NO* DIGESTIVE FUNCTION. NO DIGESTIVE ENZYMES are secreted in the COLON. (Remember, by the time the intestinal contents reach the LARGE INTESTINE, DIGESTION and ABSORPTION are almost totally complete.)

■ What little chemical digestion does occur is by BACTERIAL ACTION, not ENZYMES, in the LAST STAGES OF CHYMAL BREAK-DOWN.

■ There are also some VITAMINS synthesized by BACTERIA (B and K).

■ There is a great deal of WATER absorbed in the COLON rather than in the SMALL INTESTINE. (Thus, there is an efficient conservation of body fluids during most of digestion and absorption.)

■ Timewise, the RESIDUE OF DIGESTION enters the COLON for the longest part of the digestive journey. With the absorption of WATER and ELECTROLYTES from the CHYME into the BLOOD-STREAM, the contents of the COLON are gradually dehydrated and assume the consistency of semisolid or very hard FECES.

■ FECES FORMATION and STORAGE (See Chart #6)

■ EXPULSION OF FECES from the body via the RECTUM and muscular ANAL CANAL, regulated by the ANS.

MICROSCOPIC ANATOMY

★ The LARGE INTESTINE contains the same 4 basic tunicas as the SMALL INTESTINE, although there are many structural differences:

■ LARGE INTESTINAL MUCOSA

•There are *no* large, deep permanent folds (PLICAE CIR-CULARES) in the mucosa

•There are *no* villi (and thus no microvilli with their digestive enzymes) to be found in the mucosa

•Does contain SIMPLE COLUMNAR LINING EPITHELIUM

•Does contain interspersed GOBLET CELLS (many more than in the small intestine) which secrete MUCUS. This MUCUS lubricates the contents in the colon as they pass through.

•Does contain ABSORPTIVE CELLS that absorb WATER and ELECTROLYTE SALTS into the BLOOD STREAM (then carried by the PORTAL CIRCULATION to the LIVER before they enter into general circulation). Body fluids are conserved and FECES are dried.

•CRYPTS OF LIEBERKÜHN do not contain secretory PANETH CELLS in the bottom of their pits in the colon, as there is little need for secretory digestive enzymes. Digestive cells are present, but they lack intrinsic enzymatic activity.

4 Regions of the LARGE INTESTINE

A │ CECUM │ HANGS *BELOW* ILEOCECAL VALVE OPEN TO THE ASCENDING COLON

Ⓑ │ COLON │ MAJOR REGION

C │ RECTUM │ TERMINAL 7 1/2 in. of GI Tract

D │ ANAL CANAL │ TERMINAL INCH OF RECTUM

The CECUM: BLIND POUCH

1 │ ILEOCECAL VALVE │ (See 12.11)

2 │ VERMIFORM APPENDIX │

The COLON: 4 Regions

Ⓑ₁ │ ASCENDING COLON │

Ⓑ₂ │ TRANSVERSE COLON │

Ⓑ₃ │ DESCENDING COLON │

Ⓑ₄ │ SIGMOID (PELVIC) COLON │

The RECTUM

3 │ RECTAL TRANSVERSE FOLDS │

4 │ PELVIC PORTION │

5 │ RECTAL AMPULLA │

6 │ ANAL COLUMNS │

The ANAL CANAL

7 │ ANUS │

8 │ INTERNAL ANAL SPHINCTER │

9 │ EXTERNAL ANAL SPHINCTER │

MICROSCOPIC ANATOMY

THE 4 BASIC TUNIC LAYERS

a │ LARGE INT. MUCOSA │

b │ LARGE INT. SUBMUCOSA │

c │ LARGE INT. MUSCULARIS │

d │ LARGE INTESTINAL SEROSA │

LARGE INTESTINAL MUCOSA

a₁ │ LINING EPITHELIUM WITH ABSORPTIVE CELLS AND GOBLET CELLS │

10 │ CRYPTS OF LIEBERKÜHN │

a₂ │ LAMINA PROPRIA │

a₃ │ MUSCULARIS MUCOSA │

LARGE INTESTINAL MUSCULARIS

c₁ │ INNER CIRCULAR LAYER │

c₂ │ OUTER LONGITUDINAL LAYER │ TAENIAE COLI

Adaptations of the TUNICAS of the LARGE INTESTINE

11 │ HAUSTRA │ HAUSTRATIONS

12 │ TAENIAE COLI │ 3 MIDDLE FLAT SHEETS FROM LONGITUDINAL MUSCULARIS (OUTER LAYER)

13 │ EPIPLOIC APPENDAGES │ FAT-FILLED POUCHES ATTACHED TO TAENIAE COLI IN SEROUS LAYER

14 │ PLICAE SEMILUNARIS │ TRANSVERSE FOLDS OF MUCOSA LYING BETWEEN HAUSTRA

▪ LARGE INTESTINAL SUBMUCOSA
- Similar to the rest of the GI tract; loose fibrous tissue with BLOOD VESSELS, LYMPH VESSELS, and NERVES.

▪ LARGE INTESTINAL MUSCULARIS
- The outer longitudinal layer is incomplete and does not form a continuous sheet around the wall as it does in the rest of the alimentary canal. This smooth muscle layer is broken up into 3 flat bands or sheets, which run most of the length of the large intestine, placed equidistantly around the circumference of the organ. Since these TAENIAE COLI are not as long as the gut, their contractions gather the colon into a series of sacculated pouches called HAUSTRA (HAUSTRATIONS). These pouches and a large diameter visually distinguish the LARGE INTESTINE from the SMALL INTESTINE.

▪ LARGE INTESTINAL SEROSA
- Part of VISCERAL PERITONEUM

★ See Chart #7 for a summary of the mechanical activity of the entire alimentary canal, including the types of motility in the large intestine.

RECTAL AREA (c)

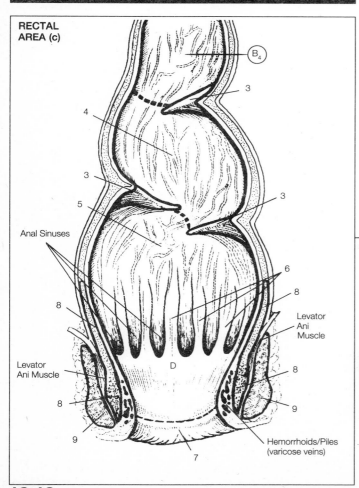

Anal Sinuses

Levator Ani Muscle

Levator Ani Muscle

Hemorrhoids/Piles (varicose veins)

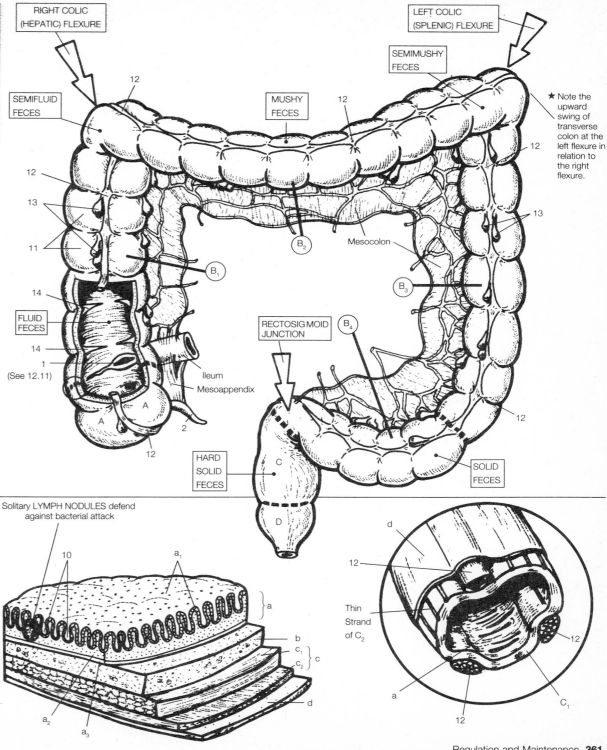

RIGHT COLIC (HEPATIC) FLEXURE

LEFT COLIC (SPLENIC) FLEXURE

SEMIMUSHY FECES

SEMIFLUID FECES

MUSHY FECES

★ Note the upward swing of transverse colon at the left flexure in relation to the right flexure.

Mesocolon

FLUID FECES

RECTOSIGMOID JUNCTION

Ileum

Mesoappendix

HARD SOLID FECES

SOLID FECES

1 (See 12.11)

Solitary LYMPH NODULES defend against bacterial attack

Thin Strand of C_2

DIGESTIVE SYSTEM: ACCESSORY ORGANS OF THE ABDOMINAL REGION
The Gallbladder and the Pancreas

★ See Chart #8, 8.7, 12.2

★ The GALLBLADDER, PANCREAS, and LIVER are the 3 accessory digestive organs in the abdominal cavity that aid in the CHEMICAL BREAKDOWN OF FOOD. The LIVER and a portion of the PANCREAS function as EXOCRINE GLANDS, essential to the digestive process in that their secretions are transported to the GI tract (DUODENUM) via a SYSTEM OF DUCTS.

★ The GALLBLADDER is a pear-shaped, saclike organ located in a fossa of the visceral (or inferior) surface of the LIVER (RIGHT LOBE).
■ 3–4 in. long
■ Contains RUGAE (similar to the stomach) that allow it to expand to its pear shape when filled with BILE (up to 2 ounces)
■ FUNCTIONS AS A STORAGE SITE FOR BILE. BILE is continuously manufactured by the LIVER. The BILE leaving the LIVER enters the GALLBLADDER via drainage of the R. and L. HEPATIC DUCTS, COMMON HEPATIC DUCT, and CYSTIC DUCT.
■ When BILE is needed (food enters GI tract) the middle muscularis of the GALLBLADDER contracts, ejecting the BILE down the DUCT SYSTEM to the DUODENUM, as ODDI'S SPHINCTER opens.
■ When the small intestine is empty of food, the SPHINCTER OF ODDI at the DUODENUM closes, and the BILE is forced back up the duct system into the GALLBLADDER for STORAGE.
■ Gallbladder also CONCENTRATES BILE through the loss of fluids absorbed by the GALLBLADDER MUCOSA. The capacity of viscid concentrated bile is 50–75 ml, equivalent to 1 1/2 pints (710 ml) of LIVER BILE. Among the functions of BILE is included the *emulsification* of undigested FAT GLOBULES into finer fat droplets in the DUODENUM (See 12.2, Chart #8)

★ The PANCREAS is a soft, lobulated, spongy, glandular organ about 6 in. long and 1 in. thick. It lies posterior to the great curvature of the STOMACH and nestles into the DUODENAL CURVE, attached to the DUODENUM by the MAIN PANCREATIC DUCT (and sometimes an AC-CESSORY PANCREATIC DUCT).
■ Has both EXOCRINE and ENDOCRINE functions:
■ The endocrine function is performed by ISLETS OF LANGER-HANS, clusters of cells (alpha, beta, delta cells), that secrete hormones directly into the blood stream. (REGULATES BLOOD SUGAR LEVEL)
■ Within the lobules around the ISLETS are numerus excretory secretory units called ACINI. Each ACINUS consists of a SINGLE LAYER of EPITHELIAL CELLS that empty PANCREATIC JUICE into a LUMEN DUCT, which carries the juice to the PANCREATIC DUCTS, and then to the DUODENUM. (COMPLETE DIGESTION of food molecules requires action of BOTH PANCREATIC ENZYMES and BRUSH BORDER EN-ZYMES.)
■ PANCREATIC JUICE contains WATER, PROTEIN, BICARBON-ATES, and mostly inactive DIGESTIVE ENZYMES which are triggered into action when released into the DUODENUM.
3 IMPORTANT ENZYMES are :

TRYPSIN ⟶ digests PROTEIN PEPTONES into AMINO ACIDS
(inactive form = TRYPSINOGEN)

AMYLASE ⟶ digest STARCH into MALTOSE and short chains of Glucose Molecules
(always active)

LIPASE ⟶ digests TRIGLYCERIDES cleaving FATTY ACIDS from GLYCEROL
(always active)

Ⓖ GALLBLADDER

Regions of the GALLBLADDER

a FUNDUS

b BODY

c NECK

d HARTMANN'S POUCH — necklike pouch; frequent reservoir for lodging of gallstones

System of ACCESSORY ORGAN DUCTS:
EXCRETORY DUCT from the GALLBLADDER
1 CYSTIC DUCT

CYSTIC DUCT Enters EXCRETORY BILE DUCT from the LIVER:
2 COMMON HEPATIC DUCT

Union of CYSTIC DUCT and COMMON HEPATIC DUCT:
3 COMMON BILE DUCT

COMMON AMPULLA where the COMMON BILE DUCT and MAIN PANCREATIC DUCT MEET:
4 AMPULLA OF VATER

ELEVATED AREA in DESCENDING POR-TION of the DUODENUM FOR EXIT of AMPULLA:
5 GREATER DUODENAL PAPILLAE

CONTROLLING SMOOTH MUSCLE at TERMINAL END of the AMPULLA for FLUID EXIT:
6 SPHINCTER OF ODDI

Ⓟ PANCREAS

Regions of the PANCREAS

e HEAD

f BODY

g TAIL

EXOCRINE Portion of the PANCREAS:
("External" Digestive Secretions Travel via DUCTS)

7 MAIN PANCREATIC DUCT
DUCT of WIRSUNG

8 ACCESSORY PANCREATIC DUCT
DUCT of SANTORINI

9 ALVEOLAR SEROUS ACINI CELLS
secrete PANCREATIC JUICE

Endocrine Portion of the PANCREAS:
(See 8.7)
(Internal Hormonal Secretions Travel via capillaries to Blood Stream)

10 ISLETS OF LANGERHANS

α ALPHA CELLS
secrete GLUCAGON HORMONE

β BETA CELLS
secrete INSULIN HORMONE

Δ DELTA CELLS
secrete HGHIF (HUMAN GROWTH HORMONE INHIBITING FACTOR)

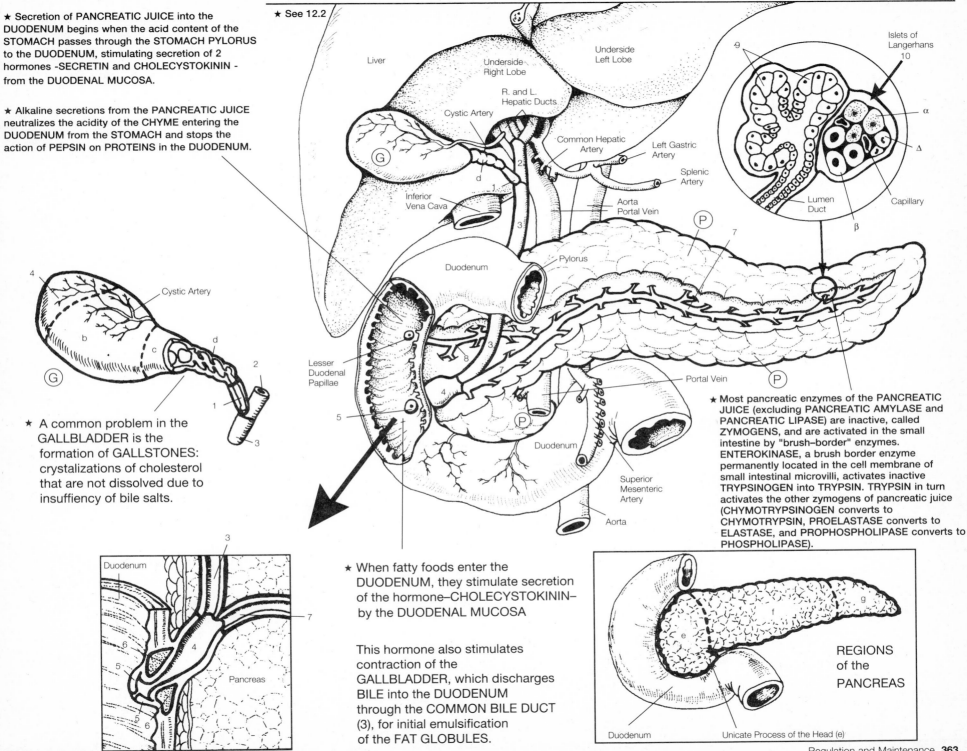

★ Secretion of PANCREATIC JUICE into the DUODENUM begins when the acid content of the STOMACH passes through the STOMACH PYLORUS to the DUODENUM, stimulating secretion of 2 hormones -SECRETIN and CHOLECYSTOKININ - from the DUODENAL MUCOSA.

★ Alkaline secretions from the PANCREATIC JUICE neutralizes the acidity of the CHYME entering the DUODENUM from the STOMACH and stops the action of PEPSIN on PROTEINS in the DUODENUM.

★ See 12.2

Liver

Underside Right Lobe

Underside Left Lobe

R. and L. Hepatic Ducts

Cystic Artery

Common Hepatic Artery

Left Gastric Artery

Splenic Artery

Aorta
Portal Vein

Inferior Vena Cava

G

d

Islets of Langerhans
10

9

α

Δ

Lumen
Duct

Capillary

β

Duodenum

Pylorus

P

7

P

Lesser Duodenal Papillae

Portal Vein

P

Duodenum

Superior Mesenteric Artery

Aorta

Cystic Artery

4

b

c

d

G

2

1

3

★ A common problem in the GALLBLADDER is the formation of GALLSTONES: crystalizations of cholesterol that are not dissolved due to insuffiency of bile salts.

★ Most pancreatic enzymes of the PANCREATIC JUICE (excluding PANCREATIC AMYLASE and PANCREATIC LIPASE) are inactive, called ZYMOGENS, and are activated in the small intestine by "brush–border" enzymes. ENTEROKINASE, a brush border enzyme permanently located in the cell membrane of small intestinal microvilli, activates inactive TRYPSINOGEN into TRYPSIN. TRYPSIN in turn activates the other zymogens of pancreatic juice (CHYMOTRYPSINOGEN converts to CHYMOTRYPSIN, PROELASTASE converts to ELASTASE, and PROPHOSPHOLIPASE converts to PHOSPHOLIPASE).

Duodenum

3

7

6

5

4

5   6

Pancreas

★ When fatty foods enter the DUODENUM, they stimulate secretion of the hormone–CHOLECYSTOKININ– by the DUODENAL MUCOSA

This hormone also stimulates contraction of the GALLBLADDER, which discharges BILE into the DUODENUM through the COMMON BILE DUCT (3), for initial emulsification of the FAT GLOBULES.

g

f

e

REGIONS
of the
PANCREAS

Duodenum

Unicate Process of the Head (e)

DIGESTIVE SYSTEM: ACCESSORY ORGANS OF THE ABDOMINAL REGION
The Liver: Gross Anatomy

Ⓛ = reddish brown A = red
IVC = blue Ⓖ = yellow-green
Ligaments = yellows and warm colors

★ See Chart #9

★ The LIVER is the largest single organ in the body, weighing from 3–4 lbs. in the adult. It is a highly complex organ that we cannot live without. The LIVER is the site of a wide variety of SYNTHETIC, STORAGE, and EXCRETORY FUNCTIONS. It is reddish brown in color, owing to its high-vascularity. (See Chart #9)

★ The LIVER is located directly beneath the DIAPHRAGM on the right side of the body occupying the RIGHT HYPO-CHONDRIAC REGION of the ABDOMINAL CAVITY, the upper portion of the EPIGASTRIC REGION, and sometimes extending into the LEFT HYPOCHONDRIAC REGION (see 1.3).
  • Level with the bottom of the STERNUM
  • Undersurface is concave
  • Covers the STOMACH, DUODENUM, RIGHT COLIC (HEPATIC) FLEXURE OF THE COLON, RIGHT KIDNEY, and SUPRARENAL (ADRENAL) CAPSULE

★ The LIVER is *almost completely* covered by VISCERAL PERI-TONEUM (exception is a BARE AREA, mainly in the posterior portion of the RIGHT LOBE). It is *completely* covered by a tough, fibrous connective tissue layer, GLISSON'S CAPSULE (not shown), that lies beneath the PERITONEUM. This capsule is thickest at the transverse fissure (hilum) where the blood vessels and hepatic duct enter the organ at the PORTA.

★ The LIVER is divided into 4 lobes, a large RIGHT LOBE, a smaller LEFT LOBE, and 2 very small lobes: QUADRATE and CAUDATE. The LIVER is connected by 5 ligaments to:
① undersurface of the DIAPHRAGM
② ANTERIOR ABDOMINAL WALL
  ■ FALCIFORM LIGAMENT
    • Divides the RIGHT and LEFT LOBE
    • Attaches to the DIAPHRAGM
  ■ ROUND LIGAMENT
    • Contained partially within FALCIFORM LIGAMENT
    • Extends to the UMBILICUS (Navel)
  ■ CORONARY LIGAMENT
    • Extends over medial surface of the RIGHT LOBE from the posterior edge of the LIVER
    • Attaches to the DIAPHRAGM
  ■ RIGHT and LEFT TRIANGULAR LIGAMENTS
    • Lie on the lateral aspects of the RIGHT and LEFT LOBES, and pass to the DIAPHRAGM
  ■ LESSER OMENTUM (See 12.4)
    • Passes from the lesser curvature of the STOMACH to the TRANSVERSE FISSURE of the LIVER
  ■ LIGAMENTUM VENOSUM
    • Lies between CAUDATE and LEFT LOBES
    • Connects left branch of the PORTAL VEIN to the INFERIOR VENA CAVA

Ⓛ **LIVER**

Area Not Invested in PERITONEUM
BA **BARE AREA**

Divided into 4 LOBES:
2 MAJOR LOBES
1 **RIGHT LOBE**

2 **LEFT LOBE**

2 MINOR LOBES   Associated with RIGHT LOBE
3 **QUADRATE LOBE** ADJACENT to GALLBLADDER

4 **CAUDATE LOBE**   near INFERIOR VENA CAVA

---

5 Connecting LIGAMENTS
5 **FALCIFORM LIGAMENT**

6 **ROUND LIGAMENT/ LIGAMENTUM TERES** DEGENERATED PRODUCT of LEFT UMBILICAL VEIN of FEMIS

7 **CORONARY LIGAMENT** BECOMES PARIETAL PERITONEUM of DIAPHRAGM

8 **R. AND L. TRIANGULAR LIGAMENTS**

Other Related Structures
9 **LESSER OMENTUM**

10 **LIGAMENTUM VENOSUM** OBLITERATED DUCTUS VENOSUS of the FETUS

Ⓖ **GALLBLADDER AND CYSTIC DUCT**

IVC **INFERIOR VENA CAVA**

A **AORTA**

UNIQUE BLOOD SUPPLY of the LIVER
Unlike any other organ, the LIVER has 2 sources of BLOOD:
■ 80% of the BLOOD is received via the PORTAL VEIN from the intestine. The LIVER is the first organ to receive blood from the intestines. This deoxygenated blood is filled with the major portion of the end products of digestion and decomposition products.
■ 20% of the BLOOD is oxygenated and is supplied to the LIVER from the AORTA via the HEPATIC ARTERY.

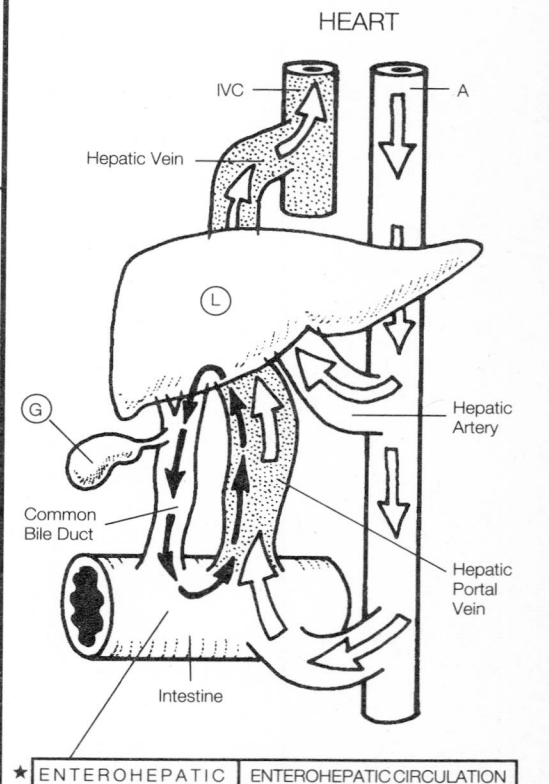

HEART

IVC
A
Hepatic Vein
Ⓛ
Ⓖ
Common Bile Duct
Hepatic Artery
Hepatic Portal Vein
Intestine

★ ENTEROHEPATIC CIRCULATION allows certain compounds excreted in the BILE to return to the LIVER via intestinal absorption and hepatic portal vein.

ENTEROHEPATIC CIRCULATION of UROGLOBIN

Bile pigment (BILIRUBIN) is converted in the intestine into UROBILINOGEN. Some is excreted in the FECES, some is recycled through the BILE, or filtered by the KIDNEYS into the URINE.

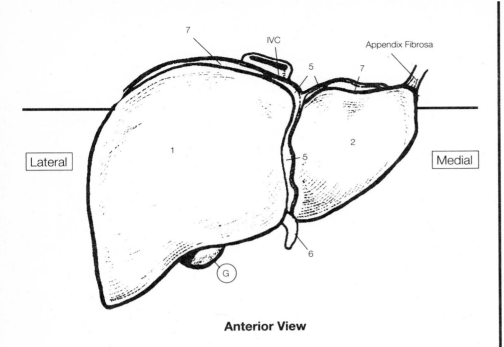

Anterior View

Lateral — Medial

7 · IVC · Appendix Fibrosa · 5 · 7 · 1 · 5 · 2 · 5 · 6 · G

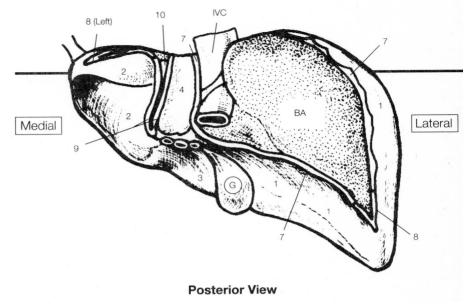

Posterior View

Medial — Lateral

8 (Left) · 10 · IVC · 7 · 7 · 2 · 4 · 2 · BA · 1 · 9 · 3 · G · 1 · 7 · 8

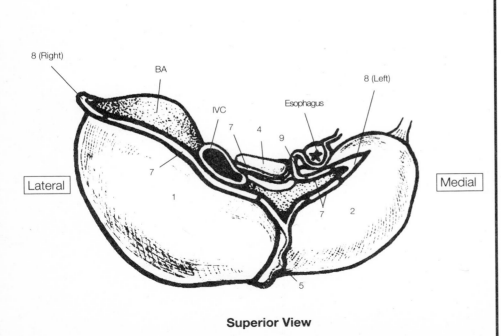

Superior View

Lateral — Medial

8 (Right) · BA · IVC · 7 · 4 · 9 · Esophagus · 8 (Left) · 7 · 1 · 7 · 2 · 5

Inferior View Exposes the 2 MINOR LOBES, the HILUS ...here the Blood Vessels and HEPATIC DUCT Enter (and the underlying VISCERAL IMPRESSIONS)

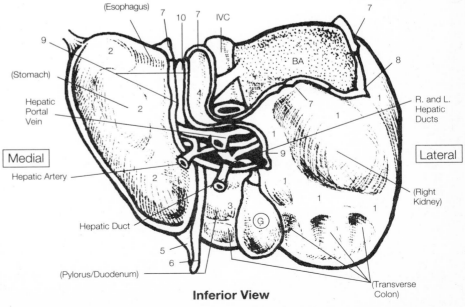

Inferior View

Medial — Lateral

(Esophagus) · 7 · 10 · 7 · IVC · 7 · 9 · 2 · BA · 8 · (Stomach) · 4 · 7 · 1 · Hepatic Portal Vein · 2 · 1 · R. and L. Hepatic Ducts · Hepatic Artery · 2 · 9 · 1 · (Right Kidney) · 3 · 1 · Hepatic Duct · 5 · G · 6 · 1 · (Pylorus/Duodenum) · (Transverse Colon)

DIGESTIVE SYSTEM: Accessory Organs of the Abdominal Region
The Liver: Microscopic Anatomy

★ Although the LIVER is divided into 4 LOBES, the internal structure and overall function of all lobes is the same. The microarchitecture of the LIVER is very complex, and STRUCTURE AND FUNCTION are CLOSELY RELATED. The structural and functional unit of the LIVER is the LIVER LOBULE:

■ Generally cylindrical, polyhedral units about 2 mm in diameter, with 5–7 sides per lobule

■ Strands of connective tissue originating from the surrounding capsule (Glisson's Capsule) of the entire LIVER enter the liver parenchyma at the hilus and form the supporting network of the liver. They separate the LOBULES from one another. These channels of interlobular connective tissue carry a *constant trinity of vessels:*

• A branch of the HEPATIC PORTAL VEIN
• A branch of the HEPATIC ARTERY
• A BILE DUCT (sometimes also lymphatic vessels)

■ Each LOBULE has at its center a blood vessel called the CENTRAL VEIN.

■ Surrounding the CENTRAL VEIN are HEPATOCYTE PLATES, or CORDS OF HEPATIC CELLS, arranged in a more or less radial fashion, which interconnect with one another crosswise. These plates are only 1–2 cell-layers thick, allowing DIRECT CONTACT OF EACH HEPATOCYTE CELL WITH THE BLOOD.

■ Located at the periphery of each lobule (in general, equally spaced around it) are the 6-8 HEPATIC TRIADS OF VESSELS, which open into SINUSOIDAL SPACES *between* the HEPATOCYTE PLATES.

■ The ARTERIAL BLOOD, and PORTAL VENOUS BLOOD (which contains the absorbed end products of digestion) *mix* as the blood flows from the periphery of the lobule to the CENTRAL VEIN.

■ The CENTRAL VEINS of all the lobules converge to form the large HEPATIC VEIN, which drains all blood from the LIVER.

■ Partly lining the large SINUSOIDAL SPACES are "fixed" or immobile phagocytic KUPFFER CELLS which help to remove old red blood cells and bacteria from the blood. There are large intercellular gaps between the KUPFFER CELLS, which permit extreme *high permeability* of the HEPATOCYTE CELLS, more so than any other capillaries.

■ *Within* each HEPATOCYTE PLATE are thin channels called BILE CANALICULI. The HEPATOCYTE CELLS produce BILE and secrete it into these channels, which are drained at the periphery of the lobule by the BILE DUCTS.

■ The BILE DUCTS drain into larger Right and Left HEPATIC DUCTS that carry BILE AWAY FROM THE LIVER (to be stored in the GALLBLADDER, through the CYSTIC DUCT).

Structural Units of the LIVER:

(LL) **LIVER LOBULES**

Structures within each LOBULE:

1 **CENTRAL VEIN of LOBULE**

2 **SINUSOIDAL SPACES** SINUSOIDS

3 **KUPFFER CELLS** RETICULOENDOTHELIAL LINING OF SINUSOIDS

4 **HEPATOCYTE PLATES**

5 **BILE CANALICULI**

6–8 TRIADS of VESSELS
Surrounding Each LOBULE:
EACH HEPATIC TRIAD (HEPATIC TRINITY)

a **BRANCH of HEPATIC PORTAL VEIN**

b **BRANCH of HEPATIC ARTERY**

c **BILE DUCT**

Main VESSEL Draining All LOBULES:
6 **HEPATIC VEIN** (BRANCHES)

CONNECTIVE TISSUE Separating LOBULES from One Another
7 **INNER EXTENDED STRANDS of GLISSON'S CAPSULE**

★ See Chart #9
The LIVER'S WIDE VARIETY of FUNCTIONS:

■ Production of bile
■ Carbohydrate metabolism
■ Lipid metabolism
■ Protein synthesis
■ Detoxification
■ Storage
■ Other activities

Portion of Right Lobe sectioned to reveal inner Lobular units interspersed with blood vessels.

Note central veins within each LOBULE

Note LOBULES are separated from each other by connective tissue (7), which carries the constant TRINITY OF VESSELS

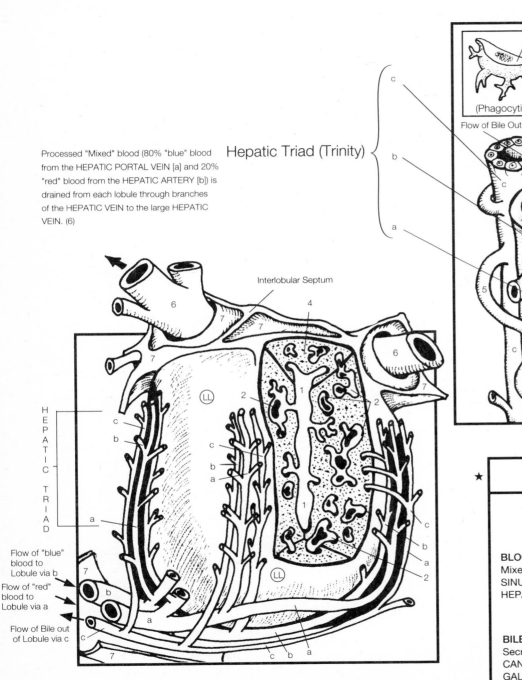

MICROARCHITECTURE within EACH LOBULE

Blood Flow to Hepatic Vein (6)

3

(Phagocytic)

Flow of Bile Out

Hepatic Triad (Trinity)

Processed "Mixed" blood (80% "blue" blood from the HEPATIC PORTAL VEIN [a] and 20% "red" blood from the HEPATIC ARTERY [b]) is drained from each lobule through branches of the HEPATIC VEIN to the large HEPATIC VEIN. (6)

Interlobular Septum

HEPATIC TRIAD

Flow of "blue" blood to Lobule via b

Flow of "red" blood to Lobule via a

Flow of Bile out of Lobule via c

CYLINDRICAL LIVER LOBULE UNIT

(Sectioned to reveal Central Vein and Sinusoids)

★ FLOW of BLOOD and BILE within EACH LIVER LOBULE

- ■ BLOOD and BILE NEVER MIX in the LIVER LOBULES
- ■ BLOOD and BILE TRAVEL in OPPOSITE DIRECTIONS
- ■ BLOOD FLOWS *BETWEEN* HEPATOCYTE PLATES
- ■ BILE FLOWS *WITHIN* HEPATOCYTE PLATES

BLOOD FLOW: From the periphery to the center of the lobule
Mixed blood from branches of HEPATIC ARTERY and PORTAL VEIN→through the SINUSOIDAL SPACES→to the CENTRAL VEIN→(then to branches of the→HEPATIC VEIN→to HEPATIC VEIN→to the INFERIOR VENA CAVA→GENERAL CIRCULATION)

BILE FLOW: From the center of the lobule to the periphery
Secreted within the HEPATOCYTE PLATES→through thin channels (BILE CANALICULI)→to the BILE DUCTS→(to the HEPATIC DUCTS→in and out of GALLBLADDER via CYSTIC DUCT→to the COMMON BILE DUCT→to DUODENUM and INTESTINES)

Chapter 12: *DIGESTIVE SYSTEM:* CHARTS

Chart #1 : The 7 Functional Processes of the Digestive System
(A One-Way Transportation System): Mouth ➡ Anus (See 12.1) (See Chart #7)

Process	Function
INGESTION	The process of taking material (particularly food) into the mouth in order to begin digestive processing.
MASTICATION	The act of CHEWING food. The comminution (pulverizing) of food and the insalivation of food is the first stage of digestion. Certain muscles close the mouth, raise and lower the mandible, tense the cheeks, and coordinate complex movements of the tongue. The teeth INCISORS *cut* the food and the MOLARS *grind* the food. The object is to get the food into as small pieces as possible since digestive enzymes work only on the surface of the food.
DEGLUTITION (SWALLOWING)	Movement of food from the mouth through the pharynx and esophagus to the stomach in such a way as to avoid food entering the trachea and respiratory system. A complicated muscular act, normally initiated voluntarily, but always completed reflexively—occurring in three stages.
PERISTALSIS	A progressive, wavelike movement (occuring in hollow tubes) that propels digesting material through the digestive tract. It is an involuntary movement induced reflexively by distention of the walls of the gastrointestinal tract. The wave consists of *contraction* of the smooth circular muscle above the distension with *relaxation* of the region (by smooth longitudinal muscle) immediately distal to the distended portion.The stomach uses the compactions to mix food with secretions as well as to empty the stomach. In the case of the small intestine, the action also spreads the contents throughout the length of the intestine. In the large intestine, mass peristalsis is a forced movement of short duration that moves contents from the middle of the transverse colon toward the rectum, occurring three or four times daily.
DIGESTION	The process by which food is broken down mechanically and chemically (hydrolytic reactions) in the alimentary canal and converted into absorbable forms. The chemical actions are brought about by a wide variety of enzymes in solution. Carbohydrates (starches), fats (lipids), and proteins are not absorbable until split into smaller molecules: CARBOHYDRATES are absorbed as MONOSACCHARIDES; FATS are absorbed as GLYCEROL and FATTY ACIDS; PROTEINS are absorbed as AMINO ACIDS. Salt, simple sugars (such as glucose), crystalloids, and water can be absorbed unchanged.
ABSORPTION	Movement of absorbable, digested nutrients mainly from the small intestine into either blood or lymph vessels. The absorbable monomer subunits of carbohydrates (monosaccharides) and the monomers of proteins (amino acids) are absorbed into the cardiovascular system, as are small aggregations of fat monomers (glycerol and fatty acids). Larger aggregations of fat monomers are absorbed into the lymph vascular system. Some substances move by active transport (glucose, amino acids, etc.) and others by diffusion (fructose). In the stomach, there is some absorption of water, alcohol, and salts. In the large intestine, water (important in the conservation of body fluids) and products of bacterial action are absorbed mainly in the ascending colon. In the lower bowel, some nutrients and drugs are absorbed.
DEFECATION	The evacuation of undigested and nonabsorbable material, bacteria, excretions, and water from the body. One does not have to eat in order to have bowel movements, because a large quantity of cellular material is desquamated from the epithelial lining of the intestinal tract each day. Once these food residues reach the rectum, they cause the urge to defecate.This sensation is related to periodic increase of pressure (to a predetermined limit established by habit) within the rectum and contracture of its musculature.

| Chart #2 | : The Four Layers of the Gastrointestinal Tract (See 12.3)

	Layer	Histology and Functions
Lumen (inside) ↓	MUCOSA	The mucosa consists of three tissue layers: (1) An inner layer of simple columnar epithelial cells with specialized mucous-secreting goblet cells interspersed; (2) An anchoring layer of connective tissue, termed the *lamina propria*; (3) The muscularis mucosa, which causes the walls to have small folds and thereby increases absorption surface area and allows for distension as new material is ingested.
	SUBMUCOSA	A layer of tissue made substantially of collagen and elastin, with blood vessels embedded in glands, and autonomic and somatic nerve branches.
	MUSCULARIS	A double layer of muscle tissue; (1) An inner circular layer that compresses food and, in some locations, is specialized into sphincters to regulate the passage of food; (2) An outer layer of longitudinal muscle that acts as the principal muscle of peristalsis. The muscular contractions are coordinated by the myenteric plexus to cause peristalsis.
Outside	SEROSA (VISCERAL PERITONEUM)	A layer of areolar connective tissue covered with simple squamous epithelial cells protecting and binding the digestive tract.

Chart #3 : Extensions and Folds of the Peritoneum: Serous Membranes of the Abdominal Cavity (See 12.4)

FALCIFORM
LIGAMENT: Attaches the (LIVER) to the (DIAPHRAGM and ANTERIOR ABDOMINAL WALL)

LESSER OMENTUM   Extended folds in the serosa of the (STOMACH and DUODENUM) suspending these organs from the (LIVER).
(2 layers, 2 folds):

GREATER OMENTUM   Extended folds in the serosa of the (STOMACH) that hang down like an apron in front of the intestines. Called
(4 layers, 2 folds):   the "fatty apron" because it contains adipose tissue and many lymph nodes.

MESENTERY   Projects from the posterior (dorsal) aspect of the abdominal cavity and supports the [GI tract].
(2 layers, 1 fold):

DORSAL MESENTERY: Outward fold of the serosa of the (SMALL INTESTINE) (portion of the MESENTERY whose tip attaches to the
                                posterior abdominal wall).

MESOCOLON: Specific portion of the MESENTERY that binds the (LARGE INTESTINE) to the posterior abdominal wall.

MESOAPPENDIX: Mesentery of the vermiform Appendix.

SEROUS
MESEMBRANES ▷
- Line the body wall of the thoracic and abdominal cavities as the PARIETAL PORTION
- Support and cover organs within these cavities as the VISCERAL PORTION
- Consist of a layer of simple squamous epithelium (mesothelium) and an underlying layer of aveolar connective tissue. (Mesothelium refers to the layer of cells derived from the Mesoderm lining the primitive body cavity in the embryo.)

Type	Location	Form and Function	Roots on Each Tooth	DECIDUOUS (1st dentition) Baby Teeth	Average Age of Eruption	Average Age of Loss	PERMANENT (2nd dentition)	Average Age of Eruption
INCISORS (Central and Lateral)	Anteriormost	Chisel-shaped for cutting and shearing	1	4 Pairs (8)	6–12 mos.	7 yrs.	4 Pairs (8)	6–9 yrs.
CANINES (Cuspids)	Anterior corners of the mouth	Cone-shaped for holding and tearing	1	2 Pairs (4)	16–20 mos.	10 yrs.	2 Pairs (4)	9–14 yrs.
PREMOLARS (1st and 2nd)	Posterior to the canines	Rounded cusps for crushing and grinding mastication)	2–3	——	——	——	4 Pairs (8)	1st: 9–13 yrs. 2nd: 10–14 yrs.
MOLARS (1st, 2nd, and 3rd)	Posterior to the Premolars at the back of the oral cavity	Rounded cusps for crushing and grinding (mastication)	2–3	4 Pairs (8) (No 3rd Molars)	12–24 mos.	11 yrs.	6 Pairs (12)	1st: 5–8 yrs. 2nd: 10–14 yrs. 3rd: 17–24 yrs.
				10 PAIRS: 5 Pairs in Upper Jaw 5 Pairs in Lower Jaw TOTAL: 20			16 PAIRS: 8 Pairs in Upper Jaw 8 Pairs in Lower Jaw TOTAL: 32	

Adapted from Kent M. Van De Graaff, *Human Anatomy*, 2d ed., p 588.

Chart #5 : Salivary Glands and Saliva (See 12.7)

MAJOR SALIVARY GLANDS

GLAND	Location	Duct	Entry into Oral Cavity	Type of Secretion
PAROTID	Below and in front of ear between skin and masseter muscle,	Parotid, or Stensen's, Duct	Opposite second upper molar	Watery serous fluid, (Salts and Amylase)
SUBMANDIBULAR (SUBMAXILLARY)	Inferior to body of mandible, midway along inside of jaw	Submandibular, or Wharton's Duct	Floor of mouth on either side of the Lingual Frenulum	Watery serous fluid (Salts, Amylase, Some Mucus)
SUBLINGUAL	Floor of mouth inside of the tongue	Several small ducts	Floor of mouth posterior to the Papilla of the Submandibular Duct	Thick mucoid salts and Amylase

Functions of Saliva	**Contents of Saliva**
•Solvent for cleansing teeth and dissolving food chemicals for taste •Enzymatic action (salivary amylase) for digestion of starch (carbohydrates) and fat •Mucus action lubricates pharynx to facilitate swallowing (deglutition) of food •After swallowing, washes out the mouth, and buffers chemical remnants of irritating substances	99.5% Water 0.5% Solutes (mostly salts) •Chloride salts activate salivary amylase •Bicarbonate and phosphate salts buffer chemicals (pH level 6.35–6.85) •Mucin protein forms mucus when dissolved in water •Urea and uric acid are waste products of the salivary glands •Lysozyme enzyme destroys bacteria, prevents tooth decay
•1000—1500 ml of saliva are secreted daily •Saliva is continuously secreted to keep the mucous membranes of the mouth moist •When food is ingested, larger amounts are secreted, even after food is swallowed	

Adapted from Kent M. Van De Graaff, *Human Anatomy*, 2d ed., p 590.

Chart #6 : **Contents of Feces (See 12.13)**

- Residues of indigestible parts of food not attacked by bacteria (e.g., fruit skins)
- Some water
- Inorganic material (mostly calcium and phosphates)
- Bile (pigments and salts)
- Shed epithelial cells from mucosa of the alimentary canal
- Large quantities of bacteria (can make up to 1/3 of the total solid weight)
- Products of bacterial decomposition
- Intestinal secretions (includes MUCUS)
- Leukocytes (migrated from the blood stream)
- Very little digestible food

Chart #7 : Mechanical Activity of the Alimentary Canal (See 12.13) (See Chart #1)

Region	Motility Type	Result
ORAL CAVITY	Mastication: cutting and grinding	Reduced particle size and mixing with saliva
ORAL CAVITY TO PHARYNX	Deglutition: swallowing action	Movement of masticated food with saliva into esophagus and not into trachea and respiratory system
ESOPHAGUS	By gravity if individual is upright and food is properly masticated. Peristalsis occurs if food becomes lodged in the esophagus.	Movement of food from upper esophagus to lower esophagus
STOMACH	Relaxation with swallowing	Aids filling of stomach, increases enzyme action;
	Mixing contractions	Allows food to mix with secretions to form chyme
	Emptying contractions	Moves chyme from stomach into small intestine
	"Hunger contractions"	Contractions develop when stomach has been habituated to mixing at regular intervals and feeding did not occur on schedule. Contractions can result in stimulating feeding behavior. Also stimulated by a low blood sugar level.
SMALL INTESTINE	Peristalsis	Movement of chyme through small intestine
	Segmental peristalsis	Peristalsis that begins and ends at a variety of locations in order to spread chyme throughout the length of the small intestine and forward toward the large intestine.
LARGE INTESTINE	Peristalsis	Movement of contents through large intestine (slow, sluggish)
	Haustral churning	Movement of material from one haustrum to another
	Mass movement	A very strong, peristaltic wave beginning in the middle of the transverse colon. Caused by food entering the stomach, initiated by the GASTROCOLIC REFLEX. Involving the action of the TAENIAE COLI, the powerful wave sweeps feces into the descending and sigmoid colons, and on into the rectum.
	"Antiperistalsis"	Tone waves moving backwards from the middle of the transverse colon into the ascending colon
	Defecation	Movement of fecal matter out of the large intestine into the environment

Adapted from Kent M. Van De Graaff, *Human Biology*, 2d ed., p 604.

Chart #8 : Bile Composition and Functions (See 12.14)

■ BILE COMPOSITION (thick, viscid; bitter taste)
- CHOLESTEROL
- LECITHIN
- MUCIN
- INORGANIC SALTS
- BILE PIGMENTS (BILIRUBIN, BILIVERDIN):

Pigments responsible for colors of BILE:
- LIVER BILE = straw color
- GALLBLADDER BILE = yellow, brown, green

■ BILE FUNCTIONS
■ EMULSIFYING ACTION (decreases SURFACE TENSION)
- BILE SALTS coat fat droplets in intestine (Separating larger FAT GLOBULES into finer fat droplets), preventing mass coalescence of fat. This makes hydrolysis of fats by PANCREATIC STEAPSIN (LIPASE) enzyme into fatty acids and glycerol easier.
■ HYDROTROPIC ACTION
- BILE SALTS combine with fats to form water soluble compounds that are more rapidly absorbed into the blood stream.

■ STIMULATES PERISTALSIS

Chart #9 : The Liver's Wide Variety of Functions (See 12.16)

■ Main chemical processing station of the body, collecting, in particular, the NEWLY ABSORBED NUTRIENTS from the intestine via the HEPATIC PORTAL VEIN

■ PRODUCTION OF BILE
- Continuous secretion of BILE SALTS for storage in the GALLBLADDER
- Conjugation and excretion of BILE PIGMENT (BILIRUBIN)

■ CARBOHYDRATE METABOLISM
Regulation of blood GLUCOSE concentration according to the body's needs. Example:
After a carbohydrate-rich meal:
- Conversion of blood GLUCOSE into GLYCOGEN and FAT (which can be stored)
During fasting:
- Production of GLUCOSE from stored liver GLYCOGEN or
- Conversion into GLUCOSE from noncarbohydrate molecules (amino acid or fatty acids) by gluconeogenesis
- Secretion of GLUCOSE into the blood

■ LIPID METABOLISM
- Synthesis of CHOLESTEROL and TRIGLYCERIDES
- Excretion of CHOLESTEROL in BILE into BLOOD
- Enzymes convert free fatty acids into KETONE BODIES (secreted during fasting)

■ PROTEIN SYNTHESIS
- Production of ALBUMIN (70% of TOTAL PLASMA PROTEIN)
- Production of PLASMA GLOBULINS (TRANSPORT PROTEINS)
- Production of CLOTTING FACTORS (FIBRINOGEN, PROTHROMBIN), and others

■ DETOXIFICATION
- Enzymes break down POISONS and/or transform them into less toxic compounds. (ex: AMMONIA is a toxic, nitrogenous by-product of the burning of amino acids for energy. AMMONIA is converted into harmless UREA.)
- Chemical alteration by biologically active molecules (drugs and hormones) and their excretion in BILE
- Phagocytosis by Kupffer cells of worn-out red and white blood cells, and bacteria

■ STORAGE
- GLYCOGEN
- Vitamins A, D, E, K, and B_{12} (antipernicious anemia factor)
- Poisons that cannot be broken down and excreted (ex: DDT)
- Storage of BLOOD (up to 1 Liter): REGULATION OF BLOOD VOLUME TO SYSTEMIC CIRCULATION

■ OTHER ACTIVITIES
- One of the MAIN SOURCES OF BODY HEAT
- Synthesis of HEPARIN (anticoagulant)
- Synthesis of HIPPURIC ACID and URIC ACID
- Synthesis of RED BLOOD CELLS in the FETUS

UNIT 4: REGULATION AND MAINTENANCE

Chapter 13 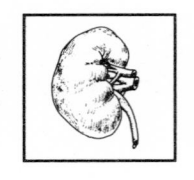 Urinary System

EXCRETORY ORGANS AIDING in WASTE ELIMINATION		
SYSTEM ORGANS	**PRIMARY ELIMINATION**	**SECONDARY ELIMINATION**
URINARY (KIDNEYS)	• H_2O • Soluble salts from protein catabolism • Nitrogenous substances (urea, uric acid, creatine, creatinine) • Inorganic minerals and salts	• CO_2 • Heat
DIGESTINE (GI TRACT)	• Solid wastes (indigestible residue) • Secretions • Bacteria	• CO_2 • Heat • H_2O • Salts
RESPIRATORY (LUNGS)	• CO_2	• H_2O vapor • Heat • Other gases
INTEGUMENTARY (SKIN)	• Heat	• H_2O • Salts • Minute quantities of urea (stimulated by Kidney inactivity)

Cellular metabolism of nutrients
produces a wide variety of by-products,
many of which can threaten BODY HOMEOSTASIS.
　　These include CARBON DIOXIDE, excess WATER
and HEAT, and AMMONIA and UREA, which are
toxic nitrogenous wastes of protein catabolism.

　　In addition, many essential ions and other
inorganic substances (and water) are ingested
through foods and fluids in greater
amounts than the body requires or needs.

　　All of the wastes and excess essential materials
must be ELIMINATED from the body in order
to maintain the body's required EQUILIBRIUM and HEALTH.
All of these substances are transported in the blood.

　　THE PRIMARY FUNCTION OF THE URINARY SYSTEM
IS TO *PRESERVE THE HOMEOSTASIS
(DYNAMIC EQUILIBRIUM) OF THE BODY
FLUIDS* THROUGH ACTION OF THE KIDNEYS BY:

- ◼ CONTROLLING THE COMPOSITION
  OF BLOOD

- ◼ CONTROLLING THE VOLUME
  OF BLOOD

This is achieved by:

- ◼ REGULATING (REMOVING AND RESTORING)
  THE BLOOD CONCENTRATIONS OF
  WATER AND SOLUTES

- ◼ EXCRETING SELECTED AMOUNTS OF
  VARIOUS SOLUTE WASTES OF METABOLISM
  (IN THE FORM OF URINE)
  (Solutes are solid substances that have
  been dissolved in a solution.)

THE KIDNEYS ARE ALSO RESPONSIBLE FOR
MAINTAINING A CONSTANT METABOLIC ACID-BASE
BALANCE.

This is achieved by:
- ◼ REGULATING THE BLOOD LEVEL OF H^+(PH) AND BICARBONATE (HCO_3^-)
  AND (NH_3) VERSUS (NH_4^+), AND ($HPO_4^=$) VERSUS ($H_2PO_4^-$)

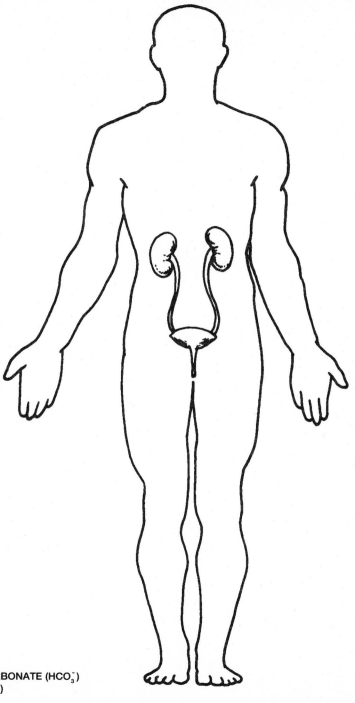

Anterior View

System Function

- **REGULATION OF BLOOD CHEMICAL
  COMPOSITION**

- **ELIMINATION OF WASTES**

- **REGULATION OF FLUID/ELECTROLYTE
  BALANCE AND VOLUME**

- **MAINTENANCE OF BODY ACID-BASE
  BALANCE**

Chapter Coloring Guidelines

- *You may want to choose your color
  palette by comparing any colored
  illustrations from your main textbook*

*Outer Cortex of Kidney = deep
                         reddish brown*

*Inner Medulla of Kidney = light
                          reddish brown*

*Arterial system = warm colors
Venous system = cool colors*

URINARY SYSTEM: GENERAL ORGANIZATION
The Urinary Tract

★ See Chart #1

The urinary system provides more functions than simply forming and discharging urine from the body.

★ The KIDNEYS do not only act as ORGANS OF EXCRETION OF UNDESIRED PRODUCTS OF METABOLISM.

The KIDNEYS also act as:
■ ORGANS OF CONSERVATION OF METABOLIC PRODUCTS

■ REGULATORS OF BLOOD CHEMICAL COMPOSITION

■ REGULATORS OF FLUID/ELECTROLYTE BALANCE AND VOLUME

■ REGULATORS OF BODY ACID-BASE BALANCE

■ CONTROL OF BLOOD PRESSURE VIA THE RENIN-ANGIOTENSIN MECHANISM (See 13.4)

■ RELEASE OF ERYTHROPROTEIN IN RED BLOOD CELL PRODUCTION

This is all achieved in a state of CONTINUOUS DYNAMIC EQUILIBRIUM, called HOMEOSTASIS, where the internal environment of the body is maintained by dynamic processes of FEEDBACK and REGULATION.

What is excreted in one instant may be deemed desirable the next instant, depending on the immediate balance required!

★ • The KIDNEYS are named for their similar shape to kidney beans.

• Because of the large size of the LIVER, the RIGHT KIDNEY is slightly lower than the LEFT.

Components of the URINARY SYSTEM

A **2 KIDNEYS**

B **2 URETERS**

C **1 URINARY BLADDER**

D **1 URETHRA**

Divisions of the URETHRA (MALE)

1 **PROSTATIC URETHRA**

2 **MEMBRANOUS URETHRA**

3 **SPONGY URETHRA**

★ The Left and Right Kidneys are supplied by the Left and Right Renal Arteries, respectively (branches of the Abdominal Aorta) and are drained by the Left and Right Renal Veins (tributaries of the Inferior Vena Cava)

★ • The KIDNEYS remove waste from the blood through the excretion of URINE from the KIDNEYS.

• The URINE is transported passively to the URETERS from the KIDNEYS.

• By PERISTALSIS and GRAVITATIONAL FORCES, the URINE is carried down the URETERS to the URINARY BLADDER.

• URINE is stored in the URINARY BLADDER until it exceeds 200–400 ml. It is then expelled (by conscious desire and an unconscious reflex: MICTURITION) through the canal of the URETHRA and discharged from the body.

(See 1.3, 12.4)

Hepatic Veins — Esophagus — Celiac Artery and Branches — Diaphragm (9) — Right Renal Artery — Right Renal Vein — Left Renal Artery — Left Renal Vein — LEFT KIDNEY (A) — 11th Rib — Superior Mesenteric Artery — 12th Rib — Inferior Mesenteric Artery — RIGHT KIDNEY (A) — Left Common Iliac Artery — ★ Right Testicular Artery and Vein (Right Testicular Vein empties into Inferior Vena Cava) — Left Common Iliac Vein — ★ Left Testicular Artery and Vein (Left Testicular Vein empties into the Left Renal Vein) — Iliacus Muscle — Rectum — Ductus (Vas) Deferens (cross over and in front of the Ureters) (See 14.1) — Symphysis Pubis — Inferior Vena Cava — Abdominal Aorta

ABDOMINAL ORGANS SURROUNDING ANTERIOR KIDNEY
(Reference numbers apply to above illustration only.)

1 - SUPRARENAL (ADRENAL) GLANDS	7 - PANCREAS
2 - LIVER	8 - JEJUNUM OF SMALL INTESTINE
3 - DUODENUM OF SMALL INTESTINE	9 - DIAPHRAGM M.
4 - COLON OF LARGE INTESTINE	10 - PSOAS MUSCLES
5 - SPLEEN	11 - QUADRATUS LUMBORUM M.
6 - STOMACH	12 - TRANSVERSUS ABDOMINUS M.

★ The KIDNEYS are exterior and posterior to the PERITONEAL LINING OF THE ABDOMINAL CAVITY (See 12.4). Therefore, they are called RETROPERITONEAL ORGANS. (The URETERS and ADRENAL GLANDS are also described as RETROPERITONEAL.) The KIDNEYS are just above the waist between the PARIETAL PERITONEUM and the POSTERIOR ABDOMINAL WALL, upon which they lie. The KIDNEYS, URETERS, and ADRENAL GLANDS, are the only PAIRED organs in the abdominal cavity.

★ • Because of the kidneys' retroperitoneal nature, many abdominal organs surround the anterior areas of the kidneys, providing a useful guide when studying locational relationships of abdominal organs. (1-8 above)

• The URINARY BLADDER is an ANTIPERITONEAL ORGAN (located in front of the peritoneal cavity).

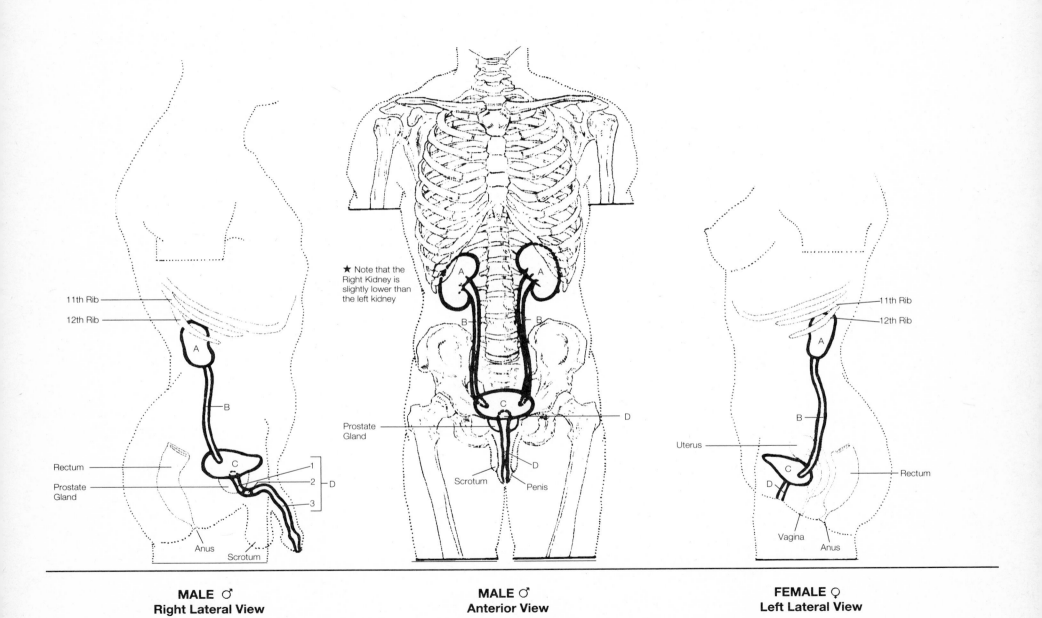

11th Rib
12th Rib

A

B

Rectum

Prostate
Gland

C

1
2 } D
3

Anus

Scrotum

MALE ♂
Right Lateral View

★ Note that the
Right Kidney is
slightly lower than
the left kidney

A   A

B   B

C

Prostate
Gland

D

D

Scrotum   Penis

MALE ♂
Anterior View

11th Rib
12th Rib

A

B

Uterus

C

D

Rectum

Vagina   Anus

FEMALE ♀
Left Lateral View

The URINARY TRACT

URINARY SYSTEM: THE KIDNEYS
Gross Anatomy (Right Kidney)

2 = yellow 8 = reddish 9 = pink
10 = light reddish brown 12 = yellow orange
13-14 = cool colors 16 = cream

★ See Chart #1, 13.3, 13.6

EXTERNAL ANATOMY: EACH KIDNEY IS EMBEDDED IN A 3-LAYERED FATTY, FIBROUS POUCH

3 SURROUNDING TISSUE LAYERS
1 | RENAL FASCIA | OUTER

2 | ADIPOSE CAPSULE | MIDDLE

3 | FIBROUS (RENAL) CAPSULE | INNER

CAVITIES
4 | HILUS (HILUM) | *

5 | RENAL SINUS | ‡

BLOOD VESSELS
6 | RENAL ARTERY |

7 | RENAL VEIN |

---

* HILUS = A depression or recess of the Kidney
where vessels (and nerves) enter and exit;
Entrance to the RENAL SINUS
  Entering vessels = → RENAL ARTERY
  Exiting vessels = → RENAL VEIN and URETER
  (also INNERVATION)

‡ RENAL SINUS = A deep concavity in the
center of the medial surface. This
inner hollow space goes halfway into
the kidney.

INTERNAL ANATOMY: Each KIDNEY

Outer Peripheral Part

8 | CORTEX | PRIMARILY VASCULAR

9 | RENAL COLUMNS | CORTICAL SUBSTANCE ENTERING MEDULLA AND SEPARATING THE RENAL PYRAMIDS

10 | RENAL MEDULLA (RENAL PYRAMIDS) |

11 | RENAL PAPILLAE | APEX OF RENAL PYRAMIDS

12 | FAT | BELOW RENAL COLUMNS

See 13.3

★ OUTER PERIPHERAL PART

■ (PARENCHYMA = CORTEX and RENAL PYRAMIDS)
  • Contains the PROCESSING CONDUCTORS,
the RENAL TUBULES. The RENAL TUBULES
consist of NEPHRONS (RENAL CORPUSCLES and
ATTACHED URINARY TUBULES), their
VASCULAR COMPONENTS and COLLECTING TUBULES.
■ CORTEX (reddish brown and granular):
  • Extends from FIBROUS CAPSULE to the
outer bases of the PYRAMIDS (and into the spaces
between them).
  • Contains the RENAL CORPUSCLES, the
CONVOLUTED PORTIONS (PROXIMAL and DISTAL) of
the URINARY TUBULES, and vascular components.
■ MEDULLA (lighter reddish and striated)
  • 8–18 PYRAMIDS
  • Contains the COLLECTING TUBULES,
and the STRAIGHT TUBULES OF NEPHRONS
(the ASCENDING and DESCENDING LOOPS OF HENLE),
the LOOPS OF HENLE, and vascular components.
  • The COLLECTING TUBULES converge to the APEX
of each PYRAMID and open into a MINOR CALYX
in the RENAL SINUS.
■ There are thousands of branches of
the RENAL ARTERY and RENAL VEIN in
both the CORTEX and MEDULLA (See 13.5).
This accounts for the reddish brown
color throughout the kidneys.

Inner Central Part

13 | MINOR CALYCES | CALYX (singular)

14 | MAJOR CALYCES |

15 | RENAL PELVIS |

16 | URETER |

★ INNER CENTRAL PART

Contains the PASSIVE CONDUCTORS,
the tubes that conduct formed urine
passively to the URETER.

■ Several MINOR CALYCES group
  together (in different areas) to
  form MAJOR CALYCES.

■ All the MAJOR CALYCES unite to
  form one funnel-shaped RENAL PELVIS.

■ The RENAL PELVIS collects urine
  from the CALYCES and funnels it
  into the URETER. (See 13.6)

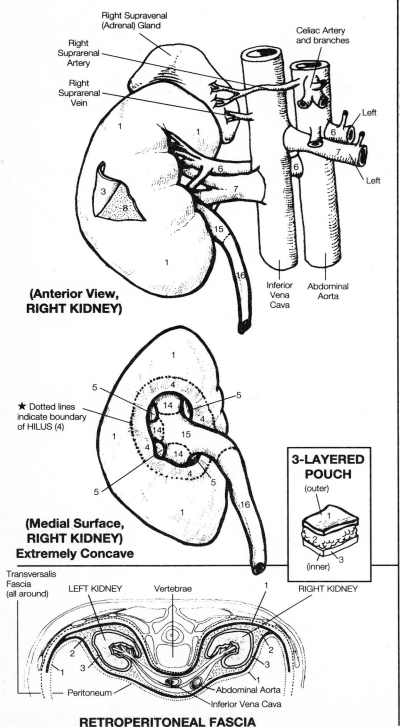

Right Supravenal
(Adrenal) Gland

Right
Suprarenal
Artery

Right
Suprarenal
Vein

Celiac Artery
and branches

Left

Left

Left

Inferior
Vena
Cava

Abdominal
Aorta

**(Anterior View,
RIGHT KIDNEY)**

★ Dotted lines
indicate boundary
of HILUS (4)

**(Medial Surface,
RIGHT KIDNEY)
Extremely Concave**

**3-LAYERED
POUCH**

(outer)

(inner)

RIGHT KIDNEY

Transversalis
Fascia
(all around)

LEFT KIDNEY

Vertebrae

Peritoneum

Abdominal Aorta

Inferior Vena Cava

RETROPERITONEAL FASCIA

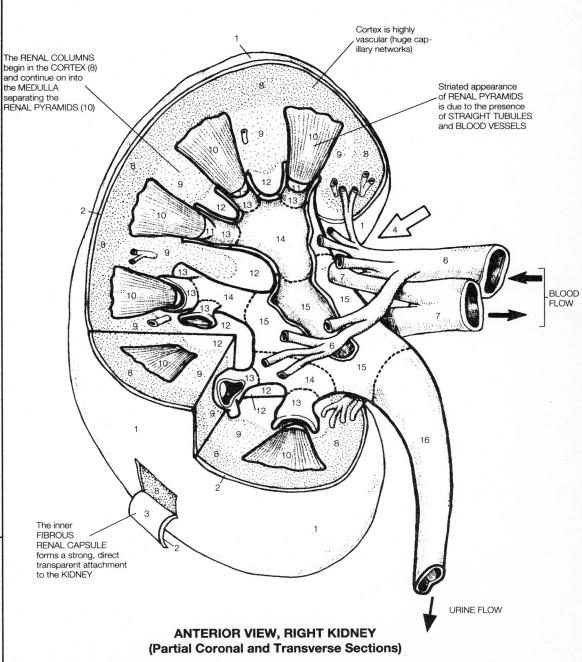

The RENAL COLUMNS
begin in the CORTEX (8)
and continue on into
the MEDULLA
separating the
RENAL PYRAMIDS (10)

Cortex is highly
vascular (huge cap-
illary networks)

Striated appearance
of RENAL PYRAMIDS
is due to the presence
of STRAIGHT TUBULES
and BLOOD VESSELS

BLOOD
FLOW

The inner
FIBROUS
RENAL CAPSULE
forms a strong, direct
transparent attachment
to the KIDNEY

URINE FLOW

**ANTERIOR VIEW, RIGHT KIDNEY
(Partial Coronal and Transverse Sections)**

13.2

13.3

URINARY SYSTEM: THE KIDNEYS
Microscopic Anatomy: **The Renal Tubule**

Regulation and Maintenance **382**

6-7 = *cool colors*
8 = *red-orange* 9 = *light blue*
11 = *warm colors*
★ See Chart #1

A RENAL CORPUSCLE = A GLOMERULUS and a BOWMAN'S CAPSULE
A NEPHRON = A RENAL CORPUSCLE and an ATTACHED URINARY TUBULE (with CONVOLUTED and STRAIGHT PORTIONS)
A RENAL TUBULE = A NEPHRON (and its VASCULAR SUPPLY) and a COLLECTING TUBULE (See Text below)

★ The NEPHRON is the structural and functional unit of the KIDNEY.

★ The NEPHRON consists of (I) a RENAL (MALPIGHIAN) CORPUSCLE (A GLOMERUS cluster of capillary blood vessels enveloped within a thin BOWMAN'S CAPSULE) and (II) its attached URINARY TUBULE (consisting of a PROXIMAL CONVOLUTED PORTION, a LOOP OF HENLE, and a DISTAL CONVOLUTED PORTION). These connect by ARCHED COLLECTING TUBULES with STRAIGHT COLLECTING TUBULES.

★ URINE is formed by filtratration in the RENAL CORPUSCLES, (Blood enters the GLOMERULUS CAPILLARY PLEXUS in each NEPHRON through the AFFERENT ARTERIOLE)
★ SELECTIVE REABSORPTION and SECRETION is achieved by the CELLS of the RENAL TUBULE.
• REABSORPTION involves the active movement of materials OUT OF the filtrate into tubule cells, and eventually into blood vessels that return the substances to the circulation.
• SECRETION involves active movement of materials INTO the FILTRATE (opposite of REABSORPTION)

There are about 1 million NEPHRONS/KIDNEY.

They are located in the CORTEX of the PARENCHYMA.

★ NEPHRONS:
    • HELP REGULATE COMPOSITION OF BLOOD AND VOLUME OF BLOOD (REMOVE SELECTED AMOUNTS OF H_2O AND SOLUTES)

    • REMOVE TOXIC WASTES FROM THE BLOOD

    • REGULATE BLOOD PH MAINTAIN ACID-BASE BALANCE

    • REABSORBS ESSENTIAL NUTRIENTS

    • CONSERVES WATER SUPPLY

A NEPHRON = (I and II)

I. | RENAL CORPUSCLE | MALPIGHIAN CORPUSCLE ⟶

II.   A URINARY TUBULE: (1-5)

1 | PROXIMAL CONVOLUTED TUBULE | BULK OF RENAL CORTEX

2 | DESCENDING LIMB of the LOOP of HENLE | STRAIGHT

3 | LOOP of HENLE |

4 | ASCENDING LIMB of the LOOP of HENLE | STRAIGHT

5 | DISTAL CONVOLUTED TUBULE |

6 | STRAIGHT COLLECTING TUBULE | DETERMINES FINAL URINE VOLUME

7 | PAPILLARY DUCT (COLLECTING DUCT) |

★ The DISTAL CONVOLUTED TUBULES (6) of *several nephrons* drain into a single COLLECTING TUBULE (7). In the MEDULLA, the STRAIGHT COLLECTING TUBULES pass through the RENAL PYRAMIDS and open at the RENAL PAPILLAE into the MINOR CALYCES through a number of short, large PAPILLARY DUCTS. (Numerous collecting tubules form a single PAPILLARY DUCT.)

I. A RENAL CORPUSCLE

8 | GLOMERULUS (CAPILLARY PLEXUS)
9 | BOWMAN'S (GLOMERULAR) CAPSULE

10 | AFFERENT ARTERIOLE
11 | EFFERENT ARTERIOLE

★ **ACTIVITIES of STRUCTURES of the RENAL TUBULE**

GLOMERULAR FILTRATION
RENAL CORPUSCLE → *Filters* blood plasma from the blood (H_2O and dissolved solutes), now called FILTRATE.

ACTIVE TRANSPORT SYSTEMS
PROXIMAL CONVOLUTED TUBULE → • *Reabsorption of solutes back into the blood*, (by surrounding capillaries preceding from the efferent arteriole) All glucose, small proteins, amino acids, salts, and some vitamins; 85% of filtrate is reabsorbed (85% solutes, followed by 85% water)
    • *Little or no secretion into tubule filtrate* (Some organic acids, antibiotics, and organic bases)
    • Wastes not reabsorbed

LOOP of HENLE → • *Absorption of H_2O into surrounding tissue cells* for H_2O conservation and increased concentration of urine solutes (high salts) in the interstitial fluid outside the tubule
    • Hypotonic fluid

DISTAL CONVOLUTED TUBULE COLLECTING TUBULE → • *Reabsorption of 10–15% of nutrients*
    • *Secretion* of hydrogen ions into tubule filtrate acidifies filtrate
    • Influenced by hormones (antidiuretic hormone permits water to pass through collecting tubule membrane to the salt-rich fluid surrounding the tubule, conserving it. Filtrate to be released becomes 2-3 times more concentrated than the blood.)

★ 99% of original filtrate is returned to tissues.
    1% of original filtrate is passed as urine to the calyces.

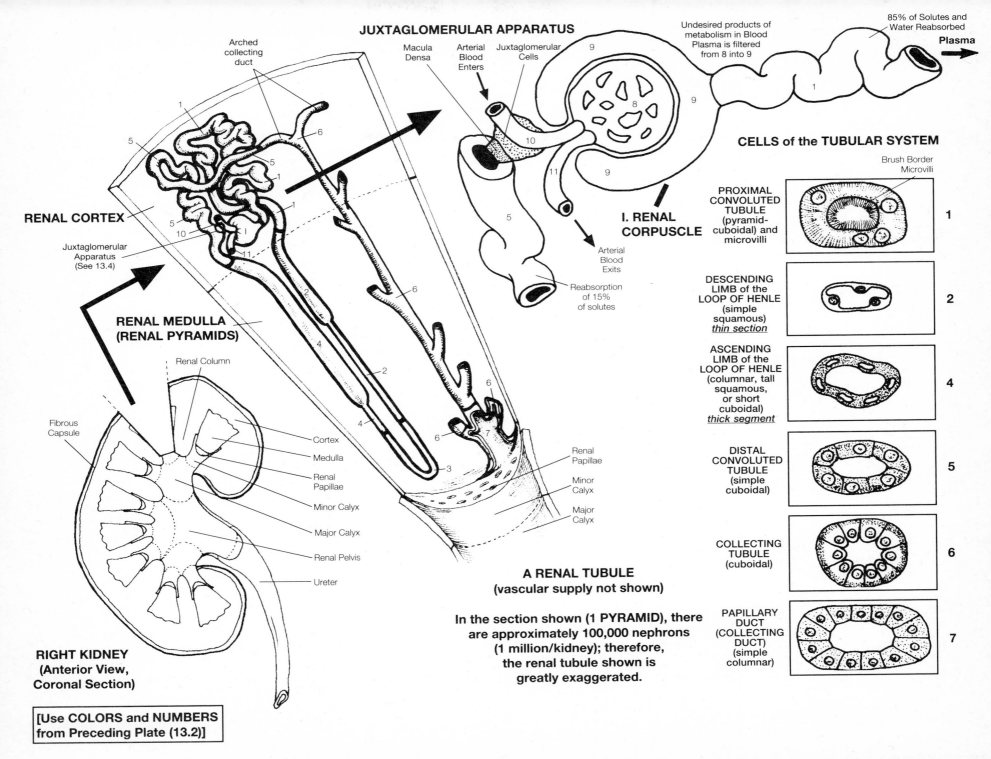

JUXTAGLOMERULAR APPARATUS

Arched collecting duct

Macula Densa

Arterial Blood Enters

Juxtaglomerular Cells

Undesired products of metabolism in Blood Plasma is filtered from 8 into 9

85% of Solutes and Water Reabsorbed

Plasma

RENAL CORTEX

Juxtaglomerular Apparatus (See 13.4)

RENAL MEDULLA (RENAL PYRAMIDS)

Arterial Blood Exits

Reabsorption of 15% of solutes

I. RENAL CORPUSCLE

CELLS of the TUBULAR SYSTEM

Brush Border Microvilli

PROXIMAL CONVOLUTED TUBULE (pyramid-cuboidal) and microvilli — **1**

DESCENDING LIMB of the LOOP OF HENLE (simple squamous) *thin section* — **2**

ASCENDING LIMB of the LOOP OF HENLE (columnar, tall squamous, or short cuboidal) *thick segment* — **4**

DISTAL CONVOLUTED TUBULE (simple cuboidal) — **5**

COLLECTING TUBULE (cuboidal) — **6**

PAPILLARY DUCT (COLLECTING DUCT) (simple columnar) — **7**

Fibrous Capsule

Renal Column

Cortex

Medulla

Renal Papillae

Minor Calyx

Major Calyx

Renal Pelvis

Ureter

Renal Papillae

Minor Calyx

Major Calyx

RIGHT KIDNEY (Anterior View, Coronal Section)

[Use COLORS and NUMBERS from Preceding Plate (13.2)]

A RENAL TUBULE (vascular supply not shown)

In the section shown (1 PYRAMID), there are approximately 100,000 nephrons (1 million/kidney); therefore, the renal tubule shown is greatly exaggerated.

13.3

URINARY SYSTEM: THE KIDNEYS: *Microscopic Anatomy:* The Renal Corpuscle and Glomerular Filtration

★ The undesired products of metabolism in the blood plasma (H_2O and dissolved solutes) are filtered through the GLOMERUS and into BOWMAN'S CAPSULE. They then travel through the tubular system.

★ The initial step in urine formation by the NEPHRONS is GLOMERULAR FILTRATION. Blood from the AFFERENT ARTERIOLE (carrying both wastes and useful nutrients) arrives at the GLOMERULUS at high pressure (60-75 mm Hg)

• The GLOMERULAR CAPILLARIES and their surrounding PODOCYTES act as a COARSE SIEVE, permitting pressure-caused passage of materials from the blood vessels into BOWMAN'S SPACE according to size or molecular weight (mw).
• Any substance with a mw of 10,000 or less passes freely through the FILTRATION SLITS between the PODOCYTE PEDICEL processes. Beyond 10,000 mw passage becomes more difficult.
• Nearly all plasma materials (Glucose, amino acids, vitamins, minerals, H_2O, and wastes) except FORMED ELEMENTS (i.e., red and white blood cells and platelets) can filter through the GLOMERULAR WALL (most plasma proteins—excluding small ALBUMIN—do not filter).

★ The open pores, or FENESTRATIONS, in the CAPILLARY ENDOTHELIUM of the GLOMERULUS allow 100–400 times the PERMEABILITY TO BLOOD PLASMA, H_2O, and DISSOLVED SOLUTES than the CAPILLARIES of SKELETAL MUSCLES.

A RENAL CORPUSCLE

1 | GLOMERULUS (CAPILLARY PLEXUS) |

2 | BOWMAN'S (GLOMERULAR) CAPSULE |

A GLOMERULUS

3 | CAPILLARY ENDOTHELIUM |

4 | BASEMENT MEMBRANE |

A BOWMAN'S CAPSULE: DOUBLE-WALLED CUP

⑤ | VISCERAL LAYER/PODOCYTES | INNER LAYER SURROUNDS CAPILLARIES

6 | BOWMAN'S SPACE | SPACE BETWEEN the 2 WALLS or LAYERS

7 | PARIETAL LAYER | OUTER LAYER

PODOCYTES: SPECIALIZED EPITHELIUM of VISCERAL LAYER of BOWMAN'S CAPSULE

8 | CELL BODY |

9 | PRIMARY BRANCH |

9a | SECONDARY BRANCH |

10 | PEDICELS | CYTOPLASMIC EXTENSIONS

11 FILTRATION SLITS SLIT PORES

JUXTAGLOMERULAR APPARATUS

★JUXTAGLOMERULAR APPARATUS is a structure consisting of myoepitheloid (contractile) cells forming a cuff around the AFFERENT ARTERIOLE .
• Regulates renal blood pressure through the production of RENIN. (Renin acts on ANGIOTENSIN to form ANGIOTENSIN I and II, a vasopressor which causes contraction of muscles of capillaries and arteries. This increases resistance to blood flow and elevates blood pressure.)
• ANGIOTENSIN II also stimulates ALDOSTERONE secretion which regulates SODIUM, CHLORIDE, and POTASSIUM METABOLISM.

12 | JUXTAGLOMERULAR CELLS | → Modified cells of the AFFERENT ARTERIOLE

13 | MACULA DENSA | → Modified cells of the DISTAL CONVOLUTED TUBULE

14 | AFFERENT ARTERIOLE |

15 | EFFERENT ARTERIOLE |

Direction of substances filtered by the kidney

★ ENDOTHELIAL-CAPSULAR MEMBRANE

■ CAPILLARY ENDOTHELIUM of the GLOMERULUS
Substances in the lumen of the capillary filter through open pores called FENESTRATIONS.
↓
■ BASEMENT MEMBRANE of the GLOMERULUS
(Has no open pores; a dialyzing membrane with myobrils. Negative charge of glycoprotein in the membrane repels large, negatively charged plasma protein molecules that could pass through the pores.)
↓
■ EPITHELIUM of VISCERAL LAYER of BOWMAN'S CAPSULE
(Contains PODOCYTES with PEDICEL extensions separated by FILTRATION SLITS. Solutes of blood plasma must pass through these slits to enter capsule interior called BOWMAN'S SPACE. Large proteins do not cross Visceral Layer.)

TUBULAR SYSTEM

★ ENDOTHELIAL-CAPSULAR MEMBRANE FILTER:
• FILTERS H_2O and SOLUTES out of the blood, which pass into BOWMAN'S SPACE and then into the RENAL TUBULE.

• Plasma protein molecules and formed blood elements DO NOT FILTER through the membrane. This UNFILTERED BLOOD circulates through the EFFERENT ARTERIOLE

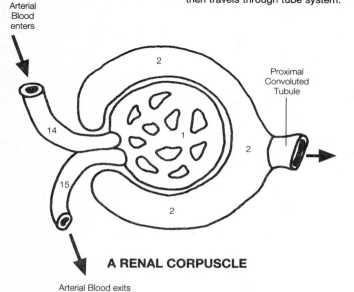

★ Undesired products of metabolism in the Blood Plasma are filtered from 1 into 2 then travels through tube system.

Arterial Blood enters

2

2

1

2

14

15

Proximal Convoluted Tubule

A RENAL CORPUSCLE

Arterial Blood exits (minus filtered plasma)

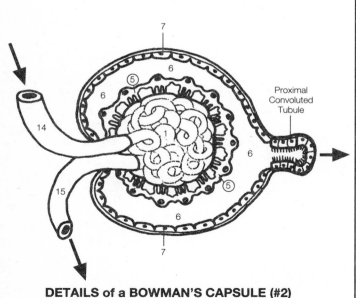

7

6

5

6

6

1

6

5

14

15

6

7

Proximal Convoluted Tubule

DETAILS of a BOWMAN'S CAPSULE (#2)

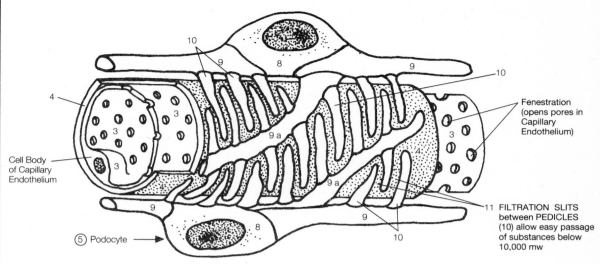

10

9

8

9

10

4

3

3

3

9a

9a

3

Cell Body of Capillary Endothelium

5 Podocyte

8

9

9

10

10

Fenestration (opens pores in Capillary Endothelium)

11 FILTRATION SLITS between PEDICLES (10) allow easy passage of substances below 10,000 mw

ENDOTHELIAL-CAPSULAR MEMBRANE: PODOCYTES ⑤ ENVELOP CAPILLARIES

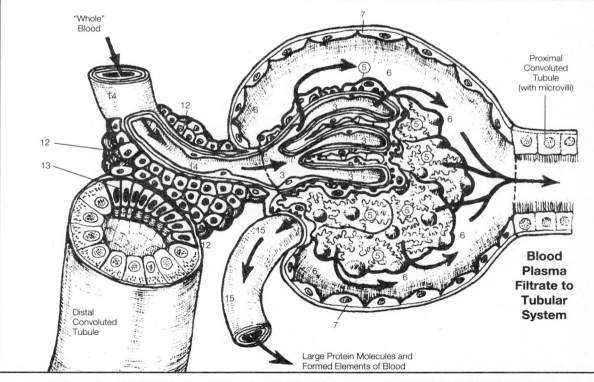

"Whole" Blood

7

12

14

12

13

14

3

3

6

5

6

6

5

3

3

5

5

5

6

6

6

5

15

15

12

7

Distal Convoluted Tubule

Large Protein Molecules and Formed Elements of Blood

Proximal Convoluted Tubule (with microvilli)

Blood Plasma Filtrate to Tubular System

SECTIONAL SCHEMATIC of RENAL CORPUSCLE and ASSOCIATED STRUCTURES

URINARY SYSTEM: THE KIDNEYS
Vascular Structure and Circulatory Pattern

1-8 = reds, warm colors
9-11 = same color for all, light purple
13-17 = blues, cool colors
★ See 13.2

★ To assist the NEPHRONS, the KIDNEYS are supplied with a vast amount of blood vessels in the outer peripheral portion (parenchyma). The L. and R. RENAL ARTERIES are very large and transport 1/4 of the total cardiac output to the kidneys at a rate of 1,200 ml/minute!

★ Each RENAL ARTERY branches to each kidney from the AORTA. Just inside the RENAL SINUS (See 13.2) the artery branches to form 2 SEGMENTAL ARTERIES, which branch to form 12 or so INTERLOBAR ARTERIES that pass between the RENAL PYRAMIDS through the RENAL COLUMNS and penetrate the CORTEX.
• The INTERLOBAR ARTERIES then branch to form ARCUATE ARTERIES that curve around the PYRAMID BASES. (The ARCUATE ARTERIES of a given INTERLOBAR ARTERY do not connect with those of any other INTERLOBAR ARTERY, thus ensuring high blood pressure to the GLOMERULI.)
• Radially directed INTERLOBULAR ARTERIES arise from the ARCUATE ARTERIES, whose side branches are the AFFERENT ARTERIOLES which become the GLOMERULI CAPILLARY PLEXUSES.

★ The AFFERENT-EFFERENT ARTERIOLE PATHWAY IS UNIQUE IN THE BODY.
(Blood from arterioles usually flows from capillaries into venules, rather than into other arterioles.)

★ An EFFERENT ARTERIOLE exits from each GLOMERULUS and forms PERITUBULAR CAPILLARY NETWORKS (around the NEPHRON TUBULES) or a looped VASA RECTI (that follows the long LOOPS of HENLE).
• INTERLOBULAR, ARCUATE, INTERLOBAR and SEGMENTAL VEINS follow the same course along the arteries of the same names.
★ However, ARCUATE VEINS do connect with one another, and can receive INTERLOBULAR VEINS or VASA RECTI directly.
• Blood is carried out of the kidney as the SEGMENTAL VEINS coalesce to form the RENAL VEIN.

★ There are two types of NEPHRONS:

■ CORTICAL NEPHRONS:
• Mainly located in the outer 2/3 of the CORTEX. Short Loops of Henle sometimes dip slightly into the medulla.

■ JUXTAMEDULLARY NEPHRONS
• The GLOMERULUS and convoluted tubules are located at the corticomedullary junction in the cortex. Long Loops of Henle penetrate deep into the medulla.

Blood flow into RENAL ARTERIES

Blood flow out of RENAL VEINS

ARTERIAL FLOW

1	**RENAL ARTERY**
2	**SEGMENTAL ARTERIES**
3	**INTERLOBAR ARTERIES**
4	**ARCUATE ARTERIES**
5	**INTERLOBULAR ARTERIES**
6	**AFFERENT ARTERIOLE**
7	**GLOMERULUS**
8	**EFFERENT ARTERIOLE**
9	**PERITUBULAR CAPILLARIES**
10	**VASA RECTI**

VENOUS FLOW

17	**RENAL VEIN**
16	**SEGMENTAL VEINS**
15	**INTERLOBAR VEINS**
14	**ARCUATE VEINS**
13	**INTERLOBULAR VEINS**
12	**PERITUBULAR CAPILLARIES**
11	**VASA RECTI**

• The FIRST SEPARATION of the BLOOD occurs in the GLOMERULUS:
■ PLASMA FILTRATE (H_2O and dissolved solutes) enter the tubular system

■ Protein MOLECULES and FORMED ELEMENTS enter the EFFERENT ARTERIOLE

★ A major portion of the PLASMA FILTRATE (glucose, small proteins, amino acids, H_2O and Na^+Cl^-) is reabsorbed by the convoluted tubules and collecting tubules, and enters the PERITUBULAR CAPILLARIES for recirculation to the tissues. The VASA RECTI share a critical relationship with the LOOPS OF HENLE in the absorption of water.

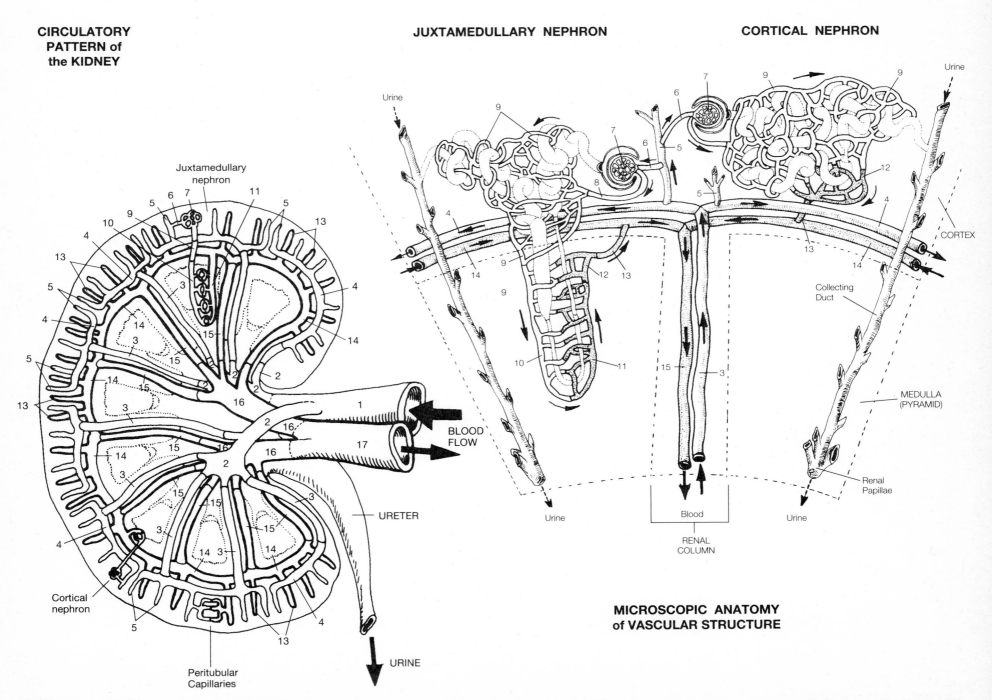

CIRCULATORY PATTERN of the KIDNEY

Juxtamedullary nephron

Cortical nephron

BLOOD FLOW

URETER

URINE

Peritubular Capillaries

JUXTAMEDULLARY NEPHRON

Urine

Urine

Blood

RENAL COLUMN

CORTICAL NEPHRON

Urine

CORTEX

Collecting Duct

MEDULLA (PYRAMID)

Renal Papillae

Urine

MICROSCOPIC ANATOMY of VASCULAR STRUCTURE

13.5

URINARY SYSTEM: THE URETERS, THE URINARY BLADDER, AND THE URETHRA

★ Each URETER is a continuation of the RENAL PELVIS of its associated kidney, and the URETERS drain the PELVES.

Ureters are only 1/2 in. in diameter at their widest point.

The URETERS extend from the beginning of each funnel-shaped PELVIS ≈ 28-34 crm to the URINARY BLADDER. (They are always retroperitoneal in position [See 3.1])

The URETERS pass into the URINARY BLADDER for several centimeters. Pressure in the bladder compresses the ureters and prevents backflow during urination.

URETER FUNCTION is to transport urine from the RENAL PELVIS into the URINARY BLADDER for storage.

Urine is transported in the URETERS by:
- Peristaltic contraction of ureteric muscular walls
- Hydrostatic pressure
- Gravity

Mucus secreted by the inner mucosa of the URETERS prevents their cells from coming in contact with urine.

B | **2 URETERS** ➡ **C** | **1 URINARY BLADDER** ➡ **D** | **1 URETHRA**

Divisions of the URETHRA (MALE) ♂

1 | **2 URETERAL ORIFICES**

2 | **INTERURETERIC FOLD**

3 | **TRIGONE**

4 | **UVULA**

11 | **PROSTATIC URETHRA**

12 | **MEMBRANOUS URETHRA**

13 | **SPONGY (CAVERNOUS) URETHRA** PENILE URETHRA

3 COATS OF THE URETERS (AND RENAL CALYCES AND RENAL PELVES)
- **MUCOSA**
  - Transitional epithelium and Lamina Propria (thicker in the PELVIS and URETER)

- **MUSCULARIS** (Smooth)
  - In the CALYCES, PELVIS, and upper 2/3 of URETERS there are 2 layers: INNER CIRCULAR and OUTER LONGITUDINAL, which become progressively thicker toward the URETER. Lower 1/3 of URETER a 3rd INNER LONGITUDINAL LAYER is added.

- **ADVENTITIA**
  - Fibrous coat blends with the Fibrous tunic of the kidney above and the SEROSA of the Bladder below.
  - This is not a Serosa (no peritoneum involved)

4 COATS

5 | **MUCOSA** TRANSITIONAL EPITHELIUM AND LAMINA PROPRIA

6 | **SUBMUCOSA** ‡

7 | **DETRUSOR MUSCLE** THICK, HEAVY CIRCULAR LAYER

8 | **SEROSA** CONTINUOUS WITH THE PERITONEUM

9 | **INTERNAL SPHINCTER** SPHINCTER VESICAE

10 | **EXTERNAL SPHINCTER** SPHINCTER URETHRAE

★ Average capacity of the URINARY BLADDER is 700–800 ml. When it exceeds 200–400 ml, urine is expelled by conscious desire and unconscious reflex (MICTURITION), both results of nervous impulses.

TRIGONE = Triangle formed by the 2 URETHRAL ORIFICES and the INTERNAL URETHRAL ORIFICE

INTERNAL SPHINCTER = Middle muscularis of the DETRUSOR MUSCLE at the area opening to the urethra (INVOLUNTARY)

EXTERNAL SPHINCTER = Skeletal muscle (VOLUNTARY)

‡ SUBMUCOSA considered by some as lacking and is really a deep loose layer of the underlying LAMINA PROPRIA of the MUCOSA.

Portions of the PROSTATIC URETHRA ♂

14 | **URETHRAL CREST** LONGITUDINAL FOLD

15 | **SEMINAL COLLICULUS**
ELEVATION OF 14, RECEIVES OPENINGS OF THE EJACULATORY DUCTS AND PROSTATIC UTRICLE

★ The URETHRA is the passageway for discharging urine from the body. ♀ In the FEMALE, it is 1 1/2 in. long, and is entirely separate from the reproductive organs.

♂ In the MALE, it is about 8 in. long, and also acts as a duct for the discharge of semen (reproductive fluid) from the body. (See 14.4) It is divided into 3 parts (11-13).

♂ URETHRA
■ Consists of basically a MUCOSA: INNER EPITHELIUM is TRANSITIONAL in (11), PSEUDOSTRATIFIED in (12) and (13), becomes STRATIFIED SQUAMOUS at the end of (13).
- A LAMINA PROPRIA connects the epithelium to surrounding tissue.

♀ URETHRA
■ Mucosal EPITHELIUM is TRANSITIONAL close to the BLADDER, and becomes STRATIFIED SQUAMOUS distally. The LAMINA PROPRIA connects epithelium to SPONGY TISSUE (and contains many veins called *erectile tissue*.)
■ *MUSCLARIS* is continuous with the bladder, forms a circular layer the entire length.

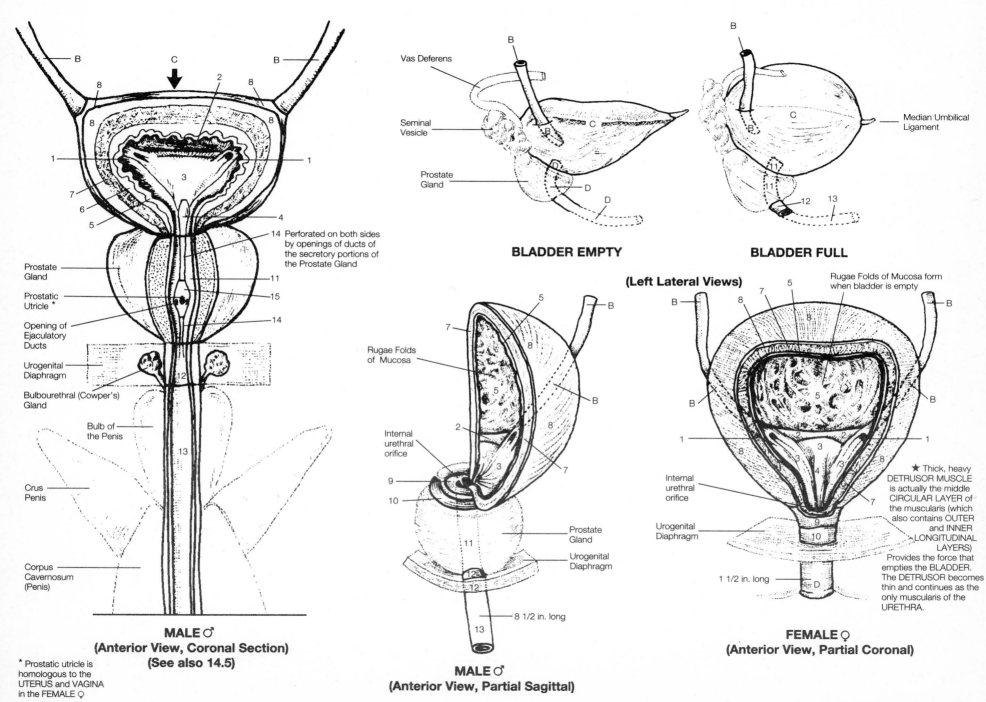

Vas Deferens

Seminal Vesicle

Prostate Gland

Median Umbilical Ligament

BLADDER EMPTY

BLADDER FULL

(Left Lateral Views)

Rugae Folds of Mucosa form when bladder is empty

Rugae Folds of Mucosa

Internal urethral orifice

Prostate Gland

Urogenital Diaphragm

8 1/2 in. long

Internal urethral orifice

Urogenital Diaphragm

1 1/2 in. long

★ Thick, heavy DETRUSOR MUSCLE is actually the middle CIRCULAR LAYER of the muscularis (which also contains OUTER and INNER LONGITUDINAL LAYERS) Provides the force that empties the BLADDER. The DETRUSOR becomes thin and continues as the only muscularis of the URETHRA.

14 Perforated on both sides by openings of ducts of the secretory portions of the Prostate Gland

Prostate Gland

Prostatic Utricle *

Opening of Ejaculatory Ducts

Urogenital Diaphragm

Bulbourethral (Cowper's) Gland

Bulb of the Penis

Crus Penis

Corpus Cavernosum (Penis)

MALE ♂
(Anterior View, Coronal Section)
(See also 14.5)

* Prostatic utricle is homologous to the UTERUS and VAGINA in the FEMALE ♀

MALE ♂
(Anterior View, Partial Sagittal)

FEMALE ♀
(Anterior View, Partial Coronal)

13.6

Chapter 13: *URINARY SYSTEM:* CHARTS

Chart #1 : Major Components of the Urinary System and Functional Summary

Structure	Function
KIDNEYS	The kidneys are responsible of ridding the body of fluid and electrolyte excesses, and conserving fluids and electrolytes when there are shortages. The kidneys also rid the body of some of the by-products of metabolism. Finally, the kidneys participate with the respiratory system and blood to maintain a constant pH level in extracellular fluid.
NEPHRON	
Renal Corpuscle	The formation of a protein-free filtrate of plasma called the glomerular filtrate. The rate of this formation is called the GLOMERULAR FILTRATION RATE (GFR).
Proximal Convoluted Tubule	The reabsorption of 70–80% of the glomerular filtrate into the peritubular capillaries. 100% reabsorption of many nutrients, such as glucose, amino acids, and fatty acids.
Loop of Henle	The establishment and maintenance of a graded, hypertonic environment for the potential excretion of a hypertonic urine. The excretion of such a urine is dictated by body fluid and electrolyte conditions.
Distal Convoluted Tubule	The reabsorption of 15–20% of the glomerular filtrate into the peritubular capillaries. The acidification of urine. The secretion of excess substances, e.g., potassium.
Collecting Ducts	Determination of final urine volume based on body fluid condition.
URETERS	Conduction of urine from the renal pelvis to the urinary bladder.
URINARY BLADDER	Retention of urine until an adequate amount of urine is available to stimulate micturition.
URETHRA	Conduction of urine from the urinary bladder to the environment. The urethra is also used to conduct semen from the body in males.

UNIT 5: CONTINUANCE OF THE HUMAN SPECIES

Chapter **14**  Reproductive System

THE REPRODUCTIVE SYSTEM IS UNIQUE IN 3 ASPECTS FROM OTHER BODY SYSTEMS	
OTHER SYSTEMS	REPRODUCTIVE SYSTEM
■ Function is to sustain the individual	■ Function is to perpetuate the species
■ Minor sexual differences	■ Major sexual differences
■ Systems functional at or close to birth	■ System is latent until puberty (development under hormonal control)

The reproductive system of the human organism sustains life by giving rise to offspring through the process of duplication.

In the human, this is accomplished by the union of GAMETES (GERM CELLS or SEX CELLS). Gametes are HAPLOID CELLS—each contains 23 single chromosomes, or a half-complement of the genetic material (each normal somatic cell contains 46 chromosomes in its nucleus). Fertilization of a female gamete (EGG, or OVUM) by a male gamete (SPERMATOZOA, or SPERM) produces a normal DIPLOID CELL (23 chromosomes of the ovum being paired with 23 chromosomes of the spermatozoon)—the ZYGOTE. Sex is determined by whether the fertilizing spermatozoon is of X-type (for a female) or of Y-type (for a male). It is the random combination of specific genes in the gametes during fusion that results in the unique individual.

THE GENERAL PROCESS OF CELL DIVISION:

■ MAINTAINS THE LIFE OF THE INDIVIDUAL
(By single cell duplication, — MITOSIS; of genetic material for growth and repair of the organism)

■ MAINTAINS CONTINUANCE OF THE SPECIES
(REPRODUCTION: By passing genetic material from generation to generation)

The structures of the male and female reproductive systems can be grouped according to FUNCTION:

① PRIMARY SEX ORGANS: GONADS

• These are MIXED GLANDS because they produce SEX HORMONES as well as GAMETES.

• Male gonads are called TESTES, which produce SPERMATOZOA through the process of SPERMATOGENESIS.

• Female gonads are called OVARIES, which produce OVA through the process of OOGENESIS. Both processes involve a special kind of cell division—MEIOSIS (see 14.10).

② SECONDARY SEX ORGANS

• They mature at puberty due to the influence of sex hormones and are essential for successful sexual reproduction in the TRANSPORTATION, AID, and STORAGE of GAMETES.

• DUCT SYSTEMS Receive, transport, and store gametes

• ACCESSORY GLANDS Produce substances that support gametes

• COPULATORY ORGANS Provide physical intercourse for gametes

③ SECONDARY SEX CHARACTERISTICS

• These are characteristics that are associated with but not essential to reproduction. Their appearance at puberty and maintenance after puberty are due to the appropriately timed activity and proper concentration of sex hormones.

MALES
• Growth of testes and penis
• Pubic hair, facial hair, underarm hair, abdominal hair
• Body growth
• Growth of larynx (voice pitch lowers, voice deepens)
• Acne from blocked glands
• Characteristic male shape of body
• Growth of bone, muscle development, and alteration of skeletal proportions

FEMALES
• Breast development
• Pattern of body hair, pubic hair, and underarm hair
• Body physique (distribution of subcutaneous fat deposits on shoulders, hips, and thighs)
• Characteristic female proportions of body
• Growth of bones, increase in height, and broader pelvis

In addition to the development of the SECONDARY SEX ORGANS and SECONDARY SEX CHARACTERISTICS at PUBERTY, PSYCHOLOGICAL CHANGES and SEXUAL INSTINCTS associated with adulthood begin to develop. (PERSONALITY CHANGES also tend to occur more in the female.)

14.0

System Components

ORGANS (TESTES AND OVARIES) FOR PRODUCTION OF REPRODUCTIVE CELLS (SPERM AND OVA) AND ORGANS FOR TRANSPORTATION AND STORAGE OF THOSE CELLS

System Function

• REPRODUCTION OF THE ORGANISM

• PERPETUATION OF THE SPECIES

• PASSAGE OF GENETIC MATERIAL FROM GENERATION TO GENERATION

Chapter Coloring Guidelines

• *You may want to choose your color palette by comparing any colored illustrations from your main textbook.*

• *Color choice is wide open, but you'll always want to color subcutaneous adipose (fat) tissue yellow and muscles in reds and oranges.*

♀
FEMALE

♂
MALE

Split Anterior View

THE MALE ♂ REPRODUCTIVE SYSTEM: GENERAL ORGANIZATION
Organs of the Male System

★ The organs of the MALE REPRODUCTIVE SYSTEM are the PRIMARY SEX ORGANS, the TESTES (MALE GONADS), and the SECONDARY SEX ORGANS, which consist of a SYSTEM OF DUCTS, ACCESSORY GLANDS, and SUPPORTIVE STRUCTURES.

FUNCTIONS OF THE MALE REPRODUCTIVE STRUCTURES
★ ■ PAIRED TESTES
   • Production of SPERM (SPERMATOZOA), MALE GERM CELLS
   • Secretion of ANDROGEN hormones (including the male sex hormone TESTOSTERONE) directly into the blood stream. At the onset of puberty, androgens cause the appearance of (and after puberty, the maintenance of):
   ■ Spermatogenesis (See 14.2)
   ■ Secondary sex characteristics
   ■ Secondary sex organ development and functioning (especially SEMINAL VESICLES, PROSTATE GLAND, PENIS, and SCROTUM enlargement)
   ■ Anabolic effects include protein synthesis and red blood cell formation.

★ ■ SYSTEM OF DUCTS
   • Storage of SPERM
   • Transportation of SPERM (and secretions contributed by glands) to the exterior
   • Technically begins inside each TESTIS: (SEMINIFEROUS TUBULES, STRAIGHT TUBULES, RETE TESTIS), then EFFERENT DUCTULES, EPIDIDYMIS, VAS DEFERENS, EJACULATORY DUCT; both EJACULATORY DUCTS empty into the single URETHRA.

★ ■ ACCESSORY GLANDS
   • Produce secretions that comprise the SEMEN (fluid that carries the SPERM).

★ ■ SUPPORTIVE STRUCTURES
   • SCROTUM—Sac of skin that houses and suspends the TESTES, EPIDIDYMIS, and a portion of the VAS DEFERENS.
   • PENIS—3 columns of erectile tissue. The median column (CORPUS SPONGIOSUM) surrounds the PENILE URETHRA.
   The penis serves as an INTROMITTENT organ (organ injecting into a cavity) during sexual intercourse (COITUS, or COPULATION).
   • UROGENITAL DIAPHRAGM—A musculofascial sheath lying superficial to the PELVIC DIAPHRAGM that surrounds the MEMBRANOUS URETHRA (See 6.11)

PRIMARY SEX ORGANS: ♂ GONADS

A | **PAIRED TESTES** | (SING. TESTIS)

SECONDARY SEX ORGANS: Mature at Puberty
SYSTEM OF DUCTS, ACCESSORY GLANDS, AND SUPPORTING STRUCTURES

SYSTEM OF DUCTS

B | **2 EPIDIDYMIDES** | (SING. EPIDIDYMIS)

C | **2 DUCTUS DEFERENTIA** | (SING. VAS, OR DUCTUS, DEFERENS)

D | **2 COMMON EJACULATORY DUCTS**

1 URETHRA:
DIVISIONS OF THE URETHRA:
CANAL FOR DISCHARGE OF SEMEN (AND URINE)

E | **PROSTATIC URETHRA**

F | **MEMBRANOUS URETHRA**

G | **PENILE (CAVERNOUS) URETHRA**

ACCESSORY GLANDS

H | **PROSTATE GLAND**

I | **2 SEMINAL VESICLES**

J | **2 BULBOURETHRAL GLANDS** COWPER'S GLANDS

SUPPORTIVE STRUCTURES
External Genital Organs

K | **SCROTUM** | 2 SCROTAL SACS

L | **PENIS**

M | **UROGENITAL DIAPHRAGM**

★ During fetal development, the TESTES arise as paired retroperitoneal organs on the posterior abdominal wall near the paired kidneys. As the body grows, the TESTES descend into the scrotum, a cutaneously derived pouch, approximately during the 28th–29th week of development. The scrotum is an outpocketing (herniation) of the abdominal wall, anterior to the PUBIC SYMPHYSIS, which progresses into the skin posterior to the PENIS. The canal through which the herniation occurs remains as the INGUINAL CANAL (See 6.10) that conveys the SPERMATIC CORD from the SCROTUM to the PELVIC CAVITY.

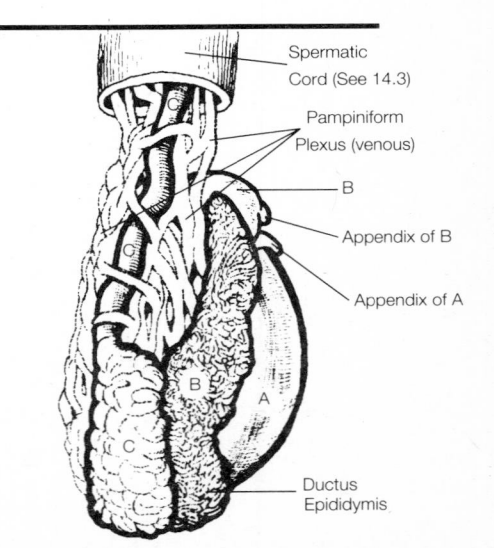

Spermatic Cord (See 14.3)
Pampiniform Plexus (venous)
B
Appendix of B
Appendix of A
Ductus Epididymis

POSTERIOR VIEW of EXPOSED TESTIS

PUBERTY in the MALE (See 8.3, 8.8)

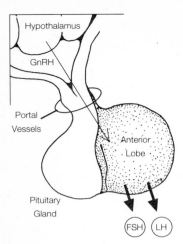

■ At the onset of PUBERTY (between the ages of 13–16)
- Maturational changes in the HYPOTHALAMUS produce an increase in regulating factor GnRH (GONADOTROPHIN-RELEASING HORMONE) and its secretion into the HYPOTHALAMO-HYPOPHYSEAL PORTAL VESSELS.

- GnRH stimulates release of GONADOTROPHIC HORMONES FSH (FOLLICLE-STIMULATING HORMONE, named for its action in the female) and LH (LUTEINIZING HORMONE) from the ANTERIOR LOBE of the PITUITARY GLAND directly into the blood stream.

- FSH and LH stimulate secretion of SEX STEROIDS (ANDROGENS and ESTROGENS) inside the TESTES. (A decrease in NEGATIVE FEEDBACK INHIBITION by the SEX STEROIDS to the sensitivity of the HYPOTHALAMUS and PITUITARY OCCURS, thus increasing FSH and LH release.) These sex steroids (especially TESTOSTERONE) cause the appearance and maintenance of SECONDARY SEX CHARACTERISTICS and SECONDARY SEX ORGAN development and functioning. (See 14.0)

- FSH stimulates the onset and regulation of SPERMATOGENESIS.
  [FSH RECEPTORS are *only* found in the NURSE, or SERTOLI, CELLS (See 14.2).]

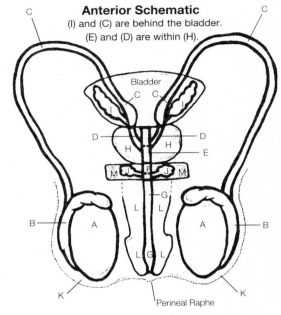

Anterior Schematic
(I) and (C) are behind the bladder.
(E) and (D) are within (H).

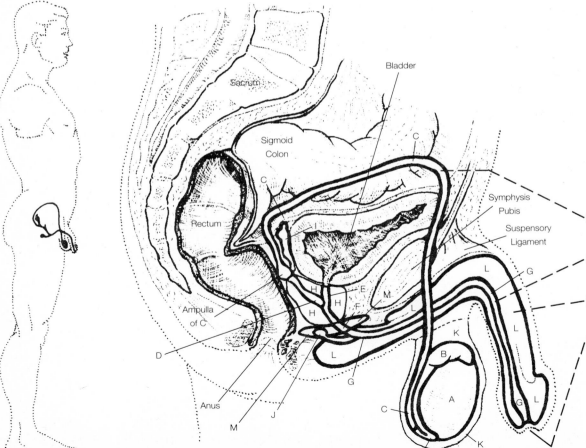

■ Smooth muscle wall of VAS DEFERENS (C) contracts and propels SPERMATOZOA into the URETHRA (E,F,G)

■ Smooth muscle of URETHRA (and striated muscle of the PERINEUM) contract and eject SEMEN from the URETHRA

■ Blood fills in the matrix of blood spaces in the PENIS (L) to render it firm and erect (INTROMITTENT)

■ In the act of EJACULATION (expulsion of semen, through the URETHRA to the exterior)—involving sympathetic innervation— 2–4 ml of SEMEN is deposited in the female VAGINA, containing from 200–500 million SPERMATOZOA . . . (approximately 100 million/ml).

THE MALE ♂ REPRODUCTIVE SYSTEM: PRIMARY SEX ORGANS
Testes and Spermatozoa

1 = light grey 2 = grey 3 = reddish-brown
4 = yellow 5 = ochre 6 = light orange 7 = light blue
B = light brown 8 = orange 9 = red C = cream

★ The TESTES are paired whitish, oval organs that measure about 4–5 cm (1.5–2 in.) in length and 2.5 cm (1 in.) in diameter. Each weighs between 10–15 gms.

★ Two tissue layers (tunicas) cover the TESTES. The TUNICA VAGINALIS (not shown, see 14.3) is a thin, serous sac derived from peritoneum during the descent of the testes. The tough, fibrous TUNICA ALBUGINEA directly surrounds each TESTIS. Inward extensions of this tunica divide each testis into a series of 200–300 internal compartments called LOBULES.

★ Each lobule contains 1–3 tightly packed, coiled SEMINIFEROUS TUBULES—THE FUNCTIONAL UNITS OF THE TESTES. (Each tube if uncoiled may exceed 2 1/3 feet.)

  • They produce SPERM through the process of SPERMATOGENESIS.

  • Endocrine cells between the seminiferous tubules called INTERSTITIAL CELLS OF LEYDIG produce and secrete the male sex hormones (androgens), and to a lesser extent the tubules do also, with a precursor.

  • Thus, the TESTES are both EXOCRINE GLANDS (SPERM) and ENDOCRINE GLANDS (ENDROGENS).

★ Once produced, the SPERM travel through the lumen of the convoluted seminiferous tubules. Sperm then converge into a partially ciliated duct network (RETE TESTIS) for further maturation within the MEDIASTINUM TESTIS (the thickened portion of the TUNICA ALBUGINEA on the posterior surface of the TESTIS).

★ From the RETE TESTIS, the SPERM are transported to the EPIDIDYMIS via the EFFERENT DUCTULES (uncoiled, the EPIDIDYMIS = 17 feet)

EPIDIDYMIS ⟶ • Final stages of SPERM MATURATION
(with tall microvilli   • Storage of SPERM (SPERMATOZOA)
or stereocilia)   • Transport SPERM to VAS DEFERENS

VAS DEFERENS ⟶ • SPERM are morphologically mature
(18 inches)   • Storage of SPERM
(4.57 cm)   • Transport SPERM to common EJACULATORYDUCT for discharge during EJACULATION (Use of smooth muscular walls.)

PRIMARY SEX ORGANS: ♂ GONADS (2 TESTES)

A SINGLE TESTIS:

2 TUNIC (TISSUE) LAYERS

  TUNICA VAGINALIS (See 14.3)

1 **TUNICA ALBUGINEA** TOUGH, FIBROUS MEMBRANE

2 **SEPTA OF THE TUNICA ALBUGINEA**

3 **LOBULES**

4 **SEMINIFEROUS TUBULES**

5 **STRAIGHT PORTIONS OF SEMINIFEROUS TUBULES**

PRELIMINARY SPERM MATURATION SITE

6 **RETE TESTIS** within #7

7 **MEDIASTINUM TESTIS**

SECONDARY SEX ORGANS
Primary Region of SPERM MATURATION

B **EPIDIDYMIS**

DUCT SYSTEM

8 **EFFERENT DUCTULES** COILED, CILIATED

9 **DUCTUS EPIDIDYMIS**

C **VAS (DUCTUS) DEFERENS**

SAGITTAL SECTIONS of a TESTIS

★ Muscular tissue of C propels sperm to prostatic urethra

★ Canal of C through which sperm are transported

★ Here the spermatozoa increase in fertilizing power. If not ejaculated, they will soon degenerate and will be absorbed in these tubules.

★ SPERMATOGENESIS

• The process of the formation of mature SPERMATOZOA (See 14.10). It starts just after puberty and normally continues until old age. Production occurs in advancing stages in the GERMINAL EPITHELIUM progressing from outer regions of the SEMINIFEROUS TUBULE toward the inner lumen of each tubule.

Through a process of specialized cell division called MEIOSIS:

EACH DIPLOID SPERMATOGONIUM (46 chromosomes each) is transformed into 4 HAPLOID SPERMATOZOA (23 chromosomes each)

★ MEIOSIS is a special nuclear division

occurring *only* in the production of gametes, which ensures that the chromosome number does not double with each generation. *Only* gametes (SPERM and OVA) are HAPLOID—having half the number of chromosomes present in other body cells.

★ SPERMATOGONIA duplicate by

MITOSIS (normal somatic cell division). One diploid daughter cell continues MITOSIS. The other diploid daughter cell becomes a PRIMARY SPERMATOCYTE. Each PRIMARY SPERMATOCYTE undergoes MEIOSIS and ultimately produces 4 *interconnected* HAPLOID SPERMATIDS.

★ SPERMIOGENESIS

• The final stage of SPERMATOGENESIS involved in the maturational changes that transform SPERMATIDS into SPERMATOZOA. Each SPERMATID embeds in a NURSE, or SERTOLI, CELL and develops a HEAD with an ACROSOME and a TAIL (FLAGELLUM). (The ACROSOME contains the enzyme HYALURONIDASE for penetration of the OVUM (See 14.6 and 14.10). It also contains the enzyme FERTILIZIN, which hardens the hyaline layer of the ovum after penetration.)

The 4 HAPLOID SPERMATIDS are disconnected with the aid of the phagocytic action of the nongerminal NURSE, or SERTOLI, CELLS.

★ SPERMIATION

• The process by which the NURSE, or SERTOLI, CELLS release the freely moving SPERM (SPERMATOZOA) into the lumen of the SEMINIFEROUS TUBULES. (Recall that FSH hormone stimulates the onset and regulation of SPERMATOGENESIS, and that FSH receptors are found only in the NURSE CELLS!)

14.2

SEMINIFEROUS TUBULE :

A BASEMENT MEMBRANE

GERMINAL EPITHELIUM

B SPERMATOGONIA
C PRIMARY SPERMATOCYTES
D SECONDARY SPERMATOCYTES
E SPERMATIDS ⎤
F SPERMATOZOA ⎦ SPERMIOGENESIS

G NURSE [SERTOLI] CELLS: SPERMIATION
H INTERSTITIAL CELLS OF LEYDIG
T Secrete ⟨TESTOSTERONE⟩

SPERMATOGENESIS

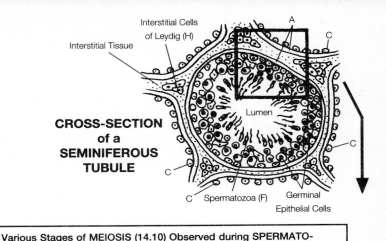

Interstitial Cells of Leydig (H)
Interstitial Tissue
Lumen
CROSS-SECTION of a SEMINIFEROUS TUBULE
Spermatozoa (F)
Germinal Epithelial Cells

A SPERMATOZOÖN (60 μm long)
MATURE SPERM CELL

a HEAD
b NECK
c BODY
d TAIL (FLAGELLUM)

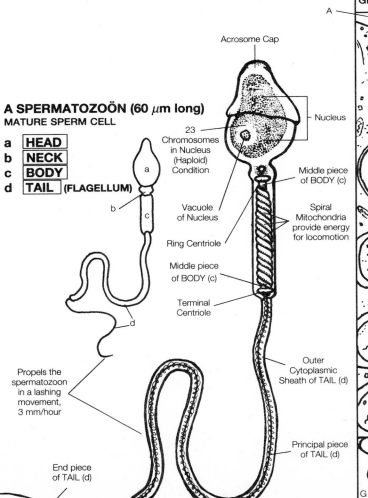

Acrosome Cap
Nucleus
23 Chromosomes in Nucleus (Haploid) Condition
Middle piece of BODY (c)
Spiral Mitochondria provide energy for locomotion
Vacuole of Nucleus
Ring Centriole
Middle piece of BODY (c)
Terminal Centriole
Outer Cytoplasmic Sheath of TAIL (d)
Propels the spermatozoon in a lashing movement, 3 mm/hour
Principal piece of TAIL (d)
End piece of TAIL (d)

Various Stages of MEIOSIS (14.10) Observed during SPERMATOGENESIS in the GERMINAL EPITHELIUM of a SEMINIFEROUS TUBULE

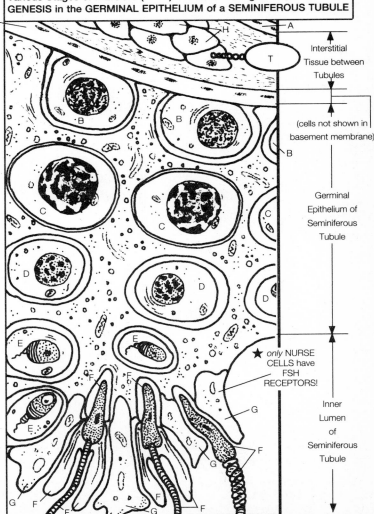

Interstitial Tissue between Tubules
(cells not shown in basement membrane)
Germinal Epithelium of Seminiferous Tubule
★ *only* NURSE CELLS have FSH RECEPTORS!
Inner Lumen of Seminiferous Tubule

THE MALE ♂ REPRODUCTIVE SYSTEM: SECONDARY SEX ORGANS:
Scrotum and Spermatic Cord

★ FUNCTIONS OF THE SCROTUM:
- ■ Support of the TESTES
- ■ Protection of the TESTES
- ■ Maintains testicle position in relation to pelvis

•The MEDIAN SEPTUM divides the SCROTUM into 2 sacs; each TESTIS is in its own separate compartment, thus increasing protection.

•The SCROTAL SACS are suspended below the PERINEUM, a region anterior to the anal opening and just behind the base of the penis.

★ The COVERINGS of the SPERMATIC CORD continue on as the inner layers of the SCROTAL SACS (SCROTUM). They are outwardly enveloped by the DARTOS muscle fibers in the subcutaneous tissue and an outer, hairy skin layer.

★ The DARTOS and CREMASTER MUSCLES maintain a constant temperature of 35∞Celsius (95∞F), the required temperature for the development and storage of the spermatozoa.

•In cold weather, the muscles contract and move the testes closer to the heat of the pelvis.

•In warm weather, the muscles relax and the testes descend, moving away from the body heat.

BLOOD CIRCULATION

•BLOOD SUPPLY to the TESTES is through the TESTICULAR ARTERIES (See 9.11, 13.1), which arise from the ABDOMINAL AORTA just below the origin of the RENAL ARTERIES.

BLOOD DRAINAGE is from the TESTICULAR VEINS—the RIGHT TESTICULAR VEIN enters directly into the INFERIOR VENA CAVA; the LEFT TESTICULAR VEIN enters into the LEFT RENAL VEIN.

The SCROTUM
SEPTAL Subdivision

1 | MEDIAN SEPTUM | (INTERNAL) a continuation of the DARTOS (4)
2 | PERINEAL RAPHE | (EXTERNAL RIDGE)

Layers of the SCROTUM

3 | OUTER SKIN
4 | DARTOS SMOOTH MUSCLE FIBERS

5 | EXTERNAL SPERMATIC FASCIA
6 | CREMASTERIC FASCIA | *
7 | CREMASTER MUSCLE | * SKELETAL MUSCLE
8 | INTERNAL SPERMATIC FASCIA

★ 9 | TUNICA VAGINALIS

This is considered the outer layer of the TESTES. It is a thin, serous sac (derived from the PERITONEUM during the descent of the TESTES). It folds over itself and forms a VISCERAL LAYER and a PARIETAL LAYER, with a CAVITY separating them.

*CREMASTERIC FASCIA and MUSCLE are continuations of the INTERNAL OBLIQUE MUSCLE and FASCIA of the ABDOMINAL WALL. (See 6.10)

The SPERMATIC CORD

★ • As the VAS (DUCTUS) DEFERENS ascends in the scrotum and passes through the abdominal wall, it travels with the TESTICULAR ARTERY, VEINS draining the TESTES (PAMPINIFORM PLEXUS), and various NERVES and LYMPHATIC VESSELS. These form the structures within the spermatic cord.

• The Spermatic Cord coverings are continuations of layers of the abdominal wall. The INTERNAL OBLIQUE MUSCLE and FASCIA of the abdominal wall become the CREMASTER MUSCLE and FASCIA.) (See 6.10)

• The INGUINAL CANAL is the slitlike passageway in the anterior abdominal wall through which the SPERMATIC CORD passes into the abdominal cavity. †

Coverings of the SPERMATIC CORD

5 | EXTERNAL SPERMATIC FASCIA
6 | CREMASTERIC FASCIA
7 | CREMASTER MUSCLE
8 | INTERNAL SPERMATIC FASCIA

Constituents of the SPERMATIC CORD

C | VAS (DUCTUS) DEFERENS
10 | TESTICULAR ARTERY
11 | PAMPINIFORM PLEXUS | VENOUS NETWORK
12 | NERVES AND LYMPHATICS

† The INGUINAL CANAL is a common site for INGUINAL HERNIAS.

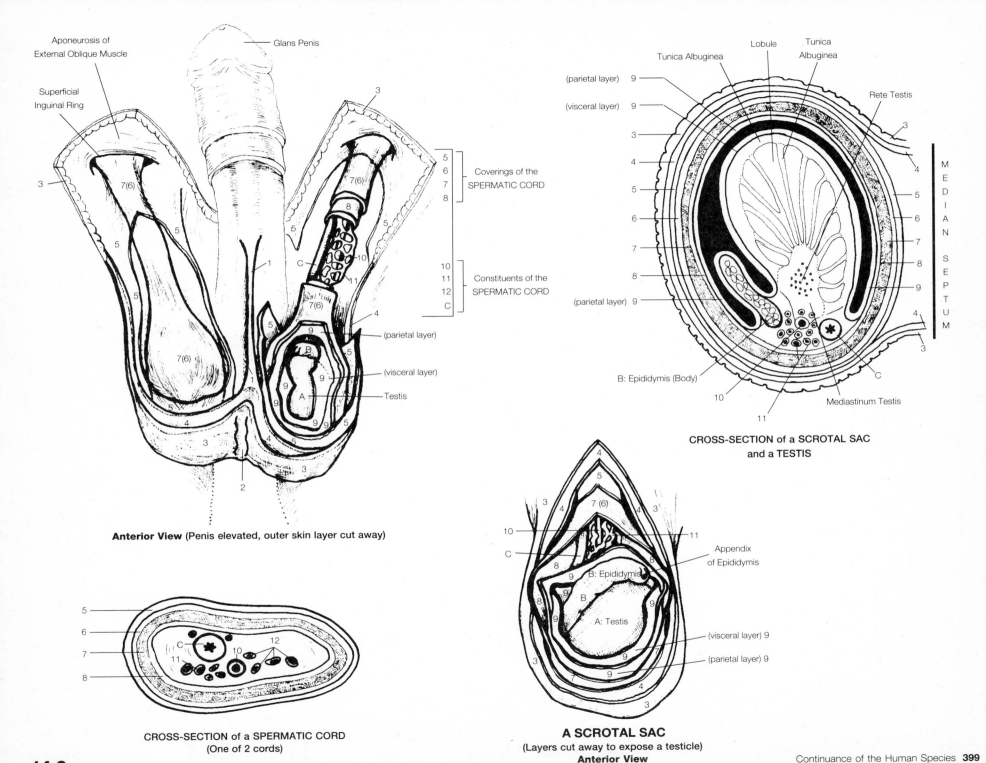

Aponeurosis of
External Oblique Muscle

Glans Penis

Superficial
Inguinal Ring

3

7(6)

3

5

5

7(6)

5

5

5

5

5

C

8

10

1

11

7(6)

4

9

(parietal layer)

5

B

9

(visceral layer)

9

A

Testis

9

7(6)

9

9

5

4

5

4

3

1

3

2

5
6
7
8
} Coverings of the
SPERMATIC CORD

10
11
12
C
} Constituents of the
SPERMATIC CORD

Anterior View (Penis elevated, outer skin layer cut away)

5
6
7
8

C

*

10

12

11

CROSS-SECTION of a SPERMATIC CORD
(One of 2 cords)

Lobule

Tunica Albuginea

Tunica
Albuginea

(parietal layer) 9

(visceral layer) 9

3

4

5

6

7

8

(parietal layer) 9

Rete Testis

3

4

5

6

7

8

9

4

3

M
E
D
I
A
N

S
E
P
T
U
M

B: Epididymis (Body)

10

11

C

Mediastinum Testis

CROSS-SECTION of a SCROTAL SAC
and a TESTIS

4

5

3 4 7 (6) 4 3

10

11

C

8

8

Appendix
of Epididymis

9 B: Epididymis

8 9 B

9 A: Testis

9

(visceral layer) 9

9

(parietal layer) 9

3

7

9

4

3

A SCROTAL SAC
(Layers cut away to expose a testicle)
Anterior View

THE MALE ♂ REPRODUCTIVE SYSTEM: SECONDARY SEX ORGANS
Urethra, Penis, and Accessory Glands

E = light green F = light blue G = light purple
C, C₁ = cream D = yellow H = pink I = blue-green
J = lime 1 = light pink 2 = red 3 = red-orange
4 = orange 7 = purple 9 = flesh
★ See 6.11, 13.6

★ The MALE URETHRA is a common tube for both the REPRODUCTIVE and URINARY SYSTEMS. Semen and urine, however, cannot pass through the urethra simultaneously since the SYMPATHETIC EJACULATION REFLEX automatically inhibits the URINARY REFLEX (MICTURITION).

• PROSTATIC URETHRA → Passes through the PROSTATE GLAND

• MEMBRANOUS URETHRA → Passes through the UROGENITAL DIAPHRAGM

PENILE (CANERNOUS) URETHRA → Passes through the SHAFT OF THE PENIS in the center of the midventral CORPUS SPONGIOSUM

★ SYSTEM OF DUCTS → Store and transport SPERM CELLS

★ ACCESSORY GLANDS → Secrete the liquid portion of semen:

• 2 SEMINAL VESICLES (60%) secrete a viscous liquid that keeps spermatozoa alive and motile.

PROSTATE GLAND (13–33%) and

• 2 BULBOURETHRAL GLANDS (7–27%) add a thin lubricant to semen.

★ •The AMPULLA of the VAS DEFERENS and the DUCT of the SEMINAL VESICLE unite to form the EJACULATORY DUCT. The Left and Right common EJACULATORY DUCTS pass through the PROSTATE GLAND and OPEN INTO THE PROSTATIC URETHRA.

•There are also many small ducts from the PROSTATE GLAND that open into the PROSTATIC URETHRA.

•The BULBOURETHRAL (COWPER'S) GLANDS are located in the UROGENITAL DIAPHRAGM. Their ducts open into the PENILE (CAVERNOUS) URETHRA.

•Mucus is secreted in response to sexual stimulation prior to ejaculation.

★ The PENIS is composed mainly of erectile tissue (arranged in 3 columns), the whole being enveloped by skin. The median (midventral)column—the CORPUS SPONGIOSUM—contains the URETHRA. When distended (filled with blood), the penis serves as the COPULATORY ORGAN of the male. When flaccid, it serves as a conduit for urine from the URINARY BLADDER.

The URETHRA
3 Main Regions

E | PROSTATIC URETHRA
F | MEMBRANOUS URETHRA
G | PENILE (CAVERNOUS) URETHRA   SPONGY URETHRA

System of Ducts

C | VAS (DUCTUS) DEFERENS (2)
C₁ | AMPULLA OF VAS DEFERENS (2)
D | COMMON EJACULATORY DUCT (2)

ACCESSORY GLANDS

H | PROSTATE GLAND (1)
I | SEMINAL VESICLES (2)
J | BULBOURETHRAL GLANDS (2)   COWPER'S GLAND

1 | LEVATOR ANI (PELVIC DIAPHRAGM)
2 | UROGENITAL DIAPHRAGM

The PENIS
3 Cylindrical Bodies of ERECTILE TISSUE

2 Lateral: **3** | CORPUS CAVERNOSUM   CORPUS CAVERNOSUM PENIS
1 Midventral **4** | CORPUS SPONGIOSUM   CORPUS CAVERNOSUM URETHRAE

Root Portion of CORPUS CAVERNOSUM PENIS

5 | CRUS OF THE PENIS

Root and Front Portions of CORPUS SPONGIOSUM (CORPUS CAVERNOSUM URETHRAE)

6 | BULB OF THE PENIS   BULBOUS URETHRAE
7 | GLANS OF THE PENIS

Skin Coverings around PENIS

8 | PREPUCE (FORESKIN)   AROUND GLANS
9 | OUTER SKIN

Inner Binding FASCIA

10 | SUPERFICIAL FASCIA   LOOSE AREOLAR
11 | DEEP FASCIA
(12) | TUNICA ALBUGINEA   (See 14.3)

(between the corpora CAVERNOSA PENIS, the Deep Fascia of the TUNICA ALBUGINEA forms a MEDIAN SEPTUM)

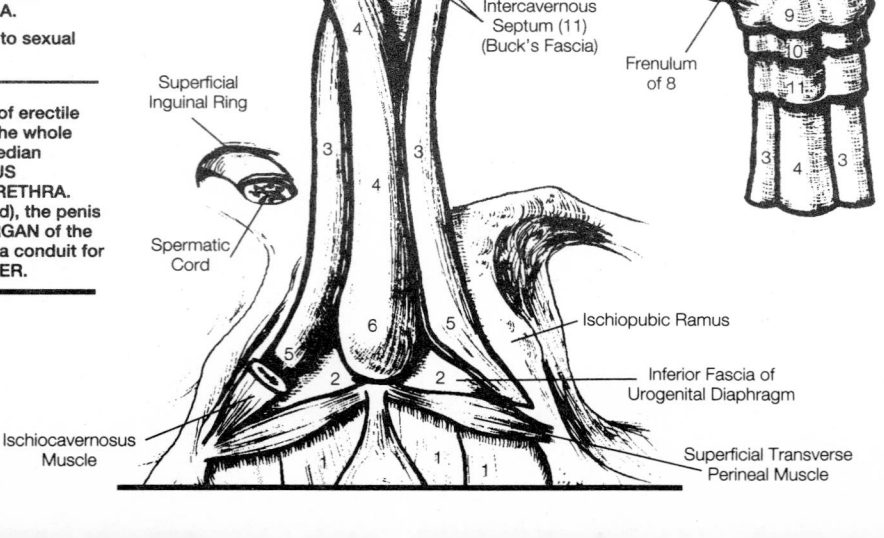

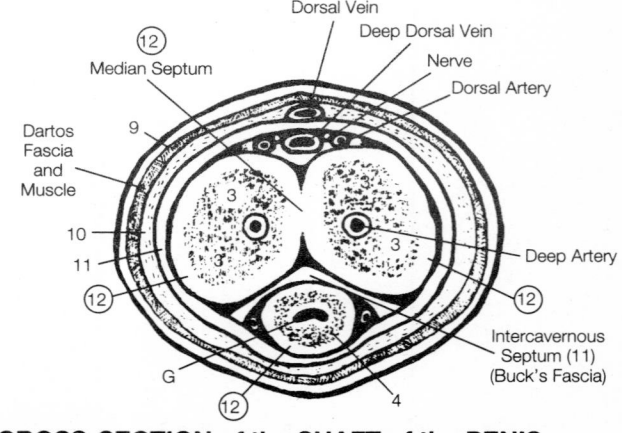

CROSS-SECTION of the SHAFT of the PENIS

(Sagittal Section, Right Lateral View)

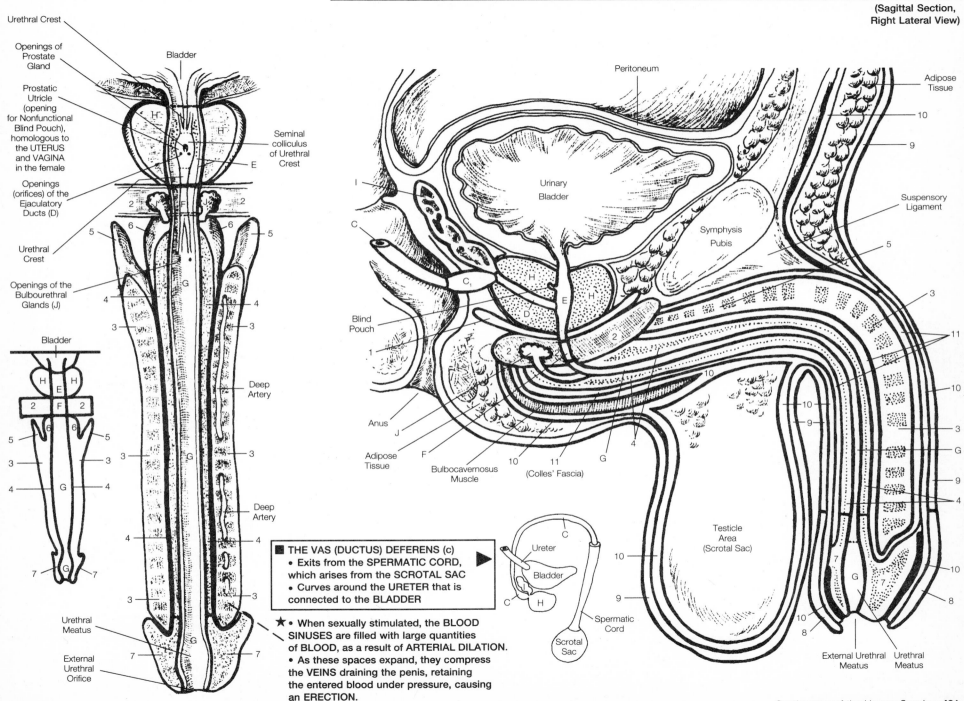

Urethral Crest

Openings of Prostate Gland

Prostatic Utricle (opening for Nonfunctional Blind Pouch), homologous to the UTERUS and VAGINA in the female

Openings (orifices) of the Ejaculatory Ducts (D)

Urethral Crest

Openings of the Bulbourethral Glands (J)

Bladder

Seminal colliculus of Urethral Crest

Bladder

Deep Artery

Deep Artery

Urethral Meatus

External Urethral Orifice

Peritoneum

Adipose Tissue

Urinary Bladder

Symphysis Pubis

Suspensory Ligament

Blind Pouch

Anus

Adipose Tissue

Bulbocavernosus Muscle

(Colles' Fascia)

Testicle Area (Scrotal Sac)

External Urethral Meatus

Urethral Meatus

Ureter

Bladder

Spermatic Cord

Scrotal Sac

■ THE VAS (DUCTUS) DEFERENS (c)
• Exits from the SPERMATIC CORD, which arises from the SCROTAL SAC
• Curves around the URETER that is connected to the BLADDER

★ • When sexually stimulated, the BLOOD SINUSES are filled with large quantities of BLOOD, as a result of ARTERIAL DILATION.
• As these spaces expand, they compress the VEINS draining the penis, retaining the entered blood under pressure, causing an ERECTION.

THE FEMALE ♀ REPRODUCTIVE SYSTEM: GENERAL ORGANIZATION
Organs of the Female System

A = reddish-brown B = light green
C = light red D = pink E = yellow
F = fleshy G = light orange H = orange
★ See Charts #1, #3, #4

The organs of the FEMALE REPRODUCTIVE SYSTEM are the *PRIMARY SEX ORGANS*—the OVARIES (FEMALE GONADS) and the *SECONDARY SEX ORGANS,* which consist of INTERNAL ACCESSORY ORGANS (2 FALLOPIAN TUBES, the UTERUS, and VAGINA), EXTERNAL STRUCTURES (the VULVA), and ACCESSORY GLANDS (the PARA-URETHRAL and VESTIBULAR GLANDS, and the BREASTS, or MAMMARY GLANDS).

FUNCTIONS OF THE FEMALE REPRODUCTIVE STRUCTURES

★ ■ **PAIRED OVARIES**

• Production, development, and eventual expulsion of OVA (EGG CELLS)—FEMALE GERM CELLS.

• Secretion of ESTROGENIC HORMONES (the female sex hormones, ESTRADIOL and ESTRONE). At the onset of puberty, estrogens are responsible for the development of:

■ Secondary sex characteristics
■ Secondary sex organs
■ Cyclical changes leading to the onset of MENSTRUATION, called MENARCHE.

★ ■ **SYSTEM OF CHANNELS (INTERNAL ORGANS)**

■ **FALLOPIAN TUBES**—Convey the ova from the ovaries to UTERUS. Also the site of FERTILIZATION (sperm travel from UTERUS toward the OVARIES).

■ **UTERUS**—Site of zygote implantation. Provides nourishment, and protects and sustains the life of the embryo and fetus during pregnancy.

• Active role in PARTURITION (birth of the baby) as muscular layer (MYOMETRIUM) produces powerful rhythmic contractions.

• Uterine mucosal lining (the ENDOMETRIUM) undergoes changes associated with the MENSTRUAL CYCLE

■ **VAGINA**—Provides housing for erect penis (and semen) during copulation (and ejaculation)

• Lower part of the BIRTH CANAL

• Excretory duct for uterine secretions and the menstrual flow

★ ■ **EXTERNAL STRUCTURES (VULVA: PUDENDUM)**

• Forms margins and protective barriers for vagina and urethra

• Clitoris associated with sexual pleasure (stimulation of sensory nerve endings)

■ **ACCESSORY GLANDS**

• Produce secretions that moisten and lubricate vestibule and vaginal opening during intercourse (vestibular glands.) Produce milk secretions for infant nourishment (mammary).

PRIMARY SEX ORGANS: ♀ GONADS

A ☐ PAIRED OVARIES ☐ [SING. OVARY]

SECONDARY SEX ORGANS

Internal ORGANS: System of CHANNELS

B ☐ 2 UTERINE (FALLOPIAN) TUBES ☐ OVIDUCTS

C ☐ 1 UTERUS ☐

D ☐ 1 VAGINA ☐

External ORGANS: VULVA (PUDENDUM)

E ☐ MONS PUBIS ☐ VENERIS

F ☐ LABIA MAJORA ☐

G ☐ LABIA MINORA ☐

H ☐ CLITORIS ☐

I VESTIBULE

Accessory GLANDS

J ☐ PARAURETHRAL GLANDS ☐ SKENE'S GLANDS

K ☐ VESTIBULAR GLANDS ☐ BARTHOLIN'S GLANDS

L ☐ MAMMARY GLANDS (BREASTS) ☐

PUBERTY IN THE FEMALE (See 8.3, 8.8), (14.1)

■ Girls attain puberty 6 months–1 year earlier than boys (usually between the age of 12–14, but sometimes even as early as 9 years or as late as 17 years).

■ The transition to puberty is more abrupt than in boys, due to the ONSET of the MENSTRUAL CYCLE (MENSTRUATION), called MENARCHE. This changes the GIRL CHILD into a WOMAN able to bear children. (average onset = 12.5 years)

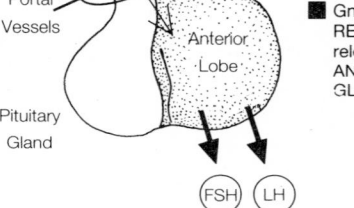

■ As in the male, maturational changes in the HYPOTHALAMUS produce an increase in GnRH (formerly known as LRH, the LUTEINIZING-RELEASING HORMONE) and its secretion into the HYPOTHALAMO-HYPOPHYSEAL PORTAL VESSELS.

■ GnRH (GONADOTROPHIC-RELEASING HORMONE) stimulates release of FSH and LH from the ANTERIOR LOBE of the PITUITARY GLAND directly into the blood stream.

■ FSH and LH stimulate secretion of SEX STEROID HORMONES (ESTROGENS) inside the OVARIES. (A decrease in NEGATIVE FEEDBACK INHIBITION BY THESEX STEROIDS to the sensitivity of the HYPOTHALAMUS and PITUITARY occurs, thus increasing FSH and LH release.)

■ These SEX STEROIDS (especially ESTRADIOL and PROGESTERONE) cause the appearance and maintenance of SECONDARY SEX CHARACTERISTICS and SECONDARY SEX ORGAN DEVELOPMENT and FUNCTIONING (See 14.0):

•FALLOPIAN TUBES LENGTHEN
•VAGINA epithelium THICKENS
•UTERINE muscle layer (MYOMETRIUM) ENLARGES
•UTERINE inner lining (ENDOMETRIUM) PROLIFERATES

★ MENARCHE (under the *control* of the HYPOTHALAMUS and PITUITARY) signals the onset of the regulation of all female reproductive activities in a *continual cyclical time pattern* (ᵃevery 28 days), ending ᵃ36 years later in MENOPAUSE.

■ This FEMALE HORMONAL CYCLE manifests itself in 2 ways, effected by the same hormonal changes:

•THE OVARIAN CYCLE (See 14.6) focuses on the changes occurring in the OVARIES.

•THE MENSTRUAL CYCLE (See 14.7) focuses on the flow or nonflow of blood from the VAGINA.

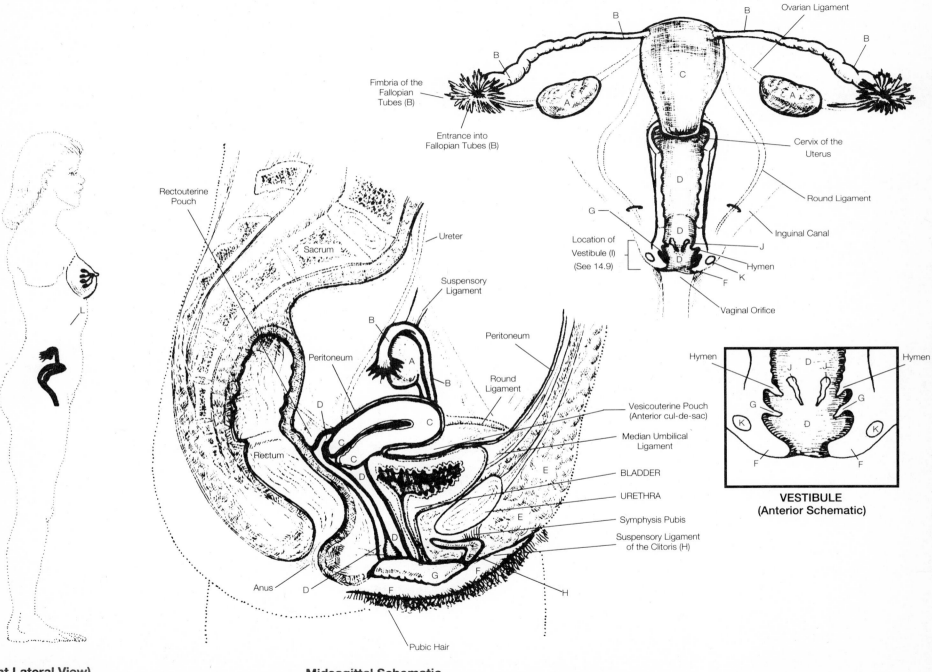

FEMALE REPRODUCTIVE ORGANS
Anterior Schematic (Coronal Section of VAGINA)

B
B
B
B
B
Ovarian Ligament

Fimbria of the Fallopian Tubes (B)

A
C
A

Entrance into Fallopian Tubes (B)

Cervix of the Uterus

D

Round Ligament

G

D

Inguinal Canal

Location of Vestibule (I) (See 14.9)

D

J

Hymen

K

F

Vaginal Orifice

Rectouterine Pouch

Sacrum

Ureter

Suspensory Ligament

Peritoneum

B

A

Peritoneum

B

D

Round Ligament

C

Vesicouterine Pouch (Anterior cul-de-sac)

C

Median Umbilical Ligament

Rectum

C

D

BLADDER

E

URETHRA

E

Symphysis Pubis

Suspensory Ligament of the Clitoris (H)

D

Anus

D

G

F

H

F

Pubic Hair

Midsagittal Schematic
(Right Lateral View)

VESTIBULE
(Anterior Schematic)

Hymen
J
D
J
Hymen
G
G
K
D
K
F
F

L

(Right Lateral View)

14.5

Continuance of the Human Species **403**

THE FEMALE ♀ REPRODUCTIVE SYSTEM: PRIMARY SEX ORGANS
Ovaries and Supporting Structures

A = reddish-brown B = light green
C = light red D = pink 1 = reddish-brown
2 = light grey 3 = light orange 4 = cream
5-7 = reds 6-8 = blues 9 = purple
Ligaments = greys, neutral colors

★ **FEMALE GONADS: OVARIES**

■ Paired glands, almond-shaped, positioned on each side of the UTERUS in the upper PELVIC CAVITY; held in position by several ligaments. Each OVARY is nestled in a depression of the posterior abdominal wall called the OVARIAN FOSSA (not shown).

★ On the MEDIAL PORTION of each OVARY is an entranceway called the HILUM. *All* vessels and nerves to each ovary enter only through the HILUM, which is supported by the OVARIAN LIGAMENT. The LATERAL PORTION of each ovary is in close contact with the open ends of each FALLOPIAN TUBE, through which the released OVUM from a ruptured GRAAFIAN FOLLICLE enters for possible fertilization from roving spermatozoa.

■ **HISTOLOGY:**
•GERMINAL EPITHELIUM—Covers the free surface, it is one layer of simple cuboidal epithelium

•TUNICA ALBUGINEA—Directly beneath the GERMINAL EPITHELIUM, it is a collagenous connective tissue capsule

•STROMA—Interior principal substance

CORTEX: outer, dense layer contains OVARIAN FOLLICLES in various stages of development

MEDULLA: inner, loose vascular layer

■ **SERIES OF LIGAMENTS MAINTAIN OVARIAN POSITION**
•BROAD LIGAMENT—The principal supporting membrane of the female reproductive tract

•MESOVARIUM—Portion of the double-layered fold of peritoneum that connects the anterior border of the ovary to the posterior layer of the BROAD LIGAMENT. Surrounds the ovary and OVARIAN LIGAMENT.

•Medial position of ovary attached to the OVARIAN LIGAMENT OF THE UTERUS

•Lateral position of ovary attached to the SUSPENSORY LIGAMENT

★ **RELATED STRUCTURES**

B | FALLOPIAN TUBE |

C | UTERUS |

D | VAGINA |

A | LEFT OVARY |

HISTOLOGICAL GENERAL STRUCTURE OF EACH OVARY: 4 LAYERS

1 | GERMINAL EPITHELIUM |

2 | TUNICA ALBUGINEA |

STROMA

3 | MEDULLA | VASCULAR

4 | CORTEX | CONTAINS OVARIAN FOLLICLES

BOUNDARY NOT DISTINCT

BLOOD CIRCULATION

• BLOOD SUPPLY is by OVARIAN ARTERIES arising from the lateral sides of the ABDOMINAL AORTA just below the origin of the RENAL ARTERIES (See 9.11, 13.1). The OVARIAN ARTERIES anastomose with OVARIAN BRANCHES of the UTERINE ARTERIES.
• BLOOD DRAINAGE is through the OVARIAN VEINS. The RIGHT OVARIAN VEIN empties directly into the INFERIOR VENA CAVA. The LEFT OVARIAN VEIN empties into the LEFT RENAL VEIN.

5 | OVARIAN ARTERY |

6 | OVARIAN VEIN |

7 | UTERINE ARTERY | (OVARIAN BRANCHES)

8 | UTERINE VEIN |

9 | CAPILLARIES IN MEDULLA |

SUPPORTIVE CONNECTING STRUCTURES ATTACHING TO EACH OVARY

13 | BROAD LIGAMENT |

■ Consists of 2 leaves of a part of the PARIETAL PERITONEUM *between* which are found:
•Remnants of the MESONEPHRIC DUCT (See 14.8 and Chart #4)
•Cellular Tissues
•Major blood vessels of PELVIS
•URETER from KIDNEY
•UTEROSACRAL LIGAMENT
•ROUND LIGAMENT

■ Attached to lateral borders of the UTERUS from insertion of the FALLOPIAN TUBE above to the PELVIC WALL below:

14 | MESOSALPINX OF THE BROAD LIGAMENT |

■ Free margin of the upper division of the BROAD LIGAMENT [(UTERINE TUBES (OVIDUCTS) lie within)]

10 | MESOVARIUM | ——→ Anchors to the MESOSALPINX of the BROAD LIGAMENT

11 | OVARIAN LIGAMENT | ——→ Anchors to the UTERUS

12 | SUSPENSORY LIGAMENT | ——→ Anchors to the POSTEROLATERAL PELVIC WALL

15 | ROUND LIGAMENT |

■ Not directly attached to the OVARY
■ One of 4 principal supporting ligaments of the UTERUS. Anchored immediately below and in front of the entrance of the FALLOPIAN TUBE into the UTERUS.
■ Each extends laterally inside the BROAD LIGAMENT to the PELVIC WALL, where it passes through the INGUINAL RING. It terminates in the LABIA MAJORA. (See 14.1, 14.8, 14.9)

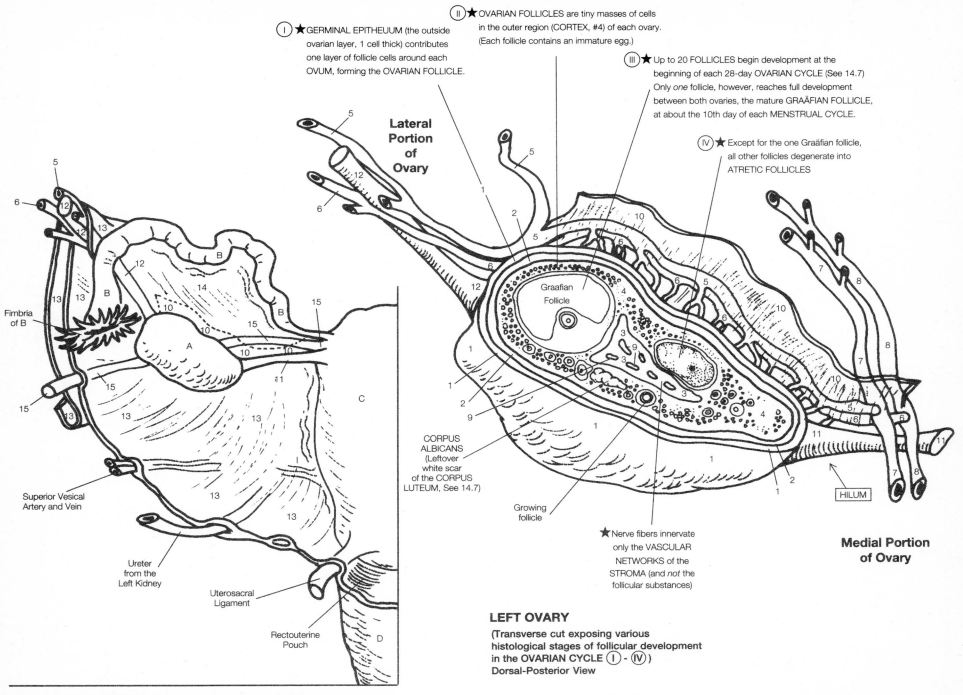

I ★ GERMINAL EPITHEUUM (the outside ovarian layer, 1 cell thick) contributes one layer of follicle cells around each OVUM, forming the OVARIAN FOLLICLE.

II ★ OVARIAN FOLLICLES are tiny masses of cells in the outer region (CORTEX, #4) of each ovary. (Each follicle contains an immature egg.)

III ★ Up to 20 FOLLICLES begin development at the beginning of each 28-day OVARIAN CYCLE (See 14.7) Only *one* follicle, however, reaches full development between both ovaries, the mature GRAÄFIAN FOLLICLE, at about the 10th day of each MENSTRUAL CYCLE.

IV ★ Except for the one Graäfian follicle, all other follicles degenerate into ATRETIC FOLLICLES

Lateral Portion of Ovary

Medial Portion of Ovary

Graafian Follicle

Fimbria of B

Superior Vesical Artery and Vein

Ureter from the Left Kidney

Uterosacral Ligament

Rectouterine Pouch

CORPUS ALBICANS (Leftover white scar of the CORPUS LUTEUM, See 14.7)

Growing follicle

★ Nerve fibers innervate only the VASCULAR NETWORKS of the STROMA (and *not* the follicular substances)

HILUM

LEFT OVARY

(Transverse cut exposing various histological stages of follicular development in the OVARIAN CYCLE I - IV)
Dorsal-Posterior View

SUPPORTIVE CONNECTING STRUCTURES of LEFT OVARY

(Dorsal-Posterior View)

THE ♀ FEMALE REPRODUCTIVE SYSTEM: PRIMARY SEX ORGANS
Ovarian Cycle (Oögenesis) and Ovulation

a = yellow b,Ⓑ = orange G = purple 9 = red
10-12 = yellow-orange, ochres c, d, e = light cool colors
f = reddish-brown g = light brown h = cream i = light pink

★ See Charts #2 and #3, 14.11

★ **THE OVARIAN CYCLE**

■ 28-DAY CYCLE in which hormonal changes effect changes in the OVARIES. OÖGENESIS (OVIGENESIS) is the process by which in each cycle one mature OVUM is prepared (from a set of up to 20 immature eggs) for fertilization.

★ ■ There are 6–7 million OÖGONIA (immature epithelial germ egg cells) produced by MITOSIS (See 14.11) in the FETAL OVARIES by the end of the 5th month of gestation, then stopped. Production of new oogonia never begins again.

■ At the end of gestation (9th month) the OÖGONIA are called PRIMARY OÖCYTES (IMMATURE OVA), as they begin MEIOSIS (See 14.11). (As in the male, genesis of the gonadal cells is arrested at PROPHASE I of the 1st MEIOTIC DIVISION.) (See Chart #2)

★ ■ Then, the PRIMARY OÖCYTES are surrounded by one layer of FOLLICULAR CELLS from the GERMINAL EPITHELIUM, forming the tiny PRIMORDIAL FOLLICLES.

■ In response to FSH stimulation from the anterior lobe of the pituitary gland:

•Some PRIMORDIAL FOLLICLES get larger (including the inner OÖCYTES)

•The surrounding FOLLICULAR CELLS divide to produce many small GRANULOSA CELLS. They fill the follicle and surround the oocyte, thus forming a PRIMARY FOLLICLE.

★ ■ Through further FSH stimulation, some PRIMARY FOLLICLES develop a fluid-filled ANTRUM, and are now called GROWING SECONDARY FOLLICLES.

•PRIMARY OOCYTE completes its 1st MEIOTIC DIVISION in a SECONDARY FOLLICLE: 2 OÖCYTES result: 1 SECONDARY OÖCYTE (gets all the cytoplasm) and 1 POLAR BODY, which degenerates.

■ The SECONDARY OÖCYTE in a SECONDARY FOLLICLE now enters the 2nd MEIOTIC DIVISION. It, too, is arrested at METAPHASE II, and is not completed unless fertilization occurs.

■ By the 10th–14th day of MENSTRUATION, only one follicle has continued to grow into a MATURE GRAÄFIAN FOLLICLE (under further FSH stimulation). All other SECONDARY FOLLICLES degenerate into ATRETIC FOLLICLES.

★ ■ In the process of OVULATION, the GRAÄFIAN FOLLICLE containing the MATURE OVUM (SECONDARY OÖCYTE) bulges from the surface, ruptures, and releases the MATURE OVUM, which finds its way into the opening of a FALLOPIAN TUBE, in search of a roving spermatozoon.

FOLLICULAR Development in the OÖGENIC EPITHELIUM of the CORTEX

GESTATING OÖGONIA NOT SHOWN

(a) PRIMARY OÖCYTE (IMMATURE OVUM)
↓
2 PRIMORDIAL FOLLICLE
↓
3 PRIMARY FOLLICLE
↓
4 SECONDARY (GROWING) FOLLICLE
↓
5 MATURE (GRAÄFIAN) FOLLICLE
↓
6 ATRETIC FOLLICLES

OVULATION
7 RUPTURED FOLLICLE
⑧ RELEASED SECONDARY OOCYTE

• Surrounded by ZONA PELLUCIDA (c)
• Surrounded by CORONA RADIATA (i)
• If not fertilized, will disintegrate in 2 days
• If fertilized by a spermatozoon, the 2nd MEIOTIC division is completed (See 14.11)

CORPUS LUTEUM

■ Following OVULATION (under the influence of LH), the empty RUPTURED FOLLICLE (7) undergoes both structural and biochemical changes in transforming into a CORPUS LUTEUM, an endocrine structure.

⑨ CORPUS HEMORRAGICUM HUGE BLOOD CLOT FORMED IN CAVITY
↓
⑩ YOUNG CORPUS LUTEUM
↓
⑪ MATURE CORPUS LUTEUM
↓
⑫ CORPUS ALBICANS REPLACES REGRESSING CORPUS LUTEUM

Types of OVA

a PRIMARY OÖCYTE (IMMATURE OVUM)

b, Ⓑ SECONDARY OÖCYTE (MATURE OVUM)

STRUCTURE OF SECONDARY (GROWING) FOLLICLE

■ Under FSH stimulation:
• ANTRUM fills with LIQUOR FOLLICULI fluid
• GRANULOSA CELLS form:
  CUMULUS OOPHORUS—Mound that supports the OVUM
  CORONA RADIATA—Ring encircling OVUM
  GRANULOSA RING—Ring around the circumference of the follicle
• ZONA PELLUCIDA, a thin gel-like layer of proteins and polysaccharides forms between the OÖCYTE and the CORONA RADIATA.

FOLLICULAR CAVITIES

c ZONA PELLUCIDA

d FOLLICULAR FLUID CAVITIES

e ANTRUM (FLUID-FILLED CAVITY)

FOLLICULAR CELL GROUPS

f GRANULOSA (FOLLICULAR) CELLS

g THECA FOLLICULI] THECA INTERNA / THECA EXTERNA

h CUMULUS OÖPHORUS

i CORONA RADIATA

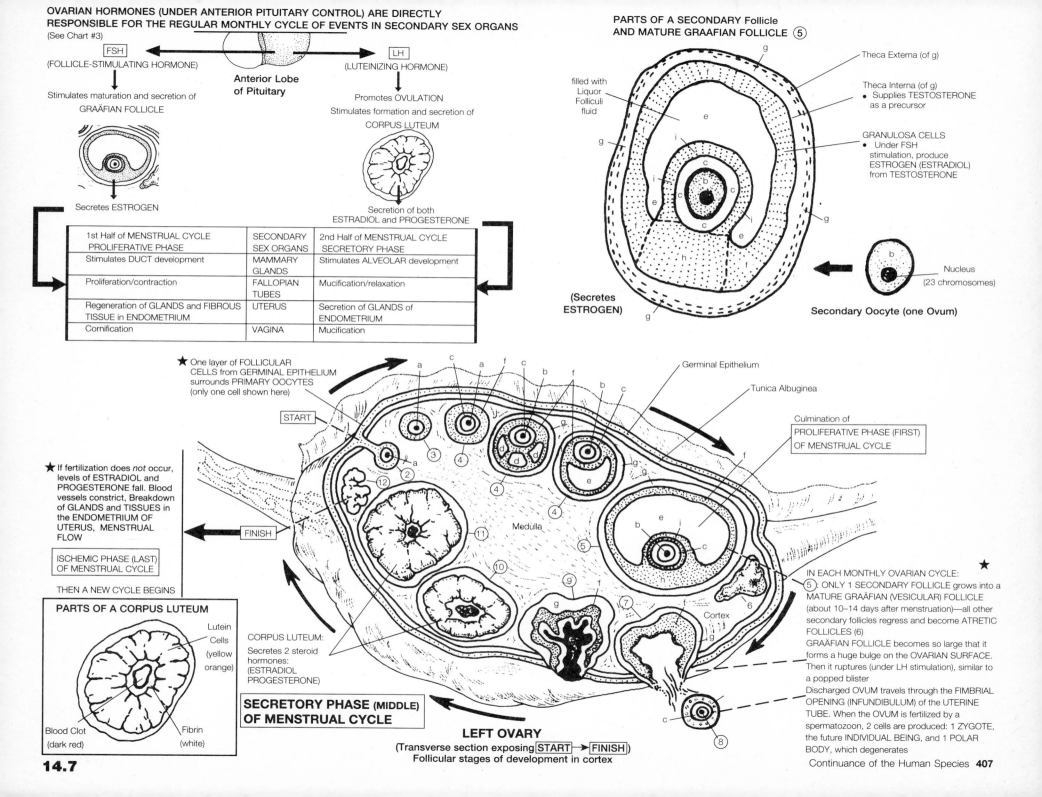

OVARIAN HORMONES (UNDER ANTERIOR PITUITARY CONTROL) ARE DIRECTLY RESPONSIBLE FOR THE REGULAR MONTHLY CYCLE OF EVENTS IN SECONDARY SEX ORGANS
(See Chart #3)

Anterior Lobe of Pituitary

FSH
(FOLLICLE-STIMULATING HORMONE)

Stimulates maturation and secretion of
GRAÄFIAN FOLLICLE

Secretes ESTROGEN

LH
(LUTEINIZING HORMONE)

Promotes OVULATION

Stimulates formation and secretion of
CORPUS LUTEUM

Secretion of both
ESTRADIOL and PROGESTERONE

1st Half of MENSTRUAL CYCLE PROLIFERATIVE PHASE	SECONDARY SEX ORGANS	2nd Half of MENSTRUAL CYCLE SECRETORY PHASE
Stimulates DUCT development	MAMMARY GLANDS	Stimulates ALVEOLAR development
Proliferation/contraction	FALLOPIAN TUBES	Mucification/relaxation
Regeneration of GLANDS and FIBROUS TISSUE in ENDOMETRIUM	UTERUS	Secretion of GLANDS of ENDOMETRIUM
Cornification	VAGINA	Mucification

PARTS OF A SECONDARY Follicle AND MATURE GRAAFIAN FOLLICLE ⑤

filled with Liquor Folliculi fluid

Theca Externa (of g)

Theca Interna (of g)
• Supplies TESTOSTERONE as a precursor

GRANULOSA CELLS
• Under FSH stimulation, produce ESTROGEN (ESTRADIOL) from TESTOSTERONE

(Secretes ESTROGEN)

Nucleus
(23 chromosomes)

Secondary Oocyte (one Ovum)

★ One layer of FOLLICULAR CELLS from GERMINAL EPITHELIUM surrounds PRIMARY OOCYTES (only one cell shown here)

START

★ If fertilization does *not* occur, levels of ESTRADIOL and PROGESTERONE fall. Blood vessels constrict, Breakdown of GLANDS and TISSUES in the ENDOMETRIUM OF UTERUS, MENSTRUAL FLOW

ISCHEMIC PHASE (LAST) OF MENSTRUAL CYCLE

THEN A NEW CYCLE BEGINS

FINISH

Germinal Epithelium

Tunica Albuginea

Culmination of
PROLIFERATIVE PHASE (FIRST) OF MENSTRUAL CYCLE

Medulla

Cortex

PARTS OF A CORPUS LUTEUM

Lutein Cells (yellow orange)

Blood Clot (dark red)

Fibrin (white)

CORPUS LUTEUM:
Secretes 2 steroid hormones:
(ESTRADIOL PROGESTERONE)

SECRETORY PHASE (MIDDLE) OF MENSTRUAL CYCLE

LEFT OVARY
(Transverse section exposing START → FINISH)
Follicular stages of development in cortex

★ IN EACH MONTHLY OVARIAN CYCLE:
⑤: ONLY 1 SECONDARY FOLLICLE grows into a MATURE GRAÄFIAN (VESICULAR) FOLLICLE (about 10–14 days after menstruation)—all other secondary follicles regress and become ATRETIC FOLLICLES (6)
GRAÄFIAN FOLLICLE becomes so large that it forms a huge bulge on the OVARIAN SURFACE. Then it ruptures (under LH stimulation), similar to a popped blister
Discharged OVUM travels through the FIMBRIAL OPENING (INFUNDIBULUM) of the UTERINE TUBE. When the OVUM is fertilized by a spermatozoon, 2 cells are produced: 1 ZYGOTE, the future INDIVIDUAL BEING, and 1 POLAR BODY, which degenerates

B = light greens and blues C = light reds
D = pink Ligaments = grays and light colors
Histological Layers = warm colors
★ See 14.6

THE FEMALE ♀ REPRODUCTIVE SYSTEM: SECONDARY SEX ORGANS:
Fallopian (Uterine) Tubes, Uterus, and Vagina

★ ■ The unfertilized OVUM (a) enters one of the FALLOPIAN TUBES (OVIDUCTS) between an umbrellalike or fingerlike FIMBRIA (sometimes touching the OVARY with a single OVARIAN FIMBRIA).

■ Ciliary action and peristalsis conduct the OVUM into the wide INFUNDIBULUM.

■ After SPERM travels up the UTERUS and into the FALLOPIAN TUBES toward the OVARY, FERTILIZATION takes place with the OVUM in the FALLOPIAN TUBE.

■ Ciliary action and peristalsis move the ZYGOTE (if FERTILIZATION occurs) to the UTERUS.

★ UTERUS:

•The UTERUS is supported by 4 pairs of ligaments derived from serosa.

•The thick, 3-layered muscular MYOMETRIUM is responsible for the muscular contractions needed during LABOR and PARTURITION (delivery).

•The zygote enters the UTERUS from the FALLOPIAN TUBE and implants in the 2–layered mucosal ENDOMETRIUM:

■ STRATUM FUNCTIONALE

Superficial layer, composed of columnar epithelium and secretory glands is shed as MENSES during menstruation, and is built up again by steroid ovarian hormones.

■ STRATUM BASALE
Regenerative deeper layer that replenishes STRATUM FUNCTIONALE after each menstruation.

★ VAGINA: •Copulation passageway

•Delivers sperm from INTROMISSION of the PENIS to CERVIX OF UTERUS

•Canal for discharge of MENSTRUAL FLOW

•Birth canal for delivery

THE 3 HISTOLOGICAL LAYERS (WALLS) OF THE UTERINE TUBES, UTERUS, AND VAGINA. ▶

B: FALLOPIAN (UTERINE) TUBES

1 INFUNDIBULUM

2 FIMBRIA

3 OVARIAN FIMBRIA

4 AMPULLA FERTILIZATION SITE

5 ISTHMUS

6 INTRAMURAL

7 MUCOSA (CILIATED COLUMNAR)

8 MUSCULARIS

9 SEROSA (part of the visceral petoneum)

C: The UTERUS

Structure

10 FUNDUS

11 BODY

12 ISTHMUS

13 CERVIX CERVIX UTERI (NECK)

Inner CAVITIES (and Their Openings †)

14 UTERINE CAVITY

15 INTERNAL OS †

16 CERVICAL CANAL

17 EXTERNAL OS †

Support of UTERUS

4 Paired LIGAMENTS:

18 BROAD LIGAMENTS PERITONEUM

19 UTEROSACRAL LIGAMENTS

20 CARDINAL LIGAMENTS LATERAL CERVICAL

21 ROUND LIGAMENTS (See 14.6)

22 ENDOMETRIUM 2-LAYERED MUCOSA (See 14.8)

23 MYOMETRIUM 3-LAYERED, THICK MUSCULARIS

24 PERIMETRIUM SEROSA

D: The VAGINA

25 FORNIX OF VAGINA
4 FORNICES

• Anterior, posterior, and 2 lateral spaces into which the upper vagina is divided

• These recesses are formed by protrusion of the CERVIX UTERI (1") into the Vagina (forming the upper horizontal crescent-shaped walls)

• The BLADDER is situated adjacent to the anterior vaginal wall

• The RECTUM is situated behind the posterior wall

• The VAGINA represents a *potential space*, the walls of which are in contact with each other (forming a vertical slit close to the VULVA)

• The stratified, squamous epithelium (mucous membrane) of the mucosa forms a series of transverse folds called VAGINA RUGAE.

MUCOSA•

MUSCULARIS

FIBROUS LAYER

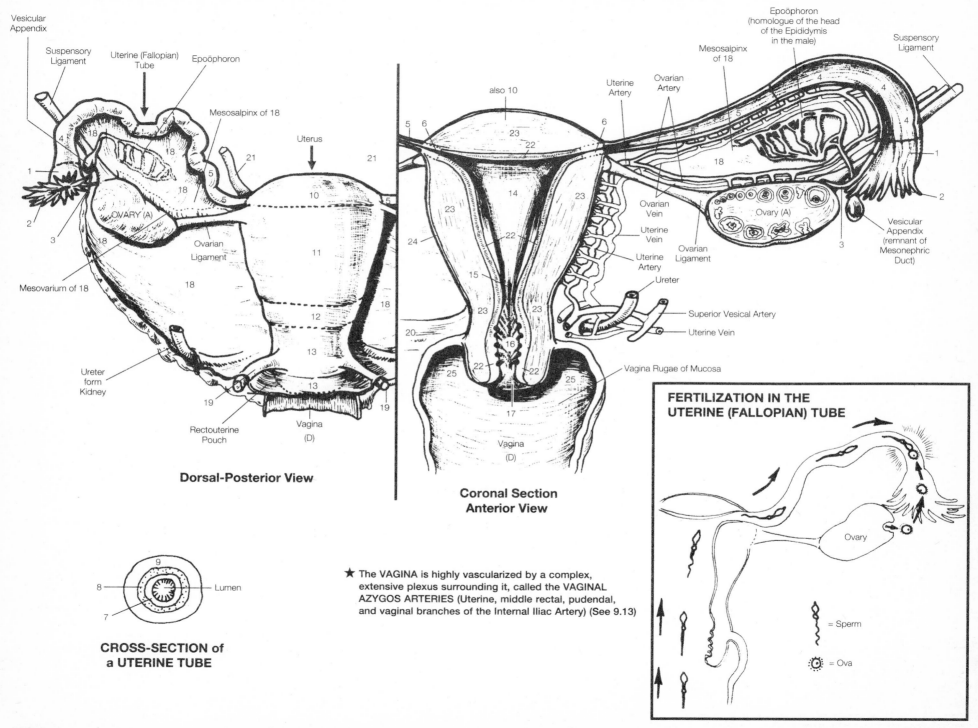

Vesicular
Appendix

Suspensory
Ligament

Uterine (Fallopian)
Tube

Epoöphoron

Mesosalpinx of 18

Uterus

21

Ovarian
Ligament

Mesovarium of 18

Ureter
form
Kidney

Rectouterine
Pouch

Vagina
(D)

OVARY (A)

18

18

18

18

18

18

18

10

11

12

13

13

19

19

5

5

5

4

1

2

3

21

Dorsal-Posterior View

also 10

Uterine
Artery

Ovarian
Artery

Mesosalpinx
of 18

Epoöphoron
(homologue of the head
of the Epididymis
in the male)

Suspensory
Ligament

5   6

23

22

6

Ovarian
Vein

Uterine
Vein

Ovarian
Ligament

Uterine
Artery

Ureter

Superior Vesical Artery

Uterine Vein

Vagina Rugae of Mucosa

14

23

23

23

24

22

15

23

23

20

16

22

22

25

25

17

Vagina
(D)

18

4

4

4

5

1

2

Ovary (A)

3

Vesicular
Appendix
(remnant of
Mesonephric
Duct)

**Coronal Section
Anterior View**

9

8 ——— Lumen

7

**CROSS-SECTION of
a UTERINE TUBE**

★ The VAGINA is highly vascularized by a complex,
extensive plexus surrounding it, called the VAGINAL
AZYGOS ARTERIES (Uterine, middle rectal, pudendal,
and vaginal branches of the Internal Iliac Artery) (See 9.13)

**FERTILIZATION IN THE
UTERINE (FALLOPIAN) TUBE**

Ovary

= Sperm

= Ova

THE FEMALE ♀ REPRODUCTIVE SYSTEM: SECONDARY SEX ORGANS
The Vulva: External Genital Structures

★ See 6.11

• MONS PUBIS: Pad of adipose tissue and coarse skin that cushions the SYMPHYSIS PUBIS and VULVA during copulation; hairy after puberty

★ The EXTERNAL GENITALIA (VULVA or PUDENDUM) serve to form margins, and to enclose and protect the VAGINAL ORIFICE (opening to the VAGINA) and other external reproductive organs. The VULVA lies posterior to the MONS PUBIS and consists of the 2 LABIA MAJORA, 2 LABIA MINORA, CLITORIS, VESTIBULE, (including VAGINAL and URETHRAL ORIFICES, and BULBS and GLANDS of the VESTIBULE).

• LABIA MAJORA: 2 thickened longitudinal folds enclose and protect VULVA (homologous to the SCROTUM); hairy. (They are separated by a spatial cleft called the RIMA PUDENDI into which the URETHRA and VAGINA opens. Their medial surfaces unite above the clitoris to form the ANTERIOR COMMISSURE.)

• LABIA MINORA: 2 small, hairless longitudinal folds protect the VAGINAL and URETHRAL OPENINGS and enclose the VESTIBULE. They lie between the LABIA MAJORA and the HYMEN. Anteriorly, they split to form the PREPUCE and FRENULUM of the CLITORIS.

• CLITORIS: Small, erectile organ
Corresponds to the origin and structure of the PENIS, although it does not have a urethra. It is richly supplied with sensory nerve endings that enhance pleasure during sexual stimulation.

• Located beneath the Anterior Commissure of the Labia Majora, and partially hidden by the cloaks of the Labia Minora
Measures 2 cm long (0.8 in.) 0.5 cm diameter (0.2 in.)

• VESTIBULE
Longitudinal spatial cleft enclosed by LABIA MINORA. (In the illustration, the VESTIBULE can be visualized as the space above the Urogenital Diaphragm.)
Contains openings of the VAGINA and URETHRA. During sexual excitement, the VAGINAL OPENING is lubricated by secretions from a pair of VESTIBULAR GLANDS located within the wall just inside the vaginal orifice.

Lateral walls of the VESTIBULE are formed   inwardly by vascular, erectile tissue: the VESTIBULAR BULBS.

★ ■ Parasympathetic nervous stimulation causes dilation of arterioles of the genital erectile tissue, and, as in the male, compress venous return.

E **MONS PUBIS**   MONS VENERIS (PUBIC EMINENCE)

The VULVA: PUDENDUM
F **2 LABIA MAJORA** † LATERAL BORDERS OF THE VULVA

G **2 LABIA MINORA** †

H **CLITORIS**

I VESTIBULE

1 **FOURCHETTE**   FRENULUM OF BOTH LABIA

2 **PERINEAL BODY** "CENTRAL TENDON" OF PERINEAL MUSCLES

† In young girls, the medial surface of the LABIA MAJORA are in contact with each other concealing the LABIA MINORA and VESTIBULE. In older women, the LABIA MINORA may protrude between the LABIA MAJORA.

3 Portions of the CLITORIS (H)
3 **GLANS** Exposed portion composed of erectile tissue

4 **BODY** Consists of 2 Fused Crura (Continuations of 5)

5 **2 CORPORA CAVERNOSA**

6 **SUSPENSORY LIGAMENT**

UNITING CLOAKS of the LABIA MINORA (G)
7 **PREPUCE OF CLITORIS** (ANTERIOR PREPUTIUM CLITORIDIS)

8 **FRENULUM OF CLITORIS** (POSTERIOR FRENULUM CLITORIDIS)

VESTIBULE: SPATIAL CLEFT between Attachments of LABIA MINORA : Receptacle of ORIFICES

9 **EXTERNAL URETHRAL ORIFICE**

10 **VAGINAL ORIFICE**

11 **HYMEN (CARUNCULAE)** *

J **PARAURETHRAL GLANDS** SKENE'S GLANDS

K **VESTIBULAR GLANDS** BARTHOLIN'S GLANDS

12 **VESTIBULAR BULBS**

* The HYMEN is a fold of mucous membrane that partially covers the entrance to the vagina. (Contrary to folklore, rupture or absence of the hymen cannot be used to prove or disprove virginity or history of sexual intercourse. Conversely, pregnancy can occur with the hymen intact.)

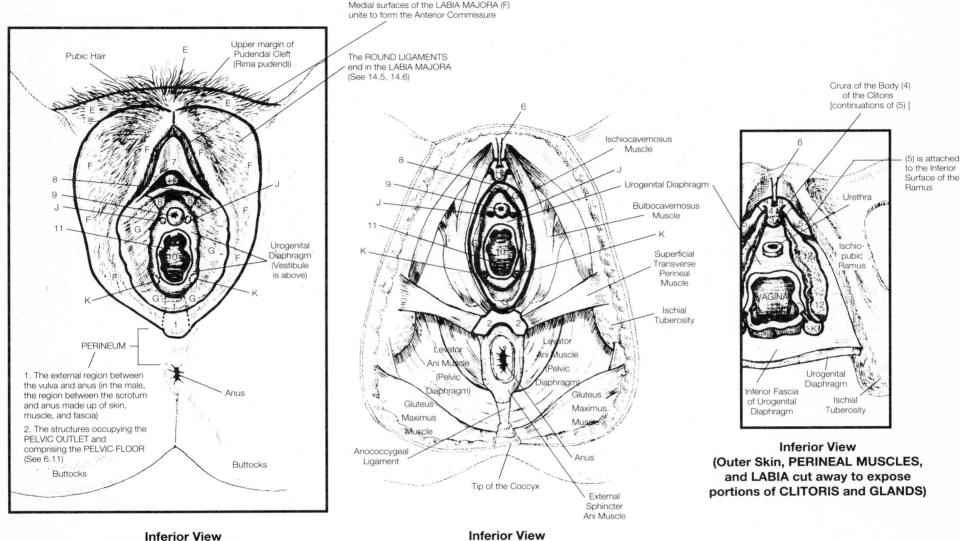

Medial surfaces of the LABIA MAJORA (F) unite to form the Anterior Commissure

The ROUND LIGAMENTS end in the LABIA MAJORA (See 14.5, 14.6)

Pubic Hair

Upper margin of Pudendal Cleft (Rima pudendi)

Urogenital Diaphragm (Vestibule is above)

PERINEUM

1. The external region between the vulva and anus (in the male, the region between the scrotum and anus made up of skin, muscle, and fascia)

2. The structures occupying the PELVIC OUTLET and comprising the PELVIC FLOOR (See 6.11)

Anus

Buttocks

Buttocks

**Inferior View
(LABIA MAJORA and MINORA
spread and folded out to
expose VESTIBULE)**

Ischiocavernosus Muscle

Urogenital Diaphragm

Bulbocavernosus Muscle

Superficial Transverse Perineal Muscle

Ischial Tuberosity

Levator Ani Muscle (Pelvic Diaphragm)

Levator Ani Muscle (Pelvic Diaphragm)

Gluteus Maximus Muscle

Gluteus Maximus Muscle

Anococcygeal Ligament

Anus

Tip of the Coccyx

External Sphincter Ani Muscle

**Inferior View
♀ PERINEUM
(Outer Skin Layer and LABIA cut away)**

Crura of the Body (4) of the Clitoris [continuations of (5)]

(5) is attached to the Inferior Surface of the Ramus

Urethra

Ischio-pubic Ramus

VAGINA

Inferior Fascia of Urogenital Diaphragm

Urogenital Diaphragm

Ischial Tuberosity

**Inferior View
(Outer Skin, PERINEAL MUSCLES,
and LABIA cut away to expose
portions of CLITORIS and GLANDS)**

THE FEMALE ♀ REPRODUCTIVE SYSTEM: SECONDARY SEX ORGANS
Accessory Glands: Mammary Glands (Breasts)

Muscles = reds and oranges
Duct Systems = cool colors
Lymph Nodes = greens 8 = reddish brown
9 = light brown 17 = yellow
★ See 14.7

★ Two compound MAMMARY GLANDS are located within the BREASTS, one in each breast.

■ STRUCTURALLY, they are part of the INTEGUMENTARY SYSTEM, being modified sweat glands embedded in superficial fascia (fat).

■ The amount of adipose tissue determines breast size and shape only.

■ FUNCTIONALLY, they are part of the REPRODUCTIVE SYSTEM, since they secrete MILK for the nourishment of the young (triggered by the hormones PROLACTIN and OXYTOCIN). (See below.)

•At PUBERTY, the OVARIES secrete ESTROGEN, which stimulates growth and development of:
• Duct system
• Mammary glands (alveoli)
• Breast adipose tissue

★ ■ Each MAMMARY GLAND is composed of between 15–20 LOBES
• Each lobe has its own drainage pathway externally. Each LOBE is subdivided into LOBULES, which contain the GLANDULAR ALVEOLI. The ALVEOLAR clusters secrete MILK only in response to secretions of LUTEOTROPIC and OXYTOCIN HORMONES.
• MILK is channeled through a series of SECONDARY TUBULES and MAMMARY DUCTS, being stored in LACTIFEROUS AMPULLAE (SINUSES) before draining at the NIPPLE.
• The NIPPLE remains pliable due to secretions of AREOLAR GLANDS within the pigmented circular AREOLA surrounding the NIPPLE.

★ ■ Support of the BREASTS is from SUSPENSORY LIGAMENTS of COOPER, which run between the LOBULES extending from the skin to the deep fascia overlying the PECTORALIS MUSCLE.

★ DEVELOPMENT DURING PREGNANCY
• PROGESTERONE secreted by the CORPUS LUTEUM in the OVARY (See 14.7) and PLACENTA acts synergistically with ESTROGENS to bring the ALVEOLI to complete development

★ DEVELOPMENT AFTER PARTURITION
• PROLACTIN (LUTEOTROPHIN) in conjunction with ADRENAL CORTICOIDS initiates LACTATION (MILK SECRETION)
• OXYTOCIN from the POSTERIOR PITUITARY GLAND induces ejection of milk
• The sucking and MILKING REFLEX restimulates milk secretion and discharge.

L The BREASTS (L. & R.)

The MAMMARY GLANDS
Glandular Structure

1 LOBE

2 LOBULES

3 GLANDULAR ALVEOLI → SECRETE MILK †

4 SECONDARY TUBULES

5 MAMMARY DUCTS

6 LACTIFEROUS AMPULLAE (SINUSES)

7 LACTIFEROUS DUCTS

Gross Anatomy

8 NIPPLE PRIMARY AREOLA

9 AREOLA

10 AREOLAR GLANDS

† In the first 2 or 3 days after birth (and before the onset of true lactation) the breasts secrete COLOSTRUM, a thin, yellowish fluid containing large quantities of proteins and calories (in addition to antibodies and lymphocytes important for protective immunity against infection).

Lymphatic Drainage

11 AXILLARY NODES

12 AXILLARY-APICAL NODES

13 PARASTERNAL NODES

14 NODES TO OPPOSITE BREAST

15 NODES TO RECTUS SHEATH AND DIAPHRAGM

Related Structures

16 SUSPENSORY LIGAMENTS OF COOPER SUPPORTS BREASTS

17 SUPERFICIAL FASCIA (FAT) ADIPOSE TISSUE

18 PECTORALIS MAJOR MUSCLE AND MINOR

19 DEEP FASCIA OF PECTORALIS MAJOR MUSCLE

20 INTERCOSTAL MUSCLES

21 SERRATUS ANTERIOR MUSCLE

22 CLAVICLE

23 RIBS

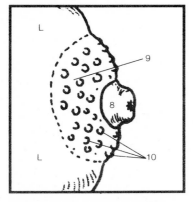

GROSS ANATOMY of BREAST

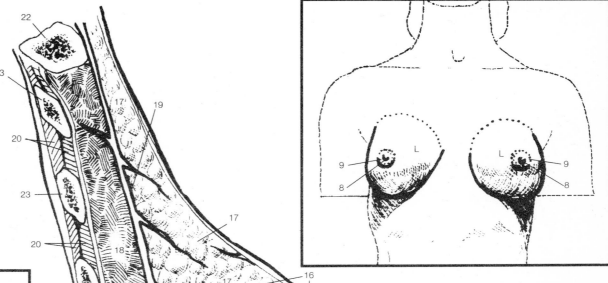

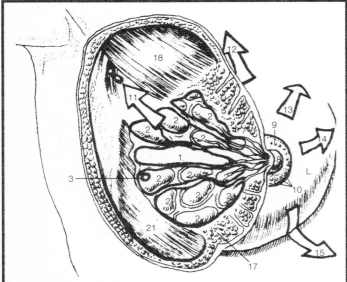

LYMPHATIC DRAINAGE CHANNELS of the BREAST
(Partially Sectioned)

★ Lateral margin of the BREAST is alongside the anterior border of the AXILLA (ARMPIT). The AXILLARY TAIL of the BREAST comes in close contact with axillary vessels, and this region is clinically associated with a high incidence of breast cancer within the lymphatic drainage

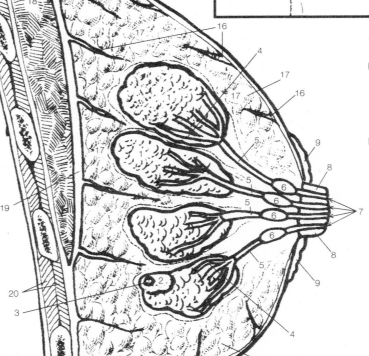

★ **BREAST DEVELOPMENT**
DURING PREGNANCY
■ First 6-12 weeks
  • Fullness
  • Tenderness
  • Erectile tissues develop in nipples
  • Pigment deposited around the nipple (the primary areola)
■ Next, 16-20 weeks
  • The secondary areola shows small whitish spots in pigmentation due to hypertrophy of the SEBACEOUS GLANDS (GLANDS of MONTGOMERY) present in the aroela surrounding the nipple.

SAGITTAL SCHEMATIC (SIMPLIFIED)

THE REPRODUCTIVE SYSTEM: SUMMARY COMPARISONS
Mitosis and Meiosis/Spermatogenesis and Oögenesis

★ See Chapter 2, Chart #1

★★ See Chart #2, and 14.2, 14.7

CELL DIVISIONS: MITOSIS and MEIOSIS

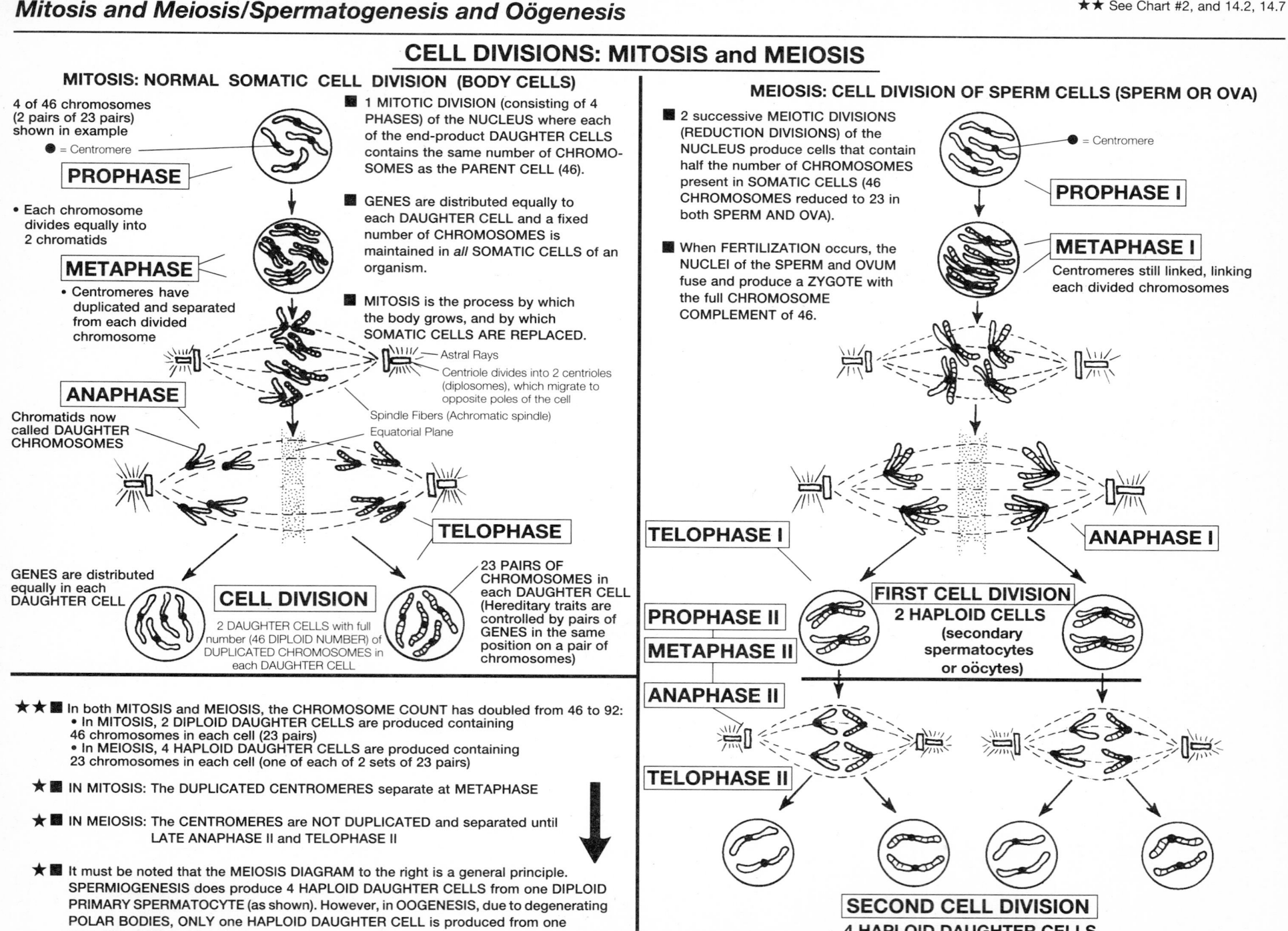

MITOSIS: NORMAL SOMATIC CELL DIVISION (BODY CELLS)

4 of 46 chromosomes
(2 pairs of 23 pairs)
shown in example

● = Centromere

PROPHASE

• Each chromosome
divides equally into
2 chromatids

METAPHASE

• Centromeres have
duplicated and separated
from each divided
chromosome

ANAPHASE

Chromatids now
called DAUGHTER
CHROMOSOMES

■ 1 MITOTIC DIVISION (consisting of 4
PHASES) of the NUCLEUS where each
of the end-product DAUGHTER CELLS
contains the same number of CHROMO-
SOMES as the PARENT CELL (46).

■ GENES are distributed equally to
each DAUGHTER CELL and a fixed
number of CHROMOSOMES is
maintained in *all* SOMATIC CELLS of an
organism.

■ MITOSIS is the process by which
the body grows, and by which
SOMATIC CELLS ARE REPLACED.

Astral Rays
Centriole divides into 2 centrioles
(diplosomes), which migrate to
opposite poles of the cell
Spindle Fibers (Achromatic spindle)
Equatorial Plane

TELOPHASE

GENES are distributed
equally in each
DAUGHTER CELL

CELL DIVISION

2 DAUGHTER CELLS with full
number (46 DIPLOID NUMBER) of
DUPLICATED CHROMOSOMES in
each DAUGHTER CELL

23 PAIRS OF
CHROMOSOMES in
each DAUGHTER CELL
(Hereditary traits are
controlled by pairs of
GENES in the same
position on a pair of
chromosomes)

MEIOSIS: CELL DIVISION OF SPERM CELLS (SPERM OR OVA)

■ 2 successive MEIOTIC DIVISIONS
(REDUCTION DIVISIONS) of the
NUCLEUS produce cells that contain
half the number of CHROMOSOMES
present in SOMATIC CELLS (46
CHROMOSOMES reduced to 23 in
both SPERM AND OVA).

■ When FERTILIZATION occurs, the
NUCLEI of the SPERM and OVUM
fuse and produce a ZYGOTE with
the full CHROMOSOME
COMPLEMENT of 46.

● = Centromere

PROPHASE I

METAPHASE I

Centromeres still linked, linking
each divided chromosomes

TELOPHASE I

ANAPHASE I

FIRST CELL DIVISION
2 HAPLOID CELLS
(secondary
spermatocytes
or oöcytes)

PROPHASE II

METAPHASE II

ANAPHASE II

TELOPHASE II

SECOND CELL DIVISION
4 HAPLOID DAUGHTER CELLS
Each cell contains 23 chromosomes

★★■ In both MITOSIS and MEIOSIS, the CHROMOSOME COUNT has doubled from 46 to 92:
• In MITOSIS, 2 DIPLOID DAUGHTER CELLS are produced containing
46 chromosomes in each cell (23 pairs)
• In MEIOSIS, 4 HAPLOID DAUGHTER CELLS are produced containing
23 chromosomes in each cell (one of each of 2 sets of 23 pairs)

★■ IN MITOSIS: The DUPLICATED CENTROMERES separate at METAPHASE

★■ IN MEIOSIS: The CENTROMERES are NOT DUPLICATED and separated until
LATE ANAPHASE II and TELOPHASE II

★■ It must be noted that the MEIOSIS DIAGRAM to the right is a general principle.
SPERMIOGENESIS does produce 4 HAPLOID DAUGHTER CELLS from one DIPLOID
PRIMARY SPERMATOCYTE (as shown). However, in OOGENESIS, due to degenerating
POLAR BODIES, ONLY one HAPLOID DAUGHTER CELL is produced from one
DIPLOID PRIMARY OOCYTE

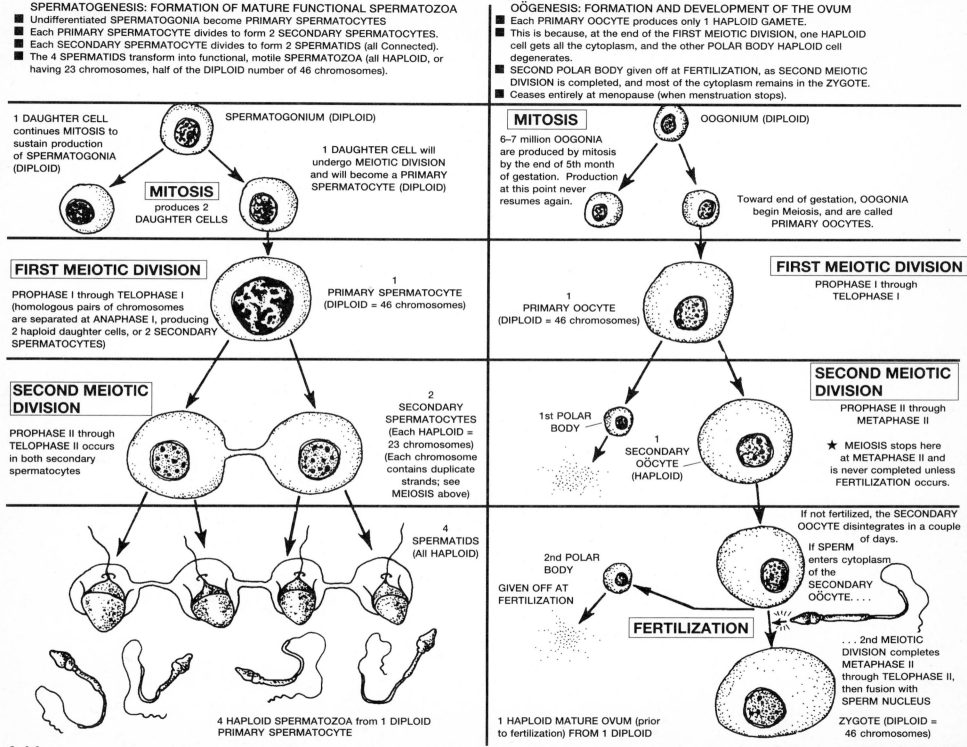

SPERMATOGENESIS: FORMATION OF MATURE FUNCTIONAL SPERMATOZOA
- Undifferentiated SPERMATOGONIA become PRIMARY SPERMATOCYTES
- Each PRIMARY SPERMATOCYTE divides to form 2 SECONDARY SPERMATOCYTES.
- Each SECONDARY SPERMATOCYTE divides to form 2 SPERMATIDS (all Connected).
- The 4 SPERMATIDS transform into functional, motile SPERMATOZOA (all HAPLOID, or having 23 chromosomes, half of the DIPLOID number of 46 chromosomes).

OÖGENESIS: FORMATION AND DEVELOPMENT OF THE OVUM
- Each PRIMARY OOCYTE produces only 1 HAPLOID GAMETE.
- This is because, at the end of the FIRST MEIOTIC DIVISION, one HAPLOID cell gets all the cytoplasm, and the other POLAR BODY HAPLOID cell degenerates.
- SECOND POLAR BODY given off at FERTILIZATION, as SECOND MEIOTIC DIVISION is completed, and most of the cytoplasm remains in the ZYGOTE.
- Ceases entirely at menopause (when menstruation stops).

1 DAUGHTER CELL continues MITOSIS to sustain production of SPERMATOGONIA (DIPLOID)

SPERMATOGONIUM (DIPLOID)

1 DAUGHTER CELL will undergo MEIOTIC DIVISION and will become a PRIMARY SPERMATOCYTE (DIPLOID)

MITOSIS
produces 2
DAUGHTER CELLS

MITOSIS

6–7 million OOGONIA are produced by mitosis by the end of 5th month of gestation. Production at this point never resumes again.

OOGONIUM (DIPLOID)

Toward end of gestation, OOGONIA begin Meiosis, and are called PRIMARY OOCYTES.

FIRST MEIOTIC DIVISION

PROPHASE I through TELOPHASE I (homologous pairs of chromosomes are separated at ANAPHASE I, producing 2 haploid daughter cells, or 2 SECONDARY SPERMATOCYTES)

1
PRIMARY SPERMATOCYTE
(DIPLOID = 46 chromosomes)

1
PRIMARY OOCYTE
(DIPLOID = 46 chromosomes)

FIRST MEIOTIC DIVISION

PROPHASE I through TELOPHASE I

SECOND MEIOTIC DIVISION

PROPHASE II through TELOPHASE II occurs in both secondary spermatocytes

2
SECONDARY SPERMATOCYTES
(Each HAPLOID = 23 chromosomes)
(Each chromosome contains duplicate strands; see MEIOSIS above)

1st POLAR BODY

1
SECONDARY OÖCYTE
(HAPLOID)

SECOND MEIOTIC DIVISION

PROPHASE II through METAPHASE II

★ MEIOSIS stops here at METAPHASE II and is never completed unless FERTILIZATION occurs.

4
SPERMATIDS
(All HAPLOID)

2nd POLAR BODY

GIVEN OFF AT FERTILIZATION

FERTILIZATION

If not fertilized, the SECONDARY OOCYTE disintegrates in a couple of days.

If SPERM enters cytoplasm of the SECONDARY OÖCYTE. . . .

. . . 2nd MEIOTIC DIVISION completes METAPHASE II through TELOPHASE II, then fusion with SPERM NUCLEUS

ZYGOTE (DIPLOID = 46 chromosomes)

4 HAPLOID SPERMATOZOA from 1 DIPLOID PRIMARY SPERMATOCYTE

1 HAPLOID MATURE OVUM (prior to fertilization) FROM 1 DIPLOID

Chapter 14: *REPRODUCTIVE SYSTEM:* CHARTS

 **Chart #1** : **Structures of the Reproductive System (See 14.1, 14.5)**

♂ Male Reproductive System

Structure	Function
TESTES	
SEMINIFEROUS TUBULES	Produce spermatozoa and male sex hormones; produce spermatozoa
INTERSTITIAL CELLS	Produce and secrete male sex hormones
EPIDIDYMIS	Storage and maturation of spermatozoa; conveys spermatozoa to Vas Deferens.
VAS DEFERENS	Stores spermatozoa; conveys spermatozoa to ejaculatory ducts.
EJACULATORY DUCTS	Receive spermatozoa and additives to produce seminal fluid
SEMINAL VESICLES	Secrete alkaline fluid containing nutrients and prostaglandins
PROSTATE GLAND	Secretes alkaline fluid that helps neutralize acidic seminal fluid and enhances motility of spermatozoa
BULBOURETHRAL (COWPER'S) GLANDS	Secrete fluid that lubricates urethra and end of penis
SCROTUM	Encloses and protects testes
PENIS	Conveys urine and seminal fluid externally; organ of copulation.

♀ Female Reproductive System

Structure	Function
OVARIES	Produce ova and female sex hormones
FALLOPIAN TUBES	Convey ovum toward uterus; site of fertilization; convey developing zygote to uterus.
UTERUS	Site of implantation; protects and sustains life of embryo and fetus during pregnancy; active role in parturition.
VAGINA	Conveys uterine secretions externally; receives erect penis and semen during copulation and ejaculation; passage for fetus during parturition.
LABIA MAJORA	Form margins of pudendal cleft (rima pudendi); enclose and protect other external reproductive organs.
LABIA MINORA	Form margins of vestibule; protect openings of vagina and urethra.
CLITORIS	Glans of the clitoris richly supplied with sensory nerve endings associated with sensations generated by pressure stimulation
VESTIBULE	Cleft between labia minora that includes vaginal and urethral openings
VESTIBULAR (BULBOURETHRAL) GLANDS (BARTHOLIN'S GLANDS)	Secrete fluid that moistens and lubricates the vestibule and vaginal opening during intercourse
MAMMARY GLANDS	Produce and secrete milk

Chart #2 : Mitosis and Meiosis (See 14.11)

Summary of Mitosis
(a continuous process divided into 4 phases:

a) *Prophase*--
• Dispersed **chromatin granules** of the nucleus become organized into 46 chromosomes (Each chromosome possesses a clear mid-region called a **centromere**, which separates it into 2 arms)
• Each chromosome is longitudinally divided into 2 identical **chromatids**, but they are still attached by the single centromere.
• The nuclear membrane and the nucleolus disappear
• The **centriole** of the cell divides into 2 daughter centrioles called **diplosomes** and they move to opposite poles of the cell (The diplosomes are connected by fine protoplasmic fibrils called **spindle fibers**)

b) *Metaphase*--
• The centromeres themselves separate and the 2 chromatids of each chromosome are now completely separated, but close together
• The homologous chromosomes (now expressed as the 2 paired chromatids of each chromosome) arrange themselves in an **equatorial plane** midway between, and perpendicular to, the 2 diplosomes, forming the **equatorial plate**

c) *Anaphase*--
• The chromatids (now called **daughter chromosomes**) diverge and move toward their respective diplosomes
• The end of the migration toward the diplosomes marks the beginning of the next phase

d) *Telophase*--
• The daughter chromosomes (chromatids) at each pole of the **achromatic spindle** undergo changes the reverse of those in Prophase-- they begin to disperse into chromatin granules
• The nuclear membrane reforms as well as the nucleolus.
• The outlines of the chromosomes disappear and chromatin appears as granules scattered throughout the nucleus connected by a net.
• By constriction in the equatorial plane, the cytoplasm becomes separated into 2 parts, resulting in 2 complete cells, each now with 46 chromosomes (23 pairs) in its nucleus.

e) *Interphase*--
• The resting period between 2 successive mitotic divisions

Summary of Spermatogenesis (♂ Meiosis)

1) Embryonic "stem cells" called spermatogonia go through a normal mitosis producing two daughter cells.

2) One of the daughter cells, called the primary spermatocyte, begins the first meiotic division.

3) First meiotic division
   a) *Prophase I*—Chromosomes appear double-stranded. Each strand, called a chromatid, contains duplicate DNA joined together by a structure known as a centromere. Homologous chromosomes pair up side by side.
   b) *Metaphase I*—Homologous chromosome pairs line up at equator. Spindle apparatus is completed.
   c) *Anaphase I*—Homologous chromosomes are separated; each member of a homologous pair moves to opposite pole.
   d) *Telophase I*—Cytoplasm divides to produce two haploid cells called secondary spermatocytes.

4) The secondary spermatocytes produced by the first meiotic division both enter the second meiotic division.

5) Second meiotic division
   a) *Prophase II*—Chromosomes appear, each containing two chromatids.
   b) *Metaphase II*—Chromosomes align along equator as spindle formation is completed.
   c) *Anaphase II*—Centromeres split and chromatids move to opposite poles.
   d) *Telophase II*—Cytoplasm divides to produce two haploid cells called spermatids.

6) **Spermiogenesis**—Spermatids produced by the second meiotic division are transformed into mature sperm with the assistance of the nurse, or Sertoli, cells.

Summary of Oogenesis (♀ Meiosis)

1) Embryonic "stem cells" in the fetal ovaries produce 6–7 million oogonia by the 5th month of gestation. Oogonia production terminates at this time, never to resume again.

2) At birth, the oogonia are called primary oocytes and number about 2 million. By puberty, the number of primary oocytes has been reduced to about 300,000–400,000.

3) Under stimulation of a reproductive hormone (Follicle-Stimulating Hormone), the primary oocyte begins the first meiotic division. *The first meiotic division is identical to the first meiotic division of the male, with one exception. One of the secondary oocytes receives all of the cytoplasm during telophase I, whereas the other becomes a small polar body that eventually degenerates.*

4) The second meiotic division begins with one secondary oocyte and continues to metaphase II. Meiosis stops at this point unless fertilization occurs.

5) If fertilization occurs then meiosis II is completed, with the fertilized oocyte receiving all of the cytoplasm and the other "cell" becomes another polar body, which again eventually degenerates.

6) The development and changes of the follicle surrounding the oocytes and the details of the entire menstrual cycle can be found in Chart #3.

Adapted from Stuart Ira Fox, *Human Physiology*, 3d ed., p. 76.

Chart #3 :**The 28-Day Menstrual Cycle: Its Relation to the Ovarian and Hormonal Cycles (See 14.7)**

The OVARIAN CYCLE

FOLLICULAR PHASE
② **PRIMORDIAL FOLLICLE**
③ **PRIMARY FOLLICLE**
④ **SECONDARY FOLLICLE**
⑤ **MATURE GRAAFIAN FOLLICLE**

OVULATORY PHASE
⑦+⑧ **OVULATION** RUPTURED GRAAFIAN FOLLICLE DISCHARGED SECONDARY OOCYTE

LUTEAL PHASE
⑩ **YOUNG CORPUS LUTEUM**
⑪ **MATURE CORPUS LUTEUM**
⑫ **CORPUS ALBICANS**

THE HORMONAL CYCLE	
PITUITARY SECRETIONS	OVARIAN SECRETIONS
← **FSH** (FOLLICLE-STIMULATING HORMONE)	E **ESTROGEN** (ESTRADIOL) →
← **LH** (LUTEINIZING HORMONE)	P **PROGESTERONE** →

The MENSTRUAL CYCLE

a **MENSTRUATION** ISCHEMIC PHASE
b **PROLIFERATIVE PHASE**

c **SECRETORY PHASE**

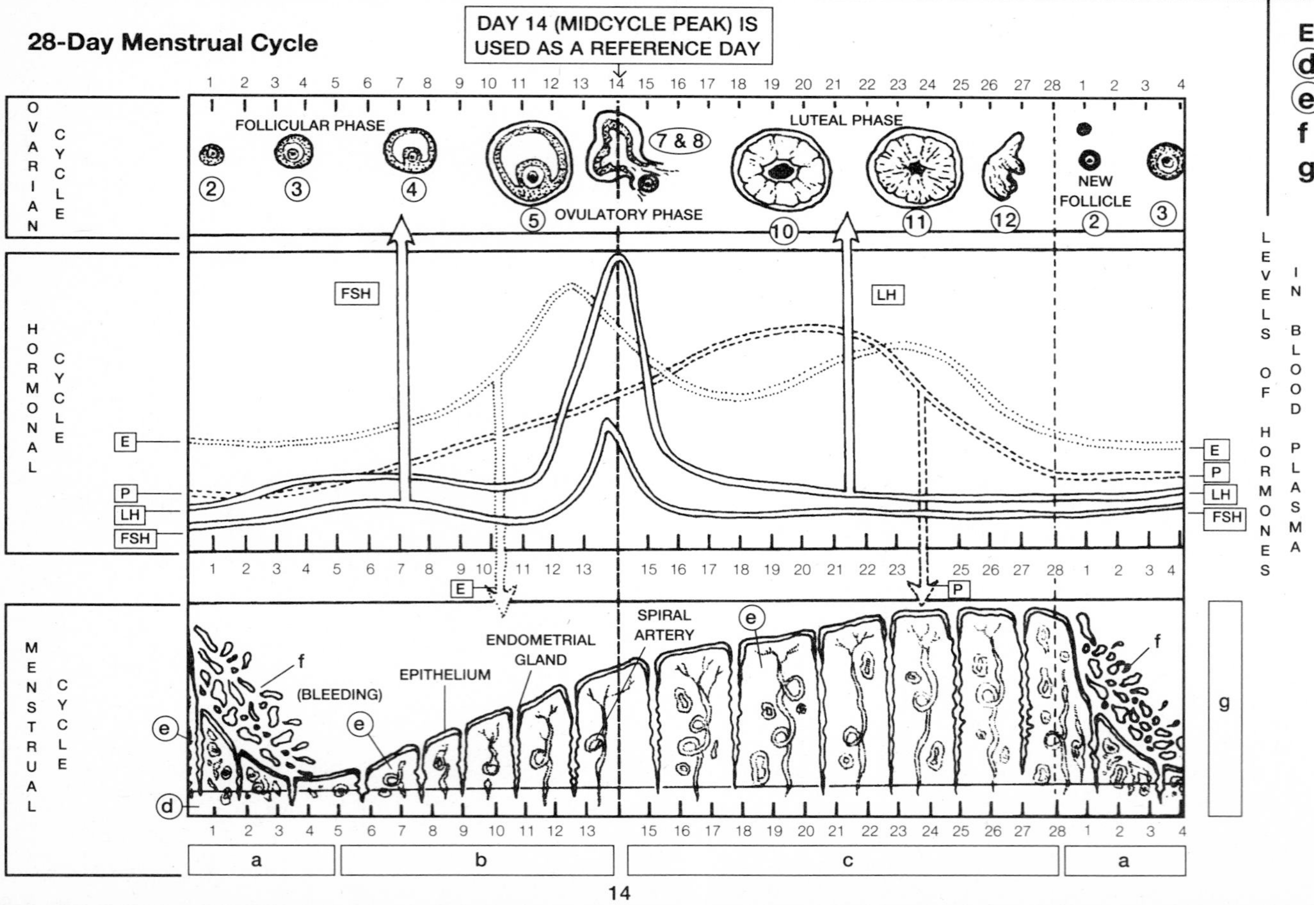

ENDOMETRIUM
d **STRATUM BASALE**
e **STRATUM FUNCTIONALE**
f **MENSES** BLEEDING
g **THICKNESS OF THE ENDOMETRIUM**

28-Day Menstrual Cycle

DAY 14 (MIDCYCLE PEAK) IS USED AS A REFERENCE DAY

OVARIAN CYCLE

FOLLICULAR PHASE   7 & 8   LUTEAL PHASE   OVULATORY PHASE

② ③ ④ ⑤ ⑩ ⑪ ⑫ NEW FOLLICLE ② ③

HORMONAL CYCLE

FSH   LH   E   P   LH   FSH

LEVELS OF HORMONES   IN BLOOD PLASMA

MENSTRUAL CYCLE

SPIRAL ARTERY   e   EPITHELIUM   ENDOMETRIAL GLAND   f   (BLEEDING)   e   e   d   g

E   P

a   b   c   a

14

Chart #4 : Some Homologies of the Male and Female Reproductive Systems (See 14.1, 14.5)

	DERIVATIVE OF STRUCTURE IN THE ♂ MALE	NAME OF STRUCTURE IN INDIFFERENT STAGE	DERIVATIVE OF STRUCTURE IN THE ♀ FEMALE	
1	TESTIS	← GONAD →	OVARY	1
2	EFFERENT DUCTULES and EPIDIDYMAL HEAD	MESONEPHRIC DUCT ← CRANIAL PORTION →	(VESTIGIAL REMNANTS) EPOOPHORON	2
3	EPIDIDYMIS and VAS DEFERENS	← REMAINDER →	DUCT OF EPOOPHORON AND GARTNER'S DUCT	3
4	EPIDIDYMAL APPENDIX	← REMAINDER →	VESICULAR APPENDIX	4
5	————	MULLERIAN DUCT ← UPPER PART →	UTERUS	5
6	————	← MIDDLE PART →	UTERUS	6
7	————	← LOWER PART →	VAGINA	7
8+	PROSTATIC URETHRA	← UROGENITAL SINUS →	VESTIBULE	8
9	PENIS (GLANS and SHAFT)	← PHALLUS →	CLITORIS (GLANS AND SHAFT)	9
10	PENOSCROTAL RAPHE	LIPS OF UROGENITAL ← GROOVE → (URETERAL FOLD)	LABIA MINORA	10
11	SCROTUM	← GENITAL TUBERCLES →	LABIA MAJORA	11

SPERMATIC CORD   ROUND LIGAMENT

+ Note that the vestibule (8) in the illustration is a space and not surrounding tissue

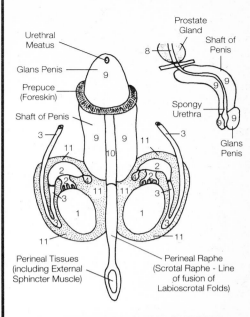

♂ MALE REPRODUCTIVE SYSTEM

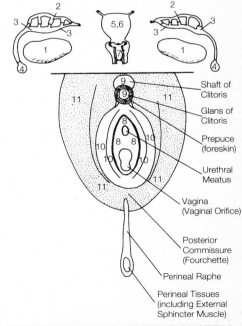

♀ FEMALE REPRODUCTIVE SYSTEM

BIBLIOGRAPHY

Dorland's Illustrated Medical Dictionary. 25th ed.
Philadelphia: W. B. Saunders & Co., 1974.

Grant, J. C. *An Atlas of Anatomy*. 4th ed. Baltimore: Williams
& Wilkens Co., 1956.

Hale, R. B., ed. *Artistic Anatomy*, P. Richer, New York:
Watson-Guptill Publications, 1971.

Hole, John W., Jr. *Human Anatomy and Physiology*. 4th ed.
Dubuque, IA: Wm. C. Brown Publishers, 1987.

McClintic, J. R. *Basic Anatomy and Physiology of the Human
Body*. 2d ed. New York: John Wiley & Sons, 1980.

McClintic, J. R. *Human Anatomy*. St. Louis: C. V. Mosby Co.,
1983.

McNaught, A. B., and R. Callander. *Illustrated Physiology*.
3d ed. New York: Churchill Livingstone, 1975.

Netter, F. H. *Ciba Collection of Medical Illustrations*. Vols. I–
III. Summit, NJ: Ciba Pharmaceutical Co., 1962.

Peck, S. R. *Atlas of Human Anatomy for the Artist*. 11th ed.
New York: Oxford University Press, 1968.

Romanes, G. J., ed. *Cunningham's Textbook of Anatomy*. 10th
ed. New York: Oxford University Press, 1964.

Spence, A. P., and E. B. Mason. *Human Anatomy and
Physiology*. 2d ed. Reading, MA: Benjamin and Cummings,
1983.

Stedman's Medical Dictionary. 25th ed. Philadelphia: W. B.
Saunders & Co., 1974.

Taber's Cyclopedic Medical Dictionary. 15th ed. Philadelphia:
F. A. Davis Co., 1985.

Tortora, G. J. *Principles of Human Anatomy*. New York:
Harper & Row, 1977.

Tortora, G. J. *Principles of Human Anatomy*. 3d ed. New
York: Harper & Row, 1983.

Truex, R. C., and M. B. Carpenter. *Human Neuroanatomy*.
8th ed. Baltimore: Williams & Wilkens Co., 1982.

Van De Graaff, K. M., and S. Fox. *Concepts of Human
Anatomy and Physiology*. Dubuque, IA: Wm. C. Brown
Publishers, 1989.

Van De Graaff, K. M., and R. W. Rhees. *Schaum's Outline of
Theory and Problems of Human Anatomy and Physiology*.
New York: McGraw-Hill Book Co., 1987.

Vannini, V., and G. Pogliani, ed. *The Color Atlas of Human
Anatomy*. New York: Harmony Books, 1980.

GLOSSARY-INDEX

Externus (eks-TER-nus) 6.17
Facies (FĀ-shē-ez) 1.2
Falciform (FAL-si-form) 12.4, 12.15
Fallopian (fah-LŌ-pē-an) 14.5, 14.8
Falx Cerebri (falks sē-REB-rē) 7.4
Fascia (FĀSH-ē-ah) 3.1, 13.2, 14.3, 14.4, 14.10
Fasciae (FASH-ē-ē) 6.17
Fascicle (FAS-i-k'l) 6.2
Fasiculata (fa-SIK-ū-la-tah) 8.6
Fauces (FAW-sez) 11.2, 12.5
Femoral (FEM-or-al) 7.14, 9.13, 9.14, 9.19, 9.21
Femoris (fem-OR-is) 6.17, 6.18, 6.19, 9.14
Femur (FĒ-mur) 4.14
Fibrinogen (fī-BRIN-ō-jen) 9.3
Fibroblast (FĪ-brō-blast) 2.4
Fibrocartilage (fī-brō-KAR-ti-lij) 2.2, 2.4
Fibula (FIB-ū-lah) 4.14
Fibular (FIB-ū-lar) 5.3
Filiform (FIL-i-form) 12.7
Filum (FĪ-lum) 7.4
Filum Terminale (FĪ-lum TER-mi-nal-ē) 7.4, 7.11
Fimbria (FĪM-brē-ah) 14.8
Fissure (FISH-ūr) 7.7, 11.7
Flexor (FLEK-sor) 6.3, 6.14, 6.21, 6.22, 6.23
Flocculonodular (FLOK-ū-lō-NOD-ū-lar) 7.10
Flocculus (FLOK-ū-lus) 7.10
Follicle (FOL-i-k'l) 8.4, 14.7
Folliculotrophin (fo'lik-ū-lō-TRŌ-fin) 8.3
Foramen (fō-RĀ-men) 7.5, 12.7
Fornix (FOR-niks) 14.8
Fossa Ovalis (FOS-ah ŌVA-lis) 9.5
Fourchette (foor-SHET) 14.9
Fovea (FŌ-vē-ah) 4.14, 7.20, 7.21
Frenulum (FREN-ū-lum) 12.5, 12.7, 14.9
Frontalis (fron-TĀ-lis) 6.4
Fundus (FUN-dus) 12.9, 14.8
Fungiform (FUN-ji-form) 7.18
Funiculi (fu-NIK-ū-lī) 7.11
Fusiform (FŪ-si-form) 6.1
Galea (GĀ-lē-ah) 6.4
Galen (GĀ-len) 9.16
Gastric (GAS-trik) 8.7, 9.11, 9.20
Gastrin (GAS-trin) 8.7
Gastrocnemius (gas'trok-NĒ-mē-us) 6.21
Gastroduodenal (gas'trō-du'ō-DĒ-nal) 9.12
Gastroepiploic (gas'trō-ep'i-PLŌ-ik) 9.12
Gemellus (jē-MEL-us) 6.17
Genicular (jē-NIK-ū-lar) 9.14, 9.21
Geniohyoid (jē'nē-ō-HĪ-oyd) 6.6
Genitofemoral (jen'i-tō-FEM-or-al) 7.14
Germinal (JUR-mi-nal) 10.3, 14.6
Germinativum (JER-mi-na-tē-vum) 3.1
Gianuzzi (jah-NOOT-zē) 12.7
Gingiva (jin-JI-vah) 12.5, 12.6
Glandular (GLAN-dū-lar) 14.10
Glans (glanz) 14.9
Globulins (GLOB-ū-lins) 9.3
Globus pallidus (GLŌ-bus pal-LID-us) 7.8
Glomerulosa (glō-MER-ū-losa) 8.6
Glomerulus (glō-MER-ū-lus) 13.3, 13.4, 13.5

Glossopharyngeal (glos'ō-fah-RIN-jē-al) 7.15, 7.17, 7.18
Glottis (GLOT-is) 11.2, 11.4
Glucagon (Glū-kah-gon) 8.7
Glucorticoids (glu'kō-KOR-ti-koyds) 8.6
Gluteal (GLŪ-tē-al) 9.13, 9.14
Gluteus (gloo-TĒ-us) 6.17
Gnostic (NOS-tik) 7.7
Golgi (GŌL-jē) 2.1
Golgi-Mazzoni (GŌL-jē mad-ZO-nē) 3.3
Gomphoses (gom-FŌ-sez) 5.1
Gonadocorticoids (gon'ah-dō-KOR-ti-koyds) 8.6
Gonadotrophin (gon'ah-dō-TRŌ-fin) 8.6, 8.8
Graafian (GRĀF-ē-an) 14.7
Gracilis (GRAS-i-lis) 6.18
Granulosa (gran'ū-LŌ-sah) 8.8, 14.7
Granulosum (gran-ū-LŌ-sum) 3.1
Gustatory (GUS-tah-tō'rē) 7.18
Gyrus (JĪ-rus) 7.7
Habenular (hah-BEN-ū-lar) 7.9
Hallucis (hah-LU-sis) 6.20, 6.21, 6.22, 6.23
Hamate (hah-MATE) 4.11
Haustra (HAWS-trah) 12.13
Haversian (ha-VER-shŭn) 4.2
Hemiazygos (hem'ē-ĀZ-i-gus) 9.19
Hemocytoblast (hē'mō-SĪ-tō-blast) 9.3
Hemorrhagicum (hem'ō-RĀ-ji-kum) 14.7
Henle (HEN-lē) 13.3
Hepatic (hē-PAT-ik) 9.1, 9.11, 9.12, 9.19, 12.2, 12.16
Hepatocyte (HEP-ah-tō-sīt) 12.16
Hepatopancreatic (he-pa-TŌ-pan-krē-ă-tik) 12.11
Hillock (HIL-ok) 7.1
Hilus (HĪ-lus) 13.2
Hippocampal (hip'ō-KAM-pal) 7.7
Histamine (HIS-tah-min) 12.9
Histiocyte (HĪS-tē-ō-sīt) 2.4
Holocrine (HOL-ō-krīn) 2.3
Humerus (HŪ-mer-us) 4.10
Hyaline (HĪ-ah-līn) 2.2, 2.4
Hydrocortisone (hī'drō-KOR-ti-son) 8.6
Hymen (HĪ-men) 14.9
Hyoglossus (hī'ō-GLOS-us) 6.6
Hyoid (HĪ-oyd) 4.4, 4.5
Hypochondriac (hī-'pō-KON-drē-ak) 1.3
Hypodermis (hī'pō-DER-mis) 3.1
Hypogastric (hī'pō-GAS-trik) 1.3, 9.19
Hypoglossal (hī'pō-GLOS-al) 7.13, 7.15
Hyponychium (hī'pō-NIK-ē-um) 3.3
Hypophyseal (hī-pof'i-ZĒ-al) 8.2
Hypophysis (hī-POF-i-sis) 8.1
Hypothalamus (hī'pō-THAL-ah-mus) 7.9, 8.2
Ileal (IL-ē-al) 9.12
Ileocecal (il-ē-ō-SĒ-kal) 12.11, 12.13
Ileocolic (il'ē-ō-KOL-ik) 9.12
Ileum (IL-ē-um) 12.11
Iliac (IL-ē-ak) 1.3, 9.14, 9.20
Iliacs (IL-ē-aks) 9.13
Iliacus (il-Ī-ah-kus) 6.9
Iliococcygeus (il'ē-ō-kok-SIJ-ē-us) 6.11
Iliocostalis (il'ē-ō-kos-TĀ-lis) 6.7

Iliohypogastric (il'ē-ō-hī'pō-GAS-trik) 7.14
Ilioinguinal (il'ē-ō-IN-gwī-nal) 7.14
Iliolumbar (il'ē-ō-LUM-bar) 9.13
Iliopectineal (il'ē-ō-pek-TIN-ē-al) 4.12
Ilium (IL-ē-um) 4.12
Incisor (in-SĪ-zer) 12.6
Incisura (in-sī-SU-rah) 12.9
Indicis (IN-di-sis) 6.15, 9.10
Infraglottic (in'frah-GLOT-ik) 11.4
Infraspinatus (in'frah-spī-NĀ-tus) 6.12
Infundibulum (in'fun-DIB-ū-lum) 8.2, 14.8
Inguinal (ING-gwi-nal) 1.3, 6.10, 10.2
Innominate (i-NOM-i-nāt) 9.8, 9.17
Insertion (in-SER-shun) 6.1
Insulin (IN-sū-lin) 8.7
Integumentary (in-teg-ū-MEN-tar-ē) 3.1
Interarytenoid (in'ter-ar'ē-TĒ-noyd) 11.4
Interatrial (in'ter-Ā-trē-al) 9.5
Intercalated (in-TER-kahl-āt-ed) 9.6
Intercondylar (in'ter-KON-dī-lar) 4.14
Intercostal (in'ter-KOS-tal) 6.9, 9.11, 14.10
Intercostalis (in'ter-KOS-tā-lis) 6.9
Interlobar (in'ter-LŌ-bar) 13.5
Interlobular (in'ter-LOB-ū-lar) 13.5
Intermedia (in'ter-MĒ-dē-ah) 8.2
Intermedius (in'ter-MĒ-de-us) 6.19
Internodal (in'ter-NŌ-dal) 9.6
Internuncial (in'ter-NUN-shē-al) 7.2
Internus (in-TER-nus) 6.17
Interossei (in'ter-OS-ē-ī) 6.16, 6.23
Interosseous (in'ter-OS-ē-us) 9.10
Interspinalis (in'ter-spī-NĀ-les) 6.7
Interstitial (in'ter-STISH-al) 8.8, 10.1
Intertransversarii (in'ter-trans'ver-SAR-i) 6.7
Intertrochanteric (in'ter-trō'kan-TER-ik) 4.14
Interureteric (in'ter-ū're-TER-ik) 13.6
Interventricular (in'ter-ven-TRIK-ū-lar) 7.5, 9.5, 9.7
Inversion (in-VER-shun) 5.4
Ipselateral (ip-SĪ-lat'er-al) 1.1
Iridica (ī-RID-i-ka) 7.21
Iris (Ī-ris) 7.20
Ischiocavernous (is'kē-ō-kav'er-NŌ-sus) 6.11
Ischiopubis (is'kē-ō-PŪ-bis) 4.12
Ischium (IS-kē-um) 4.12
Islet (Ī-let) 8.7, 12.14
Isthmus (IS-mus) 14.8
Jejunal (jē-JŪ-nal) 9.12
Jejunum (jē-JŪ-num) 12.11
Jugular (JUG-ū-lar) 9.16, 9.17
Juxta (JUKS-tah) 13.4
Krause (KROW-zez) 3.3
Kupffer (KOOP-fer) 12.16
Labia (LĀ-bē-ah) 14.5, 14.9
Labial (LĀ-bē-al) 9.17, 12.5
Labii (LA-bē-ī) 6.4
Labyrinth (LAB-i-rinth) 7.22
Lacrimal (LAK-ri-mal) 4.4, 4.5, 4.6, 4.7, 7.19
Lacteal (LAK-tē-al) 12.12
Lactiferous (lak-TIF-er-us) 14.10
Lacuna (lah-KŪ-nah) 4.2

Semilunaris (sem-ē-LŪ-nar-is) 12.13
Semimembranosus (sem'ē-mem'brah-NŌ-sus) 6.18
Seminal (SEM-i-nal) 13.6, 14.1
Seminiferous (se'mī-NIF-er-us) 14.2
Semispinalis (sem'ē-spī-NĀ-lis) 6.7
Semitendinosus (sem'ē-ten'dī-NŌ-sus) 6.18
Septa (SEP-tah) 8.5
Septal (SEP-tal) 4.7
Septi (SEP-tē) 6.4
Septum (SEP-tum) 7.8, 11.3
Septum pellucidum (SEP-tum pel-LŪ-sė-dum) 7.8
Serosa (sē-RŌ-sah) 12.3, 12.4, 12.7, 13.6
Serotonin (ser'ō-TŌ-nin) 8.1
Serous (SĒ-rus) 1.4
Serrata (ser-RĀ-tah) 7.20, 7.21
Serratus (ser-RĀ-tus) 6.7, 6.8, 14.10
Sertoli (ser-TŌ-lē) 14.2
Sheath (shēth) 5.2
Sigmoid (SIG-moyd) 9.12, 9.16, 12.13
Sinoatrial (sī'nō-Ā-trē-al) 9.6
Sinus (SĪ-nus) 10.3, 11.3, 13.2
Sinusoids (SĪ-nus-oyds) 9.20, 12.16
Soleus (SŌ-lē-us) 6.21
Somatic (sō-MAT-ik) 7.2
Somatomammotrophin (sō'mah-tō-mam-MŌ-trō-fin) 8.8
Somatotrophin (sō'mah-tō-TRŌ-fin) 8.3
Somesthetic (so'mes-THET-ik) 7.7
Spermatic (sper-MAT-ik) 14.2, 14.3
Spermatids (SPER-mah-tids) 14.2
Spermatocyte (sper-MĂT-ō-sīt) 14.2
Spermatogonia (sper'măt-ō-GŌ-nē-ah) 14.2
Spermatozoa (sper'măt-ō-ZŌ-ah) 14.2
Spermatozoon (sper'măt-ō-ZŌ-on) 14.2
Sphenoid (SFĒ-noyd) 4.4, 4.5, 4.6, 4.7
Sphenoidal (sfē-NOYD-al) 11.3
Sphincter (SFING-ter) 6.11, 9.2, 12.9, 12.14, 13.6
Spinalis (spī-NĀ-lis) 6.7
Spinosum (spi-NŌ-sum) 3.1
Splanchnic (SPLANK-nik) 7.12
Splenic (SPLEN-ik) 9.11, 9.12, 9.20
Splenius (SPLĒ-nē-us) 6.7
Spongy (SPŬN-jē) 13.1
Squamous (SKWĀ-mus) 2.3
Sternocleidomastoid (ster'nō-klī'dō-MAS-toyd) 6.6
Sternocostal (ster'nō-KOS-tal) 9.4
Sternohyoid (ster'nō-HĪ-oyd) 6.6
Sternothyroid (ster'nō-HĪ-oyd) 6.6
Sternum (STER-num) 4.9
Stratum (STRA-tum) 3.1
Stylohyoid (stī'lō-HĪ-oyd) 6.6
Stylopharyngeus (stī'lō-fah-RIN-jē-us) 6.5
Subclavian (sub-KLĀ-vē-an) 9.11, 9.17, 10.1
Subclavius (sub-KLĀ-vē-us) 6.8
Sublingual (sub-LING-gwal) 12.7
Submandibular (sub'man-DIB-ū-lar) 7.17, 12.2, 12.7

Submuscosal (sub'mū-KŌ-sal) 12.3, 12.8
Subscapularis (sub'skap-ū-LĀ-ris) 6.12
Sudoriferous (sū'dor-IF-er-us) 3.2
Sulcus (SUL-kus) 7.11, 9.5, 12.7
Superficialis (sū'per-fish'ē-Ā-lis) 6.14
Superioris (sū-PĒ-rē-ōr-is) 6.4
Supination (sū-pī-NĀ-shun) 5.4
Supinator (sù'pi-NĀ-tor) 6.13, 6.15
Supraclavicular (su'prah-klah-VIK-ū-lar) 7.13
Suprarenals (sū-prah-RĒ-nals) 8.1, 9.11, 9.19
Suprascapular (sū'prah-SKAP-ū-lar) 9.8
Supraspinatus (sū'prah-spī-NĀ-tus) 6.12
Surfactant (sur-FAK-tant) 11.6
Suspensory (sus-PEN-sō-rē) 14.6
Suture (SŪ-cher) 4.4, 4.5, 5.1
Sylvius (SIL-vē-us) 7.5
Symphysis (SIM-fi-sis) 4.12, 5.1
Synapse (SIN-aps) 7.2
Synaptic (sī-NAP-tik) 7.1
Synaptic vesicles (si-NAP-tik VES-i'kls) 7.3
Synarthroses (sin'ahr-THRŌ-sēz) 5.1
Syndesmoses (sin'dez-MŌ-sēz) 5.1
Synovial (si-NŌ-vē-al) 5.1, 5.2, 5.3
Synovium (sin-ō-vē-um) 5.2
Systemic (sis-TEM-ik) 10.1
Taenia coli (TĒ-nē-ā KŌ-lī) 12.13
Talus (TĀ-lus) 4.15
Tarsals (TAHR-sals) 9.14
Tectorial (tek-TŌ-rē-al) 7.23
Telodendria (tel'Ō-DEN-drē-ah) 7.1
Telodendrion (tel-ō-DEN-drē-on) 7.3
Temporal (TEM-pō-ral) 4.4, 4.5, 4.6, 7.7
Temporalis (tem-pō-RĀ-lis) 6.4
Tendon (TEN-don) 5.2, 6.1
Tensor (TEN-sor) 6.5, 6.17
Tentorium cerebelli (ten-TŌ-rē-um ser'ē-BEL-lī) 7.4
Teres (TĒ-rēz) 6.12, 6.13, 6.14, 12.15
Tertius (TER-shus) 6.20
Testes (TES-tēz) 8.1, 14.1
Testicular (tes-TIK-ū-lar) 9.11, 9.19, 14.3
Testis (TES-tis) 14.1
Testosterone (tes-TOS-ter-ōn) 8.8, 14.2
Tetraiodothyronine (tet'rah-ī-ō'dō-THĪ-rō-nēn) 8.4
Theca (THĒ-kah) 14.7
Thenar (THĒ-nar) 6.16
Thoracic (thō-RAS-ik) 1.4, 4.8
Thoracis (THŌ-rah-kis) 6.7
Thoracoacromial (thō'rah-kō-ah-KRŌ-mē-al) 9.10, 9.18
Thoracoepigastric (thō-rah-kō-ep'ĭ-GAS-trik) 9.19
Thorax (THŌ-raks) 1.2
Thrombocytes (THROM-bō-sīts) 9.3
Thymus (THĪ-mus) 8.1
Thyrocalcitonin (thī'rō-kal-sī-TŌ-nin) 8.4
Thyrocervical (thī'rō-SER-vi-kal) 9.8
Thyroepiglottic (thī'rō-ep'ĭ-GLOT-ik) 6.5, 11.4
Thyroglobulin (thī'rō-GLOB-ū-lin) 8.4
Thyrohyoid (thī'rō-HĪ-oyd) 6.6, 11.4

Thyroid (THĪ-royd) 8.1, 8.4, 8.5
Thyrotrophin (thī-RO-trō-fin) 8.3
Thyroxine (thī-ROK-sin) 8.4
Tibia (TIB-ē-ah) 4.14
Tibial (TIB-ē-al) 9.14, 9.21
Tibialis (tib'ē-Ā-lis) 6.21
Tonsil (TON-sil) 10.3, 12.7
Trabeculae (trah-BEK-ū-lē) 4.2, 10.3, 13.4
Trabeculae carneae (trah-BEK-ū-lē KAR-nē-ē) 9.5
Trachea (TRĀ-kē-ah) 8.5, 11.1, 11.5
Transversis (trans-VER-sis) 6.9
Transversus (trans-VER-sus) 6.4, 6.10, 6.11
Trapezium (trah-PĒ-zē-um) 4.11
Trapezius (trah-PĒ-zē-us) 6.8
Trapezoid (TRAP-ē-zoyd) 4.11
Triceps (TRĪ-seps) 6.13
Tricuspid (trī-KUS-pid) 9.5
Trigeminal (trī-JEM-i-nal) 7.15
Trigone (TRĪ-gōn) 13.6
Triiodothyronine (trī'ī-ō'dō-THĪ-rō-nēn) 8.4
Triquetral (trī-KWĒ-tral) 4.11
Trochanter (tro'KAN-ter) 4.14
Trochlear (TROK-lē-ar) 7.15
Tuberalis (TU-ber-al-is) 8.2
Tubule (TŪ-būl) 13.3
Tunica adventitia (TŪ-ni-kah ad'ven-TISH-ē-ah) 9.2
Tunica intima (TŪ-ni-kah IN-ti-mah) 9.2
Tunica media (TŪ-ni-kah MĒ-dē-ah) 9.2
Tympanic (tim-PAN-ik) 7.22
Ulna (UL-na) 4.10
Ulnar (UL-nar) 9.10
Ulnaris (ul-NĀ-ris) 6.14, 6.15
Umbilical (um-BIL-i-kal) 1.3, 9.13
Unipennate (u'nē-PEN-nāt) 6.1
Unipolar (ū'nē-PŌ-lar) 7.2
Ureter (ū-RĒ-ter) 13.1, 13.2, 13.6
Urethra (ū-RĒ-thrah) 3.1, 13.6, 14.1, 14.4
Urethral (ū-RĒ-thral) 13.6
Urogenital (ū-rō-JEN-i-tal) 14.1, 14.4
Uterine (Ū-ter-in) 9.13, 14.5
Uterosacral (ū'ter-ō-SĀ-kral) 14.8
Uterus (Ū-ter-us) 14.5
Utriculus (ū-TRIK-ū-lus) 7.23
Uvula (Ū-vū-lah) 6.5, 12.5, 13.6
Vacuole (VAK-ū-ōl) 2.1
Vagina (vah-JĪ-nah) 14.5
Vaginalis (VAJ-i-nal-is) 14.2, 14.3
Vagus (VĀ-gus) 7.15, 7.17, 7.18, 9.6
Vas deferens (vas-DEF-er-ens) 14.2, 14.3, 14.4
Vasa recti (VĀ-sah REK-tē) 13.5
Vasa vasorum (VĀ-sa vā-SŌ-rum) 9.2
Vasopressin (vas'ō-PRES-in) 8.3
Vastus (VAS-tus) 6.19
Vater (FĀ-ter) 12.11, 12.14
Veins (vāns) 9.2
Vena Cava (VĒ-nah KĀ-vah) 9.1, 9.17, 9.19, 9.21, 12.15
Venosum (vė-NŌ-sum) 12.15
Venter (VEN-ter) 1.2

Ventricle (VEN-tri-k′l) 7.5, 9.5, 11.4

Venule (VEN-ūl) 9.2, 10.1

Vermiform (VER-mi-form) 12.13

Vermis (VER-mis) 7.10

Vertebral (VER-tē-bral) 9.9

Vesicle (VES-i-k′l) 9.13

Vestibule (VES-ti-būl) 7.23, 11.2, 11.4, 12.5, 12.9, 14.5, 14.9

Vestibulocochlear (ves-tib′ū-lō-KOK-lē-ar) 7.15, 7.22

Villi (VIL-ī) 12.12

Visceral (VIS-er-al) 1.1, 12.3, 12.4, 13.4

Vitreous humor (VIT-rē-us HŪ-mor) 7.20

Vomer (VŌ-mer) 4.4, 4.5, 4.6, 4.7

Vomeronasal (vō′mer-ō-NĀ-sal) 4.7

Wirsung (VER-soong) 12.14

Xiphoid (ZIF-oyd) 4.9

Zeis (zē-IS) 7.19

Zonula adherens (ZŌN-ū-lah ad-HĒR-enz) 2.1

Zonula occludens (ZŌN-ū-lah ō-KLOOD-enz) 2.1

Zygomatic (zī′gō-MAT-ik) 4.4, 4.5, 4.6

Zygomaticus (zī-gō-MĀT-i-kus) 6.4

Zymogenic (zī′mō-JEN-ik) 12.10